Neuroendocrine Correlates of Sleep/Wakefulness

Daniel P. Cardinali
S. R. Pandi-Perumal

Editors

Neuroendocrine Correlates of Sleep/Wakefulness

With 103 Illustrations

 Springer

Daniel P. Cardinali
Department of Physiology
Faculty of Medicine
University of Buenos Aires
Argentina
Buenos Aires 1121
cardinal@mail.retina.ar

S. R. Pandi-Perumal
SUNY Downstate Medical Center
Brooklyn 11203
New York, USA

Library of Congress Control Number: 2005926929

ISBN-10: 0-387-23641-4   e-ISBN: 0-387-23692-9   Printed on acid-free paper.
ISBN-13: 978-0387-23641-4

Printed in the United States of America. (SPI/IBT)

9 8 7 6 5 4 3 2 1

springeronline.com

Dedication

To our wives and families,
Who are the reasons for any of our accomplishments
Who have taught and aided us
In much of what we know and do

In Memoriam

It is with great sadness that we announce the passing of Louis D. Van de Kar, Ph.D., Department of Pharmacology and Experimental Therapeutics, Loyola University of Chicago, USA. Luke died on September 4, 2004, after a fight with cancer. Luke was a pioneer in research on serotonergic control of stress hormone release. He then applied this pioneering research approach to study the molecular biological mechanisms responsible for the therapeutic effects of serotonin-specific reuptake inhibitors in the brain. Luke was a tremendously helpful and influential mentor to his students, colleagues, and friends. He will be greatly missed by all of us.

Preface

Many recent discoveries in both the laboratory and the clinical setting have rapidly increased our understanding of sleep medicine and neuroendocrinology. These are being continually reported in various neuroendocrine and sub-specialty journals as well as in dedicated sleep publications. Sleep medicine is thus becoming increasingly interdisciplinary while other areas of neuroscience and endocrinology are beginning to take an interest in the subject of sleep. A parallel development is that these fields are now reorganizing themselves at higher levels of complexity. The consequence of these phenomena is that for the researcher who is interested in the various facets of sleep physiology, it is becoming increasingly challenging to assimilate—let alone to master—the relevant findings in each of these fields.

Neuroendocrine Correlates of Sleep/Wakefulness summarizes and reviews many of the major new discoveries concerning neuroendocrine correlates of sleep/wakefulness. We have endeavored to select a limited number of outstanding contributions from chosen experts in their respective fields. The goal of the volume is to present the more recent developments in the fields of sleep and neuroendocrinology and to provide a context for considering them both in depth and from a multidisciplinary perspective. This volume thus brings together the knowledge and expertise of sleep specialists, neuroscientists, clinicians and basic researchers, neuroendocrinologists, and biological rhythm researchers.

We appreciate that an exhaustive review of the field would run the risk of becoming an unwieldy tome that would undermine the final objective of creating a practical and useful resource. It has been our goal to provide a concise yet comprehensive review of the expanding and increasingly interdisciplinary area of sleep medicine. It is our hope that readers will find that this effort represents a thoughtful balance of basic experimental and clinical viewpoints, and further that this will serve as a foundation for understanding and ultimately treating sleep disorders.

Inasmuch as we envision continuing updates and new editions of this volume, readers are encouraged to contact us with any thoughts or suggestions for revisions.

Daniel P. Cardinali
Department of Physiology
Faculty of Medicine
University of Buenos Aires
ARGENTINA

S.R. Pandi-Perumal
Section of Sleep Medicine
Department of Neurology
SUNY Downstate Medical Center
New York, USA

Contents

List of Contributors

S. M. Abbott
Department of Cell & Structural
  Biology
The Neuroscience Program and the
  College of Medicine
University of Illinois at Urbana-
  Champaign
Urbana, IL

M. Y. Agargun
Department of Psychiatry
Yuzuncu Yil University School of
  Medicine
Van, Turkey

J. M. Arnold
Division of Laboratory Animal
  Medicine
Department of Pharmacology
Southern Illinois University School
  of Medicine
Springfield, IL

N. Basavaraj
Department of Psychiatry
University of California, San Diego
La Jolla, CA

L. Besiroglu
Department of Psychiatry &
  Neuroscience Research Unit
Yuzuncu Yil University School of
  Medicine
Van, Turkey

D. Boivin
Center for Study and Treatment of
  Circadian Rhythms Douglas
  Hospital Research Centre
Department of Psychiatry
McGill University
Verdun (Quebec), Canada

G. Brandenberger
Laboratoire des Regulations
  Physiologiques et des Rythmes
  Biologiques chez l'Homme
Institut de Physiologie
Faculté de Médecine
France

T. Bratlid
Department of Psychiatric Research
  and Development
University Hospital of Northern
  Norway
Tromsø, Norway

R. M. Buijs
Netherlands Institute for Brain
  Research
Amsterdam, The Netherlands

D. P. Cardinali
Departamento de Fisiología,
  Facultad de Medicina
Universidad de Buenos Aires
Buenos Aires

V. M. Cassone
Department of Biology
Texas A&M University
College Station, TX

F. Chiappelli
University of California, Los Angeles
School of Dentistry,
  Psychoneuroimmunology
Los Angeles, CA

M. E. Coussons-Read
Department of Psychology
University of Colorado at Denver
Denver, CO

J. W. Crane
School of Biomedical Sciences
Queensland Brain Institute
University of Queensland
St. Lucia, Brisbane
Queensland, Australia

H. De Leersnyder
Department of Genetics
Hôpital Robert Debré
Paris, France

J. S. Emens
Department of Psychiatry
Oregon Health & Science University
Sleep and Mood Disorders
    Laboratory
Portland, OR

A. I. Esquifino
Departamento de Bioquímica y
    Biología Molecular III
Facultad de Medicina
Universidad Complutense
Madrid, Spain

P. Feng
Pulmonary and Critical Care
    Division
Department of Medicine
Case Western Reserve University
Cleveland, OH

S. Fulda
Max Planck Institute of Psychiatry
Munich, Germany

M. U. Gillette
Department of Cell & Structural
    Biology
The Neuroscience Program and the
    College of Medicine
University of Illinois at Urbana-
    Champaign
Urbana, IL

F. J. Iribarren
UCLA Neuropsychiatric Institute
Center for Community Health
Los Angeles, CA

A. Kalsbeek
Hypothalamic Integration
    Mechanisms
Netherlands Institute for Brain
    Research
Amsterdam, The Netherlands

N. Kaushal
Department of Psychiatry
Harvard Medical School/VA
    Medical
    Center
MA

L. Kayumov
Department of Psychiatry
University of Toronto
Toronto, Canada

V. M. Kumar
Department of Physiology
All India Institute of Medical
    Sciences
New Delhi, India

A. J. Lewy
Department of Psychiatry
Director, Sleep and Mood Disorders
    Laboratory, Oregon Health &
    Science University
Portland, OR

A. M. Lopez
Department of Psychiatry
University of California, San Diego
La Jolla, CA

R. Luboshitzky
Endocrine Institute
Haemek Medical Center
Afula, Israel

V. Madan, M. Sc.
School of Life Sciences
Jawaharlal Nehru University
New Delhi, India

B. N. Mallick
School of Life Sciences
Jawaharlal Nehru University
New Delhi, India

L. F. Martínez
Department of Psychiatry
University of California, San Diego
La Jolla, CA

E. L. Maurer
Department of Psychiatry
University of California, San Diego
La Jolla, CA

C. J. Meliska
Department of Psychiatry
University of California, San Diego
La Jolla, CA

N. Neagos
University of California, Los
  Angeles
School of Dentistry
Psychoneuroimmunology
Los Angeles, CA

L. P. Niles
Department of Psychiatry &
  Behavioural Neurosciences
McMaster University, Faculty of
  Health Sciences
Ontario, Canada

M. Okun
University of Colorado Health
  Sciences Center
School of Nursing
Denver, CO

D. Pal
School of Life Sciences
Jawaharlal Nehru University
New Delhi, India

S. R. Pandi-Perumal
Section of Sleep Medicine
Department of Neurology
SUNY Downstate Medical Center
Brooklyn, NY

B. L. Parry
Department of Psychiatry
University of California, San Diego
  and San Diego Veterans
  Administration Health Care
Jolla, CA

T. Partonen
Department of Mental Health and
  Alcohol Research
National Public Health Institute
Helsinki, Finland

M. F. P. Peres
Director, Sao Paulo Headache
  Center Hospital Israelita Albert
  Einstein
Instituto de Ensino e Pesquisa
Al Joaquim Eugenio de Lima
Sao Paulo SP. Brazil

O. Polo
Department of Pulmonary Diseases
Tampere University Hospital and
  Sleep Research Unit
University of Turku
Tampere, Finland

P. Prolo
University of California, Los
  Angeles
School of Dentistry
Psychoneuroimmunology
Los Angeles, CA

A. A. Putilov
Siberian Branch
Russian Academy of Medical
  Sciences
Novosibirsk, Russia

V. Ramesh
Department of Psychiatry
Harvard Medical School and VA
  Medical Center
Brockton, MA

S. A. Rivkees
Yale Child Health Research Center
Yale Department of Pediatrics
Yale University School of Medicine
New Haven, CT

M. Ruiter
Hypothalamic Integration
  Mechanisms
Netherlands Institute for Brain
  Research
Amsterdam, The Netherlands

T. Saaresranta
Department of Pulmonary Diseases
Turku University Central Hospital
University of Turku
Turku, Finland

C. M. Sinton
Department of Internal Medicine
University of Texas Southwestern
  Medical Center
Dallas, TX

M. G. Smits
Department of Neurology and Sleep
  Disorders
Gelderse Vallei Hospital
The Netherlands

D. L. Sorenson
Department of Psychiatry
University of California, San Diego
La Jolla, CA

V. Srinivasan
Department of Physiology
Pusat Pengajian Sains Perubatan
School of Medical Sciences
Universiti Sains Malaysia
Kampus Kesihatan
Kelantan, Malaysia

A. Steiger
Max Planck Institute of Psychiatry
Munich, Germany

D. Suchecki
Universidade Federal de São Paulo
  Escola Paulista de Medicina
Department of Psychobiology
Brazil

M. J. Thorpy
Sleep-Wake Disorders Center,
Montefiore Medical Center and
  Department of Neurology
Albert Einstein College of Medicine
Bronx, NY

L. A. Toth
Director, Division of Laboratory
  Animal Medicine
Department of Pharmacology
Southern Illinois University School
  of Medicine
Springfield, IL

S. Tufik
Universidade Federal de São Paulo
Escola Paulista de Medicina
Department of Psychobiology
Brazil

L. D. Van de Kar (deceased)
Department of Pharmacology
Loyola University of Chicago
Stritch School of Medicine

Antoine U. Viola
Laboratoire des Regulations
  Physiologiques et des Rythmes
  Biologiques
  chez l'Homme
Institut de Physiologie
Faculté de Médecine

T. C. Wetter
Max Planck Institute of Psychiatry
Munich, Germany

J. T. Willie
Department Molecular Genetics
UT Southwestern Medical Center
Dallas, TX

G. G. Zirpoli
Department of Psychiatry
University of California, San Diego
La Jolla, CA

N. Zisapel
Department of Neurobiochemistry
The George S. Wise Faculty of
 Life Sciences
Tel Aviv University
Tel Aviv, Israel

Basic Sciences

The Developmental Regulation of Wake/Sleep System

Pingfu Feng

Sleep associated with a high frequency of phasic activities, particularly muscle twitches, dominate the daily life of altricial mammals and human neonates. "Active sleep" is the term that has been used to describe these behavioral phenomena in the neonatal rat, cat and rabbit in comparison to the sleep state without phasic activities, which is called "quiet sleep".[1]

Rapid eye movements (REMs) are recorded during active sleep but rarely during quiet sleep. Thus, visually-scored active sleep is roughly the state of REM sleep. However, a small portion of visually-scored active sleep (i.e., behaviorally "active" sleep) may show high-amplitude EEG at postnatal day (PN) 10 or older in the rat if recorded by a polygraph. This type of sleep is first called "half-activated sleep" but is mostly scored as slow-wave sleep or non-REM (NREM) sleep.[1,2] This suggests that the term of "REM sleep" is better than that of "active sleep" in distinguishing the difference between REM sleep and non-REM sleep in matured mammals including human, and the usage of active sleep should be limited to describe the sleep state in the neonatal period.

Developmental features of sleep-wake states include: 1) very high percent (> 70%) and dramatic decrease of REM sleep (active sleep) progressively replaced by wakefulness and NREM sleep; and 2) a high frequency and progressively decrease of phasic activities. The active (voluntary) behavioral activity during the wake state significantly increases as waking time is prolonged. Simultaneously, brain plasticity indicated by the adult effect of neonatal drug treatment or environmental alteration dramatically decreases.[3]

One amazing feature of neonatal REM sleep is that it has its highest percentage shortly after birth and then decreases as maturation proceeds (Figure 1). According to Jouvet-Mounier et al., the neonatal rat spends 72% of time in REM sleep in the first ten days of life and this percentage dramatically drops to 40% or less during the second week.[4]

By the end of the second week, about one-third of time is distributed almost equally into each state of wake, REM sleep and NREM sleep according to our study using a long-term, continuous polysomnographic recording method.[5] Over the third and fourth week, REM sleep continues to decrease and reaches to the level (14%) observed in adulthood by the end of

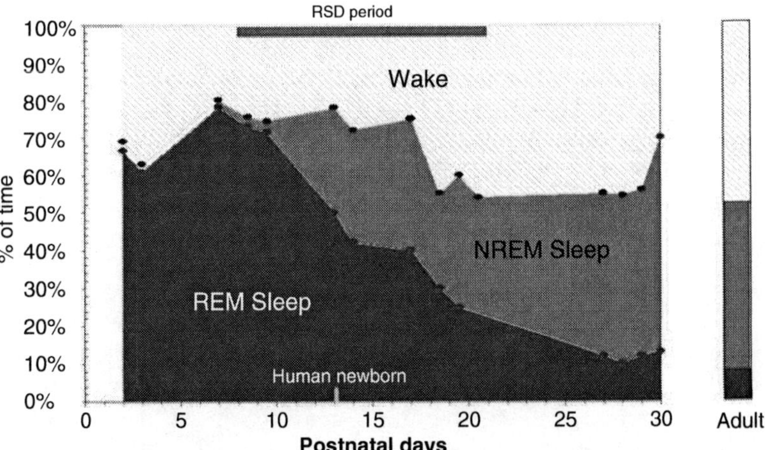

FIGURE 1. Evolution of sleep-wake states in the first month of life and the adulthood in the rat. REM sleep dominates daily life in the first postnatal week. Waking without eyes open is about 30% or less. NREM sleep, however, is almost zero during this period of time. Thereafter, REM sleep is dramatically decreased while wake and NREM sleep increase. REM sleep is about 70% in neonatal rats and about 7-8% in adult ones. Figure is re-drawn from Jouvet-Mounier D, Astic L and Lacote D. Dev Psych 1970; 2:216-239 and Feng P, Bergmann BM, Rechtschaffen A. Brain Res 1995; 703:93-9.

the fourth week.[1,5] Thus, REM sleep drops 80% over the first month of life. In addition, there is a decrease in the number of REM sleep episodes, mean REM period duration, and the number of sleep-onset REM periods. REM latency (the time from wakefulness to the first epoch of REM sleep) also dramatically increases during this period. All of these changes are closely correlated with changes in age and in brain maturation in particular. It may be important the number of REM episodes and the mean duration of each REM period decrease together. Thus, propensity to initiate REM sleep episodes and the propensity to sustain REM sleep episodes decrease monotonically from age two weeks to age four weeks.[5,6]

Quiet sleep, used to describe a sleep state without phasic movements in the first two weeks of life, is behaviorally and polysomnographically identical to adult NREM sleep.[1] The actually-scored NREM sleep, however, includes the part of half-activated sleep, which comprises a small portion of total NREM sleep.[1,2] This type of sleep is almost non-existent in the first week and increases to about 15% in the second week. By PN 12, NREM sleep is identified by high-amplitude EEG, and increases, rapidly paralleling the increase of wakefulness. Newborn rats do not open their eyes until PN 14. Before PN 14, the wake state is identified by "behavioral" criteria such as walking and eating without eyes open. This state is only 10-30% in the first two weeks of life. Similar to REM sleep development, all states develop to a near-adult level by the end of the fourth week.[1]

Phasic muscle twitches are one of the major REM activities in the neonate and also comprise the prominent neonatal behavior. A similar feature is found not only in rats but also in other rodents and the human as well. The period that shows frequent muscle twitches is the first four weeks in the rat[1], the first forty days in the kitten[2] and the first 8 months in the human newborn[7]. In humans, this feature is also typical in the premature fetus.[8] Phasic muscle twitches appear primarily during REM sleep but also in a small portion of NREM sleep, i.e., the "half-activated" sleep.[1] The rate of muscle twitches is only 1.5/min and 0.3/min during quiet sleep compared to the rate of 7.5/min and 3/min during REM sleep at the same age in PN 10 and PN 20 kittens respectively.[2] The number of phasic events dramatically decreases as animals mature.[2,3,7,9] It is of interest to note that the dramatic reduction of phasic activities in REM sleep is associated with the increase of wakefulness.

1. Neuronal Regulation of Wakefulness in Neonates

1.1. Hypocretinergic (Orexinergic) System

Recent studies in adults indicate that hypocretins are critical for the maintenance of wakefulness.[10] Microdialysis perfusion of hypocretin-1 (also called Orexin A) in both the basal forebrain[11] and brain stem increase wakefulness in freely behaving adult rats.[12] Fos expression in hypocretin neurons correlates positively with the amount of wakefulness and negatively with the amounts of non-REM and REM sleep.[13] The majority of hypocretinergic neurons express c-fos during active waking.[13] In rat brain slices of centromedial nuclei and rhomboid nuclei in the thalamus, hypocretin depolarizes and excites all neurons tested through a direct postsynaptic action.

Hypocretin-1 and Hypocretin-2 (also called orexin B) are synthesized in neurons of the perifornical region and the lateral hypothalamus. Fibers of these neurons innervate most brain regions including brain stem and basal forebrain, cortex and spinals[14,15] and act through Hypocretin-1 and Hypocretin-2 receptors. Local application of Hypocretin-1 in basal forebrain[11] or LDT/PPT[12] increases wakefulness dramatically. Hypocretin-2, which is lacking in narcoleptic dogs,[16] has strong and direct excitatory effect on the cholinergic neurons of the contiguous basal forebrain.[11] Hypocretin-1 promotes both quiet and active waking but hypocretin-2 promotes active waking only.[17]

Hypocretinergic neurons may not be functionally active in neonatal rats. The *prepro-orexin* mRNA is nearly undetectable in the lateral hypothalamic area (LHA) from PN 0 to PN 15. Hypocretin-1 and hypocretin-2-like immunopositive cells and fibers were not detected from days PN 0 to PN 10, but are observed after PN 15. The *prepro-orexin* mRNA in the LHA dramatically increased between PN 15 and PN 20, and reached adult level after three weeks.

1.2. Serotonergic System

As discussed in other chapters, serotonergic neurons are widely involved in the regulation of sleep/wake states, behavioral and basic physiological activities. These neurons are located in the raphé nucleus project widely to the entire central nervous system and act via release of 5-hydroxytryptamine (5-HT). 5-HT dorsal raphé nucleus (DRN) neurons fire at their highest rate during wakefulness, at a lower rate during non REM sleep, and completely cease to fire during REM sleep.[18-21] Extracellular level of 5-HT in DRN[22], medial medullary reticular formation,[23] PPT/LDT,[24] hippocampus[25] and frontal cortex[22] exhibited the similar pattern, i.e. the highest in waking, the lower in slow-wave sleep and the lowest in REM sleep. 5-HT DRN activity is involved more in the promoting wakefulness than in suppressing REM sleep.[26,27]

At the highest concentration tested, however, REM sleep occurred directly after waking, as in narcolepsy.[27] 5-HT DRN neurons are locomotor activity dependent.[28,29] 5-HT DRN neurons cease firing by disfacilitation during REM sleep.[30] Atonia induced by electrical stimulation of the pontine inhibitory area and gigantocellular reticular nucleus causes a reduction in the activity of the LC.[29]

At least fifteen subtypes of 5-HT receptors have been identified.[31] 5-HT1a receptor is an inhibitory autoreceptors situated on the cell body as well as the dendrites of 5-HT DRN neurons.[32] 5-HT reuptake is regulated by 5-HT transporter (5-HTT), which is widely distributed throughout the rodent brain.[33-35] 5-HT transport capacity is regulated by transporter phosphorylation and sequestration involving protein kinase C.

Phosphorylation and sequestration of the 5-HTT are substantially impacted by the occupancy of ligands, such as 5-HT and antidepressants. This occupancy prevents PKC-dependent 5-HTT phosphorylation. 5-HT plays a facilitatory role in activating the cortex through the activation of 5-HT2A receptors on thalamocortical neuronal terminals and thereby increases glutamate release, which in turn activates cortical neurons as demonstrated y an increase of c-Fos expression.[36]

Systemic administration of 5-HT2a receptor agonist, DOI, significantly decreases neocortical high-voltage spindle activity, and this effect may be blocked by different 5-HT2 receptor antagonists.[37] Thus, upregulation of 5-HT DRN neurons or an increase in systemic 5-HT is likely to activate cortical neurons via the activation of 5-HT2A receptors on the terminals of thalamocortical neurons and thereby increases glutamate release.[36]

Brain levels of 5-HT are very low in neonatal rats and increase markedly after the second postnatal week.[38] However, different brain regions show a different developmental profile of serotonin level.[39]

In the brain stem, serotonin gradually rises from embryonic day (E) 15 to levels seen in the adult by PN 32.[40] Binding of serotonin to rat brain membranes increased linearly from birth to adulthood, but newborn receptor densities were already 39% of adult levels.[41]

The circadian rhythm of hypothalamic 5-HT content exists in PN 12-, PN 23-, PN 27 and in adult rats.[42] Tryptophan hydroxylase is the rate-limiting enzyme in the synthesis of serotonin. During development, brain tryptophan hydroxylase activities increase. Tryptophan hydroxylase messenger RNA levels in the DRN increased 35-fold between E18 and PN 22, then decreased by 40% between PN 22 and 61. Tryptophan hydroxylase messenger RNA expression in the nucleus raphé obscuris increased 2.5-fold between PN 8 and 22.[43]

Thus, developmental alterations of hypocretinergic and serotonergic systems are consistent with the facts that wake percentage is much less in the early age and progressively increased as rats mature and that REM sleep deprivation leads to a significant increase of wakefulness in adult but not in neonates.[3,44] This evidence implicates that these two neural systems play an important role in promoting wakefulness in adult as well as in neonatal mammals possibly including human.

2. Neuronal Regulation of NREM Sleep

Adenosine is a ubiquitous neuromodulator that increases sleep, particularly high amplitude NREM sleep, and suppresses wakefulness. Many of these effects are mediated by A1 receptors and are thought to be functioning in the basal forebrain. Microdialysis perfusion of A1 receptor antisense in the basal forebrain significantly reduced NREM sleep with an increase in wakefulness. After 6 hours of sleep deprivation, the antisense-treated animals spent a significantly reduced amount of time in NREM sleep, and an even greater postdeprivation reduction in delta power (60-75%) and a concomitant increase in wakefulness.[45] Adenosine perfusion into the basal forebrain increased the relative power in the delta frequency band, whereas higher frequency bands (theta, alpha, beta, and gamma) showed a decrease.[46] The neuronal activity of wakefulness-active neurons in the magnocellular basal forebrain, are the highest discharge activity during wakefulness, showed marked reduction in activity just before and during the entry to NREM sleep. The adenosine concentrations in the basal forebrain were increased following prolonged wakefulness. This suggests that the reduction of wake related discharges is due to an increase in the extracellular concentration of adenosine during wakefulness. Adenosine acts via the A1 receptor to reduce the activity of wakefulness-promoting neurons.[47]

About 30% of orexin-containing neurons were labeled with adenosine A1 receptor protein. The data supports the presence of adenosine A1 receptors on orexinergic neurons and suggests a possible substrate for a functional role of adenosine in the regulation of orexinergic activity.[48] One study reports that adenosine agonist cyclohexaladenosine injected into the medial pontine reticular formation of the rat induces a long-lasting increase in rapid eye movement sleep. The A1 mechanism operates at a different locus possibly through an inhibition of GABA neurotransmission. The A2a mechanism requires the cholinergic system and may act through the increased release of acetylcholine.[49]

The availability of extracellular adenosine depends on its release by transporters or by the extracellular ATP catabolism performed by the ecto-nucleotidase pathway. Major metabolic enzymes for adenosine, adenosine deaminase, adenosine kinase, ecto- and cytosolic 5′-nucleotidase exhibited diurnal variations in their activity in most tested brain regions. Activity of adenosine deaminase increased during the active period in the ventrolateral pre-optic area but decreased significantly in the basal forebrain. Enzymatic activity of adenosine kinase and cytosolic-5′-nucleotidase was higher during the active period in all brain regions tested.[50] Significant diurnal variation in enzyme activities was noted in the cortex and the basal forebrain brain areas. Adenosine kinase and both nucleotidases showed their lowest activity in the middle of the rest phase, suggesting the level of adenosine metabolism is related with activity and may be associated with the lower level of energy metabolism during sleep compared to wakefulness.[51] In the central nervous system, an increase in neuronal activity enhances energy consumption as well as extracellular adenosine concentrations. In most brain areas high extracellular adenosine concentrations, through A1 adenosine receptors, decrease neuronal activity and thus the need for energy. Adenosine seems to act as a direct negative feed-back inhibitor of neuronal activity. In addition to the immediate effects, high extracellular adenosine concentrations also induce intracellular changes in signal transduction and transcription, e.g. increase in A1 receptor expression and NF-kappa B binding activity. These changes may at least partially mediate the long term effects of prolonged wakefulness. Adenosine may also be a common mediator of the effects of several other sleep-inducing factors.[52]

The developmental progression of the adenosine system begins in the embryo. Adenosine A receptor mRNA was detected at E 14 and receptors at E18. A1 mRNA levels increased from the level reached at E18 between PN 3 and PN 14 (maximally a doubling), whereas A receptors increased later and to a much larger extent (about 10 to 19 fold) postnatally.[53,54] Adenosine A2A receptors were low in newborn and increased four-fold to adulthood in multiple brain regions. In contrast to the large increase in A2A receptors, there was a decrease in the levels of A2A messenger RNA during the postnatal period.[54] This is consistent to the occurrence of NREM sleep during developmental period, which showed the earliest appearance at PN 10-12 and progressively increased, and reached to adult level at the end of the first month in rats.[1] However, adenosinergic activity may not be crucial for the occurrence of NREM sleep since that neonatal REM sleep deprivation is able to easily generate large amount of NREM sleep at the age that there is no detectable natural NREM sleep.[3]

3. Neuronal Regulation of REM Sleep

Cholinergic laterodorsal and pedunculopontine tegmental (LDT/PPT) neurons have been shown to be responsible for the generation of REM sleep in the adult mammal.

Within the pontine reticular formation, the signal transduction pathway activated by the muscarinic cholinergic receptor m2/m4 contributes to cholinergic REM sleep generation. Cholinergic neurons in the LDT/PPT projections to cholinoceptive medial pontine reticular formation (MPRF) neurons to provide Ach.[55-58] LDT/PPT neurons fire tonically during wakefulness and increase firing before and throughout REM sleep.[59,60] LDT/PPT neurons can be defined phenotypically based on the presence of the Ach synthetic enzyme, choline acetyltransferase (ChAT) and by the presence of vesicular acetylcholine transporter (VAChT). Muscarinic cholinergic receptors have been identified in at least 5 subtypes (M1-M5). M2 autoreceptors regulate Ach release within the MPRF and the signal transduction pathway modulated by m2/m4 receptors participate in cholinergic REM sleep generation. The content of Ach is also regulated by VAChT, a presynaptic protein located on the vesicle membrane. Normal REM sleep generation requires VAChT in presynaptic terminals of LDT/PPT neurons. Blocking VAChT binding site of vesamicol receptor by intracerebroventricular injections of vesamicol decreases the amount of Ach released from presynaptic terminals[61,62] and inhibits REM sleep in rat and cat.[63,64]

Cholinergic receptor ml/m3/m5 subtypes are coupled to stimulatory G proteins, which stimulate phospholipase C. The m2/m4 subtypes are linked to inhibitory G proteins, the major type of G protein present in mammalian brain.[65] The signal transduction cascade modulated by m2/m4 subtypes involves-inhibition of adenylate cyclase, a decrease in cyclic adenosine monophosphate (cAMP) production, and subsequent inhibition of protein kinase A (PKA).[66] Stimulation of PKA leads to phosphorylation of serine and threonine residues of various intracellular proteins such as ion channels, receptors, G proteins, and synthetic enzymes. Phosphorylation of these proteins leads to the alteration of neuronal excitability, the synthesis and release of neurotransmitters, and changes in receptor sensitivity.

Rat PPT neurons are capable of responding to stimulation and treatment with cholinergic drug when tested at the age of PN 12 and PN 21.[67] This indicates that the cholinergic system may play a role in REM sleep regulation in neonates similar to that of adulthood. Compared with REM sleep in adulthood, REM sleep in neonates is 10 times higher in rats[1] and 5 to 8 times higher in human.[68] Provided that REM sleep and its phasic activities are driven by certain neuronal or humoral systems, the functional indication of this system, such as the level of the transmitters or the density of receptors or critical enzymes involved in this mechanism, should have a developmental pattern similar to that of REM sleep, i.e., showing the highest level in the neonate and then progressively decreasing as maturation proceeds. In fact, brain levels of Ach and acetylcholinesterase (AchE) activity are low in neonates and gradually increase.

The Ach level attains adult values at PN 70 and AchE activity shows a rapid increase between the PN 7 and PN 30.[69] The time course of the increase in ChAT activity correlates negatively to the amount of REM sleep or phasic activities after birth.[70] None of M1-M5 receptors in the brain regions of cortex, hippocampus, basal forebrain and striatum exhibits the changing pattern comparable to that of REM sleep.[71-75]

M1 receptors are expressed at 31% of adult levels, M2 receptors at 32% of adult levels, M3 receptors at 36% of adult levels, and M4 receptors at 20% of adult levels in PN 3 to PN 4. The combined M2/M4 receptor density increases at a uniform rate during development from 173 to 757 fmol/mg.[72] Furthermore, application of muscarinic receptor antagonist, atropine, in neonatal rats suppressed REM sleep at PN 14 and PN 20 but not PN 11.[76] These observations suggest that cholinergic function may not be the major driving force to the high percentage of REM sleep, but rather plays a critical role in the executive generation of REM sleep during the neonatal period.

Other neural systems, such as serotonergic, dopaminergic, noradrenergic, GABAergic and hypocretinergic systems are all involved in the regulation, particularly inhibition, but not promotion of REM sleep.[77-79] Adenosinergic neurons are recently identified to promote NREM only.[80] Glutamate is one of the major excitatory neurotransmitters in the brain and exhibits cholinergic LDT/PPT exciting effect.[81] However, it does not stimulate cholinergic PPT neurons in the rat between PN 12 and PN 21.[67] Thus, there is still lack of evidence to address the drive for the high percent of REM sleep in neonatal period.

4. A Candidate Driving Force for Neonatal REM Sleep

Above discussion suggests that cholinergic neurons may not be the causal for the high percentage of REM sleep in the neonatal period, because the developmental patterns of the cholinergic neurotransmitters and receptors do not show a positive correlation to the change of REM sleep over the developmental period. One interesting piece of evidence is that corticotrophins releasing hormone (CRH) mRNA expression does exhibit such a developmental correlation.

CRH is a 41-amino acid peptide, which plays an important neuroendocrine role in the regulation of the hypothalamic-pituitary-adrenal (HPA) axis,[82] and is involved in the coordination of various responses to stress.[83-85] It acts via the secretion of ACTH, which stimulates the release of adrenal glucocorticoids.

In the brain, CRH induces different autonomic, electrophysiological and behavioral effects, indicating that this peptide can serve as a neuromodulator.[86,87] CRH neurons are located in the hypothalamic paraventricular nucleus (PVN) and have the highest activity and secretion in the beginning of the life.

Birth in most animal species is triggered by the fetus through activation of the fetal HPA axis.[88] CRH binding shows the highest level in the first week of birth and decreases significantly in the hippocampus and striatum from PN 7-21.[89] CRH receptor mRNA in hippocampal CA1, CA2 and CA3a increases to 300-600% of adult levels by PN 6 with a subsequent decline. In the amygdala, CRH receptor mRNA abundance increases steadily between PN 2 and PN 9, to levels twice higher than those in the adult. In the cortex, CRH receptor mRNA levels are high on PN 2 and are decreased to adult levels by PN 12.[90] The high levels of CRH and CRH receptor mRNA in the first two weeks of life in the rat co-occur with the extremely high percent of REM sleep (>70%) and phasic activities.[1] Adult findings indicate a relationship between elevated HPA axis and disinhibited REM propensity. The prenatally-stressed adult rat shows increased amounts of paradoxical sleep, positively correlated with plasma corticosterone levels.[91] Sleep deprivation results in a stressful reaction, including elevated HPA axis and an increased REM rebound.[92,93] Increased blood levels of ACTH and corticosterone are found in rats that exhibit higher REM propensity.[94,95] Increased cerebrospinal fluid levels of CRF, blood ACTH, and cortisol are also found in human depression,[96] which has reliable signs of increased REM propensity.[97] Furthermore, intracerebroventricular injection of CRH to rats after 72 hrs of sleep deprivation markedly increases REM sleep,[98] and a CRH antagonist blocks the sleep deprivation-induced REM sleep rebound.[99] Intracerebroventricular injection of CRH increases locomotion in the chick[100] and fish.[101] Above evidence suggests our hypothesis for the neonatal REM sleep regulation as dialogued in the Figure 2.

One potential criticism may be that in the adult model CRF is linked to the behavioral and electrophysiological wakefulness.[102] To counter this contention, we point out that the HPA axis is an open loop and that cells that secrete ACTH and corticosterone are immature in the neonate in contrast to the adult.[103] Similar situation can be seen in adult rat model of depression and human depression, in which there is an semi-open loop in the HPA axis and an elevated brain level of CRF co-occurred with increased REM sleep propensity.[95,104,105] Thus, a hypothesis is put forth that CRF may play a crucial role in driving the high percentage of REM sleep and its phasic activities in the neonate. However, empirical testing is needed.

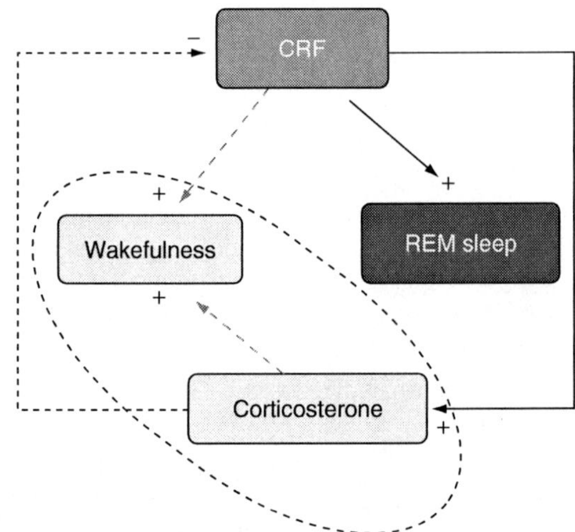

FIGURE 2. This figure illustrates the pathways in the regulation of wakefulness and REM sleep through corticotropin releasing factor (CRF). In the adult a functioning hypothalamic-pituitary-adrenal (HPA) axis that produces corticosterone along with the presence of a fully developed wake-regulation system permits CRF to have a primary function of promoting vigilance. In contrast, the neonate has both poorly functioning adrenal gland and wake-regulation system (inside the box); hence, the primary function of CRF is to promote the REM state.

5. Acknowledgments. Work was supported by the grants of NIMH, MH 069854, NHLB1 and HL64278

6. *References*

1. Jouvet-Mounier D, Astic L, Lacote D. Ontogenesis of the states of sleep in rat, cat, and guinea pig during the first postnatal month. Dev Psychobiol 1970; 2:216-39.
2. McGinty DJ, Stevenson M, Hoppenbrouwers T et al. Polygraphic studies of kitten development: sleep state patterns. Dev Psychobiol 1977; 10:455-69.
3. Feng P, Ma Y. Clomipramine suppresses postnatal REM sleep without increasing wakefulness: implications for the production of depressive behaviors. Sleep 2002; 25:177-84.
4. Komissarov IV, Talalaenko AN. Nature of the central adreno- and serotonin-sensitive structures participating in the facilitating influence of bioamines on the respiratory reflex from the upper respiratory passages. Biull Eksp Biol Med 1970; 70:52-5.
5. Vogel GW, Feng P, Kinney GG. Ontogeny of REM sleep in rats: possible implications for endogenous depression. Physiol Behav 2000; 68:453-61.

6. Feng P, Ma Y, Vogel GW. The critical window of brain development from susceptive to insusceptive. Effects of clomipramine neonatal treatment on sexual behavior. Brain Res Dev Brain Res 2001; 129:107-10.

7. Kohyama J. A quantitative assessment of the maturation of phasic motor inhibition during REM sleep. J Neurol Sci 1996; 143:150-5.

8. Dreyfus-Brisac C. Sleep Ontogenesis in early human prematurity from 24 to 27 weeks. Dev Psychol 1968; 1:162-169.

9. Denenberg VH, Thoman EB. Evidence for a functional role for active (REM) sleep in infancy. Sleep 1981; 4:185-91.

10. Kilduff TS, Peyron C. The hypocretin/orexin ligand-receptor system: implications for sleep and sleep disorders. Trends Neurosci 2000; 23:359-65.

11. Thakkar MM, Ramesh V, Strecker RE et al. Microdialysis perfusion of orexin-A in the basal forebrain increases wakefulness in freely behaving rats. Arch Ital Biol 2001; 139:313-28.

12. Xi MC, Morales FR, Chase MH. Effects on sleep and wakefulness of the injection of hypocretin-1 (orexin-A) into the laterodorsal tegmental nucleus of the cat. Brain Res 2001; 901:259-64.

13. Estabrooke IV, McCarthy MT, Ko E et al. Fos expression in orexin neurons varies with behavioral state. J Neurosci 2001; 21:1656-62.

14. Peyron C, Tighe DK, van den Pol AN et al. Neurons containing hypocretin (orexin) project to multiple neuronal systems. J Neurosci 1998; 18:9996-10015.

15. Zhang JH, Sampogna S, Morales FR et al. Orexin (hypocretin)-like immunoreactivity in the cat hypothalamus: a light and electron microscopic study. Sleep 2001; 24:67-76.

16. Lin L, Faraco J, Li R et al. The sleep disorder canine narcolepsy is caused by a mutation in the hypocretin (orexin) receptor 2 gene. Cell 1999; 98:365-76.

17. Espana RA, Baldo BA, Kelley AE et al. Wake-promoting and sleep-suppressing actions of hypocretin (orexin): basal forebrain sites of action. Neuroscience 2001; 106:699-715.

18. Hobson JA, McCarley RW, Wyzinski PW. Sleep cycle oscillation: reciprocal discharge by two brainstem neuronal groups. Science 1975; 189:55-8.

19. McGinty DJ, Harper RM. Dorsal raphe neurons: depression of firing during sleep in cats. Brain Res 1976; 101:569-75.

20. Trulson ME, Jacobs BL. Raphe unit activity in freely moving cats: correlation with level of behavioral arousal. Brain Res 1979; 163:135-50.

21. Aston-Jones G, Akaoka H, Charlety P et al. Serotonin selectively attenuates glutamate-evoked activation of noradrenergic locus coeruleus neurons. J Neurosci 1991; 11:760-9.

22. Portas CM, Bjorvatn B, Fagerland S et al. On-line detection of extracellular levels of serotonin in dorsal raphe nucleus and frontal cortex over the sleep/wake cycle in the freely moving rat. Neuroscience 1998; 83:807-14.

23. Blanco-Centurion CA, Salin-Pascual RJ. Extracellular serotonin levels in the medullary reticular formation during normal sleep and after REM sleep deprivation. Brain Res 2001; 923:128-36.

24. Strecker RE, Thakkar MM, Porkka-Heiskanen T et al. Behavioral state-related changes of extracellular serotonin concentration in the pedunculopontine tegmental nucleus: a microdialysis study in freely moving animals. Sleep Res Online 1999; 2:21-7.

25. Park SP, Lopez-Rodriguez F, Wilson CL et al. In vivo microdialysis measures of extracellular serotonin in the rat hippocampus during sleep-wakefulness. Brain Res 1999; 833:291-6.
26. Bjorvatn B, Ursin R. Changes in sleep and wakefulness following 5-HT1A ligands given systemically and locally in different brain regions. Rev Neurosci 1998; 9:265-73.
27. Sakai K, Crochet S. Role of dorsal raphe neurons in paradoxical sleep generation in the cat: no evidence for a serotonergic mechanism. Eur J Neurosci 2001; 13:103-12.
28. Jacobs BL. Single unit activity of locus coeruleus neurons in behaving animals. Prog Neurobiol 1986; 27:183-94.
29. Mileykovskiy BY, Kiyashchenko LI, Kodama T et al. Activation of pontine and medullary motor inhibitory regions reduces discharge in neurons located in the locus coeruleus and the anatomical equivalent of the midbrain locomotor region. J Neurosci 2000; 20:8551-8.
30. Sakai K, Crochet S. Serotonergic dorsal raphe neurons cease firing by disfacilitation during paradoxical sleep. Neuroreport 2000; 11:3237-41.
31. Andrade R. Regulation of membrane excitability in the central nervous system by serotonin receptor subtypes. Ann N Y Acad Sci 1998; 861:190-203.
32. Blier P, Pineyro G, el Mansari M et al. Role of somatodendritic 5-HT autoreceptors in modulating 5-HT neurotransmission. Ann N Y Acad Sci 1998; 861:204-16.
33. Graham D, Langer SZ. Advances in sodium-ion coupled biogenic amine transporters. Life Sci 1992; 51:631-45.
34. Qian Y, Melikian HE, Rye DB et al. Identification and characterization of anti-depressant-sensitive serotonin transporter proteins using site-specific antibodies. J Neurosci 1995; 15:1261-74.
35. Veasey SC, Fenik P. Pharmacological Characterization of Serotonergic Receptor Activity in the Hypoglossal Nucleus. Am J Respir Crit Care Med 2002.
36. Scruggs JL, Patel S, Bubser M et al. DOI-Induced activation of the cortex: dependence on 5-HT2A heteroceptors on thalamocortical glutamatergic neurons. J Neurosci 2000; 20:8846-52.
37. Jakala P, Sirvio J, Koivisto E et al. Modulation of rat neocortical high-voltage spindle activity by 5-HT1/5-HT2 receptor subtype specific drugs. Eur J Pharmacol 1995; 282:39-55.
38. Hedner T, Lundborg P. Serotoninergic development in the postnatal rat brain. J Neural Transm 1980; 49:257-79.
39. Cano J, Reinoso-Suarez F. Postnatal development in the serotonin content of brain visual structures. Brain Res 1982; 281:199-201.
40. Kirby ML, Mattio TG. Developmental changes in serotonin and 5-hydroxyindoleacetic acid concentrations and opiate receptor binding in rat spinal cord following neonatal 5,7-dihydroxytryptamine treatment. Dev Neurosci 1982; 5:394-402.
41. Uphouse LL, Bondy SC. The maturation of cortical serotonin binding sites. Brain Res 1981; 227:415-7.
42. Macho L, Kvetnansky R, Culman J et al. Neurotransmitter levels in the hypothalamus during postnatal development of rats. Exp Clin Endocrinol 1986; 88:142-50.
43. Rind HB, Russo AF, Whittemore SR. Developmental regulation of tryptophan hydroxylase messenger RNA expression and enzyme activity in the raphe and its target fields. Neuroscience 2000; 101:665-77.

44. Feng P, Ma Y, Vogel GW. Ontogeny of REM rebound in postnatal rats. Sleep 2001; 24:645-53.

45. Thakkar MM, Winston S, McCarley RW. A1 receptor and adenosinergic homeostatic regulation of sleep-wakefulness: effects of antisense to the A1 receptor in the cholinergic basal forebrain. J Neurosci 2003; 23:4278-87.

46. Portas CM, Thakkar M, Rainnie DG et al. Role of adenosine in behavioral state modulation: a microdialysis study in the freely moving cat. Neuroscience 1997; 79:225-35.

47. Thakkar MM, Delgiacco RA, Strecker RE et al. Adenosinergic inhibition of basal forebrain wakefulness-active neurons: a simultaneous unit recording and microdialysis study in freely behaving cats. Neuroscience 2003; 122:1107-13.

48. Thakkar MM, Winston S, McCarley RW. Orexin neurons of the hypothalamus express adenosine A1 receptors. Brain Res 2002; 944:190-4.

49. Marks GA, Shaffery JP, Speciale SG et al. Enhancement of rapid eye movement sleep in the rat by actions at A1 and A2a adenosine receptor subtypes with a differential sensitivity to atropine. Neuroscience 2003; 116:913-20.

50. Mackiewicz M, Nikonova EV, Zimmerman JE et al. Enzymes of adenosine metabolism in the brain: diurnal rhythm and the effect of sleep deprivation. J Neurochem 2003; 85:348-57.

51. Alanko L, Heiskanen S, Stenberg D et al. Adenosine kinase and 5′-nucleotidase activity after prolonged wakefulness in the cortex and the basal forebrain of rat. Neurochem Int 2003; 42:449-54.

52. Porkka-Heiskanen T, Alanko L, Kalinchuk A et al. Adenosine and sleep. Sleep Med Rev 2002; 6:321-32.

53. Aden U, Herlenius E, Tang LQ et al. Maternal caffeine intake has minor effects on adenosine receptor ontogeny in the rat brain. Pediatr Res 2000; 48:177-83.

54. Johansson B, Georgiev V, Fredholm BB. Distribution and postnatal ontogeny of adenosine A2A receptors in rat brain: comparison with dopamine receptors. Neuroscience 1997; 80:1187-207.

55. Jones SV, Barker JL, Goodman MB et al. Inositol trisphosphate mediates cloned muscarinic receptor-activated conductances in transfected mouse fibroblast A9 L cells. J Physiol 1990; 421:499-519.

56. Mitani A, Ito K, Hallanger AE, Wainer BH et al. Cholinergic projections from the laterodorsal and pedunculopontine tegmental nuclei to the pontine gigantocellular tegmental field in the cat. Brain Res 1988; 451:397-402.

57. Semba K, Reiner PB, Fibiger HC. Single cholinergic mesopontine tegmental neurons project to both the pontine reticular formation and the thalamus in the rat. Neuroscience 1990; 38:643-54.

58. Shiromani PJ, Armstrong DM, Berkowitz A et al. Distribution of choline acetyltransferase immunoreactive somata in the feline brainstem: implications for REM sleep generation. Sleep 1988; 11:1-16.

59. el Mansari M, Sakai K, Jouvet M. Unitary characteristics of presumptive cholinergic tegmental neurons during the sleep-waking cycle in freely moving cats. Exp Brain Res 1989; 76:519-29.

60. Kayama Y, Ohta M, Jodo E. Firing of `possibly' cholinergic neurons in the rat laterodorsal tegmental nucleus during sleep and wakefulness. Brain Res 1992; 569:210-20.

61. Buccafusco JJ, Wei J, Kraft KL. The effect of the acetylcholine transport blocker vesamicol on central cholinergic pressor neurons. Synapse 1991; 8:301-6.

62. Prior C, Marshall IG, Parsons SM. The pharmacology of vesamicol: an inhibitor of the vesicular acetylcholine transporter. Gen Pharmacol 1992; 23:1017-22.
63. Salin-Pascual RJ, Jimenez-Anguiano A. Vesamicol, an acetylcholine uptake blocker in presynaptic vesicles, suppresses rapid eye movement (REM) sleep in the rat. Psychopharmacology (Berl) 1995; 121:485-7.
64. Capece ML, Lydic R. cAMP and protein kinase A modulate cholinergic rapid eye movement sleep generation. Am J Physiol 1997; 273:R1430-40.
65. Spiegel AM, Shenker A, Weinstein LS. Receptor-effector coupling by G proteins: implications for normal and abnormal signal transduction. Endocr Rev 1992; 13:536-65.
66. Felder CC. Muscarinic acetylcholine receptors: signal transduction through multiple effectors. Faseb J 1995; 9:619-25.
67. Homma Y, Skinner RD, Garcia-Rill E. Effects of pedunculopontine nucleus (PPN) stimulation on caudal pontine reticular formation (PnC) neurons in vitro. J Neurophysiol 2002; 87:3033-47.
68. Roffwarg HP, Muzio JN, Dement WC. Development of the human sleep-dream cycle. Science 1966; 152:604-619.
69. Hrdina PD, Ghosh PK, Rastogi RB et al. Ontogenic pattern of dopamine, acetylcholine, and acetylcholinesterase in the brains of normal and hypothyroid rats. Can J Physiol Pharmacol 1975; 53:709-15.
70. Ninomiya Y, Koyama Y, Kayama Y. Postnatal development of choline acetyltransferase activity in the rat laterodorsal tegmental nucleus. Neurosci Lett 2001; 308:138-40.
71. Mechawar N, Watkins KC, Descarries L. Ultrastructural features of the acetylcholine innervation in the developing parietal cortex of rat. J Comp Neurol 2002; 443:250-8.
72. Wall SJ, Yasuda RP, Li M, Ciesla W et al. The ontogeny of m1-m5 muscarinic receptor subtypes in rat forebrain. Brain Res Dev Brain Res 1992; 66:181-5.
73. Buwalda B, de Groote L, Van der Zee EA et al. Immunocytochemical demonstration of developmental distribution of muscarinic acetylcholine receptors in rat parietal cortex. Brain Res Dev Brain Res 1995; 84:185-91.
74. van Huizen F, March D, Cynader MS et al. Muscarinic receptor characteristics and regulation in rat cerebral cortex: changes during development, aging and the oestrous cycle. Eur J Neurosci 1994; 6:237-43.
75. Balduini W, Cimino M, Reno F et al. Effects of postnatal or adult chronic acetylcholinesterase inhibition on muscarinic receptors, phosphoinositide turnover and m1 mRNA expression. Eur J Pharmacol 1993; 248:281-8.
76. Frank MG, Page J, Heller HC. The effects of REM sleep-inhibiting drugs in neonatal rats: evidence for a distinction between neonatal active sleep and REM sleep. Brain Res 1997; 778:64-72.
77. Taheri S, Zeitzer JM, Mignot E. The role of hypocretins (orexins) in sleep regulation and narcolepsy. Annu Rev Neurosci 2002; 25:283-313.
78. Gottesmann C. The neurochemistry of waking and sleeping mental activity: the disinhibition-dopamine hypothesis. Psychiatry Clin Neurosci 2002; 56: 345-54.
79. Lancel M, Langebartels A. Gamma-aminobutyric Acid(A) (GABA(A)) agonist 4,5,6, 7-tetrahydroisoxazolo[4,5-c]pyridin-3-ol persistently increases sleep maintenance and intensity during chronic administration to rats. J Pharmacol Exp Ther 2000; 293:1084-90.

80. Benington JH, Kodali SK, Heller HC. Stimulation of A1 adenosine receptors mimics the electroencephalographic effects of sleep deprivation. Brain Res 1995; 692:79-85.
81. Datta S. Evidence that REM sleep is controlled by the activation of brain stem pedunculopontine tegmental kainate receptor. J Neurophysiol 2002; 87:1790-8.
82. Vale W, Spiess J, Rivier C et al. Characterization of a 41-residue ovine hypothalamic peptide that stimulates secretion of corticotropin and beta-endorphin. Science 1981; 213:1394-7.
83. Laatikainen TJ. Corticotropin-releasing hormone and opioid peptides in reproduction and stress. Ann Med 1991; 23:489-96.
84. Rivier C, Rivest S. Effect of stress on the activity of the hypothalamic-pituitary-gonadal axis: peripheral and central mechanisms. Biol Reprod 1991; 45:523-32.
85. Webster EL, Grigoriadis DE, De Souza EB. Corticotropin-releasing factor receptors in the brain-pituitary-immune axis. In: McCubbin JA, Kaufmann PJ, Nemeroff CB, eds. Neuropeptides, and Systemic Disease. San Diego: Academic Press, 1991:233-260.
86. Dunn AJ, Berridge CW. Physiological and behavioral responses to corticotropin-releasing factor administration: is CRF a mediator of anxiety or stress responses? Brain Res Brain Res Rev 1990; 15:71-100.
87. Owens MJ, Nemeroff CB. Physiology and pharmacology of corticotropin-releasing factor. Pharmacol Rev 1991; 43:425-73.
88. Challis J, Sloboda D, Matthews S et al. Fetal hypothalamic-pituitary adrenal (HPA) development and activation as a determinant of the timing of birth, and of postnatal disease. Endocr Res 2000; 26:489-504.
89. Pihoker C, Cain ST, Nemeroff CB. Postnatal development of regional binding of corticotropin-releasing factor and adenylate cyclase activity in the rat brain. Prog Neuropsychopharmacol Biol Psychiatry 1992; 16:581-6.
90. Avishai-Eliner S, Yi SJ, Baram TZ. Developmental profile of messenger RNA for the corticotropin-releasing hormone receptor in the rat limbic system. Brain Res Dev Brain Res 1996; 91:159-63.
91. Dugovic C, Maccari S, Weibel L et al. High corticosterone levels in prenatally stressed rats predict persistent paradoxical sleep alterations. J Neurosci 1999; 19:8656-64.
92. Meerlo P, Koehl M, van der Borght K et al. Sleep restriction alters the hypothalamic-pituitary-adrenal response to stress. J Neuroendocrinol 2002; 14:397-402.
93. Hairston IS, Ruby NF, Brooke S et al. Sleep deprivation elevates plasma corticosterone levels in neonatal rats. Neurosci Lett 2001; 315:29-32.
94. Vogel G, Neill D, Kors D, Hagler M. REM sleep abnormalities in a new animal model of endogenous depression. Neurosci Biobehav Rev 1990; 14:77-83.
95. Prathiba J, Kumar KB, Karanth KS. Hyperactivity of hypothalamic pituitary axis in neonatal clomipramine model of depression. J Neural Transm 1998; 105:1335-9.
96. Arborelius L, Owens MJ, Plotsky PM et al. The role of corticotropin-releasing factor in depression and anxiety disorders. J Endocrinol 1999; 160:1-12.
97. Riemann D, Berger M. The effects of total sleep deprivation and subsequent treatment with clomipramine on depressive symptoms and sleep electroencephalography in patients with a major depressive disorder. Acta Psychiatr Scand 1990; 81:24-31.

98. Marrosu F, Gessa GL, Giagheddu M et al. Corticotropin-releasing factor (CRF) increases paradoxical sleep (PS) rebound in PS-deprived rats. Brain Res 1990; 515:315-8.
99. Gonzalez MM, Valatx JL. Involvement of stress in the sleep rebound mechanism induced by sleep deprivation in the rat: use of alpha-helical CRH (9-41). Behav Pharmacol 1998; 9:655-62.
100. Ohgushi A, Bungo T, Shimojo M et al. Relationships between feeding and locomotion behaviors after central administration of CRF in chicks. Physiol Behav 2001; 72:287-9.
101. Clements S, Schreck CB, Larsen DA et al. Central administration of corticotropin-releasing hormone stimulates locomotor activity in juvenile chinook salmon (Oncorhynchus tshawytscha). Gen Comp Endocrinol 2002; 125:319-27.
102. Opp MR. Corticotropin-releasing hormone involvement in stressor-induced alterations in sleep and in the regulation of waking. Adv Neuroimmunol 1995; 5:127-43.
103. Ordyan NE, Pivina SG, Rakitskaya VV et al. The neonatal glucocorticoid treatment-produced long-term changes of the pituitary-adrenal function and brain corticosteroid receptors in rats. Steroids 2001; 66:883-8.
104. Levine S. Primary social relationships influence the development of the hypothalamic–pituitary–adrenal axis in the rat. Physiol Behav 2001; 73:255-60.
105. Pariante CM, Miller AH. Glucocorticoid receptors in major depression: relevance to pathophysiology and treatment. Biol Psychiatry 2001; 49:391-404.

Basic Mechanisms of Circadian Rhythms and their Relation to the Sleep/Wake Cycle

MARTHA U. GILLETTE AND SABRA M. ABBOTT

1. The Temporal Spectrum of Life

Organisms exhibit cyclic variations in a variety of essential functions, including the sleep-wake cycle, hormonal regulation, and reproduction. A primary environmental signal regulating these functions is the daily alternation of darkness and light exerted by the rotation of the earth. Superimposed upon the daily light-dark cycle is a seasonal influence that modifies the relative durations of day and night over the course of a year. These environmental changes make it necessary for organisms to be able to modify their behavior so that they are active during times when the opportunity to acquire nutritional resources exceeds the risk of predation, and resting during times when the need for vigilance is minimized. Be they day-active or night-active, all organisms need a means of keeping time in a 24-hour world and adjusting to changes in day length or transition times that may occur.

At first, one might assume that these circadian rhythms are simply a reflection of the external day-night cycle. However, when all exogenous timing cues are removed, these rhythms persist. Every organism expresses an endogenous rhythm that varies slightly from 24 hours, making it *circadian*, or 'about a day.' Unperturbed, this circadian rhythm persists, as can be seen in Figure 1B. Due to the slight deviation of the period of the rhythm from 24 hours under these constant environmental conditions, the onset of activity drifts from its original position by a small but constant amount each day. In an aperiodic environment, the animal's activity eventually would become completely out of phase with the environment and then pass back into phase, proceeding at a regular interval with each circadian cycle. This

"

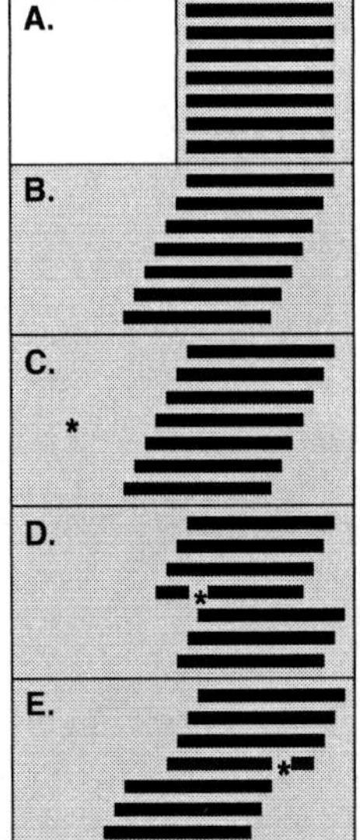

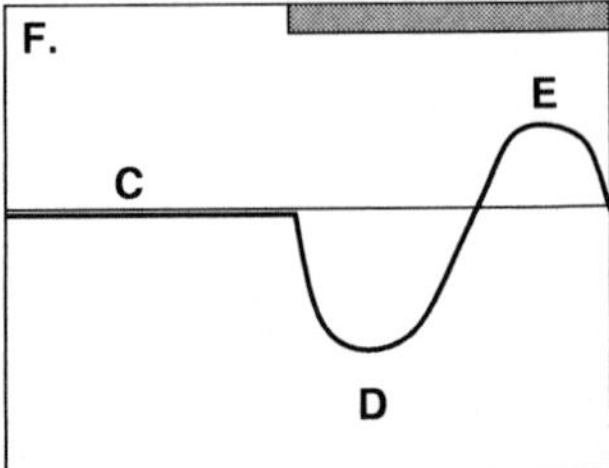

FIGURE 1. Organization of circadian rhythms of locomotor activity and its regulation by light. Panels A-E are schematic actograms from a nocturnal rodent. Dark bars in all figures indicate time when the animal is active, and grey areas indicate time when the animal is in darkness. Black bars indicate running-wheel activity. A. depicts an animal that has entrained to a light-dark cycle. Activity occurs with a regular 24-h cycle, and primarily during darkness. B. represents the same animal placed under constant conditions, showing a free-running period (*tau*) of less than 24 h. Activity begins slightly earlier each day. C. represents the effect on phase of activity onset in an animal presented with a light pulse (*) during the subjective daytime. No change is observed in the onset of activity on the following day. D. shows the response to a light pulse (*) during the subjective early evening. Activity onset is delayed on the following cycles. E. shows the response to a light pulse (*) during the subjective late evening. The onset of activity occurs earlier than predicted on the following cycles. F. is a *phase response curve* (PRC) for the response to light presented at various circadian phases for animals in constant darkness. The grey bar represents subjective night, and deflection above or below the line represents advances or delays, respectively. Times of treatment for the representative actograms (left panel) are represented by the corresponding letters on the graph.

demonstrates that the persistence of rhythmicity is not simply an after-effect of the previous environmental conditions, but rather that an endogenous rhythm is present.

This persistent endogenous rhythm is the expression of an internal circadian clock. The daily period of this biological clock differs from 24 hours by a small, but measurable amount. The circadian clock keeps time without external timing signals and, in turn, times circadian rhythms in brain and body functions.

In addition to timing behavioral rhythms, the circadian clock also can influence the levels of a variety of hormones, and many of these can be used as markers of the circadian clock. For example, plasma corticosterone shows a peak shortly before waking,[1] while plasma melatonin normally peaks shortly after sleep onset and appears to play a role in regulation of the sleep-wake cycle.[2] Circadian rhythms can also be observed in glucose metabolism; these are independent of daily food intake, although feeding and drinking behaviors are themselves circadian rhythms.[3]

Regardless of the marker of circadian phase that is being examined, however, one unique property of circadian clocks is their ability to be reset by environmental signals. The most prominent of these signals is light. When an animal in constant conditions is presented with a brief (~15 min) pulse of light, the response of the animal depends upon the time of light exposure (Figure 1C-E). During the subjective daytime, the inactive period for this animal, light has little effect. However, light exposure during subjective early evening, the beginning of the active period, delays the onset of activity on the following cycle, and subsequent rhythms continue from this new phase. On the other hand, light exposure during the subjective late night will advance the onset of activity on the following day. A plot called a *phase response curve* (PRC) can be constructed demonstrating the observed circadian response following treatments at various phases of the day/night cycle, as seen in Figure 1F.

When an animal is housed in conditions with exogenous time cues, such as a light-dark cycle, time is measured as *zeitgeber time* (ZT), because of the "time-giving' quality of light in the environment. ZT 0 is designated as the time of light onset. If an animal is housed in conditions without exogenous time cues and is relying on the endogenous clock for timing, we discuss time of day in terms of *circadian time* (CT), where CT 12 in nocturnal animals is designated as the time of onset of activity. The endogenous period of the animal is referred to as *tau* (τ). CT differs from ZT by a factor of $24/\tau$.

1.1. The Central Clock in the Suprachiasmatic Nucleus

Now that we have an idea of the many rhythms that are endogenously regulated and the fact that they can be controlled by environmental signals such as light, the question arises as to what might be controlling these rhythms. In mammals, circadian rhythms are regulated by a paired brain nucleus located at the base of the hypothalamus, directly above the optic chiasm, hence the name – the suprachiasmatic nucleus (SCN). Multiple experiments have demonstrated the role of the SCN as a central pacemaker for circadian rhythms. Lesioning studies found that selectively damaging the SCN disrupts rhythmicity in corticosterone levels, drinking activity, and wheel-running behavior.[4,5] This provided the initial evidence that the central pacemaker for the mammalian circadian clock lay within the SCN.

In later work, it was found that transplanting fetal SCN tissue into the third ventricle of animals in which the SCN had been lesioned could restore rhythmicity.[6] Furthermore, if fetal SCN tissue from a wild-type hamster was implanted into a hamster with a shortened free-running period, the new free-running period resembled that of the SCN donor rather than the host animal. This evidence suggested that not only was the SCN necessary for generating rhythms, but also that this rhythmicity was an intrinsic property of the SCN cells, which could drive the rhythms of the entire animal.[7]

In the rat, each SCN measures approximately 300 μm medial to lateral, 300 μm dorsal to ventral, and spans approximately 700 μm from rostral to caudal end. One SCN contains a total of approximately 8,000-10,000 cells, occupying a volume of approximately 0.036 mm^3.[8] Based on peptide localization, it is common to divide the rat SCN into a ventrolateral or 'core' region, and a dorsomedial or 'shell' region. The core neurons are smaller (~30 μm^2) and contain vasoactive intestinal peptide (VIP), somatostatin and gastrin-releasing peptide (GRP) colocalized with γ-amino butyric acid (GABA), while the shell neurons are larger (~45 μm^2) and contain arginine vasopressin (AVP).[9] There are topographic connections between the contralateral shells and the contralateral cores, as well as bidirectional innervation between the core and shell within each nucleus, but the core of one nucleus does not communicate directly with the shell of the contralateral nucleus.[8]

1.2.1. Outputs and the Circadian Timing System

The SCN exerts its influence on the rest of the body primarily by sending projections to the rest of the hypothalamus. Neurons from the core region project to the lateral region of the subparaventricular zone (sPVHz), the peri-suprachiasmatic area (PSCN) and the ventral tuberal area (VTU), all within the hypothalamus. The destinations of shell region projections include the medial preoptic area (MPOA), medial sPVHz and the dorsal medial hypothalamus (DMH), also all within the hypothalamus.[9] The DMH projections are particularly interesting, as many of these neurons appear to be projecting to neurons containing hypocretin/orexin, a peptide well known for its role in arousal.[10] The SCN also contains a minor set of efferents to the ventrolateral preoptic nucleus (VLPO), a region that produces prolonged reduction in sleep duration and amplitude when lesioned.[11] In addition, the SCN contains projections to the paraventricular nucleus (PVN) of the thalamus, as well as the intergeniculate leaflet (IGL). The targets of these efferents consist of endocrine neurons, autonomic neurons or intermediate neurons that potentially serve to integrate a number of hypothalamic signals.[12] Overall, the SCN appears to be uniquely situated within a network that allows it to interact closely with the regions controlling the sleep and arousal states of the animal.

One of the major outputs of the SCN appears to be an inhibitory signal for activity. Two recently discovered candidate factors for communicating such signals include transforming growth factor-α (TGF-α) and prokineticin 2 (PK2). Under normal conditions, TGF-α is expressed rhythmically in the SCN, and when infused continuously into the cerebral ventricles, it inhibits locomotor activity. Conversely, mice lacking the epidermal growth factor (EGF) receptor, making them unable to respond to TGF-α, show an excessive amount of daytime activity.[13] PK2 is also expressed rhythmically, and can inhibit locomotor activity when infused continuously.[14] This suggests a role for SCN outputs in promoting an inactive state that would be permissive for sleep.

1.2.2. Inputs and Gating by the Circadian Clock

The circadian clock in the SCN also is capable of being reset by changes in behavioral or environmental state. The changes are conveyed to the SCN by projections from a variety of different brain regions.

One of the most well-studied inputs to the SCN comes from subpopulation of retinal ganglion cells that form the retinohypothalamic tract (RHT). Lesions of the SCN disrupt the development of these neurons,[15] and disruption of the RHT results in an inability to respond to light as a circadian resetting/entraining signal.[16,17] Recent work has found that many of the retinal ganglion cells that comprise the RHT contain a photopigment, melanopsin.[18]

These melanopsin-containing cells show a response to light that parallels circadian responses to light.[19] Additionally, the terminals of the melanopsin-positive retinal ganglion cells collocalize both glutamate and pituitary adenylate cyclase-activating polypeptide (PACAP),[20] the putative neurotransmitters of the RHT.

The geniculohypothalamic tract (GHT) also appears to play a role in transmitting light information to the clock. The GHT projects from the intergeniculate leaflet (IGL) to the SCN, and releases neuropeptide Y (NPY). NPY is believed to be involved in behavior-induced phase shifts during the daytime, but also appears to be able to modulate light-induced phase shifts.[21,22] However, while the GHT pathway can transmit photic signals, disruption of this pathway does not prevent entrainment to light.[23]

The SCN also receives serotonergic input, primarily from the median raphe. Activation of the median raphe results in an increase in serotonin (5-HT) release at the SCN.[24] 5-HT release also shows a strong circadian pattern in the SCN, with 5-HT release peaking at CT 14, and 5-hydroxyindole acetic acid (5-HIAA), the major metabolite of 5-HT peaking at CT 16.[25]

Cholinergic projections to the SCN originate both in the basal forebrain and in the laterodorsal tegmental (LDT), pedunculopontine tegmental (PPT) and parabigeminal nuclei (PBg) in the brainstem.[26] The PBg is considered a satellite region of the superior colliculus, which appears to play a role in generating target-location information as part of saccadic eye-movements.[27]

The substantia innominata (SI) within the nucleus basalis magnocellularis (NBM) in the basal forebrain contributes to arousal and focused attention,[28] while the LDT and PPT both are important for regulating the sleep-wake cycle.[29] This would suggest that the cholinergic input to the SCN is providing a signal regarding the sleep-wake state of the animal, and may provide a link between the sleep-wake cycle and circadian rhythms.

2. Circadian Clock Regulators

With the anatomy in place, we now can begin looking at how the SCN responds to various inputs. One model that works particularly well for studying the signaling mechanisms that regulate circadian rhythms is to examine the electrophysiological properties of a hypothalamic brain slice preparation containing the SCN. Coronal slices of hypothalamus containing the SCN can be maintained at the interface of a warm, moist, high oxygen atmosphere and a solution containing minimal salts and glucose for at least 3 days, and will show a peak in neuronal firing activity that occurs at approximately the same time every day (Figure 2).[30,31]

Using this preparation, the pattern and rate of extracellular activity of individual neurons within the SCN can be recorded. When data from individual neurons are averaged over circadian time of recording, a peak in average firing rate of the neuronal ensemble occurs at approximately CT 7.

Using this as a baseline marker of circadian time in the brain slice, experiments can then be performed where neurotransmitters or candidate elements in a signaling pathway can be applied to the slice at different times of the day and the resultant change in time of peak can be measured to determine the effect on SCN circadian rhythms. This generates the PRC of the isolated SCN for each substance.

This system is very versatile, and it can be used to examine the effect of different agonists on resetting circadian rhythms, as well as whether antagonists of putative downstream members of the signaling pathway can block the normally observed effects. While the SCN appears to be involved in sending signals as to time of day to the rest of the body, the fact that the isolated preparation expresses differing sensitivities to resetting signals that are dependent on circadian time demonstrates that the SCN itself is intrinsically sensitive to time of day. We will now examine the various resetting pathways that are present in the SCN, focusing on the different pathways that can regulate the clock during the different times of the day.

2.1. Daytime

A number of signaling molecules appear to be important in resetting circadian rhythms during the daytime, including 5-HT, PACAP, NPY and GABA. The majority of these experiments have been performed in nocturnal rodents,

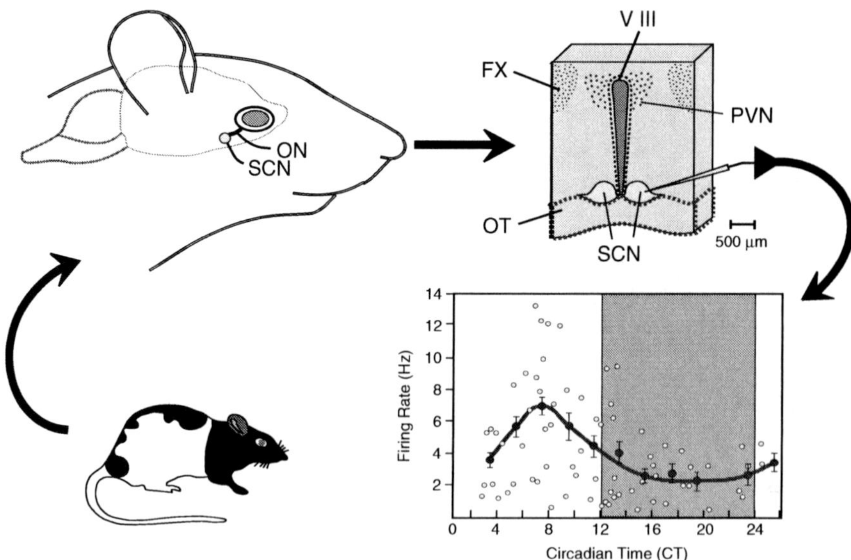

FIGURE 2. A diagrammatical representation of the hypothalamic brain slice preparation. A coronal slice of the hypothalamus is removed from the Long Evans BluGill rat, placed in culture, and extracellular neuronal activity is recorded. A representative plot of individual neuron activity along with successive 2-h means are plotted, as indicated by the line. This preparation can be maintained for at least 3 days *in vitro*, producing the characteristic peak in firing rate at ~CT 7 each day. ON/OT, optic nerve/tract; VIII, third cerebral ventricle; open circles, single unit activity; closed circles, 2-h mean ensemble activity; shading CT 12-24, subjective night.

so daytime is defined as the time in which the lights are on and/or the animal is inactive. As a result, the functional context of this regulation seems to be tied to non-photic (non-light stimulated) resetting. Non-photic resetting covers a wide variety of resetting phenomenon, including sleep deprivation, activity associated with exposure to a novel wheel and even cage changes. The unifying factor in non-photic resetting is that it involves arousal during a time when the animal would normally be inactive. Many of the neurotransmitters that have been linked experimentally with this daytime resetting also appear to modulate nighttime resetting. Daytime effects will be discussed here, while the modulation of nighttime effects will be covered in the next section.

While 5-HT is believed to play a role in activity-induced or non-photic phase shifts during the day, there is some question about whether this form of phase shifting is entirely due to 5-HT. If serotonergic agonists are applied to the hypothalamic brain slice during the daytime, the peak in electrical firing activity advances, but no change in time of peak firing rate is seen if the agonists are applied during the night (Figure 3).[32] Similar results are seen

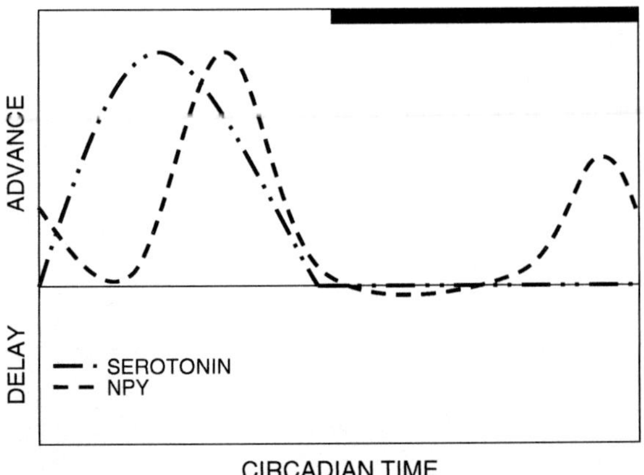

FIGURE 3. **Summary of daytime circadian resetting signals.** This figure depicts the *in vitro* response of the SCN to serotonin and neuropeptide Y, which act on the circadian clock during daytime. The circadian time of treatment is represented on the X-axis, while the magnitude of advance or delay is represented on the Y-axis. Time in which the animal would be in darkness is represented by a black bar. As can be seen, each neurotransmitter has a unique resetting profile, both in time and direction of response throughout the day. Replotted from M. Medanic and M. U. Gillette, Serotonin regulates the phase of the rat suprachiasmatic circadian pacemaker *in vitro* only during the subjective day. *J. Physiol.* **450**, 629-642 (1982) and M. Medanic and M. U. Gillette, Suprachiasmatic circadian pacemaker of rat shows two windows of sensitivity to neuropeptide Y *in vitro*. *Brain Res.* **620**:281-286 (1993).

in vivo if the dorsal or medial raphe is stimulated,[33] a paradigm that has been shown by microdialysis to increase 5-HT release at the level of the SCN.[34] During forced wheel running or sleep deprivation during the daytime, there is an increase in 5-HT release at the SCN.[35,36] This suggests a link between 5-HT and non-photic phase shifting, but evidence also exists to complicate this assertion. If 85-95% of the serotonin is depleted from the raphe projections to the SCN, animals still are capable of phase shifting in response to daytime forced activity.[37] In addition, these activity-induced phase shifts are not significantly attenuated following injection of serotonergic antagonists.[38] These data suggest that, while 5-HT may play a role in non-photic resetting, the full resetting response depends on modulation by additional neurotransmitters.

PACAP appears to play a dual role in the SCN, producing effects both during the daytime and at night, when it acts in conjunction with glutamate. PACAP is released from the RHT, and studies have found that it is colocalized with glutamate.[39] Examining levels of PACAP throughout the 24-h cycle revealed that PACAP exhibits a significant oscillation in the SCN, but not in other brain regions, and is lower during the light period than the dark period.[40] If PACAP is applied to the brain slice at different times of day, micromolar quantities will cause an advance in neuronal firing activity during the daytime, but have relatively little effect during the night.[20] However, when PACAP is injected into the SCN of the hamster between CT 4-8, transient phase advances in wheel-running activity are seen during the first day after treatment, but the long-term effects of a PACAP injection appear to produce a delay in wheel-running activity.[41] This suggests that, while PACAP has an effect on circadian rhythms during the daytime, further work is needed to determine the precise nature of this signal.

NPY also appears to play a dual role, and is effective in resetting the circadian clock both during the daytime and at night. NPY is released from the GHT, the projection from the IGL of the thalamus to the SCN. Studies have examined the effects of either injecting NPY into the SCN region of the intact animal and monitoring wheel running behavior[42,43] or applying NPY directly to the hypothalamic brain slice and examining the peak in neuronal firing activity.[21] In both cases it was found that when NPY was applied during the daytime, it induced an advance in the circadian parameter of interest (Figure 3). Additional studies implanted stimulating electrodes into the IGL, which would presumably release NPY at the SCN when activated, and examined the effects of IGL stimulation on circadian wheel-running behavior.[44] These stimulations also produced advances in circadian phase when given during the daytime. Interestingly, it has been found that exposing an animal to light[45] or applying glutamate to the brain slice[46] both were capable of blocking the response to daytime application of NPY. Co-administration of the $GABA_A$ antagonist, bicuculline, is capable of inhibiting the effects of NPY *in vivo*,[47] suggesting that the NPY signal is linked to GABAergic signaling.

One factor that daytime signaling pathways hold in common is that they all appear to be mediated by cyclic adenosine monophosphate (cAMP). In the hypothalamic brain slice, cAMP analogs applied during the daytime induce phase advances in the circadian clock, while at night they have little effect.[31,48] In addition, endogenous cAMP is high during late day, and late night,[49] suggesting a role for cAMP in the transition periods between day and night, and that induction of a rise in cAMP may move the clock to a time-point where cAMP is normally high.

2.2. Dawn and Dusk

The primary resetting signal associated with dawn and dusk is melatonin. This hormone of darkness is produced at night in the absence of light, providing a means by which the animal can measure night-length.[50] Photoperiod is an important measure for animals, such as the hamster, that are seasonally reproductive.

Melatonin is produced by the pineal, and in many vertebrates the pineal is actually the primary regulator of circadian rhythms, rather than the SCN. However, in mammals, removal of the pineal does not significantly disrupt circadian rhythms.[51]

Despite the fact that a pineal is not necessary for maintenance of mammalian circadian rhythms, it is possible to entrain free-running rats with daily injections of melatonin. Entrainment appears to work best if the melatonin injections are timed to occur shortly before the onset of the animal's active period. In addition, lesioning the SCN, but not the pineal, abolishes the ability of a rat to entrain to melatonin injections.[52]

Evidence that melatonin can reset circadian rhythms led to a number of studies looking at the direct effect of melatonin on the SCN. Using either 2-deoxy-glucose (2-DG) or neuronal activity as a marker of SCN activity, it was found that melatonin decreased both 2-DG uptake and neuronal firing activity in the rat most significantly when applied right before subjective dusk.[53,54] Melatonin-induced suppression of neuronal firing activity at the end of the day also is seen in the hamster.[55]

The SCN shows a similar window of melatonin sensitivity in terms of resetting the circadian clock. Using the hypothalamic brain slice, it was found that melatonin applied at either dawn or dusk advanced the peak in neuronal firing, but when applied at other times of day no effect was observed (Figure 4).[56,57] This resetting pattern mimicked that seen in response to activation of protein kinase C (PKC), and was blocked by inhibitors of PKC, suggesting that PKC was a downstream component of this resetting pathway.[57]

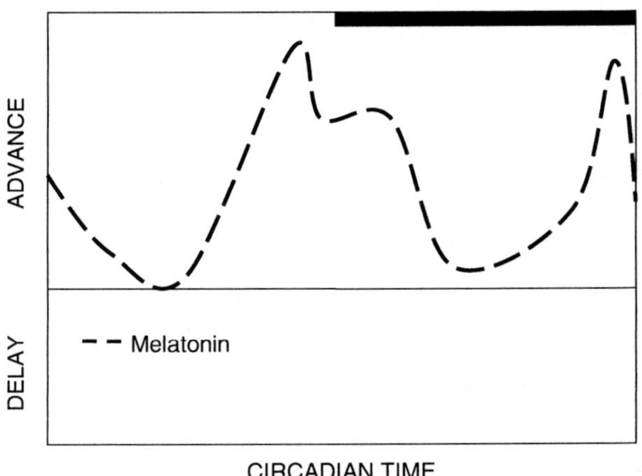

FIGURE 4. **Summary of dawn/dusk circadian resetting signals**. This figure depicts the *in vitro* response of the SCN to melatonin. The circadian time of treatment is represented on the X-axis, while the magnitude of advance or delay is represented on the Y-axis. Time in which the animal would be in darkness is represented by a black bar. As can be seen, this neurohormone has a unique resetting profile, both in time and direction of response at the subjective day- to night and night-to-day transitions. Replotted from A. J. McArthur, A. E. Hunt, and M. U. Gillette, Melatonin action and signal transduction in the rat suprachiasmatic circadian clock: activation of protein kinase C at dusk and dawn. *Endocrinology* **138**(2), 627-634 (1997).

In addition, this resetting could be inhibited with antagonists that were specific for the MT-2 type melatonin receptor.[58]

2.3. Nighttime

In the nighttime domain there are two key players, glutamate and ACh, as well as a number of substances that modulate these signals. As we have discussed previously, considerable evidence supports glutamate as the primary neurochemical signal transmitting photic stimuli from the retina to the SCN, while the functional context of the cholinergic resetting signal is still unknown.

The glutamate signaling pathway is similar to many of the pathways that have already been discussed, in that it will reset the circadian clock at a particular time of day, and in a specific direction. However, what is interesting about the glutamate signaling pathway is that it can either advance or delay the clock, depending on what time of day the signal is presented (Figure 5).[59]

This pathway has been demonstrated both *in vitro* and *in vivo* to be mediated through an N-methyl-D-aspartate (NMDA) receptor-mediated rise in

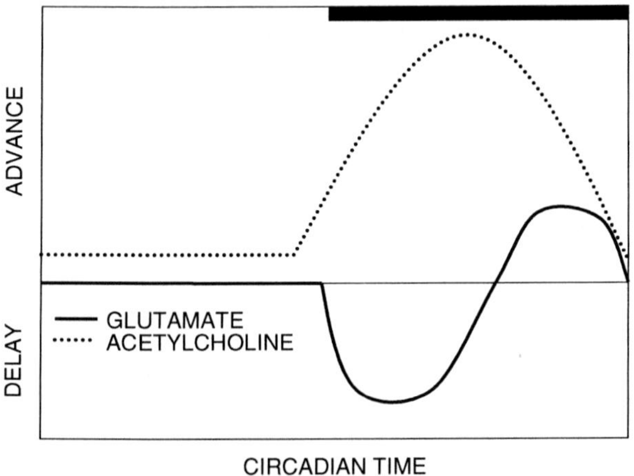

FIGURE 5. **Summary of nighttime circadian resetting signals**. This figure depicts the *in vitro* response of the SCN to glutamate and accetylcholine, neurotransmitters effective primarily at night. The circadian time of treatment is represented on the X-axis, while the magnitude of advance or delay is represented on the Y-axis. Time in which the animal would be in darkness is represented by a black bar. As can be seen, each neurotransmitter has a unique resetting profile, both in time and direction of response throughout the night. Replotted from C. Liu and M. U. Gillette, Cholinergic regulation of the suprachiasmatic nucleus circadian rhythm via a muscarinic mechanism at night. *J. Neurosci.* **16**(2), 744-751 (1996) and J. M. Ding, D. Chen, E. T. Weber, et al., Resetting the biological clock: mediation of nocturnal circadian shifts by glutamate and NO. *Science* **266**, 1713-1717 (1994).

intracellular calcium, followed by nitric oxide synthase (NOS) activation and resultant production of nitric oxide (NO).[60-64] Beyond this point, the early- and late-night pathways diverge.

During the early night, glutamate induces delays in the circadian clock. Downstream of NO, the next step in this pathway appears to be ryanodine receptor (RyR)-mediated calcium release. In the brain slice, pharmacological activation of the RyR induces a phase delay in neuronal activity during the early night, but not at any other time of day. This effect can be blocked both *in vitro* and *in vivo* with dantrolene, an RyR inhibitor.[65]

Glutamate exposure during the late night, however, advances the circadian clock. During the late night, NO appears to activate a cyclic guanosine monophosphate/protein kinase G (cGMP/PKG) signaling cascade followed by cAMP response element binding protein (CREB)-activated transcription. Inhibiting PKG in the brain slice blocks light-induced advances in neuronal activity.[65] Phospho-CREB (P-CREB) is induced in the SCN following exposure to light at night, while antagonists of putative elements of the upstream signaling pathway, such as NMDA or NOS, inhibit the induction of P-CREB.[66] Furthermore, adding an excess oligodeoxynucleotide decoy sequence that inhibits cAMP response element (CRE)-mediated transcription blocks the phase advance induced by glutamate both *in vitro* and *in vivo*.[67]

While glutamate alone is capable of resetting circadian rhythms, there are many substances that modulate this resetting. These can be divided into two categories: those that decrease the phase resetting effect of glutamate during both the early and late night, and those that have differing effects on glutamate-induced phase shifts, depending on what time of night they are applied. In the first category, both NPY and GABA, which are active in daytime phase shifting, decrease the phase-resetting effect of light or glutamate when applied during the night.[68,69] Glutamate applied during the daytime also seems to be capable of inhibiting GABA- and serotonin-induced resetting.[69,70]

Modulatory effects of serotonin and PACAP on glutamate-induced phase resetting are more complicated. If animals are depleted of serotonin, they show increased phase delays in response to light.[71] On the other hand, co-application of a PACAP antagonist, either *in vitro* or *in vivo*, decreases the phase delay seen in early night, and increases the late-night phase advance.[72,73]

When PACAP is administered in conjunction with glutamate, it increases the early-night phase delays, but decreases the late-night phase advances. This is similar to the effects seen following application of cAMP analogs to the hypothalamic brain slice, suggesting that the effects of PACAP may be mediated by a cAMP pathway.[74]

The role of ACh in resetting circadian rhythms is rather complicated, and has been unclear, because the effect varies with the site of application. The first evidence that ACh might play a role in resetting the circadian clock came in 1979, when Zatz and Brownstein examined whether pharmacological manipulation of the SCN could affect serotonin N-acetyltransferase (SNAT) activity in the pineal.[75]

They found that injections of carbachol into the lateral ventricle of Sprague-Dawley rats at CT 15 caused phase delays in SNAT activity that were similar to, but not as large as, the phase delays produced by light.[75] Carbachol injections into the lateral ventricle were also later repeated in mice[76] and hamsters,[77] where it was found that administration of carbachol during the early night caused phase delays, while late-night administration caused phase advances. This pattern of sensitivity and response is similar to that induced by light exposure. Further evidence for the involvement of ACh in the light response came from studies looking at ACh levels in the rat SCN using a radioimmunoassay (RIA).[78] Using this technique, no significant oscillation in ACh levels was found under constant conditions, but light pulses administered at CT 14 increased ACh levels in the SCN. However, no work was done to determine whether this increase was simply a response to exposure to light or if there was actually a circadian pattern to the light-stimulated ACh release. The implication of these studies was that ACh could be the primary neurotransmitter providing the signal of light to the clock.

However, significant evidence began to emerge indicating that ACh was not likely to be the primary signal of light. First of all, whereas it had previously been determined that the RHT transmitted the signal of light from the eye to the SCN, choline acetyltransferase (ChAT) was not present in this projection,[79] making it anatomically unlikely that ACh was the primary neurotransmitter involved in this signal. This evidence can be reconsidered, however, as recent studies have found an alternative splice variant of ChAT present in retinal ganglion cells that was not picked up using previous antibodies.[80] Experiments have not yet been published addressing whether this alternative form of ChAT is present in the RHT.

Additional evidence against ACh being the signal of light came from experiments that found injecting hemicholinium, which significantly depletes ACh stores, did not block the ability of the animal to phase shift in response to light.[81] There is also evidence that injecting NMDA receptor antagonists could block carbachol-induced phase shifts, suggesting that although ACh may play a role in the light response, it is not the direct cause of the observed phase shifts.[82] Finally, Liu and Gillette using extracellular recording *in vitro*, found that microdrop applications of carbachol directly to the SCN caused only phase advances, regardless of whether it was applied early or late in subjective nighttime (Figure 5).[83]

In an attempt to explain these contradicting data, the hypothesis that our lab has developed is that the dual response pattern of the SCN to cholinergic stimulation is a result of the location of application. Note that in the initial *in vivo* studies, carbachol was injected into the lateral or third ventricle, where the drug could have a diffuse effect, while in the *in vitro* studies, carbachol was applied in microdrops directly to the SCN. Emerging evidence suggests that, if the *in vivo* experiments are performed by injecting carbachol directly into the SCN rather than into the ventricle, a similar phase-response pattern is seen as resulted from the *in vitro* experiments using microdrop applications.[84] Furthermore, preliminary studies suggest that mice lacking the M_1-type muscarinic receptor (M_1AChR) do not respond to these carbachol injections,[85] but exhibit normal responses to light.[86] Together, the evidence indicates that, while ACh may contribute to the light response, as seen in earlier experiments, the direct cholinergic signal to the SCN is providing an entraining signal that is different than the signal of light, and this response is mediated by an M_1AChR. Based on the anatomical studies looking at cholinergic projections to the SCN that originate in the LDT and PPT, as well as the SI, the current hypothesis is that this cholinergic signal is involved in tying the sleep-wake and circadian cycles together.

3. Circadian and Sleep/Wake Cycle Interactions

3.1. *The Two-Process Mode of Sleep and Circadian Interactions*

The circadian and sleep-wake cycles have been known to interact. This interaction was modeled by Borbely as regulated by two processes: process S and process C. Process S is sleep-dependent; its influence becomes greater the longer one has been without sleep. Process C is circadian-dependent; it varies with circadian phase of the day or night.[87] Process S uses as its marker electroencephalographic (EEG) slow-wave activity, which may be promoted by accumulation of adenosine. During the time awake, the rising homeostatic sleep pressure from process S is compensated for by a decline in circadian process C sleep pressure.[88] The converse is true at night. The interaction between the two processes accounts for both the consolidation of sleep and wake under normal conditions, as well as the fact that there is a rhythmic variation in sleep propensity throughout the duration of sleep deprivation. The interaction between these two processes can be further demonstrated by looking at what happens to one process in the absence of the other, as will be described in the next section.

3.2. *Sleep and Circadian Interactions*

Initial studies evaluating the interaction between the circadian and sleep-wake cycles simply determined the effect of SCN lesions on circadian rhythms. Bilateral SCN lesions in rat were found to eliminate the circadian rhythm of the sleep-wake cycle.[89] While total time spent in sleep remained the same, the occurrence of sleep was now randomly distributed between the light and dark phases, rather than occurring primarily during daytime.[89] However, the ultradian rhythm of slow wave sleep (SWS) cycles was still present.[90] Later studies found a slight increase in total SWS and slight decrease in total rapid eye movement (REM) sleep in SCN-lesioned rats.[91]

The next step was to look at whether circadian inputs influenced the homeostatic response, by looking at how SCN-lesioned animals responded to periods of sleep deprivation. Initially, these animals were found to still show a homeostatic recovery response, but on further study, differences were found in the speed of recovery under different conditions.[92] Animals that contained an intact SCN and were placed under constant lighting conditions showed the fastest recovery from a period of sleep-deprivation. Animals with SCN-lesions or on a regular light-dark cycle required more time than controls to recover the lost sleep, suggesting that an intact circadian system, without the influence of outside time cues such as light, provides the most efficient means of recovering lost sleep.[93] Although squirrel monkeys exhibited slightly different patterns of response than rodents to SCN lesions followed by sleep deprivation, in that their total sleep time increases, the total time spent in NREM and REM sleep did not increase.[94]

Feedback from the sleep-wake cycle to the circadian clock was evaluated by looking at the effect of sleep deprivation and arousal on circadian rhythms. If hamsters are sleep-deprived during the latter part of their active period, either through exposure to a novel wheel, or by gentle handling, they exhibit a dramatic advance in onset of wheel-running activity that remains stable as long as they remain in constant conditions.[36,95] This sleep-deprivation paradigm has also been used in mice to demonstrate that the phase delays observed in response to an early night light pulse are actually decreased if the animal has been sleep-deprived prior to the light pulse.[96]

A recent study provided a more functional examination of this feedback loop by examining the correlation between sleep-wake state and neuronal firing rate in the SCN. Early work found that SCN cells undergo changes in firing rate that relate to arousal state of the animal.[97] More recent studies conducted a detailed analysis of SCN neuronal activity during different sleep stages, and found that the SCN firing rate exhibited the previously observed diurnal pattern, but superimposed on it was a change in firing rate dependent on sleep state of the animal. During wake or REM sleep, SCN firing rates were much higher than during NREM sleep.[98] As discussed earlier in this chapter, there are circuits in place for most sleep centers in the brain to feed back to the SCN, and these studies provide evidence that this feedback actually has functional consequences.

4. Conclusion

Circadian rhythms are an integral part of an organism's existence. In mammals, these rhythms are regulated by a circadian clock in the SCN, which is uniquely situated to receive a variety of inputs regarding the animal's environment and behavioral state. Of particular relevance for this chapter is the fact that the circadian system plays an important role in regulating when an animal is active and inactive. There is evidence that the SCN sends efferents to parts of the brain involved in arousal, and that in the absence of a circadian clock, consolidation of the sleep-wake cycle disappears. In addition, recent studies suggest that the sleep-wake state of the animal is capable of feeding back to the circadian clock. Sleep deprivation of hamsters at the appropriate time of day can reset the phase of the circadian clock, and changes in sleep-wake state appear to influence the firing rate of neurons within the SCN. Overall, evidence indicates that the circadian and sleep-wake cycles are intricately intertwined, and that study of one necessitates the study and consideration of the other.

5. Acknowledgments. Thanks to past and present members of the Gillette lab for contributions to studies on which this chapter is based, and to S. C. Baker for manuscript preparation. Supported by Public Health Service Grants NS22155, NS35859 and HL67007 (MUG) and NIH Training Grant GM07143 (SMA). Any opinions, findings, and conclusions or recommendations expressed in this publication are those of the author and do not necessarily reflect the views of the National Institutes of Neurological Diseases & Stroke, Heart, Lung, and Blood, or General Medicine.

6. References

1. R. M. Buijs, C. G. van Eden, V. D. Goncharuk and A. Kalsbeek, The biological clock tunes the organs of the body: timing by hormones and the autonomic nervous system. *J. Endocrinol.* 177, 17-26 (2003).
2. C. Cajochen, K. Krauchi and A. Wirz-Justice, Role of melatonin in the regulation of human circadian rhythms and sleep. *J. Neuroendocrinol.* 15, 432-437 (2003).
3. S. E. la Fleur, Daily rhythms in glucose metabolism: suprachiasmatic nucleus output to peripheral tissue. *J. Neuroendocrinol.* 15, 315-322 (2003).
4. F. K. Stephan and I. Zucker, Circadian rhythms in drinking behavior and locomotor activity of rats are eliminated by hypothalamic leions. *Proc. Natl. Acad. Sci. U.S.A.* 69, 1583-1586 (1972).
5. R. Y. Moore and V. B. Eichler, Loss of a circadian adrenal corticosterone rhythm following suprachiasmatic lesions in the rat. *Brain Res.* 42, 201-206 (1972).
6. R. Drucker-Colin, R. Aguilar-Roblero, F. Garcia-Hernandez, F. Fernandez-Cancino and F. B. Rattoni, Fetal suprachiasmatic nucleus transplants: diurnal rhythm recovery of lesioned rats. *Brain Res.* 311, 353-357 (1984).
7. M. R. Ralph, R. G. Foster, F. C. Davis and M. Menaker, Transplanted suprachiasmatic nucleus determines circadian period. *Science* 247, 975-978 (1990).
8. R. Y. Moore, J. C. Speh and R. K. Leak, Suprachiasmatic nucleus organization. *Cell Tissue Res.* 309, 89-98 (2002).
9. R. K. Leak and R. Y. Moore, Topographic organization of suprachiasmatic nucleus projection neurons. *J. Compara. Neurol.* 433, 312-334 (2001).

10. E. E. Abrahamson, R. K. Leak and R. Y. Moore, The suprachiasmatic nucleus projects to posterior hypothalamic arousal systems. *Neuroreport* 12(2), 435-440 (2001).
11. T. C. Chou, A. A. Bjorkum, S. E. Gaus, J. Lu, T. E. Scammell and C. B. Saper, Afferents to the ventrolateral preoptic nucleus. *J. Neurosci.* 22(3), 977-990 (2002).
12. A. Kalsbeek and R. M. Buijs, Output pathways of the mammalian suprachiasmatic nucleus: coding circadian time by transmitter selection and specific targeting. *Cell Tissue Res.* 309, 109-118 (2002).
13. A. Kramer, F.-C. Yang, P. Snodgrass, et al., Regulation of daily locomotor activity and sleep by hypothalamic EGF receptor signaling. *Science* 294, 2511-2515 (2001).
14. M. Y. Cheng, C. M. Bullock, C. Li, et al., Prokineticin 2 transmits the behavioural circadian rhythm of the suprachiasmatic nucleus. *Nature* 417, 405-410 (2002).
15. S. Mosko and R. Moore, Retinohypothalamic tract development: alteration by suprachiasmatic lesions in the neonatal rat. *Brain Res.* 164, 1-15 (1979).
16. B. Rusak, Neural mechanisms for entrainment and generation of mammalian circadian rhythms. *Fed. Proc.* 38(12), 2589-2595 (1979).
17. R. F. Johnson, R. Y. Moore and L. P. Morin, Loss of entrainment and anatomical plasticity after lesions of the hamster retinohypothalamic tract. *Brain Res.* 460, 297-313 (1988).
18. S. Hattar, R. J. Lucas, N. Mrosovsky, et al., Melanopsin and rod-cone photoreceptive systems account for all major accessory visual functions in mice. *Nature* 424, 76-81 (2003).
19. D. M. Berson, F. A. Dunn and M. Takao, Phototransduction by retinal ganglion cells that set the circadian clock. *Science* 295, 1070-1073 (2002).
20. J. Hannibal, J. M. Ding, D. Chen, et al., Pituitary adenylate cyclase-activating peptide (PACAP) in the retinohypothalamic tract: a potential daytime regulator of the biological clock. *J. Neurosci.* 17, 2637-2644 (1997).
21. M. Medanic and M. U. Gillette, Suprachiasmatic circadian pacemaker of rat shows two windows of sensitivity to neuropeptide Y *in vitro. Brain Res.* 620, 281-286 (1993).
22. P. C. Yannielli and M. E. Harrington, The neuropeptide Y Y5 receptor mediates the blockade of "photic-like" NMDA-induced phase shifts in the golden hamster. *J. Neurosci.* 21(14), 5367-5373 (2001).
23. V. Reghunandanan, R. Reghunandanan and P. I. Singh, Neurotransmitters of the suprachiasmatic nucleus: role in the regulation of circadian rhythms. *Prog. Neurobiol.* 41, 647-655 (1993).
24. A. N. Van den Pol and K. L. Tsujimoto, Neurotransmitters of the hypothalamic suprachiasmatic nucleus: immunocytochemical analysis of 25 neuronal antigens. *Neuroscience* 15, 1049-1086 (1985).
25. S. Barassin, S. Raison, M. Saboureau, et al., Circadian tryptophan hydroxylase levels and serotonin release in the suprachiasmatic nucleus of the rat. *Eur. J. Neurosci.* 15, 833-840 (2002).
26. K. G. Bina, B. Rusak and K. Semba, Localization of cholinergic neurons in the forebrain and brainstem that project to the suprachiasmatic nucleus of the hypothalamus in rat. *J. Compara. Neurol.* 335, 295-307 (1993).
27. H. Cui and J. G. Malpeli, Activity in the parabigeminal nucleus during eye movements directed at moving and stationary targets. *J. Neurophysiol.* 89(6), 3128-3142 (2003).

28. K. Semba, Multiple output pathways of the basal forebrain: organization, chemical heterogeneity, and roles in vigilance. *Behav. Brain Res.* 115, 117-141 (2000).
29. S. Deurveilher and E. Hennevin, Lesions of the pedunculopontine tegmental nucleus reduce paradoxical sleep (PS) propensity: evidence from a short-term PS deprivation study in rats. *Eur. J. Neurosci.* 13, 1963-1976 (2001).
30. D. J. Green and R. Gillette, Circadian rhythm of firing rate recorded from single cells in the rat suprachiasmatic brain slice. *Brain Res.* 245, 198-200 (1982).
31. R. A. Prosser and M. U. Gillette, The mammalian circadian clock in the suprachiasmatic nuclei is reset *in vitro* by cAMP. *J. Neurosci.* 9(3), 1073-1081 (1989).
32. M. Medanic and M. U. Gillette, Serotonin regulates the phase of the rat suprachiasmatic circadian pacemaker *in vitro* only during the subjective day. *J. Physiol.* 450, 629-642 (1992).
33. J. D. Glass, L. A. DiNardo and J. C. Ehlen, Dorsal raphe nuclear stimulation of SCN serotonin release and circadian phase-resetting. *Brain Res.* 859, 224-232 (2000).
34. T. E. Dudley, L. A. Dinardo and J. D. Glass, *In vivo* assessment of the midbrain raphe nuclear regulation of serotonin release in the hamster suprachiasmatic nucleus. *J. Neurophysiol.* 81(4), 1469-1477 (1999).
35. T. E. Dudley, L. A. DiNardo and J. D. Glass, Endogenous regulation of serotonin release in the hamster suprachiasmatic nucleus. *J. Neurosci.* 18(13), 5045-5052 (1998).
36. G. H. Grossman, R. E. Mistlberger, M. C. Antle, J. C. Ehlen and J. D. Glass, Sleep deprivation stimulates serotonin release in the suprachiasmatic nucleus. *Neuroreport* 11(9), 1929-1932 (2000).
37. K. J. Bobrzynska, N. Vrang and N. Mrosovsky, Persistence of nonphotic phase shifts in hamsters after serotonin depletion in the suprachiasmatic nucleus. *Brain Res.* 741(1-2), 205-214 (1996).
38. M. C. Antle, E. G. Marchant, L. Niel and R. E. Mistlberger, Serotonin antagonists do not attenuate activity-induced phase shifts of circadian rhythms in the Syrian hamster. *Brain Res.* 813, 139-149 (1998).
39. J. Hannibal, M. Moller, O. P. Ottersen and J. Fahrenkrug, PACAP and glutamate are co-stored in the retinohypothalamic tract. *J. Compara. Neurol.* 418, 147-155 (2000).
40. C. Fukuhara, N. Suzuki, Y. Matsumoto, et al., Day-night variation of pituitary adenylate cyclase-activating polypeptide (PACAP) level in the rat suprachiasmatic nucleus. *Neurosci. Lett.* 229, 49-52 (1997).
41. H. D. Piggins, E. G. Marchant, D. Goguen and B. Rusak, Phase-shifting effects of pituitary adenylate cyclase activating polypeptide on hamster wheel-running rhythms. *Neurosci. Lett.* 305, 25-28 (2001).
42. H. E. Albers and C. F. Ferris, Neuropeptide Y: role in light-dark cycle entrainment of hamster circadian rhythms. *Neurosci. Lett.* 50, 163-168 (1984).
43. K. L. Huhman and H. E. Albers, Neuropeptide Y microinjected into the suprachiasmatic region phase shifts circadian rhythms in constant darkness. *Peptides* 15(8), 1475-1478 (1994).
44. B. Rusak, J. H. Meijer and M. E. Harrington, Hamster circadian rhythms are phase-shifted by electrical stimulation of the geniculo-hypothalamic tract. *Brain Res.* 493, 283-291 (1989).

45. S. M. Biello and N. Mrosovsky, Blocking the phase-shifting effect of neuropeptide Y with light. *Proc. R. Soc. Lond., B, Biol. Sci.* 259, 179-187 (1995).

46. S. M. Biello, D. A. Golombek and M. E. Harrington, Neuropeptide Y and glutamate block each other's phase shifts in the suprachiasmatic nucleus in vitro. *Neuroscience* 77(4), 1049-1057 (1997).

47. K. L. Huhman, T. O. Babagbemi and H. E. Albers, Bicuculline blocks neuropeptide Y-induced phase advances when microinjected in the suprachiasmatic nucleus of syrian hamsters. *Brain Res.* 675, 333-336 (1995).

48. M. U. Gillette and R. A. Prosser, Circadian rhythm of the rat suprachiasmatic brain slice is rapidly reset by daytime application of cAMP analogs. *Brain Res.* 474, 348-352 (1988).

49. R. A. Prosser and M. U. Gillette, Cyclic changes in cAMP concentration and phosphodiesterase activity in a mammalian circadian clock studied *in vitro. Brain Res.* 568, 185-192 (1991).

50. M. U. Gillette and A. J. McArthur, Circadian actions of melatonin at the suprachiasmatic nucleus. *Behav. Brain Res.* 73, 135-139 (1996).

51. P. W. Cheung and C. E. McCormick, Failure of pinealectomy or melatonin to alter circadian activity rhythm of the rat. *Am. J. Physiol.* 242, R261-R264 (1982).

52. V. M. Cassone, M. J. Chesworth and S. M. Armstrong, Entrainment of rat circadian rhythms by daily injection of melatonin depends upon the hypothalamic suprachiasmatic nuclei. *Physiol. Behav.* 36, 1111-1121 (1986).

53. V. M. Cassone, M. H. Roberts and R. Y. Moore, Effects of melatonin on 2-deoxy-[1-^{14}C]glucose uptake within rat suprachiasmatic nucleus. *Am. J. Physiol.* 255(2), R332-R337 (1988).

54. S. Shibata, V. M. Cassone and R. Y. Moore, Effects of melatonin on neuronal activity in the rat suprachiasmatic nucleus *in vitro. Neurosci. Lett.* 97, 140-144 (1989).

55. R. R. Margraf and G. R. Lynch, An *in vitro* circadian rhythm of melatonin sensitivity in the suprachiasmatic nucleus of the Djungarian hamster, *Phodopus sungorus. Brain Res.* 609, 45-50 (1993).

56. A. J. McArthur, M. U. Gillette and R. A. Prosser, Melatonin directly resets the rat suprachiasmatic circadian clock *in vitro. Brain Res.* 565, 158-161 (1991).

57. A. J. McArthur, A. E. Hunt and M. U. Gillette, Melatonin action and signal transduction in the rat suprachiasmatic circadian clock: activation of protein kinase C at dusk and dawn. *Endocrinology* 138(2), 627-634 (1997).

58. A. E. Hunt, W. M. Al-Ghoul, M. U. Gillette and M. L. Dubocovich, Activation of MT2 melatonin receptors in rat suprachiasmatic nucleus phase advances the circadian clock. *Am. J. Physiol.* 280, C110-C118 (2001).

59. T. Shirakawa and R. Y. Moore, Glutamate shifts the phase of the circadian neuronal firing rhythm in the rat suprachiasmatic nucleus *in vitro. Neurosci. Lett.* 178, 47-50 (1994).

60. J. M. Ding, D. Chen, E. T. Weber, L. E. Faiman, M. A. Rea and M. U. Gillette, Resetting the biological clock: mediation of nocturnal circadian shifts by glutamate and NO. *Science* 266, 1713-1717 (1994).

61. S. Shibata, A. Watanabe, T. Hamada, M. Ono and S. Watanabe, N-methyl-D-aspartate induces phase shifts in circadian rhythm of neuronal activity of rat SCN *in vitro. Am. J. Physiol.* 267(2), R360-R364 (1994).

62. A. Watanabe, T. Hamada, S. Shibata and S. Watanabe, Effects of nitric oxide synthase inhibitors on N-methyl-D-aspartate-induced phase delay of circadian

rhythm of neuronal activity in the rat suprachiasmatic nucleus *in vitro. Brain Res.* 646, 161-164 (1994).

63. A. Watanabe, M. Ono, S. Shibata and S. Watanabe, Effect of a nitric oxide synthase inhibitor, N-nitro-L-arginine methylester, on light-induced phase delay of circadian rhythm of wheel-running activity in golden hamsters. *Neurosci. Lett.* 192, 25-28 (1995).

64. E. T. Weber, R. L. Gannon, A. M. Michel, M. U. Gillette and M. A. Rea, Nitric oxide synthase inhibitor blocks light-induced phase shifts of the circadian activity rhythm, but not c-fos expression in the suprachiasmatic nucleus of the Syrian hamster. *Brain Res.* 692, 137-142 (1995).

65. J. M. Ding, G. F. Buchanan, S. A. Tischkau, et al., A neuronal ryanodine receptor mediates light-induced phase delays of the circadian clock. *Nature* 394, 381-384 (1998).

66. J. M. Ding, L. E. Faiman, W. J. Hurst, L. R. Kuriashkina and M. U. Gillette, Resetting the biological clock: mediation of nocturnal CREB phosphorylation via light, glutamate, and nitric oxide. *J. Neurosci.* 17(2), 667-675 (1997).

67. S. A. Tischkau, J. W. Mitchell, S.-H. Tyan, G. F. Buchanan and M. U. Gillette, Ca2+/cAMP response element-binding protein (CREB)-dependent activation of Per1 is required for light-induced signaling in the suprachiasmatic nucleus circadian clock. *J. Biol. Chem.* 278(2), 718-723 (2003).

68. P. C. Yannielli and M. E. Harrington, Neuropeptide Y applied *in vitro* can block the phase shifts induced by light *in vivo. Neuroreport* 11(7), 1587-1591 (2000).

69. E. M. Mintz, A. M. Jasnow, C. F. Gillespie, K. L. Huhman and H. E. Albers, GABA interacts with photic signaling in the suprachiasmatic nucleus to regulate circadian phase shifts. *Neuroscience* 109(4), 773-778 (2002).

70. R. A. Prosser, Glutamate blocks serotonergic phase advances of the mammalian circadian pacemaker through AMPA and NMDA receptors. *J. Neurosci.* 21(19), 7815-7822 (2001).

71. M. J. Bradbury, W. C. Dement and D. M. Edgar, Serotonin-containing fibers in the suprachiasmatic hypothalamus attenuate light-induced phase delays in mice. *Brain Res.* 768, 125-134 (1997).

72. A. L. Bergstrom, J. Hannibal, P. Hindersson and J. Fahrenkrug, Light-induced phase shift in the Syrian hamster (Mesocricetus auratus) is attenuated by the PACAP receptor antagonist PACAP6-38 or PACAP immunoneutralization. *Eur. J. Neurosci.* 18, 2552-2562 (2003).

73. D. Chen, G. F. Buchanan, J. M. Ding, J. Hannibal and M. U. Gillette, Pituitary adenylyl cyclase-activating peptide: a pivotal modulator of glutamatergic regulation of the suprachiasmatic circadian clock. *Proc. Natl. Acad. Sci. U.S.A.* 96, 13468-13473 (1999).

74. S. A. Tischkau, E. A. Gallman, G. F. Buchanan and M. U. Gillette, Differential cAMP gating of glutamatergic signaling regulates long-term state changes in the suprachiasmatic circadian clock. *J. Neurosci.* 20(20), 7830-7837 (2000).

75. M. Zatz and M. J. Brownstein, Intraventricular carbachol mimics the effects of light on the circadian rhythm in the rat pineal gland. *Science* 203, 358-360 (1979).

76. M. Zatz and M. A. Herkenham, Intraventricular carbachol mimics the phase-shifting effect of light on the circadian rhythm of wheel-running activity. *Brain Res.* 212, 234-238 (1981).

77. D. J. Earnest and F. W. Turek, Role for acetylcholine in mediating effects of light on reproduction. *Science* 219, 77-79 (1983).

78. N. Murakami, K. Takahashi and K. Kawashima, Effect of light on the acetylcholine concentrations of the suprachiasmatic nucleus in the rat. *Brain Res.* 311, 358-360 (1984).

79. R. J. Wenthold. Glutamate and aspartate as neurotransmitters for the auditory nerve. In: *Glutamate as a Neurotransmitter,* edited by G. DiChiara and G. L. Gessa (Raven Press, New York, 1981) pp. 69-78.

80. O. Yasuhara, I. Tooyama, Y. Aimi, et al., Demonstration of choinergic ganglion cells in rat retina: expression of an alternative splice variant of choline acetyltransferase. *J. Neurosci.* 23(7), 2872-2881 (2003).

81. J. R. Pauly and N. D. Horseman, Anticholinergic agents do not block light-induced circadian phase shifts. *Brain Res.* 348, 163-167 (1985).

82. C. S. Colwell, C. M. Kaufman and M. Menaker, Phase-shifting mechanisms in the mammalian circadian system: new light on the carbachol paradox. *J. Neurosci.* 13(4), 1454-1459 (1993).

83. C. Liu and M. U. Gillette, Cholinergic regulation of the suprachiasmatic nucleus circadian rhythm via a muscarinic mechanism at night. *J. Neurosci.* 16(2), 744-751 (1996).

84. G. F. Buchanan and M. U. Gillette. Carbachol directly stimulating the SCN induces phase advances in mouse circadian rhythms throughout the night *in vitro* and *in vivo*. Abstract presented at: Society for Neuroscience, 2001.

85. G. F. Buchanan, L. R. Artinian, S. E. Hamilton, N. M. Nathanson and M. U. Gillette. The M1 muscarinic acetylcholine receptor is a necessary component in cholinergic circadian signaling. Abstract presented at: Society for Neuroscience, 2000.

86. G. F. Buchanan. Thesis: Cholinergic regulation of the mammalian circadian system: analysis of cholinergic-induced phase shifting *in vivo* and *in vitro* in wildtype and M1 knockout mice: Molecular and Integrative Physiology, University of Illinois at Urbana-Champaign, 2002.

87. A. A. Borbely, A two process model of sleep regulation. *Hum. Neurobiol.* 1(3), 195-204 (1982).

88. A. A. Borbely, Processes underlying sleep regulation. *Horm. Res.* 49, 114-117 (1998).

89. N. Ibuka and H. Kawamura, Loss of circadian rhythm in sleep-wakefulness cycle in the rat by suprachiasmatic nucleus lesions. *Brain Res.* 96, 76-81 (1975).

90. N. Ibuka, S.-I. T. Inouye and H. Kawamura, Analysis of sleep-wakefulness rhythms in male rats after suprachiasmatic nucleus lesions and ocular enucleation. *Brain Res.* 122, 33-47 (1977).

91. W. B. Mendelson, B. M. Bergmann and A. Tung, Baseline and post-deprivation recovery sleep in SCN-lesioned rats. *Brain Research* 980, 185-190 (2003).

92. I. Tobler, A. A. Borbely and G. Groos, The effect of sleep deprivation on sleep in rats with suprachiasmatic lesions. *Neurosci. Lett.* 42, 49-54 (1983).

93. R. E. Mistlberger, B. M. Bergmann, W. Waldenar and A. Rechtschaffen, Recovery sleep following sleep deprivation in intact and suprachiasmatic nuclei lesioned rats. *Sleep* 6(3), 217-233 (1983).

94. D. M. Edgar, W. C. Dement and C. A. Fuller, Effect of SCN lesions on sleep in squirrel monkeys: evidence for opponent processes in sleep-wake regulation. *J. Neurosci.* 13(3), 1065-1079 (1993).

95. M. C. Antle and R. E. Mistlberger, Circadian clock resetting by sleep deprivation without exercise in the Syrian hamster. *J. Neurosci.* 20(24), 9326-9332 (2000).

96. E. Challet, F. W. Turek, M.-A. Laute and O. V. Reeth, Sleep deprivation decreases phase-shift responses of circadian rhythms to light in the mouse: role of serotonergic and metabolic signals. *Brain Res.* 909, 81-91 (2001).
97. S. F. Glotzbach, C. M. Cornett and H. C. Heller, Activity of suprachiasmatic and hypothalamic neurons during sleep and wakefulness in the rat. *Brain Res.* 419, 279-286 (1987).
98. T. Deboer, M. J. Vansteensel, L. Detari and J. H. Meijer, Sleep states alter activity of suprachiasmatic nucleus neurons. *Nat. Neurosci.* 6(10), 1086-1090 (2003).

The Neuroendocrine Loop Model Revisited: Is It Valid or Even Relevant?

Vincent M. Cassone

1. Introduction

Biological rhythms and the biological clocks that control them are fundamental properties of most living organisms, ranging from several forms of bacteria to multicellular plants and animals. These properties share many formal and biochemical properties, and affect all aspects of physiological function, ranging from control of transcription to metabolism to cell cycle to behavior. Because these rhythms are expressed similarly in single-celled and multi-cellular organisms, and because clock function pervades all levels of biological organization, many authors have suggested that biological clocks are properties of all cells.[1]

2. Master Pacemakers for Circadian Regulation

In multi-cellular animals, however, control of overt rhythmicity is regulated hierarchically by a set of neural and neuroendocrine structures. In 1968, Gaston and Menaker[2] showed that the pineal gland is necessary for self-sustained circadian rhythms of locomotor activity of house sparrows, *Passer domesticus*. Subsequent work has shown the pineal gland is critical for behavioral rhythms in other passerine species.[3-5] Transplantation of the pineal gland into arrhythmic pinealectomized sparrows confers both rhythmicity and the circadian phase of the donors to the transplant host,[6] and explanted pineal glands contain both the circadian oscillators and photoreceptors that are sufficient to generate circadian rhythms of the biosynthesis of the hormone melatonin and entrain that rhythm to light:dark regimes.[7-10] Rhythmic administration of melatonin to pinealectomized sparrows,[11-13] pinealectomized starlings, *Sturnus vulgaris*[14] and to pinealectomized/enucleated pigeons, *Columba livia*[15] restores locomotor patterns of activity to these birds, punctuating the important role pineal melatonin plays in avian species. The role played by the pineal gland in columbiform[16] and galliform[17] is less clear than in passeriform species. However, in these species, the ocular retina synthesizes and releases melatonin as well, and when pinealectomy is combined with enucleation, pigeons, at least, become arrhythmic as in pinealectomized sparrows.[16] In the Japanese quail, whose eyes are responsible for melatonin rhythmicity in the circulation, enucleation alone abolishes locomotor rhythms.[17] Thus, the avian pineal gland (and retina in some species) is a circadian oscillator and pacemaker that regulate overt circadian rhythms via the rhythmic secretion of melatonin (Figure 1A).

This feature is not unique to birds. Many squamate reptiles and teleost fish have been shown to rely on pineal activity for overt circadian rhythms, although there is significant variability among species in this regard.[18,19] In contrast, however, the mammalian pineal gland is neither photoreceptive nor does it contain circadian oscillators to generate rhythms on melatonin.

Instead, the mammalian pineal gland is a "slave oscillator" to the master pacemaker in mammalian circadian organization, the hypothalamic suprachiasmatic nucleus (SCN).[19] Destruction of the SCN or transection/disruption of neural pathways from the SCN to the pineal gland abolishes circadian rhythms in melatonin biosynthesis. Thus, although the mammalian pineal gland does not itself contain oscillators and photoreceptors, its melatonin rhythm is nonetheless a direct output of the mammalian circadian clock. We will return to this feature below.

In mammals, the SCN is the master pacemaker regulating circadian rhythms,[20,21] at least in rodents and primates. These nuclei are characterized anatomically as recipients of a direct retinohypothalamic tract (RHT).[22-24] All of the neurons contain the inhibitory neurotransmitter γ-amino-butyric acid (GABA) and one of several neuropeptides, including arginine vasopressin, vasoactive intestinal polypeptide, neurotensin and substance P, among others, although the regional distribution of these peptides within the SCN is highly variable among mammalian taxa.[24] The SCN of rats, mice and hamsters also

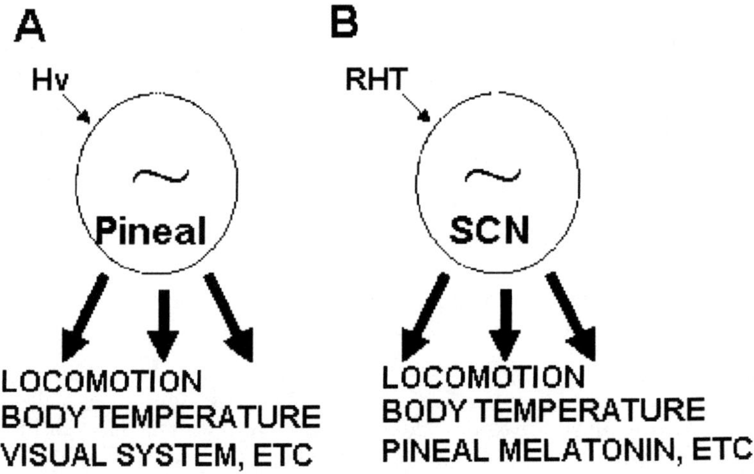

FIGURE 1. Hierarchical models for circadian organization based upon lesion studies.

contain a very high concentration of fibrous astrocytes.[25,26] Further, the nuclei consistently receive serotoninergic input from the midbrain, peptidergic (primarily NPY) input from the visual thalamus[20] and contains a high density of melatonin receptors.[27]

The evidence for the proposition that the SCN are the master pacemaker is legion (Figure 1B). Lesion of the SCN abolishes all known overt circadian rhythms ranging from activity, sleep-wake cycles, body temperature, multiple hormone secretions and cardiovascular function.[28-30] Transplantation of embryonic SCN tissue or cells derived from embryonic tissue restores circadian rhythms to rodents.[31-33] Finally, explants of SCN or dissociated cells from SCN tissue express circadian rhythms of electrical activity, metabolism, peptide release, and gene expression.[34-38] Thus, the mammalian SCN are circadian oscillators and pacemakers that regulate overt circadian rhythms.

As with the avian pineal gland, this feature is probably not unique to mammals, although the evidence is by no means clear. An RHT has been identified in several hypothalamic structures in different vertebrate taxa, and each of these contains some but not all of the anatomical characteristics of the mammalian SCN. In birds, two structures have been touted as avian homologues to the SCN. First, a structure in the medial hypothalamus, situated in the preoptic recess, has been designated medial suprachiasmatic nucleus (mSCN) by some authors[39] and periventricular preoptic nucleus (PPN) by others.[40,41] This structure does not receive RHT input, has little neuroanatomical similarities to the SCN and has not been shown to exhibit any physiological rhythmicity. However, it does express several clock genes rhythmically on a daily and circadian basis,[42-44] and lesions directed at the mSCN disrupt overt circadian rhythms.[4,45] On the other hand, a structure located caudally and laterally of the mSCN, which has been designated the visual suprachiasmatic nucleus (vSCN) by some[41] and lateral hypothalamic nucleus by others.[39] The vSCN receives direct RHT input, has a high concentration of GABA neurons, fibrous astrocytes and contains neurons that express arginine vasotocin, vasoactive intestinal polypeptide, substance P and neurotensin. It also receives a dense 5HT input, the source of which is not known, and NPY input from the visual thalamus.[41] Finally, the vSCN expresses circadian rhythms in 2DG uptake *in vivo*[46,48] electrical activity *in vitro*,[49] a high density of melatonin receptors[50-52] and clock genes, at least in house sparrows.[42]

3. The Neuroendocrine Loop Model for Avian Circadian Organization

In spite of the clear role as pacemaker played by the passerine pineal gland, there is extensive evidence that the pineal gland represents only part of a complex system of neuroendocrine structures involved in avian circadian organization.

First, pinealectomized house sparrows become arrhythmic gradually after placement in constant darkness (DD), damping over 6-10 days.[2] Using 2DG brain imaging techniques, we have found that, while most brain structures become arrhythmic in their 2DG uptake following pinealectomy and placement in DD, the vSCN persists for at least 3 days in rhythmicity.[47] However, by the tenth day in DD, when all behavioral rhythmicity is lost, all 2DG rhythmicity is lost as well.[47] Further, rhythmic administration of melatonin reestablishes both a daily activity rhythm and a daily 2DG rhythm in the vSCN.[12] Secondly, although the pineal gland produces melatonin rhythms *in vivo* indefinitely, *in vitro*, the pineal gland circadian rhythm in melatonin damps in DD after 5-7 days.[9,53] This effect can be simulated *in vivo* by sympathetic denervation of the gland.[54] Thirdly, sympathetic innervation of the avian pineal gland derives from a multi-synaptic pathway arising from the vSCN, similar to the situation in mammals.[19] This innervation releases norepinephrine (NE) on a daily and circadian basis such that NE is released during the day and subjective day,[55] which inhibits melatonin biosynthesis.[53,54] This rhythm derives from the vSCN, since lesion of the vSCN, but not the mSCN, abolishes the rhythm of NE turnover in the pineal gland.[56]

Based upon these and other studies, we[57,58] proposed the "neuroendocrine loop model" for avian circadian organization (Figure 2). At its core, the model's premise states that the system is composed of damped circadian oscillators residing within the vSCN and pineal gland, which are not capable

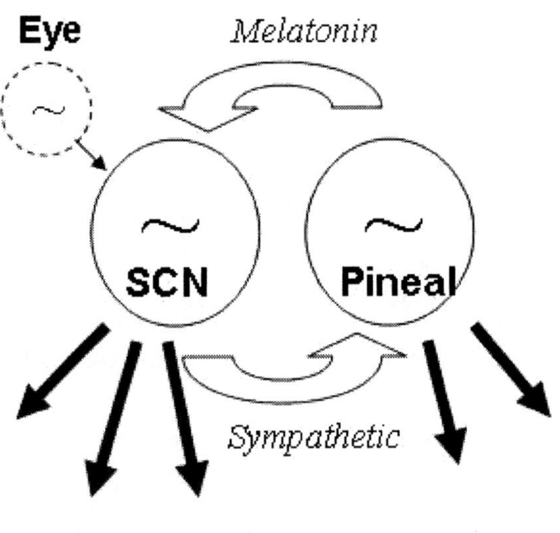

FIGURE 2. Neuroendocrine Loop Model.

of self-sustained oscillation in the absence of (1) photic input and/or (2) neural/endocrine input from the rest of the system. At this point, we do not know if pacemaker damping is due to the damping of the oscillatory mechanisms within each pacemaker cell and/or (they are not mutually exclusive) whether the population of more self-sustained oscillators within each pacemaker drift out of phase in the absence of input, resulting in the damping of the output of the population. It's best to explain the model sequentially: The vSCN is metabolically and electrically active during the subjective day, stimulating output pathways, including increasing sympathetic tone, which inhibits pineal gland output and has broad effects throughout the body. Since the vSCN is an oscillator, it spontaneously wanes in its activity as the subjective day progresses, until at subjective dusk, it is no longer active. This releases pineal oscillators from their inhibition, allowing pinealocytes to synthesize and release melatonin. Melatonin broadly affects a wide array of processes in tissues that express melatonin receptors. Among the structures expressing melatonin receptors is the vSCN, which are inhibited by melatonin. Since the pineal gland contains circadian oscillators, it too spontaneously wanes in its biosynthesis of melatonin, stopping by subjective dawn and disinhibiting vSCN output for the succeeding day. Each of the co-pacemakers in this system is directly affected by light; the pineal gland contains several photopigments and phototransduction systems that affect melatonin biosynthesis, and the vSCN receives RHT input. In addition, each of the co-pacemakers may independently affect downstream processes. The pineal gland influences CNS and peripheral sites via the secretion of melatonin during the night, and tissues that express melatonin receptors are affected by this pacemaker directly. Conversely, the SCN pacemaker (vSCN and mSCN) affects output via several pathways. A humoral output affects local hypothalamic function (at least in mammals), and a neural output via SCN afferents affects CNS and peripheral sites to which they project. Among these is a global regulation of sympathetic tone.[56]

Gwinner[61] has proposed a complementary model, the "internal resonance model", for avian circadian organization. This model states that, rather than mutual inhibition, the pineal gland and SCN interact by increasing the amplitude of the rhythmic output of the other. The model is attractive in several respects. One is that active regulation of system amplitude may explain seasonal migratory behaviors such as *Zugunruhe* or nocturnal migratory restlessness. Indeed, Gwinner and his colleagues have noted a decrease in melatonin titers in nocturnal migratory birds expressing *Zugunruhe*. The mechanism by which resonance occurs is not explicitly stated in this model, but it is not mutually exclusive from the neuroendocrine loop model, since, following inhibition by norepinephrine *in vivo*[54] or *in vitro*,[53] pineal melatonin output rebounds with a higher amplitude. Thus, internal resonance may occur through the mutually inhibitory relationship between SCN and pineal oscillators.

4. A Mammalian Neuroendocrine Loop?

The neuroendocrine loop model (and its corollary) has been extended to mammals by several authors, based upon the fact that even though the mammalian pineal gland is not an oscillator, melatonin directly affects circadian rhythms and the SCN.[61-63] Daily administration of melatonin to free-running rats, hamsters and humans entrains the circadian patterns of activity to the period of the administration cycle.[64-68] This effect requires an intact SCN, since lesions in the SCN, which make rats arrhythmic, prevent the effects of the hormone,[69] even though rats made arrhythmic with constant light (LL) are resynchronized by daily melatonin administration.[70] The effect does not require a pineal gland,[71] serotoninergic or noradrenergic afferents.[69] Melatonin affects the rat SCN directly. The SCN of most mammalian species studied contains high affinity melatonin receptors, the MT1 (Mel$_{1A}$) and MT2 (Mel$_{1B}$) receptor sub-types.[72,73]

Administration of melatonin acutely inhibits 2DG uptake in the rat SCN *in vivo*[74] and electrical activity *in vitro*[75] and phase-shifts the SCN clock[76] *in vitro*. The phase-shifting effect of melatonin appears to be mediated via the MT2 receptor, since MT2 knockout mice do not phase shift,[77] while the acute effect of the hormone is mediated via the MT1 receptor, since melatonin does not decrease SCN electrical activity in MT1 knockout mice.[78] Thus, although the circadian rhythm of pineal melatonin is a slave oscillation in mammals, melatonin nonetheless provides temporal feedback to the SCN, closing a similar neuroendocrine loop.

5. Challenges to both Circadian Hierarchy
and the Neuroendocrine Loop Models

In the past 10 years, the biological clocks field has been revolutionized by the discovery of mammalian genes that are orthologous to genes involved in clock function in *Drosophila*.[79,80] These genes, collectively called "clock genes" by the field, can be grossly categorized into "positive elements" and "negative elements". Positive elements *clock* and *Bmal1* are transcribed, translated and then dimerized to reenter the nucleus. There the *clock/Bmal1* dimer stimulates the transcription of genes with an "e-box". Among these genes are the "negative elements" *period1 (per1), period2 (per2), period3 (per3), cryptochrome1 (cry1)* and *cryptochrome2 (cry2)*. These are then transcribed, translated and form oligomers that reenter the nucleus where they inhibit the actions of the *clock/Bmal1* dimers. There are of course other "clock genes" that have been discovered. These include casein kinase 1 epsilon *(CK1E)*, which encodes a kinase that targets the PERIOD proteins for degradation, and *timeless*, which is important in flies but whose function in mammals is not known. One of the major surprises of these discoveries is the fact that these "clock genes" are expressed in many CNS and peripheral tissues and that the expression of these genes is rhythmic. Even more surprising is the fact that explanted tissues from animals expressed rhythms of at least the *period* genes when placed *in vitro* and fibroblasts could be induced to express clock gene rhythms with serum shock.[80] The prevailing view is that these peripheral rhythms represent peripheral clocks, and that the role of the SCN is to synchronize rather than drive these clocks.

As one might expect, many authors have now identified avian orthologs of these mammalian clock genes.[43,44,81-84] Indeed, we have just completed a functional genomics analysis that identifies and characterizes all of the avian orthologs of mammalian clock genes from the chick pineal gland[85] and retina.[86] As stated above, several of these authors have shown that these clock genes are expressed in the mSCN, and we corroborate this, although the vSCN also expresses them at a similar level. However, the remarkable claim is that, even though pinealectomy and exogenous melatonin affect overt clock function, these procedures have no effect on clock gene expression.[43,84,87] The authors independently and reasonably state that the data question the neuroendocrine loop model.

If the pineal gland is critical for overt rhythmicity in passerine birds, if the administration of melatonin acutely inhibits 2DG uptake in the avian vSCN, and if rhythmic administration of melatonin restores behavioral and brain 2DG rhythms in pinealectomized birds, how is it that pinealectomy and melatonin have no effect of SCN and brain clock gene expression?

Since it is incontrovertible that pineal melatonin is critical for passerine bird circadian rhythms, one could posit that the clock genes identified by many authors, including our group, are not really involved in the control of overt circadian rhythms in birds. Until knockout or other genetic manipulations in birds is possible, we cannot answer this question definitively.

Alternatively, but not necessarily exclusively, it is possible that there are at least two oscillatory systems in place within the avian circadian system; one involving transcriptional regulation of clock genes that is insensitive to melatonin and one that is relatively independent of clock gene expression but sensitive to melatonin.

Surprisingly (and similarly), the recent observation of Poirel et al.[88] showed that injection of melatonin in free-running rats at circadian time 10 (CT10; two circadian hours before activity onset), the time at which melatonin most efficaciously phase shifts locomotor activity,[71] phase shifts electrical activity,[76,77] suppresses electrical activity[75,77] and suppresses 2DG uptake[74] *has absolutely no effect on SCN clock gene expression* on the day of the injection and on the day following the injection. The authors suggest in their paper that melatonin must affect clock genes at a post-transcriptional level, but the truth is that, if these data hold up, melatonin must act on the SCN clock at a level that does not immediately include the transcription and translation of clock genes.

6. Is there Evidence for a Second Oscillatory Mechanism within the Circadian Clock?

There is growing evidence in the circadian microbiological literature that multiple oscillators can reside within single cells. In cyanobacteria, interference of transcription via knockout of sigma factors differentially affects the circadian periods of different output pathways, and single cells can express multiple periods, depending on the output measured.[89] In the single celled dinoflagellate, *Gonyaulax polyedra,* different outputs such as swarming, bioluminescence and cell division are expressed with different circadian periods.[90] In the filamentous fungus, *Neurospora crassa,* circadian oscillations in development and gene expression are regulated by at least two circadian mechanisms, one that is dependent on the expression of the transcription factor *frq* and one that persists in *frq* null strains.[91] It should not therefore be surprising if complex neuroendocrine structures involved in circadian organization house multiple molecular mechanisms. Do they exist?

At this stage, we don't really know, but several lines of evidence from our lab and others point to multiple oscillatory mechanisms underlying vertebrate clocks. Immortalized SCN2.2 cells express rhythms in 2DG uptake as well as rhythms in neurotrophins, *per1, per2, cry1, cry2* and *bmal1*.[33,60,92] These cells will also confer rhythmicity to SCN lesioned rats *in vivo*,[33] and to co-cultured cells *in vitro*[60] (Figure 3).

Thus, the SCN2.2 cells retain both oscillatory and pacemaker properties. When SCN2.2 cells are co-cultured with NIH 3T3 fibroblasts, the SCN2.2 cells induce a rhythm of 2DG uptake and clock gene expression in the co-cultured fibroblasts.[60,92] Interestingly, the induced fibroblast 2DG rhythm lags in phase the SCN2.2 2DG rhythm by approximately 4 hrs, while the *per1* and *per2* expression rhythms lag by approximately 12 hrs,[60] suggesting differential regulation. Recently, we have found that knockdown of positive element *clock* with antisense oligonucleotides in SCN2.2 cells differentially affects 2DG and clock gene expression.[92]

Anti-*clock* oligonucleotides applied to SCN2.2 cells lengthens the period of 2DG uptake in the SCN2.2 cells as well as co-cultured NIH3T cells, which never were exposed to the anti-sense molecules (Figure 4). This observation shows that the molecular clockworks in the SCN2.2 cells are responsible for NIH3T3 metabolic rhythmicity. In contrast, rhythms of *per2* and *cry1* expression are disrupted and/or abolished, punctuating the view that the positive element *clock* is critical for "clock gene" rhythms. Thus, although *clock* clearly affects both metabolic and clock gene patterns, it does so differentially, suggesting separate rhythmic processes are at work.

Astrocytes are neuroglial cells that mediate many processes within the central nervous system (CNS), including the majority of CNS intermediary

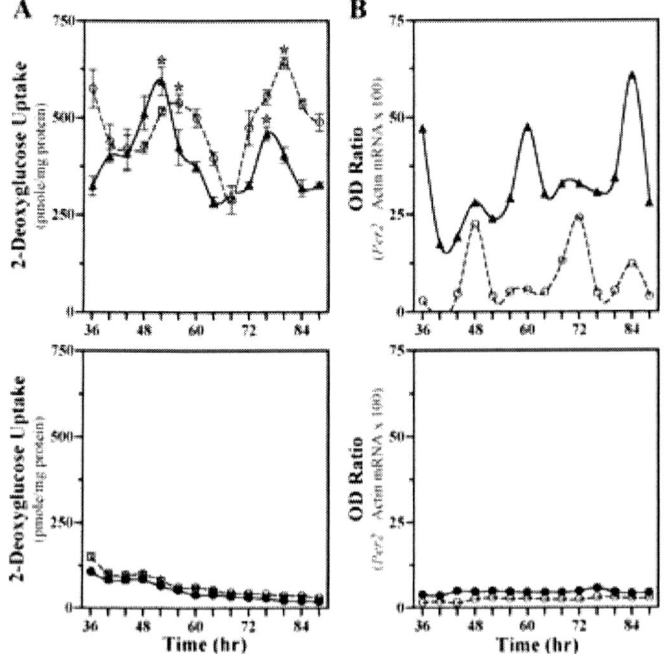

FIGURE 3. SCN 2.2 cells co-cultured with mouse NIH3T3 cells impose a rhythm of 2DG uptake (A) and clock gene expression (B) with varying phase angles. SCN2.2 (solid) with NIH 3T3 (dotted) on top, NIH 3T3 with NIH3T3 on bottom (Allen et al. 2001; ref.[60]).

metabolism.[93,94] Glucose is taken up by both neurons and glia, but astrocytes take up glucose at a higher rate than do other cell-types. Glucose is catabolized by astrocytes by at least three pathways: (1) Glucose is catabolized via glycolysis, Krebs cycle and electron transport to produce ATP for astrocytic processes. (2) When metabolic load is low, glucose is polymerized to glycogen, where it remains until metabolic load is higher. Indeed, glycogen biosynthesis has been hypothesized as a major reason for sleep.[95]

Then, presumably during waking, astrocytic glycogen is catabolized to glucose. (3) Finally, when signaled by neurons with a high metabolic activity, astrocytes will undergo lactate fermentation and release both lactate and pyruvate into the extracellular space. These are then taken up by neurons for a relatively "cheap" energy source.

When melatonin receptors were first identified in the avian brain, we[50-52] noted that specific binding and expression of the Mel$_{1C}$ receptor sub-type was present in brain areas that were bereft of neuronal cell bodies and hypothesized then that neuroglia express melatonin receptors. To continue this line of research, we developed a cell culture model for avian astrocytes, and found that chick astrocytes do indeed express functional Mel$_{1A}$ and Mel$_{1C}$ receptors.[96]

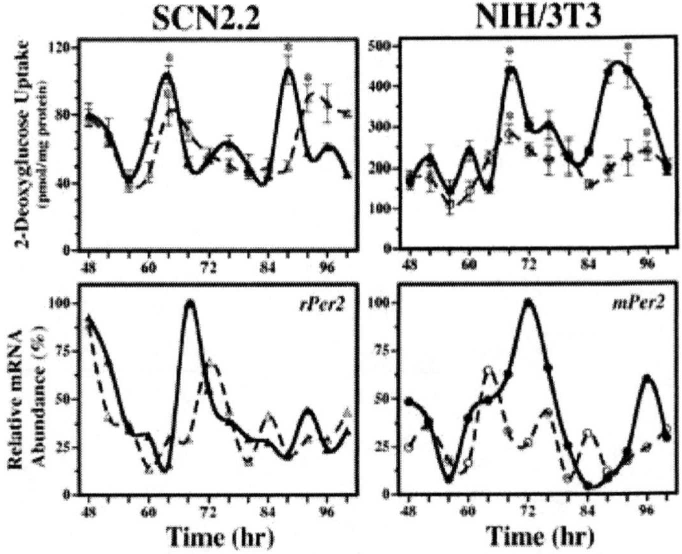

FIGURE 4. Anti-*clock* morpholino reduces amplitude and increases period of 2DG uptake in SCN 2.2 cells and co-cultured NIH3T3 cells. *Per2* expression rhythms are also disrupted but differentially (Allen et al. 2004; ref 92).

When cultured astrocytes are administered daily cycles of physiological levels of melatonin, the melatonin cycle induces a rhythm of glucose utilization, lactate and pyruvate release (Figure 5). Further, acute administration of melatonin increases glycogen biosynthesis.

Recently, we have found that chick astrocytes also express all of the known orthologs of the mammalian clock genes.[85,97] Interestingly, even though cycles of melatonin induce metabolic rhythms, the regimen has no effect on clock gene expression, corroborating the *in vivo* studies.[42-44] Thus, as with the SCN2.2 cells, metabolic and clock gene expression are differentially regulated. Might this explain the discrepancy in data and suggest a modification for the neuroendocrine loop model?

7. The Neuroendocrine Loop Model Revisited

If one were to assume that melatonin indeed affects circadian clock function and sleep in birds and mammals, as many studies indicate, and if one were to assume that Abraham et al.,[42] Yasuo et al,.[43,44] and Poirel et al.[88] are all correct in their observations that acute melatonin administration has no effect on clock gene expression, and if one were to assume that the rhythmic expression of clock genes are a primary force in circadian rhythm generation, then

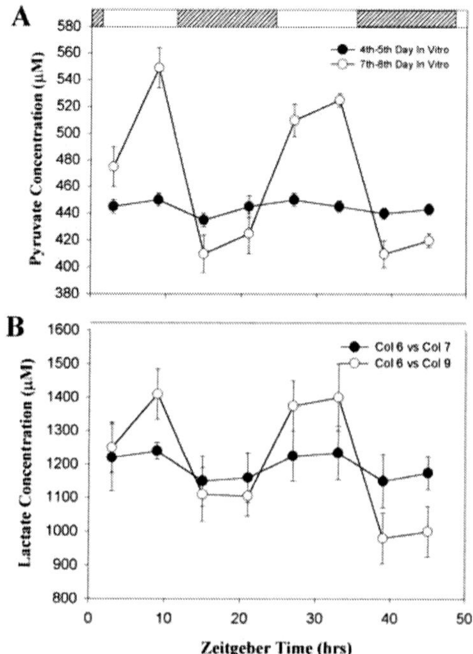

FIGURE 5. Effects of melatonin cycles on cultured chick astrocyte pyruvate and lactate release. (from Adachi et al. 2002; ref. 96).

the most parsimonious view is that metabolic and clock gene expression are separable properties of circadian clock function (Figure 6).

Perhaps, melatonin affects SCN activity via its effects on astrocytic metabolism and only indirectly affects SCN clock gene expression. In turn, clock gene expression affects metabolic rhythms in the SCN, since knockdown of *clock* decreases the amplitude and increases the period on SCN2.2 2DG uptake rhythms, but not to the extent that it affects transcriptional rhythmicity. Thus, in birds, the SCN (a composite of mSCN transcriptional activity and vSCN metabolic activity) is active during the subjective day and is activated by RHT afferents. During the subjective day, the avian SCN mediates many downstream processes via humoral and neuronal outputs, including activation of sympathetic outflow. Among the targets of sympathetic activity is the pineal gland, whose biosynthesis of melatonin in inhibited by both sympathetic norepinephrine but also directly by light. Since the SCN are oscillators, their output wanes as dusk approaches, thereby disinhibiting pineal melatonin biosynthesis. During the subjective night, this modified hypothesis suggests, melatonin is released and affects physiological functions in cells and tissues that express melatonin receptors. Among the targets of melatonin are the astrocytes within the SCN region and throughout the brain, where it inhibits glycolytic activity and increases glycogen biosynthesis. This increase in glycogen biosynthesis

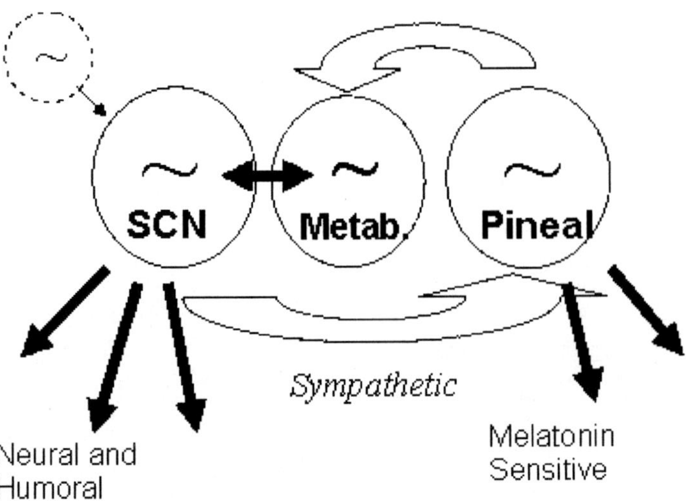

FIGURE 6. Modified Neuroendocrine Loop Model.

during avian sleep builds energy stores for brain activity during the day. How or whether this change in metabolic activity affects SCN clock gene expression is at this point not known. It is likely to occur over a longer timeframe than has been studied thus far. Since, as with the SCN, the pineal is an oscillator, its output wanes as dawn approaches, thereby disinhibiting SCN activity.

In mammals, the pineal gland is not an endogenous oscillator but still feeds back to influence SCN metabolic and electrical activity. Since most intermediary metabolism is accomplished within brain astrocytes, it is reasonable to predict that the effects of melatonin on SCN metabolism are mediated by astrocytes. The role melatonin plays in glycogen homeostasis is at this stage completely unknown. Melatonin itself inhibits SCN activity, but increases locomotion in the nocturnal rodents. One might therefore assume that the role of melatonin on sleep and glycogen biosynthesis would be opposite to that found in birds (and humans). If this is the case, research on the role of melatonin in sleep/wake cycles and the utility of melatonin as a therapeutic sleep aid should be reexamined in the context of metabolic switching. Clearly, this is an important direction for future research to take.

Thus, the neuroendocrine loop model, some 20 years old now, certainly must be modified in light of modern data, but the concept itself remains somewhat intact. The role played by the pineal in circadian rhythm generation and regulation in birds and in circadian regulation in mammals may be important for sleep research.

8. Acknowledgements. The Cassone laboratory is supported by the NIH, NSF and NASA.

9. *References*

1. C.S. Pittendrigh CS. Temporal organization: reflections of a Darwinian clock-watcher. *Annu Rev Physiol.* 55:16-54. (1993)
2. S. Gaston and M. Menaker. Pineal function: the biological clock in the sparrow? *Science* 160(832):1125-7 (1968)
3. E. Gwinner. Effects of pinealectomy on circadian locomotor activity rhythms in European starlings, *Sturnus vulgaris. J Comp Physiol* 126: 123-129 (1978)
4. Ebihara, S. and Kawamura, H. The role of the pineal organ and the suprachiasmatic nucleus in the control of circadian locomotor rhythms in the Java sparrow (Padda oryzivova). J Comp Physiol A 141:207-214. (1981).
5. Fuchs, J.L. Effects of Pinealectomy and subsequent melatonin implants on activity rhythms in the house finch (*Carpodacus* mexicanus). *J. Comp. Physiol. [A]* 153:413-19 (1983)
6. Zimmerman NH, Menaker M. The pineal gland: a pacemaker within the circadian system of the house sparrow. *Proc Natl Acad Sci U S A.* 76(2):999-1003. (1979)
7. Binkley S. A timekeeping enzyme in the pineal gland. *Sci Am.* 240(4):66-71. (1979)
8. Kasal CA, Menaker M, Perez-Polo JR. Circadian clock in culture: N-acetyltransferase activity of chick pineal glands oscillates in vitro. *Science.* 203(4381):656-8. (1979)
9. Takahashi JS, Hamm H, Menaker M. Circadian rhythms of melatonin release from individual superfused chicken pineal glands in vitro. *Proc Natl Acad Sci U S A.*;77(4):2319-22. (1980)
10. Wainwright SD, Wainwright LK. Chick pineal serotonin acetyltransferase: a diurnal cycle maintained in vitro and its regulation by light. *Can J Biochem.* 57(6):700-9 (1979)
11. V.M.Cassone, D.S. Brooks, D.B. Hodges, T.A. Kelm, J. Lu, W.S. Warren Integration of circadian and visual function in mammals and birds: brain imaging and the role of melatonin in biological clock regulation. In: *Advances in Metabolic Mapping Techniques for Brain Imaging of Behavioral and Learning Functions.* F. Gonzalez-Lima, T. Finkenstaedt and H. Scheich (eds) Kluwer Academic Publishers, Dordrecht/Boston/London, pp. 299-318. (1992)
12. J. Lu and V.M. Cassone. Daily melatonin administration synchronizes circadian patterns of brain metabolism and behavior in pinealectomized house sparrows, *Passer domesticus. J. Comp. Physiol. A* 173: 775-782 (1993)
13. S Heigl and E. Gwinner. Synchronization of circadian rhythms of house sparrows by oral melatonin: effects of changing period. *J Biol Rhythms.* 10(3): 225-33. (1995)
14. E.Gwinner and I. Benzinger . Synchronization of a circadian rhytm in pinealectomized European starlings, *Sturnus vulgaris. J. Comp. Physiol [A]w,* 127: 209-213 (1978)
15. C.C. Chabot, M. Menaker. Effects of physiological cycles of infused melatonin on circadian rhythmicity in pigeons. *J Comp Physiol [A].* 170(5):615-22 (1992)
16. Ebihara S, Uchiyama K, and Oshima I Circadian organization in the pigeon *Columbia livia:* the role of the pineal organ and the eye. *J Comp Physiol A* 154:59-69 (1984)
17. Underwood H, Siopes T. Circadian organization in Japanese quail. *J Exp Zool.* 232(3):557-66 (1984).

18. Cassone VM, Natesan AK. Time and time again: the phylogeny of melatonin as a transducer of biological time. *J Biol Rhythms*. 12(6):489-97 (1997)

19. Klein DC, Coon SL, Roseboom PH, Weller JL, Bernard M, Gastel JA, Zatz M, Iuvone PM, Rodriguez IR, Begay V, Falcon J, Cahill GM, Cassone VM, Baler R. The melatonin rhythm-generating enzyme: molecular regulation of serotonin N-acetyltransferase in the pineal gland. *Recent Prog Horm Res.*;52:307-57; (1997)

20. R.Y. Moore, R. Silver. Suprachiasmatic nucleus organization. *Chronobiol Int.* 15(5):475-87 (1998).

21. J.D.Miller, L.P. Morin, W.J. Schwartz, R.Y. Moore. New insights into the mammalian circadian clock. *Sleep.* 19(8):641-67. (1996)

22. Moore RY, Lenn NJ. A retinohypothalamic projection in the rat. J Comp *Neurol.*;146(1):1-14. (1972)

23. Hendrickson AE, Wagoner N, Cowan WM. An autoradiographic and electron microscopic study of retino-hypothalamic connections. *Z Zellforsch Mikrosk Anat.*;135(1):1-26 (1972).

24. Cassone VM, Speh JC, Card JP, Moore RY. Comparative anatomy of the mammalian hypothalamic suprachiasmatic nucleus. *J Biol Rhythms.*;3(1):71-91 (1988)

25. Botchkina GI, Morin LP. Ontogeny of radial glia, astrocytes and vasoactive intestinal peptide immunoreactive neurons in hamster suprachiasmatic nucleus. *Brain Res Dev Brain Res.* 86(1-2):48-56 (1995)

26. Lavialle M, Serviere J. Circadian fluctuations in GFAP distribution in the Syrian hamster suprachiasmatic nucleus. *Neuroreport.* 4(11):1243-6 (1993)

27. Weaver DR, Namboodiri MA, Reppert SM. Iodinated melatonin mimics melatonin action and reveals discrete binding sites in fetal brain. *FEBS Lett.* 228(1):123-7. (1988)

28. Moore RY, Eichler VB. Loss of a circadian adrenal corticosterone rhythm following suprachiasmatic lesions in the rat. *Brain Res.* 42(1):201-6.(1972)

29. Stephan FK, Zucker I. Circadian rhythms in drinking behavior and locomotor activity of rats are eliminated by hypothalamic lesions. *Proc Natl Acad Sci U S A.* 69(6):1583-6 (1972)

30. Warren WS, Champney TH, Cassone VM. The suprachiasmatic nucleus controls the circadian rhythm of heart rate via the sympathetic nervous system. *Physiol Behav.* 55(6):1091-9 (1994)

31. Sawaki Y, Nihonmatsu I, Kawamura H. Transplantation of the neonatal suprachiasmatic nuclei into rats with complete bilateral suprachiasmatic lesions. *Neurosci Res.*;1(1):67-72 (1984).

32. Ralph MR, Lehman MN. Transplantation: a new tool in the analysis of the mammalian hypothalamic circadian pacemaker. *Trends Neurosci.* 14(8):362-6 (1991).

33. Earnest DJ, Liang FQ, Ratcliff M, Cassone VM. Immortal time: circadian clock properties of rat suprachiasmatic cell lines. *Science.* 283(5402):693-5 (1999)

34. Green DJ, Gillette R. Circadian rhythm of firing rate recorded from single cells in the rat suprachiasmatic brain slice. *Brain Res*;245(1):198-200.(1982)

35. Shibata S, Oomura Y, Kita H, Hattori K. Circadian rhythmic changes of neuronal activity in the suprachiasmatic nucleus of the rat hypothalamic slice. *Brain Res.*;247(1):154-8.(1982)

36. Earnest DJ, Sladek CD. Circadian vasopressin release from perifused rat suprachiasmatic explants in vitro: effects of acute stimulation. *Brain Res.* 422(2):398-402.(1987)

37. Newman GC, Hospod FE, Patlak CS, Moore RY. Analysis of in vitro glucose utilization in a circadian pacemaker model. *J Neurosci.*;12(6):2015-21.(1992)

38. Yoo SH, Yamazaki S, Lowrey PL, Shimomura K, Ko CH, Buhr ED, Siepka SM, Hong HK, Oh WJ, Yoo OJ, Menaker M, Takahashi JS. PERIOD2::LUCIFERASE real-time reporting of circadian dynamics reveals persistent circadian oscillations in mouse peripheral tissues. *Proc Natl Acad Sci U S A.*;101(15):5339-46.(2004)

39. Brandstatter R, Abraham U. Hypothalamic circadian organization in birds. I. Anatomy, functional morphology, and terminology of the suprachiasmatic region. *Chronobiol Int.* 20(4):637-55.(2003)

40. Van Tienhoven A, Juhasz LP. The chicken telencephalon, diencephalon and mesencephalon in sterotaxic coordinates. *J Comp Neurol* 118:185-97(1962)

41. Cassone VM, Moore RY. Retinohypothalamic projection and suprachiasmatic nucleus of the house sparrow, Passer domesticus. *J Comp Neurol.* 266(2):171-82 (1987)

42. Abraham U, Albrecht U, Gwinner E, Brandstatter R. Spatial and temporal variation of passer Per2 gene expression in two distinct cell groups of the suprachiasmatic hypothalamus in the house sparrow (Passer domesticus). *Eur J Neurosci* 16(3): 429-36 (2002)

43. Yasuo S, Yoshimura T, Bartell PA, Iigo M, Makino E, Okabayashi N, Ebihara S. Effect of melatonin administration on qPer2, qPer3, and qClock gene expression in the suprachiasmatic nucleus of Japanese quail. *Eur J Neurosci.* 16(8):1541-6. (2002)

44. Yasuo S, Watanabe M, Okabayashi N, Ebihara S, Yoshimura T. Circadian clock genes and photoperiodism: Comprehensive analysis of clock gene expression in the mediobasal hypothalamus, the suprachiasmatic nucleus, and the pineal gland of Japanese Quail under various light schedules. *Endocrinology.*;144(9):3742-8 (2003)

45. Takahashi JS, Menaker M. Role of the suprachiasmatic nuclei in the circadian system of the house sparrow, Passer domesticus. *J Neurosci.* 2(6):815-28 (1982)

46. Cassone VM. Circadian variation of [14C]2-deoxyglucose uptake within the suprachiasmatic nucleus of the house sparrow, Passer domesticus. *Brain Res*459(1):178-82. (1988)

47. Lu, J. and V.M. Cassone Pineal regulation of circadian rhythms of 2-deoxy[^{14}C]-glucose uptake and 2[^{125}I]iodomelatonin binding in the visual system of the house sparrow, *Passer domesticus. J. Comp. Physiol. A* 173: 765-774 (1993)

48. Cantwell EL, Cassone VM Daily and circadian fluctuation in 2-deoxy[(14)C]-glucose uptake in circadian and visual system structures of the chick brain: effects of exogenous melatonin. *Brain Res Bull*;57(5):603-11 (2002)

49. Juss, T.V.S., Davis, I.R., Follett, B.K. and Mason, R.: Circadian rhythm in neuronal discharge activity in the quail lateral hypothalamic retinorecipient nucleus (LHRN) recording *in vitro, J. Physiol.* 475:132 (1994)

50. Rivkees SA, Cassone VM, Weaver DR, Reppert SM. Melatonin receptors in chick brain: characterization and localization. *Endocrinology.* 4: (1989)

51. Cassone VM, Brooks DS, Kelm TA. Comparative distribution of 2[125I]iodomelatonin binding in the brains of diurnal birds: outgroup analysis with turtles. *Brain Behav Evol.* 45(5):241-56 (1995)

52. Reppert SM, Weaver DR, Cassone VM, Godson C, Kolakowski LF Jr. Melatonin receptors are for the birds: molecular analysis of two receptor subtypes differentially expressed in chick brain. *Neuron* 15(5):1003-15 (1995)

53. Zatz M. Light and norepinephrine similarly prevent damping of the melatonin rhythm in cultured chick pineal cells: regulation of coupling between the pacemaker and overt rhythms? *J Biol Rhythms*. 6(2):137-47 (1991)
54. Cassone VM, Menaker M. Sympathetic regulation of chicken pineal rhythms. *Brain Res*. 272(2):311-7 (1983)
55. Cassone VM, Takahashi JS, Blaha CD, Lane RF, Menaker M. Dynamics of noradrenergic circadian input to the chicken pineal gland. *Brain Res*. 384(2):334-41 (1986)
56. Cassone VM, Forsyth AM, Woodlee GL. Hypothalamic regulation of circadian noradrenergic input to the chick pineal gland. *J Comp Physiol [A]*. 167(2):187-92 (1990)
57. Cassone VM, Menaker M. Is the avian circadian system a neuroendocrine loop? *J Exp Zool*. 232(3):539-49 (1984).
58. Cassone, V.M., and J. Lu The pineal gland and avian circadian organization: the neuroendocrine loop. *Adv. Pineal Res*. 8: 31-40(1994)
59. Silver R, LeSauter J, Tresco PA, Lehman MN. A diffusible coupling signal from the transplanted suprachiasmatic nucleus controlling circadian locomotor rhythms. *Nature*. 382(6594):810-3. (1996)
60. Allen G, Rappe J, Earnest DJ, Cassone VM. Oscillating on borrowed time: diffusible signals from immortalized suprachiasmatic nucleus cells regulate circadian rhythmicity in cultured fibroblasts. *J Neurosci*. 21(20):7937-43 (2001).
61. Cassone VM. Melatonin: time in a bottle. O*xf Rev Reprod Biol*;12:319-67 (1990).
62. Armstrong SM. Melatonin as a chronobiotic for circadian insomnia. Clinical observations and animal models. *Adv Exp Med Biol*. 460:283-97 (1999)
63. Pevet P. Melatonin: from seasonal to circadian signal. *J Neuroendocrinol*.;15(4): 422-6 (2003)
64. Redman J, Armstrong S, Ng KT. Free-running activity rhythms in the rat: entrainment by melatonin. *Science*. 219(4588):1089-91 (1983)
65. Cassone VM, Chesworth MJ, Armstrong SM. Dose-dependent entrainment of rat circadian rhythms by daily injection of melatonin. *J Biol Rhythms*. 1(3):219-29 (1986)
66. Pitrosky B, Kirsch R, Malan A, Mocaer E, Pevet P. Organization of rat circadian rhythms during daily infusion of melatonin or S20098, a melatonin agonist. *Am J Physiol*. 277(3 Pt 2):R812-28 (1999)
67. Schuhler S, Pitrosky B, Kirsch R, Pevet P. Entrainment of locomotor activity rhythm in pinealectomized adult Syrian hamsters by daily melatonin infusion. *Behav Brain Res*. 133(2):343-50 (2002)
68. Lewy AJ, Bauer VK, Hasler BP, Kendall AR, Pires ML, Sack RL. Capturing the circadian rhythms of free-running blind people with 0.5 mg melatonin. *Brain Res*. 918(1-2):96-100 (2001)
69. Cassone VM, Chesworth MJ, Armstrong SM. Entrainment of rat circadian rhythms by daily injection of melatonin depends upon the hypothalamic suprachiasmatic nuclei. *Physiol Behav*. 36(6):1111-21 (1986).
70. Chesworth MJ, Cassone VM, Armstrong SM. Effects of daily melatonin injections on activity rhythms of rats in constant light. *Am J Physiol*.;253(1 Pt 2):R101-7, (1987)
71. Warren WS, Hodges DB, Cassone VM. Pinealectomized rats entrain and phaseshift to melatonin injections in a dose-dependent manner. *J Biol Rhythms*. 8(3):233-45, (1993).

72. Dubocovich ML, Rivera-Bermudez MA, Gerdin MJ, Masana MI. Molecular pharmacology, regulation and function of Mammalian melatonin receptors. *Front Biosci.*, 8:D1093-108, (2003).
73. Reppert SM, Weaver DR, Godson C. Melatonin receptors step into the light: cloning and classification of subtypes. *Trends Pharmacol Sci.* 17(3):100-2, (1996).
74. Cassone VM, Roberts MH, Moore RY. Effects of melatonin on 2-deoxy-[1-14C]glucose uptake within rat suprachiasmatic nucleus. *Am J Physiol.*;255(2 Pt 2):R332-7 (1988).
75. Shibata S, Cassone VM, Moore RY. Effects of melatonin on neuronal activity in the rat suprachiasmatic nucleus in vitro. *Neurosci Lett.*;97(1-2):140-4, (1989).
76. Gillette MU, McArthur AJ. Circadian actions of melatonin at the suprachiasmatic nucleus. *Behav Brain Res.* 73(1-2):135-9 (1996)
77. Jin X, von Gall C, Pieschl RL, Gribkoff VK, Stehle JH, Reppert SM, Weaver DR. Targeted disruption of the mouse Mel(1b) melatonin receptor. *Mol Cell Biol.* 23(3):1054-60 (2003).
78. Liu C, Weaver DR, Jin X, Shearman LP, Pieschl RL, Gribkoff VK, Reppert SM. Molecular dissection of two distinct actions of melatonin on the suprachiasmatic circadian clock. *Neuron.* 19(1):91-102 (1997).
79. Glossop NR, Hardin PE. Central and peripheral circadian oscillator mechanisms in flies and mammals. *J Cell Sci.* 115(Pt 17):3369-77 (2002).
80. Reppert SM, Weaver DR. Coordination of circadian timing in mammals. *Nature.* 418(6901):935-41 (2002).
81. Yoshimura T, Yasuo S, Suzuki Y, Makino E, Yokota Y, Ebihara S. Identification of the suprachiasmatic nucleus in birds. *Am J Physiol Regul Integr Comp Physiol.* 280(4): R1185-9 (2001).
82. Bailey MJ, Chong NW, Xiong J, Cassone VM. Chickens' Cry2: molecular analysis of an avian cryptochrome in retinal and pineal photoreceptors. *FEBS Lett.* 513(2-3):169-74 (2002).
83. Abraham U, Albrecht U, Brandstatter R. Hypothalamic circadian organization in birds. II. Clock gene expression. *Chronobiol Int.* 20(4): 657-69 (2003).
84. Yasuo S, Yoshimura T, Bartell PA, Iigo M, Makino E, Okabayashi N, Ebihara S. Effect of melatonin administration on qPer2, qPer3, and qClock gene expression in the suprachiasmatic nucleus of Japanese quail. *Eur J Neurosci* 16(8):1541-6 (2002).
85. Bailey MJ, Beremand PD, Hammer R, Bell-Pedersen D, Thomas TL, Cassone VM. Transcriptional profiling of the chick pineal gland, a photoreceptive circadian oscillator and pacemaker. *Mol Endocrinol.* 17, 2084-95, (2003)
86. Bailey MJ, Beremand PD, Hammer R, Reidel E, Thomas TL, Cassone V. Transcriptional profiling of circadian patterns of mRNA expression in the chick retina. *J. Biol Chem* in press
87. Brandstatter R, Abraham U. Hypothalamic circadian organization in birds. I. Anatomy, functional morphology, and terminology of the suprachiasmatic region. *Chronobiol Int.* 20(4):637-55 (2003).
88. Poirel VJ, Boggio V, Dardente H, Pevet P, Masson-Pevet M, Gauer F. Contrary to other non-photic cues, acute melatonin injection does not induce immediate changes of clock gene mRNA expression in the rat suprachiasmatic nuclei. *Neuroscience.* 120(3):745-55 (2003).
89. Golden SS, Canales SR. Cyanobacterial circadian clocks—timing is everything. *Nat Rev Microbiol.* 1(3):191-9 (2003).

90. Roenneberg T, Colfax GN, Hastings JW. A circadian rhythm of population behavior in Gonyaulax polyedra. *J Biol Rhythms.*;4(2):201-16. (1999)
91. Correa A, Lewis ZA, Greene AV, March IJ, Gomer RH, Bell-Pedersen D. Multiple oscillators regulate circadian gene expression in Neurospora. *Proc Natl Acad Sci U S A.* 100(23):13597-602 (2003).
92. Allen, G., Y. Farnell, D. Bell-Pedersen, V.M. Cassone, and D.J. Earnest Effects of altered clock gene expression on the pacemaker properties of SCN2.2 cells and oscillatory properties of NIH/3T3 cells. *Neuroscience* 127: 989-999 (2004)
93. Wiesinger H, Hamprecht B, Dringen R. Metabolic pathways for glucose in astrocytes. *Glia.*;21(1):22-34 (1997).
94. Magistretti PJ. Cellular bases of functional brain imaging: insights from neuron-glia metabolic coupling. *Brain Res.* 886(1-2):108-112 (2000).
95. Benington JH, Heller HC. Restoration of brain energy metabolism as the function of sleep. *Prog Neurobiol.* 45(4):347-60 (1995).
96. Adachi A, Natesan AK, Whitfield-Rucker MG, Weigum SE, Cassone VM. Functional melatonin receptors and metabolic coupling in cultured chick astrocytes. *Glia.* 39(3):268-78 (2002).
97. Peters J., Bailey, MJ, Cassone, VM (2003) Differential regulation of clock genes and metabolic activity in astrocytes. Neurosci Abs. 16: 6

Biological Rhythms in Neuroendocrinology

DANIEL P. CARDINALI AND ANA I. ESQUIFINO

1. The Temporal Spectrum of Life

As the Earth rotates on its axis, it presents two well-defined environments, the day and the night. As the Earth's rotation axis is tilted, the relative length of day and night changes systematically along the year. As an evolutionary consequence, living creatures have responded to these two geophysical environments by developing predictive mechanisms to adapt themselves successfully. This is the origin of the 24 hour rhythms (called circadian, from Latin circa, about, and die, day) and of the seasonal or annual rhythms. The name circadian was given to these rhythms because of their property of being closed but not exactly 24 h under free running conditions, e.g., in continuous darkness. In humans this period is longer than 24 h, close to 25 h.

Therefore, when animals change between daily, nocturnal or seasonal behaviors they are not merely responding in a passive way to conditions of the environment. Rather, they are following internally generated signals produced by a master pacemaker which is in synchrony with the Earth's cycles, anticipates the transitions between day and night, and triggers behavioral and physiological changes in accordance to those transitions. In other words, a "day" and a "night" are created within the organism, a sort of a mirror of what happens in the environment. Overall, this allows adaptation to a world which changes periodically in an anticipated way.

The most conspicuous biological rhythm is the sleep-wake cycle. The sequence of transitions between the main three physiological states of our life: wakefulness, slow, non REM (NREM) sleep and rapid eye movement (REM) sleep is organized following a strict temporal order under the command of a central pacemaker (or "biological clock"). During sleep this biologic clock prepares us for the anticipated period of wakefulness by controlling the two major communication systems in the body, the endocrine and the autonomic nervous system.[1,2]

Since the "stigma" of time is present in the genes, the same mechanisms that govern the central pacemaker are detectable in every peripheral cell. Therefore, a biologic periodicity exists in association with geophysical periodicity. This association has two advantages: predictability (the speed of Earth rotation has not changed substantially in the last 4.5 billion years of life on Earth) and the possibility to detect changes of enough amplitude in the environment.

2. Reactive and Predictive Homeostasis

The term "homeostasis" was introduced more than 80 years ago by Walter Cannon to define the physiological factors that maintain the equilibrium status of the organism.[3] Following Claude Bernard, Cannon improved the concept of constancy of the internal milieu. As Cannon stated, the prefix "homeo" (alike) was selected rather than "homo" (the same) to include explicitly the normal oscillation detectable in physiological variables. The first time the term "homeostasis" was used was in reference to the mechanisms regulating blood glucose levels. Cannon used the term to describe how blood glucose was controlled with normal values oscillating between a minimum and a maximum. When these limits were trespassed, mechanisms become activated to compensate for them. Then, these processes only function in face of disturbances, never anticipating them ("reactive" homeostasis).[3]

It should be stressed, however, that most regulatory mechanisms have a latency that can be inappropriately long. For example, if a new protein is needed, latency may be as long as 1-2 h, a time which may increase if there is necessity of the synthesis of a hormone. A significant contribution of the circadian organization is that it allows the physiological response to be ready in advance to the expected perturbation of the internal milieu. This happens providing the perturbation occurs at approximately the same hour every day ("predictive homeostasis"[3]).

Let us give an example. If we consider that a small diurnally active mammal finds food at about 2 h distance of its burrow at night, it is essential for the animal to predict the beginning of the night at least 2 h in advance to survive from a nocturnal predator. Rather than looking to environmental variables, which may change in an unpredictable way, like temperature or solar position (because of clouds), it looks for endogenous phenomena which changes predictably as a function of the time of day. It is therefore essential for animal's survival to have a "built-in" time-controlled mechanism which makes time measurement independent from meteorological variations. Indeed, the circadian system is very suitable to perform this task: we can have a reasonable appraisal to time of day by paying attention to our biological functions rather than to our wristwatch. To have an inner "day" and "night" was a very important step in Evolution.

In humans such a predictive activity can be demonstrable. Body temperature and ACTH and cortisol rhythms increase some hours in advance to the major stress of morning awakening; the digestive system anticipates meal time; the cardiovascular system starts in advance during sleep the major changes in cardiovascular control to overcome the change in position during awakening. What Evolution has selected is the development of a precise "biological clock", flexible enough to be resynchronized readily to new temporal situations.

Therefore, because of the seminal influence of Cannon and Moore-Ede the term homeostasis is used today not only for describing the biological strategies to cope with changes in environment (reactive homeostasis) but also the time-related mechanisms which allow the organisms to predict the expected environmental modification by preparing to it in advance (predictive homeostasis).

Another consequence of the circadian organization is that there is no "normal" value of a given variable but a changing normal value during the 24 h cycle. If some external or internal disruptor make this variable to change by trespassing the minimal or maximal values, corrective mechanisms are triggered to return the variable to its normal limits for that precise moment in the 24 h cycle.

An attractive hypothesis (because of its therapeutic importance) is that the majority of diseases and specific pathologies of organs and systems entails a circadian alteration of the patient, which aggravates the disease. Therefore, any successful therapy must combine not only the specific treatment to control the disease, but also the restoration of the circadian activity lost as a consequence of the disease.

As a consequence of the development of the concept of homeostasis, disease can be analyzed both as the alteration of the internal environment it provokes as well as the result of the compensatory mechanisms triggered by the body to protect itself from the disease. As first defined by Hans Selye, the stress response implies a number of processes that are the consequence, not of the original noxious agent, but of the body itself in the course of the defense reaction.[4] Lately, McEwen introduced the concept of allostasis to define the price everyone pays for the stress response.[5] Even a moderate stimulus leaves its mark if repeated in time. In follow-up studies on urban populations the allostatic charge was found to be proportional to: i. arterial systolic pressure; ii. nocturnal excretion of cortisol and catecholamines; iii. body mass index; iv. glycosilated hemoglobin; v. HDL/cholesterol ratio; vi. sleep disturbance.

Levels of hormones like melatonin[6,7] or dehydroepiandrosterone[5] are inversely linked to the allostatic charge. Therefore, at least a part of the allostatic process includes modification of the circadian organization (sleep-wake cycle, melatonin secretion).

The stress reaction program, which encodes the whole level of behavioral autonomic and neuroendocrine responses, is located in the hypothalamus. It is therefore not a surprise that one of the functions of the autonomic nervous system is the synchronization of the trillions of cellular oscillators in the human body to the activity of the central pacemaker located in the suprachiasmatic nuclei (SCN) of the hypothalamus.

3. The History of the Biological Clock

Ancient men were aware of the existence of biological rhythms. Hippocrates made several references to the changes in health and disease with the hour of the day or the time of the year. However, all these phenomena (blossoming of plants, seasonal activity of animals, bird migration, hibernation) were seen as a passive reaction to an environment that changes cyclically. In accordance to this opinion, which prevailed for more than two thousand years, the environment imposes its mark on passively respondent living creatures.

The first recorded objection to such common-sense knowledge occurred in 1729 when the French astronomer Jean-Jacques d'Ortous de Mairan (who was also a physician, geologist, botanist and mathematician) reported that the leaf movements in heliotrope (Mimosa pudica) were still found after isolation of the plant in a dark box. De Mairan interpreted his results as indication that a "biologic clock" was present within the plant.[8] These results were skeptically received by de Mairan's colleagues because they were against the accepted scientific paradigm of the epoch, that is, that the biological rhythms were passive reactions to changes in the environment.

Thirty years later, two botanists, Du Monceau and Zinn, demonstrated that the movements of heliotrope leaves were independent not only from light but also from environmental temperature. In 1832, another botanist, Alphonse De Candolle, confirmed the experiments of De Mairan and established that the heliotrope did not have a 24 h rhythm but a 22-23 h endogenous rhythm. De Candolle made the first reversal experiment by putting the plants in darkness during the day and in light during the night. He found that after a transition period of some days, the plants became synchronized to the new environment by opening the leaves at night under light and closing the leaves during the day in darkness.

These botanical observations had to wait for more than 100 years before they were extended to animals and then to man. Presumably the very influential concept of Claude Bernard about the constancy of the internal milieu precluded an earlier application of biological rhythms. Indeed, the very idea of "constancy" of the internal milieu was against that of rhythmic variation. Only at the end of s. XIX some reports appeared in the scientific literature about daily rhythms of body temperature in shift workers, or in soldiers during nigh duties. In addition the first experiments using primates were published.[9] By 1930 the first scientific society devoted to the study of biologic rhythms was founded (Society for Biological Rhythms). In 1960, at Cold Spring Harbor, a Symposium was held that is considered as the initial landmark of Chronobiology as an independent discipline.

4. Rhythmometry: The Properties of the Circadian Rhythms

Generally we called a rhythm to a sequence of events that repeats in a regular way along time, that is, with the same order and at the same intervals.

Biological rhythm defines any oscillation, regular in time, observable at any level of biological organization. A given rhythm is characterized by the following parameters:

- Period: the time interval between two identical events, that is, the duration of a complete cycle
- Mesor: the mean value of the observed variable, calculated from all values in a complete cycle.
- Amplitude: the difference between the mesor and the maximal value reached in a period.
- Phase: the relationship between two or more rhythms. The term phase is also used to define a part of the cycle, e.g., ascending or descending phase, or maximum of a rhythmic phenomenon.

Occasionally, frequency is employed instead of period to define the number of cycles per time unit. Generally, frequency is used to characterize fast rhythms (like those of electroencephalogram or electrocardiogram). A frequently used frequency unit is the "hertz" (Hz), or cycles per sec.

The phase of a rhythm describes the moment along the temporal course when the biological rhythm is examined. Generally, it refers to another periodic function, either external (time of the day) or internal (another biologic rhythm). As an example of an external reference we define at what time the peak in body temperature occurs or in what season of the year a given species mates. When we say that cortisol rhythm is almost inverse in phase to that of melatonin we are referring to a phase relation between the two rhythms. Usually the phase is characterized by determining the moment at which the examined variable reaches its maximal value (or "acrophase").

The mathematical analysis of the rhythm can be performed by several techniques. A popular procedure is Cosinor that determines, by means of the least square method, the cosinoid function which best describes the variation. This method allows definition of the principal parameters of the rhythm, i.e., period, amplitude, mesor and acrophase, as well as of their confidence intervals.[10] The use of Cosinor implies the existence of a sinusoidal rhythm with Gaussian distribution of experimental errors, a situation not very common in real life. This is the reason why in many cases the preferred analysis is the spectral or Fourier analysis which does not have such a constraint. However, Fourier series analysis needs a great number of observations to be statistically significant.

Among the several biologic rhythms in Nature, those most frequently addressed as having medical application are the 24-h cycles (or circadian). The term "circadian", introduced by Franz Halberg, defines those rhythms that exhibit a period as short to 20-22 h to as long as 26-28 h. Rhythms of a shorter period are called "ultradian" whereas those of a longer period are called "infradian".

It has becomes clear that the biologic rhythms are an essential property of life. This property is transmitted genetically: animals maintained under total darkness by generations show a circadian rhythmicity of motor activity even though they have never been exposed to light.[11-13] Plants selected on the basis of different periodicities will give after cross-breeding, generations of plants that combine individuals of short, intermediate and long periods, following the rules of classical Genetics. These observations support the genetic origin of the circadian phenomena.

The Greek letter τ (tau) is used to define the endogenous period of the rhythm, i.e. that determined under free running conditions. It varies among species, and among individuals within a given species but it is the same for all the circadian rhythms of the same individual.

Another important property of the circadian rhythms is their stability in face of thermal changes. This property was discovered in plants which showed the same circadian period regardless of extreme variations of environmental temperature. Rather than following the Vant'hoff-Arrhenius law of chemical reaction (Q_{10} law, that holds that reaction velocity increases 2.5 - 3 times by increasing 10°C the reaction temperature) the period of the circadian rhythms does not follow this law showing a Q_{10} value of 0.8 - 1.2. Incidentally, if this would not be the case, the circadian structure would be a very weak mechanism to measure biological time.

An important property of the circadian oscillator is the way it varies after exposure to an external synchronizer (called also "Zeitgeber", from the German: time giver). The period in the absence of the synchronizer (τ) becomes T, the period of the oscillator after the action of the Zeitgeber. In other words, a circadian rhythm of period τ different from 24 h is constantly synchronized to a T = 24 h by the influence of environmental Zeitgebers, remarkably bright light.

Indeed, light is the more powerful ambient synchronizer, both for animals and plants. Other Zeitgebers like social factors, physical exercise, meals or temperature also contribute with bright light to entrain the human circadian rhythms.

5. The Synchronizing Effect of Light

As above mentioned, Alphonse De Candolle was the first to demonstrate the influence of light on the biological rhythms. He reported that an inversion of 180° in light-dark cycle caused a 180° change in leaf movement of heliotropes. As less as 1 h of light, appropriately applied in time, is enough to synchronize locomotor activity. Locomotor activity starts every day at the same time, in phase with beginning of scotophase. It is important to stress that light must be applied at certain times to be effective as a synchronizer, as we will discuss later. An important effect of Zeitgeber is related to its intensity. Relevant to this are the following Aschoff's rules:[14]

1. The increase of light intensity prolongs activity time and shortens the rest time in diurnal species, while it shortens activity time and prolongs rest time in nocturnal species
2. Bright light augments amplitude of activity phase in diurnally active animals and decreases it in nocturnally active animals
3. τ is longer than 24 h in diurnal species and shorter than 24 h in nocturnal species. It must be noted that this rule has numerous exceptions (e.g. the rat, a nocturnal rodent, has a τ longer than 24 h)

For other, secondary synchronizers, like social factors, temperature, food availability, similar rules are applied. Secondary synchronizers generally act as cofactors of light but under certain circumstances they can replace the principal Zeitgeber when it becomes suppressed or severely attenuated.

As indicated by the Aschoff's rules the phase relationships between the Zeitgeber and the circadian rhythm does not depends only on the values of T and τ but also on the intensity of the Zeitgeber. This "coupling force" varies with time of application, age, etc. A mathematical double oscillator model was developed by us to explain by dynamic analysis the changes in the oscillator bought about by Zeitgeber intensity.[15]

As stated, the synchronization of a rhythm entails the modification of rhythm's period (τ) to attain T. However, this does not occur homogeneously throughout the cycle but at certain sensitivity times, as described by the "phase response curve" (PRC). This was first demonstrated in unicellular algae.[16] Depending on when a synchronizer agent is applied it can phase advance, phase delay or have no effect on the circadian rhythm. That is, there is a rhythmic variation in the response of the organism to the Zeitgeber.

In man, exposure to bright light during the first part of the night, when body temperature decreases, will delay the clock while the same stimulus applied during the second part of the night, when body temperature increases, will advance the clock. At other times of the day, e.g. at noon, light exposure will not delay nor fast the clock. Melatonin, a "chemical code" of the night (the "darkness hormone") has a PRC opposite to light, with phase advances if given in the first part of the night, phase delays if given in the second part of the night and absence of phase effects when given during daytime (Figure 1).

The following are general rules for PRC's:

1. PRC's are universal and are applied to all living things
2. Similar PRC's are found in diurnal and nocturnal animals
3. Interindividual or interspecies differences in PRC's occur in the magnitude of phase advances or delays found
4. PRC's are not symmetric, with advances predominating on delays or vice versa

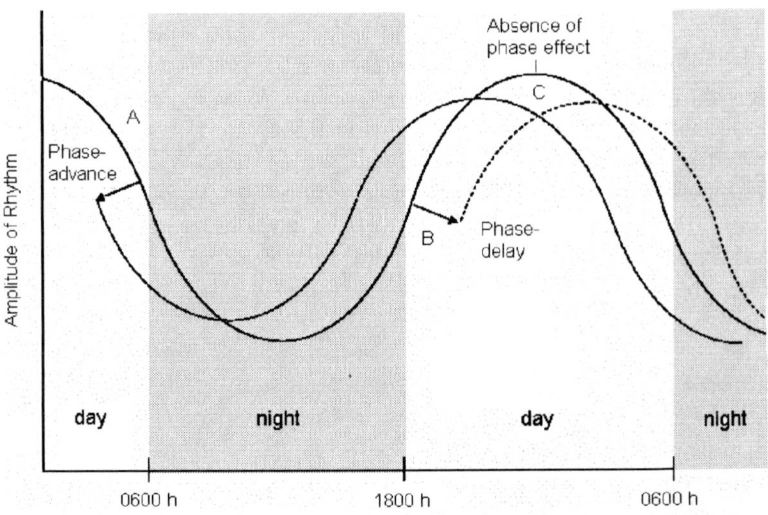

FIGURE 1. Phase-response curve for the effect of melatonin on the circadian system. Melatonin given in late subjective night phase-advances circadian rhythms (A), melatonin administration given in late subjective day delays the rhythms (B) and melatonin administration in mid-subjective day (C) has no effect on phase of rhythm.

5. When light is the Zeitgeber bright light causes greater shifts than dimmer light (1st Aschoff's rule).
6. Phase delays are produced immediately while phase advances take some days to be observed.

To have a PRC implies that during most of daily cycle the synchronizer does not affect the circadian phase. Usually the effect is found close to light-dark or dark-light transitions. Pittendrigh demonstrated that two brief expositions to light at transitions were as effective as the regular photoperiod to synchronize the circadian system (skeleton photoperiod).[16] Summarizing, the PRC is a formal prerequisite to define the action of a physical, chemical or environmental perturbation on the circadian clock.

From the analysis of the PRC for light in humans it can be concluded that bright natural light in early morning is the stimulus that phase-advances the clock every day from 25 to 24 h. This is of fundamental importance in the present "24/7 Society", working 24 h, 7 days a week, since often we are underexposed to such a synchronizing influence of morning natural light. Secondary Zeitgebers like physical exercise, diet or social interaction may also play a role when associated with environmental light.

Almost every physiologic function exhibits circadian rhythmicity, but generally their maxima occur at dissimilar times. This ordered sequence of maximal values that reveals cause-effect relations at all the organizational levels of the body constitutes the quintessence of health and are described as "phase maps". "Phase maps" can be transitorily disrupted in situations like jet lag or shift work when an organism faces a rapid change of temporal parameters. Under these circumstances, the 24 h rhythms do not adapt themselves at the same velocity and the temporal relations become distorted. Resynchronization needs several days (about 1 day/h of phase shift) and during this transition time symptoms of malaise are found (e.g., "jet-lag").

6. The Circadian Oscillator

During the 1960's it turned to be clear that the overt rhythms were not the oscillator itself but a consequence of the oscillator activity. Using the clock analogy, arms movement is not the clock mechanism but rather its manifestation. PRC allows differentiation between the oscillator and the overt rhythms. In algae individual inhibition of effector rhythms (cell division, photosynthesis, luminescence, etc.) does not affect the other rhythms, whereas if the primary genomic oscillator is suppressed, all the overt rhythms disappear.[16]

Reinberg introduced the concept of phase map, to define the temporal structure of the different biological rhythms found in a single individual and originated by a single pacemaker.[17] Identical PRC with varying acrophases are found. Therefore, the same mechanism seems to govern the complete set of biologic rhythms in a giving organism, even though different acrophases are displayed.

In unicells this is easily demonstrable. In pluricellular organisms at least two theoretical possibilities exist: i. the cells are passive responders to a rhythmic signal generated by the oscillator; ii. every cell is endowed with a circadian oscillator adjusted by the master clock to a similar pace.

Several lines of evidence obtained in the last 20 years support the second view. Almost all cell types in a pluricellular organism express intrinsic circadian properties.[11,13,18] In pluricellular organisms the host of cellular individual circadian oscillators is synchronized by a hierarchically superior, a master pacemaker located in the suprachiasmatic nuclei (SCN) of the anterior hypothalamus, whose activity is affected by Zeitgebers in a more or less direct way.

One subject of debate still open concerns to the existence of more than one master pacemaker in vertebrates, thus implying some sort of hierarchically arrangement between them. An important argument in favor of this hierarchical order is that in some cases an internal desynchronization occurs in a given individual, suggesting that the hierarchy has become disarrayed. This disorganization may derive form external causes (e.g. a transmeridian flight or the shift work) or from internal ones, like in those seen in individuals kept for long periods in a free running condition.

Two strategies have been followed to analyze the molecular mechanisms of circadian rhythmicity: i. the use of drugs of known action on intracellular metabolism: ii. the use of classic genetic techniques and more recently of molecular biology to analyze the different genes involved in circadian cell phenomena.

As far as the pharmacological strategy, a major problem was to differentiate between possible direct effects of the drug on the oscillator from those exerted on the overt rhythms. Clear-cut results were obtained with inhibitors of cell oxidative metabolism indicating the dependence of circadian oscillation from energy supply. The administration of deuterium oxide (heavy water) has the curious property to prolong the endogenous period of circadian rhythm. Inhibitors of RNA or protein synthesis were mainly employed in unicell studies. In Acetabularia, algae whose cell nucleus is easily dissected out, the circadian rhythm in photosynthesis persisted for weeks in the absence of the nucleus. If the nucleus from another Acetabularia is transplanted the chimera algae shows the circadian rhythmicity of the donor. Actinomycin D (an inhibitor of RNA polymerase) suppresses circadian rhythmicity in intact Acetabularia but not in the enucleated one. These results indicate that the RNA that codes the rhythm persists for a long time in the cytoplasm.[11]

Contrariwise, translational phenomena are absolute requirements for circadian rhythmicity. This is based on numerous experiments using pharmacological blockers of protein synthesis (e.g. antibiotics like puromycin, cycloheximide or cloramphenicol). These drugs block gene translation in 80S ribosomes (participating in the synthesis of cytoplasm proteins, like puromycin or cycloheximide) or in 70S ribosomes (involved in the synthesis of mitochondrial and chloroplast proteins, like cloramphenicol). In algae, circadian rhythms of photosynthesis and chemiluminescence are inhibited by puromycin or cycloheximide, but not by cloramphenicol. Therefore, the proteins involved in circadian rhythmicity seem to be of a cytoplasm type. In rodents, the administration of puromycin or cycloheximide caused phase shifts of circadian rhythms showing typical PRC, whereas the injection of cloramphenicol did not affect the clock.

About 20 years ago, some aspects of the early gene expression in the circadian system were uncovered. The first gene studied in the SCN was the c-fos oncogene which exhibited a circadian rhythmicity and a PRC similar to the PRC to bright light. From these results it was proposed that the c-fos product was a part of the sequence of events that transduce environmental light information in circadian response. Lately c-fos studies were extended to several other early genes in SCN.[19-23]

Several membrane events were shown to be instrumental in circadian rhythm generation. This is relevant when dealing with ultradian phenomena, like electrical activity of neurons, whose frequency can be as high 0.5 KHz. The rhythmic production of action potentials of different periods is the result of membrane processes involving ion channels, second messengers and the metabolic cascade triggered by transmitter interaction with receptors. The proposed model includes an "oscillatory" protein that becomes periodically fixed to the membrane to trigger the cyclic changes in ion permeability in the circadian oscillator. As discussed below, this model generally coincides with that derived from molecular biology studies.[13,18]

Genetic studies have been very useful to get information on oscillator's nature. They were carried out initially in Drosophila melanogaster and Neurospora crassa and more recently extended to rodents by a transgenic gene deletion methodology. Generally the objective pursued was the isolation of mutants with circadian periods shorter or longer that the wild type.[12] In Drosophila mutants occurred with modified periods for circadian rhythms like motor activity. In some cases, the mutants were arrhythmic; in some other τ values were shorter or longer than 24 h.

These genes were called per S or per L, respectively. These mutations followed the classical genetic laws, being dominant, semi-dominant or recessive. Interestingly, ultradian rhythms, like sex behavior, were also abnormal. Mutant genes in Drosophila melanogaster are located in the X chromosome. A brain transplant of a wild type fly to a per S fly will shorten τ in the expected way. In some cases, genetic mosaics with two maxima in locomotor activity occurred, as if parts of the brain alternate to rule the rhythm.[12] By means of molecular biology techniques, the per gene was cloned and sequenced. It consists of a 7.7 Kb DNA fragment. Its mRNA was used to transfect flies.

In Neurospora crassa the rhythm examined was the growth of conidia. As in the case of Drosophila, mutants with τ shorter or longer than 24 h were detected. These mutations were called frq (for "frequency") and followed the classical genetic laws, being dominant, semi-dominant or recessive.

Mutations in per gene of Drosophila or in frq gene of Neurospora include an allelic semi dominant series that codes for short or long τ as well as for arrhythmicity.There are similarities between the frq and per genes, and in some cases, identical sequences. This indicates that at least a portion of the gene remained unaltered for more than a million years, the estimated time for separation between plant and animals.[12]

The regulatory mechanisms that link gene expression with circadian phenomena have been explored with some detail. After elucidation of per structure the synthesis of the coded protein was possible. Per protein is an 1127 amino acid proteoglycan. The use of fluorescent anti-per protein antibodies allowed the study of its intracellular processing.

Per protein is a cytoplasm protein widely distributed in tissues of Drosophila. It is associated with gap-junction regulation, the most primitive mechanism of cell to cell communication.[11,12] Studies on the per gene indicated two sites of interaction with per protein. This exerts a negative feedback on its own synthesis thus giving the basis for an oscillatory synthesis control. Per protein and other immunologically related proteins are widely distributed among animals and plants. Several circadian genes have been identified in animals and plants.[13,18]

Indeed, the clock genes are a universal property of living organisms, as universal as the cell cycle. Since a striking homology is demonstrable in these genes from Drosophila to humans it can be postulated that these genes "speak" a common language, as common as the genetic code.

Figure 2 shows the proposed logical elements of the "interlocked loops" model of eukaryotic circadian clocks. Positive transcriptional complexes made up of two types of PAS-containing proteins initiate the circadian cycle.

The activity of PAS1 factors is periodic, transient and limiting, and is triggered by elements of the preceding cycle. Examples are mBmal 1 in mice, clock in Drosophila melanogaster and WC-1 in Neurospora. The second category of PAS-containing activator (PAS2) is constitutively active (mClock in mice, Cycle in the fly and WC-2 in the fungus). The transiently active heteromeric complexes drive expression of core clock genes, encoding functionally equivalent proteins.

The protein products of these genes form heterodimeric complexes that control the transcription of other clock genes, notably three Period (Per1/Per2/Per3) genes and two Cryptochrome (Cry1/Cry2) genes, which in

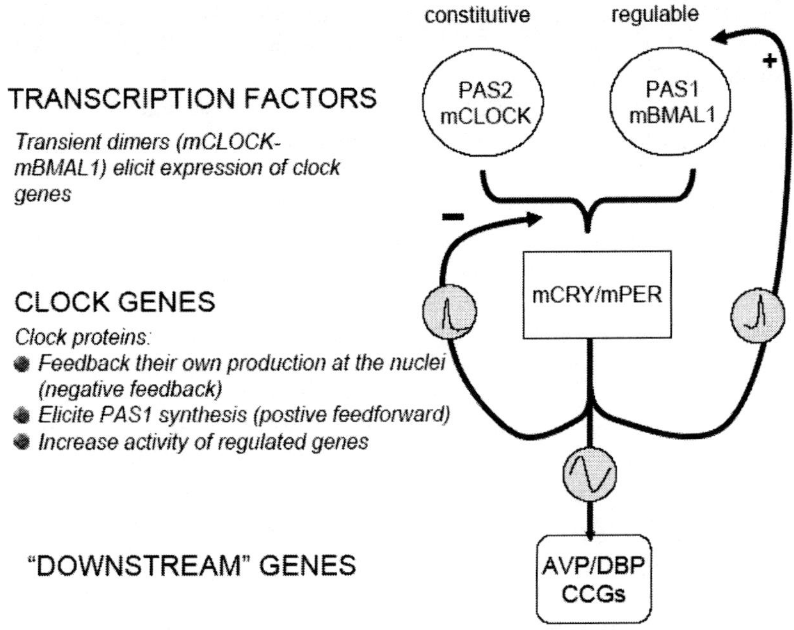

FIGURE 2. Interlocked loop model of cellular circadian clocks. Positive transcriptional complexes made up of two types of PAS-containing proteins initiate the circadian cycle. The activity of PAS1 factors is periodic, transient and limiting, and is triggered by elements of the preceding cycle. The second category of PAS-containing activator (PAS2) is constitutively active. The transiently active heteromeric complexes drive expression of core clock genes, encoding functionally equivalent proteins. "Downstream" genes include arginine vasopressin (AVP), albumin D element-binding protein (DBP) and other clock.controlled genes (CCGs). For further detail see text.

turn provide the negative feedback signal that shuts down the Clock/Bmal drive to complete the circadian cycle.

Overall these proteins fulfill three types of function. First, subject to post-translational modifications and heteromeric associations of their own, they enter the nucleus to oppose the positive drive to their cognate genes. This delayed negative feedback is facilitated by the simultaneous decline in activity of the PAS1-type factor, and could alone sustain an oscillation in gene expression. Such a rhythm would, however, tend to dampen with time.

The second function of the negative factors is to feed forward to trigger the subsequent transient availability of the PAS1-type factor responsible for initiating the next cycle. This positive feed forward is accomplished at transcriptional (flies, mammals) or post-transcriptional (Neurospora) levels and, in combination with the negative loop, enhances the stability, precision and amplitude of the cycle.

The third role of the clock factors is to direct the rhythmic expression of downstream, output genes sitting out of the core loops. These out-of-loop signals are embedded in essential routes, often transcriptional cascades, through which the clockwork communicates circadian phase to the rest of the cell.

Evidence so far indicates that it is accomplished at both transcriptional and post-transcriptional levels (Figure 2).[13,18]

7. The Central Clock

In pluricellular organisms individual genomic circadian expression needs to be synchronized by a hierarchically superior structure in order to give rise to the circadian rhythms. In mammals there is evidence that a region in the anterior hypothalamus, the SCN, is the central pacemaker for the circadian rhythms (Figure 3).[24]

These nuclei, that contain a few thousand neurons in man, have the property to generate circadian rhythms even in isolation from other brain structures. SCN integrity is required for generation and maintenance of 24 h rhythms, as well as for their synchronization by the environmental light/dark cycle. This is remarkable property since most complex behaviors like sleep or wakefulness involve operation of large areas of the brain whereas for circadian rhythmicity the major brain region participating is single and has a minimal anatomical volume.[24]

Environmental light of an appropriate intensity (like that of natural light in the morning) activates a particular type of photosensitive ganglion cells in the retina which through specific neuronal projections (the retino-hypothalamic tract) cause the genomic activation of neurons in SCN.

These photoreceptive cells are unique because: i. they are not located in the same retinal layers as the rods and cones; ii. they do not participate in visual

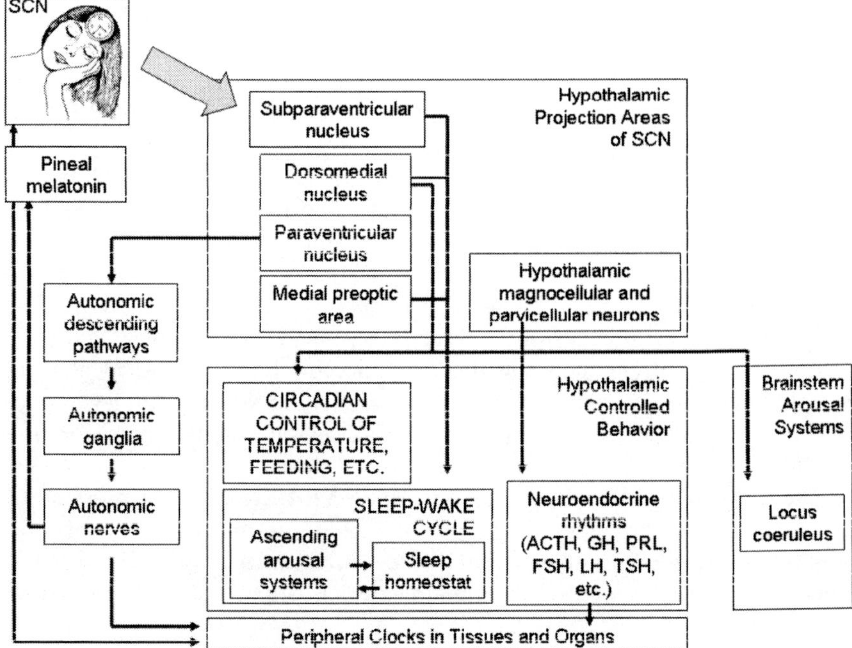

FIGURE 3. Hierarchical organization of the circadian system. The different pathways employed by the SCN are depicted.

processes; iii. they contain a unique photopigment (melanopsin). Information generated in SCN neurons is conveyed to specific areas in the basal hypothalamus that control the two major communicating systems in the body, the endocrine and the autonomic nervous systems.

Like human oligarchies, the "SCN neuronal oligarchy" exerts genomic control of the trillions of cells in the body by monopolizing both the major information from the environmental Zeitgeber (light received via the retino-hypothalamic tract) and the control of the endocrine and autonomic information network (Figure 3). The cooperation of other Zeitgebers, e.g. physical activity, to synchronize the clock is also warranted.

As already mentioned, after the first experimental demonstration of an endogenous 24 h rhythm made in plants[8] almost 200 years elapsed before similar rhythms were described in primates.[9] This set the basis for looking for the tissue responsible to maintain such rhythmicity.

Since several endocrine organs in isolation (like the adrenals) were able to keep a circadian way of secretion even in vitro it was postulated that these organs served to maintain the rhythms in other tissues of the body. However, surgical endocrine ablations, or lesion of a large number of CNS areas, did not eliminate cellular or behavioral circadian rhythms. Only the destruction of SCN brought about the complete disappearance of circadian rhythms.[25-28]

Fundamental evidence to support the clock nature of SCN is given by the following observations:

1. SCN neurons posses an endogenous rhythm of electrical activity even in vitro, with a high electrical activity during the day and low at night.
2. Concomitant to the electrical changes a higher SCN metabolic activity and glucose uptake is found during the subjective day. Per and Cry messenger RNA's peak in the SCN in mid-to-late circadian day, regardless of whether an animal is nocturnal or diurnal. This is synchronized by light.
3. Transplantation of intact SCN to animals having their own SCN destroyed reestablished the lost circadian rhythmicity.

8. Circadian Clock Effectors

SCN grafts restored normal locomotor activity in SCN lesioned hamsters, an effect seen in advance to the establishment of synaptic connection between the graft cells and the donor. Thus a humoral mechanism was postulated as the primary way in which the SCN transplant acted. However it was then established that paracrine circadian rhythm restoration was only partial (e.g. neither melatonin nor cortisol rhythms were restored) indicating that precise synaptic connections are needed for full operation of SCN.[13,24] SCN paracrine signaling includes the secretion of autacoids like transforming growth factor-α or prokineticin-2.The circadian effects of SCN are exerted through neuronal projections (Figure 3 and Table 1). They include:

1. Hypothalamic neuroendocrine parvicellular neurons
2. Central autonomic neurons located at the hypothalamic paraventricular nucleus (PVN)

TABLE 1. Functional neuroendocrine relevance of SCN projections.

SCN projections to:	Anatomical substrate of circadian effect on:
Estrogen receptor-containing neurons in rostral forebrain bundle and some GnRH neurons	Hypophysial gonadal axis
Tuberoinfundibular dopaminergic neurons	Prolactin secretion.
PVN neurons projecting to the inter-mediolateral column	Release of CRH and ACTH. Direct effects of autonomic innervation on pineal, thyroid, parathyroid, adrenal, pancreatic, hepatic, gonadal and immune function
SubPVN and dorsomedial hypothalamic nuclei	Sleep-wake cycle
Hypothalamic parvicellular neurons	TSH, GH release
Hypothalamic magnocellular neurons	Neurohypophysial hormone release

3. Hypothalamic areas (subPVN, dorsomedial nucleus, medial preoptic area) relaying information from SCN to other autonomic and neuroendocrine neurons. SCN projection of the medial preoptic area is important for sleep regulation.
4. Extrahypothalamic areas like the lateral geniculate body or the paraventricular thalamic nuclei. These projections are important for synchronization of circadian hypothalamic and behavioral responses.

Subpopulations of SCN neurons contain different neuropeptides like AVP or somatostatin, thus suggesting a functional specialization. Most of SCN neurons contain GABA together with a neuropeptide. Glutamate was demonstrated to be present in SCN neurons by electrophysiological means.

The existence of several combinations of transmitters in SCN neurons suggests that these neurons have an important degree of diversification to transmit signals.[24] For example, in the case of the daily rhythm of corticosterone, particular populations of SCN neurons exert two effects at least: inhibition via the rhythmic release of AVP and stimulation through an unidentified transmitter. These mechanisms are important for the time-of-day changes in stress response.[24]

Another example is given by sex cycle regulation. SCN AVP fibers make synaptic contacts with gabaergic interneurons that contain estrogen receptors in medial preoptic area and to a much lesser degree with gonadotropin-releasing hormone (GnRH) neurons. A normal release of AVP is needed to have a normal secretion of GnRH and thus of luteinizing hormone (LH).

It is of interest that this facilitatory effect of AVP on the GnRH-LH axis coincides with the inhibitory action of AVP on corticosterone secretion. Thus through the secretion of AVP, SCN neurons decrease the stress response and facilitate sexual receptivity.[29,30]

Recently another mechanism has been increasingly mentioned to convey SCN signals to tissues and organs in the body. This is the modulation of hormonal or immune response via the autonomic innervation to these tissues.

We proposed such a mechanisms more than 10 years ago by examining the effect of the sympathetic and parasympathetic innervation in a number of endocrine glands, describing that in a number of circumstances it is needed for normal response to physiological stimuli (circulating thyrotropin, TSH, for thyroid follicular cells; extracellular Ca^{2+} changes for thyroid C cells and parathyroid cells).[31-33]

In the case of the adrenal cortex there is evidence that besides the classic neuroendocrine command given by the hypothalamic-hypophysial axis gland's innervation also plays a role to determine the final levels of corticosterone secretion.[34,35] In view of the relevant projections of SCN neurons on autonomic descendent pathways it is clear that the final tune of corticoid secretion by SCN is also exerted via neuroendocrine and autonomic means.

This seems to be a general principle not only for the endocrine system but also for primary and secondary immune organs.[36-38] Circadian rhythmicity of lymph cell proliferation submaxillary lymph nodes is regulated by autonomic sympathetic and parasympathetic nerves. Recently similar findings were made in the liver, e.g. with the beginning of the activity phase SCN neurons augment tissue sensitivity to insulin and glucose production in the liver by a combined hormonal and neural effect.[39]

It is important to stress that anatomopathologic studies of human SCN have revealed that most of neuroanatomically obtained results in rodents can be extrapolated to humans.[40] High electrical activity of SCN during the day occurs both in nocturnal animals like the hamsters as well as in diurnal animals like primates. However, in primates corticoid secretion and initiation of the activity phase (characterized by a high sympathetic tone and higher body temperature) occur at the beginning of the light phase and not at the beginning of the night as in rodents. This means that the signal produced by the SCN is differentially interpreted in the case of nocturnal and diurnal animals by mechanisms that remain undefined. It should be noted that the only sympathetic territory stimulated in humans at night is that of the pineal gland, responsible for melatonin secretion.

9. The Neuroendocrine Rhythms

In humans both sleep and hormone secretion is controlled by a double command system given by the predictive and the reactive homeostasis processes.[2,41] Reactive homeostatic mechanisms link the depth of sleep with duration of preceding wakefulness while the circadian mechanisms play a fundamental role to determine initiation of sleep and relative duration of REM and NREM sleep phases.

The recurrent cycles of NREM and REM sleep are accompanied by major changes in all physiological systems of the body. Indeed it can be said that we live sequentially in "three different bodies": that of wakefulness, that of NREM and that of REM. For an adult living 75 years, approximately 50 years are lived in wakefulness, 19 years in NREM sleep and 6 years in REM sleep. NREM sleep duration readily decreases (about 30 min per decade) and stages III and IV or non REM sleep represent less than 10% of total sleep after 40-50 years of age. This decrease is compensated by increases in stages I and II while REM sleep or total duration of sleep remain more stable.

Striking differences between these three physiological stages have been documented. During NREM sleep there are decreases in blood pressure, heart rate, and respiratory rate, and release of anabolic hormones like growth hormone (GH) and prolactin together with a general increase of the immune function. The concomitance of these events gives credence to the notion that NREM sleep is functionally associated with anabolic and cytoprotective processes. The brain itself is hypoactive as indicated by a

20-30% reduction in oxygen consumption, resembling what is seen in a light anesthesia.

In contrast, REM sleep is associated with an "antihomeostatic" stage. The regulatory mechanisms controlling the cardiovascular, respiratory and thermoregulatory functions become grossly inefficient. Heart rate and blood pressure increase out of feedback control, and respiratory rate becomes irregular. Penile erection in males and clitoral engorgement in females accompany the brain and autonomic activation of this phase and the somatic musculature is actively inhibited. Awakening from activated or REM sleep typically yields detailed reports of hallucinoid dreaming, even in subjects who rarely or never recall dreams spontaneously. This indicates that the brain activation of this phase of sleep is sufficiently intense and organized to support complex mental processes and again argues against a rest function for most of the brain during REM sleep. Indeed, several areas of the brain, like the limbic system, are more active in REM sleep than during wakefulness.[1,2]

A significant physiological concomitant of the antihomeostatic stage of REM sleep is the loss of temperature regulation. If ambient or core temperature begins to fall, sleep is interrupted, but thermoregulatory processes cannot be brought into play during REM sleep. Thus the notion that we humans are exclusively homoeothermic animals is not longer tenable.

Two interacting processes regulate the timing, duration and depth, or intensity, of sleep: a homeostatic process that maintains the duration and intensity of sleep within certain boundaries and a circadian rhythm that determines the timing of sleep. The homeostatic process depends on immediate history: the interval since the previous sleep episode and the intensity of sleep in that episode. This drive to enter sleep increases, possibly exponentially, with the duration since the end of the previous sleep episode. It declines exponentially once sleep has been initiated. This reinforces the cyclical nature of sleep and wakefulness and equates sleep with other physiological needs such as hunger or thirst. The homeostatic sleep drive controls NREM sleep rather than REM sleep.[2]

In contrast, the phase and amplitude of the circadian rhythm are independent of the history of previous sleep but are generated by the major pacemaker, the SCN. The circadian variation of human sleep propensity is roughly the inverse of the core body temperature rhythm: maximum propensity for sleep and the highest continuity of sleep occur in proximity to the minimum of temperature.

For most adenohypophysial hormones their 24 h rhythmicity result from the interaction of the circadian clock with the sleep homeostat and include pulsatile, ultradian components of about 90 min, which is the time elapsed between every NREM and REM. These hormonal rhythms depend mostly on the sleep homeostat (e.g. GH, prolactin), on the circadian clock (e.g. cortisol, melatonin), or on both (e.g. TSH).[42-44]

9.1. Hormones Controlled by Reactive Homeostatic Mechanisms

9.1.1. Prolactin

In normal young adults the pattern of prolactin secretion is characterized by an increase immediately after sleep onset and during the first half of the night. Sleep onset, which is accompanied by a predominance of NREM sleep, has a prolactin releasing effect regardless of the time of the day sleep occurs. However, amplitude of prolactin release is less when sleep occurs at a time different from the night, indicating that a circadian component is also involved. This circadian component is more pronounced in women.[42] Presumably it involves the pineal gland and melatonin as shown by the inhibition of prolactin levels in subjects in darkness exposed to a bright light pulse (a situation in which melatonin secretion is inhibited) and by the stimulatory effect of melatonin on prolactin secretion when administered during the day.

Prolactin release is closely associated with NREM sleep. Sleep alterations (e.g. sleep apnea) show very low nocturnal prolactin levels. Aging is associated with changes in both processes of sleep (circadian and homeostatic).[43] Those of the sleep homeostat are shown by the exponential decrease of NREM sleep and of prolactin and GH secretion and are completed at 50-60 years of age. In the case of the circadian component (as evaluated by the secretory pattern of cortisol and melatonin) they become evident at a later age.[43,45]

9.1.2. GH

The effect of NREM sleep on GH release has been known for more than 30 years. In normal young adults the circulating profile of GH consists of low stable concentrations together with secretory pulses. In men, more than 70% of constitutive GH secretion occurs shortly after sleep onset. In women, some high amplitude secretory pulses also occur during the day, with a positive correlation between estradiol and daily GH levels.[42]

Changes in sleep/wake cycle are followed readily by changes in GH secretion. As in the case of prolactin, circadian influences on GH secretion are apparent. In addition, the relation of GH release to NREM sleep although obligatory is not consistent. About one/third of NREM sleep episodes are not accompanied by GH secretion. This could depend on variations in somatostatinergic tone in median eminence.

The decrease of GH levels as a function of age follows an exponential relationship, reaching at the middle of life 40% of the values found in the young in the absence of significant decreases of steroid hormone levels. In a study comprising 114 normal men, daily secretion of GH at 25-35 years of age and at 60 years of age was 50 and 30% of that seen at 16-25 years of age. These changes were independent of body mass index.[41,43,45]

Aging is associated with dramatic decrease in the circulating levels of GH and insulin-like growth factor (IGF)-1 α. This is due to reduction of peak amplitude rather than to modification of frequency of release. The decline of IGF-1 α is more gradual than that of GH. In the elderly, IGF-1 α levels are about 50% those in young, with a great interindividual variability. About 20% of normal men elder than 60 years exhibit normal IGF1 α levels.

The primary alteration that explains GH reduction with age is a concomitant increase in somatostatinergic tone of median eminence. The secretory capacity of adenohypophysial cells remains unaltered with age. These observations are the basis for the use of growth hormone releasing hormone (GHRH) in old subjects. It must be noted that it is impossible to reproduce the normal release of GH by these means, or by GH administration. Some studies indicate that NREM sleep promotion could represent a novel "secretagoge" for GH useful for somatopause treatment. Examples of this are γ-hydroxybutyrate[46] or melatonin, that promote central gabaergic activity[47] and NREM sleep.[48]

9.1.3. TSH

TSH is a hormone controlled both by the sleep wake homeostat and by the circadian clock. Frequency and amplitude modulation of TSH release rhythm occurs. During the day, TSH levels are low and relatively stable. In young adults a late afternoon elevation ensues, attaining its maximum at the initiation of sleep. The late afternoon elevation indicates the regulatory effect of the circadian clock whereas the decrease at the beginning of sleep underlines the inhibitory influence of sleep on TSH release. During sleep deprivation the nocturnal decline of TSH is not longer seen. This inhibition is related to NREM sleep and in some cases REM sleep coincides with TSH secretion.

The late afternoon increase of TSH is considered a circadian marker. On the other hand the inhibition of TSH by NREM sleep illustrates the interaction between sleep and the circadian rhythmicity. Aging is associated with a progressive decrease of TSH secretion by decreasing amplitude rather than frequency of the secretory pulses.[41]

9.2. Hormones Controlled Principally by the Circadian Clock

9.2.1. Cortisol

The 24 h changes in circulating cortisol levels are a reflection of the circadian control of ACTH secretion. This results from the periodic changes in pituitary stimulation by CRH. In young adults the maximum in plasma cortisol occurs in the morning, around 0700-0800 h, and is followed by an impending decrease with a nadir at midnight and an abrupt elevation late at the second half of sleep.[41] This profile is under the control of the circadian pacemaker through modulation of amplitude of pulses. However, participation of sleep is also demonstrable since initiation of sleep always coincides with profound inhibition of cortisol secretion by an effect attributed to NREM sleep. During the second part of the night, awakenings and particularly the last REM episode are consistently followed by bursts of cortisol secretion.

Aging is associated with changes in plasma cortisol. In a retrospective study of 90 men and 87 women (age 18 to 83 years) higher circulating cortisol levels were seen in young men as compared to age-matched women.[41] The 24 h variations are detectable even in the elder subjects. The cortisol profile after 50 years of age shows a consistent and moderate increase in both sexes. Between 20 and 80 years of age cortisol levels increases by 20-50% (more in females). Therefore mean cortisol levels tend to be similar between sexes after 60 years of age. Typically, the nadir of cortisol rhythm at 70-90 years of age is about 3-4 times higher than in a young adult. There is also a phase advance of the cortisol rhythm in the aged subjects.[41]

Several clinical and experimental studies have indicated a deleterious effect of high cortisol levels on the hippocampus, more pronounced in the nadir of the 24 h rhythms rather than in its zenith. Even a modest increase of afternoon cortisol can be associated with memory disturbances and increased insulin resistance.[5]

9.2.2. Melatonin

Plasma concentration of melatonin varies on a 24 h basis, with maximal values at night (about 200 pg/ml) and minimal during the day (10-30 pg/ml).[49] In humans the peak of melatonin secretion is at 0002 – 0006 h and does not exhibit any consistent relation with the sleep phase (Figure 4). While melatonin is produced in most organisms from algae to mammals, and its role varies considerably across the phylogenetic spectra,[50] in humans it seems to play a major function in the coordination of circadian rhythmicity, remarkable the sleep-wake cycle.[49,51]

The circadian rhythm of melatonin is generated by the SCN of the hypothalamus, and like many other circadian rhythms, it is synchronized to a 24-hour period largely by cues from the light–dark cycle received mainly via the retino-hypothalamic pathway to the SCN.

The evening increase of melatonin secretion is associated with an increase in the sleep propensity. Secretion of melatonin during the day, as seen in diverse pathologic or occupational health situations, is strongly associated with daytime sleepiness or napping, while the administration of melatonin during the day causes sleepiness.[52]

Melatonin secretion is an "arm" of the biologic clock in the sense that it responds to signals from the SCN and in that the timing of the melatonin rhythm indicates the status of the clock, both in terms of phase (i.e., internal clock time relative to external clock time) and amplitude. From another point of view, melatonin is also a chemical code of night: the longer the night, the

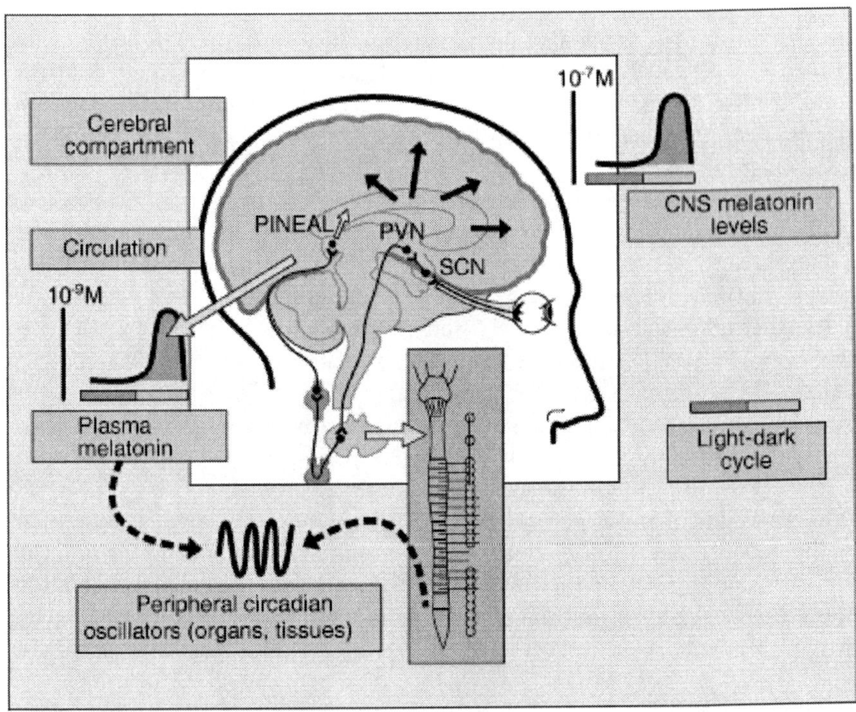

FIGURE 4. Involvement of melatonin in the transmission of photoperiodic information to peripheral clocks. Neuronal efferent pathways from the SCN directly distribute circadian information to different brain areas, including the pineal gland, that generates the melatonin rhythm. The neural route for environmental lighting control of melatonin secretion includes the intermediolateral column of the thoracic chord gray and the superior cervical ganglion. The generated melatonin rhythm is used by the SCN to distribute its rhythmic information. The possibility that there could be two compartments of melatonin affecting physiological function, with melatonin concentration in brain about 50 times greater than in plasma, has been proposed.[54]

longer the duration of its secretion. In many species, this pattern of secretion serves as a time cue for seasonal rhythms.[49]

Daily timed administration of melatonin shifts the phase of the circadian clock (Figure 1), and this phase shifting may explain melatonin effect on sleep in humans. Indirect support for this physiological role came from clinical studies on blind subjects showing free running of their circadian rhythms while a more direct support for the hypothesis was provided when the PRC for melatonin was demonstrated to be opposite (i.e. 180 degrees out of phase) to that of light.[51]

Melatonin (in a dose of 5 mg daily, timed to advance the phase of the internal clock) can maintain synchronization of the circadian rhythm to a 24-hour cycle in sighted persons who are living in conditions likely to produce a free-running rhythm, and it synchronizes the rhythm in some persons after a short period of free-running. In blind subjects with free-running rhythms, it has been possible to stabilize, or entrain, the sleep–wake cycle to a 24-hour period by giving melatonin, with resulting improvements in sleep and mood.[53] The phase shifting effects of melatonin were also sufficient to explain its utility as a treatment for circadian-related sleep disorders such as jet lag or the delayed phase sleep syndrome.

Coincidence of melatonin secretion with the dark phase of 24 h cycle is a common fact in living organisms. In higher vertebrates there are some differences in the temporal relationship of the peak in melatonin with initiation of the scotophase.

Three patterns are detectable in melatonin secretion. For Type I (e.g. hamsters, mice) there is a latency of several hours from the beginning of the dark phase and a blunting of the peak some hours in advance to the end of darkness. Type II includes those species in which melatonin levels gradually augment after light-dark transition to attain their maxima at the middle of scotophase, decreasing thereafter before dark-light transition. The rat and the human show this type of melatonin profile. In the case of type III, melatonin augments abruptly with the beginning of the dark phase and is kept high till the very end of the dark period. This is the case of sheep or cats.

The amplitude of melatonin peak is affected by several factors:

- Age: day-night differences in melatonin concentration decrease with age (i.e. levels in childhood are about 6-7 times higher than in the old age).
- Season: with phase advances in summer and phase delays in the winter.
- Menstrual cycle: with moderate decrease at the periovulatory period.
- Life style: any situation (shift work, night work, chronic disease) that affects exposure time to sunlight
- Drugs: some pharmacological agents decrease melatonin secretion (e.g. benzodiazepines, adrenergic beta-blockers). Other augment it (e.g. MAO inhibitors, tricyclic antidepressants).
- Environmental light: artificial bright light (2500 lux at least, approximately the intensity in late afternoon during the summer) blunts the nocturnal peak of melatonin after a brief exposure. The same light intensity applied at late night, close to dark-light transition, will cause a phase advance of melatonin rhythm whereas when applied at early night will cause a phase delay.
- Stress: several stressors (e.g. hypoglycemia, surgery, immobilization) modify melatonin secretion. In man physical exercise augments melatonin secretion
- Cardiovascular disease, like coronary disease or congestive heart failure.[7,55]

Although day to day melatonin secretion is remarkable constant within individuals the interindividual variation is very high. Hypotheses entertained to explain this were genetic as well as epigenetic (e.g. environmental influences during intrauterine life). Recent studies offer support to the hypothesis that the interindividual variations of melatonin production are determined genetically.[56]

9.3. Other Hormones

Most circulating hormones show a circadian rhythmicity.[57-59] For example, the renin-angiotensin-aldosterone system exhibits nocturnal peaks mainly coinciding with sympathetic activation in REM sleep.

Gonadotropins show a pulsatile pattern of release in addition to circadian variations, LH release coinciding somewhat with REM episodes during the night.[41] Hormones participating in the control of calcemia show a circadian rhythmicity.[60-62] In addition bone resorption processes are linked to the quality of sleep. A significant effect of melatonin on bone calcification has been recently described.[63]

Neurohypophysial hormones show nocturnal maxima that are dependent on the circadian clock via melatonin secretion.[64] In humans melatonin administration increases the release of oxytocin and vasopressin while in the rat an opposite effects is found.

Pinealectomy affects both the circulating oxytocin and vasopressin levels as well as the electrical activity of supraoptic and paraventricular magnocellular neurons. Pinealectomized rats exhibit a reduced homeostatic response to hypovolemia or hypernatremia. Interestingly, not only release of neurohypophysial hormones but their renal effects show circadian rhythmicity possibly by the activation of autonomic mechanisms above mentioned.[64]

10. Acknowledgments. Work in authors' laboratories was supported in part by Agencia Nacional de Promoción Científica y Tecnológica, Argentina, the University of Buenos Aires, Consejo Nacional de Investigaciones Científicas y Técnicas (CONICET), Argentina, Fundación Bunge y Born, Buenos Aires and DGES, Spain.

11. References

1. J. A. Hobson and E. F. Pace-Schott, The cognitive neuroscience of sleep: neuronal systems, consciousness and learning, *Nat. Rev. Neurosci.* 3, 679-693(2002).
2. E. F. Pace-Schott and J. A. Hobson, The neurobiology of sleep: genetics, cellular physiology and subcortical networks, *Nat. Rev. Neurosci.* 3, 591-605 (2002).
3. M. C. Moore-Ede, F. M. Sulzman, and C. A. Fuller. *The Clock that Times Us: Physiology of the Circadian System* (Harvard University Press; Cambridge, 1982).
4. H. Selye, Syndrome produced by diverse nocuous agents, *Nature* 138, 32-38 (1936).
5. B. McEwen, Protective and damaging effects of stress mediators, *N. Engl. J. Med.* 338, 171-179 (1998).

6. L. Girotti, M. Lago, O. Ianovsky et al, Low urinary 6-sulphatoxymelatonin levels in patients with coronary artery disease, J. *Pineal Res.* 29, 138-142 (2000).

7. L. Girotti, M. Lago, O. Ianovsky et al, Low urinary 6-sulphatoxymelatonin levels in patients with severe congestive heart failure, *Endocrine* 22, 245-248 (2003).

8. J. De Mairan, Observation botanique, *Histoire de L'Academie Royale des Sciences* 35-36 (1729).

9. S. Simpson and J. J. Galbraith, Observations on the normal temperature of the monkey and its diurnal variation, and on the effect of changes in the daily routine on this variation, *Trans. R. Soc. Edinb.* 45, 65-106 (1906).

10. D. P. Cardinali, J Jorda Catala, and E J Sanchez Barcelo. *Introducción a la Cronobiología. Fisiología de los Ritmos Biológicos* (Editorial Universidad de Cantabria; Caja Cantabria, Santander, 1994).

11. J. C. Dunlap, Molecular bases for circadian clocks, *Cel.* 96, 271-290 (1999).

12. K. Wager-Smith and S. A. Kay, Circadian rhythm genetics: from flies to mice to humans, *Nat. Genet.* 26, 23-27 (2000).

13. M. H. Hastings, A. B. Reddy, and E. S. Maywood, A clockwork web: circadian timing in brain and periphery, in health and disease, *Nat. Rev. Neurosci.* 4, 649-661 (2003).

14. J. Aschoff, Human circadian rhythms in activity, body temperature and other functions, *Life Sci. Space Res.* 5, 159-173 (1967).

15. G. Ortega, D. A. Golombek, D. Otero, L. Romanelli, and D. P. Cardinali, Effect of zeitgeber intensity reduction on a simulated dual-oscillator human circadian system: Classical and dynamical analysis, *Chronobiol. Int.* 9, 137-147 (1992).

16. C. S. Pittendrigh, Temporal organization: reflections of a Darwinian clock-watcher, *Annu. Rev. Physiol.* 55, 16-54 (1993).

17. A. Reinberg and F. Halberg, Circadian chronopharmacology, *Annu. Rev. Pharmacol.* 11, 455-492 (1971).

18. R. M. Buijs, C. G. Van Eden, V. D. Goncharuk, and A. Kalsbeek, The biological clock tunes the organs of the body: timing by hormones and the autonomic nervous system, *J Endocrinol.* 177, 17-26 (2003).

19. G. A. Groos and J. Hendriks, Circadian rhythms in electrical discharge of rat suprachiasmatic neurones recorded in vitro, *Neurosci. Lett.* 34, 283-288 (1982).

20. R. Silver, J. Lesauter, P. A. Tresco, and M. Lehman, A diffusible coupling signal from the transplanted suprachiasmatic nucleus controlling circadian locomotor rhythms, *Nature* 382, 810-813 (1996).

21. J. M. Ding, L. E. Faiman, W. J. Hurst, L. R. Kuriashkina, and M. U. Gillette, Resetting the biological clock: mediation of nocturnal CREB phosphorylation via light, glutamate, and nitric oxide, J. *Neurosci.* 17, 667-675 (1997).

22. S. Yamazaki, R. Numano, M. Abe et al, Resetting central and peripheral circadian oscillators in transgenic rats, *Science* 88, 682-685 (2000).

23. M. Y. Cheng, C. M. Bullock, C. Li et al, Prokineticin 2 transmits the behavioural circadian rhythm of the suprachiasmatic nucleus, *Nature* 417, 405-410 (2002).

24. R. M. Buijs and A. Kalsbeek, Hypothalamic integration of central and peripheral clocks, *Nat. Rev. Neurosci.* 2, 521-526 (2001).

25. R. Y. Moore and V. B. Eichler, Loss of a circadian adrenal corticosterone rhythm following suprachiasmatic lesions in the rat, *Brain Res.* 42, 201-206 (1972).

26. F. K. Stephan and I. Zucker, Circadian rhythms in drinking behavior and locomotor activity of rats are eliminated by hypothalamic lesions, *Proc. Natl. Acad. Sci. USA* 69, 1583-1586 (1972).

27. S. I. T. Inouye, H. Kawamura, D. J. Green, and R. Gillette, Persistence of circadian rhythmicity in a mammalian hypothalamic 'island' containing the suprachiasmatic nucleus, *Proc. Natl. Acad. Sci. USA* 76, 5962-5966 (1979).
28. M. Kaneko, K. Kaneko, J. Shinsako, and M. F. Dallman, Adrenal sensitivity to adrenocorticotropin varies diurnally, *Endocrinology.* 109, 70-75 (1981).
29. A. Kalsbeek, J. J. Van Heerikhuize, J. Wortel, and R. M. Buijs, A diurnal rhythm of stimulatory input to the hypothalamo-pituitary-adrenal system as revealed by timed intrahypothalamic administration of the vasopressin V1 antagonist, *J. Neurosci.* 16, 5555-5565 (1996).
30. I. F. Palm, E. M. Van der Beek, V. M. Wiegant, R. M. Buijs, and A. Kalsbeek, Vasopressin induces an LH surge in ovariectomized, estradiol-treated rats with lesion of the suprachiasmatic nucleus, *Neuroscience.* 93, 659-666 (1999).
31. D. P. Cardinali and H. E. Romeo, Peripheral neuroendocrine interrelationships in the cervical region, *News Physiol. Sci.* 5, 100-104 (1990).
32. D. P. Cardinali and H. E. Romeo, The autonomic nervous system of the cervical region as a channel of neuroendocrine communication, *Front. Neuroendocrinol.* 12, 278-297 (1991).
33. D. P. Cardinali and J. E. Stern, Peripheral neuroendocrinology of the cervical autonomic nervous system, *Bra. J. Med. Biol. Res.* 27, 573-599 (1994).
34. M. S. Jasper and W. C. Engeland, Splanchnic neural activity modulates ultradian and circadian rhythms in adrenocortical secretion in awake rats, *Neuroendocrinology.* 59, 97-109 (1994).
35. R. M. Buijs, J. Wortel, J. J. Van Heerikhuize et al, Anatomical and functional demonstration of a multisynaptic suprachiasmatic nucleus adrenal(cortex) pathway, *Eur.J. Neurosci.* 11, 1535-1544 (1999).
36. A. I. Esquifino and D. P. Cardinali, Local regulation of the immune response by the autonomic nervous system, *Neuroimmunomodulation.* 1, 265-273 (1994).
37. D. P. Cardinali and A. I. Esquifino, Neuroimmunoendocrinology of the cervical autonomic nervous system, *Biomed. Rev.* 9, 47-59 (1998).
38. D. P. Cardinali, R. A. Cutrera, and A. I. Esquifino, Psychoimmune neuroendocrine integrative mechanisms revisited, *Biol. Signals Recept.* 9, 215-230 (2000).
39. S. E. La Fleur, A. Kalsbeek, J. Wortel, M. L. Fekkes, and R. M. Buijs, A daily rhythm in glucose tolerance. A role for the suprachiasmatic nucleus in insulin-independent glucose uptake?, *Diabetes.* 50, 1237-1243 (2001).
40. J. P. Dai, D. F. Swaab, J. Van der Vliet, and R. M. Buijs, Postmortem tracing reveals the organization of hypothalamic projections of the suprachiasmatic nucleus in the human brain, *J. Comp. Neurol.* 400, 87-102 (1998).
41. G. Copinschi, K. Spiegel, R. Leproult, and E. Van Cauter, Pathophysiology of human circadian rhythms, *Novartis Found. Symp.* 227, 143-157 (2000).
42. E. Van Cauter and G. Copinschi, Interrelationships between growth hormone and sleep, *Growth Horm. IGF Res.* 10 Suppl B, S57-S62 (2000).
43. Y. Touitou and E. Haus, Alterations with aging of the endocrine and neuroendocrine circadian system in humans, *Chronobiol. Int.* 17, 369-390 (2000).
44. P. Penev, K. Spiegel, M. L'Hermite-Baleriaux, R. Schneider, and E. Van Cauter, Relationship between REM sleep and testosterone secretion in older men, *Ann. Endocrinol. (Paris).* 64, 157 (2003).

45. E. Van Cauter, L. Plat, R. Leproult, and G. Copinschi, Alterations of circadian rhythmicity and sleep in aging: endocrine consequences, *Horm. Res.* 49, 147-152 (1998).

46. E. Van Cauter, L. Plat, M. B. Scharf et al, Simultaneous stimulation of slow-wave sleep and growth hormone secretion by gamma-hydroxybutyrate in normal young Men, *J. Clin. Invest.* 100, 745-753 (1997).

47. D. A. Golombek, P. Pevet, and D. P. Cardinali, Melatonin effect on behavior: Possible mediation by the central GABAergic system, *Neurosci. Biobehav. Rev.* 20, 403-412 (1996).

48. J. M. Monti, F. Alvarino, D. P. Cardinali, I. Savio, and A. Pintos, Polysomnographic study of the effect of melatonin on sleep in elderly patients with chronic primary insomnia, *Arch. Gerontol. Geriatr.* 28, 85-98 (1999).

49. D. P. Cardinali and P. Pevet, Basic aspects of melatonin action, *Sleep Med. Rev.* 2, 175-190 (1998).

50. R. J. Reiter, D. X. Tan, S. Burkhardt, and L. C. Manchester, Melatonin in plants, *Nutr. Rev.* 59, 286-290 (2001).

51. D. J. Kennaway and H. Wright, Melatonin and circadian rhythms, *Curr. Top. Med. Chem.* 2, 199-209 (2002).

52. I. V. Zhdanova, R. J. Wurtman, M. M. Regan, J. A. Taylor, J. P. Shi, and O. U. Leclair, Melatonin treatment for age-related insomnia, *J. Clin. Endocrinol. Metab.* 86, 4727-4730 (2001).

53. D. J. Skene, S. W. Lockley, and J. Arendt, Melatonin in circadian sleep disorders in the blind, *Biol. Signals Recept.* 8, 90-95 (1999).

54. J. M. Monti and D. P. Cardinali, A critical assessment of the melatonin effect on sleep in humans, *Biol. Signals Recept.* 9, 328-339 (2000).

55. L. Girotti, M. Lago, O. Yanovsky et al, Low urinary 6-sulphatoxymelatonin levels in patients with coronary artery disease, *J Pineal Res.* 29, 138-142 (2000).

56. B. Griefahn, P. Brode, T. Remer, and M. Blaszkewicz, Excretion of 6-hydroxymelatonin sulfate (6-OHMS) in siblings during childhood and adolescence, *Neuroendocrinology.* 78, 241-243 (2003).

57. L. Q. Qin, J. Li, Y. Wang, J. Wang, J. Y. Xu, and T. Kaneko, The effects of nocturnal life on endocrine circadian patterns in healthy adults, *Life Sci.* 73, 2467-2475 (2003).

58. L. Morgan, S. Hampton, M. Gibbs, and J. Arendt, Circadian aspects of postprandial metabolism, *Chronobiol. Int.* 20, 795-808 (2003).

59. S. Andersen, N. H. Bruun, K. M. Pedersen, and P. Laurberg, Biologic variation is important for interpretation of thyroid function tests, *Thyroid.* 13, 1069-1078 (2003).

60. A. Schlemmer, P. Ravn, C. Hassager, and C. Christiansen, Morning or evening administration of nasal calcitonin? Effects on biochemical markers of bone turnover, *Bone.* 20, 63-67 (1997).

61. P. Shao, M. Ohtsuka-Isoya, and H. Shinoda, Circadian rhythms in serum bone markers and their relation to the effect of etidronate in rats, *Chronobiol. Int.* 20, 325-336 (2003).

62. L. Rejnmark, A. L. Lauridsen, P. Vestergaard, L. Heickendorff, F. Andreasen, and L. Mosekilde, Diurnal rhythm of plasma 1,25-dihydroxyvitamin D and vitamin

D-binding protein in postmenopausal women: relationship to plasma parathyroid hormone and calcium and phosphate metabolism, *Eur. J. Endocrinol.* 146, 635-642 (2002).

63. D. P. Cardinali, M. G. Ladizesky, V. Boggio, R. A. Cutrera, and C. A. Mautalen, Melatonin effects on bone: Experimental facts and clinical perspectives, J. *Pineal Res.* 34, 81-87 (2003).

64. M. L. Forsling, Diurnal rhythms in neurohypophysial function, *Exp Physiol.* 85, 179S-186S (2000).

Biological Clock Control of Glucose Metabolism

Timing Metabolic Homeostasis

MARIEKE RUITER, RUUD M. BUIJS AND ANDRIES KALSBEEK

1. Homeostasis

The development of the concept of homeostasis began in the mid 19th century, when Claude Bernard (1813-1878) claimed that 'la fixeté du milieu interieur' is essential for higher organisms to survive in an ever changing environment. "Standing or staying the same" is the literal meaning of the term homeostasis, which was introduced later by Walter B. Cannon (1871-1945) and derived from the Greek words *homeios* (the same) and *stasis* (staying). However, as Cannon emphasized, homeostasis does not mean something set and immobile that stays exactly the same all the time. In his words, homeostasis "means a condition that may vary, but which is relatively constant". This definition was refined by Donald C. Jackson (1987): "Homeostasis is not a single optimal control condition but rather a variety or continuum that varies with the animal's circumstances. Set points or regulated values are not fixed, but may change depending on ambient conditions or because of changing physiological conditions or demands."

A very clear example of a changing ambient condition with consequent changes in physiological demands is the daily cycle of light and dark. As the earth rotates around its own axis, light and dark periods occur intermittently and in a regular pattern. Most animals have therefore adopted a certain daily activity pattern, depending on their specific habitat, body composition or feeding preferences. This has a profound impact on their energy requirements throughout the daily cycle of light and dark. Fortunately, as the light-dark cycles keep repeating, this variation in energy need is very predictable. This offers the possibility not only of maintaining homeostasis by keeping the physiological parameters at a preferred level, but also of *preventing disturbance* of homeostasis by *anticipation* of the regularly recurring changes in energy need. This means that the set-point itself changes over the 24-hour period, thus providing an organism with the major advantage of having a timing system that guarantees storage and mobilization of energy resources at the right moments of the day.

2. The Timing System

The mammalian biological clock in the brain, harbored in the suprachiasmatic nucleus (SCN), is located in the anterior hypothalamus, on top of the optic chiasm.[1-3] It oscillates endogenously[4,5] and synchronizes these oscillations to the light-dark cycle by means of the light-dark information it receives directly through the eyes[6-9]. At the basis of the daily rhythms generated by the SCN are a set of genes known as the 'clock genes'. Rhythmic transcription of two of these genes, named *clock* and *bmal1*, drives negative and positive feedback loops, involving the expression of the other clock genes: *period (per1-per3), cryptochrome (cry1* and *cry2), Rev-Erbα* (reviewed by Reppert and Wea-ver[10]) and *dec1* and *dec2*.[11] The molecular clock drives the expression of many other genes in the SCN, thus composing a timing signal, which is sent to other areas within the brain and the rest of the body via several types of projections. The phase and function of this molecular clock in the various subpopulations of SCN-neurons may differ.[12]

Anterograde tracer injections into the SCN have revealed a very restricted range of projection sites, mainly within the medial hypothalamus.[13] Outside the hypothalamus, only the intergeniculate leaflet (IGL) and the paraventricular nucleus of the thalamus (PVT) receive direct SCN projections. Retrograde tracing from these target areas has revealed distinct subpopulations of SCN cells, each sending their signal to different target areas.[14] Direct projections to neuroendocrine neurons[15-19] and autonomic paraventricular (PVN) neurons[20-23] enable the SCN to impose its rhythmic signal on the physiology. Many PVN neurons project to the median eminence and the pituitary and regulate pituitary hormone secretion.[23] Moreover, the preautonomic PVN neurons can change the sensitivity of peripheral organs to the pituitary hormones via the autonomic nervous system.[24]

Other intra-hypothalamic target areas, such as the medial preoptic area (MPO), dorsomedial hypothalamus (DMH), lateral hypothalamus (LHA), ventromedial hypo-thalamus (VMH) and subPVN probably serve as integration areas, since they are involved in a wide variety of physiological functions. For instance, the MPO is involved in the regulation of body temperature and gonadal hormone secretion,[25] the DMH is important for sleep/wake regulation,[26,27] ACTH and corticosterone secretion[28] and the regulation of food intake,[29-31] the LHA is involved in sleep-wake regulation and ingestive behavior[32] and the VMH in regulation of energy metabolism.[33,34]

By projecting to, and closely cooperating with, these nuclei, the SCN is able to modify the autonomic output of the hypothalamus and induce daily hormone rhythms, thus integrating its daily signal with mammalian physiology. Furthermore, the SCN neurons and their projections express a wide range of different neurotransmitters and peptides that have inhibitory or stimulatory effects on their target neurons. These substances are present in many different combinations,[35,36] offering the possibility of an even more differentiated SCN output.

3. Peripheral Clocks

The molecular clock machinery described above is present not only in most of the SCN neurons but also in many peripheral tissues. Clock genes have been identified in liver,[37,38] lung, and skeletal muscle tissue,[39] as well as in the heart[38,40] and pancreas.[41]

A discussion about the role of the SCN as a 'master clock' continues, as conflicting evidence is being gained on this subject. Evidence in favor of the 'master clock' theory is abundant (reviewed by Reppert and Weaver[10]). Blockade and stimulation of sympathetic activity showed that the activity of the autonomic nervous system may be an important factor in the regulation of peripheral clocks in the liver.[42] Furthermore, lesions of the SCN abolish or dampen rhythms in peripheral *per1* gene expression,[42-44] which can be reinstated by adrenaline injections.[42] It has been proposed that the SCN is able to sustain endogenous oscillations because of its paracrine neuropeptide (vasoactive intestinal peptide, VIP) signaling, a feature that peripheral oscillating cells do not possess.[36] Until recently, it was widely accepted that *per1* gene expression in organ explants *in vitro* oscillates no longer than 7 days.[5,37,39,43,44] However, recent results, gained from *per2* transgenic mice, do show persistent *per2* expression in mice with SCN lesions and in cultured peripheral tissue.[45] Without the SCN, the phases of *per2* expression in different peripheral tissues seem to drift apart, instead of dampening. These findings indicate that the SCN acts as a 'synchronizer' rather than a 'driver' of circadian gene expression.

Besides light as the main Zeitgeber, other time cues, such as food availability, have been shown to have a major impact on the synchronization of peripheral clock gene expression to the environment. Evidence for a "food entrainable oscillator" was first ob-tained by Stephan and colleagues,[46] who have shown that rats with an SCN lesion can still be entrained to a feeding schedule. Furthermore, intact rats can be entrained to a feeding schedule, thereby uncoupling peripheral clock gene rhythms from the rhythm in the SCN.[47,48] Besides affecting molecular clocks, the effect of restricted feeding can also be seen in liver physiology and hormonal profiles.[49,50] Recently, carbohydrate avail-ability, which affects the redox state of cells, was proposed to act as an input regulator for the peripheral clock genes Clock, Bmal1, and Npas2,[51,52] providing the first specific mechanism proposed for feeding-mediated entrainment of circadian clocks in mammals.

The exact function of the peripheral molecular clock has thus not yet been eluci-dated. We suggest that it may serve a purpose in situations where food availability is not in synchrony with an animal's usual activity pattern, which may create the need to un-couple peripheral oscillations from the central timing signal of the SCN in order to use available energy resources as efficiently as possible. Furthermore, peripheral clocks ena-ble peripheral organs to oscillate without the need of a continuous SCN input.

In the present chapter, we will discuss the daily glucose metabolism in more detail, to illustrate the importance of the circadian timing system for mammalian physiology and the mechanisms used by the SCN to control general physiology.

4. Glucose Metabolism

Glucose is a very important energy source, in fact for some organs, such as the brain, it is in normal circumstances the only energy source. Because the brain is unable to store glucose as glycogen, it depends completely on glucose that enters the brain via the circulation. Sufficient amounts of glucose must therefore be available at all times, and controlling other organs, that do have the capacity to store or produce glucose, is very important.

The first evidence that the SCN is directly involved in the regulation of glucose metabolism came from experiments by Nagai and coworkers, who showed that electrical stimulation of the SCN resulted in augmented plasma glucose concentrations.[53] Intraperitoneal administration of α- and β-blockers prevented this effect.[54] In the following years, more studies were performed, that tried to elucidate how the SCN is involved in the control of the many different aspects of glucose homeostasis regulation.

Plasma glucose concentrations in both humans and rats were shown to rise during the sleep period, and to peak at the onset of the activity period, independent of food intake and locomotor activity (Figure 1B).[55,56] In humans, this morning glucose peak has been termed the 'dawn phenomenon'. Furthermore, the speed at which an injected bolus of glucose is cleared from the plasma (glucose tolerance, Figure 1A) varies at different moments of the light-dark cycle, as does the capability of insulin to stimulate glucose uptake (insulin sensitivity). Both are highest at the onset of the activity period.[57-59]

Increases or decreases of plasma glucose levels can be caused by a number of processes involving several organs in the body and several stimulatory and inhibitory mechanisms. Plasma glucose concentrations decrease when glucose is taken up into cells for oxidation (e.g. by skeletal muscle or brain) or storage (e.g. in muscle, liver or adipose tissue).

Plasma glucose concentrations increase due to absorption of glucose from the gastrointestinal tract after feeding, or due to glucose production via the breakdown of glycogen (glycogenolysis in the liver), or *de novo* glucose synthesis (gluconeogenesis in liver, kidneys[60] and small intestines[61]). Suprachiasmatic modulation of these processes may, as argued above, occur via hormonal or neuronal pathways, or a combination of both.

The control of glucose homeostasis involves several well-known hormones. Elevated plasma glucose levels stimulate the pancreatic β-cells to

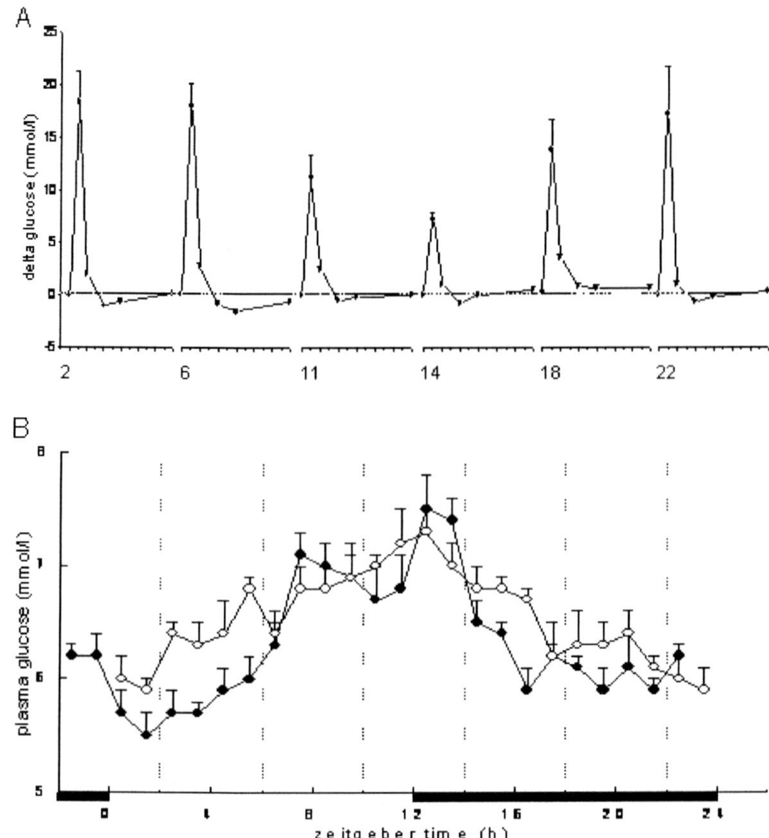

FIGURE 1. In rats, both glucose tolerance[57] (A) and plasma glucose concentrations[55] (B) show peak levels at the onset of the dark (activity) period. A: A glucose bolus was injected at different zeitgeber times and blood samples were taken at t = 0, 2, 10, 20, 30 and 60 min. Values are changes in glucose concentration from baseline at t = 0. Small peaks indicate rapid clearance of the injected glucose from plasma into tissue. B: 24-hour plasma glucose profiles in freely moving ad libitum (black circles) and meal-fed (open circles) rats. 10-minute meals were given every four hours, depicted here by vertical dotted lines. Black bars indicate the dark period. Values are given as mean ± SEM.

secrete insulin. Subsequently, insulin stimulates the uptake of glucose into cells, e.g. after a meal or, in a more experimental situation, after injection of a glucose bolus (glucose tolerance test). Furthermore, it decreases the plasma glucose concentration by inhibiting hepatic glucose production.[62] Glucagon, produced in the pancreatic α-cells, is known for its stimulatory effect on hepatic glucose production[63,64] and its release is elicited by a rapid decrease in plasma glucose levels. Glucagon secretion is inhibited by insulin[65,66] and glucagon stimulates insulin release via both indirect (glucose-mediated) and direct pathways.[67]

Apart from glucagon, several other hormones have a stimulatory influence on hepatic glucose output. Epinephrine, produced in the adrenal medulla, stimulates glucose production in case of hypoglycemia, but also during exercise. When infused, it first stimulates glycogenolysis. Later on, it enhances gluconeogenesis.[68]

Another hormone known for its stimulatory effect on glucose mobilization is growth hormone, secreted from the pituitary. It induces hepatic insulin resistance and thus promotes hepatic glucose output.[69] Corticosterone (or cortisol in humans), produced in the adrenal cortex, also stimulates gluconeogenesis and induces insulin resistance.[70]

Besides a hormonal control of glucose homeostasis, a multitude of studies has provided evidence of a neuronal control of several aspects of glucose metabolism. As early as 1855, Claude Bernard showed that stimulating the floor of the fourth ventricle in the brain caused hyperglycemia.[71]

Furthermore, in tracing studies, the central origins of the autonomic innervation of peripheral organs have been identified. By use of the transsynaptic retrograde tracer Pseudorabies virus (PRV), direct multisynaptic connections were shown between the hypothalamus and many organs that are in some way important for glucose homeostasis, such as the liver,[72,73] the pancreas,[20] adipose tissue,[74,75] the heart,[76] kidneys,[77] adrenals[24,73] and pineal gland.[22] Via these neuronal pathways, the brain may be able to control organs involved in glucose production and utilization directly, as well as the secretion of the above-mentioned hormones, that in turn influence glucose metabolism.

Thus, there are hormonal and neuronal pathways that the SCN can use to send its timing signal to the periphery and modulate daily glucose homeostasis (Figure 1). Furthermore, SCN control of the availability and utilization of other nutrients may have an impact on the way different aspects of glucose metabolism are regulated. In the following we will discuss how, i.e. by means of which output pathways, the SCN might control the daily plasma glucose rhythm.

5. The Liver

The liver is a central organ in glucose metabolism. When glucose is taken up from the intestinal tract, it reaches the liver via the portal vein. There, it is taken up mostly via insulin-independent pathways, via the high-capacity glucose transporter GLUT2.[78] It is then either transformed into the polysaccharide glycogen via the glycogenic enzymes glucokinase and glycogen synthase, to store it with reduced osmotic value, or directed to glycolysis. Thus, glucose can be stored in the liver and released when needed, e.g. during sleep when no food intake occurs. In rats, the glycogen content of the liver shows a circadian rhythm, with peak levels at the end of the dark period, as shown in several studies.[79-82] The liver is not only a glucose storage organ; it is also capable of producing glucose. This was already recognized by Claude Bernard in 1848. Glucose release from the liver into the circulation occurs via the breakdown of stored glycogen (glycogenolysis) and through *de novo* glucose synthesis from non-glucose precursors like lactate, alanine, glycerol and glutamine, in a process called gluconeogenesis. In a normal situation, both processes occur continuously, at basal rates, but during fasting the contribution of gluconeogenesis increases gradually. Glycogen stores are mostly depleted after one night of fasting, and although the rate of glycogenesis is decreased, both glycogenesis and glycogenolysis continue at a basal rate until extreme starvation.[83-85]

In rats, both plasma glucose concentrations and glucose tolerance peak at the onset of the activity period,[55,57] i.e. at the time when rats have not started feeding yet. Therefore, the increase in plasma glucose can only occur if glucose production by the liver is high at that moment as well and it has therefore been suggested that hepatic glucose production is stimulated at the onset of the activity period.[86] Indeed, gluconeogenesis,[87] and more specifically the presence and activity of the gluconeogenic enzyme phosphoenolpyruvate carboxykinase (PEPCK) is rhythmic in the liver, with high levels at times when the animals are not feeding. Although the abundance of this enzyme is strongly correlated with feeding, the rhythm persists even in the fasted state, indicating that it is a feeding-independent circadian rhythm.[88,89]

The liver receives projections from both parasympathetic and sympathetic preautonomic neurons,[72,90] and is connected to the SCN via both branches of the autonomic nervous system.[91] It has been known for decades that sympathetic nerve activity in general stimulates hepatic glucose production.[92-95] More specifically, electrical stimulation of the VMH causes an increase in glucose production via activation of sympathetic nerves innervating the liver.[96] Furthermore, electrical stimulation of the SCN causes hyperglycemia, which is prevented by the administration of autonomic blockers.[53,54]

Hepatic glucose uptake caused by an increase of glycogenesis and inhibition of gluconeogenesis is stimulated by activation of the LHA.[97] Blockade of sympathetic hepatic nerves also results in an increase of glucose uptake. Activation of parasympathetic nerves innervating the liver leads to increased glycogen storage and an inhibition of hepatic glucose output.[98,99] Conversely, parasympathetic hepatic denervation impairs glycogen storage in the liver[100] and it increases hepatic glucose production.[101] Moreover, a diurnal rhythm in hepatic glycogen content is likely to be induced via the autonomic nervous system, as this rhythm is independent of circulating glucose and insulin.[102,103]

Indeed, selective hepatic denervation disrupts the rhythmic properties of plasma glucose concentrations. Both hepatic sympathectomy as well as hepatic vagotomy abolished the daily glucose rhythm published previously,[55] and resulted in glucose levels comparable to the daily peak measured in intact rats.[104] This effect was somewhat surprising since, based on the literature about sympathetic blockade presented above, decreased glucose concentrations were expected after a sympathetic denervation. However, after selective denervation of e.g. the liver, pancreas, heart and colon, PRV transport in the remaining nerves had slowed down compared to intact controls.[20,105,106] The speed of PRV transport is suggested to depend on nerve activity.[105] Thus, these tracing data indicate that dissection of the sympathetic nerve may induce a (paradoxal) decrease of activity in the vagal nerve. Nonetheless, these studies indicate that both the sympathetic and parasympathetic input of the autonomic nervous system to the liver are important for the SCN timing signal to be conveyed.

Besides parasympathetic and sympathetic nerve activity towards the liver, the insulin to glucagon ratio influences the balance between hepatic glucose uptake or output; i.e. the presence of either more insulin or more glucagon in the portal vein will induce glucose uptake in, or glucose release from the liver, respectively.[107-109] However, combining daily insulin and glucagon patterns measured in our separately, but very similarly conducted studies[55,110] does not suggest involvement of this ratio in the daily glucose rhythm measured in the same studies.

6. The Pancreas

The pancreas has both an exocrine and endocrine function. Its exocrine function is to produce a multitude of enzymes that are released in the gut, where they facilitate digestion of carbohydrates, proteins and lipids. However, in this chapter, the endocrine function of the pancreas is the most important. Both insulin and glucagon are produced by the endocrine pancreas, in the islets of Langerhans (by the β-cells and α-cells, respectively). They are released into the circulation, to act on many other organs, although they also have a paracrine effect on neighboring cells. The impact of the hypothalamus on the stimulation of pancreatic hormone secretion has been known for a long time.[111-114] More recently, anatomical evidence of a multisynaptic connection between the SCN and the pancreas has been obtained through studies using PRV as a retrograde tracer.[20,115] Both the sympathetic and parasympathetic nervous system are involved in the control of pancreatic islet function.[116] Apart from the classical cholinergic and adrenergic neuro-transmitters, a number of parasympathetic and sympathetic neuropeptides has been identified in the pancreas. Galanin and neuropeptide Y (NPY) have been implied in sympathetic control of insulin secretion, but results are conflicting, probably due to differences in experimental conditions. Therefore, an exact role for these peptides could thus far not be elucidated. Furthermore, parasympathetic peptides such as gastrin releasing peptide (GRP), pituitary adenylate cyclase activating polypeptide (PACAP) and vasoactive intestinal polypeptide (VIP) are expressed in pancreatic ganglia and islets, and released upon vagal electrical stimulation. They all stimulate release of insulin and glucagon (reviewed in detail by Ahren[117]). Functional and electrophysiological studies have provided evidence that neuropeptide Y (NPY) and substance P stimulate, whereas somatostatin and calcitonin gene–related peptide inhibit the dorsal motor nucleus of the vagus (DMV) neurons that project to the pancreas, when injected in the lateral ventricle.[118,119] The origin of these peptides is as yet unknown, but retrograde tracing has revealed a range of hypothalamic nuclei projecting to the pancreas via either sympathetic or parasympathetic pathways[20]. One of these areas is the LHA,[20,32,120] that projects to the DMV. These projections contain the neuropeptide orexin.[20] Insulin-induced hypoglycemia induces increased electrical activity in the LHA and c-fos in orexin neurons.[121] In fact, glucose injection into the LHA induces increased firing rate in the pancreatic nerves.[122]

Insulin secretion from the β-cells in the pancreas is in large part regulated by the autonomic nervous system. Electrical activation of the vagus nerve was shown to stimu-late insulin secretion both *in vivo* and *in vitro* in several different species including humans.[116,123] Several studies have provided evidence for acetylcholine (Ach) to stimulate insulin release, and for the muscarinic antagonist atropine to inhibit this response[116,124,125]. Bereiter and colleagues have shown that these vagal branches that stimulate insulin release originate in the nucleus ambiguus (AMB) in the brainstem.[126] Stimulation of this area induced vagally mediated (blocked by vagotomy or atropine treatment) insulin release. However, AMB output seems to be of dual nature, as it also induced a sympathetically mediated (blocked by phentolamine) inhibition of insulin secretion.[126] Thus, the AMB may be an integrative area, its output depending on a combination of stimuli.

As shown here, many studies report involvement of parasympathetic stimulation of insulin secretion. However, insulin release was found to be mediated by both branches of the autonomic nervous system. The effect of sympathetic stimulation seems to depend on the activation of either α-adrenergic (stimulates insulin release) or β-adrenergic (inhibits insulin release) receptors in the pancreas.[124,127-129]

Several studies have been performed to investigate circadian rhythms in insulin secretion and sensitivity. Although glucose tolerance in humans changes throughout the 24-hour period (glucose tolerance decreases towards the evening and is lowest in the middle of the night), the insulin secretion pattern does not parallel these changes in glucose.[130,131] In rats, glucose tolerance shows a pattern similar to that in humans; it decreases towards the end of the dark (activity) period and is highest at the onset of the activity period.[57] A daily rhythm in basal plasma insulin concentrations in the rat has been described previously, although results are somewhat conflicting.[55,132] This rhythm could not be correlated completely to the daily rhythm in plasma glucose concentrations. Furthermore, the amount of insulin released in response to an intravenous glucose bolus did not show differences relating to the time-of-day at which it was injected. Insulin sensitivity, however, does seem to vary over the 24-hours, and can be correlated to the rhythm in glucose tolerance.[57] Differences in glucose tolerance between the different times of the day are quite large, whereas insulin sensitivity only shows small (but significant) differences. This indicates that, besides insulin sensitivity, other factors possibly also mediate the SCN control of glucose tolerance. Isolated pancreatic islets secrete insulin in a circadian way, with a tau between 22 and 26 hours.[133] To our knowledge, however, this has not been investigated in the *in vivo* situation. Although it is relatively easy to measure plasma insulin concentrations, the liver metabolizes insulin (and glucagon) and thus plasma insulin concentrations only give a partial view on insulin release, i.e. one should measure these hormone levels in the portal vein.

Glucagon, known for its stimulatory effect on hepatic glucose production[63,64] is released by the α-cells of the pancreas after stimulation of either the sympathetic[134-137] or parasympathetic branch of the autonomic nervous system[138-141] as well as by circulating epinephrine.[142] These three pathways are all active in case of hypoglycemia, when glucagon is released to stimulate glucose production and restore normal plasma glucose concentrations (counterregulation).[143-146] These counterregulatory glucagon responses display a diurnal variation in rats, with highest responses at the onset and end of the dark period,[147] showing that the SCN modulates at least one of the three ways in which glucagon release is stimulated.

Glycogenolysis and gluconeogenesis are stimulated by glucagon. Because both plasma glucose concentrations and glucose tolerance in rats peak at the onset of the activity period,[55,57] glucose production by the liver is suggested to be high at that moment as well. This may be mediated by glucagon and, indeed, several studies have indicated the presence of a daily rhythm in basal plasma glucagon concentrations.[80,107,110,148-150] Results of these studies were inconsistent with respect to phase and amplitude of the rhythm. In our own study, however, we have found no correlation between daily glucagon and glucose rhythms measured in the same animals[110]. Furthermore, the glucose rhythm is independent of food intake, whereas glucagon does respond to feeding in a scheduled feeding regimen. These meal-responses last 30-90 min after the initiation of a meal and are possibly mediated via ingested amino acids.[151] Total glucagon release in response to meals, as measured by the area under the curve (AUC) did not show any daily variation. These meal responses are unlike the well-known cephalic-phase glucagon (and insulin) release, occurring very quickly (within a minute) after the initiation of food intake, before nutrients are taken up in the blood via the gastrointestinal tract, and last no more than 10 minutes. These responses are induced by taste and texture properties of the ingested food and probably mediated via the autonomic nervous system.[152-155] To our knowledge, it is unknown whether cephalic-phase (glucagon or insulin) release shows diurnal variation.

As described above for the liver, a complex balance between parasympathetic and sympathetic activity also seems to be very important for a correct modulation of the islets of Langerhans in the pancreas. Both branches seem to be able to modify the α- and β-cell sensitivity for several factors (such as glucose, insulin and glucagon) and possibly the sympathetic branch influences pancreatic sensitivity for parasympathetic input, and vice versa, as well.

This offers the possibility of very precise control of pancreatic secretion, depending on the metabolic state of the body.

7. Skeletal Muscle

Besides the brain, skeletal muscle is the tissue that uses most glucose for oxidation. Glucose is stored in the form of glycogen, as it is in the liver, and skeletal muscle contains the largest glycogen store in the body. Until recently it was believed that skeletal muscle does not express the final enzyme in the glucose production pathway, glucose-6-phos-phatase. This means that glucose, once taken up, cannot be released again from the muscle, and can be used only as a local energy supply. However, recently this enzyme has also been shown in the muscle, indicating a possible new role for muscle tissue in glucose metabolism.[156] The skeletal muscle glycogen content in the rat displays a clear daily rhythm with a peak at the end of the activity period and a dip at the onset of the activity period, which seems to be related to the timing of food intake.[79,80,82] Because skeletal muscle uses such a large amount of nutrients for energy expenditure, we suggest that the rhythm in glucose tolerance shown previously[57] is in large part attributable to skeletal muscle glucose uptake. At least part of this is accounted for by a daily variation in insulin sensitivity.[55] Furthermore, the VMH seems to have an important role in the regulation of glucose uptake in skeletal muscle. Electrical stimulation of the VMH and leptin injections in the VMH lead to increased glucose disposal in skeletal muscle.[157,158] Glucose is taken up in cells by means of active transport, via glucose transporters. Several glucose transporter subtypes have been identified, with different functions.[159] Glucose transporter 4 (GLUT4) is the one that acts in response to insulin. Under basal circumstances, GLUT4 is present in vesicles inside the cell, where they are not actively involved in the transport of glucose. Insulin recruits these vesicles towards the cell membrane, via the insulin receptor and an intracellular second messenger cascade.[160] Once present in the plasma membrane, GLUT4 facilitates glucose uptake. Another glucose transporter expressed in muscle is GLUT1, which is permanently present in the cell membrane and does not respond to insulin. Possibly, the daily rhythm in glucose uptake in tissue can be explained by a rhythmic expression of these glucose transporters, or by a rhythm in the recruitment of GLUT4 to the plasma membrane.

8. Adipose Tissue

In white adipose tissue, insulin stimulates glucose uptake GLUT4 as well.[159] Although glucose uptake occurs in adipose tissue, and glucose can be used in lipogenesis, this is an energy-requiring process. Glucose and lipid metabolism profoundly influence each other. Adipose tissue blood flow can increase up to four-fold after a glucose load, which has been attributed to the resulting hyperinsulinemia.[161] Furthermore, plasma concentrations of free fatty acids, the building bricks of lipids, released by adipose tissue or taken up from the intestines, have a profound influence on glucose metabolism. Both substances compete for oxidation (known as the Randle cycle[162]). High FFA concentrations cause insulin resistance,[163,164] i.e. increased plasma glucose levels due to a decreased tissue uptake. Both high glucose and FFA concentrations will cause increased storage of lipids in adipose tissue.[164,165] Thus, lipid and glucose metabolism are in close relation, and rhythmic properties of lipid metabolism may have an impact on daily glucose metabolism. Indeed, FFA and triglyceride concentrations in plasma show daily rhythms.[148,150] but their relation with daily glucose metabolism could not yet be shown (Ruiter *et al.* unpublished data).

Recently, it was shown that next to the well-known sympathetic innervation,[75] white adipose tissue also receives a parasympathetic input,[74] Both inputs are closely linked to the metabolic activity of white adipose tissue. Sympathetic denervation leads to an increased fat pad size in hamsters, due to a lack of fat-mobilizing capacity.[166] Parasympathetic denervation results in a decrease of both lipid and glucose uptake.[74] This indicates that the balance in autonomic outflow to white adipose tissue may determine whether a net build up or break down of nutrient stores occurs.

Furthermore, adipose tissue is nowadays considered an important endocrine organ rather than just a lipid storage compartment[167-169] and the new substances isolated from it have a putative role in glucose metabolism. Leptin, for example, shows a pronounced circadian rhythm[170] and has been suggested to influence pancreatic function[171] and energy metabolism.[172-174] Another example is adiponectin, a hormone that reverses obesity-related insulin resistance by stimulating glucose and FFA utilization,[175,176] and which, in humans, also shows a diurnal pattern in plasma concentrations.[177]

9. The Kidneys

The kidneys are nowadays considered very important in glucose metabolism, as they produce glucose during the post-absorptive phase and are proposed to account for ~20% of total gluconeogenesis.[60,178-180] Other than the liver, the kidneys do not contain a substantial amount of glycogen, and thus renal glucose production is mainly accounted for by gluconeogenesis, which is stimulated by epinephrine and inhibited by insulin.[181] The presence of a circadian rhythm in renal gluconeogenesis has not been clearly shown. A daily rhythm in the gluconeogenic enzyme PEPCK was found,[182] but no rhythm in gluconeogenic substrate specificity could be measured,[183] and the authors suggested that renal gluconeogenesis serves merely to meet the minimal glucose requirement of the brain.

10. The Pituitary, Adrenals and Pineal Gland

As stated earlier in this chapter, many PVN neurons project to the pituitary and regulate pituitary hormone secretion.[23] As some of these hormones have an impact on glucose metabolism, this may be an extra way of control for the SCN. Growth hormone (GH) has been suggested to play a role in the dawn phenomenon in humans,[184] since it promotes hepatic glucose output.[69] Rats, however, show an ultradian rather than circadian pattern of GH release,[185-187] which makes it unlikely for GH to have a significant role in the dawn phenomenon, at least in rats.

Furthermore, the hypothalamus-pituitary-adrenal axis has been suggested to play a role in daily glucose metabolism. Cortisol and corticosterone enhance glucose produc-tion[70] and show a clear daily rhythm with peak values just before the onset of the activity period in humans[184,188] and rats.[189-192] In fact, the daily rhythm of plasma corticosterone levels correlates very well with the glucose patterns observed in mammals[55,56,193] and would therefore be a good candidate for regulating this glucose pattern.[194,195] However, a rise in blood glucose levels still occurs if the morning rise of cortisol levels in humans is prevented,[196,197] suggesting that a major role for corticosteroids in the modulation of daily glucose metabolism is not likely.

Epinephrine, an important hormone that counteracts hypoglycemia, is released from the adrenal medulla and is known to have a circadian release pattern in humans, with a trough in the middle of the night and a peak in the late morning.[198,199] This peak occurs somewhat later than the peak in plasma glucose concentrations, which occurs just at the onset of the activity period.[55,56] Thus far, no animal data are available either that would suggest a role of epinephrine in the control of daily glucose homeostasis by the SCN.

Finally, a hormone which is very clearly and exclusively controlled by the SCN is melatonin. It is produced by the pineal gland, and its release is controlled by an autonomic input via multisynaptic connections originating in the SCN.[22] Although several studies have suggested a role for melatonin in the control of glucose metabo-lism,[200,201] its exact function is not clear. The liver and pancreatic β-cells contain melatonin receptors[202] and an effect of melatonin on insulin secretion and sensitivity has been suggested[203,204]. Furthermore, pinealectomy affects glucose homeostasis, but again, results were not consistent.[205-207] Since melatonin is secreted only during the night in both nocturnal and diurnal species, whereas the basal daily rhythms in plasma glucose concentrations, glucose tolerance and insulin sensitivity show a 12-hour shift in these species, it is unlikely that melatonin has a direct effect on aspects of glucose metabolism such as tested in these studies. It is more likely that melatonin influences glucose metabolism in an indirect way, by affecting the SCN, and by acting as a general 'night signal' to the body and the brain.

11. Peripheral Feedback – Glucose Sensing

A very important aspect of keeping homeostasis is the integration of peripheral feedback information about the metabolic state of the body with other sensory information, before the final output signal is generated. The major part of this integration takes place within the hypothalamus. Several mechanisms for detecting glucose levels in the body have evolved, and it occurs at many different locations in the body. Glucose can be sensed and adjusted at the organ level, e.g. by altered secretion of glucagon or insulin after GLUT-2 mediated glucose-sensing by the pancreatic α- or β-cells.[208-210] Furthermore, in the hepatoportal vein, GLUT2-dependent vagal afferents sense blood glucose levels in a way that is very similar to that in the pancreatic β-cells.[78,211,212] The vagus may also conduct information from glucose sensors present in the splenic vein, in the renal tube, and in muscle and adipose tissue.[213] Glucose sensing cells are also present in the carotid body.[214]

Glucose sensing also occurs in the central nervous system. In the brainstem nucleus of the solitary tract (NTS) and DMV, the firing rate of glucose responsive neurons can either increase or decrease during hyper- and hypoglycemia.[215] This may depend on their specific target areas.[216,217] Furthermore, several hypothalamic neurons respond to either high or low glucose levels by increasing or decreasing their firing rate.[218] Several of these hypothalamic nuclei, such as the arcuate[219] and PVN[220-222] as well as the VMH[223] and LHA[32,218] contain glucose-sensing neurons. Even the SCN itself seems to be glucose-sensitive, as the phase of the spontaneous activity of SCN cells in a slice preparation is altered in response to different glucose concentrations in the bathing medium.[224] Therefore, it seems that information about the glycemic state of the body and the time of day is integrated in the hypothalamus. Several studies indicate that the PVN and VMH are likely candidates for the mediation of autonomic responses to hypoglycemia,[225,226] whereas the DMH is more likely to mediate hormonal (HPA-axis) responses to counteract hypoglycemia.[227] Since the SCN projects to both of these areas, as discussed above, it is likely that these connections mediate the daily variation in counterregulatory responses.[147]

12. Energy Restriction

The mammalian biological clock has been ticking over thousands and thousands of years, and during a large portion of this time, adequate food resources were not always available. Therefore, the mammalian body is highly specialized in spending, in times of scarcity, only the absolute necessary minimum amount of energy. A major advantage of having a circadian system, when saving energy, is the ability to adjust the metabolic rate even to the specific time of the day. Indeed, metabolic rhythms persist during starvation.[228] The metabolic rate can be lowered during inactivity, while it is kept unchanged when really needed, i.e., at the onset of the activity period. Fasted rats become more active at the onset of the dark period, because this is the time to search for food. This is reflected by heart rate, body temperature and locomotor activity, that are decreased during inactivity, but remain comparable to *ad libitum* fed rats at the onset of the activity period.[229] Furthermore, the catabolic hormone glucagon, which stimulates glucose output from the liver, shows normal plasma levels at the onset of the activity period, but is decreased during the light phase.[110] Perhaps, the decreased plasma glucagon level also contributes to lower metabolism. This phenomenon has also been shown for triglyceride metabolism in fasted rats.[103] Thus, the modulating effect of food restriction on the SCN results in an increase in the amplitude of many physiological rhythms, especially by a decrease of its trough values.

13. Energy Surplus

As already mentioned above, during thousands and thousands of years of evolution, food has usually been scarce. Most animals, including humans, have therefore adopted a strategy to store as much of the available nutrients as possible (known as Neel's "thrifty" gene hypothesis[230]), to be able to survive the periods of scarcity. In the present time, however, food has become abundant in many parts of the world and industrialization has decreased our daily load of physical activity. So, although our body does not need it anymore, our genes are still set to store energy reserves. This combination of factors causes more and more people to become obese. Constant high levels of carbohydrates and fat continuously stimulate insulin release, and cells become insulin-resistant, leading to diabetes mellitus.

Apart from the amount of food intake, the timing of food intake has also changed. Instead of consuming a few regular meals during daytime, people now have access to snacks at all times. Daily activity is not restricted to the light period anymore either. Thus, the amplitude of daily variation in many of our environmental conditions has diminished, i.e. the difference between night and day in the amount of activity, light and food intake is decreased. Although not very many people actually work during the night, the general activity pattern has shifted towards evening hours. Consequently, also the intake of meals and snacks during evening and night time has increased. Since light,[7] food intake[46-48] and physical activity[231,232] are powerful Zeitgebers, a shift in their presence may have profound impact on physiological parameters. A flattening in the rhythm of these Zeitgebers will result in a more constant environmental input to the brain.

As has been hypothesized previously,[233] such a disturbance may have profound effects on metabolism and may even cause disease. Food intake in the late evening, a time-of-day when insulin sensitivity is low and fewer nutrients are cleared from the plasma, may result in high plasma FFA and glucose values. This causes increased fat storage instead of oxidation by muscle tissue.[164] In fact, Qin *et al.* showed that people with a nocturnal life style have decreased glucose tolerance and a higher risk for diabetes, cardiovascular disease and the metabolic syndrome.[234]

This has also been shown in shift workers,[235-237] another group of people with disturbed daily rhythms. Moreover, more and more evidence is being obtained for a disturbance of daily physiological rhythms in obesity and Diabetes Mellitus[58,59,238-241]. Therefore, it seems that the 24-hour society that has evolved during the last 50 years causes disturbances of the daily activity pattern that metabolism, regulated by a circadian system that has evolved during a time span of thousands of years, cannot cope with.

14. Conclusion

As shown above, biological rhythms are found throughout the various components of glucose homeostasis. By sending the signal of day and night to the rest of the body, the SCN synchronizes the different organs in the body to the presence of light and adapts physiological processes to the specific energy requirement of that moment. As said, homeostasis is thus protected, not only by responding to acute changes in the internal or external environment, but also by anticipating daily changes in energy requirement. Considering the importance of sufficient glucose supply, it is logical that many organs act simultaneously and in a possibly redundant way.

Thus, stimulation of glucose uptake may be necessary in one part of the body, but at the same time it may need to be inhibited in another. A central mechanism that orchestrates the differentiated needs for these processes according to the time of day is therefore essential. The biological clock fine-tunes the various simultaneous processes in the body that are part of energy metabolism, such as glucose production, utilization and storage in several organs, but also feeding behavior and counterregulatory responses to acute (metabolic) stressors.

Although glucose metabolism in general is controlled via both hormonal and neuronal pathways, the results obtained so far provide evidence that biological clock control mainly occurs via the autonomic nervous system. Secretion of the pancreatic hormones insulin and glucagon is stimulated via autonomic nerves but although these hormones show circadian patterns of release, no evidence has been found for these rhythms to have an essential role in the daily rhythm of plasma glucose.

On the other hand, it is likely that the SCN connections to the endocrine pancreas are involved in the changes of the insulin and glucagon responses to feeding and hypoglycemia, respectively, over the day-night cycle.[147,242] Furthermore, also other hormones, such as corticosteroids or growth hormone, do not seem to have an important role in the control of the daily plasma glucose rhythm.

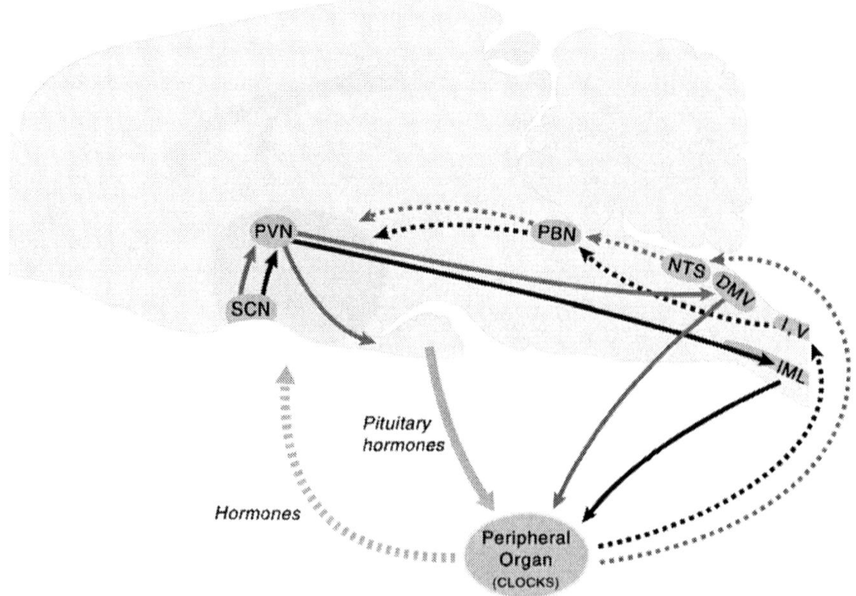

FIGURE 2. The suprachiasmatic nucleus (SCN) sends its signal of day and night to the rest of the body via several different pathways. Both "parasympathetic" (light gray) and "sympathetic" (dark grey) neurons project to cells in the PVN (and other hypothalamic nuclei involved in glucose metabolism). The pre-autonomic PVN cells project to the dorsal motor nucleus of the vagus (DMV, parasympathetic) in the brainstem and the intermediolateral column of the spinal cord (IML, sympathetic) that relay the signal to organs in the periphery. Furthermore, the pituitary receives the SCN signal through the release of "releasingfactors" from the PVN and rhythmically secretes hormones that may have an effect on glucose production. Peripheral feedback signals concerning glucose availability, energy reserves in the body and hormone levels and are received by the hypothalamus via different sensory pathways.

However, there is strong evidence of an important role of the autonomic innervation of organs involved in glucose production. Multisynaptic connections between the SCN and the liver[72] have been shown. Selective (sympathetic or parasympathetic) denervation of the liver leads to disruption of the daily plasma glucose rhythm.[104] Furthermore, electrical stimulation of the SCN induces hyperglycemia,[53] a response that is blocked by α- and β-adrenergic antagonists.[54]

It is well-known that many SCN neurons projecting to the PVN express the inhibitory neurotransmitter gamma-aminobutyric acid (GABA).[243,244] Experiments investigating the SCN control of the daily melatonin rhythm have shown a circadian rhythm in the GABA-ergic output signal from the SCN to the PVN.[245,246] Recent results obtained by our group show that this GABA signal to the PVN may also be important in the mediation of the

SCN signal to the liver. Infusion of the GABA-antagonist bicuculline (BIC) into the PVN induced a rise in plasma glucose concentration, an effect that could be prevented by sympathetic denervation of the liver. Furthermore, the SCN-DMH pathway may also play a role in the regulation of glucose output, as blocking synaptic transmission by infusion of tetrodotoxin (TTX) into either the SCN or the DMH induced a similar rise in the plasma glucose concentration. Therefore, the daily anticipatory rise in plasma glucose levels may be induced by the SCN, through a withdrawal of the inhibitory inputs to the sympathetic pre-autonomic neurons in the PVN, and subsequent stimulation of glucose production by the liver.[90]

The VMH may be an important relay in the hypothalamic control of glucose uptake. Electrical stimulation or leptin injection into the VMH (but not LH) increases glucose uptake in skeletal muscle and the heart, as well as in brown adipose tissue (BAT).[157,158] Surgical denervation of BAT prevented this increase, indicating that this effect is indeed mediated via the central nervous system. Direct inputs from the SCN to the VMH, however, are scarce. Glucose uptake in white adipose tissue was not affected in these experiments, suggesting that this may be mediated via other hypothalamic nuclei. Thus, also the regulation of glucose uptake in the liver, muscle and adipose tissue, either insulin-dependent or –independent, is probably mediated via direct autonomic innervation, as glucose uptake is impaired in autonomically denervated tissues.[74,100,101]

In conclusion, there is a delicate balance in the control of the many target areas within the brain and in the rest of the body that are involved in glucose homeostasis. Many hypothalamic nuclei are involved in the control of glucose metabolism, and many of these nuclei receive input from the SCN in order to adjust their outputs to the daily physiological needs of the body. Based on the papers reviewed here, we suggest that it is likely that the SCN signal that synchronizes glucose homeostasis to the day-night cycle is conveyed via the autonomic nervous system rather than via hormonal pathways.

15. References

1. Moore, R. Y. & Eichler, V. B. Loss of a circadian adrenal corticosterone rhythm following suprachiasmatic lesions in the rat. *Brain Res* 42, 201-6 (1972).
2. Stephan, F. K. & Zucker, I. Circadian rhythms in drinking behavior and locomotor activity of rats are eliminated by hypothalamic lesions. *Proc Natl Acad Sci U S A* 69, 1583-6 (1972).
3. Reppert, S. M. et al. Effects of damage to the suprachiasmatic area of the anterior hypothalamus on the daily melatonin and cortisol rhythms in the rhesus monkey. *J Neurosci* 1, 1414-25 (1981).
4. Aschoff, J. & Wever, R. Spontanperiodik des Menschen bei ausschluss aller Zeitgeber. *Naturwissenschaften* 49, 337-342 (1962).
5. Abe, M. et al. Circadian rhythms in isolated brain regions. *J Neurosci* 22, 350-6 (2002).

6. Groos, G. A. & Meijer, J. H. Effects of illumination on suprachiasmatic nucleus electrical discharge. *Ann N Y Acad Sci* 453, 134-46 (1985).

7. Meijer, J. H. et al. Light entrainment of the mammalian biological clock. *Prog Brain Res* 111, 175-90 (1996).

8. Hattar, S., Liao, H. W., Takao, M., Berson, D. M. & Yau, K. W. Melanopsin-containing retinal ganglion cells: architecture, projections, and intrinsic photosensitivity. *Science* 295, 1065-70 (2002).

9. Freedman, M. S. et al. Regulation of mammalian circadian behavior by non-rod, non-cone, ocular photoreceptors. *Science* 284, 502-4 (1999).

10. Reppert, S. M. & Weaver, D. R. Coordination of circadian timing in mammals. *Nature* 418, 935-41 (2002).

11. Noshiro, M. et al. Rhythmic expression of DEC1 and DEC2 in peripheral tissues: DEC2 is a potent suppressor for hepatic cytochrome P450s opposing DBP. *Genes Cells* 9, 317-29 (2004).

12. Honma, S., Shirakawa, T., Katsuno, Y., Namihira, M. & Honma, K. Circadian periods of single suprachiasmatic neurons in rats. *Neurosci Lett* 250, 157-60 (1998).

13. Watts, A. G., Swanson, L. W. & Sanchez-Watts, G. Efferent projections of the suprachiasmatic nucleus: I. Studies using anterograde transport of Phaseolus vulgaris leucoagglutinin in the rat. *J Comp Neurol* 258, 204-29 (1987).

14. Leak, R. K. & Moore, R. Y. Topographic organization of suprachiasmatic nucleus projection neurons. *J Comp Neurol* 433, 312-34 (2001).

15. Van der Beek, E. M., Horvath, T. L., Wiegant, V. M., Van den Hurk, R. & Buijs, R. M. Evidence for a direct neuronal pathway from the suprachiasmatic nucleus to the gonadotropin-releasing hormone system: combined tracing and light and electron microscopic immunocytochemical studies. *J Comp Neurol* 384, 569-79 (1997).

16. De la Iglesia, H. O., Blaustein, J. D. & Bittman, E. L. The suprachiasmatic area in the female hamster projects to neurons containing estrogen receptors and GnRH. *Neuroreport* 6, 1715-22 (1995).

17. Choi, S., Wong, L. S., Yamat, C. & Dallman, M. F. Hypothalamic ventromedial nuclei amplify circadian rhythms: do they contain a food-entrained endogenous oscillator? *J Neurosci* 18, 3843-52 (1998).

18. Buijs, R. M., Markman, M., Nunes-Cardoso, B., Hou, Y. X. & Shinn, S. Projections of the suprachiasmatic nucleus to stress-related areas in the rat hypothalamus: a light and electron microscopic study. *J Comp Neurol* 335, 42-54 (1993).

19. Kalsbeek, A., Fliers, E., Franke, A. N., Wortel, J. & Buijs, R. M. Functional connections between the suprachiasmatic nucleus and the thyroid gland as revealed by lesioning and viral tracing techniques in the rat. *Endocrinology* 141, 3832-41 (2000).

20. Buijs, R. M., Chun, S. J., Niijima, A., Romijn, H. J. & Nagai, K. Parasympathetic and sympathetic control of the pancreas: a role for the suprachiasmatic nucleus and other hypothalamic centers that are involved in the regulation of food intake. *J Comp Neurol* 431, 405-23 (2001).

21. Gerendai, I., Toth, I. E., Boldogkoi, Z., Medveczky, I. & Halasz, B. CNS structures presumably involved in vagal control of ovarian function. *J Auton Nerv Syst* 80, 40-5 (2000).

22. Larsen, P. J., Enquist, L. W. & Card, J. P. Characterization of the multisynaptic neuronal control of the rat pineal gland using viral transneuronal tracing. *Eur J Neurosci* 10, 128-45 (1998).

23. Swanson, L. W. & Kuypers, H. G. The paraventricular nucleus of the hypothalamus: cytoarchitectonic subdivisions and organization of projections to the

pituitary, dorsal vagal complex, and spinal cord as demonstrated by retrograde fluorescence double-labeling methods. *J Comp Neurol* 194, 555-70 (1980).

24. Buijs, R. M. et al. Anatomical and functional demonstration of a multisynaptic suprachiasmatic nucleus adrenal (cortex) pathway. *Eur J Neurosci* 11, 1535-44 (1999).

25. Sakurada, S. et al. Autonomic and behavioural thermoregulation in starved rats. *J Physiol* 526 Pt 2, 417-24 (2000).

26. Aston-Jones, G., Chen, S., Zhu, Y. & Oshinsky, M. L. A neural circuit for circadian regulation of arousal. *Nat Neurosci* 4, 732-8 (2001).

27. Chou, T. C. et al. Critical role of dorsomedial hypothalamic nucleus in a wide range of behavioral circadian rhythms. *J Neurosci* 23, 10691-702 (2003).

28. Kalsbeek, A. et al. GABA receptors in the region of the dorsomedial hypothalamus of rats are implicated in the control of melatonin and corticosterone release. *Neuroendocrinology* 63, 69-78 (1996).

29. Brugger, M. Fresstrieb als hypothalamisches symptom. *Helv Physiol Acta* 1, 183-189 (1943).

30. Larsson, S. On the hypothalamic organisation of the nervous mechanism regulating food intake. *Acta Physiol Scand Suppl* 32, 7-63 (1954).

31. Bellinger, L. L. & Bernardis, L. L. The dorsomedial hypothalamic nucleus and its role in ingestive behavior and body weight regulation: lessons learned from lesioning studies. *Physiol Behav* 76, 431-42 (2002).

32. Bernardis, L. L. & Bellinger, L. L. The lateral hypothalamic area revisited: ingestive behavior. *Neurosci Biobehav Rev* 20, 189-287 (1996).

33. Ono, T., Sasaki, K. & Shibata, R. Diurnal- and behaviour-related activity of ventromedial hypothalamic neurones in freely behaving rats. *J Physiol* 394, 201-20 (1987).

34. Ono, T., Sasaki, K. & Shibata, R. Feeding- and chemical-related activity of ventromedial hypothalamic neurones in freely behaving rats. *J Physiol* 394, 221-37 (1987).

35. Buijs, R. M., Van Eden, C. G., Goncharuk, V. D. & Kalsbeek, A. The biological clock tunes the organs of the body: timing by hormones and the autonomic nervous system. *J Endocrinol* 177, 17-26 (2003).

36. Harmar, A. J. An essential role for peptidergic signalling in the control of circadian rhythms in the suprachiasmatic nuclei. *J Neuroendocrinol* 15, 335-8 (2003).

37. Stokkan, K. A., Yamazaki, S., Tei, H., Sakaki, Y. & Menaker, M. Entrainment of the circadian clock in the liver by feeding. *Science* 291, 490-3 (2001).

38. Storch, K. F. et al. Extensive and divergent circadian gene expression in liver and heart. *Nature* 417, 78-83 (2002).

39. Yamazaki, S. et al. Resetting central and peripheral circadian oscillators in transgenic rats. *Science* 288, 682-5 (2000).

40. Young, M. E., Razeghi, P. & Taegtmeyer, H. Clock genes in the heart: characterization and attenuation with hypertrophy. *Circ Res* 88, 1142-50 (2001).

41. Muhlbauer, E., Wolgast, S., Finckh, U., Peschke, D. & Peschke, E. Indication of circadian oscillations in the rat pancreas. *FEBS Lett* 564, 91-6 (2004).

42. Terazono, H. et al. Adrenergic regulation of clock gene expression in mouse liver. *Proc Natl Acad Sci U S A* 100, 6795-800 (2003).

43. Sakamoto, K. et al. Multitissue circadian expression of rat period homolog (rPer2) mRNA is governed by the mammalian circadian clock, the suprachiasmatic nucleus in the brain. *J Biol Chem* 273, 27039-42 (1998).

44. Akhtar, R. A. et al. Circadian cycling of the mouse liver transcriptome, as revealed by cDNA microarray, is driven by the suprachiasmatic nucleus. *Curr Biol* 12, 540-50 (2002).

45. Yoo, S. H. et al. PERIOD2::LUCIFERASE real-time reporting of circadian dynamics reveals persistent circadian oscillations in mouse peripheral tissues. *Proc Natl Acad Sci U S A* 101, 5339-46 (2004).

46. Stephan, F. K., Swann, J. M. & Sisk, C. L. Anticipation of 24-hr feeding schedules in rats with lesions of the suprachiasmatic nucleus. *Behav Neural Biol* 25, 346-63 (1979).

47. Damiola, F. et al. Restricted feeding uncouples circadian oscillators in peripheral tissues from the central pacemaker in the suprachiasmatic nucleus. *Genes Dev* 14, 2950-2961 (2000).

48. Schibler, U., Ripperger, J. & Brown, S. A. Peripheral circadian oscillators in mammals: time and food. *J Biol Rhythms* 18, 250-60 (2003).

49. Diaz-Munoz, M., Vazquez-Martinez, O., Aguilar-Roblero, R. & Escobar, C. Anticipatory changes in liver metabolism and entrainment of insulin, glucagon, and corticosterone in food-restricted rats. *Am J Physiol Regul Integr Comp Physiol* 279, R2048-56 (2000).

50. Davidson, A. J. & Stephan, F. K. Plasma glucagon, glucose, insulin, and motilin in rats anticipating daily meals. *Physiol Behav* 66, 309-15 (1999).

51. Rutter, J., Reick, M., Wu, L. C. & McKnight, S. L. Regulation of clock and NPAS2 DNA binding by the redox state of NAD cofactors. *Science* 293, 510-4 (2001).

52. Stephan, F. K. & Davidson, A. J. Glucose, but not fat, phase shifts the feeding-entrained circadian clock. *Physiol Behav* 65, 277-88 (1998).

53. Nagai, K., Fujii, T., Inoue, S., Takamura, Y. & Nakagawa, H. Electrical stimulation of the suprachiasmatic nucleus of the hypothalamus causes hyperglycemia. *Horm Metab Res* 20, 37-9. (1988).

54. Fujii, T., Inoue, S., Nagai, K. & Nakagawa, H. Involvement of adrenergic mechanism in hyperglycemia due to SCN stimulation. *Horm Metab Res* 21, 643-5 (1989).

55. La Fleur, S. E., Kalsbeek, A., Wortel, J. & Buijs, R. M. A suprachiasmatic nucleus generated rhythm in basal glucose concentrations. *J Neuroendocrinol* 11, 643-52 (1999).

56. Bolli, G. B. et al. Demonstration of a dawn phenomenon in normal human volunteers. *Diabetes* 33, 1150-3 (1984).

57. La Fleur, S. E., Kalsbeek, A., Wortel, J., Fekkes, M. L. & Buijs, R. M. A daily rhythm in glucose tolerance: a role for the suprachiasmatic nucleus. *Diabetes* 50, 1237-43 (2001).

58. Lee, A., Ader, M., Bray, G. A. & Bergman, R. N. Diurnal variation in glucose tolerance. Cyclic suppression of insulin action and insulin secretion in normal-weight, but not obese, subjects. *Diabetes* 41, 742-9 (1992).

59. Van Cauter, E. V. et al. Abnormal temporal patterns of glucose tolerance in obesity: relationship to sleep-related growth hormone secretion and circadian cortisol rhythmicity. *J Clin Endocrinol Metab* 79, 1797-805 (1994).

60. Gerich, J. E., Meyer, C., Woerle, H. J. & Stumvoll, M. Renal gluconeogenesis: its importance in human glucose homeostasis. *Diabetes Care* 24, 382-91 (2001).

61. Croset, M. et al. Rat small intestine is an insulin-sensitive gluconeogenic organ. *Diabetes* 50, 740-6 (2001).

62. Matsuura, N., Cheng, J. S. & Kalant, N. Insulin control of hepatic glucose production. *Can J Biochem* 53, 28-36 (1975).

63. Gerich, J., Cryer, P. & Rizza, R. Hormonal mechanisms in acute glucose counterregulation: the relative roles of glucagon, epinephrine, norepinephrine, growth hormone, and cortisol. *Metabolism* 29, 1164-75 (1980).
64. Chiasson, J. L., Cherrington, A.D. Glucagon and Liver Glucose Output. *in:* *Handbook of Experimental Pharmacology. Lefèbre (ed)* 1, 361-381 (1983).
65. Samols, E., Bonner-Weir, S. & Weir, G. C. Intra-islet insulin-glucagon-somatostatin relationships. *Clin Endocrinol Metab* 15, 33-58 (1986).
66. Banarer, S., McGregor, V. P. & Cryer, P. E. Intraislet hyperinsulinemia prevents the glucagon response to hypoglycemia despite an intact autonomic response. *Diabetes* 51, 958-65 (2002).
67. Ahren, B., Nobin, A. & Schersten, B. Insulin and C-peptide secretory responses to glucagon in man: studies on the dose-response relationships. *Acta Med Scand* 221, 185-90 (1987).
68. Sacca, L., Vigorito, C., Cicala, M., Corso, G. & Sherwin, R. S. Role of gluconeogenesis in epinephrine-stimulated hepatic glucose production in humans. *Am J Physiol* 245, E294-302 (1983).
69. Kahn, C. R., Goldfine, I. D., Neville, D. M., Jr. & De Meyts, P. Alterations in insulin binding induced by changes in vivo in the levels of glucocorticoids and growth hormone. *Endocrinology* 103, 1054-66 (1978).
70. Plaschke, K., Muller, D. & Hoyer, S. Effect of adrenalectomy and corticosterone substitution on glucose and glycogen metabolism in rat brain. *J Neural Transm* 103, 89-100 (1996).
71. Bernard, C. *Lecons de Physiologie Appliquee a la Medicine* (Bailiere, Paris, 1855).
72. La Fleur, S. E., Kalsbeek, A., Wortel, J. & Buijs, R. M. Polysynaptic neural pathways between the hypothalamus, including the suprachiasmatic nucleus, and the liver. *Brain Res* 871, 50-6 (2000).
73. Card, J. P. Pseudorabies virus and the functional architecture of the circadian timing system. *J Biol Rhythms* 15, 453-61 (2000).
74. Kreier, F. et al. Selective parasympathetic innervation of subcutaneous and intra-abdominal fat -functional implications. *J Clin Invest* 110, 1243-50 (2002).
75. Bamshad, M., Aoki, V. T., Adkison, M. G., Warren, W. S. & Bartness, T. J. Central nervous system origins of the sympathetic nervous system outflow to white adipose tissue. *Am J Physiol* 275, R291-9 (1998).
76. Scheer, F. A., Ter Horst, G. J., van Der Vliet, J. & Buijs, R. M. Physiological and anatomic evidence for regulation of the heart by suprachiasmatic nucleus in rats. *Am J Physiol Heart Circ Physiol* 280, H1391-9 (2001).
77. Sly, J. D., Colvill, L., McKinley, J. M. & Oldfield, J. B. Identification of neural projections from the forebrain to the kidney, using the virus pseudorabies. *J Auton Nerv Syst* 77, 73-82 (1999).
78. Burcelin, R., Dolci, W. & Thorens, B. Glucose sensing by the hepatoportal sensor is GLUT2-dependent: in vivo analysis in GLUT2-null mice. *Diabetes* 49, 1643-8 (2000).
79. Conlee, R. K., Rennie, M. J. & Winder, W. W. Skeletal muscle glycogen content: diurnal variation and effects of fasting. *Am J Physiol* 231, 614-8 (1976).
80. Gagliardino, J. J., Pessacq, M. T., Hernandez, R. E. & Rebolledo, O. R. Circadian variations in serum glucagon and hepatic glycogen and cyclic amp concentrations. *J Endocrinol* 78, 297-8 (1978).
81. Sollberger. The control of circadian glycogen rhythms. *Ann. N.Y.Acad.Sci* 117, 519-554 (1964).

82. Peret, J., Chanez, M. & Pascal, G. Schedule of protein ingestion and circadian rhythm of certain hepatic enzyme activities involved in glucose metabolism in the rat. *Nutr Metab* 20, 143-57 (1976).
83. Corssmit, E. P., Romijn, J. A. & Sauerwein, H. P. Review article: Regulation of glucose production with special attention to nonclassical regulatory mechanisms: a review. *Metabolism* 50, 742-55 (2001).
84. Gelfand, R. A., Sherwin, R.S. Glucagon and starvation. *IN: Handbook of Experimental Pharmacology. Lefèbre (ed)* 1, 223-237 (1983).
85. Margolis, R. N. & Curnow, R. T. The role of insulin and glucocorticoids in the regulation of hepatic glycogen metabolism: effect of fasting, refeeding, and adrenalectomy. *Endocrinology* 113, 2113-9 (1983).
86. Boden, G., Chen, X. & Urbain, J. L. Evidence for a circadian rhythm of insulin sensitivity in patients with NIDDM caused by cyclic changes in hepatic glucose production. *Diabetes* 45, 1044-50 (1996).
87. Kida, K. et al. The circadian change of gluconeogenesis in the liver in vivo in fed rats. *J Biochem (Tokyo)* 88, 1009-13 (1980).
88. Mlekusch, W., Paletta, B., Truppe, W., Paschke, E. & Grimus, R. Plasma concentrations of glucose, corticosterone, glucagon and insulin and liver content of metabolic substrates and enzymes during starvation and additional hypoxia in the rat. *Horm Metab Res* 13, 612-4 (1981).
89. Phillips, L. J. & Berry, L. J. Circadian rhythm of mouse liver phosphoenolpyruvate carboxykinase. *Am J Physiol* 218, 1440-4 (1970).
90. Kalsbeek, A., La Fleur, S. E., van Heijningen, C. & Buijs, R. Suprachiasmatic GABA-ergic inputs to the PVN control plasma glucose concentrations in the rat via the sympathetic innervation of the liver. *J Neurosci in press* (2004).
91. Buijs, R. M. et al. The suprachiasmatic nucleus balances sympathetic and parasympathetic output to peripheral organs through separate preautonomic neurons. *J Comp Neurol* 464, 36-48 (2003).
92. Shimazu, T. & Fukuda, A. Increased activities of glycogenolytic enzymes in liver after splanchnic-nerve stimulation. *Science* 150, 1607-8 (1965).
93. Shimazu, T. & Amakawa, A. Regulation of glycogen metabolism in liver by the autonomic nervous system. 3. Differential effects of sympathetic-nerve stimulation and of catecholamines on liver phosphorylase. *Biochim Biophys Acta* 165, 349-56 (1968).
94. Seydoux, J., Brunsmann, M. J., Jeanrenaud, B. & Girardier, L. alpha-Sympathetic control of glucose output of mouse liver perfused in situ. *Am J Physiol* 236, E323-7 (1979).
95. Pascoe, W. S., Smythe, G. A. & Storlien, L. H. 2-deoxy-D-glucose-induced hyperglycemia: role for direct sympathetic nervous system activation of liver glucose output. *Brain Res* 505, 23-8 (1989).
96. Takahashi, A., Ishimaru, H., Ikarashi, Y., Kishi, E. & Maruyama, Y. Effects of ventromedial hypothalamus stimulation on glycogenolysis in rat liver using in vivo microdialysis. *Metabolism* 46, 897-901 (1997).
97. Shimazu, T., Fukuda, A. & Ban, T. Reciprocal influences of the ventromedial and lateral hypothalamic nuclei on blood glucose level and liver glycogen content. *Nature* 210, 1178-9 (1966).
98. Lautt, W. W. & Wong, C. Hepatic parasympathetic neural effect on glucose balance in the intact liver. *Can J Physiol Pharmacol* 56, 679-82 (1978).

99. Shimazu, T. Regulation of glycogen metabolism in liver by the autonomic nervous system. V. Activation of glycogen synthetase by vagal stimulation. *Biochim Biophys Acta* 252, 28-38 (1971).

100. Xue, C. et al. Isolated hepatic cholinergic denervation impairs glucose and glycogen metabolism. *J Surg Res* 90, 19-25 (2000).

101. Matsuhisa, M. et al. Important role of the hepatic vagus nerve in glucose uptake and production by the liver. *Metabolism* 49, 11-6 (2000).

102. Chen, C., Williams, P. F., Cooney, G. J. & Caterson, I. D. Diurnal rhythms of glycogen metabolism in the liver and skeletal muscle in gold thioglucose induced-obese mice with developing insulin resistance. *Int J Obes Relat Metab Disord* 16, 913-21 (1992).

103. Marrino, P., Gavish, D., Shafrir, E. & Eisenberg, S. Diurnal variations of plasma lipids, tissue and plasma lipoprotein lipase, and VLDL secretion rates in the rat. A model for studies of VLDL metabolism. *Biochim Biophys Acta* 920, 277-84 (1987).

104. La Fleur, S. E. et al. Autonomic projections to the liver are essential for the SCN-generated rhythm in plasma glucose concentrations. *PHd thesis* chapter 5 (2001).

105. Vizzard, M. A., Brisson, M. & de Groat, W. C. Transneuronal labeling of neurons in the adult rat central nervous system following inoculation of pseudorabies virus into the colon. *Cell Tissue Res* 299, 9-26 (2000).

106. Standish, A., Enquist, L. W., Escardo, J. A. & Schwaber, J. S. Central neuronal circuit innervating the rat heart defined by transneuronal transport of pseudorabies virus. *J Neurosci* 15, 1998-2012 (1995).

107. Tiedgen, M. & Seitz, H. J. Dietary control of circadian variations in serum insulin, glucagon and hepatic cyclic AMP. *J Nutr* 110, 876-82 (1980).

108. Vansant, G., Van Gaal, L., Van Acker, K. & De Leeuw, I. Importance of glucagon as a determinant of resting metabolic rate and glucose-induced thermogenesis in obese women. *Metabolism* 40, 672-5 (1991).

109. Moore, M. C. & Cherrington, A. D. Regulation of net hepatic glucose uptake: interaction of neural and pancreatic mechanisms. *Reprod Nutr Dev* 36, 399-406 (1996).

110. Ruiter, M. et al. The daily rhythm in plasma glucagon concentrations in the rat is modulated by the biological clock and by feeding behavior. *Diabetes* 52, 1709-15 (2003).

111. Luiten, P. G., ter Horst, G. J., Buijs, R. M. & Steffens, A. B. Autonomic innervation of the pancreas in diabetic and non-diabetic rats. A new view on intramural sympathetic structural organization. *J Auton Nerv Syst* 15, 33-44 (1986).

112. De Jong, A., Strubbe, J. H. & Steffens, A. B. Hypothalamic influence on insulin and glucagon release in the rat. *Am J Physiol* 233, E380-8 (1977).

113. Kaneto, A., Kosaka, K. & Nakao, K. Effects of stimulation of the vagus nerve on insulin secretion. *Endocrinology* 80, 530-6 (1967).

114. Louis-Sylvestre, J. Preabsorptive insulin release and hypoglycemia in rats. *Am J Physiol* 230, 56-60 (1976).

115. Jansen, A. S., Hoffman, J. L. & Loewy, A. D. CNS sites involved in sympathetic and parasympathetic control of the pancreas: a viral tracing study. *Brain Res* 766, 29-38 (1997).

116. Brunicardi, F. C., Shavelle, D. M. & Andersen, D. K. Neural regulation of the endocrine pancreas. *Int J Pancreatol* 18, 177-95 (1995).

117. Ahren, B. Autonomic regulation of islet hormone secretion–implications for health and disease. *Diabetologia* 43, 393-410 (2000).

118. Li, Y., Jiang, Y. C. & Owyang, C. Central CGRP inhibits pancreatic enzyme secretion by modulation of vagal parasympathetic outflow. *Am J Physiol* 275, G957-63 (1998).

119. Li, Y. & Owyang, C. Somatostatin inhibits pancreatic enzyme secretion at a central vagal site. *Am J Physiol* 265, G251-7 (1993).

120. Williams, G. et al. The hypothalamus and the control of energy homeostasis: different circuits, different purposes. *Physiol Behav* 74, 683-701 (2001).

121. Moriguchi, T., Sakurai, T., Nambu, T., Yanagisawa, M. & Goto, K. Neurons containing orexin in the lateral hypothalamic area of the adult rat brain are activated by insulin-induced acute hypoglycemia. *Neurosci Lett* 264, 101-4 (1999).

122. Niijima, A., Kannan, H. & Yamashita, H. Neural control of blood glucose homeostasis; effect of microinjection of glucose into hypothalamic nuclei on efferent activity of pancreatic branch of vagus nerve in the rat. *Brain Res Bull* 20, 811-5 (1988).

123. Ahren, B. & Taborsky, G. J., Jr. The mechanism of vagal nerve stimulation of glucagon and insulin secretion in the dog. *Endocrinology* 118, 1551-7 (1986).

124. Woods, S. C. & Porte, D., Jr. Neural control of the endocrine pancreas. *Physiol Rev* 54, 596-619 (1974).

125. Ahren, B., Taborsky, G. J., Jr. & Porte, D., Jr. Neuropeptidergic versus cholinergic and adrenergic regulation of islet hormone secretion. *Diabetologia* 29, 827-36 (1986).

126. Bereiter, D. A., Berthoud, H. R., Brunsmann, M. & Jeanrenaud, B. Nucleus ambiguus stimulation increases plasma insulin levels in the rat. *Am J Physiol* 241, E22-7 (1981).

127. Strubbe, J. H. & Steffens, A. B. Neural control of insulin secretion. *Horm Metab Res* 25, 507-12 (1993).

128. Samols, E. & Weir, G. C. Adrenergic modulation of pancreatic A, B, and D cells alpha-Adrenergic suppression and beta-adrenergic stimulation of somatostatin secretion, alpha-adrenergic stimulation of glucagon secretion in the perfused dog pancreas. *J Clin Invest* 63, 230-8 (1979).

129. Scheurink, A. J. et al. Sympathoadrenal influence on glucose, FFA, and insulin levels in exercising rats. *Am J Physiol* 256, R161-8 (1989).

130. Van Cauter, E., Desir, D., Decoster, C., Fery, F. & Balasse, E. O. Nocturnal decrease in glucose tolerance during constant glucose infusion. *J Clin Endocrinol Metab* 69, 604-11 (1989).

131. Shapiro, E. T. et al. Oscillations in insulin secretion during constant glucose infusion in normal man: relationship to changes in plasma glucose. *J Clin Endocrinol Metab* 67, 307-14 (1988).

132. Jolin, T. & Montes, A. Daily rhythm of plasma glucose and insulin levels in rats. *Horm Res* 4, 153-6 (1973).

133. Peschke, E. & Peschke, D. Evidence for a circadian rhythm of insulin release from perifused rat pancreatic islets. *Diabetologia* 41, 1085-92 (1998).

134. Ahren, B., Veith, R. C. & Taborsky, G. J., Jr. Sympathetic nerve stimulation versus pancreatic norepinephrine infusion in the dog: 1). Effects on basal release of insulin and glucagon. *Endocrinology* 121, 323-31 (1987).

135. Palmer, J. P., Porte Jr., D. Neural control of glucagon secretion. *IN: Handbook of Experimental Pharmacology. Lefèbre (ed)* 1, 115-131 (1983).

136. Bobbioni, E., Marre, M., Helman, A. & Assan, R. The nervous control of rat glucagon secretion in vivo. *Horm Metab Res* 15, 133-8 (1983).

137. Steffens, A. B. & Strubbe, J. H. CNS regulation of glucagon secretion. *Adv Metab Disord* 10, 221-57 (1983).
138. Kaneto, A., Miki, E. & Kosaka, K. Effects of vagal stimulation on glucagon and insulin secretion. *Endocrinology* 95, 1005-10 (1974).
139. Bloom, S. R. & Edwards, A. V. Pancreatic endocrine responses to stimulation of the peripheral ends of the vagus nerves in conscious calves. *J Physiol* 315, 31-41 (1981).
140. Jensen, S. L., Fahrenkrug, J., Holst, J. J., Nielsen, O. V. & Schaffalitzky de Muckadell, O. B. Secretory effects of VIP on isolated perfused porcine pancreas. *Am J Physiol* 235, E387-91 (1978).
141. Havel, P. J. et al. Evidence that vasoactive intestinal polypeptide is a parasympathetic neurotransmitter in the endocrine pancreas in dogs. *Regul Pept* 71, 163-70 (1997).
142. Taborsky, G. J., Jr., Ahren, B. & Havel, P. J. Autonomic mediation of glucagon secretion during hypoglycemia: implications for impaired alpha-cell responses in type 1 diabetes. *Diabetes* 47, 995-1005 (1998).
143. Havel, P. J., Mundinger, T. O., Veith, R. C., Dunning, B. E. & Taborsky, G. J., Jr. Corelease of galanin and NE from pancreatic sympathetic nerves during severe hypoglycemia in dogs. *Am J Physiol* 263, E8-16 (1992).
144. Dunning, B. E., Scott, M. F., Neal, D. W. & Cherrington, A. D. Direct quantification of norepinephrine spillover and hormone output from the pancreas of the conscious dog. *Am J Physiol* 272, E746-55 (1997).
145. Havel, P. J. & Taborsky, G. J., Jr. The contribution of the autonomic nervous system to changes of glucagon and insulin secretion during hypoglycemic stress. *Endocr Rev* 10, 332-50 (1989).
146. Schwartz, T. W. Pancreatic polypeptide: a hormone under vagal control. *Gastroenterology* 85, 1411-25 (1983).
147. Kalsbeek, A., Ruiter, M., La Fleur, S. E., Van Heijningen, C. & Buijs, R. M. The diurnal modulation of hormonal responses in the rat varies with different stimuli. *J Neuroendocrinol* 15, 1144-55 (2003).
148. Hansen, A. P. & Johansen, K. Diurnal patterns of blood glucose, serum free fatty acids, insulin, glucagon and growth hormone in normals and juvenile diabetics. *Diabetologia* 6, 27-33 (1970).
149. Tasaka, Y., Inoue, S., Maruno, K. & Hirata, Y. Twenty-four-hour variations of plasma pancreatic polypeptide, insulin and glucagon in normal human subjects. *Endocrinol Jpn* 27, 495-8 (1980).
150. Yamamoto, H., Nagai, K. & Nakagawa, H. Role of SCN in daily rhythms of plasma glucose, FFA, insulin and glucagon. *Chronobiol Int* 4, 483-91 (1987).
151. Assan, R., Marre, M., Gormley, M. The amino acid-induced secretion of glucagon. *IN: Handbook of Experimental Pharmacology. Lefèbre (ed)* 2, 19-29 (1983).
152. Benthem, L., Mundinger, T. O. & Taborsky, G. J., Jr. Meal-induced insulin secretion in dogs is mediated by both branches of the autonomic nervous system. *Am J Physiol Endocrinol Metab* 278, E603-10 (2000).
153. Abdallah, L., Chabert, M. & Louis-Sylvestre, J. Cephalic phase responses to sweet taste. *Am J Clin Nutr* 65, 737-43 (1997).
154. Berthoud, H. R., Bereiter, D. A., Trimble, E. R., Siegel, E. G. & Jeanrenaud, B. Cephalic phase, reflex insulin secretion. Neuroanatomical and physiological characterization. *Diabetologia* 20 Suppl, 393-401 (1981).
155. Berthoud, H. R. & Jeanrenaud, B. Sham feeding-induced cephalic phase insulin release in the rat. *Am J Physiol* 242, E280-5 (1982).

156. Shieh, J. J., Pan, C. J., Mansfield, B. C. & Chou, J. Y. A potential new role for muscle in blood glucose homeostasis. *J Biol Chem* 279, 26215-9 (2004).

157. Minokoshi, Y., Haque, M. S. & Shimazu, T. Microinjection of leptin into the ventromedial hypothalamus increases glucose uptake in peripheral tissues in rats. *Diabetes* 48, 287-91 (1999).

158. Sudo, M., Minokoshi, Y. & Shimazu, T. Ventromedial hypothalamic stimulation enhances peripheral glucose uptake in anesthetized rats. *Am J Physiol* 261, E298-303 (1991).

159. Wood, I. S. & Trayhurn, P. Glucose transporters (GLUT and SGLT): expanded families of sugar transport proteins. *Br J Nutr* 89, 3-9 (2003).

160. Bryant, N. J., Govers, R. & James, D. E. Regulated transport of the glucose transporter GLUT4. *Nat Rev Mol Cell Biol* 3, 267-77 (2002).

161. Frayn, K. N., Karpe, F., Fielding, B. A., Macdonald, I. A. & Coppack, S. W. Integrative physiology of human adipose tissue. *Int J Obes Relat Metab Disord* 27, 875-88 (2003).

162. Randle, P. J., PV, G., Hales, C. & Newsholme, E. A. The glucose fatty acid cycle: its role in insulin sensititvy and the metabolic disturbances of diabetes mellitus. *The Lancet* 1, 785-789 (1963).

163. Mason, T. M. et al. Prolonged elevation of plasma free fatty acids desensitizes the insulin secretory response to glucose in vivo in rats. *Diabetes* 48, 524-30 (1999).

164. Fabris, R. et al. Preferential channeling of energy fuels toward fat rather than muscle during high free fatty acid availability in rats. *Diabetes* 50, 601-8 (2001).

165. Carlson, M. G., Snead, W. L., Hill, J. O., Nurjhan, N. & Campbell, P. J. Glucose regulation of lipid metabolism in humans. *Am J Physiol* 261, E815-20. (1991).

166. Youngstrom, T. G. & Bartness, T. J. White adipose tissue sympathetic nervous system denervation increases fat pad mass and fat cell number. *Am J Physiol* 275, R1488-93 (1998).

167. Ahima, R. S. & Flier, J. S. Adipose Tissue as an Endocrine Organ. *Trends Endocrinol Metab* 11, 327-332 (2000).

168. Kershaw, E. E. & Flier, J. S. Adipose tissue as an endocrine organ. *J Clin Endocrinol Metab* 89, 2548-56 (2004).

169. Fliers, E. et al. White adipose tissue: getting nervous. *J Neuroendocrinol* 15, 1005-10 (2003).

170. Kalsbeek, A. et al. The suprachiasmatic nucleus generates the diurnal changes in plasma leptin levels. *Endocrinology* 142, 2677-85 (2001).

171. Seufert, J. Leptin effects on pancreatic beta-cell gene expression and function. *Diabetes* 53 Suppl 1, S152-8 (2004).

172. Sandoval, D. A. & Davis, S. N. Leptin: metabolic control and regulation. *J Diabetes Complications* 17, 108-13 (2003).

173. Petersen, K. F. et al. Leptin reverses insulin resistance and hepatic steatosis in patients with severe lipodystrophy. *J Clin Invest* 109, 1345-50 (2002).

174. Minokoshi, Y. et al. Leptin stimulates fatty-acid oxidation by activating AMP-activated protein kinase. *Nature* 415, 339-43 (2002).

175. Yamauchi, T. et al. The fat-derived hormone adiponectin reverses insulin resistance associated with both lipoatrophy and obesity. *Nat Med* 7, 941-6 (2001).

176. Yamauchi, T. et al. Adiponectin stimulates glucose utilization and fatty-acid oxidation by activating AMP-activated protein kinase. *Nat Med* 8, 1288-95 (2002).

177. Gavrila, A. et al. Diurnal and ultradian dynamics of serum adiponectin in healthy men: comparison with leptin, circulating soluble leptin receptor, and cortisol patterns. *J Clin Endocrinol Metab* 88, 2838-43 (2003).

178. Cano, N. Bench-to-bedside review: glucose production from the kidney. *Crit Care* 6, 317-21 (2002).

179. Stumvoll, M., Meyer, C., Mitrakou, A., Nadkarni, V. & Gerich, J. E. Renal glucose production and utilization: new aspects in humans. *Diabetologia* 40, 749-57 (1997).

180. Stumvoll, M. et al. Uptake and release of glucose by the human kidney. Postabsorptive rates and responses to epinephrine. *J Clin Invest* 96, 2528-33 (1995).

181. Meyer, C. et al. Relative importance of liver, kidney, and substrates in epinephrine-induced increased gluconeogenesis in humans. *Am J Physiol Endocrinol Metab* 285, E819-26 (2003).

182. Nagai, K., Suda, M., Yamagishi, O., Toyama, Y. & Nakagawa, H. Studies on the circadian rhythm of phosphoenolpyruvate carboxykinase. III. Circadian rhythm in the kidney. *J Biochem (Tokyo)* 77, 1249-54 (1975).

183. Kida, K., Nishio, T., Nagai, K., Matsuda, H. & Nakagawa, H. Gluconeogenesis in the kidney in vivo in fed rats. Circadian change and substrate specificity. *J Biochem (Tokyo)* 91, 755-60 (1982).

184. Trumper, B. G., Reschke, K. & Molling, J. Circadian variation of insulin requirement in insulin dependent diabetes mellitus the relationship between circadian change in insulin demand and diurnal patterns of growth hormone, cortisol and glucagon during euglycemia. *Horm Metab Res* 27, 141-7 (1995).

185. Clark, R. G., Chambers, G., Lewin, J. & Robinson, I. C. Automated repetitive microsampling of blood: growth hormone profiles in conscious male rats. *J.Endocrinol.* 111, 27-35 (1986).

186. Kimura, F. & Tsai, C. W. Ultradian rhythm of growth hormone secretion and sleep in the adult male rat. *J Physiol* 353, 305-15 (1984).

187. Pincus, S. M. et al. Females secrete growth hormone with more process irregularity than males in both humans and rats. *Am J Physiol* 270, E107-15 (1996).

188. Van Cauter, E. Diurnal and ultradian rhythms in human endocrine function: a minireview. *Horm Res* 34, 45-53 (1990).

189. De Boer, S. F. & Van der Gugten, J. Daily variations in plasma noradrenaline, adrenaline and corticosterone concentrations in rats. *Physiol Behav* 40, 323-328 (1987).

190. Dallman, M. F. et al. Feast and famine: critical role of glucocorticoids with insulin in daily energy flow. *Front Neuroendocrinol* 14, 303-47 (1993).

191. Kalsbeek, A., van der Vliet, J. & Buijs, R. M. Decrease of endogenous vasopressin release necessary for expression of the circadian rise in plasma corticosterone: a reverse microdialysis study. *J Neuroendocrinol* 8, 299-307 (1996).

192. D'Agostino, J., Vaeth, G. F. & Henning, S. J. Diurnal rhythm of total and free concentrations of serum corticosterone in the rat. *Acta Endocrinol (Copenh)* 100, 85-90 (1982).

193. Van Cauter, E. et al. Modulation of glucose regulation and insulin secretion by circadian rhythmicity and sleep. *J Clin Invest* 88, 934-42 (1991).

194. Plat, L. et al. Effects of morning cortisol elevation on insulin secretion and glucose regulation in humans. *Am J Physiol* 270, E36-42 (1996).

195. Van Cauter, E., Shapiro, E. T., Tillil, H. & Polonsky, K. S. Circadian modulation of glucose and insulin responses to meals: relationship to cortisol rhythm. *Am J Physiol* 262, E467-75 (1992).
196. Bright, G. M., Melton, T. W., Rogol, A. D. & Clarke, W. L. Failure of cortisol blockade to inhibit early morning increases in basal insulin requirements in fasting insulin-dependent diabetics. *Diabetes* 29, 662-4 (1980).
197. Shamoon, H., Hendler, R. & Sherwin, R. S. Altered responsiveness to cortisol, epinephrine, and glucagon in insulin-infused juvenile-onset diabetics. A mechanism for diabetic instability. *Diabetes* 29, 284-291 (1980).
198. Prinz, P. N., Halter, J., Benedetti, C. & Raskind, M. Circadian variation of plasma catecholamines in young and old men: relation to rapid eye movement and slow wave sleep. *J Clin Endocrinol Metab* 49, 300-4 (1979).
199. Linsell, C. R., Lightman, S. L., Mullen, P. E., Brown, M. J. & Causon, R. C. Circadian rhythms of epinephrine and norepinephrine in man. *J Clin Endocrinol Metab* 60, 1210-5 (1985).
200. Peschke, E., Peschke, D., Hammer, T. & Csernus, V. Influence of melatonin and serotonin on glucose-stimulated insulin release from perifused rat pancreatic islets in vitro. *J Pineal Res* 23, 156-63 (1997).
201. Shima, T. et al. Melatonin suppresses hyperglycemia caused by intracerebroventricular injection of 2-deoxy-D-glucose in rats. *Neurosci Lett* 226, 119-22 (1997).
202. Acuna-Castroviejo, D., Reiter, R. J., Menendez-Pelaez, A., Pablos, M. I. & Burgos, A. Characterization of high-affinity melatonin binding sites in purified cell nuclei of rat liver. *J Pineal Res* 16, 100-12 (1994).
203. Picinato, M. C. et al. Melatonin inhibits insulin secretion and decreases PKA levels without interfering with glucose metabolism in rat pancreatic islets. *J Pineal Res* 33, 156-60 (2002).
204. Cagnacci, A. et al. Influence of melatonin administration on glucose tolerance and insulin sensitivity of postmenopausal women. *Clin Endocrinol (Oxf)* 54, 339-46 (2001).
205. Lima, F. B. et al. Pinealectomy causes glucose intolerance and decreases adipose cell responsiveness to insulin in rats. *Am J Physiol* 275, E934-41 (1998).
206. La Fleur, S. E., Kalsbeek, A., Wortel, J., van der Vliet, J. & Buijs, R. M. Role for the pineal and melatonin in glucose homeostasis: pinealectomy increases nighttime glucose concentrations. *J Neuroendocrinol* 13, 1025-32 (2001).
207. Picinato, M. C., Haber, E. P., Carpinelli, A. R. & Cipolla-Neto, J. Daily rhythm of glucose-induced insulin secretion by isolated islets from intact and pinealectomized rat. *J Pineal Res* 33, 172-7 (2002).
208. Matschinsky, F. M., Glaser, B. & Magnuson, M. A. Pancreatic beta-cell glucokinase: closing the gap between theoretical concepts and experimental realities. *Diabetes* 47, 307-15 (1998).
209. Matschinsky, F. M. Glucokinase as glucose sensor and metabolic signal generator in pancreatic beta-cells and hepatocytes. *Diabetes* 39, 647-52 (1990).
210. Newgard, C. B. & McGarry, J. D. Metabolic coupling factors in pancreatic beta-cell signal transduction. *Annu Rev Biochem* 64, 689-719 (1995).
211. Thorens, B. GLUT2 in pancreatic and extra-pancreatic gluco-detection (review). *Mol Membr Biol* 18, 265-73 (2001).
212. Burcelin, R., Da Costa, A., Drucker, D. & Thorens, B. Glucose competence of the hepatoportal vein sensor requires the presence of an activated glucagon-like peptide-1 receptor. *Diabetes* 50, 1720-8 (2001).

213. Peters, A., Schweiger, U., Fruhwald-Schultes, B., Born, J. & Fehm, H. L. The neuroendocrine control of glucose allocation. *Exp Clin Endocrinol Diabetes* 110, 199-211 (2002).

214. Pardal, R. & Lopez-Barneo, J. Low glucose-sensing cells in the carotid body. *Nat Neurosci* 5, 197-8 (2002).

215. Adachi, A., Kobashi, M. & Funahashi, M. Glucose-responsive neurons in the brainstem. *Obes Res* 3 Suppl 5, 735S-740S (1995).

216. Wu, X., Gao, J., Yan, J., Owyang, C. & Li, Y. Hypothalamus-brain stem circuitry responsible for vagal efferent signaling to the pancreas evoked by hypoglycemia in rat. *J Neurophysiol* 91, 1734-47 (2004).

217. Ferreira, M., Jr. et al. Glucose effects on gastric motility and tone evoked from the rat dorsal vagal complex. *J Physiol* 536, 141-52 (2001).

218. Oomura, Y., Ooyama, H., Sugimori, M., Nakamura, T. & Yamada, Y. Glucose inhibition of the glucose-sensitive neurone in the rat lateral hypothalamus. *Nature* 247, 284-6 (1974).

219. Leloup, C., Orosco, M., Serradas, P., Nicolaidis, S. & Penicaud, L. Specific inhibition of GLUT2 in arcuate nucleus by antisense oligonucleotides suppresses nervous control of insulin secretion. *Brain Res Mol Brain Res* 57, 275-80 (1998).

220. Schuit, F. C., Huypens, P., Heimberg, H. & Pipeleers, D. G. Glucose sensing in pancreatic beta-cells: a model for the study of other glucose-regulated cells in gut, pancreas, and hypothalamus. *Diabetes* 50, 1-11 (2001).

221. Leloup, C. et al. Glucose transporter 2 (GLUT 2): expression in specific brain nuclei. *Brain Res* 638, 221-6 (1994).

222. Levin, B. E., Dunn-Meynell, A. A. & Routh, V. H. Brain glucose sensing and body energy homeostasis: role in obesity and diabetes. *Am J Physiol* 276, R1223-31 (1999).

223. Routh, V. H. Glucosensing neurons in the ventromedial hypothalamic nucleus (VMN) and hypoglycemia-associated autonomic failure (HAAF). *Diabetes Metab Res Rev* 19, 348-56 (2003).

224. Hall, A. C., Hoffmaster, R. M., Stern, E. L., Harrington, M. E. & Bickar, D. Suprachiasmatic nucleus neurons are glucose sensitive. *J Biol Rhythms* 12, 388-400 (1997).

225. Evans, S. B. et al. Inactivation of the PVN during hypoglycemia partially simulates hypoglycemia-associated autonomic failure. *Am J Physiol Regul Integr Comp Physiol* 284, R57-65 (2003).

226. De Vries, M. G., Lawson, M. A. & Beverly, J. L. Dissociation of hypothalamic noradrenergic activity and sympathoadrenal responses to recurrent hypoglycemia. *Am J Physiol Regul Integr Comp Physiol* 286, R910-5 (2004).

227. Evans, S. B. et al. Inactivation of the DMH selectively inhibits the ACTH and corticosterone responses to hypoglycemia. *Am J Physiol Regul Integr Comp Physiol* 286, R123-8 (2004).

228. Escobar, C., Diaz-Munoz, M., Encinas, F. & Aguilar-Roblero, R. Persistence of metabolic rhythmicity during fasting and its entrainment by restricted feeding schedules in rats. *Am J Physiol* 274, R1309-16 (1998).

229. Scheer, F. A., van Heijningen, C. & Buijs, R. Light and biological clock interact in the regulation of body temperature. *PhD thesis* Chapter 3 (2003).

230. Neel, J. V. Diabetes mellitus: a "thrifty" genotype rendered detrimental by "progress"? *Am J Hum Genet* 14, 353-62 (1962).

231. Marchant, E. G. & Mistlberger, R. E. Entrainment and phase shifting of circadian rhythms in mice by forced treadmill running. *Physiol Behav* 60, 657-63 (1996).

232. Kas, M. J. & Edgar, D. M. Scheduled voluntary wheel running activity modulates free-running circadian body temperature rhythms in Octodon degus. *J Biol Rhythms* 16, 66-75 (2001).

233. Kreier, F. et al. Hypothesis: shifting the equilibrium from activity to food leads to autonomic unbalance and the metabolic syndrome. *Diabetes* 52, 2652-6 (2003).

234. Qin, L. Q. et al. The effects of nocturnal life on endocrine circadian patterns in healthy adults. *Life Sci* 73, 2467-75 (2003).

235. Knutsson, A. Health disorders of shift workers. *Occup Med (Lond)* 53, 103-8 (2003).

236. Karlsson, B. H., Knutsson, A. K., Lindahl, B. O. & Alfredsson, L. S. Metabolic disturbances in male workers with rotating three-shift work. Results of the WOLF study. *Int Arch Occup Environ Health* 76, 424-30 (2003).

237. Kuhn, G. Circadian rhythm, shift work, and emergency medicine. *Ann Emerg Med* 37, 88-98 (2001).

238. O'Brien, I. A., Lewin, I. G., O'Hare, J. P., Arendt, J. & Corrall, R. J. Abnormal circadian rhythm of melatonin in diabetic autonomic neuropathy. *Clin Endocrinol (Oxf)* 24, 359-64 (1986).

239. Rana, J. S., Mukamal, K. J., Morgan, J. P., Muller, J. E. & Mittleman, M. A. Circadian variation in the onset of myocardial infarction: effect of duration of diabetes. *Diabetes* 52, 1464-8 (2003).

240. Ichikawa, M., Kanai, S., Ichimaru, Y., Funakoshi, A. & Miyasaka, K. The diurnal rhythm of energy expenditure differs between obese and glucose-intolerant rats and streptozotocin-induced diabetic rats. *J Nutr* 130, 2562-7 (2000).

241. Hotta, K. et al. Circulating concentrations of the adipocyte protein adiponectin are decreased in parallel with reduced insulin sensitivity during the progression to type 2 diabetes in rhesus monkeys. *Diabetes* 50, 1126-33 (2001).

242. Kalsbeek, A. & Strubbe, J. H. Circadian control of insulin secretion is independent of the temporal distribution of feeding. *Physiol Behav* 63, 553-8 (1998).

243. Cui, L. N., Coderre, E. & Renaud, L. P. Glutamate and GABA mediate suprachiasmatic nucleus inputs to spinal-projecting paraventricular neurons. *Am J Physiol Regul Integr Comp Physiol* 281, R1283-9 (2001).

244. Hermes, M. L., Coderre, E. M., Buijs, R. M. & Renaud, L. P. GABA and glutamate mediate rapid neurotransmission from suprachiasmatic nucleus to hypothalamic paraventricular nucleus in rat. *J Physiol* 496 (Pt 3), 749-57 (1996).

245. Perreau-Lenz, S. et al. Suprachiasmatic control of melatonin synthesis in rats: inhibitory and stimulatory mechanisms. *Eur J Neurosci* 17, 221-8 (2003).

246. Kalsbeek, A. et al. Melatonin sees the light: blocking GABA-ergic transmission in the paraventricular nucleus induces daytime secretion of melatonin [In Process Citation]. *Eur J Neurosci* 12, 3146-54 (2000).

Molecular Mechanisms of Melatonin Action: Targets in Sleep Regulation

Lennard P. Niles

1. Introduction

Melatonin, the principal hormone secreted by the pineal gland, influences the function of diverse neuroendocrine and other systems in mammals. This indoleamine hormone is also involved in maintaining brain homeostasis, entraining biological rhythms and coordinating reproductive function to changes in photoperiod in seasonal breeders.[1,2] Other studies indicate a potentially important immunomodulatory role for melatonin which binds with high-affinity to T lymphocytes[3] and is synthesized by human lymphocytes.[4]

There is considerable evidence that in high pharmacological concentrations, melatonin is a potent free radical scavenger,[5] whereas its well documented antioxidant effects may involve either pharmacological or physiological concentrations.[6] Several reports have demonstrated that melatonin can exert neuroprotective effects in various models of neurodegeneration.[7,8] These studies have often utilized pharmacological doses of melatonin, which can interact with both central-type and peripheral-type benzodiazepine receptors,[9] making it difficult to differentiate between the physiological and pharmacological effects of melatonin as a neuroprotective agent. Recent evidence that picomolar-nanomolar concentrations of melatonin can upregulate neurotrophic factor expression in vitro,[10] suggests a role for this hormone in physiological neuroprotection.

In addition to the foregoing, melatonin has long been implicated in the modulation of sleep but several questions remain regarding the targets and mechanisms involved. In a study comparing the effects of varying doses of melatonin on daytime sleepiness in young adults, lower doses, which produced typical nocturnal plasma hormone levels, were found to be ineffective, while a high dose which produced supraphysiologic hormone levels had a soporific effect during the first half hour of administration.[11] As noted by the authors, the lack of significant effects following treatment with lower doses of melatonin may be related to the time of administration. This is supported by the effectiveness of low-dose melatonin in promoting sleep, when administered 2-4 hours before habitual bedtime.[12] Why this apparent temporal sensitivity to the soporific effects of melatonin? It is now thought that the physiological effects of this hormone are primarily mediated by G protein-coupled receptors (GPCRs) in the brain and elsewhere.[13,14]

Emerging evidence that these receptors exhibit dynamic circadian-related changes in expression and sensitivity,[15-17] need to be taken into account in studies aimed at assessing the somnogenic effects of melatonin. This chapter is focused on the physiological targets and possible mechanisms underlying sleep modulation by melatonin.

2. Melatonin Receptors

2.1. G Protein-Coupled Receptors

Three melatonin receptor subtypes, which are expressed by three distinct genes, have been cloned and characterized. The Mel_{1a} receptor is present in the mammalian hypothalamus especially the suprachiasmatic nuclei (SCN), other brain areas and peripheral tissues.[18,19] A second subtype, the Mel_{1b}, is relatively enriched in the retina and hippocampus, but also present in other brain areas and peripheral tissues.[19,20] A third melatonin receptor, the Mel_{1c} subtype, has been identified in the chick brain.[21] The Mel_{1a} and Mel_{1b} receptors, now officially designated as MT_1 and MT_2 subtypes,[22] are present in mammals including humans, but to date, there is no evidence that the Mel_{1c} receptor is expressed in mammals.

RT-PCR studies indicate that the MT_1 transcript is widely distributed in the human brain, with highest levels in the cerebellum.[23] In situ hybridization has revealed a distinct pattern of melatonin receptor localization in the human cerebellar cortex. The MT_1 mRNA is expressed in cerebellar granule and basket-stellate cells, whereas the MT_2 is found in Bergmann glia and astrocytes.[24] This apparent differential expression of the MT_1 and MT_2 receptors on neurons or glial cells may not occur in other brain areas, such as the SCN, where both subtypes have been detected on neuronal-like cells.[25]

2.2. Molecular Structure

Sequences of the frog, sheep and two human cDNAs revealed that they encode proteins of 420, 366, 350 and 360 amino acids, with predicted molecular weights of 47.4, 40.4, 39.4 and 40.2 kDa, respectively.[18,20,26]

Hydropathy analysis of the Xenopus receptor revealed seven hydrophobic transmembrane segments, which is a well known characteristic of GPCRs. The amino terminus contains one consensus site for N-linked glycosylation in the frog receptor, and two such sites in each mammalian receptor. Several consensus sites for protein phosphorylation are present within the carboxyl-terminal tail, which is 119 amino acids long in the frog receptor.[26]

Partial cDNAs from the hamster and rat brain were found to be 94% identical and to have an 86% identity with the corresponding region of the sheep and human clones, indicating that they belong to the MT_1 subtype. In contrast, the human MT_1 and MT_2 receptors exhibit only a 60% homology at the amino acid level,[20] consistent with their expression by different genes.

2.3. Regulation of Melatonin Receptors

The marked diurnal rhythm in circulating levels of melatonin provides an excellent physiological system for studies of melatonin receptor regulation. Receptor binding studies utilizing basal hypothalamic membranes containing the SCN, have detected an inverse correlation between the density of high-affinity sites and the levels of circulating melatonin.[27] Thus, high-affinity binding was highest late in the light phase, following prolonged depletion of the agonist, and lowest during darkness when exposure to elevated melatonin concentrations presumably down-regulated the high-affinity receptor.[27] Similarly, a significant increase in high-affinity binding has been observed in the pars tuberalis/median eminence of rats killed at the end of the light phase as compared with animals killed in the morning.[28] In keeping with the foregoing, suppression or depletion of circulating melatonin levels by exposure to constant light or pinealectomy, caused a significant increase in the density of high-affinity sites in the pars tuberalis of the rat and hamster.[29] Conversely, a single injection of melatonin reversed the effect of constant light or pinealectomy on high-affinity binding in the rat pars tuberalis and SCN.[30] Moreover, preincubation of cultured ovine PT cells in the presence of melatonin (100 pM or 1fM) for 24 hours, resulted in a significant decrease in $[^{125}I]$MEL binding in crude PT membranes.[31] Recent studies, utilizing in vitro transfection of human melatonin receptors, indicate that following an 8 hour exposure to a physiological concentration of melatonin, there is a time-dependent increase in expression of the MT_1 receptor subtype, but similar treatment did not affect MT_2 expression.[17] Taken together, these reports suggest that melatonin is involved in the differential regulation of its GPCR subtypes in brain targets, such as the hypothalamus/SCN.

3. Melatonin Signalling

3.1. G Protein-Mediated Effects

Earlier studies, which preceded cloning of the melatonin GPCRs, had demonstrated that physiological concentrations of this indoleamine inhibited the adenylate cyclase -cAMP pathway in the hypothalamus and other tissues, via pertussis toxin (PTX)-sensitive G-proteins.[32-34] More recently, coprecipitation analysis in MT_1-transfected human embryonic kidney (HEK) 293 cells indicate that this receptor couples to G_{i2} and G_{i3} in an agonist-dependent and guanine nucleotide-sensitive manner, suggesting involvement of the α-subunits of these inhibitory G protein isoforms in melatonin-induced suppression of cAMP production.[35] The cloned MT_1 was also found to mediate phospholipase C activation by melatonin, via PTX-insensitive $G_{q/11}$ proteins.[35] Consistent with the foregoing, studies with transfected Chinese hamster ovary (CHO) cells showed that both the MT_1 and MT_2 receptors are coupled to phosphoinositide hydrolysis,[36] but based on EC_{50} determinations, the MT_2 appears to be significantly more potent than the MT_1 receptor in activating this PLC-PKC pathway. Other recent studies indicate that melatonin activates the mitogen-activated protein kinase (MAPK)-extracellular regulated kinase (ERK) pathway in a mouse neuronal cell line.[37] In addition, melatonin activates inwardly rectifying potassium (Kir3) channels via a G protein-coupled and PTX-sensitive mechanism,[38] resulting in hyperpolarization of target SCN neurons.[39] Thus, in addition to its well-documented inhibition of the adenylate cyclase (AC)-cAMP-protein kinase A (PKA) pathway, melatonin can interact with multiple G proteins to produce its diverse physiological effects.[13,14] Indeed, the MT_1 and MT_2 receptors have been found to mediate stimulation or inhibition of GABAergic activity in the rat brain, respectively.[40] Since both of these receptors are coupled to inhibition of the AC-cAMP-PKA pathway, their differential effects presumably involve interaction with additional divergent signalling pathways in the central nervous system. In keeping with this view, transfection studies with human melatonin receptors in HEK cells indicate that the MT_2 but not the MT_1 receptor is coupled to inhibition of the cGMP pathway.[41,42] Given the potential coupling of these receptors to numerous G proteins, as supported by recent chimeric experiments,[43] it is likely that more evidence of their divergent signalling will appear in the future.

3.2. Direct Effects on Intracellular Effectors

In vitro studies have shown an interaction between melatonin and calmodulin, a small Ca^{2+} binding protein which is involved in several intracellular processes including cyclic nucleotide metabolism, ion transport, protein phosphorylation, cytoskeletal function and cell proliferation.[44,45] Incubation with melatonin at a physiological concentration of 1nM, resulted in an increase of calmodulin levels in MDCK and NIE-115 cells after 3 days but the levels of this protein were decreased after 6 days. In addition to this apparently time-dependent biphasic effect on calmodulin concentrations, melatonin was also found to inhibit Ca^{2+}-calmodulin stimulated bovine heart cAMP-phosphodiesterase (PDE) activity with an IC_{50} value of 1nM.[46] Electrophoretic studies indicated that, in the presence of Ca^{2+}, tritiated melatonin ([3H]melatonin) co-migrated with calmodulin suggesting a direct interaction between melatonin and calmodulin.[46] Interestingly, saturation and kinetic studies with liposome-incorporated calmodulin, revealed that [3H] melatonin binds to a single site on the calmodulin molecule with high (picomolar) affinity. This binding is selective for melatonin as indicated by the significantly lower affinities of related analogs.[47] On the basis of these findings, it was suggested that blockade of Ca^{2+}-calmodulin function by melatonin could account for several of its effects on cellular physiology.[48]

In addition to the foregoing, there is evidence that melatonin can directly interact with calcium-dependent PKC in the absence of receptors.[49] In vitro studies have shown a concentration-dependent stimulation of PKC activity by melatonin, with an EC_{50} of 1nM. In the presence of calcium, melatonin enhanced phorbol ester-stimulated PKC activity and [^{3}H]Phorbol-12, 13-dibutyrate binding to this enzyme, suggesting that its binding site on PKC is different from that of diacylglycerol and phorbol esters.[49] In view of the ability of an inhibitor of the α, β and γ isoforms of PKC to block melatonin-induced activation of PKC, it is possible that one or more of these isoforms may be intracellular targets for melatonin.[49]

4. Pharmacological Effects of Melatonin on Sleep

Several studies have demonstrated that pharmacological doses of melatonin produce sedative/hypnotic effects and can alleviate the symptoms of jet lag in humans.[50-52] These high doses of melatonin can directly interact with benzo-diazepine (BZ) receptors, which are allosterically linked to modulation of GABAergic activity via the BZ-GABA$_A$ receptor complex.[9,53] Since GABAergic mechanisms play an important role in sleep modulation,[54] it is very likely that the acute sedative effects of pharmacological doses of mela-tonin involve its interaction with BZ-GABA$_A$ receptors. This view is sup-ported by evidence that BZ-GABA$_A$ antagonists block the sleep inducing effect of high doses of melatonin in experimental animals.[55] In contrast to the foregoing, there is evidence that relatively low doses of melatonin can also produce sedative effects in humans.[56] Moreover, the effects of these lower doses are not mediated by BZ receptors, as shown by the lack of antagonism by the central-type BZ antagonist flumazenil.[57] Interestingly, although these lower doses of melatonin produced higher than normal circulating levels,[57] recent evidence that melatonin concentrations in the mammalian third ventri-cle are 20-fold higher than nocturnal plasma levels,[58] suggests that the higher circulating levels observed may be reflective of an upper physiological range in the central nervous system (CNS). It is important to understand the physi-ological mechanisms underlying sleep modulation by melatonin, in order to fully utilize its therapeutic potential in diverse sleep-related disorders.

5. Physiological Modulation of Sleep by Melatonin

5.1. Acute Sedative Effects

In assessing the possible mechanisms involved in the effects of melatonin on sleep, it should be borne in mind that both acute soporific effects as well as longer-term reentrainment of sleep-wake cycles have been attributed to this hormone. With regard to the acute sleep inducing effect, it is likely that the

GPCRs for melatonin are involved. There is *in vivo* electrophysiological evidence that melatonin potentiates GABAergic inhibition of neuronal activity in the mammalian cortex.[59] More recently, *in vitro* studies have indicated that the MT_1 receptor is coupled to stimulation of GABAergic activity in the hypothalamus, whereas the MT_2 receptor mediates an opposite effect in the hippocampus.[40] In view of the dominant inhibitory role of GABA in the CNS, the modulation of $GABA_A$ receptor function by melatonin, via its GPCRs, might well be the cornerstone underlying its effects on neuronal activity in the SCN and other brain areas. The predominant effect of melatonin in the rat SCN appears to be inhibition of neuronal activity,[60,61] which is consistent with the relatively high expression of the MT_1 subtype in the circadian clock and the fact that this receptor is linked to enhancement of GABAergic activity.

Immunohistochemical and electrophysiological studies of the hypothalamic preoptic area (POA), which plays a major role in sleep promotion, have identified a subset of sleep-active ventrolateral POA (VLPO) neurons.[62,63] A tightly clustered group of VLPO neurons appear to promote non-REM sleep, by suppression of the histaminergic arousal system, which originates in the tuberomammillary nucleus of the posterior hypothalamus. Whereas a diffuse subgroup of VLPO neurons are thought to promote REM sleep through their inhibitory projection to monoaminergic dorsal raphe (serotonergic) and locus coeruleus (noradrenergic) nuclei in the brainstem.[64] Interestingly, sleep-active and arousal-related neurons exhibit reciprocal suppression of each other across the wake-NREM-REM cycle, providing a basis for the sleep-wake switch.[65] The inhibitory projections from the VLPO, to the histaminergic, serotonergic and noradrenergic components of the arousal system, utilize GABA and galanin, which are present in nearly 80% of VLPO neurons.[66,67] Given melatonin's ability to modulate GABAergic activity, could its elevated nocturnal levels directly activate VLPO neurons with concomitant suppression of arousal systems and sleep induction? This is certainly possible providing that the G protein-coupled melatonin receptors which potentiate GABAergic activity are expressed on VLPO neurons. Studies with MT_1 receptor-deficient mice have demonstrated that this receptor subtype mediates acute physiological inhibition of spontaneous neuronal firing in the SCN.[68] In addition, as mentioned earlier, the MT_1 receptor is coupled to stimulation of GABAergic activity in the SCN.[40]

Although these studies focused on the SCN where MT_1 expression is relatively enriched, it is now known that this widely expressed receptor is present in hypothalamic areas outside the SCN and also in other human CNS regions.[23,24] Therefore, the possibility that melatonin acts directly on VLPO neurons cannot be ruled out at present. Nonetheless, current evidence clearly indicates that the SCN is the primary hypothalamic target for melatonin, which can act on the predominantly expressed MT_1 receptor to acutely inhibit neuronal activity via enhancement of GABAergic activity and/or activation of potassium channels.

How does this inhibitory effect of melatonin on SCN activity induce sleep? Possible clues to answering this question may come from an examination of the targets reached by SCN efferents in various diencephalic regions, including the medial preoptic area (MPA), the dorsomedial hypothalamic nucleus (DMH) and the subparaventricular zone.[69,70] These SCN targets, especially the MPA and DMH project to the VLPO, allowing the SCN to indirectly regulate sleep promotion by the VLPO. In addition, the SCN is thought to have direct monosynaptic projections to the VLPO,[71] although these are fewer than its efferents to the MPA and DMH.[72]

Since the predominant effect of melatonin in the SCN is suppression of neuronal (GABAergic) activity, this action could result in disinhibition of the MPA and DMH, which can then activate sleep-promoting neurons in the VLPO. This indirect activation of VLPO neurons may be augmented by further melatonin-induced disinhibition in the direct SCN-VLPO projection. The consequent activation of GABAergic projections from the VLPO would then induce sleep by inhibiting activity in the monoaminergic arousal system.[73] Of course, other mechanisms, which do not require GPCR transduction, may also be involved in the sleep promoting effects of melatonin. For example, its antagonistic effects on intracellular targets such as calmodulin, as decribed earlier, could lead to suppression of neuronal activity in histaminergic, serotonergic or noradrenergic neurons in the arousal system, further contributing to sleep induction. Another interesting possibility is that melatonin may act directly on GPCRs or intracellular targets to modulate neuronal activity in the hypocretin/orexin system located in the posterior hypothalamus, which is involved in arousal state control and sleep-wake regulation.[74]

5.2. Phase Shifting Effects

It has long been established that the circadian clock or pacemaker in mammals resides in the SCN of the hypothalamus. Recent advances utilizing cellular and molecular approaches have identified several clock genes within the molecular oscillatory core of the clock. These genes, which include *Clock, Bmal1, Period1, Period2, Crytochrome1 and Crytochrome2*, together with their protein products, interact via transcription-translation autoregulatory feedback loops which underlie circadian rhythmicity.[75] The marked diurnal rhythm in pineal melatonin production (and secretion) is regulated by the SCN pacemaker, which receives photic information via a direct retinohypothalamic tract, and transmits this information to the pineal via a multisynaptic pathway.[2] Melatonin, in turn, is thought to influence circadian rhythmicity by acting directly on receptor sites within the SCN.[76] Neurophysiological studies indicate that melatonin alters the electrical activity of SCN neurons in vitro,[60] and phase-shifts neuronal firing in this circadian oscillator,[77] which supports a direct role for this indoleamine in synchronizing circadian rhythms. When applied to SCN-containing hypothalamic slices in vitro, melatonin, (at a physiological concentration of 1 nM), induced a significant advance in the electrical activity rhythm in the SCN, only when added late in the subjective day or early subjective night.[77] It is noteworthy that the phase-shifting effect of melatonin was most pronounced late in the day, before the day-night transition, when the high-affinity G_i-coupled receptors for this hormone are at their maximal density and sensitivity in the SCN.[27] Since these receptors mediate the inhibitory effect of melatonin on cAMP production, it seems reasonable to assume that a decrease in the intracellular concentration of this second messenger is involved in the above phase-shifting effect. In accordance with the foregoing, cAMP and its analogs have been shown to reset the SCN circadian clock in vitro.[78,79] Another second messenger, cGMP, which can be modulated by melatonin,[42] has also been implicated in regulation of the circadian pacemaker. For example, cGMP analogs can reset the phase of SCN oscillation, *in vitro,* only when applied during the subjective night of the circadian cycle.[80] A comparison of the *in vitro* effects of cAMP and cGMP analogs, indicated a difference of about 12 hours in the periods of SCN sensitivity to these second messengers, which appear to utilize distinct biochemical pathways in modulating the SCN clock.[80] More recent studies indicate that the phase-shifting effect of melatonin on the firing rate rhythm in the SCN involves activation of protein kinase C (PKC), via a pertussis toxin sensitive pathway.[81] Although the MT_1 receptor had long been thought to mediate this effect, studies with animals subjected to targeted disruption of this receptor subtype, indicate that the elusive MT_2 receptor is in fact the subtype involved in melatonin-induced phase shifts in SCN neuronal activity.[68] Therefore, it is reasonable to assume that the MT_2 plays a role in phase shifting circadian rhythmicity, via activation of the PLC-PKC pathway. This action could underlie the re-entrainment of disordered sleep cycles associated with blindness, shift work, jetlag and other conditions, following treatment with physiological doses of melatonin.[82-85] It is not known how activation of the melatonin-MT_2-PLC-PKC cascade is linked to phase shifts in SCN activity. However, the ability of GABA to phase shift neuronal activity in the mouse SCN[86] or circadian locomotor activity in a diurnal rodent,[87] suggests a potential link to the entraining effects of melatonin, which interacts with GABAergic systems. As noted earlier, the MT_2 receptor mediates inhibition of $GABA_A$ receptor activity in the hippocampus.[40] Since the MT_2 is also expressed in the SCN, it can presumably alter $GABA_A$ receptor function in the clock, at specific times in the 24-hour cycle (eg dusk and dawn), to modulate biological rhythms including sleep-wake cycles.

5.3. *Molecular Mechanisms and Conclusions*

GABA, the major inhibitory neurotransmitter in the CNS, is also the predominant neurotransmitter in the SCN,[88] where there is a widespread distribution of $GABA_A$ receptors both pre- and postsynaptically.[89] Therefore, it is not surprising that internal clock activity as well as clock output neurons utilize this transmitter, as previously suggested.[90]

Moreover, the acute as well as the circadian effects of melatonin on sleep appear to involve modulation of GABAergic activity in the clock, although it is now possible to envisage separate mechanisms converging on this crucial effector system. The $GABA_A$ receptor which mediates the hyperpolarizing effects of GABA in the CNS, is a pentameric complex of subunits assembled from seven subunit classes consisting of numerous isoforms.[91] It is now thought that the majority of central $GABA_A$ receptor subtypes consist of various pentameric arrangements of α, β and γ subunits.[92] Immunohistochemical and molecular studies have detected the α2,3,5,β1,3 and γ1,2 subunits in the SCN.[93,94] Electrophysiological studies have shown that $GABA_A$ activity can be differentially modulated by phosphorylation, depending on the subunits present in the pentameric complex.[95] For example, PKA-induced phosphorylation of serine residues on β1 or β3 subunits inhibits or enhances GABA-activated currents, respectively.[96] Therefore, MT_1-$G_{i\alpha}$ mediated suppression of the cAMP-PKA pathway in the SCN, with decreased phosphorylation of serine residues, could result in activation of β1-expressing $GABA_A$ receptors, while not affecting the normal activity of β3-expressing receptors. The consequent overall enhancement of GABAergic function, potentially supplemented by MT_1-$G_{i\beta\gamma}$ activation of inwardly rectifying potassium channels, could suppress neuronal activity in the SCN, with sleep induction via the pathways described earlier. The possible mechanisms involved in this acute sleep induction by melatonin are shown in Figure 1.

Regarding the issue of rhythm entrainment by melatonin, it is now known that this action is mediated by the MT_2 receptor and involves activation of

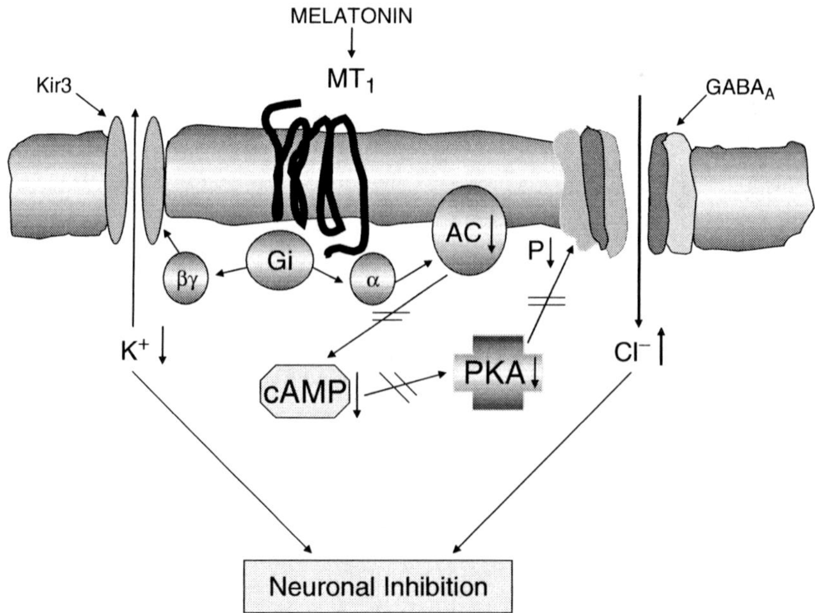

FIGURE 1. MT_1 mediated signal transduction pathways in the SCN underlying the acute sedative effects of melatonin.

PKC. While this indicates progress in understanding how melatonin phase shifts clock activity, the molecular mechanisms linking its activation of PKC to circadian modulation await clarification. Like other protein kinases, PKC can selectively modulate $GABA_A$ receptor activity with enhancement observed in hippocampal dentate gyrus granule cells, no effects found in CA1 pyramidal cells,[95] and inhibition seen in cultured cortical neurons.[97] Other studies indicate that PKC can phosphorylate γ2 subunits on the $GABA_A$ receptor, resulting in suppression of GABAergic activity in the brain.[98] Since γ2 subunits are expressed in the SCN, it is possible that activation of the MT_2-PLC-PKC pathway inhibits GABAergic transmission within the clock, resulting in phase shifts in neuronal activity and associated sleep-wake rhythms. How does this entraining signal get through to the clock at dusk and dawn, when melatonin levels are low? Recent studies indicate that nadirs in the rhythmic PKC activity of the SCN coincide with its windows of sensitivity to melatonin.[99] The relatively higher efficacy of the MT_2 receptor in driving the PLC-PKC cascade,[36] presumably allows this receptor to mediate shifts in neuronal activity in the sensitized clock, with consequent changes in sleep behaviour. The involvement of a PTX-sensitive protein in melatonin-induced activation of PKC in the SCN,[81] suggests transduction by the β and γ subunits of G_i, as illustrated in Figure 2. However, the relatively high expression

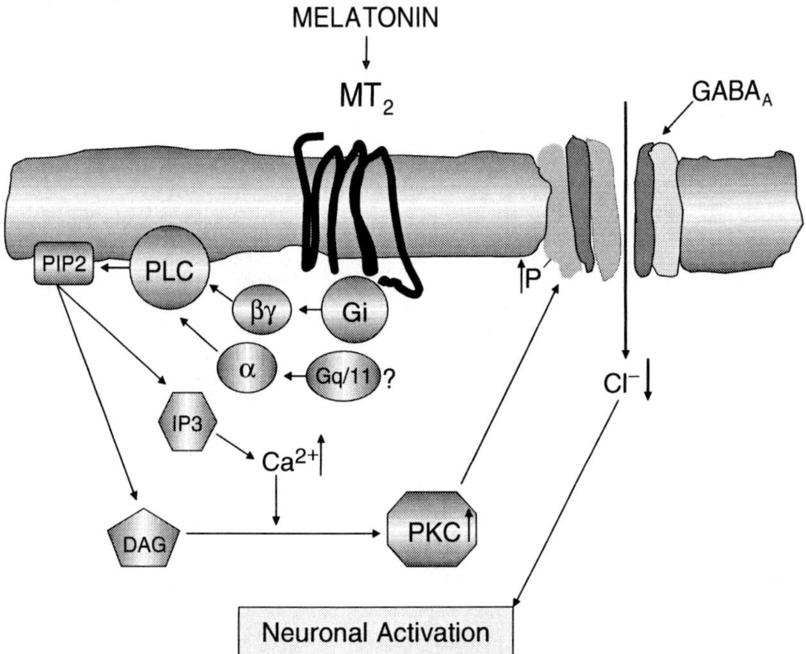

FIGURE 2. MT_2 mediated signal transduction pathways in the SCN underlying the phase shifting effects of melatonin.

of $G_{q/11}$ proteins in the hypothalamus[100] suggests that these transducers could also be utilized by the MT_2 in activating PKC. The focus here has been on possible GPCR-mediated mechanisms underlying the physiological effects of melatonin on sleep. Although GABAergic modulation appears to be a dominant theme, the multiplicity of targets for this hormone makes it likely that other supplementary mechanisms are involved. Future studies utilizing diverse cellular and molecular strategies will continue to clarify the role of melatonin in sleep, thus moving toward a full unleashing of its therapeutic potential in sleep dysfunction.

6. References

1. R.J. Reiter, Pineal melatonin: cell biology of its synthesis and of its physiological interactions, *Endocrine Rev.* 12(2), 151-180 (1991).
2. V. Simonneaux, and C. Ribelayga, Generation of the melatonin endocrine message in mammals: a review of the complex regulation of melatonin synthesis by norepinephrine, peptides, and other pineal transmitters, *Pharmacol. Rev.* 55(2), 325-395 (2003).
3. M.G. Gonzalez-Haba, S. Garcia-Maurino, J.R. Calvo, R. Goberna, and J.M. Guerrero, High-affinity binding of melatonin by human circulating T lymphocytes (CD4$^+$), *FASEB J.* 9(13), 1331-1335 (1995).
4. Carrillo-Vico, J.R. Calvo, P. Abreu, P.J. Lardone, S. Garcia-Maurino, R.J. Reiter, and J.M. Guerrero, Evidence of melatonin synthesis by human lymphocytes and its physiological significance: possible role as intracrine, autocrine, and/or paracrine substance, *FASEB J.* 18(3), 537-539 (2004).
5. R. Hardeland, R.J.Reiter, B. Poeggeler, and, D.X. Tan, The significance of the metabolism of the neurohormone melatonin: antioxidative protection and formation of bioactive substances, *Neurosci. Biobehav. Rev.* 17(3), 347-357 (1993).
6. R.J. Reiter, and D.X. Tan, Melatonin: a novel protective agent against oxidative injury of the Ischemic/reperfused heart, *Cardiovasc. Res.* 58(1), 10-19 (2003).
7. C.V. Borlongan, M. Yamamoto, N. Takei, M. Kumazaki, C. Ungsuparkorn, H. Hida, P.R. Sanberg, and H.Nishino, Glial cell survival is enhanced during melatonin-induced neuroprotection against cerebral ischemia, *FASEB J.* 14(10), 1307-1317 (2000).
8. H. Khaldy, G. Escames, J. Leon, L. Bikjdaouene, and D. Acuna-Castroviejo, Synergistic effects of melatonin and deprenyl against MPTP-induced mitochondrial damage and DA depletion, *Neurobiol. Aging.* 24(3), 491-500 (2003).
9. C.C. Tenn, and L.P. Niles, Central-type benzodiazepine receptors mediate the anti-dopaminergic effect of clonazepam and melatonin in 6-hydroxydopamine lesioned rats: involvement of a GABAergic mechanism, *J. Pharmacol. Exp. Ther.* 274(1), 84-89 (1995).
10. K.J. Armstrong, and L.P. Niles, Induction of GDNF mRNA expression by melatonin in rat C6 glioma cells, *NeuroReport.* 13(4), 473-475 (2002).
11. C.J. Van Den Heuvel, D.J. Kennaway, and D. Dawson, Effects of daytime melatonin infusion in young adults, *Am. J. Physiol.* 275(1Pt1), E19-E26 (1998).
12. I.V. Zhdanova, R.J.Wurtman, C. Morabito, V.R. Piotrovska, and H.J. Lynch, Effects of low oral doses of melatonin, given 2-4 hours before habitual bedtime, on sleep in normal young humans, *Sleep.* 19(5), 423-431(1996).

13. L.P. Niles, 1997, in: *G Protein Methods and Protocols: Role of G proteins in Psychiatric and Neurological Disorders*, edited by R.K. Mishra, G.B.Baker and A.A. Boulton (Humana Press, New Jersey, 1997), pp. 223-281.

14. J. Vanecek, Cellular mechanisms of melatonin action, *Physiol. Rev.* 78(3), 687-721 (1998).

15. M.U. Gillette, and J.W., Mitchell, Signaling in the suprachiasmatic nucleus: selectively responsive and integrative, *Cell Tissue Res.* 309(1), 99-107 (2002).

16. M.J. Gerdin, M.I. Masana, D. Ren, R.J. Miller, and M.L. Dubocovich, Short-term exposure to melatonin differentially affects the functional sensitivity and trafficking of the hMT1 and hMT2 melatonin receptors, *J. Pharmacol. Exp. Ther.* 304(3), 931-939 (2003).

17. M.I. Masana, P.A. Witt-Enderby, and M.L. Dubocovich Melatonin differentially modulate the expression and function of the hMT1 and hMT2 melatonin receptors upon prolonged withdrawal, *Biochem. Pharmacol.* 65(5), 731-739 (2003).

18. S.M. Reppert, D.R. Weaver, and T. Ebisawa, Cloning and characterization of a mammalian melatonin receptor that mediates reproductive and circadian responses, *Neuron.* 13(5), 1177-1185 (1994).

19. M.L. Dubocovich, M.A. Rivera-Bermudez, M.J. Gerdin, M.I. Masana, Molecular pharmacology, regulation and function of mammalian melatonin receptors, *Front Biosci.* 8, d1093-1108 (2003).

20. S.M. Reppert, C. Godson, C.D. Mahle, D.R. Weaver, S.A. Slaugenhaupt, and J.F. Gusella, Molecular characterization of a second melatonin receptor expressed in human retina and brain: The MEL_{1b} melatonin receptor, *Proc. Natl. Acad. Sci.* 92(19), 8734-8738 (1995).

21. S.M. Reppert, D.R. Weaver, V.M. Cassone, C. Godson, and L.F. Kolakowski, Melatonin receptors are for the birds: Molecular analysis of two receptor subtypes differentially expressed in chick brain, *Neuron.* 15(5), 1003-1015 (1995).

22. M.L. Dubocovich, D.P. Cardinali, B. Guardiola-Lemaitre, R.M. Hagan, D.N. Krause, D. Sugden, P.M. Vanhoutte, and F.D. Yocca, in: *The IUPHAR Compendium of Receptor Characterization and Classification* (IUPHAR Media, London, U.K., 1998), pp. 187-193.

23. C. Mazzucchelli, M. Pannacci, R. Nonno, V. Lucini, F. Fraschini, and B.M. Stankov, The melatonin receptor in the human brain: cloning experiments and distribution studies, *Brain Res. Mol. Brain Res.* 39(1-2), 117-126 (1996).

24. W.M. Al-Ghoul, M.D Herman, and M.L. Dubocovich, Melatonin receptor subtype expression in human cerebellum, *NeuroReport.* 9(18), 4063-4068 (1998).

25. M.A. Rivera-Bermudez, M.I. Masana, G.M., Brown, D.J. Earnest, and M.L. Dubocovich, Immortalized cells from the rat suprachiasmatic nucleus express functional melatonin receptors, *Brain Res.* 1002(1-2), 21-27 (2004).

26. T. Ebisawa, S. Karne, M.R. Lerner, and S.M. Reppert, Expression cloning of a high-affinity melatonin receptor from Xenopus dermal melanophores, *Proc. Natl. Acad. Sci.* 91(13), 6133-6137 (1994).

27. C. Tenn, and L.P. Niles, Physiological regulation of melatonin receptors in rat suprachiasmatic nuclei: Diurnal rhythmicity and effects of stress, *Mol. Cell. Endocrinol.* 98(1), 43-48 (1993).

28. J. Vanecek, E. Kosar, and J. Vorlicek, Daily changes in melatonin binding sites and the effect of castration, *Mol Cell Endocrinol.* 73(2-3), 165-170 (1990).

29. F. Gauer, M. Masson-Pévet, and P. Pévet, Pinealectomy and constant illumination increase the density of melatonin binding sites in the pars tuberalis of rodents, *Brain Res.* 575(1), 32-38 (1992).

30. F. Gauer, M. Masson-Pévet, and P. Pévet, Melatonin receptor density is regulated in rat pars tuberalis and suprachiasmatic nuclei by melatonin itself, *Brain Res.* 602(1), 153-156 (1993).

31. D.G. Hazlerrigg, A. Gonzalez-Brito, W. Lawson, M.H. Hastings, and P.J. Morgan, Prolonged exposure to melatonin leads to time-dependent sensitization of adenylate cyclase and down-regulates melatonin receptors in pars tuberalis cells from ovine pituitary, *Endocrinology.* 132(1), 285-292 (1993).

32. L.L. Carlson, D.R. Weaver, and S.M. Reppert, Melatonin signal transduction in hamster brain: Inhibition of adenylyl cyclase by a pertussis toxin-sensitive G protein, *Endocrinology.* 125(5), 2670-2676 (1989).

33. L.P. Niles, and F. Hashemi, Picomolar-affinity binding and inhibition of adenylate cyclase activity by melatonin in Syrian hamster hypothalamus, *Cell. Mol. Neurobiol.* 10(4), 553-558 (1990).

34. L.P. Niles, M. Ye, D.S. Pickering, and S-W. Ying, Pertussis toxin blocks melatonin-induced inhibition of forskolin-stimulated adenylate cyclase activity in the chick brain, *Biochem. Biophys. Res. Commun.* 178(2), 786-792 (1991).

35. L. Brydon, F. Roka, L. Petit, P. de Coppet, M. Tissot, P. Barrett, P.J. Morgan, C. Nanoff, A.D. Strosberg, and R. Jockers, Dual signaling of human Mel1a melatonin receptors via G (i2), G (i3), and G (q/11) proteins, *Mol. Endocrinol.* 13(12), 2025-2038 (1999).

36. R.S. MacKenzie, M.A. Melan, D.K. Passey, and P.A. Witt-Enderby, Dual coupling of MT (1) and MT(2) melatonin receptors to cyclic AMP and phosphoinositide signal transduction cascades and their regulation following melatonin exposure, Biochem. Pharmacol. 63(4), 587-595 (2002).

37. D. Roy, and D.D. Belsham, Melatonin receptor activation regulates GnRH gene expression and secretion in GT1-7 GnRH neurons, *J. Biol. Chem.* 277(1), 251-258(2002).

38. C.S. Nelson, J.L.Marino, and C.N. Allen, Melatonin receptors activate heteromeric G-protein coupled Kir3 channels, *Neuroreport.* 7(3), 717-720 (1996).

39. M. van den Top, R.M. Bujis, J.M. Ruijiter, P. Delagrange, D. Spanswick, and M.L. Hermes, Melatonin generates an outward potassium current in rat suprachiasmatic nucleus neurons in vitro independent of their circadian rhythm, *Neuroscience.* 107(1), 99-108 (2001).

40. Q. Wan, H.Y. Man, F. Liu, J. Braunton, H.B. Niznik, S.F. Pang, Brown, G.M. and Y.T. Wang, Differential modulation of $GABA_A$ receptor function by Mel_{1a} and Mel_{1b} receptors, *Nat. Neurosci.* 2(5), 401-403 (1999).

41. L. Petit, I. Lacroix, P. de Coppet, A.D. Strosberg, and R. Jockers, Differential signaling of human Mel1a and Mel1b melatonin receptors through the cyclic guanosine 3'-5'monophosphate pathway, *Biochem. Pharmacol.* 58(4), 633-639 (1999).

42. L. Brydon, L. Petit, P. Delagrange, A.D. Strosberg, and R. Jockers, Functional expression of MT2 (Mel1b) melatonin receptors in human PAZ6 adipocytes, *Endocrinology.* 142(10), 4264-4271 (2001).

43. J.E. Drew, P. Barrett, S. Conway, P. Delagrange, and P.J. Morgan, Differential coupling of the extreme C-terminus of G protein alpha subunits to the G protein-coupled melatonin receptors, *Biochim. Biophys. Acta.* 1592(2), 185-192 (2002).

44. M.E. Gnegy, Calmodulin in neurotransmitter and hormone action, *Annu. Rev. Pharmacol. Toxicol.* 33, 45-70 (1993).
45. H. Wang, and D.R. Storm, Calmodulin-regulated adenylyl cyclases: cross-talk and plasticity in the central nervous system, *Mol. Pharmacol.* 63(3), 463-468 (2003).
46. G. Benitez-King, L. Huerto-Delgadillo, and F. Anton-Tay, Melatonin modifies calmodulin cell levels in MDCK and N1E-115 cell lines and inhibits phosphodiesterase activity in vitro, *Brain Res.* 557(1-2), 289-292 (1991).
47. G. Benítez-King, L. Huerto-Delgadillo, and F. Antón-Tay, Binding of ^{3}H-melatonin to Calmodulin, *Life Sci.* 53(3), 201-207 (1993).
48. G. Benitez-King, and F. Anton-Tay, Calmodulin mediates melatonin cytoskeletal effects, *Experientia.* 49(8), 635-641(1993).
49. F. Anton-Tay, G. Ramirez, I. Martinez, and G. Benitez-King, In vitro stimulation of protein kinase C by melatonin, *Neurochem. Res.* 23(5), 601-606 (1998).
50. J.G. MacFarlane, J.M. Cleghorn, G.M. Brown, and D.L. Streiner, The effects of exogenous melatonin on the total sleep time and daytime alertness of chronic insomniacs: A preliminary study, *Biol. Psychiatry.* 30(4), 371-376 (1991).
51. A.B. Dollins, H.J. Lynch, R.J. Wurtman, M.H. Deng, K.U. Kischka, R.E. Gleason, and H.R. Lieberman, Effect of pharmacological daytime doses of melatonin on human mood and performance, *Psychopharmacol.* 112(4), 490-496 (1993).
52. J. Arendt, S. Deacon, J. English, S. Hampton, and L. Morgan, Melatonin and adjustment to phase shift, *J. Sleep Res.* 4(52), 74-79 (1995).
53. C.C. Tenn, and L.P. Niles, Modulation of dopaminergic activity in the striatum by benzodiazepines and melatonin, *Pharmacol. Rev. Commun.* 12(3), 171-178 (2002).
54. E. Mignot, S. Taheri, and S. Nishino, Sleeping with the hypothalamus: emerging therapeutic targets for sleep disorders, *Nat. Neurosci.* 5(S1), 1071-1075 (2002).
55. F. Wang, J. Li, C. Wu, J. Yang, F. Xu, and Q. Zhao The GABA (A) receptor mediates the hypnotic activity of melatonin in rats, *Pharmacol. Biochem. Behav.* 74(3), 573-578 (2003).
56. A.B. Dollins, I.V. Zhdanova, R.J. Wurtman, H.J. Lynch, and M.H. Deng, Effect of inducing noctural serum melatonin concentrations in daytime on sleep, mood, body temperature, and performance, *Proc. Natl. Acad. Sci.* 91(5), 1824-1828 (1994).
57. R. Nave, P. Herer, I. Haimov, A. Shlitner and P. Lavie, Hypnotic and hypothermic effects of melatonin on daytime sleep in humans: lack of antagonism by flumazenil, *Neurosci. Lett.* 214(2-3), 123-126 (1996).
58. D.C. Skinner, and B. Malpaux, High melatonin concentrations in third ventricular cerebrospinal fluid are not due to Galen vein blood recirculating through the choroid plexus, *Endocrinology.* 140(10), 4399-4405 (1999).
59. B. Stankov, G. Biella, C. Panara, V. Lucini,, S. Capsoni, J. Fauteck, B. Cozzi, and F. Fraschini, Melatonin signal transduction and mechanism of action in the central nervous system: Using the rabbit cortex as a model, *Endocrinology.* 130(4), 2152-2159 (1992).
60. S. Shibata, V.M. Cassone, and R.Y. Moore, Effects of melatonin on neuronal activity in the rat suprachiasmatic nucleus in vitro, *Neurosci. Lett.* 97(1-2), 140-144 (1989).
61. J. Stehle, J. Vanecek, and L. Vollrath, Effects of melatonin on spontaneous electrical activity of neurons in rat suprachiasmatic nuclei: an in vitro iontophoretic study, *J. Neural Transm.* 78(2), 173-177 (1989).

62. J.E. Sherin, P.J. Shiromani, R.W. McCarley, and C.B. Saper, Activation of ventrolateral preoptic neurons during sleep, *Science.* 271(5246), 216-219 (1996).
63. R Szymusiak, N. Alam, T.L. Steininger, and D. McGinty, Sleep-waking discharge patterns of ventrolateral preoptic/anterior hypothalamic neurons in rats, *Brain Res.* 803(1-2), 178-188 (1998).
64. J. Lu, M.A. Greco, P. Shiromani, C.B. Saper, Effect of lesions of the ventrolateral preoptic nucleus on NREM and REM sleep, *J. Neurosci.* 20(10), 3830-3842 (2000).
65. D. McGinty, and R. Szymusiak, Brain structures and mechanisms involved in the generation of NREM sleep: focus on the preoptic hypothalamus, *Sleep Med. Rev.* 5(4), 323-342 (2001).
66. J.E. Sherin, J.K. Elmquist, F. Torrealba, and C.B. Saper, Innervation of histaminergic tuberomammillary neurons by GABAergic and galaninergic neurons in the ventrolateral preoptic nucleus of the rat, *J. Neurosci.* 18(12), 4705-4721 (1998).
67. C.B. Saper, T.C. Chou, and T.E. Scammell, The sleep switch: hypothalamic control of sleep and wakefulness, *Trends Neurosci.* 24(12), 726-731 (2001).
68. C. Liu, D.R. Weaver, X. Jin, L.P. Shearman, R.L. Pieschl, V.K. Gribkoff, and S.M. Reppert, Molecular dissection of two distinct actions of melatonin on the suprachiasmatic circadian clock, *Neuron.* 19(1), 91-102 (1997).
69. E.F. Pace-Schott, and J.A. Hobson, The neurobiology of sleep: genetics, cellular physiology and subcortical networks, *Nat. Rev. Neurosci.* 3(8), 591-605 (2002).
70. S. Deurveilher, and K. Semba, Indirect projections from the suprachiasmatic nucleus to the median preoptic nucleus in rat, *Brain Res.* 987(1), 100-106 (2003).
71. X. Sun, B. Rusak, and K. Semba, Electrophysiology and pharmacology of projections from the suprachiasmatic nucleus to the ventromedial preoptic area in rat, *Neuroscience.* 98(4), 715-728 (2000).
72. T.C. Chou, A.A. Bjorkum, S.E. Gaus, Lu J. Scammell and C.B. Saper, Afferents to the ventrolateral preoptic nucleus, *J. Neurosci.* 22(3), 977-990 (2002).
73. D. Gervasoni, L. Darracq, P. Fort, F. Souliere, G. Chouvet, and P.H. Luppi, Electrophysiological evidence that noradrenergic neurons of the rat locus coeruleus are tonically inhibited by GABA during sleep, *Eur. J. Neurosci.* 10(3), 964-970 (1998).
74. T.S. Kilduff, and C. Peyron,, The hypocretin/orexin ligand-receptor system: implications for sleep and sleep disorders, *Trends Neurosci.* 23(8), 359-365 (2000).
75. S. Honma, and K. Honma, The biological clock: Ca2+ links the pendulum to the hands, *Trends Neurosci.* 26(12), 650-653 (2003).
76. D.N. Krause, and M.L. Dubocovich, Melatonin receptors, *Annu. Rev. Pharmacol. Toxicol.* 31, 549-568 (1991).
77. A.J. McArthur, M.U. Gillette, and R.A. Prosser, Melatonin directly resets the rat suprachiasmatic circadian clock in vitro, *Brain Res.* 565(1), 158-161 (1991).
78. M.U. Gillette, and R.A. Prosser, Circadian rhythm of the rat suprachiasmatic brain slice is rapidly reset by daytime application of cAMP analogs, *Brain Res.* 474(2), 348-352 (1988).
79. R.A. Prosser, and M.U. Gillette, The mammalian circadian clock in the suprachiasmatic nuclei is reset in vitro by cAMP, *J. Neurosci.* 9(3), 1073-1081 (1989).
80. R.A. Prosser, A.J. McArthur, and M.U. Gillette, cGMP induces phase shifts of a mammalian circadian pacemaker at night, in antiphase to cAMP effects, *Proc. Natl. Acad. Sci.* 86(17), 6812-6815 (1989).

81. A.J. McArthur, A.E. Hunt, and M.U. Gillette, Melatonin action and signal transduction in the rat suprachiasmatic circadian clock: activation of protein kinase C at dusk and dawn, *Endocrinology.* 138(2), 627-634 (1997).

82. D.J. Skene, S.W. Lockley, and J. Arendt, Use of melatonin in the treatment of phase shift and sleep disorders, *Adv. Exp. Med. Biol.* 467, 79-84 (1999).

83. K.M. Sharkey, and C.I. Eastman, Melatonin phase shifts human circadian rhythms in a placebo-controlled simulated night-work study, *Am. J. Physiol. Regul. Integr. Comp. Physiol.* 282(2), R454-R463 (2002).

84. L.M. Hack, S.W. Lockley, J. Arendt, and D.J. Skene, The effects of low-dose 0.5mg.melatonin on the free-running circadian rhythms of blind subjects, *J. Biol. Rhythms.* 18(5), 420-429 (2003).

85. M. Crasson, S. Kjiri, A. Colin, K. Kjiri, M. L'Hermite-Baleriaux, M. Ansseau, and J.J. Legros, Serum melatonin and urinary 6-sulfatoxymelatonin in major depression, *Psychoneuroendocrinol.* 29(1), 1-12 (2004).

86. C. Liu, and S.M. Reppert, GABA synchronizes clock cells within the suprachiasmatic circadian clock, *Neuron.* 25(1), 123-128 (2000).

87. C.M. Novak, and H.E. Albers, Novel phase-shifting effects of GABAA receptor activation in the suprachiasmatic nucleus of a diurnal rodent, *Am. J. Physiol. Regul. Integr. Comp. Physiol.* 286(5), R820-R825 (2004).

88. R.Y. Moore, and J.C. Speh, GABA is the principal neurotransmitter of the circadian system, *Neurosci. Lett.* 150(1), 112-116 (1993).

89. M.A. Belenky, N. Sagiv, J.M. Fritschy, and Y. Yarom, Presynaptic and postsynaptic GABAA receptors in rat suprachiasmatic nucleus, *Neuroscience.* 118(4), 909-923 (2003).

90. R.M. Buijs, Y.X. Hou, S. Shinn, and L.P. Renaud, Ultrastructural evidence for intra- and extranuclear projections of GABAergic neurons of the suprachiasmatic nucleus, *J. Comp. Neurol.* 340(3), 381-391 (1994).

91. N. Brandon, J. Jovanovic, and S. Moss, Multiple roles of protein kinases in the modulation of gamma-aminobutyric acid (A) receptor function and cell surface expression, *Pharmacol. Ther.* 94(1-2), 113-122 (2002).

92. J.T. Kittler, and S.J. Moss, Modulation of GABAA receptor activity by phosphorylation and receptor trafficking: implications for the efficacy of synaptic inhibition, *Curr. Opin. Neurobiol.* 13(3), 341-347 (2003).

93. B. Gao, J.M. Fritschy, and R.Y. Moore, GABAA-receptor subunit composition in the circadian timing system, *Brain Res.* 700(1-2), 142-156 (1995).

94. B.F. O'Hara, R. Andretic, H.C. Heller, D.B. Carter and T.S. Kilduff, GABAA, GABAC, and NMDA receptor subunit expression in the suprachiasmatic nucleus and other brain regions, *Brain Res. Mol. Brain Res.* 28(2), 239-250 (1995).

95. P. Poisbeau, M.C. Cheney, M.D. Browning, and I. Mody, Modulation of synaptic GABAA receptor function by PKA and PKC in adult hippocampal neurons, J. Neurosci. 19(2), 674-683 (1999).

96. McDonald, B.J., Amato, A., Connolly, C.N., Benke, D., Moss, S.J. and Smart, T.G., Adjacent phosphorylation sites on GABAA receptor beta subunits determine regulation by cAMP-dependent protein kinase, *Nat. Neurosci.* 1(1), 23-28 (1998).

97. N.J. Brandon, P. Delmas, J.T. Kittler, B.J. McDonald, W. Sieghart, D.A. Brown, T.G. Smart, and S.J. Moss, GABAA receptor phosphorylation and functional modulation in cortical neurons by a protein kinase C-dependent pathway, *J. Biol. Chem.* 275(49), 38856-38862 (2000).

98. J. Feng, X. Cai, J. Zhao, and Z. Yan, Serotonin receptors modulate GABA (A) receptor channels through activation of anchored protein kinase C in prefrontal cortical neurons, *J. Neurosci.* 21(17), 6502-6511 (2001).
99. M.A. Rivera-Bermudez, M.J. Gerdin, D.J. Earnest and M.L. Dubocovich, Regulation of basal rhythmicity in protein kinase C activity by melatonin in immortalized rat suprachiasmatic nucleus cells, *Neurosci. Lett.* 346(1-2), 37-40 (2003).
100. G. Milligan, Regional distribution and quantitative measurement of the phosphoinositidase C-linked guanine nucleotide binding proteins G11 alpha and Gq alpha in rat brain, *J. Neurochem.* 61(3), 845-851 (1993).

Serotonin and Neuroendocrine Regulation

Relevance to the Sleep/Wake Cycle

James W. Crane and Louis D. Van de Kar

1. Introduction

Serotonin was first isolated from serum by Rapport et al.[1,2] and was subsequently found to be present in the brain and to function as a neurotransmitter.[3-5] Over the course of the last 50 years there has been an explosion of knowledge of the serotonergic system. The relatively recent development of selective serotonin re-uptake inhibitors (SSRIs) and their effectiveness in treating a vast array of conditions (e.g. depression, anxiety, obesity, bulimia, aggression, obsessive compulsive disorder and post-traumatic stress disorder) has only served to heighten interest into the multitude of functions that serotonin influences via its actions within the central nervous system.[6-14] Two of these functions of serotonin—the activation of various neuroendocrine systems and modulation of the sleep/wake cycle—will be dealt with in this review. Particular emphasis will be placed on the possible role that neuroendocrine systems may play in mediating some of the effects of serotonin on the three major components of the sleep/wake cycle: wakefulness, slow wave sleep (SWS) and rapid eye-movement (REM) sleep.

2. Serotonin Neurons and Serotonin Receptors

The first detailed description of serotonergic neurons in the brain was pre-formed by Dahlstom and Fuxe.[15] This study combined with that of Steinbusch et al.[16] determined that the vast majority of serotonergic neurons are located within midbrain raphe nuclei. These serotonergic neurons have been classified into nine separate cell groups (B1-9). Of these cell groups, it is those located in the dorsal (B6 and B7) and the median raphe (B5 and B8) that have been found to provide the majority of the serotonergic input to the forebrain.[17,18] These serotonergic cell groups have also been implicated in the control of neuroendocrine function. However, control of neural function occurs through the action of serotonin on specific serotonin receptors located both pre- and post-synaptically.

At present, serotonin receptors are divided into 7 families (5-HT$_{1-7}$), with a total of 14 distinct receptor subtypes recognized (see refs.[19,20]). The 5-HT$_1$

family of receptors consists of 5-HT_{1A}, 5-HT_{1B}, 5-HT_{1D}, 5-HT_{1E}, and 5-HT_{1F}. The 5-HT_2 family contains 5-HT_{2A}, 5-HT_{2B}, 5-HT_{2C}, and the 5-HT_5 family contains 5-HT_{5A} and 5-HT_{5B}. At present 5-HT_3, 5-HT_4, 5-HT_6 and 5-HT_7 receptors are the only member of their respective families. With the exception of the 5-HT_3 receptor, a ligand-gated ion channel, all of these receptors are classical 7-transmembrane G protein coupled receptors. Activation of these receptors causes the disassociation of the trimeric G proteins into the Gα subunit and the βγ-subunit, both of which are capable of influencing neuronal function. In this regard, the type of G-protein that each serotonin receptor is coupled to determines its effect. The members of the 5-HT_1 family are coupled to $G_{i/o}$ and $_z$ proteins which act to inhibit adenylyl cyclase; the members of the 5-HT_2 family are coupled to the $G_{q/11}$ proteins to activate phospolipase C, increase intracellular Ca^{++} and activate protein kinase C; and $5\text{-HT}_{4,\,6}$, and $_7$ receptors couple to G_s proteins which act to increase adenylyl cyclase activity.[21] Currently, the G-protein involved in 5-HT_5 signaling has not been determined. It must be noted that, in all cases, activation of these receptors has a diverse array of effects on the neuronal function. For instance, 5-HT_{1A}, and 5-HT_7 receptors also activate mitogen-activated protein kinase.[22-24] As such, the consequence of serotonin receptor activation within the brain is still an area of intense investigation. Given that members of the 5-HT_1 and 5-HT_2 receptor families have been the focus of the majority of investigations into the serotonergic regulation of the sleep/wake cycle, this review will confine itself to the neuroendocrine effects of activation of these receptor subtypes. Hence, it is important at this point to discuss 5-HT_{1A} receptors in a little more detail.

The 5-HT_{1A} receptor subtype is the most widely studied of the 5-HT_1 receptors and possibly of all serotonin receptors. This is in no small part because these receptors have been suggested to play a role in a number of neuropsychiatric disorders, such as depression and anxiety.[25-28] Importantly, these receptors are found at both pre- and post-synaptic locations. Pre-synaptically they are found on the cell bodies of raphe neurons, where they function as autoreceptors,[29,30] such that stimulation of these receptors within the dorsal raphe decreases neuronal firing and prevents serotonin release in the forebrain.[31,32] Within the forebrain serotonin receptors are located within a number of brain regions including, importantly, a number of hypothalamic nuclei involved in the regulation of neuroendocrine function.[30,33] As will be discussed later, activation of both pre- and post-synaptic 5-HT_{1A} receptors appears to play a role in both neuroendocrine function and the sleep/wake cycle.

3. Serotonin and the Sleep/Wake Cycle: A Brief Overview

The relationship between the activity of the serotonergic system and the sleep/wake cycle has been the focus of excellent reviews in recent years (see refs.[34,35]). Therefore, we will provide only a brief outline of this field before dealing in more detail with serotonergic modulation of neuroendocrine systems and the possible influence this may have on the sleep/wake cycle.

For many years serotonin was considered to be a sleep inducing substance.[34] This changed, however, with the recording of dorsal raphe activity over the length of the sleep/wake cycle. It was found that dorsal raphe activity was at its highest while the animal was awake, with this activity dropping at the onset of sleep and falling to almost zero during periods of REM sleep.[36-38] These results were consistent with reports that cooling of raphe neurons induced sleep.[39] More recently, treatments that increase synaptic serotonin levels, such as SSRI administration, have been found to have a suppressive effect on REM sleep.[35] As such, it is now thought that serotonin acts within the brain to suppress REM sleep and promote waking.

With regards to the serotonin receptor subtypes involved in sleep/wake cycle regulation, to date the 5-HT_{1A} receptors have been the most widely investigated. Systemic administration of the selective 5-HT_{1A} receptor agonist 8-hydroxy-2-(di-n-propy-lamino) tetralin (8-OH-DPAT) increases waking, and reduces SWS and REM sleep.[34,40,41] The 5-HT_{1A} receptor partial-agonists buspirone, gepirone and ipsapirone also reduce SWS and REM and increase waking in rats,[42] and administration of ipsapirone has also been shown to reduce REM sleep in humans.[43-45] In rats, the effects of 8-OH-DPAT can be blocked by the selective 5-HT_{1A} receptor antagonist 4-(2′-methoxy-phenyl)-1-[2′-(n-2″-pyridinyl)-p-iodobenzamido]-ethyl-piperazine (p-MPPI), indicating that 5-HT_{1A} receptors mediate the sleep/wake effects of 8-OH-DPAT.[46] In contrast to the effects of systemic administration, the delivery of 8-OH-DPAT directly into the dorsal raphe (which presumably activates 5-HT_{1A} autoreceptors and inhibits serotonin neurons in this region) increases REM sleep in both cats and rats.[41,47,48] This result is consistent with the observation that serotonin neurons in the dorsal raphe are silent during REM sleep. Therefore, it appears that both pre- and post-synaptic 5-HT_{1A} receptors are capable of influencing the sleep/wake cycle: activation of 5-HT_{1A} autoreceptors increasing REM sleep, and activation of post-synaptic 5-HT_{1A} receptors reducing REM sleep. Interestingly, knockout mice lacking the 5-HT_{1A} receptor display an increase in REM sleep, suggesting that post-synaptic 5-HT_{1A} receptors exert more of an influence over the sleep/wake cycle than 5-HT_{1A} autoreceptors.[49]

The sleep/wake cycle is also influenced by receptors of the 5-HT_2 family, although much less is known than for the 5-HT_{1A} receptor. Agonists of the $5\text{-HT}_{2A/2C}$ receptors such as 1-(2,5-dimethoxy-4-methylphenyl)-2-aminopropane (DOM) and 1-(2,5-dimethoxy-4-iodophenyl)-2-aminopropane (DOI) increase waking and decrease SWS and REM sleep.[50,51] All but the reduction in REM sleep can be prevented by pre-treatment with selective $5\text{-HT}_{2A/2C}$ antagonists such as ritanserin.[34,50,51] Consistent with these findings, systemic administration of 5-HT_2 receptor antagonists alone increases SWS in both rats and humans.[50,52-54] However, reports on the influence of 5-HT$_2$ receptor antagonists on REM sleep are less conclusive.[34] All in all, however, the lack of highly selective antagonists has limited research into this area.

The influence exerted by serotonin and its receptors is largely thought to be mediated by effects on sleep/wake regulatory structures of the brain. For example, dorsal raphe projections to the lateral dorsal tegmental nuclei, pedunculopontine tegmental nuclei, and the medial pontine reticular formation are thought to influence REM sleep. Under this proposal, activation of 5-HT$_{1A}$ autoreceptors would reduce serotonin release in these brain regions and increase REM sleep. In contrast, activation of post-synaptic 5-HT$_{1A}$ receptors in these brain regions would act to inhibit REM sleep.[55,56] However, 5-HT$_{1A}$ and 5-HT$_2$ receptors are known to have effects on a number of neuroendocrine systems, some of which are known to influence the sleep/wake cycle.

4. The Hypothalamic-Pituitary-Adrenal Axis

As a result of considerable investigation during the last half of the 20th century the details of the hypothalamic-pituitary-adrenal (HPA) axis are now well known.[57] The apex of this axis is located within the medial parvocellular zone of the paraventricular nucleus of the hypothalamus (mpPVN). It is within this region that neurons containing corticotrophin-releasing-hormone (CRH) that project to the median eminence are located.[58,59] First isolated by Vale et al.,[60] CRH is released by neurons of the mpPVN into the median eminence and transported, via a portal circulation, to the anterior pituitary. On exposure to CRH, corticotropin cells within the anterior pituitary release adrenocorticotropin hormone (ACTH) into the general circulation. However, the activation of corticotropin cells is also regulated by arginine vasopressin and oxytocin. Arginine vasopressin is also release into median eminence by CRH-containing neurons of the mpPVN.[61] At the level of the pituitary arginine vasopressin acts synergistically with CRH to enhance the release of ACTH into the general circulation.[62-64] Similarly, oxytocin acts within the pituitary to influence the release of ACTH into the general circulation. Magnocellular oxytocin-containing neurons within the paraventricular nucleus (PVN) and the supraoptic nucleus largely project to the posterior pituitary where they release oxytocin into the general circulation. However, there is also evidence that they provide terminals that release oxytocin into the portal circulation of the median eminence,[65,66] and that the presence of oxytocin in the anterior pituitary potentiates the CRH-induced release of ACTH into the general circulation.[65,67-69] The adrenal cortex then responds to the increased presence of ACTH by releasing glucocorticoids (cortisol in humans; corticosterone in rats). Glucocorticoids have an incredibly diverse range of actions within the body. However, they are principally thought of as stress hormones crucial to the survival of the organism during exposure to stressful stimuli.

Perhaps because of the importance of the HPA axis in an organism survival, the activity of this axis is under the control of a multitude of factors. The neural regulation of the HPA axis is an area of intense investigation, with the major focus being on inputs from areas traditionally viewed as limbic structures. In particular, a number of structures within the amygdala areas have been shown to regulate HPA axis activity,[70-74] as have regions of the bed nucleus of the stria terminalis, medial prefrontal cortex and hippocampus.[75-84] Inputs from brainstem and midbrain nuclei are also critical regulators of HPA axis function. For instance, noradrenergic neurons of the ventrolateral medulla and nucleus tractus solitarius influence HPA axis responses to stressful stimuli.[85,86] Similarly, serotonergic neurons are known to regulate HPA axis activity (see next section). As well as synaptically released neurotransmitters, a number of humoral factors influence the HPA axis. Probably the most well characterized of these are the glucocorticoids themselves. It has been known for sometime that glucocorticoids provide a negative-feedback regulation of HPA axis activity. This occurs via two glucocorticoid receptor subtypes. Type I receptors are high affinity receptors that are thought to be largely occupied by basal levels of glucocorticoids. Conversely, Type II receptors have a relatively low affinity for glucocorticoids and are thought to only be occupied during the high levels of glucocorticoids released as a result of stressful stimuli. Type I receptors appear to be important in maintaining the level of ACTH secretion seen during the circadian trough, whereas both Type I and II receptors are involved reining in the HPA axis during the circadian peak and during stressor exposure.[61,87,88]

4.1. Serotonergic Regulation of the HPA Axis

Serotonergic neurons of the dorsal and median raphe nuclei project to the mpPVN region of the hypothalamus, the location of the CRH cells at the apex of the HPA axis.[89,90] Indeed, serotonin-containing terminals make direct synaptic contact with CRH cells in the mpPVN.[91] A number of studies have demonstrated that increasing serotonin synthesis, release or synaptic levels results in activation of the HPA axis. For example, acute administration of SSRIs results in the HPA axis activation in rats and humans.[92-94] Similarly, the administration of the serotonin releasing drugs 3,4-methylenedioxymethamphetamine (MDMA) and dexfenfluramine results in increases in ACTH and cortisol in humans.[95,96] With regards to the particular serotonin receptor subtypes that might be involved in HPA axis regulation, the focus of research has again been on 5-HT_{1A} and $5\text{-HT}_{2A/2C}$ receptors.

It has now been demonstrated by *in-situ* hybridization, autoradiography and immunohistochemistry that 5-HT_{1A} receptors are located within the PVN,[33,97-100] and activation of 5-HT_{1A} receptors stimulates the HPA axis. In rats, systemic administration of the selective 5-HT_{1A} receptor agonist 8-OH-DPAT increases the release of ACTH and corticosterone into the plasma, an effect that can be blocked by the specific 5-HT_{1A} receptor antagonist WAY 100635.[101,102] Systemic administration of partial 5-HT_{1A} receptor agonists such as buspirone, gepirone and ipsiparone also result in HPA axis activation in both rats and humans.[103-106] Of course systemically administered 5-HT_{1A} receptor agonists act on both the somatodendritic autoreceptors in the raphe and post-synaptic receptors. However, there is evidence to suggest that the HPA axis stimulatory effect is due to activation of post-synaptic 5-HT_{1A} receptors located with the PVN. Lesion studies have demonstrated that the PVN is crucial to the 8-OH-DPAT-induced ACTH and corticosterone release.[107,108] Recently, it has been reported that delivery of the 5-HT_{1A} selective antagonist WAY 100635 into the PVN prevented the ACTH response to systemic 8-OH-DPAT administration.[109] Also, the administration of 8-OH-DPAT directly into the PVN increases plasma ACTH and corticosterone levels.[110] Interestingly, it now appears that glutamate receptors within the PVN may be the involved in the HPA axis activation elicited by activation of 5-HT_{1A} receptors within the PVN.[110] However, regardless of the exact nature of the circuitry within the PVN, it is clear that 5-HT_{1A} receptor activation within this region stimulates the HPA axis. Another aspect of the 5-HT_{1A} receptor activation of the HPA axis is that it involves the pertussis toxin-insensitive G_z-protein.[111] When rats were pre-treated with pertussis toxin the ACTH response to 8-OH-DPAT was not affected. However, the injections of G_z antisense oligonucleotides into the third ventricle of the rat reduced the 8-OH-DPAT-induced increase in plasma ACTH.[111] This finding could have considerable implications for future pharamacological manipulations of the HPA axis.

The PVN also contains 5-HT_{2A} and 5-HT_{2C} receptors[97,112,113] and activation of these receptor-subtypes stimulates the HPA axis.[114-118] In fact, systemic administration of the 5-HT_2 receptor agonist DOI activates CRH cells of the mpPVN, as determined by dual-immunohistochemical detection of the protein product of the immediate-early gene *c-fos* and CRH peptide.[119] Historically, the lack of suitably selective agonists and antagonists has limited investigation into the relative contribution of 5-HT_{2A} and 5-HT_{2C} receptors to HPA axis stimulation. However, we have recently demonstrated that the ACTH response to DOI can be blocked by intra-PVN injection of the selective 5-HT_{2A} receptor antagonist MDL 100907, suggesting that 5-HT_{2A} receptors in the PVN (and not 5-HT_{2C} receptors) are involved in the regulation of the HPA axis response by 5-HT.[113]

4.1.1. Relevance to the Sleep/Wake Cycle

Deepening of sleep and increases in REM sleep are associated with a decrease in cortisol levels. Conversely, lightening of sleep and waking are associated with increases in cortisol secretion.[120] This pattern of cortisol secretion is similar to that displayed by the neurons of the dorsal raphe i.e. quiet during REM sleep and active during waking, and these neurons are known to influence the activity of the HPA axis (see above). Also, a number of studies have found that components of the HPA axis (namely CRH, ACTH, and corticosterone/cortisol) are capable of influencing the sleep/wake cycle. Repetitive intravenous injections of CRH resulted in an increase in waking and a decrease in SWS and REM sleep,[121] and intracerebroventricular administration of CRH has been reported to decrease SWS in rats.[122] Consistent with these results, intracerebroventricular administration of CRH antagonists or antisense oligonucleotides directed against CRH mRNA decreases waking and increase SWS, but do not significantly affect REM sleep.[123,124] It must be noted, however, that there exists in the brain an extensive network of CRH neurons and terminals that is not related (at least directly) to the release of ACTH from the pituitary. As such, it is possible that the sleep/wake effects of CRH, CRH antagonists and CRH antisense oligonucleotides are not due to alterations in HPA axis function.

Investigations into the effect of exogenous ACTH and cortisol/corticosterone have more conclusively determined the influence exerted by the HPA axis on the sleep/wake cycle. Continuous 8-hour infusion of ACTH has been reported to decrease REM sleep regardless of when the infusions started, but reduced SWS only when infusion began at 0800.[125] However, the influence of ACTH appears to be related to the central actions of ACTH rather than stimulation of cortisol release, as systemic administration of ebiratide (a synthetic ACTH analogue with behavioural effects but lacking the ability to stimulate cortisol secretion) reduced SWS and increased wakefulness.[126] Similar to the effects of ACTH, continuous and pulsatile cortisol administration have been shown to reduce REM sleep but increase SWS in healthy males[127,128] As exogenous CRH and cortisol exert opposite effects on SWS, it has been proposed that the influence of CRH on SWS is not mediated via its stimulation of cortisol release. In contrast, it appears that the decrease in REM sleep seen after CRH, ACTH and cortisol administration is due to the effects of increased plasma cortisol concentrations seen after administration of each of these hormones.[129]

At least some effects of the serotonergic system on the sleep/wake cycle are likely to be due to its influence on the HPA axis. The activity of serotonergic neurons of the dorsal raphe is at its highest during waking and almost nonexistent during REM sleep. Also, serotonin levels are elevated in the PVN when the animal is awake.[130] In addition to this, activation of 5-HT_{1A} and $5\text{-HT}_{2A/2C}$ receptors decreases SWS and REM sleep, and stimulates the HPA axis, as exemplified by increased plasma ACTH and cortisol/corticosterone levels. As detailed above, increased cortisol/corticosterone levels act to reduce REM sleep and could, therefore, contribute to the REM sleep reduction seen after activation of 5-HT_{1A} and $5\text{-HT}_{2A/2C}$ receptors. However, the decrease in SWS seen in response to activation of 5-HT_{1A} and $5\text{-HT}_{2A/2C}$ receptors is more likely to be due to the central effects of CRH and ACTH, as both of these peptides appear to act centrally to decrease SWS. The increase in REM sleep seen after inhibition of dorsal raphe 5-HT neurons by direct administration of the 5-HT$_{1A}$ receptor agonist 8-OH-DPAT could also be due to reduced serotonin release in the PVN and a subsequent reduction in the activity of the HPA axis. Of course further investigation will need to be performed to determine the exact involvement of the HPA axis in mediating aspects of serotonergic regulation of the sleep/wake cycle.

5. Oxytocin

Magnocellular oxytocin-containing neurons are located within both the PVN and the supraoptic nucleus of the hypothalamus, and these neurons project to the posterior pituitary where, upon activation, they release oxytocin into the general circulation.[131] As mentioned above, magnocellular neurons also release oxytocin into the portal circulation of the median eminence, a release that appears to influence the CRH-induced release of ACTH.[65-69] There is also evidence that oxytocin is released from the dendrites of magnocellular neurons and can influence the activity of surrounding neurons. Lastly, it appears that parvocellular neurons of the PVN contain oxytocin and use it as a neurotransmitter/neuromodulator in a number of extrahypothalamic areas.[131-133] Although, these neural actions of oxytocin are not technically neuroendocrine in nature, they are actions that we will deal with since they have been found to regulate the sleep/wake cycle.

5.1. Serotonergic Regulation of Oxytocin Release

It is well accepted that the release of oxytocin into the general circulation by magnocellular oxytocin neurons is controlled, at least in part, by the serotonergic system. Both the PVN and supraoptic nucleus contain serotonin-positive terminals[134] and stimulation of the dorsal raphe increases plasma oxytocin levels.[110,135] However, lesion studies suggest that the ability of serotonin to increase plasma oxytocin is largely dependent on oxytocin neurons of the PVN, rather than those of the supraoptic nucleus.[136] Serotonergic neurons also appear to be involved in maintaining basal levels of oxytocin in magnocellular neurons, as selective lesions of serotonin terminals within the PVN reduced basal levels of oxytocin in the plasma and the pituitary.[136,137] However, to date there have been no studies on the effect of the serotonergic system on oxytocin release from dendrites release or from centrally-projecting parvocellular neurons.

Similar to serotonergic regulation of the HPA axis, 5-HT$_{1A}$ and 5-HT$_{2A/2C}$ receptors are the most investigated with regards to regulation of oxytocin secretion. The systemic administration of the 5-HT$_{1A}$ selective agonist 8-OH-DPAT increases plasma oxytocin levels, and similar results have been found with other 5-HT$_{1A}$ receptor agonists.[101,102,107,108,138] This oxytocin release is due to 5-HT$_{1A}$ receptor activation as it can be blocked by pre-treatment with the selective 5-HT$_{1A}$ receptor antagonist WAY 100635.[101] Numerous pieces of evidence suggest that post-synaptic 5-HT$_{1A}$ receptors in the PVN are responsible for the oxytocin release: 5-HT$_{1A}$ receptors are present on magnocellular oxytocin neurons of the PVN;[33] lesions of the PVN reduce the oxytocin response to systemic administration of 5-HT$_{1A}$ receptor agonists;[107,108] and intra-PVN injection of WAY 100635 blocks the oxytocin response to systemic 8-OH-DPAT administration.[109] Similar to the situation with the HPA axis, 5-HT$_{1A}$ receptor-mediated regulation of oxytocin release from the PVN appears to involve the pertussis toxin-insensitive G-protein, G$_z$.[111]

A number of studies have demonstrated that 5-HT$_{2A}$ and 5-HT$_{2C}$ receptors are located within the PVN,[97,112,113,139] and that systemic administration of the 5-HT$_{2A/2C}$ receptor agonists increases plasma oxytocin concentrations.[33,107,113,119] Consistent with this increase, systemic administration of the 5-HT$_2$ receptor agonist DOI activates, as determined by c-*fos* expression, oxytocin neurons of the PVN.[119] Both the increase in c-*fos* expression and the release of oxytocin in response to systemic DOI can be blocked by systemic injection of the 5-HT$_{2A}$ selective antagonist MDL 100907.[119] Further to this, we have recently demonstrated that the DOI-induced release of oxytocin can be suppressed by intra-PVN injections of MDL 100907, indicating that it is 5-HT$_{2A}$ receptors within the PVN that are critical for the oxytocin response to DOI.[113]

5.1.1. Relevance to the Sleep/Wake Cycle

To date there appear to be no reports on the effect of systemic oxytocin administration on the sleep/wake cycle. However, it has recently been reported that retrodialysis of oxytocin into the lateral cerebral ventricle of rats increases wakefulness and decreases REM sleep.[140] Given that intra-PVN oxytocin appears to facilitate the release of ACTH in response to stress, the authors suggested that the sleep/wake effects following administration of oxytocin into the lateral cerebral ventricle might be due to its actions within the PVN to influence HPA axis activity.[141] While it is acknowledged that intra-nuclear release of oxytocin from soma and dendrites can occur in the PVN,[131-133] it is yet to be determined whether serotonin is capable of inducing this release. However, given that serotonin and serotonin receptors are known to stimulate magnocellular oxytocin neurons within the PVN, it seems highly likely that serotonergic mechanisms are also involved in the regulation of intra-nuclear release of oxytocin. If this is indeed the case, it would provide another avenue by which the serotonergic system might influence the sleep/wake cycle.

6. Prolactin

Prolactin is synthesized by lactotrophs located within the anterior pituitary, and its release into the general circulation is under the control of both prolactin-inhibiting factors and prolactin-releasing factors located in the hypothalamus.[142,143] The primary prolactin-inhibitory factor is dopamine, which is released into the median eminence by tuberoinfundibular neurons located in the arcuate nucleus of the hypothalamus. Once in the portal circulation of the anterior pituitary it binds to dopamine receptors (D_2) located on the lactotrophs and inhibits prolactin release into the general circulation.[144] As lactotrophs have a particularly high spontaneous activity, leading to prolactin release, the inhibitory action of dopamine was long considered to be the primary regulator of prolactin secretion.[144] However, a number of prolactin-releasing factors have recently been identified. Amongst these is oxytocin, which stimulates prolactin release *in vitro* and *in vivo*.[145] Consistent with this, lactotrophs possess oxytocin receptors and (as mentioned above) oxytocin is found in high concentration in the portal circulation.[65,146] However, the influence of oxytocin on prolactin secretion only seems to be apparent in the absence of dopamine.[143,145] Other prolactin-releasing factors that are believed to act directly on the lactotrophs of the anterior pituitary include thyrotropin-releasing hormone, vasoactive intestinal peptide, and the recently identified prolactin releasing peptide.[142-144] Neurotransmitters and neuromodulators have also been demonstrated to influence the release of prolactin, included in this list is serotonin.

6.1. Serotonergic Regulation of Prolactin Release

Systemic administration of serotonin and serotonin releasing agents (such as *p*-chloroamphetamine) cause a rapid increase in plasma prolactin concentrations,[147,148] and there is evidence that neurons of the dorsal raphe contribute to the serotonergic regulation of the prolactin release.[149-151] Also, serotonergic release of prolactin is mediated by PVN neurons, since lesions of this area prevent the prolactin release seen following systemic administration of serotonin receptor agonists and serotonin releasing agents.[107,148,152]

In contrast to the regulation of the HPA axis and oxytocin secretion, 5-HT_{1A} receptors do not regulate prolactin secretion. Prolactin release in humans following systemic administration of agonists selective for the 5-HT_{1A} receptor is inconsistent, with buspirone and gepirone (but not ipsapirone or tandospirone) increasing plasma prolactin levels.[153-157] In rats, systemic administration of 8-OH-DPAT and ipsapirone produce a highly transient increase in prolactin release.[158-162] However, pre-treatment with the 5-HT_{1A} receptor antagonist WAY 100635 does not prevent the prolactin release seen after systemic 8-OH-DPAT administration.[101] Consistent with this, we have found that reductions in the levels of G_z protein in the PVN, produced by local administration of G_z antisense oligonucleotides, do not alter the prolactin release seen after systemic 8-OH-DPAT administration.[111] Put together with reports that the prolactin response to 8-OH-DPAT can be blocked by pre-treatment with pertussis toxin,[111] these results suggest that receptors other than the 5-HT_{1A} receptor are utilizing the pertussis toxin-sensitive G-proteins to mediate the 8-OH-DPAT-induced increases in plasma prolactin levels.

It has been known for some time that the 5-HT$_{2A/2C}$ agonists, such as DOI, dose dependently increases plasma prolactin levels;[102,107,148,152] however, the lack of a suitably selective antagonist prevented further delineation of the relative contribution of 5-HT$_{2A}$ and 5-HT$_{2C}$ receptors in this response. Recently, however, we used the selective 5-HT$_{2A}$ receptor antagonist MDL 100907 to investigate the role of this subtype in DOI-induced prolactin secretion. Pretreatment, systemically, with MDL 100907 dose dependently reduced the increase in plasma prolactin levels seen after systemic DOI administration.[119] Further to this, injection of MDL 100907 directly into the PVN also dose dependently reduced the prolactin release seen in response to systemic administration of DOI.[113] Based on these results we conclude that 5-HT$_{2A}$ receptors, but not 5-HT$_{2C}$ receptors, in the PVN play an integral role in the control of prolactin secretion by serotonin. It should be noted, however, that at present the neural substrates responsible for 5-HT$_{2A}$ receptor-induced prolactin release are unknown. As detailed above, 5-HT$_{2A}$ receptors also activate oxytocin neurons in the PVN, suggesting that this prolactin-releasing factor may mediate some of the effects. It is also possible that 5-HT$_{2A}$ receptors stimulate the release of other prolactin-releasing factors or act to inhibit dopamine release into the median eminence, possibly through the activation of inhibitory inter-neurons present in the hypothalamus.[142]

6.1.1. Relevance to the Sleep/Wake Cycle

A number of studies indicate that the predominant effect of prolactin is to increase REM sleep. Systemic administration of prolactin results in an increase in REM sleep in both rat and rabbits.[163,164] Increased endogenous prolactin levels, resulting from systemic administration of vasoactive intestinal peptide, increase REM sleep in rats.[165] Chronic elevations in plasma prolactin levels, as a result of anterior pituitary graft under the capsule of the kidney, result in a large increases in REM sleep.[166] Also, there is a correlation between increases in REM sleep and plasma prolactin levels following central infusion of prolactin-releasing peptide,[167] and systemic administration of antibodies to prolactin produced a small reduction in REM sleep.[168] Interestingly, central administration of prolactin also increases REM sleep,[169] and the central administration of antibodies to prolactin decreases spontaneous and stress-induced REM sleep.[170] These results are consistent with the fact that circulating prolactin gains access the CNS, via a receptor-mediated transport mechanism located in the choroids plexus,[171] where it is thought to interact with specific prolactin receptors distributed throughout the forebrain to regulate REM sleep.[172]

It is difficult to reconcile the REM sleep promoting effect of prolactin with the fact that systemic administration of 5-HT$_{2A}$ receptor agonists increase prolactin release but decrease REM sleep. It must be noted, however, that prolactin does reduce REM sleep in rats when administered during the night period.[164] This suggests that the behavioral state of the animal might influence the effect of prolactin on REM sleep. Also, increases in REM sleep seen after systemic prolactin administration take 1-2 hours to manifest.[163] As such, more rapid changes in the sleep/wake cycle resulting from 5-HT$_{2A}$ receptor activation could prime the system to respond to prolactin by reducing REM sleep. Of course, it is possible that the REM sleep promoting effects of prolactin are simply overcome by the REM sleep suppressing effects of glucocorticoids that are also released following activation of 5-HT$_{2A}$ receptors in the hypothalamus.

7. Growth Hormone

The synthesis and release of growth hormone from somatotrophs of the anterior pituitary is primarily regulated by two hypothalamic peptides: growth hormone releasing hormone (GHRH) and somatostatin.[173,174] A number of hypothalamic nuclei contain GHRH; however, only GHRH-containing neurons of the arcuate nucleus project to the median eminence.[173,175,176] Upon reaching the somatotrophs, via the portal ciculation of the pituitary, GHRH acts on specific receptors to stimulate both the synthesis and release of growth hormone.[173] This is in contrast to the actions of somatostatin which, once released from neurons of the periventricular-region of the PVN into the median eminence, acts via specific receptors on somatotrophs to suppress growth hormone release.[173,177] In addition to this, somatostatin containing neurons in the arcuate nucleus are also thought to influence the activity of GHRH-containing neurons to influence the release of growth hormone from the anterior pituitary into the general circulation.[173]

7.1. Serotonergic Regulation of Growth Hormone Release

Serotonin has a stimulatory effect on growth hormone secretion. Administration of the serotonin precursors, L-tryptophan or 5-hydroxytryptophan, stimulates growth hormone release.[178-181] Similarly, administration of the SSRI fluoxetine to rat pups increases plasma growth hormone levels.[182]

In humans, pre-treatment with the 5-HT$_{1A}$ receptor antagonist pindolol attenuates the growth hormone response to an L-tryptophan challenge.[178] Similarly, systemic administration of a number of 5-HT$_{1A}$ selective agonists (such as buspirone, ipsapirone, tandospirone, and flesinoxan) produce rapid increases in plasma growth hormone concentrations in humans; increases that can be blocked by pre-treatment with pindolol.[183-187] In rats, however, results suggest that 5-HT$_{1A}$ receptor activation does not affect plasma growth hormone levels to any great degree.[158,188] To date the influence of serotonin and serotonin receptors on GHRH has not been investigated.

However, GHRH does appear to play a role in the growth hormone release seen after systemic administration of 5-hydroxytryptophan.[189] Also, the influence of 5-HT$_{1A}$ receptors on somatostatin concentrations is yet to be determined, with both a decrease and no change being reported to occur after systemic 8-OH-DPAT administration to.[190,191]

7.1.1. Relevance to the Sleep/Wake Cycle

Both growth hormone and GHRH influence the sleep/wake cycle in animals and humans (see refs.[192,193]). Growth hormone increases REM sleep in humans and rats[194] and systemic administration of GHRH to humans and animals increases both SWS and REM sleep.[122,195-198] The REM-promoting effects of GHRH appear to be due to the stimulation of growth hormone release, as GHRH-induced increases in REM sleep were not seen in hypophysectomized rats.[199] Similarly, reductions in REM sleep seen in mice with non-functional GHRH-receptors can be reversed by subcutaneous infusion of growth hormone,[200] whereas the reductions in SWS remain unaltered. These results also suggest that the GHRH acts centrally to increase SWS.

Given the limited amount of information regarding serotonergic control of growth hormone secretion, it is hard to make any strong claims about the involvement of growth hormone release in the effect of serotonin on the sleep/wake cycle. Activation of 5-HT$_{1A}$ receptors decreases SWS and REM sleep, and increases the release of growth hormone. Yet, growth hormone administration increases REM sleep.

As such, the 5-HT$_{1A}$ receptors are unlikely to influence REM sleep via the release of growth hormone. However, as there are no reports on the influence of 5-HT$_{1A}$ receptors on GHRH release, it is impossible to determine whether or not this peptide plays a role in the SWS reduction seen in response to 5-HT$_{1A}$ receptor activation.

8. Conclusion

Serotonin is known to regulate the sleep/wake cycle. While this is thought to be primarily due to its influence over sleep/wake centers within the brain, there is (as detailed above) a considerable amount of circumstantial evidence suggesting that serotonergic regulation of neuroendocrine systems may also play a role in modulating the sleep/wake cycle.

However, this possibility has yet to be directly examined. This is no doubt due in part to the paucity of specific serotonin receptors agonists and antagonists that are available. Hopefully future studies directly investigating the role that the activation of neuroendocrine systems by serotonin plays in its regulation of the sleep/wake cycle will soon be performed.

9. References

1. Rapport MM. Serum vasoconstrictor (serotonin). V. The presence of creatine in the complex: a proposed study of the vasocontrictor principle. J Biol Chem 1949;180:961-969.
2. Rapport MM, Green AA, Page IH. Serum vasoconstrictor (serotonin). IV. Isolation and characterization. J Biol Chem 1948;176:1243-1251.
3. Amin AH, Crawford TBB. The distribution of substance P and 5-hydroxytryptamine in the central nervous system of the dog. J physiol (Lond) 1954;126:596-618.
4. Twarog BM, Page IH. Serotonin content of some mammalian tissues and urine and a method for its determination. Am J Physiol 1953;175:157-161.
5. Welsh JH. Serotonin as a possible neurohumoral agent; evidence obtained in lower animals. Ann N Y Acad Sci 1957;66(3):618-30.
6. Coccaro EF, Kavoussi RJ, Hauger RL. Serotonin function and antiaggressive response to fluoxetine: a pilot study. Biol Psychiatry 1997;42(7):546-52.
7. Coplan JD, Papp LA, Pine D, et al. Clinical improvement with fluoxetine therapy and noradrenergic function in patients with panic disorder. Arch Gen Psychiatry 1997;54(7):643-8.
8. Pearlstein T, Rosen K, Stone AB. Mood disorders and menopause. Endocrinol Metab Clin North Am 1997;26(2):279-94.
9. Perez V, Gilaberte I, Faries D, et al. Randomised, double-blind, placebo-controlled trial of pindolol in combination with fluoxetine antidepressant treatment. Lancet 1997;349(9065):1594-7.
10. Yonkers KA. Antidepressants in the treatment of premenstrual dysphoric disorder. J Clin Psychiatry 1997;58 Suppl 14:4-10; discussion 11-3.
11. Dodman NH, Donnelly R, Shuster L, et al. Use of fluoxetine to treat dominance aggression in dogs. J Am Vet Med Assoc 1996;209(9):1585-7.
12. Greeno CG, Wing RR. A double-blind, placebo-controlled trial of the effect of fluoxetine on dietary intake in overweight women with and without binge-eating disorder. Am J Clin Nutr 1996;64(3):267-73.
13. Goldstein DJ, Wilson MG, Thompson VL, et al. Long-term fluoxetine treatment of bulimia nervosa. Fluoxetine Bulimia Nervosa Research Group. Br J Psychiatry 1995;166(5):660-6.
14. Barr LC, Goodman WK, McDougle CJ, et al. Tryptophan depletion in patients with obsessive-compulsive disorder who respond to serotonin reuptake inhibitors. Arch Gen Psychiatry 1994;51(4):309-17.
15. Dahlstrom A, Fuxe K. Evidence for the existence of monoamine-containing neurons in the central nervous system. I. Demonstration of monoamines in the cell bodies of brain stem neurons. Acta Physiol Scand 1964;62(Suppl 232):1-55.
16. Steinbusch HW, Verhofstad AA, Joosten HW. Localization of serotonin in the central nervous system by immunohistochemistry: description of a specific and sensitive technique and some applications. Neuroscience 1978;3(9):811-9.
17. Jacobs BL, Azmitia EC. Structure and function of the brain serotonin system. Physiol Rev 1992;72(1):165-229.
18. Baumgarten HG, Grozdanovic Z. Anatomy of central serotoninergic projection systems. In: Baumgarten HG, Gothert M, editors. Serotoninergic neurons and 5-HT receptors in the CNS. Berlin: Springer-Verlag; 1997. p. 41-90.
19. Barnes NM, Sharp T. A review of central 5-HT receptors and their function. Neuropharmacology 1999;38(8):1083-152.

20. Raymond JR, Mukhin YV, Gelasco A, et al. Multiplicity of mechanisms of serotonin receptor signal transduction. Pharmacol Ther 2001;92(2-3):179-212.

21. Hanley NR, Van de Kar LD. Serotonin and the neuroendocrine regulation of the hypothalamic-pituitary-adrenal axis in health and disease. Vitam Horm 2003; 66:189-255.

22. Garnovskaya MN, van Biesen T, Hawe B, et al. Ras-dependent activation of fibroblast mitogen-activated protein kinase by 5-HT1A receptor via a G protein beta gamma-subunit-initiated pathway. Biochemistry 1996;35(43):13716-22.

23. Millan MJ, Newman-Tancredi A, Duqueyroix D, et al. Agonist properties of pindolol at h5-HT1A receptors coupled to mitogen-activated protein kinase. Eur J Pharmacol 2001;424(1):13-7.

24. Mukhin YV, Garnovskaya MN, Collinsworth G, et al. 5-Hydroxytryptamine1A receptor/Gibetagamma stimulates mitogen-activated protein kinase via NAD(P)H oxidase and reactive oxygen species upstream of src in chinese hamster ovary fibroblasts. Biochem J 2000;347 Pt 1:61-7.

25. Schramek H. MAP kinases: from intracellular signals to physiology and disease. News Physiol Sci 2002;17:62-67.

26. Gorman JM. Treating generalized anxiety disorder. J Clin Psychiatry 2003; 64(Suppl 2):24-9.

27. Gross C, Zhuang X, Stark K, et al. Serotonin1A receptor acts during development to establish normal anxiety-like behaviour in the adult. Nature 2002; 416(6879):396-400.

28. Artigas F, Romero L, de Montigny C, et al. Acceleration of the effect of selected antidepressant drugs in major depression by 5-HT1A antagonists. Trends Neurosci 1996;19(9):378-83.

29. Sotelo C, Cholley B, El Mestikawy S, et al. Direct Immunohistochemical Evidence of the Existence of 5-HT1A Autoreceptors on Serotoninergic Neurons in the Midbrain Raphe Nuclei. Eur J Neurosci 1990;2(12):1144-1154.

30. Kia HK, Miquel MC, Brisorgueil MJ, et al. Immunocytochemical localization of serotonin1A receptors in the rat central nervous system. J Comp Neurol 1996;365(2):289-305.

31. Verge D, Calas A. Serotoninergic neurons and serotonin receptors: gains from cytochemical approaches. J Chem Neuroanat 2000;18(1-2):41-56.

32. Stamford JA, Davidson C, McLaughlin DP, et al. Control of dorsal raphe 5-HT function by multiple 5-HT(1) autoreceptors: parallel purposes or pointless plurality? Trends Neurosci 2000;23(10):459-65.

33. Zhang Y, Dudas B, Muma NA, et al. Co-localization of 5-HT_{1A} and 5-HT_{2A} serotonin receptors on oxytocin cells in the hypothalamic paraventricular nucleus. Neuroscience Abst. 2001;17(136):265.12.

34. Portas CM, Bjorvatn B, Ursin R. Serotonin and the sleep/wake cycle: special emphasis on microdialysis studies. Prog Neurobiol 2000;60(1):13-35.

35. Ursin R. Serotonin and sleep. Sleep Med Rev 2002;6(1):55-69.

36. Cespuglio R, Faradji H, Gomez ME, et al. Single unit recordings in the nuclei raphe dorsalis and magnus during the sleep-waking cycle of semi-chronic prepared cats. Neurosci Lett 1981;24(2):133-8.

37. McGinty DJ, Harper RM. Dorsal raphe neurons: depression of firing during sleep in cats. Brain Res 1976;101(3):569-75.

38. Trulson ME, Jacobs BL. Raphe unit activity in freely moving cats: correlation with level of behavioral arousal. Brain Res 1979;163(1):135-50.

39. Cespuglio R, Walker E, Gomez ME, et al. Cooling of the raphe nucleus induces sleep in the cat. Neurosci Lett 1976;3:221-227.
40. Monti JM, Jantos H. Dose-dependent effects of the 5-HT1A receptor agonist 8-OH-DPAT on sleep and wakefulness in the rat. J Sleep Res 1992;1(3): 169-175.
41. Bjorvatn B, Fagerland S, Eid T, et al. Sleep/waking effects of a selective 5-HT1A receptor agonist given systemically as well as perfused in the dorsal raphe nucleus in rats. Brain Res 1997;770(1-2):81-8.
42. Monti JM, Jantos H, Silveira R, et al. Sleep and waking in 5,7-DHT-lesioned or (-)-pindolol-pretreated rats after administration of buspirone, ipsapirone, or gepirone. Pharmacol Biochem Behav 1995;52(2):305-12.
43. Gillin JC, Sohn JW, Stahl SM, et al. Ipsapirone, a 5-HT1A agonist, suppresses REM sleep equally in unmedicated depressed patients and normal controls. Neuropsychopharmacology 1996;15(2):109-15.
44. Driver HS, Flanigan MJ, Bentley AJ, et al. The influence of ipsapirone, a 5-HT1A agonist, on sleep patterns of healthy subjects. Psychopharmacology (Berl) 1995;117(2):186-92.
45. Seifritz E, Moore P, Trachsel L, et al. The 5-HT1A agonist ipsapirone enhances EEG slow wave activity in human sleep and produces a power spectrum similar to 5-HT2 blockade. Neurosci Lett 1996;209(1):41-4.
46. Sorensen E, Gronli J, Bjorvatn B, et al. The selective 5-HT(1A) receptor antagonist p-MPPI antagonizes sleep–waking and behavioural effects of 8-OH-DPAT in rats. Behav Brain Res 2001;121(1-2):181-7.
47. Portas CM, Thakkar M, Rainnie D, et al. Microdialysis perfusion of 8-hydroxy-2-(di-n-propylamino)tetralin (8-OH-DPAT) in the dorsal raphe nucleus decreases serotonin release and increases rapid eye movement sleep in the freely moving cat. J Neurosci 1996;16(8):2820-8.
48. Monti JM, Jantos H, Monti D. Increased REM sleep after intra-dorsal raphe nucleus injection of flesinoxan or 8-OHDPAT: prevention with WAY 100635. Eur Neuropsychopharmacol 2002;12(1):47-55.
49. Boutrel B, Monaca C, Hen R, et al. Involvement of 5-HT1A receptors in homeostatic and stress-induced adaptive regulations of paradoxical sleep: studies in 5-HT1A knock-out mice. J Neurosci 2002;22(11):4686-92.
50. Dugovic C. Functional activity of 5-HT2 receptors in the modulation of the sleep/wakefulness states. J Sleep Res 1992;1(3):163-168.
51. Dugovic C, Wauquier A, Leysen JE, et al. Functional role of 5-HT2 receptors in the regulation of sleep and wakefulness in the rat. Psychopharmacology (Berl) 1989;97(4):436-42.
52. Dugovic C, Wauquier A. 5-HT2 receptors could be primarily involved in the regulation of slow-wave sleep in the rat. Eur J Pharmacol 1987;137(1):145-6.
53. Landolt HP, Meier V, Burgess HJ, et al. Serotonin-2 receptors and human sleep: effect of a selective antagonist on EEG power spectra. Neuropsychopharmacology 1999;21(3):455-66.
54. Sharpley AL, Elliott JM, Attenburrow MJ, et al. Slow wave sleep in humans: role of 5-HT2A and 5-HT2C receptors. Neuropharmacology 1994;33(3-4):467-71.
55. Sakai K, Crochet S. A neural mechanism of sleep and wakefulness. Sleep Biol Rhyth 2003;1:29-42.
56. Monti JM, Monti D. Role of dorsal raphe nucleus serotonin 5-HT1A receptor in the regulation of REM sleep. Life Sci 2000;66(21):1999-2012.

57. Plotsky PM, Thrivikraman KV, Meaney MJ. Central and feedback regulation of hypothalamic corticotropin-releasing factor secretion. Ciba Foundation symposium 1993;172:59-84.

58. Swanson LW, Sawchenko PE, Rivier J, et al. Organization of ovine corticotropin-releasing factor immunoreactive cells and fibres in the rat brain: an immunohistochemical study. Neuroendocrinology 1983;36:165-186.

59. Sawchenko PE, Brown ER, Chan RK, et al. The paraventricular nucleus of the hypothalamus and the funtional neuroanatomy of visceromotor responses to stress. Prog Brain Res 1996;107:201-22.

60. Vale W, Spiess J, Rivier C, et al. Characterization of a 41-residue ovine hypothalamic peptide that stimulates secretion of corticotropin and beta-endorphin. Science 1981;213(4514):1394-7.

61. Whitnall MH. Regulation of the hypothalamic corticotropin-releasing hormone neurosecretory system. Prog Neurobiol 1993;40:573-629.

62. Gillies GE, Linton EA, Lowry PJ. Corticotropin releasing activity of the new CRF is potentiated several times by vasopressin. Nature 1982;299(5881):355-7.

63. Rivier C, Vale W. Modulation of stress-induced ACTH release by corticotropin-releasing factor, catecholamines and vasopressin. Nature 1983;305(5932):325-7.

64. Rivier C, Vale W. Interaction of corticotropin-releasing factor and arginine vasopressin on adrenocorticotropin secretion in vivo. Endocrinology 1983;113(3):939-42.

65. Gibbs DM. High concentrations of oxytocin in hypophysial portal plasma. Endocrinology 1984;114(4):1216-8.

66. Buma P, Nieuwenhuys R. Ultrastructural demonstration of oxytocin and vasopressin release sites in the neural lobe and median eminence of the rat by tannic acid and immunogold methods. Neurosci Lett 1987;74(2):151-7.

67. Gibbs DM, Vale W, Rivier J, et al. Oxytocin potentiates the ACTH-releasing activity of CRF(41) but not vasopressin. Life Sci 1984;34(23):2245-9.

68. Gibbs DM. Stress-specific modulation of ACTH secretion by oxytocin. Neuroendocrinology 1986;42(6):456-8.

69. Miyata S, Itoh T, Lin SH, et al. Temporal changes of c-fos expression in oxytocinergic magnocellular neuroendocrine cells of the rat hypothalamus with restraint stress. Brain Res Bull 1995;37(4):391-5.

70. Beaulieu S, Di Paolo T, Barden N. Control of ACTH secretion by the central nucleus of the amygdala: implication of the serotoninergic system and its relevance to the glucocorticoid delayed negative feedback mechanism. Neuroendocrinology 1986;44(2):247-54.

71. Roozendaal B, Koolhaas JM, Bohus B. Attenuated cardiovascular, neuroendocrine, and behavioral responses after a single footshock in central amygdaloid lesioned male rats. Physiol Behav 1991;50(4):771-5.

72. Van de Kar LD, Piechowski RA, Rittenhouse PA, et al. Amygdaloid lesions: differential effect on conditioned stress and immobilization-induced increases in corticosterone and renin secretion. Neuroendocrinology 1991;54(2):89-95.

73. Xu Y, Day TA, Buller KM. The Central amygdala modulates hypothalamic-pituitary-adrenal axis responses to systemic interleukin-1b administration. Neuroscience 1999;94(1):175-183.

74. Dayas CV, Buller KM, Day TA. Neuroendocrine responses to an emotional stressor: evidence for involvement of the medial but not the central amygdala. Euro. J. Neurosci. 1999;11:2312-2322.

75. Dunn JD. Plasma corticosterone responses to electrical stimulation of the bed nucleus of the stria terminalis. Brain Research 1987;407:327-331.
76. Crane JW, Buller KM, Day TA. Evidence that the bed nucleus of the stria terminalis contributes to the modulation of hypophysiotropic corticotropin-releasing factor cell responses to systemic interleukin-1beta. J Comp Neurol 2003;467(2):232-42.
77. Crane JW, Ebner K, Day TA. Medial prefrontal cortex suppression of the hypothalamic-pituitary-adrenal axis response to a physical stressor, systemic delivery of interleukin-1beta. Eur J Neurosci 2003;17(7):1473-81.
78. Gray TS, Piechowski RA, Yracheta JM, et al. Ibotenic acid lesions in the bed nucleus of the stria terminalis attenuate conditioned stress-induced increases in prolactin, ACTH and corticosterone. Neuroendocrinology 1993;57:517-524.
79. Feldman S, Conforti N. Subcortical pathways involved in the mediation of adrenocortical responses following frontal cortex stimulation. Neurosci Lett 1987;73(1):85-9.
80. Dunn JD. Plasma corticosterone responses to electrical stimulation of the medial frontal cortex. Neurosci Res Communic 1990;7(1):1-7.
81. Diorio D, Viau V, Meaney MJ. The role of the medial prefrontal cortex (cingulate gyrus) in the regulation of hypothalamic-pituitary-adrenal responses to stress. J Neurosci 1993;13(9):3839-47.
82. Herman JP, Cullinan WE, Watson SJ. Involvement of the Bed Nucleus of the Stria Terminalis in Tonic Regulation of Paraventricular Hypothalamic CRH and AVP mRNA Expression. J. Neuroendocrinology 1994;6:433-442.
83. Herman JP, Cullinan WE. Neurocircuitry of stress: central control of the hypothalamo-pituitary-adrenocortical axis. Trends Neurosci 1997;20(2):78-84.
84. Feldman S, Weidenfeld J. Neural mechanisms involved in the corticosteroid feedback effects on the hypothalamo-pituitary-adrenocortical axis. Prog. Neurobiol. 1995;45:129-141.
85. Buller K, Xu Y, Dayas C, et al. Dorsal and ventral medullary catecholamine cell groups contribute differentially to systemic interleukin-1beta-induced hypothalamic-pituitary-adrenal axis responses. Neuroendocrinology 2001;73(2):129-38.
86. Dayas CV, Buller KM, Crane JW, et al. Stressor categorization: acute physical and psychological stressors elicit distinctive recruitment patterns in the amygdala and in medullary noradrenergic cell groups. Eur J Neurosci 2001;14(7):1143-52.
87. Keller-Wood M. Fast feedback control of canine corticotrophin by cortisol. Endocrinology 1990;126:1959-1966.
88. Dallman MF, Akana SF, Cascio CS, et al. regulation of ACTH secretion: variations on a theme of B. Rec. Prog. Horm. Res. 1987;41:113-173.
89. Petrov T, Krukoff TL, Jhamandas JH. The hypothalamic paraventricular and lateral parabrachial nuclei receive collaterals from raphe nucleus neurons: a combined double retrograde and immunocytochemical study. J Comp Neurol 1992;318(1):18-26.
90. Petrov T, Krukoff TL, Jhamandas JH. Chemically defined collateral projections from the pons to the central nucleus of the amygdala and hypothalamic paraventricular nucleus in the rat. Cell Tissue Res 1994;277(2):289-95.
91. Liposits Z, Phelix C, Paull WK. Synaptic interaction of serotonergic axons and corticotropin releasing factor (CRF) synthesizing neurons in the hypothalamic paraventricular nucleus of the rat. A light and electron microscopic immunocytochemical study. Histochemistry 1987;86(6):541-9.

92. Bianchi M, Sacerdote P, Panerai AE. Fluoxetine reduces inflammatory edema in the rat: involvement of the pituitary-adrenal axis. Eur J Pharmacol 1994; 263(1-2):81-4.

93. von Bardeleben U, Steiger A, Gerken A, et al. Effects of fluoxetine upon pharmacoendocrine and sleep-EEG parameters in normal controls. Int Clin Psychopharmacol 1989;4 Suppl 1:1-5.

94. Reist C, Helmeste D, Albers L, et al. Serotonin indices and impulsivity in normal volunteers. Psychiatry Res 1996;60(2-3):177-84.

95. Mas M, Farre M, de la Torre R, et al. Cardiovascular and neuroendocrine effects and pharmacokinetics of 3, 4-methylenedioxymethamphetamine in humans. J Pharmacol Exp Ther 1999;290(1):136-45.

96. Grob CS, Poland RE, Chang L, et al. Psychobiologic effects of 3,4-methylenedioxymethamphetamine in humans: methodological considerations and preliminary observations. Behav Brain Res 1996;73(1-2):103-7.

97. Gundlah C, Pecins-Thompson M, Schutzer WE, et al. Ovarian steroid effects on serotonin 1A, 2A and 2C receptor mRNA in macaque hypothalamus. Brain Res Mol Brain Res 1999;63(2):325-39.

98. Li Q, Wichems C, Heils A, et al. Reduction in the density and expression, but not G-protein coupling, of serotonin receptors (5-HT1A) in 5-HT transporter knockout mice: gender and brain region differences. J Neurosci 2000;20(21):7888-95.

99. Li Q, Battaglia G, Van de Kar LD. Autoradiographic evidence for differential G-protein coupling of 5-HT1A receptors in rat brain: lack of effect of repeated injections of fluoxetine. Brain Res 1997;769(1):141-51.

100. Li Q, Muma NA, Battaglia G, et al. A desensitization of hypothalamic 5-HT1A receptors by repeated injections of paroxetine: reduction in the levels of G(i) and G(o) proteins and neuroendocrine responses, but not in the density of 5-HT1A receptors. J Pharmacol Exp Ther 1997;282(3):1581-90.

101. Vicentic A, Li Q, Battaglia G, et al. WAY-100635 inhibits 8-OH-DPAT-stimulated oxytocin, ACTH and corticosterone, but not prolactin secretion. Eur J Pharmacol 1998;346(2-3):261-6.

102. Li Q, Levy AD, Cabrera TM, et al. Long-term fluoxetine, but not desipramine, inhibits the ACTH and oxytocin responses to the 5-HT1A agonist, 8-OH-DPAT, in male rats. Brain Res 1993;630(1-2):148-56.

103. Lesch KP, Sohnle K, Poten B, et al. Corticotropin and cortisol secretion after central 5-hydroxytryptamine-1A (5-HT1A) receptor activation: effects of 5-HT receptor and beta-adrenoceptor antagonists. J Clin Endocrinol Metab 1990;70(3):670-4.

104. Lesch KP, Mayer S, Disselkamp-Tietze J, et al. 5-HT1A receptor responsivity in unipolar depression. Evaluation of ipsapirone-induced ACTH and cortisol secretion in patients and controls. Biol Psychiatry 1990;28(7):620-8.

105. Sargent P, Williamson DJ, Pearson G, et al. Effect of paroxetine and nefazodone on 5-HT1A receptor sensitivity. Psychopharmacology (Berl) 1997;132(3): 296-302.

106. Urban JH, Van de Kar LD, Lorens SA, et al. Effect of the anxiolytic drug buspirone on prolactin and corticosterone secretion in stressed and unstressed rats. Pharmacol Biochem Behav 1986;25(2):457-62.

107. Bagdy G. Role of the hypothalamic paraventricular nucleus in 5-HT$_{1A}$, 5-HT$_{2A}$ and 5-HT$_{2C}$ receptor-mediated oxytocin, prolactin and ACTH/corticosterone responses. Behav Brain Res 1996;73:277-280.

108. Bagdy G. Studies on the sites and mechanisms of 5-HT1A receptor-mediated in vivo actions. Acta Physiol Hung 1996;84(4):399-401.
109. Osei-Owusu P, Sullivan NR, Van de Kar LD, et al. 5-HT1A receptors in the paraventricular nucleus of the hypothalamus mediate endocrine and behavioral but not cardiovascular responses to 8-OH-DPAT. FASEB J 2003;17:A445.
110. Feldman S, Weidenfeld J. Involvement of endogeneous glutamate in the stimulatory effect of norepinephrine and serotonin on the hypothalamo-pituitary-adrenocortical axis. Neuroendocrinology 2004;79(1):43-53.
111. Serres F, Li Q, Garcia F, et al. Evidence that G(z)-proteins couple to hypothalamic 5-HT(1A) receptors in vivo. J Neurosci 2000;20(9):3095-103.
112. Appel NM, Mitchell WM, Garlick RK, et al. Autoradiographic characterization of (+−)-1-(2,5-dimethoxy-4-[125I] iodophenyl)-2-aminopropane ([125I]DOI) binding to 5-HT2 and 5-HT1c receptors in rat brain. J Pharmacol Exp Ther 1990;255(2):843-57.
113. Zhang Y, Damjanoska KJ, Carrasco GA, et al. Evidence that 5-HT2A receptors in the hypothalamic paraventricular nucleus mediate neuroendocrine responses to (-)DOI. J Neurosci 2002;22(21):9635-42.
114. Owens MJ, Knight DL, Ritchie JC, et al. The 5-hydroxytryptamine2 agonist, (+−)-1-(2,5-dimethoxy-4-bromophenyl)-2-aminopropane stimulates the hypothalamic-pituitary-adrenal (HPA) axis. I. Acute effects on HPA axis activity and corticotropin-releasing factor-containing neurons in the rat brain. J Pharmacol Exp Ther 1991;256(2):787-94.
115. Koenig JI, Gudelsky GA, Meltzer HY. Stimulation of corticosterone and beta-endorphin secretion in the rat by selective 5-HT receptor subtype activation. Eur J Pharmacol 1987;137(1):1-8.
116. King BH, Brazell C, Dourish CT, et al. MK-212 increases rat plasma ACTH concentration by activation of the 5-HT1C receptor subtype. Neurosci Lett 1989;105(1-2):174-6.
117. Fuller RW, Snoddy HD. Serotonin receptor subtypes involved in the elevation of serum corticosterone concentration in rats by direct- and indirect-acting serotonin agonists. Neuroendocrinology 1990;52(2):206-11.
118. Rittenhouse PA, Bakkum EA, Levy AD, et al. Evidence that ACTH secretion is regulated by serotonin2A/2C (5-HT2A/2C) receptors. J Pharmacol Exp Ther 1994;271(3):1647-55.
119. Van de Kar LD, Javed A, Zhang Y, et al. 5-HT2A receptors stimulate ACTH, corticosterone, oxytocin, renin, and prolactin release and activate hypothalamic CRF and oxytocin-expressing cells. J Neurosci 2001;21(10):3572-9.
120. Gronfier C, Simon C, Piquard F, et al. Neuroendocrine processes underlying ultradian sleep regulation in man. J Clin Endocrinol Metab 1999;84(8):2686-90.
121. Holsboer F, von Bardeleben U, Steiger A. Effects of intravenous corticotropin-releasing hormone upon sleep-related growth hormone surge and sleep EEG in man. Neuroendocrinology 1988;48(1):32-8.
122. Ehlers CL, Reed TK, Henriksen SJ. Effects of corticotropin-releasing factor and growth hormone-releasing factor on sleep and activity in rats. Neuroendocrinology 1986;42(6):467-74.
123. Chang FC, Opp MR. Blockade of corticotropin-releasing hormone receptors reduces spontaneous waking in the rat. Am J Physiol 1998;275(3 Pt 2):R793-802.
124. Chang FC, Opp MR. A corticotropin-releasing hormone antisense oligodeoxynucleotide reduces spontaneous waking in the rat. Regul Pept 2004;117(1):43-52.

125. Gillin JC, Jacobs LS, Snyder F, et al. Effects of ACTH on the sleep of normal subjects and patients with Addison's disease. Neuroendocrinology 1974;15(1): 21-31.
126. Steiger A, Guldner J, Knisatschek H, et al. Effects of an ACTH/MSH(4-9) analog (HOE 427) on the sleep EEG and nocturnal hormonal secretion in humans. Peptides 1991;12(5):1007-10.
127. Born J, DeKloet ER, Wenz H, et al. Gluco- and antimineralocorticoid effects on human sleep: a role of central corticosteroid receptors. Am J Physiol 1991;260 (2 Pt 1):E183-8.
128. Friess E, U VB, Wiedemann K, et al. Effects of pulsatile cortisol infusion on sleep-EEG and nocturnal growth hormone release in healthy men. J Sleep Res 1994;3(2):73-79.
129. Steiger A. Sleep and the hypothalamo-pituitary-adrenocortical system. Sleep Med Rev 2002;6(2):125-38.
130. Orosco M, Rouch C, De Saint-Hilaire Z, et al. Dynamic changes in hypothalamic monoamines during sleep/wake cycles assessed by parallel EEG and microdialysis in the rat. J Sleep Res 1995;4(3):144-149.
131. Ludwig M. Dendritic release of vasopressin and oxytocin. J Neuroendocrinol 1998;10(12):881-95.
132. Sabatier N, Caquineau C, Douglas AJ, et al. Oxytocin released from magnocellular dendrites: a potential modulator of alpha-melanocyte-stimulating hormone behavioral actions? Ann N Y Acad Sci 2003;994:218-24.
133. Bealer SL, Crowley WR. Neurotransmitter interaction in release of intranuclear oxytocin in magnocellular nuclei of the hypothalamus. Ann N Y Acad Sci 1999; 897:182-91.
134. Sawchenko PE, Swanson LW, Steinbusch HW, et al. The distribution and cells of origin of serotonergic inputs to the paraventricular and supraoptic nuclei of the rat. Brain Res 1983;277(2):355-60.
135. Saphier D. Paraventricular nucleus magnocellular neuronal responses following electrical stimulation of the midbrain dorsal raphe. Exp Brain Res 1991;85(2): 359-63.
136. Van de Kar LD, Rittenhouse PA, Li Q, et al. Hypothalamic paraventricular, but not supraoptic neurons, mediate the serotonergic stimulation of oxytocin secretion. Brain Res Bull 1995;36(1):45-50.
137. Saydoff JA, Carnes M, Brownfield MS. The role of serotonergic neurons in intravenous hypertonic saline-induced secretion of vasopressin, oxytocin, and ACTH. Brain Res Bull 1993;32:567-572.
138. Bagdy G, Kalogeras KT. Stimulation of 5-HT1A and 5-HT2/5-HT1C receptors induce oxytocin release in the male rat. Brain Res 1993;611(2):330-2.
139. Wright DE, Seroogy KB, Lundgren KH, et al. Comparative localization of serotonin$_{1A,\ 1C}$ and $_2$receptor subtypes mRNA in rat brain. J Comp Neurol 1995; 351:357-373.
140. Lancel M, Kromer S, Neumann ID. Intracerebral oxytocin modulates sleep-wake behaviour in male rats. Regul Pept 2003;114(2-3):145-52.
141. Neumann ID, Kromer SA, Toschi N, et al. Brain oxytocin inhibits the (re)activity of the hypothalamo-pituitary-adrenal axis in male rats: involvement of hypothalamic and limbic brain regions. Regul Pept 2000;96(1-2):31-8.
142. Emiliano AB, Fudge JL. From galactorrhea to osteopenia: rethinking serotonin-prolactin interactions. Neuropsychopharmacology 2004;29(5):833-46.

143. Samson WK, Taylor MM, Baker JR. Prolactin-releasing peptides. Regul Pept 2003;114(1):1-5.
144. Freeman ME, Kanyicska B, Lerant A, et al. Prolactin: structure, function, and regulation of secretion. Physiol Rev 2000;80(4):1523-631.
145. Mogg RJ, Samson WK. Interactions of dopaminergic and peptidergic factors in the control of prolactin release. Endocrinology 1990;126(2):728-35.
146. Breton C, Pechoux C, Morel G, et al. Oxytocin receptor messenger ribonucleic acid: characterization, regulation, and cellular localization in the rat pituitary gland. Endocrinology 1995;136(7):2928-36.
147. Kamberi IA, Mical RS, Porter JC. Effects of melatonin and serotonin on the release of FSH and prolactin. Endocrinology 1971;88(6):1288-93.
148. Rittenhouse PA, Levy AD, Li Q, et al. Neurons in the hypothalamic paraventricular nucleus mediate the serotonergic stimulation of prolactin secretion via 5-HT1c/2 receptors. Endocrinology 1993;133(2):661-7.
149. Quattrone A, Schettini G, Di Renzo G, et al. Effect of midbrain raphe lesion or 5,7-dihydroxytryptamine treatment on the prolactin-releasing action of quipazine and D-fenfluramine in rats. Brain Res 1979;174:71-79.
150. Van de Kar LD, Bethea CL. Pharmacological evidence that serotonergic stimulation of prolactin secretion is mediated via the dorsal raphe nucleus. Neuroendocrinology 1982;35(4):225-30.
151. Barofsky AL, Nash JF, Meltzer HY. Dorsal raphe-hypothalamic projections provide the stimulatory serotonergic input to suckling-induced prolactin release. Endocrinology 1983;113:1894-1903.
152. Bagdy G, Makara GB. Hypothalamic paraventricular nucleus lesions differentially affect serotonin-1A (5-HT1A) and 5-HT2 receptor agonist-induced oxytocin, prolactin, and corticosterone responses. Endocrinology 1994;134(3):1127-31.
153. Lesch KP. 5-HT1A receptor responsivity in anxiety disorders and depression. Prog Neuropsychopharmacol Biol Psychiatry 1991;15(6):723-33.
154. Lesch KP, Rupprecht R, Poten B, et al. Endocrine responses to 5-hydroxytryptamine-1A receptor activation by ipsapirone in humans. Biol Psychiatry 1989;26(2):203-5.
155. Dinan TG, Barry S, Yatham LN, et al. The reproducibility of the prolactin response to buspirone: relationship to the menstrual cycle. Int Clin Psychopharmacol 1990;5(2):119-23.
156. Cowen PJ, Anderson IM, Grahame-Smith DG. Neuroendocrine effects of azapirones. J Clin Psychopharmacol 1990;10(3 Suppl):21S-25S.
157. Anderson IM, Cowen PJ. Effect of pindolol on endocrine and temperature responses to buspirone in healthy volunteers. Psychopharmacology (Berl) 1992;106(3):428-32.
158. Di Sciullo A, Bluet-Pajot MT, Mounier F, et al. Changes in anterior pituitary hormone levels after serotonin 1A receptor stimulation. Endocrinology 1990;127(2):567-72.
159. Kellar KJ, Hulihan-Giblin BA, Mulroney SE, et al. Stimulation of serotonin1A receptors increases release of prolactin in the rat. Neuropharmacology 1992;31(7):643-7.
160. Nash JF, Meltzer HY. Effect of gepirone and ipsapirone on the stimulated and unstimulated secretion of prolactin in the rat. J Pharmacol Exp Ther 1989;249(1):236-41.

161. Simonovic M, Gudelsky GA, Meltzer HY. Effect of 8-hydroxy-2-(di-n-propylamino) tetralin on rat prolactin secretion. J Neural Transm 1984;59(2):143-9.

162. Van de Kar LD, Lorens SA, Urban JH, et al. Effect of selective serotonin (5-HT) agonists and 5-HT2 antagonist on prolactin secretion. Neuropharmacology 1989;28(3):299-305.

163. Obal F, Jr., Opp M, Cady AB, et al. Prolactin, vasoactive intestinal peptide, and peptide histidine methionine elicit selective increases in REM sleep in rabbits. Brain Res 1989;490(2):292-300.

164. Roky R, Valatx JL, Jouvet M. Effect of prolactin on the sleep-wake cycle in the rat. Neurosci Lett 1993;156(1-2):117-20.

165. Obal F, Jr., Payne L, Kacsoh B, et al. Involvement of prolactin in the REM sleep-promoting activity of systemic vasoactive intestinal peptide (VIP). Brain Res 1994;645(1-2):143-9.

166. Obal F, Jr., Kacsoh B, Bredow S, et al. Sleep in rats rendered chronically hyperprolactinemic with anterior pituitary grafts. Brain Res 1997;755(1):130-6.

167. Zhang SQ, Kimura M, Inoue S. Effects of prolactin-releasing peptide (PrRP) on sleep regulation in rats. Psychiatry Clin Neurosci 2000;54(3):262-4.

168. Obal F, Jr., Kacsoh B, Alfoldi P, et al. Antiserum to prolactin decreases rapid eye movement sleep (REM sleep) in the male rat. Physiol Behav 1992;52(6):1063-8.

169. Roky R, Valatx JL, Paut-Pagano L, et al. Hypothalamic injection of prolactin or its antibody alters the rat sleep-wake cycle. Physiol Behav 1994;55(6):1015-9.

170. Bodosi B, Obal F, Jr., Gardi J, et al. An ether stressor increases REM sleep in rats: possible role of prolactin. Am J Physiol Regul Integr Comp Physiol 2000;279(5):R1590-8.

171. Walsh RJ, Slaby FJ, Posner BI. A receptor-mediated mechanism for the transport of prolactin from blood to cerebrospinal fluid. Endocrinology 1987; 120(5):1846-50.

172. Roky R, Obal F, Jr., Valatx JL, et al. Prolactin and rapid eye movement sleep regulation. Sleep 1995;18(7):536-42.

173. Muller EE, Locatelli V, Cocchi D. Neuroendocrine control of growth hormone secretion. Physiol Rev 1999;79(2):511-607.

174. Lin-Su K, Wajnrajch MP. Growth Hormone Releasing Hormone (GHRH) and the GHRH Receptor. Rev Endocr Metab Disord 2002;3(4):313-23.

175. Romero MI, Phelps CJ. Identification of growth hormone-releasing hormone and somatostatin neurons projecting to the median eminence in normal and growth hormone-deficient Ames dwarf mice. Neuroendocrinology 1997;65(2): 107-16.

176. Merchenthaler I, Vigh S, Schally AV, et al. Immunocytochemical localization of growth hormone-releasing factor in the rat hypothalamus. Endocrinology 1984; 114(4):1082-5.

177. Kawano H, Daikoku S. Somatostatin-containing neuron systems in the rat hypothalamus: retrograde tracing and immunohistochemical studies. J Comp Neurol 1988;271(2):293-9.

178. Smith CE, Ware CJ, Cowen PJ. Pindolol decreases prolactin and growth hormone responses to intravenous L-tryptophan. Psychopharmacology (Berl) 1991; 103(1):140-2.

179. Smythe GA, Lazarus L. Growth hormone regulation by melatonin and serotonin. Nature 1973;244(5413):230-1.

180. Vijayan E, Krulich L, McCann SM. Stimulation of growth hormone release by intraventricular administration of 5HT or quipazine in unanesthetized male rats. Proc Soc Exp Biol Med 1978;159(2):210-2.
181. Arnold MA, Fernstrom JD. L-Tryptophan injection enhances pulsatile growth hormone secretion in the rat. Endocrinology 1981;108(1):331-5.
182. Katz LM, Nathan L, Kuhn CM, et al. Inhibition of GH in maternal separation may be mediated through altered serotonergic activity at 5-HT2A and 5-HT2C receptors. Psychoneuroendocrinology 1996;21(2):219-35.
183. Nakayama T, Suhara T, Okubo Y, et al. In vivo drug action of tandospirone at 5-HT1A receptor examined using positron emission tomography and neuroendocrine response. Psychopharmacology (Berl) 2002;165(1):37-42.
184. Newman ME, Li Q, Gelfin Y, et al. Low doses of ipsapirone increase growth hormone but not oxytocin secretion in normal male and female subjects. Psychopharmacology (Berl) 1999;145(1):99-104.
185. Pitchot W, Wauthy J, Hansenne M, et al. Hormonal and temperature responses to the 5-HT1A receptor agonist flesinoxan in normal volunteers. Psychopharmacology (Berl) 2002;164(1):27-32.
186. Pitchot W, Wauthy J, Legros JJ, et al. Hormonal and temperature responses to flesinoxan in normal volunteers: an antagonist study. Eur Neuropsychopharmacol 2004;14(2):151-5.
187. McAllister-Williams RH, Massey AE. EEG effects of buspirone and pindolol: a method of examining 5-HT1A receptor function in humans. Psychopharmacology (Berl) 2003;166(3):284-93.
188. Meller E, Bohmaker K. Differential receptor reserve for 5-HT1A receptor-mediated regulation of plasma neuroendocrine hormones. J Pharmacol Exp Ther 1994;271(3):1246-52.
189. Murakami Y, Kato Y, Kabayama Y, et al. Involvement of growth hormone (GH)-releasing factor in GH secretion induced by serotoninergic mechanisms in conscious rats. Endocrinology 1986;119(3):1089-92.
190. Uvnas-Moberg K, Hillegaart V, Alster P, et al. Effects of 5-HT agonists, selective for different receptor subtypes, on oxytocin, CCK, gastrin and somatostatin plasma levels in the rat. Neuropharmacology 1996;35(11):1635-40.
191. Bjorkstrand E, Ahlenius S, Smedh U, et al. The oxytocin receptor antagonist 1-deamino-2-D-Tyr-(OEt)-4-Thr-8-Orn-oxytocin inhibits effects of the 5-HT1A receptor agonist 8-OH-DPAT on plasma levels of insulin, cholecystokinin and somatostatin. Regul Pept 1996;63(1):47-52.
192. Steiger A. Sleep and endocrinology. J Intern Med 2003;254(1):13-22.
193. Steiger A, Holsboer F. Neuropeptides and human sleep. Sleep 1997;20(11): 1038-52.
194. Krueger JM, Fang J, Hansen MK, et al. Humoral Regulation of Sleep. News Physiol Sci 1998;13:189-194.
195. Perras B, Marshall L, Kohler G, et al. Sleep and endocrine changes after intranasal administration of growth hormone-releasing hormone in young and aged humans. Psychoneuroendocrinology 1999;24(7):743-57.
196. Steiger A, Guldner J, Hemmeter U, et al. Effects of growth hormone-releasing hormone and somatostatin on sleep EEG and nocturnal hormone secretion in male controls. Neuroendocrinology 1992;56(4):566-73.

197. Obal F, Jr., Alfoldi P, Cady AB, et al. Growth hormone-releasing factor enhances sleep in rats and rabbits. Am J Physiol 1988;255(2 Pt 2):R310-6.
198. Kerkhofs M, Van Cauter E, Van Onderbergen A, et al. Sleep-promoting effects of growth hormone-releasing hormone in normal men. Am J Physiol 1993;264 (4 Pt 1):E594-8.
199. Obal F, Jr., Floyd R, Kapas L, et al. Effects of systemic GHRH on sleep in intact and hypophysectomized rats. Am J Physiol 1996;270(2 Pt 1):E230-7.
200. Obal F, Jr., Alt J, Taishi P, et al. Sleep in mice with nonfunctional growth hormone-releasing hormone receptors. Am J Physiol Regul Integr Comp Physiol 2003;284(1):R131-9.

Locus Coeruleus and Adrenergic Modulation of Rapid Eye Movement Sleep

BIRENDRA NATH MALLICK, VIBHA MADAN AND DINESH PAL

1. Introduction

Neurons in the brain regulate different functions of the body through the release of a variety of chemicals, the neurotransmitters. These neurotransmitters and neurochemicals may be broadly divided into two main classes. The small-molecule neurotransmitters such as acetylcholine, biogenic amines, amino acids viz. gamma aminobutyric acid (GABA), glycine, glutamate and also several neuroactive peptides which are short chains of amino acids that may act like neurohormones. The term biogenic amine has been used for certain neurotransmitters that include the catecholamines, for example, dopamine, epinephrine, norepinephrine, derived from the amino acid tyrosine and serotonin, derived from the amino acid tryptophan.

Norepinephrine (NE) was first identified in the 1930s, as a neurotransmitter whose actions resembled that of the stress hormone epinephrine. It plays an important role in the well known "fight-or-flight response", that accompanies the adrenaline rush one experiences when the brain senses a stressful situation and mobilizes the body for quicker reaction. Deficiencies and excesses of NE have also been linked to mood disorders, the former with depression while the latter with agitated, maniac states. NE-dependent modulation of long-term alterations in synaptic strength, gene transcription and other processes suggest a potentially critical role of this neurotransmitter system in experience-dependent alterations in neural function and behavior.

The locus coeruleus (LC), located in the dorso-lateral pontine central gray region in the brainstem, is the primary site possessing, depending on species, exclusively or predominantly NE-ergic neurons. Through a widespread efferent projection system, the LC-NE-ergic system supplies a major portion (more than 60%) of NE throughout the central nervous system. Initial studies provided critical insights into the basic organization and properties of this system. More recent work identifies a complicated array of behavioral and electrophysiological actions that have in common the facilitation of processing of relevant or salient information. This involves two basic levels of action. First, the system contributes to the initiation and maintenance of behavioral and forebrain neuronal activity states appropriate for the collection of sensory information (e.g. waking). Second, within the waking state, this system modulates the collection and processing of salient sensory information through diverse concentration-dependent actions within cortical and subcortical sensory, attention and memory circuits. The ability of a given stimulus to increase discharge of LC-NE-ergic neurons appears independent of affective value (appetitive vs. aversive). Combined, these observations suggest that the LC-noradrenergic system is a critical component of the central neural architecture supporting interaction with and navigation through the complex world.

2. Modulation of Sleep-Wakefulness and Corresponding Changes in the Eeg by Adrenoceptor Agonist and Antagonist

Sleep-wakefulness has been objectively identified and classified into different states (Figure 1) by the presence of associated characteristic signals in the cortical electroencephalogram (EEG), electrooculogram (EOG), electromyogram (EMG) and hippocampal (Hipp) waves. Our understanding of the effects and mechanism of action of different neurotransmitters on sleep-wakefulness has largely come from the use of agonist and antagonist of neurotransmitter(s) at least on the EEG.

The presence of adrenergic receptors in the brain has been shown since long.[1,2,3,4] The $\alpha2$-agonist, clonidine, when injected intraperitoneally, was found to reduce REM sleep in rats and cats.[5,6] A similar decrease in REM sleep has been observed in man with a dose about five times smaller than that used in the rat.[7] The $\alpha2$-antagonist, yohimbine, increased active wakefulness immediately after administration but did not affect REM sleep.[8] Systemic application of $\alpha1$-adrenergic agonist, methoxamine, increased active wakefulness and decreased sleep as well as REM sleep,[9] while $\alpha1$-antagonist, prazosin, increased REM sleep.[8] The $\beta1$-antagonist increased waking and decreased REM sleep while β-1 agonist increased REM sleep.[10] In another study prazosin administration was found to slightly modify waking and sleep while REM sleep was significantly reduced in a dose dependent manner.[9] Oral administration of prazosin in rats was found to shorten quiet waking and REM sleep while it increased active waking and slow wave sleep.[11] Although systemic application of the drugs suggested that noradrenergic agonist and antagonist might modulate sleep-wakefulness and REM sleep, their central site of action could not be commented upon. Also, systemic injections might affect sleep-waking secondary to peripheral changes e.g. blood pressure. However, the drugs are likely to act at the central site may be inferred from the observations that the EEG desynchronization induced by stimulation of the brainstem reticular formation in cats could be prevented by i.p. injection of adrenoceptor and cholinergic antagonists.[12]

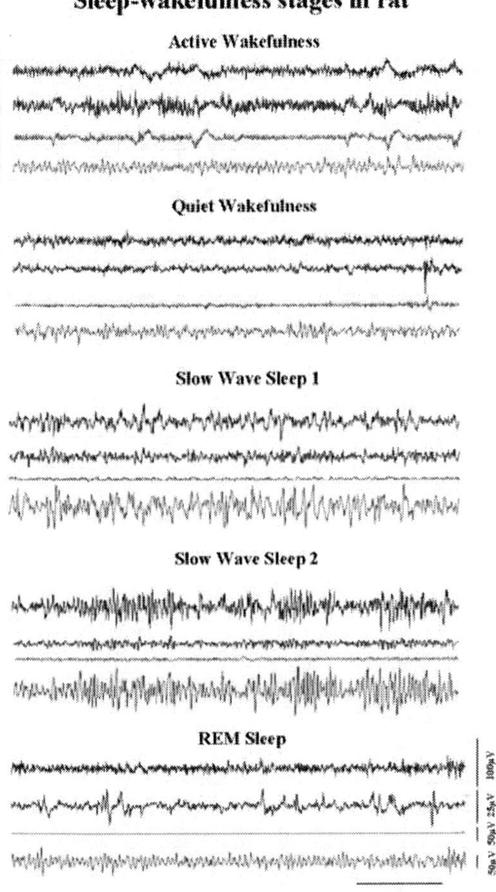

FIGURE 1. This figure shows polygraphic traces of simultaneous recording of electroencephalogram (EEG), electrooculogram (EOG), electromyogram (EMG) and hippocampal theta waves (hipp) in a freely moving rat. The electrophysiological signals are used to classify the sleep-wakefulness into five stages.

Localized bilateral injections of the β-agonist isoproterenol into the dorsal pontine tegmentum of cat reduced REM sleep while the β-antagonist propranolol consistently enhanced it, largely because of an increased number of REM sleep episodes.[13] Bilateral injections of α2-agonist clonidine into the dorsal pontine tegmentum of cat produced an almost complete suppression of REM sleep.[14] The α1-agonist methoxamine, when injected into the dorsal pontine tegmentum of cat produced significant decrease in REM sleep.

The decrease in the total REM sleep was found to be due to both an increase in REM sleep latency and a reduction in number and duration of

REM sleep episodes.[15] Local application of noradrenergic agonist and antagonist into the medial preoptic area also modulated the sleep-waking pattern in freely behaving rats. It was concluded that in the medial preoptic area the noradrenaline modulated sleep and wakefulness by acting on $\alpha 2$- and β-adrenoceptors, respectively.[16] It was further supported by the fact that microinjection of β-agonist isoproterenol into medial septal region of basal forebrain significantly increased the time spent in wakefulness and a near complete suppression of REM sleep.[17]

Microdialysis application of both epinephrine and NE into the caudal part of peri-LCα produced a marked decrease in REM sleep.[18] Both epinephrine and NE caused a dose dependent inhibition of REM sleep and induction of REM sleep without atonia when applied to the caudal part of the peri-LCα. These effects were also produced by clonidine, an $\alpha 2$-agonist while $\alpha 2$-antagonists were found to block the effect of NE. When co-applied with carbachol into the caudal peri-LCα, clonidine completely blocked the marked REM sleep inducing effect of carbachol.[19] Thus, NE agonist and antagonist infused systemically or locally into specific regions in the brain in general increased wakefulness, reduced sleep and REM sleep, however, to various degrees. One of the complexities in such studies is that sleep and wakefulness stages have been divided into multiple sub-stages leading to variations in results. However, the effect on REM sleep is relatively more consistent than that on other sleep-waking states and also depended on the extent of LC getting affected. As mentioned above since the brain receives maximum amount of NE from the neurons located in the LC, their behavior and regulation across sleep-waking, which might modulate the levels of NE in the brain, would be discussed subsequently with emphasis on REM sleep.

3. Behavior and Modulation of Locus Coeruleus Neurons across Sleep-Wakefulness

Electrolytic lesions of the dorsal noradrenergic bundle that ascends from the LC[20] resulted in increased slow wave sleep as well as REM sleep.[21] Destruction of the ventral part of LC, the LCα and peri-LCα was followed by irreversible disappearance of REM sleep atonia.[22-25] Destruction of extended area involving LC principal, LCα and peri-LCα suppressed REM sleep completely during the two post-lesion months,[26] however, a similar destruction of the dorsal part of LC did not suppress the occurrence of REM sleep.[27] Carbachol injections in the region of the LC or subcoeruleus induced a cataplectic-like behavior.[28,29,30] Further, reversible inactivation of the LC proper, the anterior part of the LC, by localized cooling induced slow wave sleep followed by REM sleep.[31] These studies showed that loss of LC neurons lead to an increase in both non-REM sleep as well as REM sleep. However, the intensity of effects differed which could be due to the extent of damage to the LC. Since maximum NE is released from the LC neurons, the results suggested that in normal circumstances NE released from the LC neurons might inhibit sleep and REM sleep.

4. Specificity of NE-Ergic Neurons in LC to REM Sleep

To confirm the behavior of the LC neurons during sleep-waking states, single neuronal activities from freely behaving cats and rats were correlated with behavioral as well as electrophysiological correlates of sleep-wakefulness. It was observed that the LC-neurons were active throughout waking as well as sleeping except during REM sleep when they ceased firing.[25,32-34] Because of such REM sleep related behavior these NE-ergic neurons in the LC were termed as REM-OFF neurons. The specificity of these neurons to REM sleep was further confirmed by the fact that they continue to be active during REM sleep deprivation.[35] Also, these neurons ceased firing during carbachol induced REM sleep, however, they behaved similar to spontaneous wakefulness if wakefulness was induced by stimulation of the midbrain reticular formation.[36]

These results suggest that at least in the brain there would be reduced NE release during REM sleep (due to cessation of the REM-OFF neurons) and during REM sleep deprivation, due to non-cessation of those neurons, there would be increased levels of NE in the brain. Also, the increased levels of NE in the brain during REM sleep deprivation might be responsible for at least REM sleep deprivation induced changes in the brain and behavior. The possible mechanism of regulation of the REM-OFF neurons and their role in REM sleep related physiological functions would be further discussed.

5. Modulation of REM-OFF Neuronal Activity

As mentioned above the LC-neurons cease firing during REM sleep and they continue firing during REM sleep deprivation. Based on these observations it was hypothesized that cessation of the LC-neurons is possibly a pre-requisite for generation of REM sleep.

To confirm, the neurons in the LC were kept active for extended period by bilateral electrical stimulation of the LC with low intensity and low frequency electrical stimulation in freely moving normally behaving rats and the effects on sleep-waking were studied. It was observed that there was significant reduction in REM sleep during the period of stimulation, however, during the post-stimulation period there was a rebound increase in REM sleep.[37]

The finding was similar to that of instrumental REM sleep deprivation where after REM sleep deprivation, during the recovery phase, a rebound increase in REM sleep has been reported.[38] The advantage of this study, unlike that of lesion study, was that the recovery function of inactivation of the LC could be studied. Although the study suggested that the LC-neurons must cease firing for the generation of REM sleep, the mechanism and neurotransmitter involved for such action needed further investigations.

6. Mechanism of Cessation of Locus Coeruleus Neurons

In addition to that of the REM-OFF neurons, isolated independent studies identified another group of neurons in the brainstem, whose firing rate increase during REM sleep. These neurons are presumably cholinergic and have been termed as REM-ON neurons.[39,40] It was proposed that these REM-ON neurons are likely to keep the REM-OFF neurons inhibited during REM sleep,[39,41] however, the mechanism of inhibition was unknown.

Since the propositions were based on isolated and independent studies it was confirmed by recording both the REM-ON and REM-OFF neurons simultaneously in the same animal. It was observed that respective increase and decrease in firing rate of these two types of neurons temporally matched fairly well supporting the reciprocal inhibition hypothesis,[42] however, the mechanism of inhibition still remained to be investigated.

7. Role of GABA in Inhibiting the REM-OFF Neuronal Activity

Based on reciprocal inhibition models[39-41] acetylcholine released from the REM-ON neurons should inhibit the NE-ergic REM-OFF neurons for the generation of REM sleep. However, cholinergic agonist, carbachol, was found to depolarise and not hyperpolarise the LC neurons.[43] Hence, it is unlikely that acetylcholine would inhibit the REM-OFF neurons. Therefore, it was hypothesised that the excitatory cholinergic input, presumably from the REM-ON neurons, might be converted into an inhibitory signal by an intermediate inhibitory neurotransmitter, GABA.[44] Subsequently, it was confirmed by double microinjection neuro-microanatomo-pharmaco-physio- behavioral studies in freely moving normally behaving rats.[45] In short, microinjection of either cholinergic or GABA-ergic agonist into the LC increased, while their respective antagonist decreased REM sleep. Further, if injection of GABA was preceded by cholinergic antagonist, REM sleep was increased, however, if microinjection of cholinergic agonist was preceded by that of GABA-ergic antagonist, REM sleep was reduced and microinjection of both the antagonists also reduced REM sleep. These findings suggested that in the LC cholinergic input is likely to mediate its effect through GABA interneurons for the regulation of REM sleep.[45,46] Further studies showed that, in addition to possible local interneurons, the LC may receive GABA-ergic inputs from prepositus hypoglossus nucleus in the brain stem as well[47] which receives inputs from cholinergic LDT/PPT area, the site of REM-ON neurons.[48] Thus, the studies showed that the NE-ergic REM-OFF neurons in the LC are neurochemically controlled by GABA-ergic neurons, which in turn are likely to be modulated by the cholinergic REM-ON neurons.

8. Role of Sleep and Waking Areas in Controlling REM-OFF Neuronal Activity

Normally the REM sleep is expressed after a subject has spent sometime in sleep, and it does not follow wakefulness. This suggests that the REM sleep generating mechanism is likely to be modulated by the sleep and waking areas in the brain. Hence, the influence of sleep and waking areas in the brain on REM-ON and REM-OFF neurons was studied. The sleep and waking areas in the brainstem and in the basal forebrain were electrophysiologically identified in freely moving normally behaving cats. Subsequently, single neuronal activity including REM-ON and REM-OFF neurons as well as electrophysiological signals characterizing sleep-wakefulness and REM sleep were recorded simultaneously in the same animal. Thereafter, those sleep and waking areas were stimulated and the influence studied on the pre-identified REM-ON and REM-OFF neurons. It was observed that the brainstem waking area excited the REM-OFF neurons and inhibited the REM-ON neurons.[49] On the other hand, the brainstem hypnogenic area stimulated the REM-ON neurons while the hypothalamic hypnogenic area was not very effective.[50]

The results of the studies mentioned so far may be summarized as follows. Normally during wakefulness the brainstem wake active neurons directly activate the REM-OFF neurons and inhibit the REM-ON neurons directly or indirectly through the activation of the REM-OFF neurons. The firing of the REM-OFF neurons keeps releasing NE in the brain. Gradually during sleep the wake active neurons reduce firing causing withdrawal of excitation from the REM-OFF neurons resulting in slowing of those neurons. This slowing also withdraws inhibition (disinhibition) from the REM-ON neurons. Finally, after certain length of sleep, when possibly certain unknown conditions are fulfilled, the brainstem sleep area excites the REM-ON neurons, which in turn activate GABA-ergic neurons that shut off the REM-OFF neurons by actively inhibiting them to initiate REM sleep (Figure 2). Since the REM-OFF neurons cease firing there should be reduced NE levels in the brain during spontaneous REM sleep, however, if REM sleep is not allowed, REM-OFF neurons do not cease firing or vice versa.

Thus, during REM sleep deprivation there is increased levels of NE in the brain. Increased NE in the brain may also be supported by the fact that after REM sleep deprivation there is increased tyrosine hydroxylase activity[51] and decreased monoamine oxidase A activity.[52]

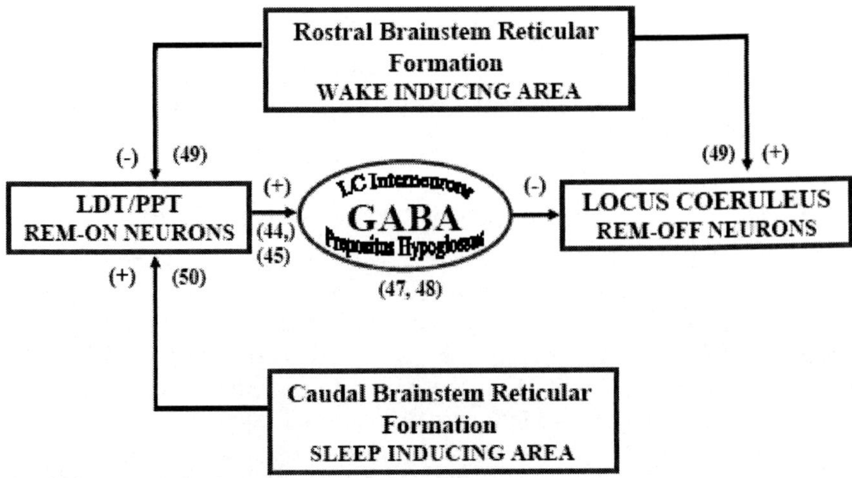

FIGURE 2. Based on the findings from isolated studies, as represented by the reference numbers in parenthesis, the connections between neurons for the regulation of REM sleep have been schematically shown in this figure. Abb: GABA – gamma amino butyric acid, LC – locus coeruleus, LDT/PPT – laterodorsal tegmentum/pedunculopontine tegmentum, REM-OFF – REM-OFF neurons, REM-ON,-REM ON neurons, (+) – excitation, (–) – inhibition.

Also, there is increased level of NE,[53] and increased synthesis of tyrosine hydroxylase-mRNA[54] as well as tyrosine hydroxylase enzyme within the NE-ergic neurons[55] after REM sleep deprivation.

9. Increased Norepinephrine might be Responsible for REM Sleep Deprivation-Induced Changes in Altered Functions

It has been shown above that after REM sleep deprivation there is likely to be an increase in the levels of NE in the brain. On the other hand, REM sleep deprivation is known to affect several physiological functions. It is therefore possible, that REM sleep deprivation induced alterations in functions could be due to increased levels of NE in the brain.

One of the ways to confirm the same would be to study if the action of NE per se or the mechanism responsible for the increase in NE after REM sleep deprivation could be prevented, the effects of REM sleep deprivation should be minimized or abolished and application of NE would further enhance the effect(s). These would be discussed with the help of a few examples.

9.1. Sleep Deprivation-Associated Increase in Na^+-K^+ATPase Activity and Role of NE

It has been reported that REM sleep deprivation increases Na^+-K^+ATPase activity in the rat brain.[56] This effect was mediated by increased NE because it could be prevented by α1-adrenoceptor blocker, prazosin, both *in vivo* as well as *in vitro*.[57]

Also, it was shown that the increase in NE after REM sleep loss was likely due to non-cessation of the LC-REM-OFF neurons because if the GABA in the LC was prevented to act by continuous microinfusion of $GABA_A$ receptor antagonist, picrotoxin, there was significant reduction in REM sleep associated with simultaneous increase in Na^+-K^+ATPase activity in the brain, an effect similar to that of instrumental REM sleep loss.[58]

This REM sleep deprivation induced increase in Na^+-K^+ATPase activity was due to withdrawal of uncompetitive inhibition[59] suggesting that there was both allosteric stimulation as well as increased number of molecules of the enzyme after REM sleep deprivation. The former is likely to be due to withdrawl of calcium mediated[60] inhibition of enzyme activity while the latter due to increased synthesis of the enzyme molecules.[61]

9.2. REM Sleep Deprivation and Brain Excitability

REM sleep deprivation is known to increase excitability, confusion, irritability, aggressiveness.[62,63] It is also known that Na^+-K^+ATPase plays a significant role in maintenance of neuronal and brain excitability.[64]

Since REM sleep deprivation increases NE and that is the cause of REM sleep deprivation induced increase in Na^+-K^+ ATPase activity, it is likely that increased NE would be the cause or at least an important factor for the REM sleep deprivation induced increase in brain excitability.

Therefore, we proposed that one of the functions of REM sleep is to maintain brain excitability.[46,65] As a corollary, it may be said that, the alterations in the brain excitability would lead to REM sleep deprivation induced changes in behavior and diseased conditions (Figure 3).

9.3. REM Sleep Deprivation and Regulation of Body Temperature

Independent and isolated studies have reported that the body temperature (Tb) falls during sleep and it fluctuates during REM sleep.[66,67] However, during REM sleep deprivation there is a fall in Tb although there is increase in food intake.[67,68] Both these phenomena could be due to REM sleep deprivation induced increased levels of NE in the brain.

The preoptico-anetrior hypothalamic area (POAH) possesses the thermosensitive neurons and is the primary site responsible for thermoregulation.[69,70] There are evidences from isolated independent studies that

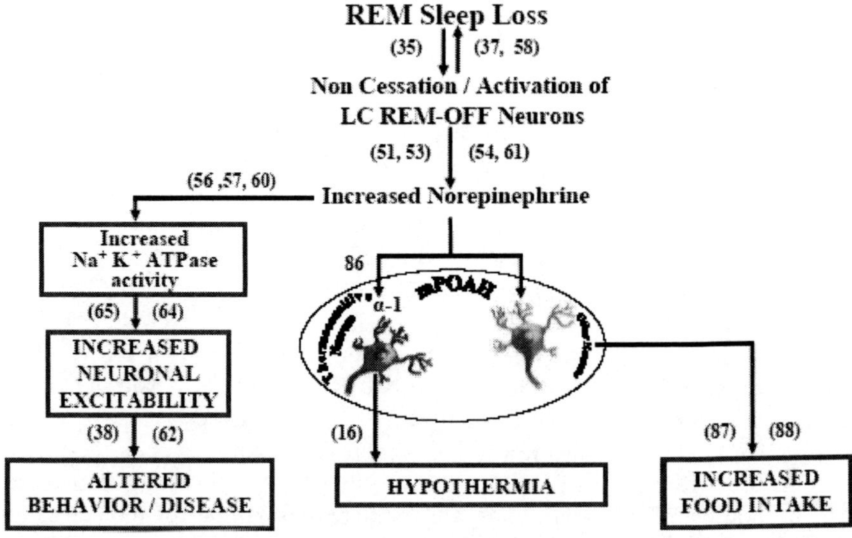

FIGURE 3. Schematic representation showing increase in norepinephrine in the brain after REM sleep deprivation and its effect in precipitating such deprivation induced effects.

aminergic,[71,72] cholinergic[73-77] as well as GABA-ergic[78,79] modulation of the POAH area neurons alter sleep-waking responses as well as thermoregulation. Heat exposure in different mammalian species has been reported to cause an increased concentration of NE in the POAH.[80-83] On the other hand, it has been observed that after 24 hours of cold exposure, there was significant decrease in the levels of NE in the paraventricular nuclei of the hypothalamus, although the contents of other monoamines were not altered.[84] Electrical stimulation of A1 region increased the NE and GABA level in the medial POAH.[85] Local microinjection of agonist and antagonist of NE individually and in combination showed that in the medial POAH NE modulates Tb by acting on α1-adrenoceptors and causes hypothermia.[16] It has further been confirmed that the thermosensitive neurons in the medial POAH possess α1-adrenoceptors.[86] Thus, it is likely that increased NE during REM sleep deprivation act on the α1-adrenoceptors located on the thermosensitive neurons in the POAH and induces hypothermia (Figure 3).

9.4. REM Sleep Deprivation and Food Intake

REM sleep deprivation has been reported to cause increased food intake[67, 68] associated with a decrease in Tb as mentioned above. On the other hand NE in the hypothalamus has also been shown to increase food intake.[87,88] Thus, it is possible that the increased levels of NE in the brain after REM sleep deprivation might be the likely cause for REM sleep deprivation associated increased food consumption (Figure 3).

10. Conclusions

The LC possesses NE-ergic REM-OFF neurons and they are inhibited, possibly by GABA, during spontaneous REM sleep. These neurons continue firing during REM sleep deprivation. Also, if these neurons are kept active by electrical stimulation or by preventing the GABA to act on them, REM sleep is significantly reduced and some effects similar to instrumental REM sleep deprivation are expressed, however, adrenoceptor antagonist could prevent such effects. Continuous activation (without periodic cessation) of these neurons would cause increased levels of NE in the brain. This elevated level of NE in the brain is the primary or at least a very important factor for REM sleep deprivation associated alteration in several functions including brain excitability, thermoregulation and food intake. Therefore, any factor, (external or from within) which would not allow the neurons in the LC (especially the REM-OFF neurons) to cease firing, would cause loss of REM sleep and associated increase in NE in the brain. The increased levels of NE in the brain would increase food intake, reduce body temperature and increase Na^+-K^+ ATPase activity. The increase in the Na^+-K^+ATPase activity in turn would affect neuronal excitability leading to REM sleep deprivation associated disorders. It may also be proposed that effects of loss of REM sleep through life possibly get gradually accumulated and get expressed as various diseases in the later phase of life. It is also possible that accumulation of such factor(s) due to REM sleep deprivation or, induced changes thereof, may predispose a person to altered physiological states e.g., increased blood pressure, heart attack, Alzheimer's disease, etc. Thus, REM sleep deprivation may have a direct, indirect, cumulative as well as compounding effect(s) on physiological functions.

11. References

1. U' Prichard DC, Greenberg DA, Sheehan P et al. Regional distribution of alpha noradrenergic receptor binding in calf brain. Brain Res 1977; 138:151-158.
2. U' Prichard DC, Charness ME, Robertson D et al. Prazosin: differential affinities for two populations of noradrenergic receptor binding sites. Eur J Pharmacol 1978; 50:87-89.
3. Palacois JM, Kuhar MJ. Beta-adrenergic-receptor localization by light microscopic autoradiography. Science 1980; 208:1378-1380.
4. Aghajanian GK, Cedarbaum JM, Wang RY. Evidence for norepinephrine-mediated collateral inhibition of locus coeruleus neurons. Brain Res 1977; 136:570-577.
5. Kleinlogel H, Scholtysik G, Sayers V. Effects of clonidine and BS 100-141 on the EEG sleep patterns in rats. Eur J Pharmacol 1975; 33:159-163.

6. Leppavuori A, Putkonen PTS. Alpha-adrenergic influences on the control of the sleep-waking cycle in the cat. Brain Res 1980; 193:95-116.

7. Gallard JM, Kafi S. Involvement of pre- and postsynaptic receptors in cate-cholaminergic control of paradoxical sleep in man. Eur J Clin Pharmacol 1979; 15:83-89.

8. Makela JP, Hilakivi I. Effect of alpha-adrenoceptor blockade on sleep and wakefulness in the rat. Pharmacol Biochem Behav 1986; 24:613-616.

9. Pellejero T, Monti JM, Baglietto J et al. Effects of methoxamine and alpha-adrenoceptor antagonists, prazosin and yohimbine, on the sleep-waking cycle of the rat. Sleep 1984; 7:365-372.

10. Hilakivi I. The role of beta and alpha-adrenoceptors in the regulation of the sleep-waking cycle in the cats. Brain Res 1983; 277:109-118.

11. Kleinlogel H. Effects of selective alpha1 adrenoceptor blocker prazosin on EEG sleep and waking stages in the rat. Neuropsychobiology 1989; 21:100-103.

12. Thankachan S, Islam F, Mallick BN. Adrenergic and cholinergic modulation of spontaneous and brain stem reticular formation stimulation induced desynchronization of the cortical EEG in freely moving cats. Sleep and Hypnosis 1999; 1:14-21.

13. Tononi G, Pompeiano M, Pompeiano O. Modulation of desynchronized sleep through microinjection of beta-adrenergic agonists and antagonists in the dorsolateral pontine tegmentum of the cat. Pflugers Arch 1989; 415:142-149.

14. Tononi G, Pompeiano M, Cirelli C. Effects of local pontine injection of noradrenergic agents on desynchronized sleep of the cat. Prog Brain Res 1991; 88:545-553.

15. Cirelli C, Tononi G, Pompeiano M et al. Modulation of desynchronized sleep through microinjection of alpha-1 adrenergic agonists and antagonists in the dorsal pontine tegmentum of the cat. Eur J Physiol 1992; 422:273-279.

16. Mallick BN, Alam MN. Different types of norepinephrinergic receptors are involved in preoptic area mediated independednt modulation of sleep – wakefulness and body temperature. Brain Res 1992; 591:8–19.

17. Berridge CW, Foote SL. Enhancement of behavioral and electroencephalographic indices of waking following stimulation of noradrenergic beta-receptors within the medial septal region of the basal forebrain. J Neurosci 1996; 16:6999-7009.

18. Crochet S, Sakai K. Effects of microdialysis application of monoamines on the EEG and behavioural states in the cat mesopontine tegmentum. Eur J Neurophysiol 1999; 11:3738-3752.

19. Crochet S, Sakai K. Alpha-2 adrenoceptor mediated paradoxical (REM) sleep inhibition in the cat. NeuroReport 1999; 10:2199–2204.

20. Maeda T, Pin C, Salvert D et al. Les neurones contenant des catecholamines du tegmentum pontique et leurs voies de projections chez le chat. Brain Res 1973, 57:119–152.

21. Petitjean F, Sakai K, Blondaux Ch et al. Hypersomnie par lesion isthmique chez le chat. II. Etude neurophysiologique et pharmacologique. Brain Res 1975; 88:439–453.

22. Jouvet M, Delorme F. Locus coeruleus et sommeil paradoxal. Comp Ren Soc de Biol 1965; 159:95-899.

23. Henley K, Morrison AR. A re-evaluation of the effects of lesions of the pontine tegmentum and locus coeruleus on phenomena of paradoxical sleep in the cat. Acta Neurobiol Exp(Warsz) 1974; 34:215-232

24. Sastre, JP, Jouvet, M. Les schemes moteurs du sommeil paradoxal. In : Proc Europ Congr Sleep Res, Karger: Basel, 1977; 18-23.

25. Sakai K. Some anatomical and physiological properties of ponto-mesencephalic tegmental neurons with special reference to the PGO waves and postural atonia during paradoxical sleep in cat. In : Hobson JA, Brazier MAB, eds. The reticular formation revisited. New York: Raven Press, 1980:27-447.

26. Sastre, JP, Sakai K, Jouvet M. Bilateral lesions of the dorsolateral pontine tegmentum. II. Effect upon muscle atonia. Sleep Res 1978; 7:44.

27. Jones BE, Harper ST, Halaris AE. Effects of locus coeruleus lesions upon cerebral monoamine content, sleep wakefulness states and the response to amphetamine in the cat. Brain Res 1977; 124:473-496

28. Mitler MM, Dement WC. Cataplectic–like behavior in cats after micro-injections of carbachol in pontine reticular formation. Brain Res 1974; 68:335-343.

29. Van Dongen PAM, Broekkamp CLE, Cools AR. Atonia after carbachol microinjections near the locus coeruleus in cats. Pharmacol Biochem Behav 1978; 8:527–532.

30. Silberman EK, Vivaldi E, Garfield J et al. Carbachol trigerring of desynchronized sleep phenomena : enhancement via small volume infusions. Brain Res 1980; 191:215–224.

31. Cespuglio R, Gomez ME, Faradji H et al. Alterations in the sleep-waking cycle induced by cooling of the locus coeruleus area. EEG Clin Neurophysiol 1982; 54:570-578

32. Chu NS, Bloom FE. Norepinephrine-containing neurons: changes in spontaneous discharge patterns during sleeping and waking. Science 1973; 179: 908-910.

33. Aston-Jones G, Bloom FE. Activity of norepinephrine containing locus coeruleus neurons in behaving rats anticipates fluctuations in the sleep-waking cycle. J Neurosci 1981; 13:876–886.

34. Jacobs BL. Single unit activity of locus coeruleus neurons in the behaving animals. Prog Neurobiol 1986; 27:183–194.

35. Mallick BN, Siegel JM, Fahringer H. Changes in pontine unit activity with REM sleep deprivation. Brain Res 1989; 515:94-98.

36. Thankachan S, Mallick BN, Islam F. Behavior of brain stem neurons during spontaneous and induced EEG desynchronization during wakefulness and rapid eye movement sleep in free moving cats. Sleep Res 1997; 26:54.

37. Singh S, Mallick BN. Mild electrical stimulation of pontine tegmentum around locus coeruleus reduces rapid eye movement sleep in rats. Neurosci Res 1996; 24:227-235.

38. Dement WC, The effect of dream deprivation. Science 1960; 131:1705-1707.

39. Hobson JA, McCarley RW and Wyzinski PW (1975) Sleep cycle oscillation: Reciprocal discharge by two brainstem neuronal groups. Science 189: 55–58.

40. Hobson JA, Lydic R, Baghdoyan HA. Evolving concepts of sleep cycle generation: from brain centers to neuronal population (review and commentaries). Behav Brain Sci 1986; 9:371-448.

41. Sakai K. Executive mechanisms of paradoxical sleep. Arch Ital Biol 1988; 126:239-257.

42. Mallick BN, Thankachan S, Islam F. Differential responses of brain stem neurons during spontaneous and stimulation induced desynchronization of the cortical EEG in freely moving cats. Sleep Res Online 1998; 1:132-146.

43. Egan TM, North RA. Acetylcholine acts on M2-muscarinic receptors to excite rat locus coeruleus neurons. Br J Pharmacol 1985; 85:733-735.

44. Alam MN, Kumari S and Mallick BN (1993) Role of GABA in acetylcholine induced locus coeruleus mediated increase in REM sleep. Sleep Res 22: 541.
45. Mallick BN, Kaur S, Saxena RN. Interactions between cholinergic and GABA-ergic neurotransmitters in and around the locus coeruleus for the induction and maintenance of rapid eye movement sleep in rats. Neuroscience 2001; 104:467-485.
46. Mallick BN, Majumdar S, Faisal M, Yadav V, Madan V, Pal D. Role of norepinephrine in the regulation of rapid eye movement sleep. J Biosci 2002; 27:539-551.
47. Kaur S, Saxena RN and Mallick BN GABAergic neurons in prepositus hypoglossi regulate REM sleep by its action on locus coeruleus in freely moving rats. Synapse 2001; 42:141–150.
48. Higo S, Ito K, Fuch D et al. Anatomical interconnections of the pedunculopontine tegmental nucleus and the nucleus prepositus hypoglossi in the cat. Brain Res 1990; 536:79-85.
49. Thankachan S, Islam F, Mallick B.N. Role of wake inducing brain stem area on rapid eye movement sleep regulation in freely moving cats. Brain Res Bull 2001; 55:43-49.
50. Mallick BN, Thankachan S, Islam F. Influence of hypnogenic brain areas on wakefulness and REM sleep related neurons in the brain stem of freely moving cats. J Neurosci Res 2004; 75:133-142.
51. Sinha AK, Ciaranello RD, Dement WC et al. Tyrosine hydroxylase activity in rat brain following REM sleep deprivation. J Neurochem 1973; 20:1289-1290.
52. Thakkar M, Mallick BN. Effect of rapid eye movement sleep deprivation on rat brain monoamine oxidases. Neuroscience 1993; 55:677-683.
53. Porkka-Heiskanen T, Smith SE, Taira T et al. Noradrenergic activity in rat brain during rapid eye movement sleep deprivation and rebound sleep. Am J Physiol 1995; 268:R1456-R1463.
54. Basheer R, Magner M, McCarley RW et al. REM sleep deprivation increases the levels of tyrosine hydroxylase and norepinephrine transporter mRNA in the locus coeruleus. Brain Res Mol Brain Res 1998; 57:235-240.
55. Majumdar S, Mallick BN. Increased Levels of Tyrosine Hydroxylase and Glutamic Acid Decarboxylase in locus coeruleus neurons after rapid eye movement sleep deprivation in rats. Neurosci Lett 2003; 338:193-6.
56. Gulyani S, Mallick BN. Effect of rapid eye movement sleep deprivation on rat brain Na-K ATPase activity. J Sleep Res 1993; 2:45-50.
57. Gulyani S, Mallick BN. Possible mechanism of REM sleep deprivation induced increase in Na-K ATPase activity. Neuroscience 1995; 64:255-260.
58. Kaur S, Panchal M, Faisal M et al. Long term blocking of GABA-A receptor in locus coeruleus by bilateral microinfusion of picrotoxin reduced rapid eye movement sleep and increased brain Na-K ATPase activity in freely moving normally behaving rats. Behav Brain Res 2004; 151:185-190.
59. Adya HVA, Mallick BN. Uncompetitive stimualtion of rat brain Na-K ATPase activity by rapid eye movement sleep deprivation. Neurochemistry Int 2000; 36:249–253.
60. Mallick BN, Adya HVA. Norepinephrine induced alpha-adrenoceptor mediated increase in rat brain Na-K ATPase activity is dependent on calcium ion. Neurochemistry Int 1999; 34:499–507.
61. Majumdar S, Faisal M, Madan V et al. Increased turnover of Na-K ATPase molecules in the rat brain after rapid eye movement sleep deprivation. J Neurosci Res 2003; 73:870-875.

62. Vogel GW. A review of REM sleep deprivation. Arch Gen Psychiat 1975; 32: 749-761.
63. Gulyani S, Majumdar S, Mallick BN. Rapid Eye Movement Sleep and significance of its deprivation studies-A review. Sleep and Hypnosis 2000; 2:49-68.
64. Trachtenberg MC, Packey DJ, Sweeney T. In vivo functioning of Na, K, activated ATPase. Curr Topics Cell Reg 1982; 19:159-217.
65. Mallick BN, Adya HVA, Thankachan S. REM sleep deprivation alters factors affecting neuronal excitability: Role of norepinephrine and its possible mechanism of action. In: Mallick BN and Inoue S eds. Rapid Eye Movement Sleep. Marcel Dekker: New York, 1999: 338-354.
66. Obal FJr. Thermoregulation during sleep. Exp Brain Res 1984; suppl8:157-172. 1984.
67. Rechtschaffen A, Bergmann BM, Everson CA et al. Sleep deprivation in the rat: X. Integration and discussion of the findings 1989. Sleep 1989; 12:68-87.
68. Rechtschaffen A, Bergmann BM, Everson CA et al. Sleep deprivation in the rat: X. Integration and discussion of the findings 1989. Sleep 2002; 25:68-87.
69. Fox RH Temperature regulation with special reference to man. In: Linden RJ, ed. Recent Advances in Physiology. Churchill Livingstone: London, 1974:340-405.
70. Boulant JA, Dean JB. Temperature receptors in the central nervous system. Annu Rev Physiol 1986; 48:639-654.
71. Day TA, Willoughby JO, Geffen LB. Thermoregulatory effects of preoptic area injections of noradrenaline in restrained and unrestrained rats. Brain Res 1979; 174:175-179.
72. Poole S, Stephenson JD. Effects of noradrenaline and carbachol on temperature regulation of rats. Br J Pharmacol 1979; 65:43-51.
73. Hernandez-Peon R. Sleep induced by local electrical or chemical stimulation of the forebrain. Electroencephalogr Clin Neurophysiol 1962; 14:423-424.
74. Hernandez-Peon R, Chavez-Ibarra G, Morgane PJ et al. Limbic cholinergic pathways involved in sleep and emotional behavior. Exp Neurol 1963; 8:93-111.
75. Yamaguchi N, Marczynski TJ, Ling GM. The effects of electrical and chemical stimulation of the preoptic region and some non-specific thalamic nuclei in unrestrained, waking animals. Electroencephalogr Clin Neurophysiol 1963; 15:154.
76. Tsoucaris-Kupfer D, Scmitt H. Hypothermic effect of alpha-sympathomimetic agents and their antagonism by adrenergic and cholinergic blocking drugs. Neuropharmacology 1972; 11:625-635.
77. Mallick BN, Joseph MM. Adrenergic and cholinergic inputs in preoptic area of rats interact for sleep-wake thermoregulation. Pharmacol Biochem Behav 1998; 61:193-199.
78. Ali M, Jha SK, Kaur S et al. Role of GABA-A receptor in the preoptic area in the regulation of sleep-wakefulness and rapid eye movement sleep. Neurosci Res 1999; 33:245-250.
79. Jha SK, Islam F, Mallick BN. GABA exerts opposite influence on warm and cold sensitive neurons in medial preoptic area in rats. J Neurobiol 2001; 48:291-300.
80. Myers RD, Beleslin DB. Changes in serotonin release in hypothalamus during cooling or warming of the monkey. Am J Physiol 1971; 220:1746-1754.
81. Myers RD, Chinn C. Evoked release of hypothalamic norepinephrine during thermoregulation in the cat. Am J Physiol 1971; 224:230-236.
82. Metcalf G, Myers RD. Precise location within the preoptic area where noradrenaline produces hypothermia. Eur J Pharmacol 1978; 51:47-53.

83. Myers RD. Hypothalamic control of thermoregulation: Neurochemical Mechanisms, In: Morgane PJ, Panksepp J, eds. Handbook of the hypothalamus. Marcel dekker: New York, 1980:83-210.

84. Ohtani NT, Sugano T, Ohta M. Alterations in monoamines and GABA in the ventromedial and paraventricular nuclei of the hypothalamus following cold exposure: a reduction in noradrenaline induces hyperphagia. Brain Res 1999; 842:6-14.

85. Herbison AE, Heavens RP, Dyer RG. Oestrogen modulation of excitatory A1 noradrenergic input to rat medial preoptic gamma aminobutyric acid neurones demonstrated by microdialysis. Neuroendocrinology 1990; 52(2):161-8.

86. Mallick BN, Jha SK and Islam F. Presence of alpha-1 adrenoreceptors on thermosensitive neurons in the medial preoptico-anterior hypothalamic area in rats. Neuropharmacology 2002; 42:697-705.

87. Chafetz MD, Parko K, Diaz S et al. Relationships between medial hypothalamic alpha 2-receptor binding, norepinephrine, and circulating glucose. Brain Res 1986; 384:404-408.

88. Leibowitz SF. Brain monoamines and peptides: role in the control of eating behavior. Fed Proc 1986; 45:1396-1403.

Neuroendocrine Outcomes of Sleep Deprivation in Humans and Animals

DEBORAH SUCHECKI AND SERGIO TUFIK

1. Abstract

Hormones can modulate and be modulated by sleep. This close relationship has been recognised for many decades. Sleep deprivation is an adverse condition that can alter the functioning of the neuroendocrine system, inasmuch as concentrations of hormones involved in anabolic processes, such as growth hormone (GH)[¥] are reduced, whilst levels of hormones involved in anabolic processes, such as glucocorticoids (GC) are increased. Therefore, prolonged periods of sleep deprivation, either internally or externally imposed, may lead to a wear and tear phenomenon, much similar to prolonged stressful conditions. In human beings, the vicious circle composed by sleep deprivation, stress and obesity has been claimed to be a major contributor to type II diabetes, cardiovascular diseases and ultimately, death.

2. Introduction

The important relationship between hormones and sleep has been well documented throughout the past decades.[1,2,3] This is a bidirectional relationship, for not only do hormones modulate sleep, but their secretion is also altered by different states of the sleep-wake cycle. The reader will have the opportunity to appreciate these relationships in several chapters of the present book.

Throughout this chapter the reader will get in contact with different human conditions that ultimately lead to chronic sleep deprivation and their repercussion on neuroendocrinology. In line with modern society, which has stimulated the round-the-clock activities, this chapter will also call the attention to the neuroendocrine consequences of shift-work and voluntary sleep curtailment.

[¥] Abbreviations used in this chapter: **ACTH**: adrenocorticotropic hormone; **ADR**: adrenaline; **Arc. N.**: arcuate nucleus; **AVP**: arginin vasopressin; **CRH**: corticotropin-releasing factor; **CSF**: cerebrospinal fluid; **GC**: glucocorticoid; **GH**: growth hormone; **GHRH**: GH-releasing hormone; **HPA**: hypothalamus-pituitary-adrenal; **LHA**: lateral hypothalamic area; **MMPM**: modified multiple platform method; **mRNA**: messenger RNA; **OT**: oxytocin; **PVN**: paraventricular nucleus; **REM**: rapid eye movements; **TSH**: thyrotropin-stimulating hormone.

And finally, animal models of prolonged sleep deprivation will show how deleterious this condition can be on hormones, stress response and metabolism.

The main purpose of the first and, in many ways, of the subsequent studies with sleep deprivation was to determine the function of sleep. Despite the numerous known consequences of sleep loss, its functions still remain a mystery. And this is probably so because sleep is involved in many functions.

Currently, many groups are interested in determining the outcomes of sleep loss, for good reasons. First, because the general population appears to be sleep-deprived, or at least, people are sleeping less than needed,[4,5] and this, of course, will have neurobehavioural, endocrine and metabolic consequences that deserve attention.[6,7] Second, because sleep deprivation or sleep loss is a common feature of many sleep pathologies, including sleep apnoea and insomnia, to mention but a few of these pathologies. The reader will have the opportunity to find the deleterious neuroendocrine consequences of these pathologies throughout this book.

The importance of sleep in general and of REM[†] sleep in particular, both in humans and in animals can be attested by its suppression. Studies of total or REM sleep deprivation show two major results. First, there is a continuous increase on the number of interventions to prevent the subject from entering sleep. In humans, for instance, is almost impossible to continue the experiment after five days of REM sleep deprivation, because the subject initiates sleep in REM.[8] Morden and co-workers[9] reported similar results in rats. From the first to the third days of REM sleep deprivation, the number of interventions raised from 134 to 350. After the third day of deprivation slow wave sleep was also interrupted.[9] Thus, the sleep pressure is so strongly built that it is impossible to prevent sleep from occurring.

The second consequence is known as *sleep rebound*, and is a compensation for the sleep that was lost during sleep deprivation or sleep restriction. The net result is that the individual or the rat will spend more time sleeping on the recovery night than on the baseline night. In addition, the latency to sleep is also shortened, indicating that a certain amount of sleep is necessary and any loss must be corrected.[9-15]

2.1. Methods to Induce Sleep Deprivation in Human Beings

The sleep of young adult humans is monophasic. The average sleep time is 7 to 8 h per night and during this period there is an alternating pattern among the different sleep stages. Let us say that an individual goes to bed at 23:00 h

---

[†] The term REM sleep is not the most adequate for rats; we prefer to use paradoxical sleep instead. However, for the purpose of standardization, REM sleep will be used as a synonym of paradoxical sleep throughout this chapter.

and wakes up at 7:00 h. If we record this individual's sleep, we will see a predominance of delta sleep during the first and a predominance of REM sleep during the second half of the night.[16] Rats, which are polyphasic and nocturnal animals, sleep predominantly during the day and are predominantly awake during the night.

But is still possible to see that both high amplitude slow wave sleep (that corresponds to delta sleep) and REM sleep are more concentrated in the period between 10:00 h and 13:00 h (Figure 1). Therefore, if we want to deprive a human being of REM sleep, we have to maintain the volunteer or patient awake from approximately 3:00 h to 7:00 h. If we want to be even more specific, it is possible to record the volunteer's sleep and wake him/her up every time there is a REM onset.

Some studies simply use total sleep deprivation and the volunteers are maintained awake for the entire night and subsequent day or for the entire experimental period, being allowed to sleep only on the following night.

Currently, a new approach to investigate the neuroendocrine and neurobehavioural outcomes of sleep loss utilises a paradigm of sleep debt. This paradigm is based on the evidence that population of industrialised countries engage in voluntary sleep curtailment. Despite all the technological amenities, the population sleeps approximately 90 minutes less than it did in the beginning of the XX century, leading to a condition of chronic sleep deprivation.[4] In this paradigm, healthy young men are allowed to sleep for 4 h per night, for one week. The sleep window ranges from 01:00 h to 5:00 h. On the following week, the volunteers are allowed to stay in bed for 12 h per night, so they can compensate for the sleep loss and the comparisons are made between these two opposite periods, sleep debt and sleep extension.

2.2. *Methods to Induce REM Sleep Deprivation in Rats*

There are several manners by which REM sleep can be suppressed. These approaches may employ drugs that stimulate the noradrenergic or serotonergic systems, or drugs that block the cholinergic system.[17,18] Lesions of nuclei involved in sleep generation or stimulation of nuclei involved in waking maintenance are also valid approaches to suppress sleep.[19-21] The focus of this chapter, however, will be the neuroendocrine effects of instrumental techniques developed to produce REM sleep deprivation.

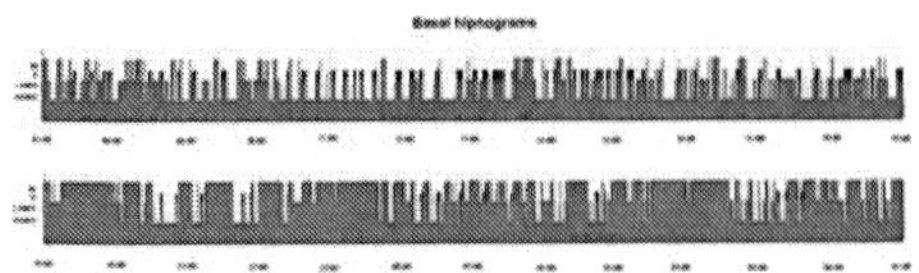

FIGURE 1. Typical sleep architecture of a rat throughout a 24h period.

2.2.1. The Disk-Over-Water Method

Dr. Allan Rechtschaffen in Chicago developed this technique in 1983. In this method the animals are connected to a polygraph that detects sleep onset and REM sleep onset. Depending on the purpose of the study, whether total or REM sleep deprivation, a computer connected to both polygraph and the disk-over-water, triggers the disk when the animal initiates sleep or REM sleep, so that the disk rotates and the animal is forced to walk in order to avoid falling in the water pan, located immediately below the disk.

The 'so-called' control or yoked rat is placed on the same disk as the sleep deprived rat and is wakened every time the disk rotates.[27] This method has generated numerous interesting data about changes in the physiology of sleep-deprived rats. Many of these findings will be discussed in the following sessions.

2.2.2. The Single Platform Method

In 1964, Jouvet and colleagues elaborated the very first method of REM sleep deprivation. The method, though very simple, was an ingenious one, because it was based on a very unique characteristic of REM sleep: the muscle atonia. With this concept in mind, the researchers placed a cat on top of an inverted flower-pot, immersed in water. Every time the cat entered REM sleep, it lost its muscle tone and balance, falling in the water.[22] In 1965, Cohen and Dement[23] adapted this method for rats, using a platform of 6.5 cm in diameter (Figure 2). Sleep recording of the animals during this procedure certified that REM sleep was completely suppressed and that it produced a major sleep rebound during the recovery period.[9,15,22,24]

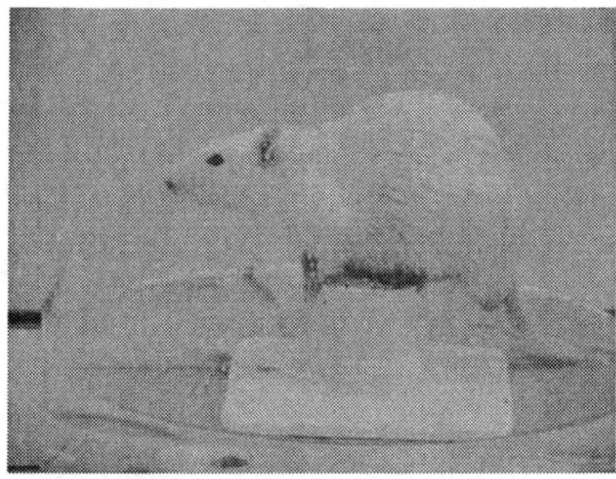

FIGURE 2. Photo of the inverted flower-pot technique, in which one rat is individually placed onto a platform deepened in water until 1 cm of the upper surface. With the onset of REM sleep, the rat loses its muscle tone and wakes up when it touches the water.

2.2.3. The Multiple Platform Method

The principle of this method is the same as the previous one, except that in the multiple platform method, the animals are allowed to move among several platforms.[25]

2.2.4. The Modified Multiple Platform Method

Rats are social animals that establish social hierarchy in the home-cage. We thought, therefore, that we could use a method to deprived 10 rats, raised together since puberty, using a large tiled water-tank (145.0 cm long × 44.0 cm wide × 45.0 cm high). Inside this water-tank we put 14 platforms, so the rats can move from one platform to the other[26] (Figure 3). This method has the advantage of preventing social isolation and movement restriction, two additional intervening variables that are present in the single platform method.

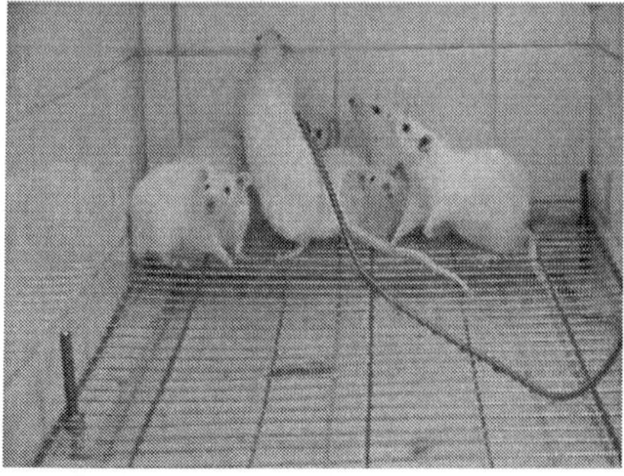

FIGURE 3. Upper panel: rats raised as a large group and placed together on narrow platform of 6.5 cm immersed in water. Lower panel: the proposed control group, placed onto a grid, located on the floor of the water-tank, also filled with water.

3. Effects on Physiological Systems

3.1. Body Temperature

Short-term (four days) REM sleep deprivation by the single platform method results in hyperthermia.[28,29] This increase seems to be partially mediated by prostaglandins, since inhibitors of prostaglandin synthesis reverses sleep deprivation-induced hyperthermia.[30] However, as REM sleep deprivation progresses, the rats become hypothermic.[31] Total sleep deprivation, on the other hand, leads to sustained hyperthermia.[32] These differential effects of REM sleep or total sleep deprivation on body temperature led to the question of whether the increased energy requirement is a consequence of the excessive heat loss or a change in the set-point of the central thermostat. To answer this question, rats were exposed to an alley with a thermal gradient, varying from $0°$ to $60°C$. Control rats sleep at the ambient temperature (T_a) corresponding to the rat's thermoneutral body temperature (T_b), i.e., $28°$ to $29°C$. Rats deprived of total sleep, however, choose to sleep at a T_a of approximately $45°C$, even though they are hyperthermic, suggesting that total sleep deprivation results in loss of the thermostat set-point.[32] On the other hand, REM sleep deprived rats also choose high T_a to sleep, which is not a surprise since they are hypothermic, suggesting that REM sleep deprivation results in loss of T_b maintenance.[31]

3.2. Metabolism and Energy Expenditure

Some of the studies that explore the metabolic effects of prolonged sleep deprivation in human beings involve extreme situations, such as military training with sleep and food deprivation,[33] or measurement of metabolic rate in the cold after sleep deprivation,[34] etc. A recent study evaluated the effect of 88 h of sleep deprivation on the levels of plasma leptin of volunteers maintained in laboratory conditions.

Leptin is an adipocyte-derived hormone that signals energy stored in adipose tissue and is involved in the regulation of feeding. Minimum levels are observed during the day, and a rise of leptin concentrations occur during early to mid sleep. Prolonged sleep deprivation results in a sustained reduction of the amplitude of leptin secretion, with smaller peaks, but no apparent change in the nadir. This change in leptin secretion may be a consequence of the increased activity of the adrenocortical axis and the sympathetic nervous system, which also influence the secretion of insulin.[35]

Many studies evaluate the metabolic outcomes in cases of sleep fragmentation, such as sleep apnoea and insomnia, or in sleep restriction paradigms. Insomnia patients, for instance, show increased metabolic rate both during the day as during the night, reflected by augmented maximum oxygen consume.[36] Likewise, sleep apnoea results in increased energy expenditure during the night, together with autonomic hyperactivity, reflected by elevated levels of urinary noradrenaline.[37,38]

Despite the chronic sleep fragmentation that sleep apnoea patients undergo, they present high leptin levels that correlate positively to the severity of the pathology. Still, sleep apnoea patients are often obese, and obesity too is positively correlated with apnoea severity.[39]

Studies using the sleep restriction paradigm in humans have shown that healthy young volunteers exhibit reduced glucose tolerance and a glucose clearance 40% slower than during the recovery period. In addition, the acute insulin response to a glucose load is 30% lower during the sleep restriction than during the sleep extension period.[7]

Thyrotropin-stimulating hormone (TSH) follows a circadian rhythm with low levels during the day and a quick increase of plasma concentrations in the beginning of the night, reaching a peak during sleep onset. TSH amplitude is virtually doubled during sleep deprivation, with increased secretion until the middle of the night followed by spontaneous reduction to basal levels early in the morning.[40] Peak TSH concentration coincides with the hypersomnolence reported by healthy volunteers subjected to 43 h of sleep deprivation.[41]

In rats, one of the most replicable effects of sleep deprivation is the hyperphagia concomitant to body weight loss. This phenomenon is observed with any kind of instrumental method used to induce sleep deprivation.[13,42-45] In general, there is a 3.5 to 9% loss of body weight,[13,45,46] reflecting the increased metabolic rate and energy expenditure that take place during this manipulation. When deprived rats are close to death, they show intense hypothermia and energy expenditure 2 to 3 times higher than baseline.[47,48]

What is interesting about this increased metabolic rate induced by sleep deprivation is that it cannot be reversed by either a supply of fat-rich diet[49] or by a 5% sucrose solution.[45] In addition, a fat-rich diet does not result in increased cholesterol, tryglicerides or glucose plasma levels, indicating accelerated turnover of nutrients.[49]

Recently, a study reported that prolonged sleep deprivation (15 days or more) results in approximately 50% lower levels of leptin compared to home-cage controls. These sleep-deprived animals eat more, but showed body weight loss.[50] Low leptin levels are probable sign of reduced adipose tissue. In fact, studies from our laboratory showed that 4 days of sleep deprivation by the flower-pot technique results in a significant loss of fat tissue, without changes in protein and water muscle content (Luz, unpublished data).

There is a close relationship between food intake and the sleep-wake cycle. In rats, for instance, fasting increases waking and reduces both slow wave sleep and REM sleep,[51] whereas sleep deprivation increases food intake. A new family of neuropeptides, the orexins, was described recently, which are concomitantly involved in feeding and waking.[52]

The levels of orexin in the cerebrospinal fluid are elevated after sleep deprivation in dogs[53] and rats,[54] suggesting that elevated levels of orexin serve as an adaptive process to forced waking.

Recently, however, several studies have indicated that orexin may, in fact, be related to activity, rather than wakefulness, since c-fos immunoreactivity was observed in cats during active, but not quiet waking[55] and during carbachol-induced REM sleep, when motor systems are activated and cats show eye movements and muscle twitches.[56]

In agreement with this idea, a recent paper showed that the orexin concentration in the CSF of rats is increased by 30 min of forced swimming, or by 30 min of sleep deprivation by gentle handling, but is reduced by 8h of immobilization.[57] Figure 4 illustrates a proposed mechanism for sleep deprivation-induced hyperphagia and body weight loss in rats.

3.3. Hormones

3.3.1. The Gonadal Axis

Prolonged sleep deprivation in rats results in augmented sexual activity. This behavioural effect seems to be in contradiction to the fact that stress, in general,

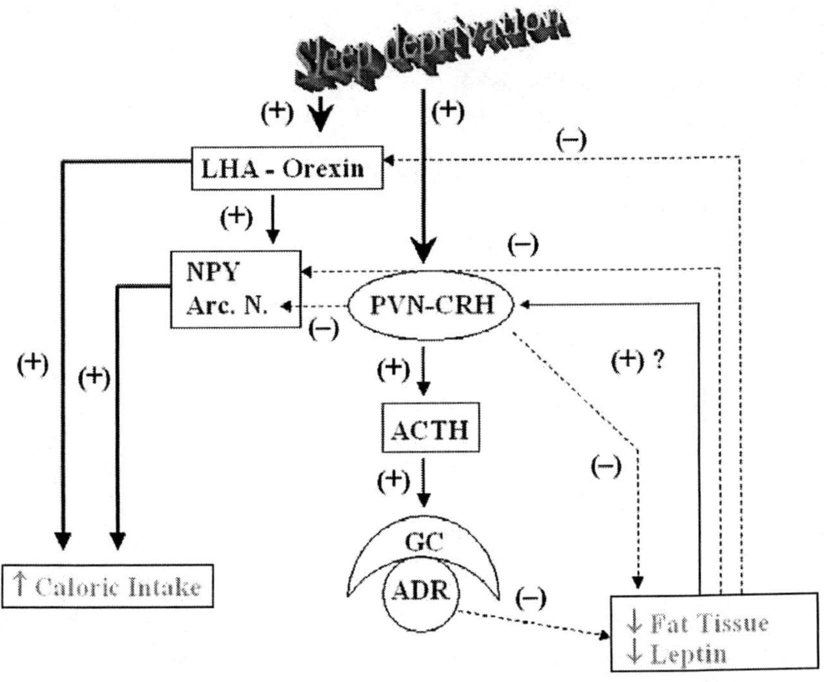

FIGURE 4. A model proposed to explain the mechanism by which sleep deprivation induces hyperphagia and body weight loss in rats. Orexins, which are involved in both feeding and sleep, are negatively regulated by leptin.[52] Sleep deprivation induces loss of fat tissue, possibly by activating the HPA axis, thus reducing the leptin signal on the lateral hypothalamic area (LHA). Full lines: stimulation (+); dashed lines: inhibition (−).

inhibits the hypothalamic-pituitary-gonadal axis. Thus, CRH, GC and β-endorphin inhibit gonadotropin-releasing hormone and, consequently, luteinizing and follicle stimulating hormones and testosterone.[58]

In fact, four days of REM sleep deprivation in rats culminate in low levels of testosterone, estrone and estradiol, but high levels of progesterone and corticosterone.[59,60] Similar hormonal alterations are also observed when animals are submitted to footshock, indicating that particular types of stress produce a deviation of the regular synthesis of sexual hormones.[60]

Despite the lowered levels of testosterone in sleep-deprived male rats, there is a marked increase in genital reflexes, indicated by penile erection and ejaculation, both in young and aged rats.[61,62] Castrated REM sleep-deprived rats treated with progesterone, but not with testosterone, display more genital reflexes than vehicle-treated rats.[63]

To restore male sexual behaviour that includes mounting, intromission and ejaculation in REM sleep-deprived rats, testosterone is essential.[64] Thus, it seems that progesterone is essential to initiate genital reflexes, whereas testosterone is vital to produce the full range of male reproductive behaviour.

3.3.2. The Somatotropic Axis

Secretion of GH is tightly linked to sleep. This phenomenon has been recognized for more than 30 years in human beings,[65,66] but it also occurs in many other species including primates, dogs, lambs[1] and rats.[67]

One of the pieces of strongest evidence for the close relationship between sleep onset and GH peak is the suppression of GH secretion during one night of sleep deprivation and the occurrence of a peak when subjects are allowed to sleep during the subsequent afternoon.[1]

The conclusion is that regardless of the time of day, the major GH pulse will take place after sleep onset, when sleep stages 3 and 4 (delta sleep) are predominant, indicating that GH secretion is driven more strongly by sleep than by circadian rhythm. Nonetheless, there is some influence of circadian rhythm on GH secretion, since GHRH induced GH secretion is more intense in the early night, coinciding with the nadir of somatostatinergic tone.[1]

GH secretion is suppressed by awakenings. This inhibition is seen even following the administration of growth hormone releasing hormone (GHRH) and it seems to be mediated by increased secretion of somatostatin, which is stimulated by corticotropin-releasing factor (CRH).[68] CRH triggers a cascade of neuroendocrine events in response to adverse situations (see next session) and it also inhibits GHRH.[69]

Brandenberg and co-workers[70] reported that during a regular night of sleep, the GH pulse that occurs after sleep onset accounts for 58% of the 24-h levels of the hormone. However, the distribution of GH pulses in 24-h sleep deprived subjects follows a much different pattern so that larger pulses are released during the day; the net result is that the total amount of GH secreted in 24-h sleep deprived individuals is not different from that of volunteers allowed to sleep.

A different picture emerges from the sleep debt condition, where volunteers sleep from 01:00 to 05:00 a.m., i.e., there sleep is delayed by 2h and awaking is advanced by 2 h. In this case two GH pulses are observed, one that takes place at the usual bedtime, 2 h before sleep onset, whereas the second pulse occurs immediately after sleep onset. Thus, one can see a pulse of GH even in the absence of sleep, at a time of strongest inhibition of somatostatin.[71] Despite these two pulses, the percentage of GH released throughout 24 h is smaller in the period of sleep debt than in the sleep extension, regardless of the impressive rebound of delta sleep and of delta activity that occurs during the short sleep period.[71]

Other examples of pathologies that result in sleep deprivation, such as sleep apnoea, support the influence of inadequate sleep on GH secretion. For instance, nocturnal GH secretion is inhibited in sleep apnoea patients. This inhibition is corrected by treating these patients with continuous positive airway pressure, resulting in two peaks of GH together with an impressive rebound of delta and REM sleep (124% and 253%, respectively).[72,73]

In rats, the relationship between GH secretion and sleep is similar to that described for humans, even though rats are polyphasic animals with very short sleep cycles. In these animals the ultradian GH rhythm is associated with the ultradian rhythm of sleep-wake cycle, and a temporal relationship between corticosterone, GH and sleep is established.[67] GH secretion is regulated positively by GHRH and negatively by somatostatin. GHRH administered in the evening increases REM sleep in intact, but not in hypophysectomised rats, whereas it increases slow wave sleep in both intact and hypophysectomised rats, indicating a circadian influence on the somatotropic axis similar to that described in humans.[74]

In fact, GHRH and somatostatin hypothalamic contents follow a well-determined circadian rhythm, with one peak for both neuropeptides occurring about 2h after lights on (at 10:00h), and a second major peak between 18:00h and 20:00h for GHRH and at 22:00h for somatostatin.

The nadirs occur at the end of lights off (at 8:00h) for both hormones and at 13:00h and 16:00h for GHRH and somatostatin, respectively.[75] A significant reduction of hypothalamic GHRH is observed after 8 h of sleep deprivation. One or two hours of sleep recovery are still not sufficient to regain basal levels of hypothalamic GHRH. In contrast, somatostatin levels only show a trend to increase with 4h and 8h of sleep deprivation.[75]

Investigation of the effects of short- and long-term sleep deprivation on GHRH and somatostatin mRNA levels in the hypothalamus show that short-term sleep deprivation (8h and 12h) induces increased GHRH mRNA levels, which are normalised by 2h of sleep recovery. Somatostatin mRNA levels, on the contrary are reduced by 8h of sleep deprivation, and 1h of sleep recovery is sufficient to return levels to basal.[76]

In distinct parts of the hypothalamus, the arcuate (Arc. N.) and the paraventricular nuclei (PVN), long-term sleep deprivation results in different alterations on mRNA levels. Using a multiple flower-pot method, Toppila and co-workers[77] showed that 24 h of REM sleep deprivation led to increased mRNA levels of somatostatin in the Arc. N. and that these levels were augmented in the PVN only after 72 h of REM sleep deprivation.

This increase may reflect the augmented REM sleep pressure observed in REM sleep deprived rats, since somatostatin has been implicated as a REM sleep inductor, both in young[78] and aged rats.[79] GHRH mRNA levels are augmented in the PVN by 6h of sleep deprivation[80] and reduced in both Arc. N. and PVN after 24 h or 72 h of REM sleep deprivation.

It is possible then that increased GHRH hypothalamic levels[76,80] regulate the slow wave sleep rebound observed after short-term sleep deprivation. The reduction of GHRH mRNA levels in the PVN of long-term REM sleep deprived rats, may, in turn, be a consequence of the increased levels of somatostatin mRNA in these conditions.[80]

3.3.2. The HPA Axis and Stress Response

Before reporting the effects of sleep deprivation on the hypothalamic-pituitary-adrenal (HPA) axis, an explanation of how this axis is regulated and how it is related to wake and sleep is in order.

The HPA axis is the main neuroendocrine transducer of the stress response. As show in Figure 5, the axis works in a concerted, coordinated manner, in which a cascade of events involving the paraventricular nucleus of the hypothalamus (PVN), the adenohypophysis and the adrenal cortex takes place, ultimately resulting in the secretion of glucocorticoids (GC). The activity of the HPA axis is governed by both stress and/or circadian rhythm, with the nadir occurring before the resting period and the peak, right before the active period.[81]

Corticotropin-releasing hormone (CRH) is synthesised in the parvocellular region of the paraventricular nucleus (PVN), whereas vasopressina (AVP) and oxytocin (OT) are synthesised in magnocellular region of the PVN and in the supraoptic nucleus of the hypothalamus.[82] These neuropeptides are stored in the pituitary stalk, a region that connects the hypothalamus to the pituitary, more specifically in the external zone of the median eminence. In response to numerous stimuli, these neuropeptides are released into the hypophyseal portal blood circulation and reach the corticotropes in the adenohypophysis[83], stimulating the secretion of adrenocorticotropin hormone (ACTH) into the systemic circulation. ACTH, in turn, stimulates the synthesis and secretion of GC (the main GC in human beings is cortisol, and in rats is corticosterone).

Regulation of the HPA axis activity is accomplished by GC, which by acting at specific receptors, type I and, mainly, type II receptors, feeds back in the pre-frontal cortex, hippocampus, hypothalamus and pituitary to shut down the system.[81] Numerous studies have shown that type I-related GC action is involved with control of circadian changes, whilst GC actions mediated by

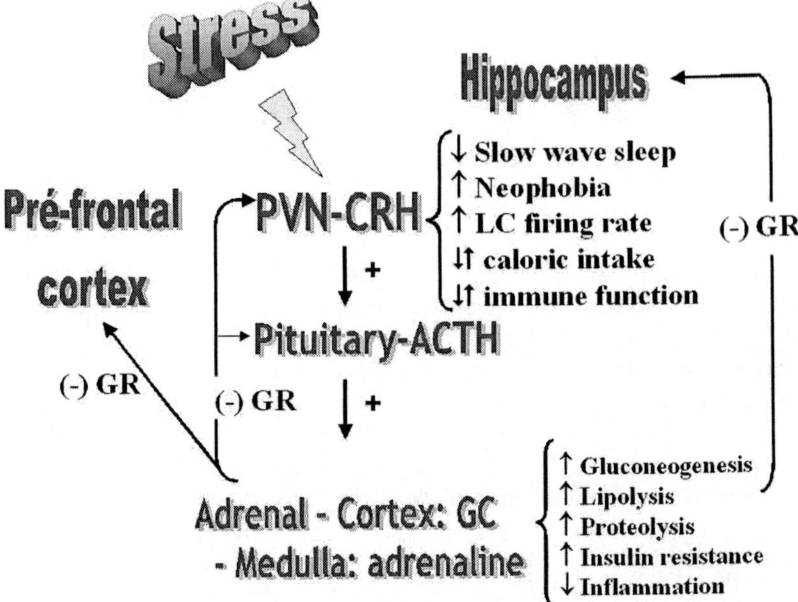

FIGURE 5. Schematic representation of the endocrine, behavioural, autonomic and immune responses to stress, mediated by CRH. In this figure, both activation (+) and inactivation (−) of the HPA axis are represented. Inactivation or glucocorticoid negative feedback is mainly accomplished by glucocorticoid receptors (GR), located in the pre-frontal cortex, hippocampus, hypothalamus and pituitary (adapted from[88]).

type II receptors involve the restoration of homeostatic disruptions that result from exposure to stressful stimuli.[81,84,85] These receptors work in a coordinated fashion to control behavioural and neuroendocrinological states.[86]

The profile of ACTH and cortisol secretion during a normal night of sleep in human beings show lowest levels of the hormones during the first half of the night, when delta sleep predominates.

During the second part of the night, when delta sleep is almost suppressed and REM sleep predominates, ACTH and cortisol begin to rise reaching the highest levels before morning awakening.[2]

Empirically is not difficult to realize that hormones of the HPA axis are related to wakefulness[87] because nobody feels asleep during a stressful situation. Quite the contrary, humans and animals are alert and fully attentive toward the stressful stimulus, ready to fight or fly. This is typical behavioural response to stress, being mediated by CRH, ACTH, cortisol and adrenaline.[88]

Pathological conditions characterised by hypercortisolemia (classical examples are melancholic and psychotic depression, and Cushing's syndrome) lead to sleep alterations, which include higher percentage of wakefulness and stage 1 of sleep, short REM sleep latency[89] and increased REM density.[90] Interestingly, healthy subjects whose first degree relatives were diagnosed with major or bipolar depression, exhibit a hyperactive HPA axis, shorter slow wave sleep during the first non-REM cycle and higher REM density in the first REM period.[91] Similar alterations were also reported in animal models of depression, in which rats were stressed in the prenatal period and exhibited an impaired corticosterone negative feedback mechanism, increased sleep fragmentation and percentage of superficial slow wave sleep, and shorter REM sleep latency.[92]

Sleep deprivation activates the HPA axis, both in human beings and rats. This alteration in humans is observed in pathological conditions such as insomnia,[93] or in experimental protocols with healthy volunteers.[94]

Insomnia patients present higher levels of ACTH than healthy volunteers almost throughout the whole day and night, whereas cortisol levels are higher in the period when activity of the HPA axis should be low, i.e., approximately from 20:00 h to 01:30 h. Not surprisingly, these patients exhibit poor quality of sleep, reflected by longer sleep latency, longer waking time and smaller sleep efficiency.[93] This shows clearly the opposite relationship between sleep and activity of the HPA axis, but does not resolve the question of causality regarding whether hyperactivity of the HPA axis leads to insomnia or the other way around.

In healthy volunteers, several protocols were used to investigate the dynamics of cortisol release during sleep compared to lack or fragmentation of sleep. Apparently, both conditions activate the adrenocortical axis. When healthy volunteers are awakened from the second REM sleep period on or aroused every time they enter sleep stages 1 or 2, ACTH and cortisol levels are immediately increased. This increase did not persist throughout the night, indicating that feedback mechanism within the HPA axis is functional even with frequent arousals.[94] During a daytime period of sleep, following one night of sleep deprivation, cortisol levels are as high as they should be for the period, even if the volunteers also slept during nightime strengthening the circadian influence, rather than the influence of sleep, on the activity of the adrenocortical axis. Moreover, during 4 h of sleep deprivation, cortisol levels were nearly 60% higher than at the same period of night sleep. The pulse-by-pulse analysis of cortisol release showed that 91% of delta sleep bouts occurred when cortisol secretory rate was low, whilst 88% of arousals were associated with increased cortisol secretory rate. Therefore, reduction of cortisol levels facilitates the occurrence of delta sleep.[95]

Total (24 h) or partial (4 h) sleep deprivation leads to increased levels of cortisol (45% and 37%, respectively) during the quiescent period of the HPA axis activity, i.e., from 18:00 h to 23:00 h, resulting in prolonged elevation of cortisol levels, suggesting that sleep loss may impair the feedback mechanism that regulates HPA activity.[96] However, not all studies report a 24 h sleep deprivation-induced change in cortisol levels. At least one study showed that 24 h average cortisol levels following a similar period of sleep deprivation is lower than the levels measured before sleep deprivation. These results, in fact are in agreement with the rebound of delta sleep observed in the post-sleep deprivation night and could be a feasible explanation for the antidepressant effect of sleep deprivation.[97] Methodological differences between these studies are the likely explanations for the divergent results reported.

Prolonged sleep debt in humans also result in increased plasma levels of cortisol, mainly during the quiescent period, i.e., late afternoon to early night. Similar elevations were observed in free salivary cortisol levels.[7] Throughout an 8-day sleep restriction protocol or a 48-h sleep deprivation in rats, both ACTH and corticosterone plasma levels are increased. Nevertheless, the HPA axis response to restraint stress, imposed to these animals 4 h after the end of each deprivation protocol, resulted in lower ACTH plasma levels and similar corticosterone responses as in control rats. In this case, the findings lead to the conclusion that the GC negative feedback mechanism is well preserved in animals.[98] Preliminary data from our laboratory, using a much longer sleep restriction protocol (21 days) in rats, indicate that sleep debt induced more intense increments of ACTH than of corticosterone secretion, suggesting an adrenal insensitivity to its trophic factor (Machado, unpublished data). This reduced sensitivity could represent an adaptive mechanism against chronic stimulation of cortisol release.

Numerous alterations of stress-related parameters are induced by prolonged sleep deprivation with the flower-pot method. These include reduced binding of the CRH receptors in the striatum and pituitary in addition to a reduction of hypothalamic content of CRH and increased CRH content in striatal, limbic and pituitary regions, suggesting that REM sleep deprivation induces a global CRH release.[99] Moreover, post-deprivation basal and ACTH-induced corticosterone levels are increased; immediately after REM sleep deprivation, ACTH plasma levels are augmented, adrenal glands are heavier, and thymus is atrophied.[26,100-103]

The HPA axis is sensitive to the negative feedback action of corticosterone. Thus, previously elevated levels of the steroid may inhibit further responses to stress. However, in chronic stress conditions, the HPA axis exhibits a facilitatory response to subsequent stressors. Therefore, "basal" corticosterone levels in chronically stressed animals increase from 5 μg/dl to 8μg/dl, but ACTH "basal" levels are generally unaltered.[104] These findings are basically the same as we reported previously, using the modified multiple platform method (MMPM).[26] Since the properties of the HPA axis, such as activation and feedback inhibition, show adaptation to chronic stress, we sought to determine whether prolonged REM sleep deprivation in the presence or absence of conspecifics could influence the hormonal responses to a mild stressor. Thus, animals previously submitted to 96h of REM sleep deprivation, either by the flower-pot method (individual deprivation) or by the MMPM (group deprivation), were challenged with a mild stressor (saline + novelty). All groups showed peak corticosterone levels 5 min after the stress, and sleep-deprived animals show higher peak levels than control rats. Twenty min after the stress, however, these levels remained elevated in control groups,

13. Suchecki D, Palma BD, Tufik S. Sleep rebound in animals deprived of paradoxical sleep by the multiple platform method. Brain Res 2000; 875:14-22.

14. Werth E, Cote KA, Gallmann E, et al. Selective REM sleep deprivation during daytime I. Time course of interventions and recovery sleep. Am J Physiol 2002; 283:R521-R526.

15. Machado RB, Hipólide DC, Benedito-Silva AA, et al. Sleep deprivation induced by the modified multiple platform technique: quantification of sleep loss and recovery. Brain Res 2004; 1004:45-51.

16. Sleep Syllabus. www.sleephomepages.org/sleepsyllabus/sleephome.html. 1995.

17. Salin-Pascual RJ, Jimenez-Anguiano A, Granados-Fuentes D, et al. Effects of biperiden on sleep at baseline and after 72 h of REM sleep deprivation in the cat. Psychopharmacology 1992; 106:540-542.

18. Mavanji V, Datta S. Sleep-wake effects of yohimbine and atropine in rats with a clomipramine-based model of depression. Neuroreport 2002; 13(13):1603-1606.

19. Kovalzon VM, Tsibulsky VL. REM-sleep deprivation, stress and emotional behavior in rats. Behav Brain Res 1984; 14:235-245.

20. Pacheco-Cano MT, García-Hernández F, Próspero-Garcia O, et al. Vasoactive intestinal polypeptide induces REM recovery in insomniac forebrain lesioned cats. Sleep 1990; 13:297-303.

21. Singh S, Mallick BN. Mild electrical stimulation of pontine tegmentum around locus coeruleus reduces rapid eye movement sleep in rats. Neurosci Res 1996; 24:227-235.

22. Jouvet D, Vilmont P, Delorme F, et al. Etude de la privation sélective de la phase paradoxale de sommeil chez le chat. Compt Rend Soc Biol 1964; 158:756-759.

23. Cohen HB, Dement WC. Sleep: changes in threshold to electroconvulsive shock in rats after deprivation of "paradoxical" phase. Science 1965; 150:1318-1319.

24. Landis C. Altered sleep patterns with the platform method of REM sleep deprivation in rats Sleep Res 1996; 25:469.

25. Coenen AML, van Luijtelaar ELJM. Stress induced by three procedures of deprivation of paradoxical sleep. Physiol Behav 1985; 35:501-504.

26. Suchecki D, Tufik S. Social stability attenuates the stress in the modified multiple platform method for paradoxical sleep deprivation in the rat. Physiol Behav 2000; 68:309-316.

27. Rechtschaffen A, Gilliland MA, Bergmann BM, et al. Physiological correlates of prolonged sleep deprivation in rats. Science 1983; 221:182-184.

28. Hoshino K. Food deprivation and hypothermia in desynchronized sleep-deprived rats. Braz J Med Biol Res 1996; 29:41-46.

29. Salin-Pascual RJ, Franco M, Garcia-Ferreiro R, et al. Differences in sleep variables, blood adenosine, and body temperature between hypothyroid and euthyroid rats before and after REM sleep deprivation. Sleep 1997; 20:957-962.

30. Seabra MLV, Tufik S. Sodium diclofenac inhibits hyperthermia induced by paradoxical sleep deprivation: the possible participation of prostaglandins. Physiol Behav 1993; 54:923-926.

31. Landis CA, Bergmann BM, Ismail MM, et al. Sleep deprivation in the rat: XV. Ambient temperature choice in paradoxical sleep-deprived rats, Sleep 1992; 15:13-20.

32. Prete FR, Bergmann BM, Holtzman P, et al. Sleep deprivation in the rat: XII. Effect on ambient temperature choice. Sleep 1991; 14:109-115.

33. Gomez-Merino D, Chennaoui M, Drogou C, et al. Decrease in serum leptin after prolonged physical activity in men. Med Sci Sports Exerc 2002; 34(10):1594-1599.

34. Young AJ, Castellani JW, O'Brien C, et al. Exertional fatigue, sleep loss, and negative energy balance increase susceptibility to hypothermia. J Appl Physiol 1998; 85(4):1210-1217.
35. Mullington JM, Chan JL, Van Dongen HP, et al. Sleep loss reduces diurnal rhythm amplitude of leptin in healthy men. J Neuroendocrinol 2003; 15(9):851-854.
36. Bonnet MH, Arand DL. Insomnia, metabolic rate and sleep restoration. J Intern Med 2003; 254:23-31.
37. Ryan CF, Love LL, Buckley PA. Energy expenditure in obstructive sleep apnea. Sleep 1995; 18:180-187.
38. Stenlof K, Grunstein R, Hedner J, et al. Energy expenditure in obstructive sleep apnea: effects of treatment with continuous positive airway pressure. Am J Physiol 1996; 271:E1036-E1043.
39. Ozturk L, Unal M, Tamer L, et al. The association of the severity of obstructive sleep apnea with plasma leptin levels. Arch Otolaryngol Head Neck Surg 2003; 129(5):538-540.
40. Allan J, Czeisler C. Persistence of the circadian thyrotropin rhythm under constant conditions and after light-induced shifts of circadian phase. J Clin Endocrinol Metab 1994; 79:508-512.
41. Leproult R, Van Reeth O, Byrne MM, et al. Sleepiness, performance, and neuroendocrine function during sleep deprivation: effects of exposure to bright light or exercise. J Biol Rhythms 1997; 23:245-258.
42. Elomaa E. The cuff pedestal: an alternative to flowerpots? Physiol Behav 1979; 23:669-672.
43. Kushida CA, Bergmann BM, Rechtschaffen A. Sleep deprivation in the rat: IV. Paradoxical sleep deprivation. Sleep 1989, 12:22-30.
44. Brock JW, Farooqui SM, Ross KD, et al. Stress-related behavior and central norepinephrine concentrations in the REM sleep-deprived rat. Physiol Behav 1994; 55:997-1003.
45. Suchecki D, Antunes J, Tufik S. Palatable solutions during paradoxical sleep deprivation: reduction of hypothalamic-pituitary-adrenal axis activity and lack of effect on energy imbalance. J Neuroendocrinol 2003; 15:815-821.
46. Andersen ML, Martins PJF, D'Almeida V, et al. Effects of paradoxical sleep deprivation on blood parameters associated with cardiovascular risk in aged rats. Exp. Gerontol., in press.
47. Everson C, Bergmann BM, Rechtshaffen A. Sleep deprivation in the rat: III. Total sleep deprivation. Sleep 1989; 12:12-21.
48. Everson CA. Functional consequences of sustained sleep deprivation in the rat. Behav Brain Res 1995; 69:43-54
49. Everson CA, Wehr TA. Nutritional and metabolic adaptations to prolonged sleep deprivation in the rat. Am J Physiol 1993; 264:R376-R387.
50. Everson CA, Crowley WR. Reductions in circulating anabolic hormones induced by sustained sleep deprivation in rats. Am J Physiol Endocrinol Metab, in press.
51. Borbely AA. Sleep in the rat during food deprivation and subsequent restitution of food. Brain Res 1997; 124:457-461.
52. Willie JT, Chemelli RM, Sinton CM, et al. To eat or to sleep? Orexin in the regulation of feeding and wakefulness. Ann Rev Neurosci 2001; 24:429-458.
53. Wu MF, John J, Maidment N, et al. Hypocretin release in normal and narcoleptic dogs after food and sleep deprivation, eating, and movement. Am J Physiol 2002; 283:R1079-R1086.

54. Pedrazzoli M, D'Almeida V, Martins PJ, et al. Increased hypocretin-1 levels in cerebrospinal fluid after REM sleep deprivation. Brain Res. 2004; 995:1-6.
55. Torterolo P, Yamuy J, Sampogna S, et al. Hypocretinergic neurons are primarily involved in activation of the somatomotor system. Sleep 2003; 26:25-28.
56. Torterolo P, Yamuy J, Sampogna S, et al. Hypothalamic neurons that contain hypocretin (orexin) express c-fos during active wakefulness and carbachol-induced active sleep. Sleep Res Online 2001; 4:25-32.
57. Martins PJ, D'Almeida V, Pedrazzoli M, et al. Increased hypocretin-1 (orexin-a) levels in cerebrospinal fluid of rats after short-term forced activity. Regul Pept 2004; 117:155-158.
58. Stratakis CA, Chrousos GP. Neuroendocrinology and pathophysiology of the stress system. Ann NY Acad Sci 1995; 771:1-18.
59. Andersen ML, Bignotto M, Tufik S. Facilitation of ejaculation after metamphetamine administration in paradoxical sleep deprived rats. Brain Res 2003; 978:31-37.
60. Andersen ML, Bignotto M, Machado RB, et al. Effects of chronic stress on steroid hormones secretion in male rats. Braz J Med Biol Res, in press.
61. Andersen ML, Bignotto M, Papale LA, et al. Age-related effects on genital reflexes induced by paradoxical sleep deprivation and cocaine in rats. Exp Gerontol 2004; 39(2):233-237.
62. Andersen ML, Bignotto M, Machado RB, et al. Does sleep deprivation and cocaine induce penile erection and ejaculation in old rats? Addict Biol 2002; 7:285-290.
63. Andersen ML, Bignotto M, Tufik S. Hormone treatment facilitates penile erection in castrated rats after sleep deprivation and cocaine. J Neuroendocrinol 2004;16(2):154-159
64. Velazquez-Moctezuma J, Monroy E, Cruz ML Facilitation of the effect testosterone on male sexual behavior in rats deprived of REM sleep. Behav Neural Biol 1989; 51:46-53.
65. Sassin JF, Parker DC, Johnson LC, et al. Effects of slow wave sleep deprivation on human growth hormone release in sleep: preliminary study. Life Sci 1969; 8(23):1299-1307.
66. Sassin JF, Parker DC, Mace JW, et al. Human growth hormone release: relation to slow-wave sleep and sleep-walking cycles. Science 1969; 165(892):513-515.
67. Mitsugi N, Kimura F. Simultaneous determination of blood levels of corticosterone and growth hormone in the male rat: relation to sleep-wakefulness cycle. Neuroendocrinology 1985; 41(2):125-130.
68. Spath-Schwalbe E, Hundenborn C, Kern W, et al. Nocturnal wakefulness inhibits growth hormone (GH)-releasing hormone-induced GH secretion. J Clin Endocrinol Metab 1995; 80:214-219.
69. Barbarino A, Corsello SM, Della Casa S, et al. Corticotropin-releasing hormone inhibition of growth-hormone releasing hormone-induced growth hormone release in man. J Clin Endocrinol Metab 1990; 71:1368-1374.
70. Brandenberger G, Gronfier C, Chapotot F, et al. Effect of sleep deprivation on overall 24 h growth-hormone secretion. Lancet 2000; 56:1408.
71. Spiegel K, Leproult R, Colecchia EF, et al. Adaptation of the 24-h growth hormone profile to a state of sleep debt Am J Physiol 2000; 279:R874-R883.
72. Cooper BG, White JS, Ashworth LA, et al. Hormonal and metabolic profiles in subjects with obstructive sleep apnea syndrome and the acute effects of

nasal continuous positive airway pressure (CPAP) treatment. Sleep 1995; 18:172-179.

73. Grunstein RR. Metabolic aspects of sleep apnea. Sleep 1996; 19:S218-S220.

74. Obál F Jr, Floyd R, Kapas L, et al. Effects of systemic GHRH on sleep in intact and hypophysectomized rats. Am J Physiol. 1996; 270(2 Pt 1):E230-237.

75. Gardi J, Obál Jr F, Fang J, et al. Diurnal variations and sleep deprivation-induced changes in rat hypothalamic GHRH and somatostatin contents. Am J Physiol 1999; 277(46):R1339-R1344.

76. Zhang J, Chen Z, Taishi P, et al. Sleep deprivation increases rat hypothalamic growth hormone-releasing hormone mRNA. Am J Physiol 1998; 275(44):R1755-R1761.

77. Toppila J, Asikainen M, Alanko L, et al. The effect of REM sleep deprivation on somatostatin and growth hormone-releasing hormone expression in the rat hypothalamus. J Sleep Res 1996; 5(2):115-122.

78. Danguir J. Intracerebroventricular infusion of somatostatin selectively increases paradoxical sleep in rats. Brain Res 1986; 367(1-2):26-30.

79. Danguir J. The somatostatin analogue SMS 201-995 promotes paradoxical sleep in aged rats. Neurobiol Aging 1989; 10(4):367-369.

80. Toppila J, Alanko L, Asikainen M, et al. Sleep deprivation increases somatostatin and growth hormone-releasing hormone messenger RNA in the rat hypothalamus. J Sleep Res 1997; 6(3):171-178.

81. Dallman MF, Akana SF, Cascio CS, Darlington DN, Jacobson L, Levin N. Regulation of ACTH secretion: Variations on a theme of B. Recent Prog Horm Res 1987; 43:113-173.

82. Plotsky PM. Hypophyseotropic regulation of adenohypophyseal adrenocorticotropin secretion. Fed Proc 1985; 44:207-213.

83. Antoni FA. Hypothalamic control of adrenocorticotropin secretion: advances since the discovery of 41-residue corticotropin-releasing factor. Endocr Rev 1986: 7(4):351-378.

84. De Kloet ER, Reul JMHM. Feedback action and tonic influence of corticosteroids on brain function: A concept arising from the heterogeneity of brain receptor systems. Psychoneuroendocrinology 1987; 2:83-105.

85. Ratka A, Sutanto W, De Kloet ER. On the role of brain mineralocorticoid (Type I) and glucocorticoid (Type II) receptors in neuroendocrine regulation. Neuroendocrinology 1989; 50:117-123.

86. De Kloet ER. Brain corticosteroid receptor balance and homeostatic control. Front Endocrinol 1991; 12:95-164.

87. Chang FC, Opp MR. Corticotropin-releasing hormone (CRH) as a regulator of waking. Neurosci Biobehav Rev. 2001; 25(5):445-453.

88. Arborelius L, Owens MJ, Plotsky PM, et al. The role of corticotropin-releasing factor in depression and anxiety disorders. J Endocrinol 1999; 160:1-12.

89. Stefos G, Staner L, Kerkhofs M, et al. Shortened REM latency as a psychobiological marker for psychotic depression? An age-, gender-, and polarity-controlled study. Biol Psychiatry 1998; 44(12):1314-1320.

90. Shipley JE, Schteingart DE, Tandon R, et al. EEG sleep in Cushing's disease and Cushing's syndrome: comparison with patients with major depressive disorder. Biol Psychiatry 1992; 32(2):146-155.

91. Krieg JC, Lauer CJ, Schreiber W, et al. Neuroendocrine, polysomnographic and psychometric observations in healthy subjects at high familial risk for affective

disorders: the current state of the 'Munich vulnerability study'. J Affect Disord 2001; 62(1-2):33-37.

92. Dugovic C, Maccari S, Weibel L, et al. High corticosterone levels in prenatally stressed rats predict persistent paradoxical sleep alterations. J Neurosci 1999; 19(19):8656-8664.

93. Vgontzas AN, Bixler EO, Lin HM, et al. Chronic insomnia is associated with nyctohemeral activation of the hypothalamic-pituitary-adrenal axis: clinical implications. J Clin Endocrinol Metab 2001; 86(8):3787-3794.

94. Späth-Schwalbe E, Gofferje M, Kern W, et al. Sleep disruption alters nocturnal ACTH and cortisol secretory patterns. Biol Psychiatry 1991; 29(6):575-584.

95. Weibel L, Follenius M, Spiegel K, et al. Comparative effect of night and daytime sleep on the 24-hour cortisol secretory profile. Sleep 1995; 18(7):549-556.

96. Leproult R, Copinschi G, Buxton O, et al. Sleep loss results in an elevation of cortisol levels the next evening. Sleep 1997; 20(10):865-870.

97. Vgontzas AN, Mastorakos G, Bixler EO, et al. Sleep deprivation effects on the activity of the hypothalamic-pituitary-adrenal and growth axes: potential clinical implications. Clin Endocrinol 1999; 51(2):205-215.

98. Meerlo P, Koehl M, van der Borght K, et al. Sleep restriction alters the hypothalamic-pituitary-adrenal response to stress. J Neuroendocrinol 2002; 14(5):397-402.

99. Fadda P, Fratta W. Stress-induced sleep deprivation modifies corticotropin releasing factor (CRF) levels and CRF binding in rat brain and pituitary. Pharmacol Biochem Behav 1997; 35(5):443-446.

100. Brock JW, Farooqui SM, Ross KD, et al. Stress-related behavior and central norepinephrine concentrations in the REM sleep-deprived rat. Physiol Behav 1994; 55:997-1003.

101. Patchev V, Felszeghy K, Korányi L. Neuroendocrine and neurochemical consequences of a long-term sleep deprivation in rats: similarities to some features of depression. Homeostasis 1991; 33:97-108.

102. Suchecki D, Lobo LL, Hipólide DC, et al. Increased ACTH and corticosterone secretion induced by different methods of paradoxical sleep deprivation. J Sleep Res 1998; 7:276-281.

103. Coenen AML, van Luijtelaar ELJM. Stress induced by three procedures of deprivation of paradoxical sleep. Physiol Behav 1985; 35:501-504.

104. Dallman MF, Akana SF, Scribner KA, et al. Stress, feedback and facilitation in the hypothalamo-pituitary-adrenal axis. J Neuroendocrinol 1991; 4: 517-526.

105. Suchecki D, Tiba PA, Tufik S. Paradoxical sleep deprivation facilitates subsequent corticosterone response to a mild stressor in rats. Neurosci Lett 2002; 320:45-48.

106. Levine S, Ursin H. What is stress? In: Brown MR, Koob GF, Rivier C, eds. Stress, Neurobiology and Neuroendocrinology. New York: Marcel Dekker, Inc., 1991, 3-21.

107. Plaut SM, Friedman SB. Stress, coping behavior and resistance to disease. Psychother Psychosom 1982; 38:274-283.

108. Suchecki D, Tiba PA, Tufik S. Hormonal and behavioural responses of paradoxical sleep-deprived rats to the elevated plus maze. J Neuroendocr 2002; 14:549-554.

109. Dallman MF, Pecoraro N, Akana SF, et al. Chronic stress and obesity: a new view of "comfort food". Proc Natl Acad Sci 2003; 100(20): 11696-11701.

Sleep-Inducing Substances
in the Regulation of Sleep

VIJAY RAMESH, NAVITA KAUSHAL
AND VELAYUDHAN MOHAN KUMAR

1. Introduction

We spend almost one-third of our life sleeping, yet very little is understood as to why we need sleep or how do we sleep. The extrinsic and intrinsic controlling mechanisms of sleep have fascinated scientists for generations and many different theories, networks and endogenous compounds have been proposed. Although various substances are labeled 'sleep-inducing substances' for example, delta sleep inducing peptide, prostaglandin etc. we still lack definitive knowledge on how these chemicals bring about a balance in regulating sleep and wakefulness. However, as the biochemical mechanisms underlying sleep control are now slowly emerging, the major question perhaps is whether these humoral mediators seem to have some relation to sleep by, for example, affecting circadian rhythms or arousal states thereby actively governing the sleep pattern or are they just responding to the sleep homeostasis.

The first ever study probably began in November 1906 when two French psychophysiologists, Legendre and Pieron,[62] deprived dogs of sleep in order to obtain an endogenously occurring sleep-inducing substance. During the same period, in July 1907, a Japanese physiochemist, Ishimori, performed similar experiments on dogs. The French scientist published a series of short reports that culminated in a full paper in 1913, while Ishimori[42] published a full article in Japanese in 1909. Both of them independently reached the conclusion that prolonged state of wakefulness eventually accumulated some chemical substance in the brain that may trigger sleep.

However there was no report in these lines of study until 1961, when Kornmuller et al.[52] by their technically advanced methodology showed these effects in cats. They induced sleep in one cat by stimulating the thalamus by chronically implanted electrodes and went on to show that blood transfusion from the latter soon caused the other to sleep. He concluded that some kind of hormone carried by the blood of sleeping cats might be involved in induction of sleep.

Monnier and his team[66] started working on isolating sleep inducing substances and obtained DSIP in 1977. In the mean time, similar systemic research was undertaken in the USA by Pappenheimer and associates[76] and in Japan by Uchizono and colleagues.[68] These events triggered the hope of finding the ideal humoral substance that can control sleep. Consequently, a number of endogenous factors have been nominated as putative sleep substances, although their particular roles in the regulatory mechanisms of sleep remain unknown, for reviews.[8,15,25,26,40,41,54,96,99] For a substance to be ultimately termed as a putative sleep regulatory substance, it should fulfill certain criteria.

1. The substance should induce physiological sleep either directly exciting the sleep regulatory mechanisms or indirectly suppressing wake regulatory mechanism.
2. The substance and its receptors should naturally be present in the animal.
3. Administration of the substance into the animal through different routes should induce injection bound sleep irrespective of circadian rhythmicity.
4. Antagonizing the effect of the substance or blocking its receptor should reduce spontaneous sleep and also reduce induced sleep brought by sleep deprivation, forced locomotion and changes in ambient temperature.
5. The concentration or turnover of the substance or its receptor should vary with the sleep-wake cycle or with prolonged wakefulness.

The main techniques in evaluating these types of drugs involve at the least a setup for long-term behavioral state recording in animals. For example,

1. Monitoring sleep-wake behavior involving multi-channel EEG set up and preparing the animals for sleep bioassays.
2. Continuous systemic, intracerebroventricular (icv) or intracerebral (ic) infusion techniques, which should enable either to administer drugs or to withdraw blood or CSF for, sleep bioassay with minimal disturbance to the animals.
3. Mathematical modeling and computer simulation techniques for understanding the total regulatory system of sleep mechanisms.
4. Techniques for the analysis of ultradian, circadian and infradian rhythms of locomotor activity, body temperature and behavioral state in animals.

There are many substances that have at least partially met the status of sleep-inducing factor, like human chorionic gonadotropin,[95] growth hormone-releasing hormone, interleukin-1, tumor necrosis factor,[72] component B of sleep-promoting substance, delta sleep-inducing peptide (DSIP), uridine, muramyl dipeptide (MDP) and prostaglandin D_2 (PGD_2).[39] The main purpose of this chapter is to introduce the subject of sleep-inducing substances. Hence we will be focusing on a few examples like DSIP, uridine, PGD_2 and MDP (Figure 1). Other putative sleep-inducing substances are beyond the scope of this chapter.

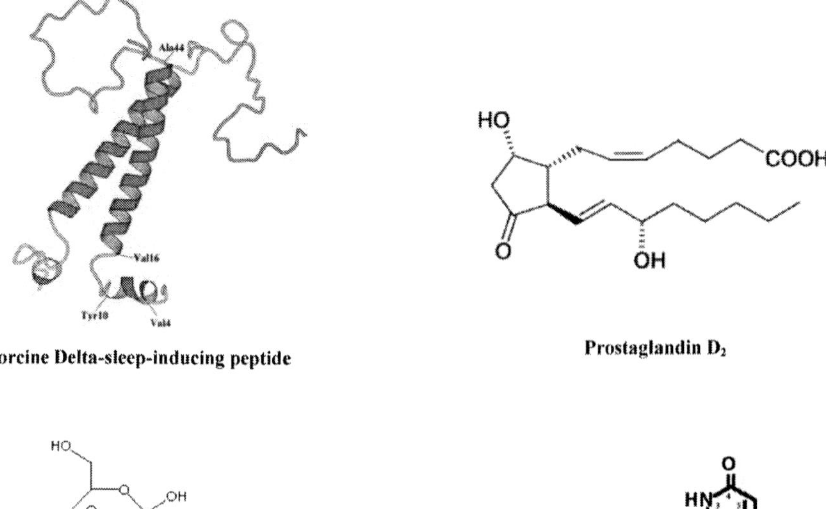

Porcine Delta-sleep-inducing peptide

Prostaglandin D₂

Muramyl dipeptide

Uridine

FIGURE 1. Structures of some sleep-inducing substances.

2. Delta Sleep-Inducing Peptide

DSIP is a naturally occurring substance and in 1977 it was for the first time isolated from rabbit brain.[25,26,79,88] As its name suggests DSIP promotes sleep, especially a particular type of sleep that is characterized by an increase in the delta rhythm of the EEG. It is a nonapeptide that is normally synthesized in the hypothalamus and targets multiple sites including the brain itself[26] and has been detected in rats, mice, rabbits, cats and human beings.[1,48,93] Immunopositive neurons were found in the neurons of rostral-caudal band extending from the primary olfactory cortex to the lateral hypothalamus, basal ganglia, amygdala, septum and the thalamus.[18] Labeled neurons were found at higher concentrations in the areas known to be active during SWS and REM sleep for example, the reticular formation and the raphe nuclei.

However the functional significance of the presence of DSIP in these areas is not clearly known, even though it is safe to assume that DSIP may have a modulatory role in these areas especially because they deal with sensory systems, such as somatic sensation, audition, and vision, and visceral brain systems.

DSIP is normally present in minute amounts in the blood. It can readily be absorbed from the gut without being denatured by enzymes.[2,45] It is degraded in blood through the amino-peptidase pathway.[71] Brain and plasma DSIP concentrations exhibit a marked diurnal variation[4] and there has been shown to be a correlation between DSIP plasma concentrations and circadian rhythm in human beings. Concentrations are higher in the afternoons than mornings and it is reported that the plasma concentrations of DSIP are influenced by the initiation of sleep.[89] DSIP is present in relatively high concentrations in human milk (10-30 ng mL^{-1})[7,23] and urine (100 pg-6 ng/ml).[15,22] It is also present in the CSF (30-560 pg/ml)[22,46] suggesting that crossing over the BBB by DSIP is not competitive and probably non-specific.[3,5]

A number of studies have examined the use of DSIP with varied success.[7,67] It is reported that subjects injected with small doses of DSIP increased the 'pressure to sleep' and to a certain extent induced delta-wave sleep. Hence, DSIP has been described as a sleep-promoting substance rather than a sedative. DSIP does not act as a hypnotic drug that needs to be given just before retiring in acute conditions. On the other hand, a dose of DSIP given during the course of the day will promote improved sleep on the next night and may be for several successive nights. Long-term infusion of DSIP resulted in rapid increase in SWS[39], the amount of hourly SWS increased to 30-40 min (Figure 2). However, such an SWS-inducing effect did not last long. It is also known to enhance delta and sigma activity in rabbits.[1,65,105] One study reported significant increase in both SWS and REM sleep at the expense of waking when DSIP was injected into the lateral ventricle during the dark period[106]. Another study observed enhanced total sleep in REM sleep deprived cats.[94] Despite these clear short-term results the beneficial efficacy of DSIP in long-term management of insomnia has still to be proven. So, it seems that the physiological actions of the peptide mostly depend on the behavioral state of the animal; they are stronger under the disturbed than normal behavioral conditions. DSIP may, paradoxically, be of use in the treatment of narcolepsy and it is possible that it exerts its effect by restoring circadian rhythms.[87] Unfortunately, no difference in DSIP concentrations has been found between narcoleptic patients and normal subjects.[100] However patients suffering from certain psychiatric disorders, such as schizophrenia and depression have lower concentrations of DSIP both in plasma and cerebrospinal fluid.[64] In these patients the concentrations were inversely correlated with sleep disturbance. DSIP when administered to healthy volunteers did not have any effect on growth hormone or prolactin.[21]

There are published reports, where no sleep inducing effect (in some case wake inducing) of DSIP was found in rabbits after IV injection,[27] in cats after

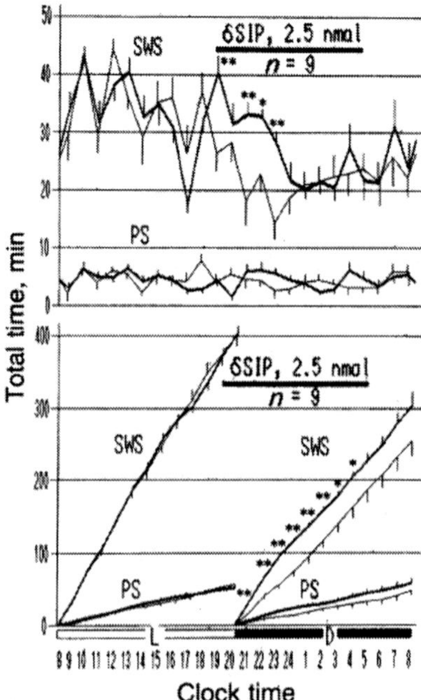

FIGURE 2. Sleep-promoting effects of a 10-hr intraventricular infusion of DSIP at 2.5 nmol from 19:00 to 05:00 (indicated by a solid bar) in otherwise saline infused rats. The upper graph shows the hourly integrated sleep amounts, while the bottom graph illustrates the cumulative values in the light (L) and the dark (D) period. Thin and thick curves represent the baseline and the experimental day, respectively. Vertical lines on each hourly value indicate the range of SEM, * and**, values on the experimental day that were significantly different from that of the baseline at $P < 0.05$ and $P < 0.01$, respectively. Adapted with permission from Inoue et al., Proc Natl Acad Sci, USA, 1984.

IP injection[92] and in rats after ICV injection.[73] Brain temperature in the rats was not affected and an analog, ω-amino-caprilyl DSIP, also failed to change any of these parameters. Rather an increase in wakefulness was observed after the injection of either peptide. Kovalzon[53] also showed a negative effect of DSIP. However, other analogs like D-Trp[1]-DSIP, D-Tyr[1]-DSIP, and D-Ala[2]-DSIP found to induce a marked increase of mainly SWS. Based on their observation they concluded that prevention of rapid metabolism of the peptide was critical in achieving the desired effect. It would appear therefore that there might be dual action within these analogues for not only sleep but also sleep reversal.

Nevertheless, with contrast results obtained with DSIP from different laboratories, it seems reasonable to conclude that the peptide can affect sleep in many animal species, although there has not been consistent reproducibility.

DSIP has also shown to produce changes in other functions apart from sleep. Previous studies have shown that DSIP has an anticonvulsant action in the rat by increasing the threshold to NMDA- and picrotoxin-induced convulsions.[11,90] DSIP has an antinociceptive action in mice, an effect which is blocked by naloxone,[60] neuroprotective effect in rats subjected to bilateral carotid ligation,[91] antihypertensive effect[24] and antimetastatic activity.[80] DSIP also reduced brain swelling in a model of toxic cerebral edema in the rat.[78] As might be expected of any substance that is naturally occurring, DSIP does not have any significant side effects. In some human studies, transient headache, nausea and vertigo have been reported. DSIP, (which has a half-life of 15 min) actually appears to be incredibly safe as its LD_{50} has never been determined because it has never so far proved possible to kill an animal whatever the dose of DSIP administered.[38] It may also reduce amphetamine-induced hyperthermia and may be beneficial in some chronic pain conditions.[60]

A mechanism of action for DISP could be due to the modulation of noradrenergic system. Modulation of POA neurons by noradrenergic system have shown to induce sleep.[58,59,82,83] Benson et al.[6] in 1985 reported a markedly reduced turnover norepinephrine and alpha-methylparatyrosine in hypothalamic median eminence after pretreatment of rats with DSIP, thus establishing a direct relation between the adrenergic neurotransmitter system and DSIP. In vitro, DSIP inhibited the effect of isoproteronol alone or in a combination with phenylephrine, in rat pineal culture for two days, but not with phenylephrine alone. Although it does not fully explain the paradoxical results of its action, it is suggested that the action of DSIP is mediated through α_1 adrenergic postsynaptic receptor. It is not known by what mechanism alpha-adrenergic receptors act synergistically with beta-adrenergic receptors in this system,[49] so that it is possible that DSIP could influence the 'link' between alpha and beta-adrenergic receptors.[24] This modulatory effect on the noradrenergic system can be used to explain the sleep effects observed after DSIP administration. It is well known that noradrenergic transmission can affect sleep stages, both SWS and REM sleep. Administration of specific adrenergic agonists and antagonists revealed that the stimulation of α_1 adrenergic receptors decrease PS while drowsy waking or light sleep is increased.[81] Stimulation of α_2 receptors also increases SWS.[50,63,82,83]

3. Uridine

Uridine, a nucleic acid which is ubiquitously distributed in many tissues, including the brain is involved in a variety of biochemical processes. Uridine consists of uracil as base moiety and D-ribose as sugar moiety on its structure. This compound has the same oxopyrimidine ring in the structure as that of barbiturates, nonbarbiturate anesthetics, and antiepileptics. The basic structure of the barbiturates is similar to that of uracil. It is suggested that uridine is released from steps of nucleic acid-nucleic protein biosynthesis (catabolism), and reaches the binding sites in the areas of the brain which regulate natural sleep. Komoda et al.[51] reported that uridine was one of the endogenous sleep-promoting substances which was isolated from brainstems of 24-hour sleep-deprived rats. The infusion of small amounts of uridine caused the enhancement of natural sleep in rats.[39]

Uridine increased the frequency of the sleep without affecting both quality and quantity of the sleep. Although uridine is shown to promote natural sleep in rats, it does not have any hypnotic activity. The effect of uridine can be effectively studied by the use of systematically synthesized nucleoside derivatives. Thus, studies showed that the hypnotic activity of the derivative, N^3-phenacyluridine was 16-fold higher than that of N^3-benzyluridine. It also induced motor incoordination,[103] had anti anxiety effects and shortened the swimming time, indicating this also acted on the CNS depressant system. However, N^3-Benzyluridine and N^3-phenacyluridine showed a dose-dependent hypnotic activity in mice indicating its possible role in physiological sleep. The hypnotic effect of N^3-benzyluridine was three-fold higher than that of thiopental at the same dose.[102]

Infusion of 1 pmol N^3-benzyluridine resulted in a significant increase in the total time and frequency of PS in the light period of the recovery. In contrast, infusion of N^3-benzyluridine at 10 pmol, significantly enhanced frequency and the total time of SWS without affecting the duration of episodes. A 10 h infusion of 10 pmol uridine induced a rather mild but continued increase in both SWS and REM sleep (Figure 3). There was a significant increase in the cumulative value of SWS and PS from baseline only after 7 and 12 h after the dark onset, respectively. In contrast, uridine did not enhance PS and SWS in the light period. Uridine did not produce any effect on the brain temperature.[39] Although the sleep-promoting effects of uridine were still evident during recovery period, the potentiation of sleep was limited in the dark period. So it is safer to say, that the mechanisms of sleep-inducing effect may be different. Kimura et al.[47] suggests that N^3-benzyluridine may play a role in sleep regulation as an intermediate between natural and artificial sleep.

It is well known that benzodiazepine binding site is coupled to $GABA_A$ receptor complex in the brain[30,74] which includes γ-aminobutyric acid (GABA) binding site, chloride channel, picrotoxin binding site and barbiturate active site. Certain barbiturates affect benzodiazepine and GABA bindings.[74,101] Furthermore, it is known that pentobarbital increased [^{3}H]diazepam binding to synaptic membranes in the presence of chloride ion. Therefore, Kimura et al.[47] predicts that uridine derivatives also affect the GABA-benzodiazepine receptor-chloride channel complex because of the structural similarity to barbiturates. Guarneri et al.[28,29] and Yamamoto et al.,[104] showed the interaction of uridine with GABA binding site in cerebellar membranes of rats, although much higher concentration of uridine is needed to compete with GABA at its binding sites. However, Kimura et al.[47] suggests that uridine binding site is also located closely to adenosine binding site, and may functionally act together to produce CNS depressant activity including promoting sleep. They further

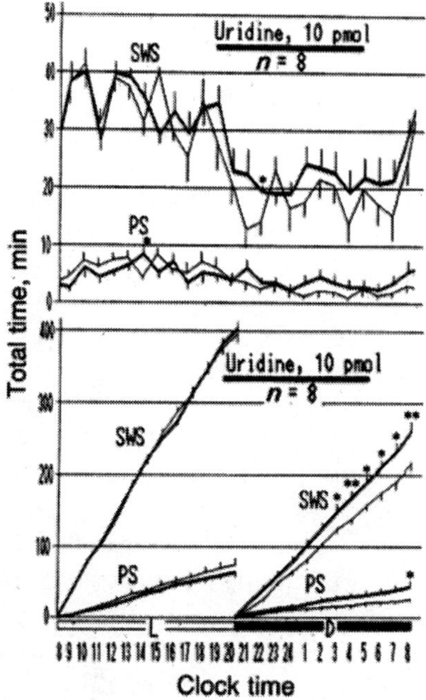

FIGURE 3. Sleep-promoting effects of uridine. For details, see figure 2 caption.

propose that the phenomenon of sleep and wakefulness is a reversible physical action; endogenous substances such as uridine might be utilized in de novo syntheses of this kind of nucleic acid and hypothesize that the uridine receptor mainly mediates the hypnotic action of derivatives of oxopyrimidine nucleoside (Figure 4).

4. Prostaglandin D_2

The prostaglandins (PGs) are a group of C_{20} polyunsaturated fatty acids that contain a unique five-membered ring structure. PGs exhibits diverse but important role in on a wide variety of physiological and pathological activities. It is widely distributed in all types of mammalian cells. In 1982, Narumiya et al.[70] showed that PGD_2 is the major prostanoid in the CNS of rats including humans and have some important neural function. Microinjection of PGD_2 into the preoptic area of the rat brain induces behavioral sleep. Such sleep-inducing effects are seen also in mice and monkeys. Roberts and co-workers[84] have shown increased levels of PGD_2 in patients with systemic mastocytosis during deep sleep episodes. A highly significant increase has also been reported in the CSF of patients with African sleeping sickness.[77]

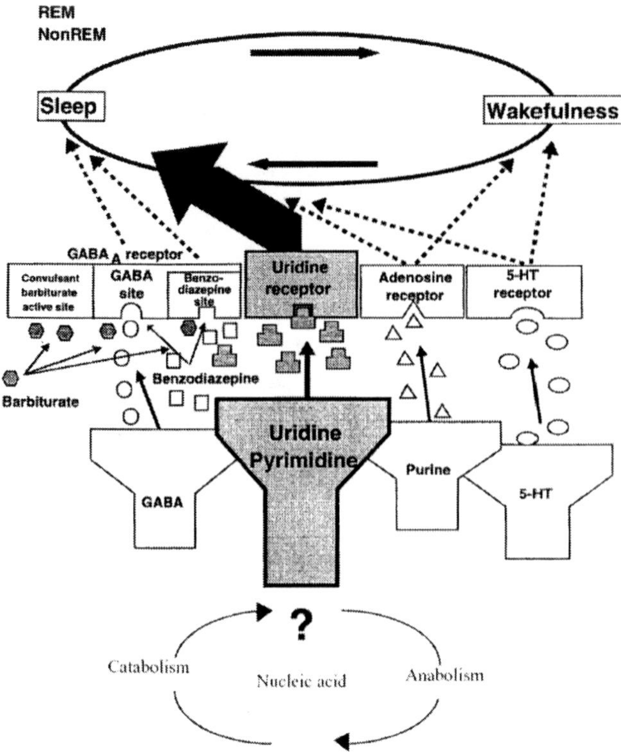

FIGURE 4. Postulated mechanisms of hypnotic activity of oxopyrimidine nucleoside derivatives. For details see Kimura et al., Sleep 2001; 24: 251-260. Adapted with permission.

A 10 h infusion of PGD_2 resulted in a rapid increase in SWS and a gradual increase in REM sleep, significant effect seen in the second half of the infusion period[39] (Figure 5). When PGD_2 was infused during the night, SWS and REM sleep showed a significant dose dependent increase, which was indistinguishable from physiological sleep as judged by EEG, EMG, food and water intake, brain temperature, heart rate, and other general behavior of the rat[97].

The cumulative increment of SWS and REM sleep in the dark period was about 64 min and 13 min respectively. This increase was entirely due to more frequency in SWS and REM sleep episodes. PGD synthase (PDGS), an enzyme that produces PGD_2, and the PGD_2 concentration in the CSF exhibited a circadian fluctuation in parallel with the sleep-wake cycle,[75] for reviews.[31-36]

PGD_2 is a monometric glycoprotein with a molecular weight of ~27 000 Da and belong to be a member of the lipocalin superfamily. It is produced in the trabecular cells of the arachnoid membrane and choroids plexus and circulates in the CSF. These proteins are widely distributed in nature and are

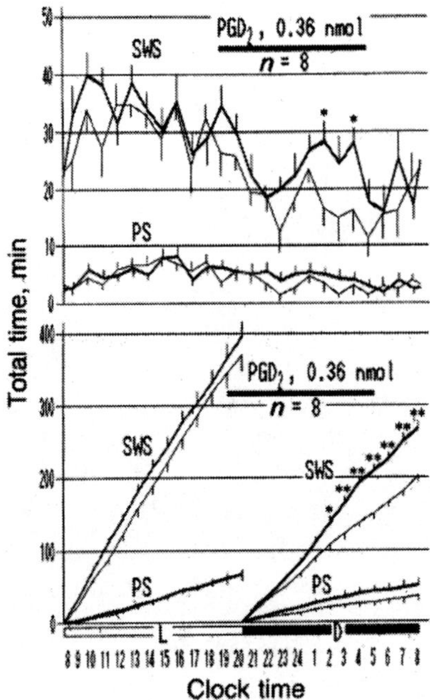

FIGURE 5. Sleep-promoting effects of PGD_2. For details, see figure 2 caption.

small secretory proteins present in the membrane system in the cell, namely, ectoproteins sharing a common feature of binding and transport of small lipophilic molecules with an exception of PGDS, which is an enzyme rather than a lipid transporter. PGDS catalyses the isomerization of the substrate PGH_2 to produce PGD_2.

When selenium chloride, which is a potent, specific, noncompetitive, and reversible inhibitors of brain PGDS[43] was infused into the third ventricle of a rat during the day, sleep was inhibited in a time and dose dependent manner after about 2 h. It could be because of its interaction with the free sulfhydryl group in the active center. This effect was reversible in that, when the infusion was interrupted, sleep was restored. These results therefore clearly showed that PGD_2 is an endogenous sleep-promoting substance. In recent studies on transgenic mice in which the human PGDS gene was incorporated sleep was significantly increased. In situ hybridization to detect mRNA and immunohistochemical staining studies showed PGDS mRNA in the rat brain and spinal cord, more significantly in the leptomeninges, namely the pia mater and arachnoid membrane, and in the choroid plexus in the ventricles than the brain parenchyma, the corpus callosum.[98] The neurons did not express the mRNA although it was expressed the oligodendrocytes. Moreover, the PGDS enzyme activity in the choroid plexus and arachnoid membrane was much higher than the whole brain both in rats and humans. The most effective site of action was the subarachnoid space in the surface area of the medial ventral region of the rostral basal forebrain. SWS was increased more than 120% indicating the presence of PGD_2 receptors in the surface of the leptomeninges somewhat rostral to the preoptic area and below the diagonal band of Broca. As PGDS is a member of the lipocalin superfamily, whose members are all lipid transporters and secretory proteins, it is reasonable to assume that PGDS is mainly, if not exclusively, produced in the membrane system surrounding the brain, namely the arachnoid membrane and choroid plexus, and is then secreted into the CSF to circulate in the CSF in the ventricular and subarachnoid spaces.

Subsequently, the mouse PGD_2 receptor was cloned by Naruymiya and coworkers, and its structure was delineated.[37] Adenosine A_{2a} agonists, 2-[p-(2-carboxyethyl) phenylethylamino]-5′-N-ethylcarboxamidoadenosine (CGS21680), when administered to the PGD_2-sensitive zone, both SWS and REM sleep increased indicating that the stimulation of A_{2a}-adenosine receptors in the rostral basal forebrain promotes SWS and REM sleep.[85] Infusion of PGD_2 into the subarachnoid space just anterior to the preoptic area induced Fos-IR in the ventrolateral preoptic area (VLPO) in association with an increase in SWS, along with a decrease in Fos in putative wake-active neurons like the tubomammilary nucleus (TMN). These observations suggest that PGD_2 may induce sleep via meningeal PGD_2 receptors with subsequent activation of VLPO neurons.[86] Thus, PGD_2 may promote sleep by inducing meningeal cells to release paracrine signaling molecules such as adenosine, which subsequently excite nearby sleep-active VLPO neurons (Figure 6). The VLPO whose axons contain GABA and galanin, on the other hand, inhibits wake-promoting TMN neurons to induce sleep.

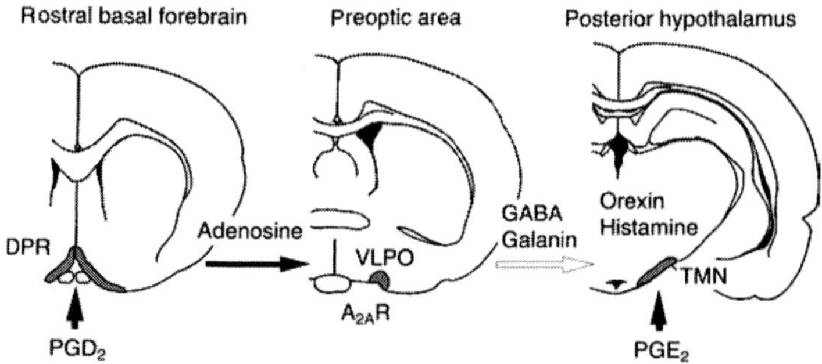

FIGURE 6. Schematic representation of the molecular mechanisms of sleep-wake regulation by PGD_2. solid arrow represent excitatory; open horizontal arrow represents inhibitory. Adapted with permission from Hayaishi, J Appl Physiol, 2002.

5. Muramyl Peptides

'Factor S' also called sleep-inducing substances are derived from bacteria that inhabit the bowel and was identified as a structurally related group of muramyl peptides (MPs) which constitute the cell wall component of both Gram positive and Gram negative bacteria.[12,19,55-57]

Muramic acid and diaminopimelic acid are constituents of polymeric peptidoglycans in the cell walls of bacteria and the subunits are immunostimulants and pyrogens[61]. It is contrary to the idea that the body synthesized its own sleep factors. Apart from its influence in regulating sleep, MPs are known to exert potent effects on the immune system by stimulating many sites on both the cellular and humoral arms of the immune system.[12] The gastrointestinal tract of the neonatal is sterile at birth[20] and no conclusive report has been published to establish a link between maternal supply of MPs to the fetus. This could be one of the reasons why neonates and babies in utero have a very large amount of REM sleep apart from the argument that immaturity of the brain prevents SWS in neonates.[44] Davenne and Krueger[13] in 1987, however, showed that administration of the synthetic MP (N-acetylmuramyl-L-alanyl-D-isoglutamine (AcMur-Ala-iGln), muramyl dipeptide (MDP; Ellouz et al.[16]) to neonatal rabbits did in fact stimulate SWS and reduce REM sleep. Microinjection studies indicate the site of action of this peptide is localized between the basal forebrain and the mesodiencephalic junction. Thus it can be safely concluded that the availability of MPs may well regulate the ontogeny of sleep in the neonate[10]; the development of the gut flora in significant numbers in growing babies slowly increases SWS. Clinically the gut flora/sleep ontogeny relationship could be important in situations where infants are placed on long-term or large doses of antibiotics, where diet is inadequate, or even in the difference between breast and bottle fed babies.[9] All of these situations produce qualitative and/or quantitative changes in gut flora composition.

A 10 h infusion of muramyl dipeptide into the third ventricle resulted in a rather slow increase in SWS.[39] Not only did it have an effect on SWS, but also it increased brain temperature by 1.0-1.5°C. The cumulative data showed the SWS significantly increased from 5 to 12 h after the onset of dark period (Figure 7). MDP did not induce any change in the REM sleep pattern. The increase in SWS was due to longer episode duration and not due to an increase in episode frequency. In another study however, icv infusion of MDP increased the hourly percentage of SWS in rabbits for 6 h or more after the infusion.[55] The amplitudes of slow waves were high during these episodes and there was an increase of 1-2°C in body temperature indicating that it may act directly on temperature-regulating neurons or indirectly through centrally generated endogenous pyrogens.

In contrast, the administration of MDP in cats showed a more complex behavior. The rise in body temperature was associated with lethargy and inability to enter deep sleep for 2-4 h after the infusion. There was no REM

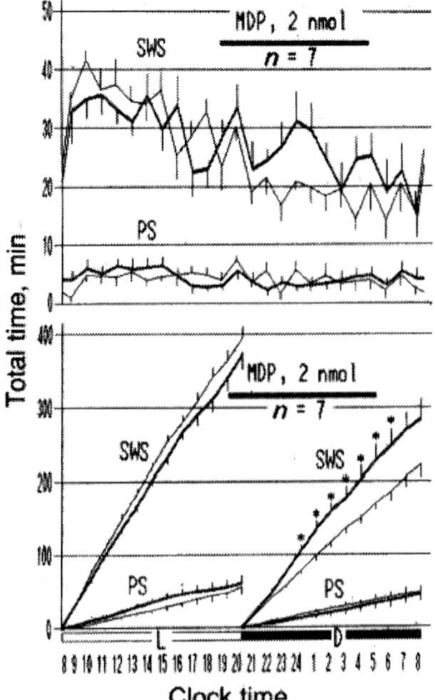

FIGURE 7. Sleep-promoting effects of MDP. For details, see figure 2 caption.

sleep during this period. However, all cats which received the drug underwent an abnormally long period of deep sleep after the initial period of sleep disturbance.[55] The somnogenic effects can also be elicited by intravenous and intraperitoneal injection or by oral administration via stomach tube.

As we mentioned in the beginning, this chapter is just an introduction to the influence of sleep-inducing substances on sleep. Although many significant strides have taken place in recent years, a unified mechanism on how we sleep and wake still remains elusive. Yet, as of today, studies to decipher the science of sleep fascinate researchers worldwide.

6. References

1. Alyautdin RN, Kalikhevich VN, Churkina VI. Hypnogenic properties of the delta-sleep peptides and its structural analogs. Farmacol Toksikol 1984; 47: 26-30.
2. Banks WA, Kastin AJ, Coy DH, Angulo E. Entry of DSIP peptides into dog CSF, role of physiochemical and pharmacokinetic parameters. Brain Res Bull 1986; 17:155-158.
3. Banks WA, Kastin AJ, Coy DH. Evidence that 125I-N-Try-delta-sleep-inducing-peptide crosses the blood brain barrier by a noncompetitive mechanism. Brain Res 1984; 301: 201-207.

4. Banks WA, Kastin AJ, Selznik JK. Modulation of immunoactive levels of DSIP and blood-brain permeability by lighting and diurnal rhythm. J Neurosci Res 1985; 14: 347-355.

5. Banks WA, Kastin AJ. CSF-plasma relationship for DSIP and some other neuropeptides. Pharmacol Biochem Behav 1983; 19: 1037-1040.

6. Benson B, Gregory VM, Schoenenberger GA. Acute reduction of LH and hypothalamic norepinephrine turnover by delta sleep-inducing peptide. Endocrinology 1985; 116: Sympos Suppl, 257.

7. Bes F, Hofman W, Van Schuur J, BC. Effect of delta sleep-inducing peptide on sleep of chronic insomniac patients. A double-blind study. Neuropsychobiology 1992; 26: 193-197.

8. Borbley A. Endogenous sleep-promoting substances. Trends Pharmacol Sci 1982; 3: 350.

9. Brown R, Andren J, Andrews K, Brown L, Chidgey J, Geary N, King MG and Roberts TK. Muramyl peptides and the functions of sleep. Behav Brain Res 1995; 69: 85-90.

10. Brown R. Pang G, Husband AJ, King MG, Bull DF. Sleep deprivation and the immune response to pathogenic and non pathogenic antigens. In A.J. Husband (Ed.), Behaviour and Immunity, CRC Press, Boca Raton, 1992, 127-133 pp.

11. Brusentsov AI, Moroz VV, Suprun SA, Pomazanova TN, Shandra AA, Godleviskii. [Changes in the blastogenic lumphocyte transformation during kindling induced by picrotoxin in rats] Ross Fiziol Zh Im I M Sechenova 1998; 84: 233-237.

12. Chedid L, Synthetic muramyl peptides: their origin, present status and future prospects. Fed Proc. 1986; 45: 2545-2548.

13. Davenne D, Krueger J. Enhancement of quiet sleep in rabbit neonates by muramyl dipeptide. Am J Physiol. 1987; 253: R646-R654.

14. Drucker-Collin R, Valvarde RC. Endocrine and peptide functions in the sleep-waking cycle. In: Ganten D, Pfaff D, eds. Sleep-clinical and experimental aspects. Berlin: Springer-Verlag, 1982; 37-81.

15. Ekman R, Larson I, Malmquist M, Thorell JI. Radioimmunoassay of delta sleep-inducing peptide using an iodinated p-hydroxyphenylpropionic acid derivative as tracer. Regul Pept 1983; 6: 371-378.

16. Ellouz F, Adam A, Ciorbaru R, Lederer E. Minimal structural requirements for adjuvant activity of bacterial peptidoglycan derivatives. Biochem Biophys Res Commun. 1974; 59: 1317-25.

17. Ernst A, Monti JC, Schoenenberger GA. Delta-sleep-inducing peptide (DSIP) in human and cow milk. In: Koella WP, Ruether R, Schulz H, eds. Sleep. Stuttgart: Fisher Verlag, 1984; 215-217.

18. Feldman SC, Kastin AJ. Localization of neurons containing immunoreactive delta sleep-inducing peptide in the rat brain: an immunocytochemical study. Neuroscience 1984; 11: 303-317.

19. Fencl V, Koski G, Pappenheimer JR. Factors in cerebrospinal fluid from goats that affect sleep and activity in rats. J Physiol. 1971; 216: 565-589.

20. Foo MC, Lee A, Cooper GN. Natural antibodies and the intestinal flora of rodents. Aus J Exp Biol.1974; 52: 321-330.

21. Giusti M, Carraro A, Porcella E. Delta sleep-inducing peptide administration does not influence growth hormone and prolactin secretion in normal woman. Psychoneuroendocrinology 1993; 18: 79-84.

22. Graf MV, Hunter CA, Fishman AJ. DSIP occurs in free form in mammalian plasma, human CSF and urine. Pharmacol Biochem Behav 1984b; 21: 761-766.
23. Graf MV, Hunter CA, Kastin AJ. Presence of delta sleep-inducing peptide-like material in human milk. J Clin Endocrinol Metab 1984a; 59: 127-132.
24. Graf MV, Kastin AJ, Schoenenberger GA. Delta sleep-inducing peptide in spontaneously hypertensive rats. Pharmacol Biochem Behav 1986; 24: 1797-1799.
25. Graf MV, Kastin AJ. Delta-sleep-inducing peptide (DSIP): a review. Neurosci Biobehav Rev 1984; 8: 83-93.
26. Graf MV, Kastin AJ. Delta-sleep-inducing peptide (DSIP): an update. Peptides 1986; 7: 1165-1187.
27. Grinjavichus KKA, Milashus AM. Study of the influence of the delta sleep-inducing peptide (DSIP) on slow sleep in rabbits. Zh Vissh Nerv Deiat 1982; 32: 1084-1089.
28. Guarneri P, Guarneri R, Bella VL, Piccoli F. Interaction between uridine and GABA-mediated inhibitory transmission: studies in vivo and in vitro. Epilepsia 1985; 26: 666-671.
29. Guarneri P, Guarneri R, Mocciaro C, Piccoli F. Interaction with uridine with GABA binding sites in cerebellar membranes of the rat. Neurochem Res 1983; 8: 1537-1545.
30. Haefely W. Antagonists of benzodiazepine: functional aspects. In: Biggio G, Costa E, eds. Benzodiazepine recognition site legends. Biochemistry and Pharmacology. New York: Raven Press, 1983: 73-93.
31. Hayaishi O. Molecular mechanisms of sleep-wake regulation: roles of prostaglandins D_2 and E_2. FASEB J. 1991; 5: 2575-2581.
32. Hayaishi O. Prostaglandin D_2 and sleep. In: Hayaishi, O. and Inoué, S. (ed.) Sleep and Sleep Disorders: From Molecule to Behaviour, Academic Press, New York, 1997: 3 10.
33. Hayaishi O. Prostaglandins and sleep. In: Meier-Ewert, K. and Okawa, M. (eds) Sleep-Wake Disorders, Plenum Press, New York, 1998: 1 10.
34. Hayaishi O. Sleep-wake regulation by prostaglandin D_2 and E_2. J Biol Chem., 1988; 263: 14593-14596.
35. Hayaishi, O. Molecular genetics studies on sleep-wake regulation , with special emphasis on the prostaglandin D_2 system. J Appl Physiol, 2002, 92, 863-868.
36. Hayaishi, O. Prostaglandin D_2 and sleep – a molecular genetic approach. J Sleep Res., 1999, 8 (suppl. 1): 60-64.
37. Hirata M, Kakizuka A, Aizawa M, Ushikubi F, Narumiya S. Molecular characterization of a mouse prostaglandin D receptor and functional expression of the cloned gene. Proc Natl Acad Sci, USA 1994; 91: 11192-11196.
38. Huang JT, Lajtha A. The degradation of a nonpeptide, sleep-inducing peptide, in rat brain: comparison with enkephlin breakdown. Res Commun Chem Pathol Pharmacol 1978; 19: 191-199.
39. Inoue S, Honda K, Komoda Y, Uchizono K, Ueno R, Hayaishi O. Differential sleep-promoting effects of five sleep substances nocturnally infused in unrestrained rats. Proc Natl Acad Sci , USA 1984; 81: 6240-6244.
40. Inoue S, Uchizono K, Nagasaki H. Endogenous sleep promoting factors. Trends Neurosci 1982; 5: 218-220.
41. Inoue S. Sleep substances: their roles and evolution (in Japanese). Seikagaku (Tokyo) 1983; 55: 445-460.

42. Ishimori K. True cause of sleep-A hypnogenic substance as evidenced in the brain of sleep deprived animals (in Japanese). Tokyo Igakkai Zasshi 1909; 23: 429-57.

43. Islam F, Watanabe, Ya, Morii H, Hayaishi O. Inhibition of rat brain prostaglandin D synthase by inorganic selenocompounds. Arch Biochem Biophys 1991; 289: 161-166.

44. Jouvet-Mounier D, Astic L, Lacote D. Ontogenesis of the states of sleep in rat, cat and guinea pig during the first postnatal month. Dev. Psychobiol.1970; 2:16-239.

45. Kastin AJ, Banks WA, Castellanos PF, Nissen C, Coy DH. Differential penetration of DSIP peptides into rat brain. Pharmacol Biochem Behav 1982; 17: 1187-1191.

46. Kato N, Nagaki S, Takahashi Y, Namura I, Saito Y. DSIP-like material in rat brain, human cerebrospinal fluid, and plasma as determined by enzyme immunossay. In: Inoue S, Boebley AA, eds. Endogenous sleep substances and sleep regulation. Utrecht: VNU science press, 1985; 141-153

47. Kimura T, Ho IK, Yamamoto I. Uridine receptor: Discovery and its involvement in sleep mechanism. Sleep 2001; 24: 251-260.

48. Kimura M, Honda K, Komoda Y, Inoue S. Interacting sleep-modulatory effects of simultaneously administered delta-sleep-inducing peptide, muramyl dipeptide and uridine in unrestrained rats. Neurosci Res 1987; 5: 157-166.

49. Klein DC, Sugden D, Weller JL. Postsynaptic alpha adrenergic receptors potentiate the beta adrenergic stimulation of pineal serotonin N-acetyltransferase. Proc Natl Acad Sci USA 1983; 80: 599-603.

50. Kleinlogl H, Scholtysik G, Sayers AC. Effects of clonidine and BS 100-141 on the EEG pattern in rats. Eur J Pharmacol 1975; 33: 159-162.

51. Komoda Y, Ishikawa M, Nagasaki H, Iriki M, Honda K, Inoue S, Higashi A, Uchizono K. Uridine, a sleep-promoting substance from brainstems of sleep-deprived rats. Biomed Res 1983; 4: 223-228.

52. Kornmuller AE, Lux HD, Winkel K, KleeM. Neurohumoral ausgeloste Schlafzustande an Tieren mit gekreuztem Kreislauf unter der Kontrolle von EEG-Ableitungen. Naturwissenschaften 1961; 48: 503-505.

53. Kovalzon VM. Search for a sleep peptide. Internat Congress APSS, Bologna, 1983; 208.

54. Krueger JM, Obal FJ, Fang J, Kubota T, Taishi P. The role of cytokines in physiological sleep regulation. Ann N Y Acad Sci 2001; 933: 211-221.

55. Krueger JM, Pappenheimer JR, Karnovsky ML. Sleep-promotig effects of muramyl peptides. Proc Natl Acad Sci, USA. 1982; 79: 6102-6106.

56. Krueger JM, Pappenheimer JR, Karnovsky ML. The composition of sleep -promoting factor isolated from human urine. J Biol Chem.1982; 257: 1664-1669.

57. Krueger JM. Somnogenic activity of muramyl peptides. TIPS.1985; 6:218-221.

58. Kumar VM, Sharma R, Wadhwa S, Manchanda SK. Sleep-inducing function of noradrenergic fibers in the medial preoptic area. Brain Res Bull 1993; 32: 153-158.

59. Kumar VM, Datta S, Chhina GS, Singh B. Alpha adrenergic system in medial preoptic area involved in sleep-wakefulness in rats. Brain Res Bull 1986; 16: 463-468.

60. Larbig W, Gerber WD, Kluck M, Schoenenberger GA. Theraputic effects of delta-sleep-inducing peptide (DSIP) inpatients with chronic, pronounced pain episodes. A clinical pilot study. Eur Neurol 1984; 23: 372-385.

61. Lederer E. Synthetic immunostimulants derived from the bacterial cell wall. J Med Chem. 1980; 23: 819-25.

62. Legendre R, Pieron H. Recherches sur le besoin de sommeil consecutif a une veille prolongee. Z Allgem Physiol 1913; 14: 235-262.

63. Leppavouri A, Putkonen PTS, Sternberg D. Sedation and suppression of paradoxical sleep in cat after clonidine. Acta Physiol Scand (suppl) 1976; 440: 60.

64. Lindstrom LH, Ekman R, Walleus H, Widerlov E. Delta-sleep inducing peptide cerebrospinal fluid from schizophrenics, depressives and healthy volunteers. Prog Neuropsychopharmacol Biol Psychiatry 1985; 9: 83-90.

65. Mikhaleva II. Delta-sleep-inducing peptide (DSIP): Novel biological properties and an active cyclic analog. Internat Congress APSS, Bologna, 1983: 136.

66. Monnier M, Koller T, Graver S, Humoral influences of induced sleep and arousal upon electrical brain activity of animals with crossed circulation. Exp Neurol 1963; 8: 264-277.

67. Monti JM, Debellis J Alterwain P, Pellejero T, Monti D. Study of delta sleep-inducing peptide efficiency in improving sleep on short term administration to chronic insomniacs. Int J Clin Pharmacol Res 1987; 7:105-110.

68. Nagasaki H, Iriki M, Inoue S, Uchizono K. The presence of sleep promoting material in the brain of sleep-deprived rats. Proc Jpn Acad 1974; 50: 241-246.

69. Nakamura A, Nakashima M, Sugao T, Kanemoto H, Fukumura Y, Shiomi H. Potent antinociceptive effect of centrally administered delta-sleep-inducing peptide (DSIP). Eur J Pharmacol 1988; 155: 247-253.

70. Narumiya, S., Ogorochi, T., Nakao, K., Hayaishi, O. Prostaglandin D_2 in rat brain, spinal cord and pituitary: basal level and regional distribution. *Life Sci.*, 1982, 31: 2093 2103.

71. Nyberg F, Thornwall M, Hetta J. Aminopeptides in human CSF which degrades delta-sleep inducing peptide (DSIP). Biochem Biophys Res Commun 1990; 167: 1256-1262.

72. Obal F Jr, Krueger JM. Biochemical regulation of non-rapid eye movement sleep. Front Biosci 2003; 8: d520-550.

73. Obal F, Torok A, Alfoldi P, Sary G, Hajos M, Penke B. Effect of intracerebroventricular injection of delta sleep-inducing peptide (DSIP) on sleep and brain temperature in rats at night. Pharmacol Biochem Behav 1985; 23: 953-957.

74. Olsen RW. Gaba-benzodiazepine-barbiturate receptor interactions. J Neurochem 1981; 37: 1-13.

75. Pandey HP, Ram A, Matsumura H and Hayaishi O. Concentration of prostaglandin D_2 in cerebrospinal fluid exhibits a circadian alteration in conscious rats, Biochem Mol Biol Int, 1995 37: 431-437.

76. Pappenheimer JR, Miller TB, Goodrich CA. Sleep promoting effects of cerebrospinal fluid from sleep-deprived goats. Proc Natl Acad Sci USA 1967; 58: 513-517.

77. Pentreath VW, Rees K, Owolabi OA, Philip KA, Duoa F. The somnogenic T lymphocyte suppressor prostaglandin D_2 is selectively elevated in cerebrospinal fluid of advanced sleeping sickness patients. Trans R Soc Trop Med Hyg 1990; 84: 795-799.

78. Platonov IA, Iasnetsov VV. The effect of the delta sleep-inducing peptide on the development of toxic brain edema-swelling. Biull Eksp Biol Med 1992; 114: 463-464.

79. Pollard BJ, Pomfrett CJD. Delta sleep-inducing peptide. Eur J Anaes 2001; 18: 419-422.

80. Prudchenko IA, Stashevskaia LV, Shepel EN et al. Synthesis and biological properties of delta-sleep peptide analogs. 2. Antimetastatic effect. Bioorg Khim 1993; 19: 1177-1190.

81. Putkonen PTS. Alpha and beta adrenergic mechanisms in the control of sleep stages. In: Priest RG, Pletcher A, Ward J, eds. Sleep Research. Preston (UK): MTP press Ltd 1979; 19-34.

82. Ramesh V, Kumar VM. The role of alpha-2 receptors in the medial preoptic area in the regulation of sleep-wakefulness and body temperature. Neuroscience 1998; 85: 807-817.

83. Ramesh V, Kumar VM, John J, Mallick H. Medial preoptic alpha-2 adrenoceptors in the regulation of sleep-wakefulness. Physiol Behav 1995; 57: 171-175.

84. Roberts JL II, Sweetman BJ, Lewis RA, Austen KF, Oates JA. Increased production of prostaglandin D_2 in patients with mastocytosis. N Engl J Med 1980; 303: 1400-1404.

85. Satoh S, Matsumura H, Suzuki F, Hayaishi O. Promotion of sleep mediated by the A_{2a}-adenosine receptor and possible involvement of this receptor in the sleep induced by prostaglandin D_2 in rats. Proc Natl Acad Sci, USA 1996; 93: 5980-5984.

86. Scammell T, Gerashchenko D, Urade Y, Onoe H, Saper C, Hayaishi O. Activation of ventrolateral preoptic neurons by the somnogen prostaglandin D_2. Proc Natl Acad Sci, USA 1998; 95: 7754-7759.

87. Schneider-Helmert D. Effects of DSIP on narcolepsy. Eur Neurol 1984; 23: 353-357.

88. Schoenenberger GA, Monnier M. Characterization of a delta-electroencephlogram (-sleep)-inducing peptide. Proc Natl Acad Sci USA 1977; 74: 1282-1286.

89. Seifritz E, Muller MJ, Schonenberger GA et al. Human plasma DSIP decreases at the initiation of sleep at different circadian times. Peptides 1995; 16: 1475-1481

90. Shandra AA, Godlevskii LS, Brusentsov AI, Karlyuga VA. Effects of delta-sleep-inducing peptide on NMDA induced convulsive activity in rats. Neuroscibehavphysiol 1998; 28: 694-697.

91. Shandra AA, Godlevskii LS, Brusentsov AI. Effects of delta-sleep-inducing peptide in cerebral ischemia in rats. Neurosci Behav Physiol 1998; 28: 443-446.

92. Sommerfelt L. Reduced sleep in cats after interaperitoneal injection of delta-sleep-inducing peptide (DSIP). Neurosci Lett 1985; 58: 73-77.

93. Susic V. The effect of subcutaneous administration of delta sleep-inducing peptide (DSIP) on some parameters of sleep in the cat. Physiol Behav 1987; 40: 569-572.

94. Susic V, Masirevic G. Effect of delta sleep-inducing peptide (DSIP) on sleep in cats. ESRS Congress, Munich, 1984; 296.

95. Toth P, Lukacs H, Hiatt ES, Reid KH, Iyer V, Roa CV. Administration of human chorionic gonadotropin affects sleep-wake phases and other associated behaviors in cycling female rats. Brain Res. 1994; 654: 181-190.

96. Uchizono K. Studies on sleep substances (in Japanese). Nihon Rinsho (Osaka) 1984; 42: 989-1006.

97. Ueno R, Honda K, Inoue S, Hayaishi, O. Prostaglandin D_2, a cerebral sleep-inducing substance in rats. Proc Natl Acad Sci, USA 1983; 80: 1735-1737.

98. Urade Y, Kitahama K, Ohishi H, Kaneko T, Mizuno N, Hayaishi O. Dominant expression of mRNA for prostaglandin D synthase in leptomeninges, choroid plexus, and oligodendrocytes of the adult rat brain. Proc Natl Acad Sci, USA 1993; 90: 9070-9074.

99. Ursin R. Endogenous sleep factors. Exp Brain Res 1984; Suppl 8: 118-132.

100. Vgontzas AN, Friedman TC, Chorousos GP, Bixler EO, Vela-Beuna A, Kales A. Delta sleep-inducing peptide in normal humans and in patients with sleep apnea and narcolepsy. Peptides 1995; 16: 1153-1156.

101. Willow M, Johnston GAR. Dual action of pentobarbitone on GABA binding: role of binding site integrity. J Neurochem 1981; 37: 1291-1294.

102. Yamamoto I, Kimura T, Tateoka Y, Watanabe K, Ho IK. N^3-Benzyluridine exerts hypnotic activity in mice. Chem Pharm Bull 1985; 33: 4088-4090.

103. Yamamoto I, Kuze J, Kimura T, Watanabe K, Kondo S, Ho IK. The potent depressant effects of N^3-phenacyluridine in mice. Biol Pharm Bull 1994; 17: 514-516.

104. Yamamoto I, Kimura T, Watanabe K, Tateoka Y, Ho IK. Action mechanism for hypnotic activity of N^3-Benzyluridine and related compounds. In: Inoue S, Krueger JM, eds. Endogenous sleep factors. The Netherlands: SPB academic Publishing bv, 1990: 133-142.

105. Ye YH, Lin Y, Lu YJ, Li CX, Ji AX, Xing QY, Liu SY, Zhang WY, Wang ZS, Dai XJ. The application of a new active amide method to the synthesis of delta sleep-inducing peptide (DSIP). Acta Sci Natl Univ Pek 1984; 4: 63-68.

106. Young AMJ, Key BJ. Antagonism of the effect of delta sleep-inducing peptide by naloxone in the rat. Neuropharmacology 1984; 23: 1347-1350.

The Influence of Orexin on Sleep and Wakefulness

CHRISTOPHER M. SINTON AND JON T. WILLIE

1. Introduction

An exciting recent development in sleep research was the discovery of the importance for vigilance state control and narcolepsy-cataplexy of neurons containing orexin (also known as hypocretin) neuropeptides. Narcolepsy-cataplexy is a chronic, debilitating sleep disorder that is characterized by excessive daytime sleepiness, manifested as attacks of daytime somnolence at inappropriate times.[1-3] Narcoleptics also show symptoms that are considered indicative of abnormal REM (Rapid Eye Movement) sleep expression. These latter symptoms include cataplexy, hypnogogic hallucinations, sleep-onset REM periods, and sleep paralysis. In contrast to daytime sleepiness, the nighttime sleep of narcoleptics is fragmented and of poor quality, typically demonstrating lengthy periods of wakefulness after sleep onset.

The autosomal recessive form of canine narcolepsy-cataplexy was found to be caused by a mutation in the orexin receptor-2 gene.[4,5] In a concurrent discovery, mice with deletion of the *prepro-orexin* gene (i.e., *orexin*[-/-] mice) exhibited a phenotype similar to human narcolepsy-cataplexy with increased REM sleep and cataplexy-like episodes entered directly from states of active movement.[6] Subsequently, to model the human disorder more closely with post-natal loss of orexin neurons, an *orexin/ataxin-3* transgenic mouse was created in which orexin neurons expressed a cytotoxic gene product (a truncated form of human *ataxin-3* containing a polyglutamine repeat) under control of the human *prepro-orexin* promoter.[7] This resulted in apoptotic cell death and complete degeneration of orexin-containing neurons by 8 weeks of age. Even by 6 weeks of age, these mice demonstrated the vigilance state changes and episodes of cataplexy that indicated a narcoleptic phenotype.[7] This work in the mouse was then extended to the rat, which also demonstrated narcolepsy-cataplexy after degeneration of orexin projections following expression of the *orexin/ataxin-3* transgene.[8] Cataplexy in canines and rodents consists of attacks of sudden bilateral atonia in postural muscles, with consequent collapse. The episodes last from a few seconds to a few minutes and are often provoked by emotion or excitement, such as food presentation to dogs.[6,9,10]

This significant series of findings provided important insights into the pathophysiology of the narcolepsy-cataplexy syndrome, though mutations of the *prepro-orexin* or orexin receptor genes appear to be extremely rare in the human.[11,12] However, undetectable to very low levels of orexin neuropeptide in the cerebrospinal fluid (CSF) were described in most patients with narcolepsy-cataplexy, while CSF orexin levels of patients presenting with other disorders were comparable to those of controls.[13-16] The number of orexin neurons is therefore likely to be diminished in narcoleptic patients[11,17] though the cause for this apparent neuronal degeneration remains undetermined.[18,19] Recently, rescue of the narcoleptic phenotype in *orexin/ataxin-3* transgenic mice was shown to follow intracerebroventricular administration of orexin or by expression of the *prepro-orexin* gene under the control of the β-actin-cytomegalovirus (CAG) hybrid promotor.[20] Taken together, these findings leave no doubt as to the critical involvement of orexin neurons in the human disorder.

In addition to the role of orexin in the control of sleep and wakefulness, which was first identified from the narcolepsy-cataplexy phenotype and is the subject of this review, the neuropeptide may have a neuromodulatory role in several neuroendocrine and homeostatic functions such as food intake, body temperature regulation, and blood pressure regulation. The reader interested in other functional aspects of orexin is referred to recent studies and reviews.[6,21-25]

2. The Orexin System

Orexin was identified in 1997 by two independent groups. De Lecea and colleages[21] first predicted the sequence of two related peptides, hypocretin-1 and −2, using a direct tag PCR subtraction technique to isolate mRNA from hypothalamic tissue. Independently, Sakurai and colleages,[26] using a systematic biochemical search based on calcium imaging in HEK293 cells expressing G protein-coupled cell surface orphan receptors, purified the two actual peptide ligands, which they termed orexin-A (hypocretin-1) and orexin-B (hypocretin-2). These first two reports indicated that neurons containing the orexin neuropeptides were found exclusively in the dorsal and lateral hypothalamic areas,[21,26] and also that orexin was likely to function as a neurotransmitter since the peptides were localized in synaptic vesicles and had neuroexcitatory effects on hypothalamic neurons.[21] Sakurai and colleages[26] reported that orexin-A and orexin-B, containing 33 and 28 amino acids, respectively, were derived from a single precursor *prepro-orexin* gene, and acted at two G-protein coupled receptors, designated orexin receptor-1 and receptor-2. Studies are continuing as to the relevant roles of the two orexin neuropeptides in the normal animal. In this review, therefore, the two sequences are not differentiated and the term "orexin" as used here indicates either, or both, orexin-A and orexin-B.

Subsequent anatomical studies defined a highly delimited population of orexin-expressing neurons in the perifornical region of the lateral hypothalamus (LH).[22,27,28] Figure 1 shows the location of orexin-containing neurons and the distribution of their processes in a schematic sagittal section of the brain. Immunohistochemical studies revealed a widespread distribution of orexin projections that was remarkable for the targeting of several distinct brain regions that were known to be involved in the regulation of sleep and wakefulness, including both brainstem and forebrain areas.[22,27,29-31] Notable orexin projections to the forebrain include innervation of the cholinergic basal forebrain (BF) (in the rat this includes the horizontal limb of the diagonal band of Broca, the magnocellular preoptic nucleus, and the substantia innominata) and the histaminergic tuberomammillary nucleus (TMN). Major brainstem targets include the pontine and medullary pontine reticular formation (PRF), the cholinergic mesopontine tegmental nuclei (including the laterodorsal tegmental nucleus, LDT), the locus coeruleus (LC), and the dorsal raphe nucleus (DRN).

Two orexin receptors have been identified.[26] Orexin-A is a high-affinity ligand for the orexin receptor 1 (OX_1R), the affinity of which for orexin-B is 1–2 orders of magnitude lower. The orexin receptor 2 (OX_2R) exhibits equally high affinity for both neuropeptides. Currently no ligands are available with sufficient specificity for OX_1R and OX_2R receptors to define their distribution. However, *in situ* hybridization studies of orexin receptor mRNA[32,33] have shown a diffuse pattern, consistent with the widespread nature of orexin projections, although a marked differential distribution of OX_1R and OX_2R

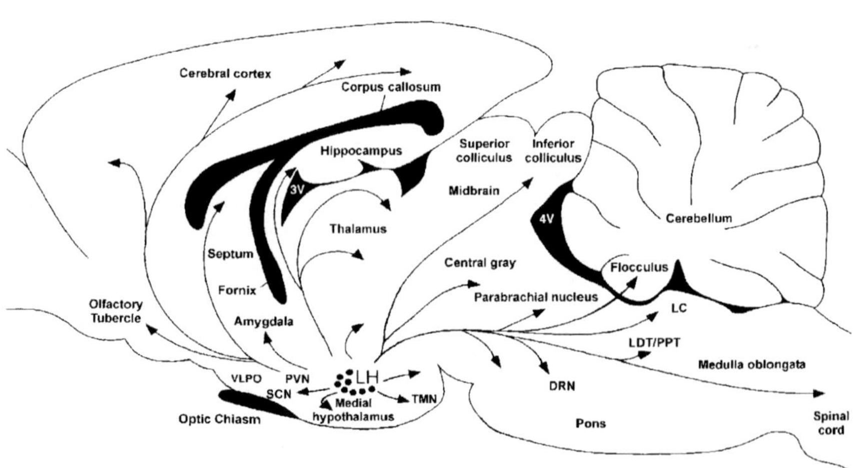

FIGURE 1. Schematic showing the location of orexin-containing neurons (dots) in the lateral hypothalamus (LH) and their widely distributed projections to all levels of the neuraxis. DRN, dorsal raphe nucleus; LC, locus coeruleus; LDT/PPT, laterodorsal tegmental and pedunculopontine tegmental nuclei; PVN, paraventricular nucleus; SCN, suprachiasmatic nucleus; TMN, tuberomammillary nucleus; VLPO, ventrolateral preoptic area; 3V/4V, third/fourth ventricle.

mRNA has also been noted. Of the brain regions involved in vigilance state control, pontine nuclei, including the DRN and LC appear to show a predominance of mRNA for OX_1R receptors, and electrophysiological studies have shown that the response of orexin in the LDT is also likely to be primarily mediated though OX_1R.[34] In contrast, forebrain regions predominately express the OX_2R receptor.

3. Rapid Eye Movement (REM) Sleep

Since its discovery, orexin has remained the focus of intense research as the mechanisms involved in its action are elucidated and the roles that it might play in the normal brain are determined. The intrusion of REM sleep signs into wakefulness in narcoleptic humans suggested that orexin might inhibit REM sleep, or specifically prevent direct shifts from wakefulness into REM sleep. In this review, reference is made to the neurophysiology and neurotransmitters important for the generation of REM sleep and the interested reader is referred to recent reviews of this important topic for more detailed discussion.[35-37] In brief, sub-populations of mesopontine cholinergic cells, including those located in the LDT, discharge preferentially just before and during REM sleep. These LDT cholinergic "REM-on" neurons act as promoters of REM sleep signs, via their excitatory projections to several populations of cells in the PRF.[38] These PRF effector cells in turn control the cardinal signs of REM sleep, such as muscle atonia and rapid eye movements. In contrast, monoaminergic neurons in the noradrenergic LC and serotonergic DRN exhibit a pattern of discharge activity that is nearly opposite: their discharge rate is greatest during waking, declines during non-REM sleep and virtually ceases prior to and during REM sleep.[39,40] This inverse correlation with REM sleep led to the hypothesis that the "Wake-on/REM-off" monoaminergic neurons inhibit the REM-promoting, REM-on neurons, and considerable evidence is now available in support of this hypothesis.[41] Figure 2 is a schematic representation of the location of the cell groups that are important for vigilance state control, especially those that control REM sleep. In this regard, the histamine-containing, TMN neurons also show a Wake-on/REM-off discharge pattern,[42,43] and these neurons are conceptualized as part of the wakefulness-promoting system, in agreement with drowsiness as a common side effect of antihistamine drugs. Other pontine cholinergic cells in the LDT are considered part of the ascending arousal system because they discharge preferentially during wakefulness and REM sleep but show reduced activity during non-REM sleep.[44]

3.1. Orexin Inhibits REM Sleep Onset

Orexin excites LC noradrenergic neurons, thus providing one mechanism by which orexin might promote wakefulness and suppress REM sleep.[45,46]

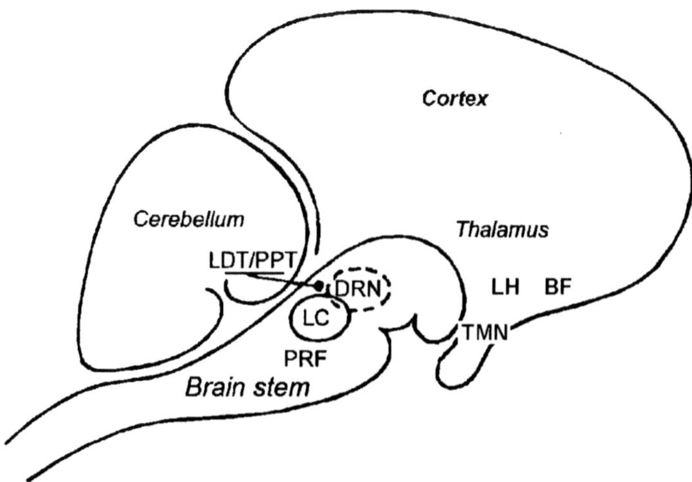

FIGURE 2. Schematic of a sagittal section of a mammalian brain showing the location of nuclei important for vigilance state control, particularly REM sleep. BF, basal forebrain; PRF, pontine reticular formation. Other abbreviations as in Figure 1.

Similarly, orexin has been shown in the *in vitro* slice preparation to excite serotoninergic neurons in the DRN,[47] a second mechanism by which orexin could suppress REM sleep by activating another population of Wake-on/REM-off neurons. However, orexin neurons also project to the brainstem cholinergic REM-on cells, including those in the LDT and the PRF. In an apparent paradoxical action, orexin excites cholinergic LDT neurons.[48] This result would indicate that REM sleep should be disfacilitated in the absence of orexin, rather than the reverse. However, the vigilant state abnormalities observed in the absence of orexin suggest that the lack of excitatory drive on the monoaminergic REM-off cells is the dominant influence. Thus, in the absence of orexin, this disinhibition from the monoaminergic projections at the level of the LDT is apparently greater than the direct disfacilatory action. Under normal circumstances, orexin promotes a tension between the activity of the Wake-on/REM-off and REM-on neurons, with the balance of the excitation being on the monoaminergic Wake-on/REM-off cells most of the time. This balance of effect, and the importance of the response of the monoaminergic Wake-on/REM-off cells, is also underlined by the action of the neuropeptide in the PRF brainstem nuclei that modulate the expression of REM sleep signs. Cholinergic stimulation of the PRF rapidly induces REM sleep,[49] an effect that is mimicked by the local antagonism of orexin. The latter was shown in a study that applied, in the absence of specific receptor antagonists, antisense oligodeoxynucleotides against the mRNA for OX_2R by microdialysis perfusion into the PRF of rats.[50] This treatment increased REM sleep time during both the light and dark phases and, significantly, also produced behavioral cataplexy. One explanation of this result is that inhibition of orexinergic neurotransmission in the PRF results in feedback disfacilitation of Wake-on/REM-off cells and hence the shift in balance towards REM sleep.

Taken together, these results indicate that an important function of orexin is to inhibit the appearance of REM sleep episodes. In confirmation of this conclusion, orexin injected into the LDT in the cat decreased the number of REM sleep episodes.[51] This study is also important, however, because this treatment did not influence the mean duration of REM sleep episodes. In a concordant study, after the post-natal ablation of orexin neurons in the *orexin/ataxin-3* transgenic rat, the number of bouts of REM sleep was increased but the average duration of a REM sleep episode was not changed.[8] These data suggested that orexin affected the switch into REM sleep but, once initiated, it did not influence the expression of the state itself. Kiyashchenko and colleagues[52] reported that orexin release is highest during both active wakefulness and REM sleep, but this result left open the possibility that orexin might also be involved in the generation of the REM sleep state. However, since orexin has now been shown to affect the switch into REM sleep but not the state itself, it is unlikely that orexin is involved in the maintenance, generation, or switch out of REM sleep. Thus the correlation between orexin release in both active wakefulness and REM sleep[52] is likely to be through ongoing motor patterns, whether they are expressed, as in active wakefulness, or inhibited, as in REM sleep. Furthermore, this pattern of release of orexin may be one way in which the inhibition of REM sleep is maintained at a higher level during active wakefulness when the brain state, including cortical activation, has the greatest similarity to REM sleep. For this reason, when orexin is absent, the switching mechanism into REM sleep appears biased[53,54] and direct transitions from wakefulness to REM sleep can occur.

The predominantly inhibitory influence of orexin on the REM sleep switch during wakefulness is consistent with the distribution of vigilance states across the nycthemeron in narcoleptic rodents. For example, comparison of the hourly distribution of REM sleep time between the *orexin/ataxin-3* transgenic and wild-type rats showed an apparently continuous effect of the loss of orexin on the expression of REM sleep (Figure 3).[8] Thus the absence of orexin in the transgenic rats resulted in increased REM sleep throughout the normally active phase and, consequentially, a reduced homeostatic drive for REM sleep during the light phase. In the wild-type rat, the homeostatic drive for REM sleep during the normal sleep phase is expressed as a gradual increase in the hourly times spent in REM sleep. Significantly, this distribution of REM sleep in the wild-type rats is inversely correlated with the diurnal variation of the orexin signal, which remains high throughout the dark period but reaches a minimum towards the end of the light phase in this species.[55-57] Together, these data support the hypothesis that the release of orexin has an inhibitory influence on the appearance of REM sleep episodes. However, once an episode of REM sleep has been initiated, it proceeds normally in the absence of orexin.

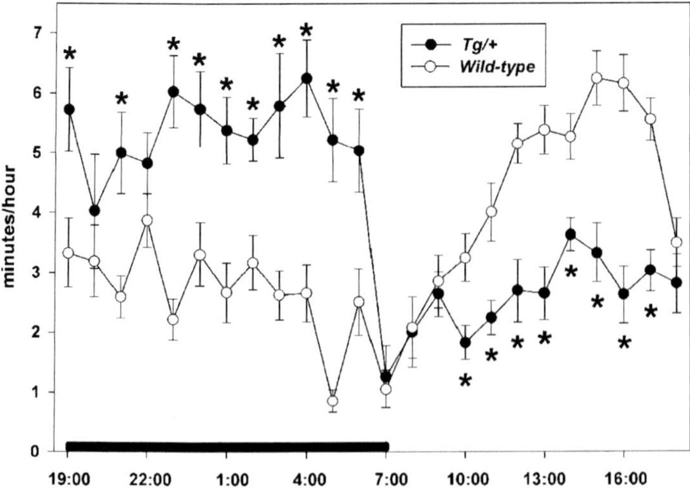

FIGURE 3. Hourly distribution of REM sleep time over the nycthemeron (in minutes; mean ± SEM) for wild-type rats and their *orexin/ataxin-3* transgenic littermates (Tg/+). Significant differences between the genotypes are marked by an asterick (t-test, P < 0.05). The dark phase is noted by the horizontal black bar. Note that REM sleep time remains elevated throughout the dark phase in the Tg/+ rats but is reduced in the light phase relative to the wild-type rats. The latter demonstrate gradually increasing REM sleep time during the light phase, with a peak towards the end of the phase.

3.2. *The Progression of non-REM Sleep to REM Sleep*

In noting the excitatory influence of orexin projections on components of the central activating system, it is clear that the neuropeptide must also play a significant role in central arousal and cortical activation. Importantly, the neuropeptide provides a balance of excitation to the cholinergic REM-on cells of the LDT and PRF and the monoaminergic Wake-on/REM-off cell groups, which are important for the maintenance of wakefulness in addition to their role in the REM sleep switch. However, since orexin is critical for initiating a REM sleep episode, but then has no apparent influence on characteristics of the state itself, such as the duration, electroencephalogram (EEG) spectral characteristics etc., the action of orexin on pontine cholinergic REM-on cells is probably most important while REM sleep is not being expressed.[54] This is especially true during wakefulness, but also to some extent during non-REM sleep, in which state the pontine cholinergic cells show reduced discharge but do not cease firing completely.[44] As non-REM sleep progresses, therefore, the balance of an excitatory influence on the Wake-on/REM-off cells, which inhibit the pontine cholinergic REM-on cells, gradually changes to a net excitation of REM-on cells. In this way, orexin will inhibit the onset of an episode of REM sleep, an effect that is strongest during wakefulness but continues during non-REM sleep. The neuropeptide can thus be considered as an important influence on the normal ordered appearance of the vigilance states from wakefulness through non-REM sleep to REM sleep.

4. Behavioral Arrests and Cataplexy

The definition and clinical diagnosis of cataplexy, in both humans and animals, has historically been derived from observation of collapsed posture with maintenance of consciousness. Willie and colleagues[10] provided a detailed discussion of the behavioral and pharmacological criteria used to define cataplexy across species. Summarized briefly, a cataplectic episode in the mouse can be characterized as a sudden behavioral arrest during ongoing motor activity, followed by an abrupt end to the episode and a return to purposeful activity. Most of these episodes are accompanied by an EEG pattern that is identical to normal REM sleep with atonia, though some are associated with a non-REM sleep pattern. The non-REM sleep episodes are generally behavioral distinct and show a more gradual onset, having the characteristics of a sleep attack as clinically defined in the human. Occasionally a brief period of apparent consciousness is observed at the beginning of a REM sleep associated episode.[10] This finding provided the basis for the important conceptualization that cataplectic episodes can be initiated in the mouse as muscle atonia during wakefulness but that these episodes then rapidly transition to REM sleep. This transition is clearly very rapid and so complete in the mouse that the EEG and electromyogram (EMG) of essentially all such episodes of behavioral arrest are indistinguishable from those recorded during REM sleep. For this reason, the time spent in such episodes is classified as REM sleep.

However in the rat,[8] this transition can be delayed and some episodes of behavioral arrest were observed, lasting for 10–20 sec, with an EEG that was indistinguishable from that of wakefulness, combined with muscle atonia. These periods therefore were more similar to episodes of cataplexy as described clinically. One preliminary conclusion that can be drawn from these data, therefore, is that the duration of the cataplectic component of these episodes varies with species. In comparison with the rodent, episodes apparently last longer in the human and canine with evidence of a more prolonged duration of consciousness, but frequently these episodes will also end as REM sleep in these species. Hence the suggestion by Hishikawa and Shimizu[58] that cataplexy in the human is a transitional state between wakefulness and REM sleep, complementing an earlier accepted definition of cataplexy as a fragmentary manifestation of REM sleep.

The non-REM sleep associated episodes of collapse in the mouse are behaviorally distinct because they show a gradual onset and a delayed (i.e., after at least 60 sec) tendency to progress to REM sleep. Noted particularly in the $OX_2R^{-/-}$ mouse,[10] these are the episodes that are most similar to sleep attacks in the human and are most likely related to a deficit in arousal (i.e., the overwhelming sleepiness as typically described by the narcoleptic patient). An alternative explanation of these episodes has recently been proposed,[59] as a characteristic of the instability of vigilance states in narcoleptic mice.

An additional type of abrupt cataplectic-type arrest, first identified behaviorally in the *orexin*[−/−] mouse,[6] was also found to be associated with non-REM sleep. However, these differed from the gradual arrest type, and the EEG showed the presence of spindles as this type of arrest begun. EEG spindles, which appear specifically during stage 2 non-REM sleep in the human, are episodic bursts of rhythmic, 14- to 16-Hz waves. They also occur at sleep onset and during pre-REM sleep in the rodent.[60] These episodes of abrupt arrest, behaviorally identified as cataplectic in nature, were therefore considered as being initiated just prior to REM sleep as indicated by the EEG.[6] Some of these episodes progress rapidly to REM sleep.[10]

The frequent tendency to transition very rapidly to REM sleep in the mouse made classification of cataplectic episodes difficult in this species initially.[6] However, subsequent careful behavioral observation,[10] supported by studies in the rat,[8] now provide a clearer picture. Figure 4 displays a schema of the different forms of behavioral arrest that have been described and classified in the mouse. In summary, the behaviorally characterized abrupt arrest, which is typically associated with REM sleep, can be initiated either with consciousness and atonia, or into a pre-REM spindling stage, or directly into

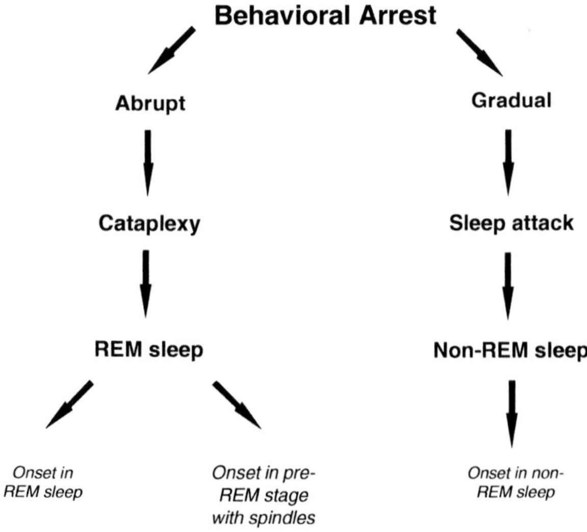

FIGURE 4. Classification of episodes of behavioral arrest as observed in the rodent.[6,8,10] Abrupt episodes are those that commence in REM sleep or a pre-REM spindling stage. Gradual arrests commence in non-REM sleep and progress, if at all, more slowly to REM sleep (i.e., after at least 60 sec).

REM sleep. This type of arrest, even if it does not begin in REM sleep, can transition to this stage in all species, and, in the mouse this transition occurs in nearly all instances and very rapidly. A second type of arrest, behaviorally more gradual in onset and associated with non-REM sleep, which in some instances shows a delayed onset (i.e., after at least 60 sec) of REM sleep, can clinically be likened to a sleep attack,[61] and is probably caused by an arousal deficit with associated rapid onset of sleepiness. This is the type of arrest that would be clinically described from a multiple sleep latency test (MSLT) as a sleep-onset REM period (i.e., SOREM).[58,62]

4.1. Direct Transitions to REM Sleep and Cataplexy

Recently, Mochizuki and colleagues[59] chose to exclude from REM sleep in *orexin*[−/−] mice all episodes during which the EEG/EMG could be considered as indicating cataplexy. For this purpose, these authors defined cataplexy as all periods which began and ended in wakefulness and which demonstrated the EEG of REM sleep combined with muscle atonia. In previous studies in the mouse, for the reasons noted above, all such episodes were treated as REM sleep. As a result of this difference in methodology, REM sleep times in this study[59] are notably different from those observed in previous studies in *orexin*[−/−] mice.[6,10] Cataplexy was therefore not defined by Mochizuki and colleagues[59] using behavioral criteria based on clinical observations, but entirely from electrophysiological criteria (i.e., direct transitions from wakefulness to REM sleep). While this method is convenient, use of this operational definition to distinguish cataplexy from normal REM sleep produces apparent differences in REM sleep times and may also produce erroneous results. For example, this method assumes that periods of REM sleep that appear to be associated with cataplexy do not contribute to REM sleep homeostasis. This assumption has not been proven, and indeed, an apparent homeostatic compensatory decrease has been observed in light phase REM sleep as a consequence of increased REM sleep, including that associated with cataplexy, during the dark phase.[10] The data from the *orexin/ataxin-3* transgenic rat also show a compensatory decrease in REM sleep during the light phase following a dark phase increase in REM sleep.[8] Furthermore, EEG/EMG recordings obtained from the narcoleptic rat[8] showed transitions to REM sleep following more evident periods of cataplexy, which appear to meet the clinical definition of cataplexy as a wakeful EEG combined with muscle atonia. In the mouse, as noted above, such episodes have been characterized but appear to be very limited in duration.[10]

Agreement on the method of analysis for vigilant state classification and cataplexy in rodents must await resolution of these issues. However, the result obtained by Mochizuki and colleagues[59] concerning the distribution of REM sleep time across the nycthemeron in narcoleptic mice impacts conclusions drawn here about the function of orexin in the normal animal. As noted above, data from the *orexin/ataxin-3* transgenic rat[8] indicate that orexin is important for suppression of the switch into REM sleep. In the absence of orexin, REM sleep episodes are triggered from active wakefulness and thus, during the active (i.e., dark) phase, the time spent in REM sleep is increased. REM sleep homeostasis is maintained, however, so that REM sleep time during the sleep (i.e., light) phase is less in the transgenic rat. It is important to note that the inhibiting influence of orexin on REM sleep episode onset is also apparent in the wild-type rat, in which the peak in REM sleep time occurs when orexin levels are at a minimum.[55-57] The primary REM sleep effect of orexin is thus to inhibit initiation of the state, but REM sleep homeostasis remains relatively unaffected in the absence of orexin. However, by excluding time spent in REM sleep following cataplexy onset from the total REM sleep time, Mochizuki and colleagues[59] found that the remaining REM sleep time was only slightly elevated in the *orexin*[−/−] mice to that recorded in the wild-type animals. This result allowed these authors[59] to suggest that sleep in the orexin-deficient mouse differs only in the stability of vigilance states. While murine narcolepsy-cataplexy is indeed associated with vigilance state instability, an independent and significant role in gating REM sleep episodes should not be discounted. Hence the absence of the neuropeptide has a net effect on the distribution of REM sleep across the nycthemeron.[6,8]

Even when cataplexy is operationally defined as direct transitions to REM sleep, different results are obtained from different animal models. For example, the data from the *orexin/ataxin-3* transgenic rats[8] were reanalyzed by the same method as Mochizuki and colleagues[59] to examine this definition in another species. A state of "cataplexy" was therefore introduced and classified whenever an epoch had the EEG/EMG characteristics of REM sleep and was preceded and followed directly by wakefulness. The results from this analysis (cf. Table 1) showed that, even when time spent in cataplexy was defined in such a broad manner and discounted from REM sleep time in the *orexin/ataxin-3* transgenic rat, the conclusions of the previous study[8] with respect to REM sleep abnormalities were not changed. That is, the rats still exhibit a significant elevation of remaining REM sleep time, even over 24 hr, and in particular, REM sleep time was significantly increased during the dark phase and significantly decreased during the light phase.

5. The Maintenance of Wakefulness

Excessive sleepiness during the daytime is another important symptom of human narcolepsy-cataplexy. Since orexin was implicated in the disorder, several authors have proposed that the neuropeptide plays a major role in maintaining wakefulness and promoting vigilant attention.[18,54,63,64] In this section, we review currently available data that support this conclusion and consider the functional importance of orexin under normal circumstances with regard to wakefulness.

TABLE 1. REM sleep data from Beuckmann and colleagues[8] reanalyzed according to Mochizuki and colleagues[59] to introduce cataplexy as a separate state.

	REM sleep		Cataplexy[b]	
	+/+	Tg/+	+/+	Tg/+
24 hr				
Total time (min)	83.3 ± 2.9	89.4 ± 3.6[a]	0	5.4 ± 4.6
Episode duration (sec)	78.2 ± 2.6	80.5 ± 2.8	–	71.3 ± 48.1
Number of episodes	65.5 ± 3.4	67.3 ± 3.9	–	3.7 ± 3.1
Light phase				
Total time (min)	50.4 ± 2.6	30.6 ± 2.6[a]	0	0
Episode duration (sec)	73.9 ± 2.8	67.1 ± 4.8	–	–
Number of episodes	41.8 ± 2.4	27.9 ± 2.6[a]	–	–
Dark phase				
Total time (min)	32.9 ± 2.6	58.8 ± 2.9[a]	0	5.4 ± 4.6
Episode duration (sec)	86.4 ± 4.8	90.7 ± 2.2	–	71.3 ± 48.1
Number of episodes	23.4 ± 1.9	39.2 ± 2.8[a]	–	3.7 ± 3.1

[a] Significant difference between wild-type (+/+) and *orexin/ataxin-3* transgenic (Tg/+) rats (P < 0.01). Data are expressed as mean ± SEM.
[b] No episodes of cataplexy according to these criteria were recorded in (+/+) rats and no episodes were recorded in (Tg/+) rats during the light phase.

Orexin fibers and receptors are found in wake-promoting regions throughout the neuraxis, including the cholinergic and monoaminergic brainstem arousal systems and the cholinergic BF. Chemelli and colleagues[6] first noted a dense concentration of orexin-containing fibers around the somata of BF cholinergic neurons. This indicated that orexin might act not only on REM sleep-related phenomena but also on the maintenance of wakefulness and vigilance. Cholinergic BF neurons, however, are not the sole substrate for such a wakefulness promoting action of orexin, and orexin-containing fibers[22,29,30] and orexin receptors[6,32] were also found in other areas known to be important for the maintenance of wakefulness. These include the noradrenergic LC, the histaminergic neurons of the TMN and the cholinergic mesopontine cell groups. Orexin has been shown to be excitatory on many of these systems[48,65-68] and the neuropeptide promotes wakefulness when administered locally in these areas.[51,69,70] Recently, Bayer and colleagues[71] have described a direct excitatory effect of orexin on cortical neurons of sublayer 6b, which project widely throughout the cortex.[72,73] Thus, in addition to the effect of orexin on subcortical arousal systems, it is now apparent that the neuropeptide has an activating influence on an important corticocortical system.

5.1. Circadian Factors and Wakefulness

The distribution of sleep and wakefulness throughout the nycthemeron depends on both the homoeostatic drive for sleep and the circadian phase.[74]

The *homeostatic process* is determined by the extent of prior wakefulness and reflects the propensity for sleep. It builds up during wakefulness and dissipates during non-REM sleep. Under normal circumstances this factor reaches a maximum in the human during the evening before the beginning of the sleep period. The second, or *circadian* factor, varies with a 24 hour periodicity and is independent of the amount of preceding sleep or wakefulness. Together, these processes modulate the need for sleep and influence the balance between central alertness and sleepiness. Interestingly, the circadian influence on sleep normally acts against the homeostatic drive at the beginning and end of the sleep period to consolidate sleep. The circadian pacemaker achieves this consolidation by a mechanism which, at first sight, seems to be in a paradoxical phase relationship to the normal timing of the sleep period in humans.[75] This follows from the fact that the circadian drive for wakefulness is greatest in the evening hours, just prior to the normal time of sleep onset. Conversely, the circadian drive for sleepiness is highest in the morning hours, just prior to the usual waking time in humans. These circadian signals promoting wakefulness and sleep thus consolidate sleep by acting, respectively, against the homeostatic drive for sleep in the evening and against the homeostatic drive for wakefulness in the morning.[76] The former signal has been termed the "circadian alertness signal".

In a study of narcoleptic patients, Dantz and colleagues[77] found that both the homeostatic drive for sleep and the circadian pacemaker were normal, but the circadian alertness signal, which should peak late in the active portion of the daily cycle, was insufficient to maintain wakefulness in these patients. This result implied that orexin might be integral to the circadian alertness signal, and the study in the *orexin/ataxin-3* transgenic rat[8] provided some confirmatory data (Figure 5). When the hourly distribution of wakefulness times in the transgenic rat was compared with the corresponding data from the wild-type animals, the wakefulness deficit in the transgenic rats during the dark, or active, period was found to be concentrated at the photoperiod boundaries and in particular at the end of this period. This time corresponds closely to the timing, in humans, of the maximal circadian alertness signal.[75] With a polyphasic sleep pattern and no single consolidated wakefulness bout, a similar alertness signal had not previously been investigated in the rodent. Indeed, no evidence has been found to date in studies of vigilance data in mice of variations in wakefulness that could be linked to the photoperiod boundaries and showed differences in *orexin*[-/-] mice.[10] However, these results in the rat[8] showed that the wild-type animals demonstrated, just prior to the beginning of the sleep period, a significant increase in wakefulness, which was absent after the loss of orexin in the transgenic rats. It is known that the circadian variation in the orexin signal in rat brain peaks at the same time as this increase in wakefulness, that is late in the active portion of the daily cycle.[55-57] These findings therefore support the proposal that orexin contributes to the daily variation in wakefulness at the end of the active period.[78] A recent study in the squirrel monkey[79] is also important, especially as this species, unlike the

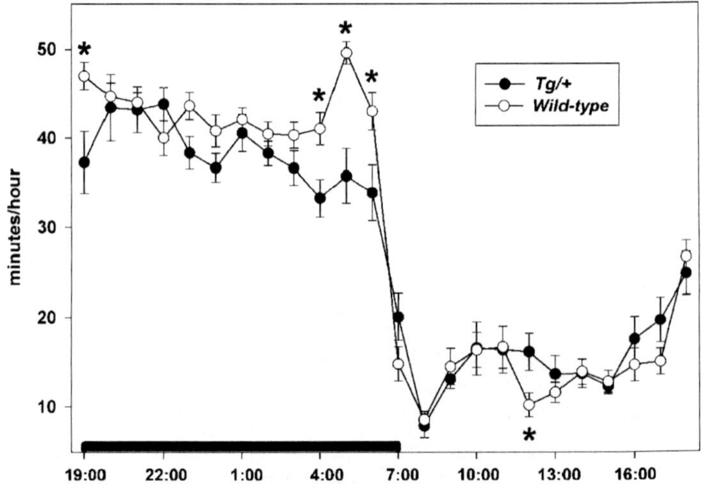

FIGURE 5. Hourly distribution of wakefulness time over the nycthemeron (in minutes; mean ± SEM) for wild-type rats and their *orexin/ataxin-3* transgenic littermates (Tg/+). Significant differences between the genotypes are marked by an asterisk. The dark phase is noted by the horizontal black bar. Note that wakefulness time is similar in both genotypes except that increased wakefulness is seen in the wild-type rats at the dark phase photoperiod boundaries, particularly at the end of this phase, at a time when orexin levels are high.[55-57]

rat, normally shows an active phase characterized by a consolidated period of wakefulness. As in the rat, the CSF orexin levels of this primate were found to peak towards the end of the active phase and reach a minimum at wake onset. Across species therefore, central orexin levels correlate closely with a circadian wake-promoting or sleep-enhancing signal, suggesting that a role of varying orexin levels might be to oppose the accumulating or dissipating homeostatic sleep drive at these times. Interestingly, Zeitzer and colleagues[79] also reported that orexin levels remained elevated during a 4 hour forced extension of the wake period in the squirrel monkey. This could be taken as evidence for a homeostatic component to orexin levels, but an alternative explanation is the coupling of orexin release with ongoing motor activity that would occur during enforced wakefulness.

As noted above, orexin has widespread interactions with all components of central arousal, both cortical and subcortical. To some extent, these are redundant parallel systems because of the critical importance of wakefulness and vigilant attention, but the various components of central arousal are also likely to play different roles under different circumstances. For example, under conditions of food deprivation, orexin appears to play a role in increasing wakefulness.[24] Mieda and colleagues[80] have also recently described a specific action of orexin in the food anticipatory increase in wakefulness that occurs during a restricted feeding schedule. However, under normal and homeostatically balanced requirements for sleep and wakefulness, as the data reviewed here show, the function of orexin appears to be linked with the circadian alertness signal.

6. Summary and Conclusions

In the few years since the discovery of the importance of orexin in narcolepsy-cataplexy, significant insights have been gained into the mechanisms by which orexin influences vigilance state control. Inhibiting the onset of REM sleep episodes appears to be one important function. In a related effect, through the balance and coordination of excitatory inputs to both the REM-on and Wake-on/REM-off cell populations, the neuropeptide may ensure rapid and complete switching into REM sleep.[81] As is evident from the symptoms of narcolepsy-cataplexy, controlling the onset of REM sleep episodes is particularly critical for inhibiting direct transitions from wakefulness into REM sleep. The mechanism for preventing these transitions may be related to coupling orexin release to ongoing motor programs, since orexin release is greatest during active wakefulness.[52] This ensures that the neuropeptide is maximally effective at this time, when the brain states are most similar, in preventing the switch into REM sleep. In agreement with this suggestion, the findings in narcoleptic rats[8] indicate that the absence of orexin creates a susceptibility to transition from active wakefulness to REM sleep via cataplexy. However, once a REM sleep episode is initiated, the absence of orexin does not appear to have a significant effect on the stability of the episode by inducing or preventing the switch from REM sleep. This follows from the fact that the mean duration of REM sleep episodes is unchanged in the narcoleptic animal,[8] despite the high levels of orexin in the normal animal during REM sleep.[52]

A second function for the neuropeptide implicates orexin in the circadian wake-promoting signal that acts in opposition to the sleep drive at the end of the active phase of the daily cycle.[8,77-79] As shown by narcoleptics who are unable to maintain wakefulness without medication, orexin is therefore necessary to maintain normal wakefulness later in the active phase of the diurnal cycle. In this regard, the circadian pacemaker itself is apparently normal in narcoleptics,[77] suggesting that the effect of the neuropeptide is as an effector pathway for the wakefulness that is linked to specific aspects of the circadian rhythm.

Over the 24 hr cycle, however, the coupling of orexin release to wakefulness and REM sleep,[52] and thus to ongoing motor activity, would be sufficient to ensure the diurnal variation in orexin levels.[55-57] A possible corollary is that orexin acts as a positive feedback on arousal during motor activity in active wakefulness. Furthermore, orexin levels *per se* can provide a basis for the variation in the expression of REM sleep over 24 hr. In this regard, it is interesting to note that McCarley and Massequoi,[82] in developing a theoretical model of the REM sleep oscillator, postulated the existence of a circadian control factor that excited monoaminergic Wake-on/REM-off cells to prevent the occurrence of REM sleep and so influence the diurnal distribution of the state. Current data thus indicate that orexin could be such a factor.

Orexin is involved in other functions, such as the wakefulness that follows a period of enforced food deprivation.[24] Other roles for the neuropeptide, so far unidentified, are perhaps also possible under normal conditions in a normal animal. But even on the basis of the current data, questions now remain as to why such apparently disparate functions, involving the onset of REM sleep episodes and the circadian alertness signal, are performed by orexin and what this might inform us about the function of REM sleep. In 1939, Nathaniel Kleitman asserted that "when we know the exact mechanism of narcolepsy and cataplexy, we shall have the correct explanation of the operation of the mechanism of the physiological sleep-and-wakefulness cycle".[83] In later writings, Kleitman[83] questioned his own assertion, but it may yet turn out to be one of the most prescient suggestions in sleep research.

7. References

1. M.S. Aldrich, Diagnostic aspects of narcolepsy, *Neurology* 50, S2-S7 (1988).
2. C.M. Sinton, and R.W. McCarley, Neuroanatomical and neurophysiological aspects of sleep: Basic science and clinical relevance, *Sem. Clin. Neuropsychiatry* 5, 6-19 (2000).
3. C.M. Sinton, and R.W. McCarley, Sleep Disorders (Article 1483), in: *Encyclopedia of Life Sciences* (Macmillan, London, 2001); *http://www.els.net.*
4. L. Lin, J. Faraco, R. Li et al, The sleep disorder canine narcolepsy is caused by a mutation in the hypocretin (orexin) receptor 2 gene, *Cell* 98, 365-376 (1999).
5. M. Hungs, J. Fan, L. Lin, X. Lin, R.A. Maki, and E. Mignot, Identification and functional analysis of mutations in the hypocretin (orexin) genes of narcoleptic canines, *Genome Res.* 11 531-539 (2001).
6. R.M. Chemelli, J.T. Willie, C.M. Sinton, et al, Narcolepsy in orexin knockout mice: Molecular genetics of sleep regulation, *Cell* 98, 437-451 (1999).
7. J. Hara, C.T. Beuckmann, T. Nambu et al, Genetic ablation of orexin neurons in mice results in narcolepsy, hypophagia, and obesity, *Neuron* 30, 345-354 (2001).
8. C.T. Beuckmann, C.M. Sinton, S.C. Williams et al, Expression of a poly-glutamine-ataxin-3 transgene in orexin neurons induces narcolepsy-cataplexy in the rat, *J. Neurosci.* 24, 4469-4477 (2004).
9. E. Mignot, Genetic and familial aspects of narcolepsy, *Neurology* 50, S16-S22 (1998).
10. J.T. Willie, R.M. Chemelli, C.M. Sinton et al, Distinct narcolepsy syndromes in orexin receptor-2 and orexin null mice: molecular genetic dissection of non-REM and REM sleep regulatory processes, *Neuron* 38, 715-730 (2003).
11. C. Peyron, J. Faraco, W. Rogers et al, A mutation in a case of early onset narcolepsy and a generalized absence of hypocretin peptides in human narcoleptic brains, *Nature Med.* 6, 991-997 (2000).
12. M. Gencik, N. Dahmen, S. Wieczorek et al, A prepro-orexin gene polymorphism is associated with narcolepsy, *Neurology* 56, 115-117 (2001).
13. S. Nishino, B. Ripley, S. Overeem, G.J. Lammers, and E. Mignot, Hypocretin (orexin) deficiency in human narcolepsy, *Lancet* 355, 39-40 (2000).

14. S. Nishino, B. Ripley, S. Overeem et al, Low cerebrospinal fluid hypocretin (orexin) and altered energy homeostasis in human narcolepsy, *Ann. Neurol.* 50, 381-388 (2001).

15. B. Ripley, S. Overeem, N. Fujiki et al, CSF hypocretin/orexin levels in narcolepsy and other neurological conditions, *Neurology* 57, 2253-2258 (2001).

16. E. Mignot, G.J. Lammers, B. Ripley et al, The role of cerebrospinal fluid hypocretin measurement in the diagnosis of narcolepsy and other hypersomnias, *Arch. Neurol.* 59, 1553-1562 (2002).

17. T.C. Thannickal, R.Y. Moore, R. Nienhuis et al, Reduced number of hypocretin neurons in human narcolepsy, *Neuron* 27, 469-474 (2000).

18. S. Taheri, J.M. Zeitzer, and E. Mignot, The role of hypocretins (orexins) in sleep regulation and narcolepsy, *Ann. Rev. Neurosci.* 25, 283-313 (2002).

19. L. Lin, M. Hungs, and E. Mignot, Narcolepsy and the HLA region, *J. Neuroimmunol.* 117, 9-20 (2001).

20. M. Mieda, J.T. Willie, J. Hara, C.M. Sinton, T. Sakurai, and M. Yanagisawa, Orexin peptides prevent cataplexy and improve wakefulness in an orexin-ablated model of narcolepsy in mice, *Proc. Natl. Acad. Sci. USA* 101, 4649-4654 (2004).

21. L. de Lecea, T.S. Kilduff, C. Peyron et al, The hypocretins: hypothalamus-specific peptides with neuroexcitatory activity, *Proc. Natl. Acad. Sci. USA* 95, 322-327 (1998).

22. C. Peyron, D. Tighe, A.N. van den Pol et al, Neurons containing hypocretin (orexin) project to multiple neuronal systems, *J. Neurosci.* 18, 9996-10015 (1998).

23. A.N. van den Pol, Hypothalamic hypocretin (orexin): robust innervation of the spinal cord, *J. Neurosci.* 19, 3171-3182 (1999).

24. J.T. Willie, R.M. Chemelli, C.M. Sinton, and M. Yanagisawa, To eat or to sleep? Orexin in the regulation of feeding and wakefulness, *Ann. Rev. Neurosci.* 24, 429-458 (2001).

25. A.V. Ferguson, and W.K. Samson, The orexin/hypocretin system: a critical regulator of neuroendocrine and autonomic function, *Front. Neuroendocrinol.* 24, 141-150 (2003).

26. T. Sakurai, A. Amemiya, M. Ishii et al, Orexins and orexin receptors: a family of hypothalamic neuropeptides and G protein-coupled receptors that regulate feeding behavior, *Cell* 92, 573-585 (1998).

27. T. Nambu, T. Sakurai, K. Mizukami, Y. Hosoya, M. Yanagisawa, and K. Goto, Distribution of orexin neurons in the adult rat brain, *Brain Res.* 827, 243-260 (1999).

28. T.C. Chou, C.E. Lee, J. Lu et al, Orexin (hypocretin) neurons contain dynorphin, *J. Neurosci.* 21, RC168 (2001).

29. C.F. Elias, C.B. Saper, E. Maratos-Flier et al, Chemically defined projections linking the mediobasal hypothalamus and the lateral hypothalamic area, *J. Comp. Neurol.* 402, 442-459 (1998).

30. Y. Date, Y. Ueta, H. Yamashita et al, Orexins, orexigenic hypothalamic peptides, interact with autonomic, neuroendocrine and neuroregulatory systems, *Proc. Natl. Acad. Sci. USA* 96, 748-753 (1999).

31. T.L. Horvath, S. Diano, and A.N. van Den Pol, Synaptic interaction between hypocretin (orexin) and neuropeptide Y cells in the rodent and primate hypothalamus: a novel circuit implicated in metabolic and endocrine regulations, *J. Neurosci.* 19, 1072-1087 (1999).

32. P. Trivedi, H. Yu, D.J. MacNeil, L.H.T. Van der Ploeg, and X.M. Guan, Distribution of orexin receptor mRNA in the rat brain, *F.E.B.S. Lett.* 438, 71-75 (1998).

33. J.N. Marcus, C.J. Aschkenasi, C.E. Lee et al, Differential expression of orexin receptors 1 and 2 in the rat brain, *J. Comp. Neurol.* 435, 6-25 (2001).
34. C.J. Tyler, K.A. Kohlmeier, J.T. Willie et al, Orexin receptor-1 mediates multiple hypocretin/orexin (H/O) actions in laterodorsal tegmentum (LDT) and dorsal raphe (DR), in: *Abstract Viewer/Itinerary Planner* (Soc. Neurosci., Washington DC, 2002), program no. 776.12.
35. R.W. McCarley, Sleep neurophysiology: basic mechanisms underlying control of wakefulness and sleep, in: *Sleep disorders medicine: basic science, technical considerations, and clinical aspects, 2nd ed.*, edited by S. Chokroverty (Butterworth-Heinemann, Woburn, MA, 1999), pp. 21-50.
36. R.W. McCarley, Human electrophysiology and basic sleep mechanisms, in: *The American Psychiatric Publishing Textbook of Neuropsychiatry and Clinical Neuroscience, 4th ed.*, edited by S.C. Yudofsky and R.E. Hales (American Psychiatric Press, Washington DC, 2002), pp. 43-70.
37. J.M. Siegel, Brainstem mechanisms generating REM sleep, in: *Principles and practice of sleep medicine, 3rd ed.*, edited by M.H. Kryger, T. Roth and W.C. Dement (Saunders, Philadelphia, PA, 2000), pp. 112-133.
38. A. Mitani, K. Ito, A.H. Hallanger, B.H. Wainer, K. Kataoka, and R.W. McCarley, Cholinergic projections from the laterodorsal and pedunculopontine tegmental nuclei to the pontine gigantocellular tegmental field in the cat, *Brain Res.* 451, 397-402 (1988).
39. D.J. McGinty, and R.M. Harper, Dorsal raphe neurons: Depression of firing during sleep in cats, *Brain Res.* 101, 569-575 (1976).
40. G. Aston-Jones, and F.E. Bloom, Activity of norepinephrine containing locus coeruleus neurons in behaving rats anticipates fluctuations in the sleep-waking cycle, *J. Neurosci.* 1, 876-886 (1981).
41. R.W. McCarley, R.W. Greene, D. Rainnie, and C.M. Portas, Brain stem neuromodulation and REM sleep, *Sem. Neurosciences* 7, 341-354 (1995)
42. G. Vanni-Mercier, K. Sakai, and M. Jouvet, "Waking-state specific" neurons in the caudal hypothalamus of the cat, *C. R. Acad. Sci. Biol.* 298, 195-200 (1984).
43. J.M. Monti, Involvement of histamine in the control of the waking state, *Life Sciences* 53, 1331-1338 (1993).
44. M. Steriade, and R.W. McCarley, *Brainstem Control of Wakefulness and Sleep* (Plenum Press, New York, 1990).
45. T.L. Horvath, C. Peyron, S. Diano et al, Hypocretin (orexin) activation and synaptic innervation of the locus coeruleus noradrenergic system, *J. Comp. Neurol.* 415, 145-159 (1999).
46. J.J. Hagan, R.A. Leslie, S. Patel et al, Orexin A activates locus coeruleus cell firing and increases arousal in the rat, *Proc. Natl. Acad. Sci. USA* 96, 10911-10916 (1999).
47. R.E. Brown, O. Sergeeva, K.S. Eriksson, and H.L. Haas HL, Orexin A excites serotoninergic neurons in the dorsal raphe nucleus of the rat, *Neuropharmacology* 40, 457-459 (2001).
48. S. Burlet, C.J. Tyler, and C.S. Leonard, Direct and indirect excitation of laterodorsal tegmental neurons by hypocretin/orexin peptides: implications for wakefulness and narcolepsy, *J. Neurosci.* 22, 2862-2872 (2002).
49. R.W. Greene, and R.W. McCarley, Cholinergic neurotransmission in the brainstem: implications for behavioral state control, in: *Brain Cholinergic Systems*, edited by M. Steriade and D. Biesold (Oxford University Press, New York, 1990) pp. 224-235.

50. M.M. Thakkar, V. Ramesh, E.G. Cape, S. Winston, R.E. Strecker, and R.W. McCarley, REM sleep enhancement and behavioral cataplexy following orexin (hypocretin) II receptor antisense perfusion in the pontine reticular formation, *Sleep Res. Online* 2, 112-120 (1999).

51. M.C. Xi, F.R. Morales, and M.H. Chase, Effects on sleep and wakefulness of the injection of hypocretin-1 (orexin-A) into the laterodorsal tegmental nucleus of the cat, *Brain Res.* 901, 259-264 (2001).

52. L.I. Kiyashchenko, B.Y. Mileykovskiy, N. Maidment et al, Release of hypocretin (orexin) during waking and sleep states, *J. Neurosci.* 22, 5282-5286 (2002).

53. C.B. Saper, T.C. Chou, and T.E. Scammell, The sleep switch: Hypothalamic control of sleep and wakefulness, *Trends Neurosci.* 24, 726-731 (2001).

54. J.G. Sutcliffe, and L. de Lecea, The hypocretins: setting the arousal threshold, *Nature Rev. Neurosci.* 3, 339-349 (2002).

55. S. Taheri, D. Sunter, C. Dakin et al, Diurnal variation in orexin A immunoreactivity and prepro-orexin mRNA in the rat central nervous system, *Neurosci. Lett.* 279, 109-112 (2000).

56. N. Fujiki, Y. Yoshida, B. Ripley, K. Honda, E. Mignot, and S. Nishino, Changes in CSF hypocretin-1 (orexin A) levels in rats across 24 hours and in response to food deprivation, *Neuroreport* 12, 993-997 (2001).

57. Y. Yoshida, N. Fujiki, T. Nakajima et al, Fluctuation of extracellular hypocretin-1 (orexin A0 levels in the rat in relation to the light-dark cycle and sleep-wake activities, *Eur. J. Neurosci.* 14, 1075-1081 (2001).

58. Y. Hishikawa, and T. Shimizu, Physiology of REM sleep, cataplexy and sleep paralysis, *Adv. Neurol.* 67, 245-271 (1995).

59. T. Mochizuki, A. Crocker, S. McCormick, M. Yanagisawa, T. Sakurai, and T.E. Scammell, Behavioral state instability in orexin knock-out mice, *J. Neurosci.* 24, 6291-6300 (2004).

60. C. Gottesmann, The transition from slow-wave sleep to paradoxical sleep: evolving facts and concepts of the neurophysiological processes underlying the intermediate stage of sleep, *Neurosci. Biobehav. Rev.* 20, 367–387 (1996).

61. F.J. Zorick, P.J. Salis, T. Roth, and M. Kramer, Narcolepsy and automatic behavior: a case report, *J. Clin. Psychiatry* 40, 194-197 (1979).

62. S. Nishino, and E. Mignot, Pharmacological aspects of human and canine narcolepsy, *Prog. Neurobiol.* 52, 27-78 (1997).

63. J.M. Siegel, R. Moore, T. Thannickal, and R. Nienhuis, A brief history of hypocretin/orexin and narcolepsy, *Neuropsychopharmacology* 25, S14-S20 (2001).

64. C.T. Beuckmann, and M. Yanagisawa, Orexins: from neuropeptides to energy homeostasis and sleep/wake regulation, *J. Mol. Med.* 80, 329-342 (2002).

65. L. Bayer, E. Eggermann, M. Serafin et al, Orexins (hypocretins) directly excite tuberomammillary neurons, *Eur. J. Neurosci.* 14, 1571-1575 (2001).

66. T.L. Horvath, C. Peyron, S. Diano et al, Hypocretin (orexin) activation and synaptic innervation of the locus coeruleus noradrenergic system, *J. Comp. Neurol.* 415, 145-159 (1999).

67. E. Eggermann, M. Serafin, L. Bayer L et al, Orexins/hypocretins excite basal forebrain cholinergic neurones, *Neuroscience* 108, 177-181 (2001).

68. K.S. Eriksson, O. Sergeeva, R.E. Brown, and H.L. Haas, Orexin/hypocretin excites the histaminergic neurons of the tuberomammillary nucleus, *J. Neurosci.* 21, 9273-9279 (2001).

69. M.M. Thakkar, V. Ramesh, R.E. Strecker, and R.W. McCarley, Microdialysis perfusion of orexin-A in the basal forebrain increases wakefulness in freely behaving rats, *Arch. Ital. Biol.* 139, 313-328 (2001).

70. Z.L. Huang, W.M. Qu, W.D. Li et al, Arousal effect of orexin A depends on activation of the system, *Proc. Natl. Acad. Sci. USA* 98, 9965-9970 (2001).

71. L. Bayer, M. Serafin, E. Eggermann et al, Exclusive postsynaptic action of hypocretin-orexin on sublayer 6B cortical neurons, *J. Neurosci.* 24, 6760-6764 (2004).

72. B. Clancy, and L.J. Cauller, Widespread projections from subgriseal neurons (layer VII) to layer I in adult rat cortex, *J. Comp. Neurol.* 407, 275-286 (1999).

73. R.L. Reep, Cortical layer VII and persistent subplate cells in mammalian brains, *Brain Behav. Evol.* 56, 212-234 (2000).

74. D.J. Dijk, and C.A. Czeisler, Contribution of the circadian pacemaker and the sleep homeostat to sleep propensity, sleep structure, electroencephalographic slow waves, and sleep spindle activity in humans, *J. Neurosci.* 15, 3526-353 (1995).

75. D.J. Dijk, and C.A. Czeisler, Paradoxical timing of the circadian rhythm of sleep propensity serves to consolidate sleep and wakefulness in humans, *Neurosci. Lett.* 166, 63-68 (1994).

76. D.J. Dijk, and J.F. Duffy, A circadian perspective on human sleep-wake regulation and ageing, in: *The Regulation of Sleep*, edited by A.A. Borbély, O. Hayaishi, T.J. Sejnowski and J.S. Altman (Human Frontier Science Program, Strasbourg, 2000), pp. 212-222.

77. B. Dantz, D.M. Edgar, and W.C. Dement, Circadian rhythms in narcolepsy: studies on a 90 minute day, *Electroenceph. Clin. Neurophysiol.* 90, 24-35 (1994).

78. E. Mignot, A commentary on the neurobiology of the hypocretin/orexin system, *Neuropsychopharmacology* 25, S5-S13 (2001).

79. J.M. Zeitzer, C.L. Buckmaster, K.J. Parker, C.M. Hauck, D.M. Lyons, and E. Mignot, Circadian and homeostatic regulation of hypocretin in a primate model: implications for the consolidation of wakefulness, *J. Neurosci.* 23, 3555-3560 (2003).

80. M. Mieda, S.C. Williams, C.M. Sinton, T. Sakurai, and M. Yanagisawa, Orexin Neurons Function in an Efferent Pathway of a Food-Entrainable Circadian Oscillator in Eliciting Food-Anticipatory Activity and Wakefulness, *J. Neurosci.* (in press) (2004).

81. A.B. Kirillov, C.D. Myre, and D.J. Woodward, Bistability, switches and working memory in a two-neuron inhibitory-feedback model, *Biol. Cybern.* 68, 441-449 (1993).

82. R.W. McCarley, and S.G. Massaquoi, Neurobiological structure of the revised limit cycle reciprocal interaction model of REM sleep cycle control, *J. Sleep Res.* 1, 132-137 (1992).

83. N. Kleitman, *Sleep and Wakefulness*, (University Chicago Press, Chicago, 1963), p.242.

Clinical Sciences

The Development and Regulation of Expressed Rhythmicity in Infants

SCOTT A. RIVKEES

1. Introduction

Circadian rhythms are endogenously generated rhythms with a period length of about 24-hrs. Evidence gathered over the past decade indicates that the circadian timing system develops prenatally and the suprachiasmatic nuclei, the site of a circadian clock, is present by mid-gestation in primates. Recent evidence also shows that the circadian system of primate infants is responsive to light at very premature stages and that low intensity lighting can regulate the developing clock. After birth, there is progressive maturation of the circadian system outputs, with pronounced rhythms in sleep-wake and hormone secretion generally developing after two months of age. Showing the importance of photic regulation of circadian phase in infants, exposure of premature infants to low-intensity cycled lighting results in the early establishment of rest-activity patterns that are in phase with the 24-hour light-dark cycle. With the continued elucidation of circadian system development and influences on human physiology and illness, it is anticipated that consideration of circadian biology will become an increasingly important component of neonatal care.

2. The Circadian Timing System

Circadian rhythms are endogenously driven rhythms with a period length of about 24-hrs.[1] Notable examples of circadian rhythms include the sleep-wake cycle and daily rhythms in hormone production. Circadian rhythms are also involved in the pathogenesis of illnesses, such as reactive airway disease and myocardial infarction.[2,3] The system responsible for the generation and regulation of circadian rhythms is the circadian timing system. This neural system consists of a biological clock, input pathways, and output pathways.[1] The paired suprachiasmatic nuclei (SCN) in the anterior hypothalamus are the site of a biological clock. The SCN are located above the optic chiasm at the base of the third ventricle.[4] The SCN exhibit endogenous rhythmicity and have a period of oscillation close to 24-hrs.

Lesion studies in rodents provided the initial evidence that the SCN are the site of a circadian pacemaker.[4] *In vivo* and *in vitro* studies have since shown day-night rhythms in electrical activity, metabolic activity, and gene expression. Transplantation of fetal SCN cells into SCN-lesioned animals restores rhythmicity to the recipient further supporting that the SCN contain a biological clock.[4] Circadian oscillations have been seen in individual rodent SCN cells, and expressed rhythmicity reflects the collective oscillations of many SCN cells.[5,6]

Because SCN oscillations are not exactly 24-hrs, it is necessary to reset the circadian pacemaker each day to prevent endogenous clock oscillations from drifting (or free-running) out of phase with the external light-dark cycle. Input pathways relay photic information from the retina to the SCN to synchronize (or entrain) the oscillations of the clock to the 24-hr light-dark cycle.[7] A direct pathway from the retina to the SCN, the retinohypothalamic tract (RHT), has been shown to be both necessary and sufficient for photic entrainment.[7] The raphe nucleus also influences SCN function via serotinergic projections.[7]

Output pathways are responsible for the overt expression of circadian rhythms. Several discrete neural pathways projecting from the SCN to several hypothalamic and non hypothalamic sites have been defined.[8-10]

Via these pathways, the circadian system acts to broadly influence neural physiology. Output pathways of the circadian system also regulate the rhythmic production of several hormones including melatonin and cortisol.[2,3,8-10]

3. The Primate Circadian System

Several lines of evidence support that the paired SCN are the site of a biological clock in primates. Similar to rodents, the primate SCN are located above the optic chiasm at the base of the third ventricle.[11] In contrast to rodents, human SCN cells are not densely clustered making the nuclei less visually apparent.[11-13]

However, using probes for melatonin receptors and SCN peptides, the human SCN can be identified.[11-13] Using DG, day-night oscillations in SCN metabolic activity have been detected in squirrel monkeys and baboons.[14-16]

Lesion studies performed in the early 1980s suggested the presence of a circadian pacemaker outside of the SCN in monkeys.[17] However, analysis of these reports revealed that either the completeness of the lesions was not verified, or monkeys were not studied in constant conditions.[18] Reexamination of this issue challenges the existence of primate circadian pacemakers outside the SCN.

Squirrel monkeys with total SCN lesions show a complete absence of circadian rhythmicity when animals are monitored in constant conditions.[18] Supporting that the SCN are the site of a circadian pacemaker in humans, tumors and congenital lesions in the SCN region result in the loss of temperature rhythms and organized sleep-wake patterns.[19,20]

The RHT has been anatomically characterized in prosimian (lemurs, shrews) and simian (squirrel monkeys, rhesus macaques, baboons, chimpanzees and apes) species.[11]

This tract also has been identified in studies of postmortem human specimens using techniques that label degenerating retinal axons.[21,22] Although it was suggested that cutaneous light exposure can influence circadian function,[23] there is little support for the notion that there is extraretinal photoreception in mammals.[24-26] Furthermore, other investigators have failed to reproduce phase shifting effects of cutaneous light exposure.[14]

Outputs of the primate circadian system have been widely characterized in human clinical studies. Many day-night rhythms have been documented.[2,3,5] Several of these rhythms have been shown to persist in constant conditions indicating that they are true endogenously generated circadian rhythms. Notable examples of circadian rhythms include the sleep-wake cycle, daily rhythms in body temperature, and day-night rhythms in cortisol and melatonin production.[2,3,25] Day-night differences in gonadotropin, testosterone, growth hormone and thyrotropin secretion are also present.[27,28]

4. Development of the Primate SCN

Although rodent studies have led to our understanding of developmental circadian physiology,[27-30] notable differences between rodents and primates have prevented the extension of rodent data to clinical care. In general, rodents are more immature at birth than humans. Differences in the sensitivity to light and other aspects of circadian physiology between humans and rodents also have been observed. However, based on evidence gathered over the past decade, it appears that the circadian clock in the SCN forms and begins oscillating *in utero* in primates.

In squirrel monkeys, SCN neurogenesis occurs early in gestation over days 27-48.[31] Because monkey and human embryonic development are very similar over the first 100 days of gestation,[32] it is therefore likely that the human SCN neurons form early in gestation.

It is not currently known when the primate SCN are first apparent morphologically. Yet, using [^{125}I]melatonin and [^{125}I]SKF38393 to label the nuclei, the human SCN have been detected at gestation week 18[33,34] (Figure 1).

Functional studies suggest that the primate SCN oscillate prenatally. Studies of squirrel monkeys reveal day-night differences in SCN metabolic activity at the end of gestation.[35] It is not known if SCN oscillations are present at earlier ages. The physiologic processes influenced by the fetal clock have yet to be elucidated in primates.

Similar to rodents, the timing of the onset of labor and birth in humans is influenced by the circadian cycle with peak incidences between midnight and the early morning.[36] However, we do not know if the fetal clock plays a role in the circadian gating of birth in humans.

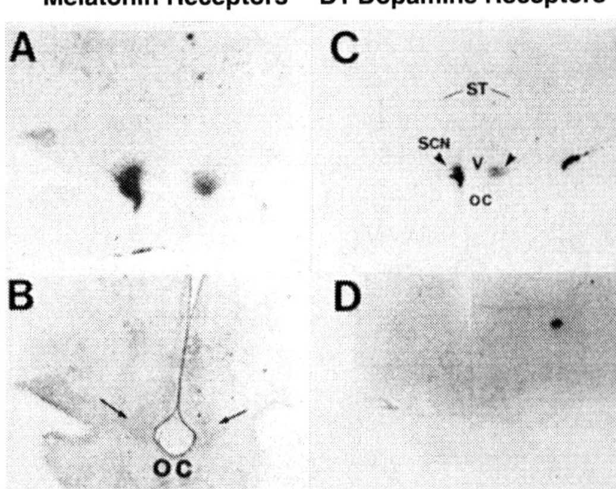

FIGURE 1. A. Localization of [125I]melatonin binding to the SCN of an 18-week gestation human fetus. Specific labeling is shown in black. B. The stained section used to generate the autoradiograph in A. Reproduced by permission from ref[33]. C. Localization of [125I]SKF38393 binding to D1 dopamine receptors in the SCN of a 20-week post conceptual human infant. Specific labeling is shown in black. D. Non-specific labeling. Reproduced by permission from ref[34]. OC, optic chiasm; ST, striatum. Arrows identify the SCN.

Immunocytochemistry studies show that SCN maturation continues after birth.[37] The SCN contain distinct populations of neurons that express arginine vasopressin or vasoactive intestinal polypeptide.[37] In term infants, the number of vasopressinergic neurons is 20% of the number present in adults.[37] It is not until one year of age that infants and adults have comparable vasopressin neuron numbers.[37] The number of vasoactive intestinal polypeptide containing SCN cells also increase after birth.[37]

5. Development of Primate Photic Entrainment

A critical issue in knowing if environmental cycles need to be considered in the care of infants is knowing when the primate circadian system becomes functionally responsive to light.

The RHT has been identified in a 36 week gestation human newborn.[38] However, because of human study limitations, has not been possible to determine if the circadian clock of human infants is functionally responsive to light at birth.

Non-invasive methods used to examine regional changes in brain activity, such as function magnetic resonance (fMRI) imaging or positron emission tomography (PET), hold promise in being able to directly examine SCN

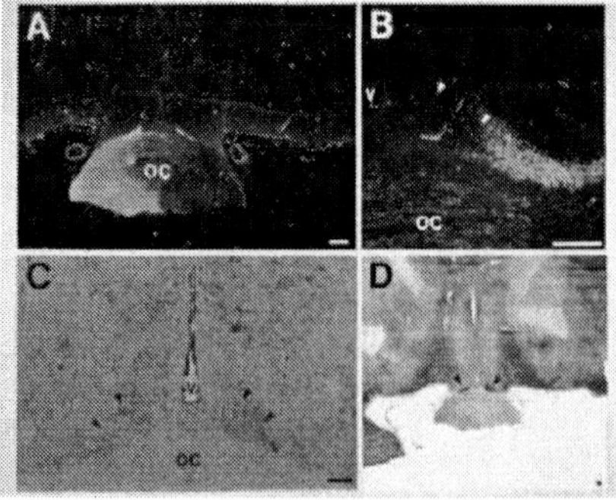

FIGURE 2. Innervation of the SCN by the retinohypothalamic tract (RHT) in a newborn baboon infant. A. Low-power image showing labeling of retinal fibers in the optic chiasm by horseradish peroxidase. B. Adjacent tissue section showing the location of the SCN. C. High power image showing projections of the RHT into the right SCN. D. Autoradiographic image generated from [^{14}C]2-deoxyglocose uptake studies showing that light exposure at night induces increases in SCN metabolic activity. Areas of increased uptake are dark. Arrows identify the SCN. Scale bar = 5 mm. Reproduced by permission from ref.[16]

function. In human adults, we have been able to observe acute increases in SCN metabolic activity after light exposure at night using ^{18}F-DG in PET studies.[39] However, because of the small size of the SCN, consistent visualization of SCN activity is difficult to achieve and these methods have not been applied to infants.

Because of human study limitations, we have studied baboons, which are excellent models for human infants, to provide insights into the developing human clock. By monitoring changes in SCN metabolic activity and gene expression (Figure 2), light responsiveness can be demonstrated at birth in term baboon infants.[16] The presence of the RHT can also be demonstrated.[16]

By monitoring the effects of different lighting conditions on newborn baboon activity patterns, we have been able to show that newborn baboons are entrained by low intensity (200 lux) lighting.[16] These findings are similar to those seen in human adults showing that circadian phase can be regulated by low intensity (ca. 180 lux) lighting.[40,41]

Thus, it is likely that low intensity lighting, similar to that found indoors, can regulate the developing primate clock. To determine when photic responsiveness first occurs in primates, we have also studied premature baboon infants.[42] To our surprise, we find that the SCN are functionally innervated by the retina at stages equivalent to 25 wks post-conception human infants[42] (Figure 3).

The primate circadian system is therefore sensitive to light in very premature infants when postnatal survival with intensive support becomes possible.

Light   Blindfolded

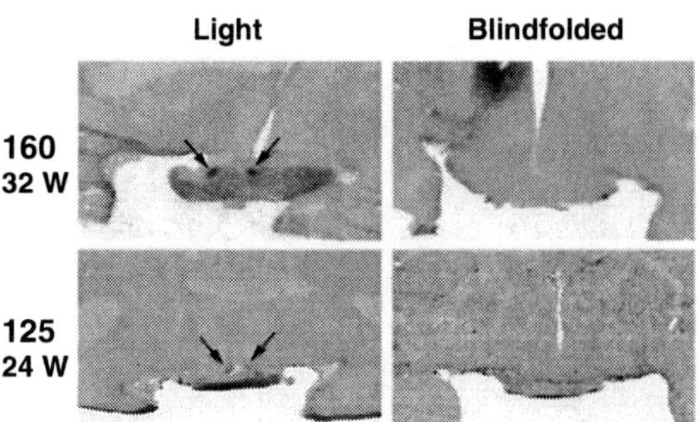

FIGURE 3. Autoradiograph images of preterm baboon brain sections showing SCN DG uptake after light exposure at night. Animals shown were PC 160 or 125 and were either blindfolded or directly exposed to light. The images are obtained from mid-SCN levels. Arrows identify the SCN image. Reproduced by permission from ref[42].

6. Development of Expressed Rhythmicity

The development of expressed rhythmicity has received attention in both human and nonhuman primates. During pregnancy, day-night rhythms are observed for a variety of hormones (esterone and progesterone) and physiological parameters (uterine contractility) in mothers.[43,44] In human fetuses, day-night rhythms in heart rate, respiratory rate, and adrenal steroidogenesis have been detected.[43,44] However, these rhythms appear to be driven by the mother.

When term human infants are examined, day-night rhythms are difficult to detect in the neonatal period.[39,45-47] Consolidated periods of activity and rest are not generally observed until after the first or second month of life. Activity plots of human newborns reveal that sleep is generally distributed over the 24-hr day during the first few weeks of life (Figure 3). At 6 wks of age, infants are awake more during the daytime than at night. By 12 wks of age, daytime sleep duration decreases further and much more sleep occurs at night. Importantly, although consolidated periods of rest and activity are not apparent until more than one to two months after birth, day-night differences in activity can be detected as early as one week of age in some babies.

At the age when day-night differences in infant activity become clearly apparent, day-night rhythms in hormone production are observed. Day-night rhythms in melatonin production can be detected at 12 weeks of age.[48,49] Circadian variation in cortisol levels appears between after 3-6 months of age.[50-52] With advancing age, circadian rhythms have been detected for a variety of other hormones and circulating factors.[53]

Because infant care influences activity patterns, it is possible that patterns of developing circadian rhythmicity in human infants reflect influences of caregivers rather than endogenous rhythmicity. Thus, to characterize the development of expressed rhythmicity in primates, we have examined the development of expressed rhythmicity in newborn baboons raised in constant conditions (continuous dim lighting, evenly spaced care)[16]. Similar to human infants, baboon infants do not manifest clear day-night differences in activity patterns in the early neonatal period (Figure 4). Yet, at one month of age, day-night differences in activity patterns are observed. Developing primate rhythmicity thus reflects maturation from a state of relative arrhythmicity to rhythmicity over the first few months of life.

As in rodents, it appears that infant circadian phase is synchronous with that of the mother in baboon and human infants. However, in some humans and baboons.[16,54] infant phase may be out of synchrony with that of the mother at birth. Thus, whereas there is maternal-infant synchrony of circadian phase in most primates, it may not be universal.

7. Rhythmicity in Premature Infants

The large number of premature infants hospitalized for extended periods has greatly facilitated studies of rhythmicity in preterm infants. Over the past decade, studies of patterns of infant activity, heart rate, temperature, and sleep state have not surprisingly flourished.[55]

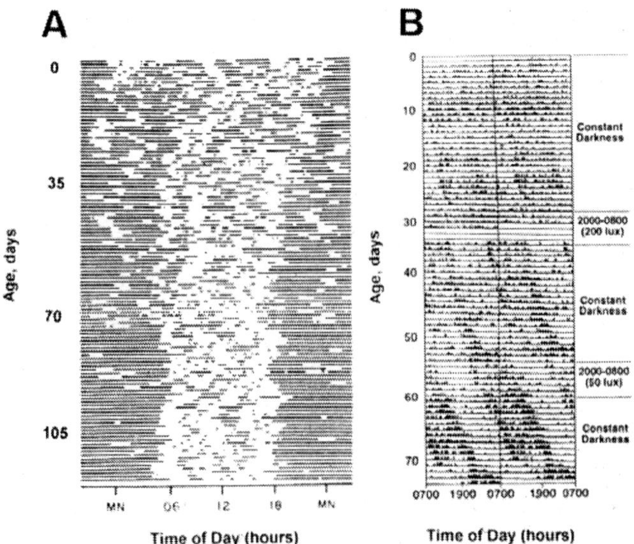

FIGURE 4. Rest-activity patterns of a newborn human (A) and a baboon (B) infant (right). Double-plotted actograms are shown. In A, dark bars represent sleep. In B, dark bars represent activity. Please note that the circadian phase of the baboons infant was shifted by exposure to a 200 lux reversed light-dark cycle at 30 days of age, and much less so by 50 lux of exposure at 55 days of age. Reproduced by permission from ref.[16]

Several of these studies have revealed the presence of ultradian rhythms (rhythms with period lengths of much less than 24 hrs). Endogenously driven circadian rhythms, however, are not clearly apparent.

When temperature and heart rate are studied beginning at a postconceptual (PC) age 24-29, circadian rhythmicity is generally not apparent even at 17 weeks after birth.[56] Studies of preterm infants at PC 32 weeks, have failed to detect day-night differences in sleep patterns whereas some differences are noted in term infants[57].

Analysis of temperature, heart rate, and activity patterns at PC 35 weeks have revealed ultradian rhythms, but no clear cut circadian rhythms.[57-59] Because feeding and physical contact influence infant temperature, heart rates and activity patterns, it is likely that infant care schedules drive the ultradian rhythms seen in preterm infants. These interventions may also mask the detection of circadian rhythms.

8. The Yale Neonatal Entrainment Study

Following the discovery that the primate circadian clock is responsive to light in very premature infants, we next assessed the effects of photic entrainment on premature infants.[60] In these studies, the development of rest-activity patterns was examined in human preterm infants exposed to continuous dim lighting or low-intensity cycled lighting before discharge from hospital to home.

In general, day/night differences in rest and activity are not apparent in hospitalized control infants (Figure 5), whereas day/night differences in rest and activity are seen in experimental infants. Over the first ten days at home, distinct day/night differences in activity are not seen in controls, but experimental infants are more active during the day than at night. It was not until 21-30 days after discharge that day/night activity ratios in control infants match those seen in experimental infants shortly after discharge. Yet, even at this age, experimental infants are considerably more active during the day as compared to control infants. Despite the differences in rest-activity patterns among groups, no differences in weight gain or change in head circumference are seen.

These observations show that exposure to low-intensity cycled lighting for 10 days before discharge induces distinct patterns of rest/activity in preterm infants that are in synchrony with the light-dark cycle that they will encounter at home. These effects are even more pronounced as soon as the child is discharged to home. In contrast, the appearance of rest/activity patterns in synchrony with the solar light-dark cycle is delayed in infants that have been exposed to continuous dim lighting in the hospital.

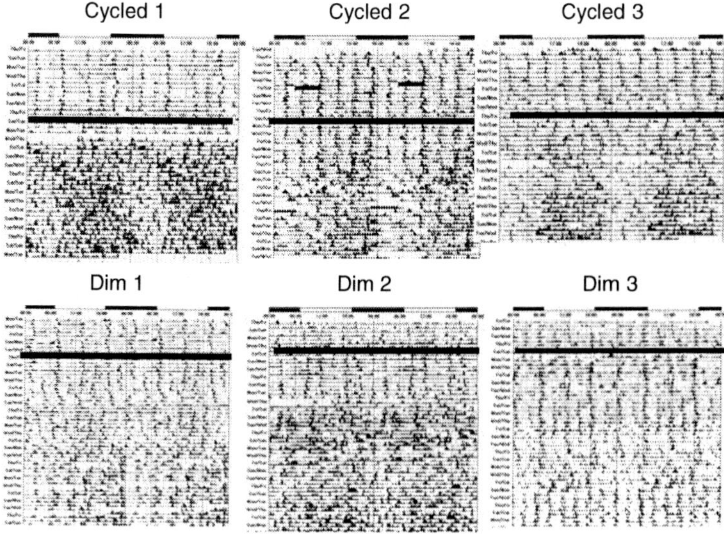

FIGURE 5. Actograms of rest-activity in representative infants exposed to cycled lighting in the (top panels) or constant dim light (bottom panels). Dark bars represent activity; the same activity scale is used in each plot. The time of day is shown on top. The thick dark line in middle of plots depicts the date of discharge. Note that distinct patterns of rest and activity in the infants are more apparent after discharge in infants exposed to cycled lighting than dim lighting before discharge from the hospital. In infants exposed to dim lighting, day-night differences in rest-activity patterns in synchrony with the light-dark cycle are generally not apparent until about 20 days after discharge from the hospital.

9. Other Studies of Lighting and Infants

Potential influences of cycled lighting on premature infants have been the subject of a few previous studies. In the Stanford Cycled Lighting Study, differences in circadian rhythms in temperature were not detected among infants exposed to either continuous dim lighting or cycled lighting before discharge.[58,59,61] These infants were studied 1 and 3 months after discharge. Because we observe that infants in both groups manifest similar circadian phase by 30 days of age, treatment effects on the rhythm of core body temperature may no longer be distinct after one month of age.

Other investigators have suggested that exposing infants to light/dark cycles improves infant weight gain. Mann and co-workers found that exposure to light/dark cycles before discharge resulted in better weight gain and more sleep over the 24-hour day than did chaotic lighting patterns.[62] These effects were seen 6 weeks after discharge and not sooner[62]. Because of this lag period, it has been suggested that the observed effects were not a direct result of cycled lighting exposure on the infant.[61] More recently, it has been suggested that exposing infants to light/dark cycles improves the in-hospital growth of babies if exposure occurs before 36 weeks of age.[63] Yet, the infants in near-darkness group in this study appeared more ill than the other groups. Considering that it is difficult to detect circadian activity in premature infants.[39,64] the potential mechanisms by which lighting could directly influence the growth of premature infants is not clear. By studying infants that were closely matched at enrollment, we failed to observe influences of lighting on growth either in-hospital or at home.

Previous studies have suggested that day/night rhythmicity is not apparent in prematurely born babies until nearly one month after term-birth age equivalency is reached (>42 weeks postmenstrual age).[58,59,61,65] These conclusions have been based on 24-48 hour assessments of rectal temperature and/or sleep patterns. However, using actigraphy to continuously monitor rest-activity patterns, we find that circadian phase can be detected in infants exposed to cycled lighting as early as a postmenstrual age of 34 weeks.

In our previous studies of non-human primate infants reared in constant conditions, we also found that day/night differences in rest and activity were apparent shortly after term birth.[16] Most importantly, we find that day/night differences in activity could be detected several weeks before it was possible to detect circadian rhythms in core temperature using internal telemetry devices.[16] Thus, analysis of rest activity patterns may provide the earliest index of developing circadian rhythmicity in infants.

10. Nursery Lighting Practices

The practice of nursery lighting has changed over the past several decades without a clear basis. Cycled lighting was often used in the hospital nurseries in the fifties and sixties. Yet, continuous bright lighting became favored when isolettes and neonatal intensive care units were introduced in the seventies. In reaction to continuous bright light, continuous dim light was introduced in the eighties and nineties, along with covering infant isolettes with blankets and quilts.

Although continuous dim lighting is the current practice in most nurseries in the United States, the scientific basis for this practice is not clear.[66] It has been suggested that ambient lighting may contribute to eye disease in premature infants.[67] Yet, rigorous clinical studies have failed to show adverse effects of low intensity lighting on premature infants.[68-70]

Investigators who propose a NIDCAP (Neonatal Individualized Developmental Care Assessment Program) have suggested that since the womb is dark, infants should be dark-reared.[71] This approach overlooks the fact that prenatally the infant is exposed to maternal time-of-day cues that synchronize the fetal clock with the external light/dark cycle.[29] Rearing premature infants in the dark thus deprives babies of the time-of-day

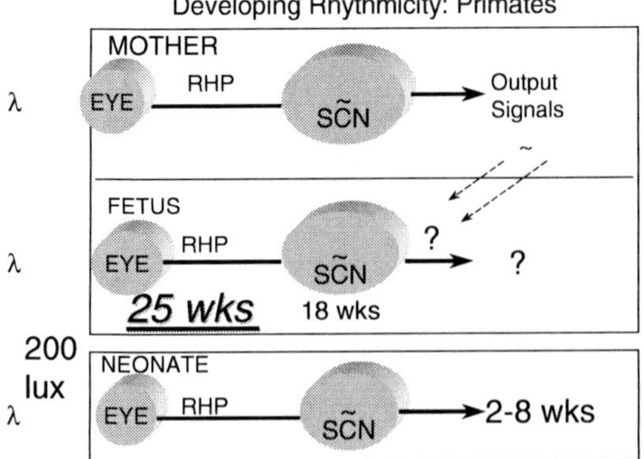

FIGURE 6. Schematic representation of primate circadian system development based on studies of non-human primates. Estimated human ages are given.

information that they would have received during full gestation. Data also show that the NIDCAP approach does not improve developmental outcome or sleep of premature infants[72]. Thus, a rational approach considering the importance of circadian rhythmicity and environmental lighting cycles is needed in the care of hospitalized infants.

11. Summary

Increasing evidence indicates that the circadian timing system is a fundamental homeostatic system that potently influences human behavior and physiology throughout development (Figure 6).

After birth there is progressive maturation of the circadian system with day-night rhythms in activity and hormone secretion developing between one and three months of age. Recent evidence shows that the circadian system of primate infants is responsive to light at very premature stages and that low intensity lighting can regulate the developing clock. With the continued elucidation of circadian system development and influences on human physiology and illness, it is anticipated that consideration of circadian biology will become an increasingly important component of neonatal care.

12. References

1. S. Panda, J. B. Hogenesch, S. A. Kay. Circadian rhythms from flies to human. *Nature*. 16;417(6886):329-335.(2002).
2. M. C. Moore-Ede, C. A. Czeisler, G. S. Richardson. Circadian timekeeping in health and disease. Part 1. Basic properties of circadian pacemakers. *N. Engl. J. Med.*; 309(8):469-476.(1983).

3. M. C. Moore-Ede, C. A. Czeisler, G. S. Richardson. Circadian timekeeping in health and disease. Part 2. Clinical implications of circadian rhythmicity. *N.Engl.J.Med.*; 309(9):530-536.(1983).

4. D. R. Weaver. The suprachiasmatic nucleus: a 25-year retrospective. *J Biol Rhythms*. 13(2):100-112.(1998).

5. D. K. Welsh, D. E. Logothetis, M. Meister, S. M. Reppert. Individual neurons dissociated from rat suprachiasmatic nucleus express independently phased circadian firing rhythms. *Neuron*. 14(4):697-706.(1995).

6. C. Liu, D. R. Weaver, S. H. Strogatz, S. M. Reppert. Cellular construction of a circadian clock: period determination in the suprachiasmatic nuclei. *Cell*. 91(6):855-860.(1997).

7. L. P. Morin. The circadian visual system. *Brain Res.Brain Res.Rev.*; 19(1): 102-127.(1994).

8. J. M. Kornhauser, K. E. Mayo, J. S. Takahashi. Light, immediate-early genes, and circadian rhythms. *Behav Genet*. 26(3):221-240.(1996).

9. A. G. Watts, L. W. Swanson, G. Sanchez-Watts. Efferent projections of the suprachiasmatic nucleus: I. Studies using anterograde transport of Phaseolus vulgaris leucoagglutinin in the rat. *J.Comp.Neurol.*; 258(2):204-229.(1987).

10. A. G. Watts, L. W. Swanson. Efferent projections of the suprachiasmatic nucleus: II. Studies using retrograde transport of fluorescent dyes and simultaneous peptide immunohistochemistry in the rat. *J.Comp.Neurol.*; 258(2): 230-252.(1987).

11. R. Y. Moore. Organization of the primate circadian system. *J.Biol.Rhythms*. 8 Suppl:S3-9:S3-9.(1993).

12. R. Lydic, W. C. Schoene, C. A. Czeisler, M. C. Moore-Ede. Suprachiasmatic region of the human hypothalamus: homolog to the primate circadian pacemaker? *Sleep.*; 2(3):355-361.(1980).

13. R. Lydic, H. E. Albers, B. Tepper, M. C. Moore-Ede. Three-dimensional structure of the mammalian suprachiasmatic nuclei: a comparative study of five species. *J.Comp.Neurol.*;204(3):225-237.(1982).

14. K. Lushington, R. Galka, L. N. Sassi, D. J. Kennaway, D. Dawson. Extraocular light exposure does not phase shift saliva melatonin rhythms in sleeping subjects. *J Biol Rhythms.*;17(4):377-386.(2002).

15. W. J. Schwartz, S. M. Reppert, S. M. Eagan, M. C. Moore-Ede. In vivo metabolic activity of the suprachiasmatic nuclei: a comparative study. *Brain Res.*; 274(1):184-187.(1983).

16. S. A. Rivkees, P. L. Hofman, J. Fortman. Newborn primate infants are entrained by low intensity lighting. *Proc.Natl.Acad.Sci.U.S.A.*;94(1):292-297.(1997).

17. S. M. Reppert, M. J. Perlow, L. G. Ungerleider, et al. Effects of damage to the suprachiasmatic area of the anterior hypothalamus on the daily melatonin and cortisol rhythms in the rhesus monkey. *J Neurosci*. 1(12):1414-1425.(1981).

18. D. M. Edgar, W. C. Dement, C. A. Fuller. Effect of SCN lesions on sleep in squirrel monkeys: evidence for opponent processes in sleep-wake regulation. *J Neurosci*. 13(3):1065-1079.(1993).

19. W. J. Schwartz, N. A. Busis, E. T. Hedley-Whyte. A discrete lesion of ventral hypothalamus and optic chiasm that disturbed the daily temperature rhythm. *J.Neurol.*; 233(1):1-4.(1986).

20. S. Rivkees. Arrhythmicity in a child with septo-optic dysplasia and establishment of sleep-wake cyclicity with melatonin. *J Pediatrics*. 139:463-465.(2001)

21. D. I. Friedman, J. K. Johnson, R. L. Chorsky, E. G. Stopa. Labeling of human retinohypothalamic tract with the carbocyanine dye, DiI. *Brain Res.*;560(1-2): 297-302.(1991).
22. A. A. Sadun, J. D. Schaechter, L. E. Smith. A retinohypothalamic pathway in man: light mediation of circadian rhythms. *Brain Res.*;302(2):371-377.(1984).
23. S. S. Campbell, P. J. Murphy. Extraocular circadian phototransduction in humans. *Science.* 279(5349):396-399.(1998).
24. M. Hebert, S. K. Martin, C. I. Eastman. Nocturnal melatonin secretion is not suppressed by light exposure behind the knee in humans. *Neurosci Lett.* 22;274(2):127-130.(1999).
25. R. G. Foster. Shedding light on the biological clock. *Neuron.* 20(5):829-832.(1998).
26. G. Jean-Louis, D. F. Kripke, R. J. Cole, J. A. Elliot. No melatonin suppression by illumination of popliteal fossae or eyelids. *J Biol Rhythms.* 15(3):265-269. (2000).
27. E. D. Weitzman, C. A. Czeisler, J. C. Zimmerman, M. C. Moore-Ede. Biological rhythms in man: relationship of sleep-wake, cortisol, growth hormone, and temperature during temporal isolation. *Adv Biochem Psychopharmacol.* 28:475-499. (1981).
28. C. A. Czeisler, E. B. Klerman. Circadian and sleep-dependent regulation of hormone release in humans. *Recent Prog Horm Res.* 54:97-130.(1999).
29. S. M. Reppert, D. R. Weaver, S. A. Rivkees. Maternal communication of circadian phase to the developing mammal. *Psychoneuroendocrinology.*;13(1-2):63-78. (1988).
30. S. M. Reppert. Interaction between the circadian clocks of mother and fetus. *Ciba.Found.Symp.*; 183:198-207(1995).
31. F. J. Van Eerdenburg, P. Rakic. Early neurogenesis in the anterior hypothalamus of the rhesus monkey. *Brain Res.Dev.Brain Res.*;79(2):290-296.(1994).
32. L. Nowell-Morris, C. E. Faherenbruch. *Practical and evolutionary considerations for the use of nonhuman primate model in prenatal research.* New York: Alan R. Liss;1985.
33. S. M. Reppert, D. R. Weaver, S. A. Rivkees, E. G. Stopa. Putative melatonin receptors in a human biological clock. *Science.* 242(4875):78-81.(1988).
34. S. A. Rivkees, J. E. Lachowicz. Functional D1 and D5 dopamine receptors are expressed in the suprachiasmatic, supraoptic, and paraventricular nuclei of primates. *Synapse.* 26(1):1-10.(1997).
35. S. M. Reppert, W. J. Schwartz. Functional activity of the suprachiasmatic nuclei in the fetal primate. *Neurosci.Lett.*;46(2):145-149.(1984).
36. A. Jolly. Hour of birth in primates and man. *Folia Primatol.* 18(1):108-121.(1972).
37. D. F. Swaab. Development of the human hypothalamus. *Neurochem Res.* 20(5):509-519.(1995).
38. S. F. Glotzbach, P. Sollars, M. Pagano. Development of the human retinohypothalamic tract. *Soc. Neurosci.*;18:857.(1992).
39. S. A. Rivkees, H. Hao. Developing circadian rhythmicity. *Semin Perinatol.* 24(4):232-242.(2000).
40. D. B. Boivin, J. F. Duffy, R. E. Kronauer, C. A. Czeisler. Dose-response relationships for resetting of human circadian clock by light. *Nature.* 379(6565): 540-542.(1996).
41. T. L. Shanahan, C. A. Czeisler. Physiological effects of light on the human circadian pacemaker. *Semin Perinatol.* 24(4):299-320.(2000).

42. H. Hao, S. A. Rivkees. The biological clock of very premature primate infants is responsive to light. *Proc Natl Acad Sci U S A.* 96(5):2426-2429.(1999).

43. M. Seron-Ferre, C. A. Ducsay, G. J. Valenzuela. Circadian rhythms during pregnancy. *Endocr. Rev.*;14(5):594-609.(1993).

44. M. Seron-Ferre, C. Torres-Farfan, M. L. Forcelledo, G. J. Valenzuela. The development of circadian rhythms in the fetus and neonate. *Semin Perinatol.* 25(6):363-370.(2001).

45. A. H. Parmelee, Jr. Sleep cycles in infants. *Dev. Med. Child Neurol.*;11(6):794-795.(1969).

46. A. Meier-Koll, U. Hall, U. Hellwig, G. Kott, V. Meier-Koll. A biological oscillator system and the development of sleep-waking behavior during early infancy. *Chronobiologia.* 5(4):425-440.(1978).

47. J. Kleitman, Engelman. Sleep characteristics of infants. *J Appl Physiolol.* 6:127-134.(1953).

48. D. J. Kennaway, G. E. Stamp, F. C. Goble. Development of melatonin production in infants and the impact of prematurity. *J. Clin. Endocrinol. Metab.*;75(2):367-369.(1992).

49. D. J. Kennaway, F. C. Goble, G. E. Stamp. Factors influencing the development of melatonin rhythmicity in humans. *J. Clin. Endocrinol. Metab.*;81(4):1525-1532.(1996).

50. I. Z. Beitins, A. Kowarski, C. J. Migeon, G. G. Graham. Adrenal function in normal infants and in marasmus and kwashiorkor. Cortisol secretion, diurnal variation of plasma cortisol, and urinary excretion of 17-hydroxycorticoids, free corticoids, and cortisol. *J Pediatr.* 86(2):302-308.(1975).

51. S. Onishi, G. Miyazawa, Y. Nishimura, et al. Postnatal development of circadian rhythm in serum cortisol levels in children. *Pediatrics.* 72(3):399-404.(1983).

52. D. A. Price, G. C. Close, B. A. Fielding. Age of appearance of circadian rhythm in salivary cortisol values in infancy. *Arch. Dis. Child.* 58(6):454-456.(1983).

53. T. Hellbrugge, J. E. Lange, J. Rutenfranz. Circadian periodicity of physiological functions in different stages of infancy and childhood. *Ann N Y Acad Sci.* 117:361-373.(1964).

54. A. H. Parmelee, Jr., W. H. Wenner, Y. Akiyama, M. Schultz, E. Stern. Sleep states in premature infants. *Dev. Med. Child Neurol.*;9(1):70-77.(1967).

55. S. A. Rivkees. Developing circadian rhythmicity. Basic and clinical aspects. *Pediatr. Clin. North Am.*;44(2):467-487.(1997).

56. S. W. D'souza, S. Tenreiro, D. Minors, M. L. Chiswick, D. G. Sims, J. Waterhouse. Skin temperature and heart rate rhythms in infants of extreme prematurity. *Arch. Dis. Child.* 67(7 Spec No):784-788.(1992).

57. T. F. Anders, M. A. Keener, H. Kraemer. Sleep-wake state organization, neonatal assessment and development in premature infants during the first year of life. II. *Sleep.*;8(3):193-206.(1985).

58. S. F. Glotzbach, D. M. Edgar, M. Boeddiker, R. L. Ariagno. Biological rhythmicity in normal infants during the first 3 months of life. *Pediatrics.* 94(4 Pt 1):482-488.(1994).

59. S. F. Glotzbach, D. M. Edgar, R. L. Ariagno. Biological rhythmicity in preterm infants prior to discharge from neonatal intensive care. *Pediatrics.* 95(2):231-237.(1995).

60. S. A. Rivkees, L. Mayes, H. Jacobs, I. Gross. Rest-activity patterns of premature infants are regulated by cycled lighting. *Pediatrics.*:in press.(2004).

61. M. Mirmiran, R. L. Ariagno. Influence of light in the NICU on the development of circadian rhythms in preterm infants. *Semin Perinatol*. 24(4):247-257.(2000).

62. N. P. Mann, R. Haddow, L. Stokes, S. Goodley, N. Rutter. Effect of night and day on preterm infants in a newborn nursery: randomised trial. *Br Med J (Clin Res Ed)*. 293(6557):1265-1267.(1986).

63. D. H. Brandon, D. Holditch-Davis, M. Belyea. Preterm infants born at less than 31 weeks' gestation have improved growth in cycled light compared with continuous near darkness. *J Pediatr*. Feb;140(2):192-199.(2002).

64. M. Mirmiran, S. Lunshof. Perinatal development of human circadian rhythms. *Prog.Brain Res*.;111:217-26:217-226.(1996).

65. K. Mcgraw, R. Hoffmann, C. Harker, J. H. Herman. The development of circadian rhythms in a human infant. *Sleep*. May 1;22(3):303-310.(1999).

66. R. L. Ariagno, M. Mirmiran. Shedding light on the very low birth weight infant. *J Pediatr*. Oct;139(4):476-477.(2001).

67. P. Glass, G. B. Avery, K. N. Subramanian, M. P. Keys, A. M. Sostek, D. S. Friendly. Effect of bright light in the hospital nursery on the incidence of retinopathy of prematurity. *N.Engl.J.Med*.;313(7):401-404.(1985).

68. A. R. Fielder, M. J. Moseley. Environmental light and the preterm infant. *Semin Perinatol*. 24(4):291-298.(2000).

69. J. D. Reynolds, R. J. Hardy, K. A. Kennedy, R. Spencer, W. A. Van Heuven, A. R. Fielder. Lack of efficacy of light reduction in preventing retinopathy of prematurity. Light Reduction in Retinopathy of Prematurity (LIGHT-ROP) Cooperative Group. *N.Engl.J.Med*.; 338(22):1572-1576.(1998).

70. K. A. Kennedy, A. R. Fielder, R. J. Hardy, B. Tung, D. C. Gordon, J. D. Reynolds. Reduced lighting does not improve medical outcomes in very low birth weight infants. *J Pediatr*.;139(4):527-531.(2001).

71. H. Als, G. Lawhon, F. H. Duffy, G. B. Mcanulty, R. Gibes-Grossman, J. G. Blickman. Individualized developmental care for the very low-birth-weight preterm infant. Medical and neurofunctional effects. *Jama*. 21;272(11):853-858.(1994).

72. R. L. Ariagno, E. B. Thoman, M. A. Boeddiker, et al. Developmental care does not alter sleep and development of premature infants. *Pediatrics*. 100(6):E9.(1997).

Sleep and Circadian Rhythm
of Melatonin in Smith-Magenis
Syndrome

HÉLÈNE DE LEERSNYDER

1. Introduction

Smith Magenis syndrome (SMS), is a rare genetic disease emblematic of neu-
rodevelopmental disorder. Clinical features include mild dysmorphism, short
stature, developmental delay and abnormal behavior. Severe sleep disturbances and maladaptive daytime behavior were linked to abnormal circadian
rhythm of melatonin. SMS is the demonstration of biological basis for sleep
disorder in a genetic disease.

First described by Ann Smith et al in 1982[1,2], SMS is a contiguous gene
deletion syndrome ascribed to interstitial deletion of chromosome 17
(17p11.2).[3,4] Its prevalence is estimated to 1/25,000 live births, however, due
to recent delineation, most persons with SMS having been identified in the
last 10 years, the prevalence could reflects under ascertainment. All cases
occur de novo, there is no parental imprinting.

2. Clinical Features and Diagnosis Criteria

Several distinctive features characterize the phenotype of Smith-Magenis syndrome,[5] including brachycephaly with a characteristic craniofacial appearance (midface hypoplasia, mouth characteristic with cupid's bow,
prognatism), ocular abnormalities (myopia and strabismus, iris anomalies),
speech delay with or without hearing loss, hoarse deep voice, short stature
with history of failure to thrive, brachydactyly, peripheral neuropathy (pes
cavus or planus, depressed deep tendon reflexes, insensitivity to pain), scolio-
sis. Other variable features include cardiac defects, renal abnormalities,
seizures, cleft palate, low immunoglobulins and thyroxin defect. All patients
have some degree of developmental delay and mental retardation. IQ scores
range between 20 and 78, most falling in the moderate range of 45-55.
Behavioral problems[6] include consistently aggression, self injurious behaviors, temper tantrums, impulsivity, repetitive behavior, hyperactivity with
attention deficit. Low sensitivity to pain could be consistant with bilateral
lenticulo-insular anomalies detected in MRI and PET.[7] Severe sleep disturbances[8] and unusual circadian rhythms are constant features of the syndrome.

The diagnosis is based on clinical features and confirmed on high resolution karyotype with detectable deletion of 17p11.2 and by fluorescence in situ hybridization (FISH) probe specific for SMS.[9,10] Most patients have a common deletion interval of 4-5 megabases. However, deletions have ranged from <2 to >9 megabases, and mutations in *RAI1 (Retinoic Acid-Induced gene)* were shown[11] in individuals who have phenotypic features consistent with SMS.

3. Night and Day Sleep Disturbances in Smith-Magenis Syndrome

Significant symptoms of sleep disturbance are seen in all persons with SMS and have a major impact on the child and other family members, many of whom become sleep deprived themselves. Questionnaires, sleep consultations, sleep diaries and actimetries revealed sleep disturbance in most SMS subjects. All patients go to bed easily after a short bed-time ritual. Bedtime is similar at 8-9 p.m. (range 6.45 pm to 10 pm) regardless of age and sex. The duration of night sleep average 7.50 hrs (range: 5-9 hrs), decline with age and is slightly shorter than that of aged matched patients. All SMS persons consistently wake up 1-3 times per night and fell back asleep within 30 minutes or more. Once awaken, they are hyperactive. This behavior force parents or care keeping to constantly look after them and to devise artifices to keep them in the bedroom by night (lock the door, switch light, remove furniture and objects). The mean wake-up time is 5.30 a.m. (range 2:00 am to 7:00 am). Behavioral problems correlate with night sleep insufficiency. Most patients exhibit morning tiredness when normally circadian vigilance is high. They have temper tantrums when tired and naps (more than 30 min) during the day, regardless of age. Most interestingly, they consistently have "sleep attacks" at the end of the day, as they suddenly fall asleep during evening meals even with fulfilled mouth.

The 24 hours-polysomnography, correlated with actimetry and sleep diaries, reveal a reduced total sleep time in 57% of the patients[8]. All sleep stages are present but 3-4 non rapid eye movement (NREM) sleep is reduced. REM sleep is disrupted and arousal with increased tonic EMG activity is frequent. Awakenings (more than 15 minutes) occurred in 75% of cases.

Interestingly, all SMS patients display a phase shift in their circadian rhythm of melatonin[12-14] (Figure 1). Indeed, time at onset of melatonin secretion in SMS is 6 am±2 (controls : 9 pm±2). Peak time is at 12 pm±1 (controls : 3.30 am±1.30) and melatonin offset is at 8 pm±1 (controls : 6 am±1). Melatonin peak value rises 94±pg/ml (controls : 76 pg/ml). Irregular levels of melatonin are noted during the day with a second peak between 6 and 8 p.m. (45 pg/ml±32) and the total duration of melatonin secretion is protracted

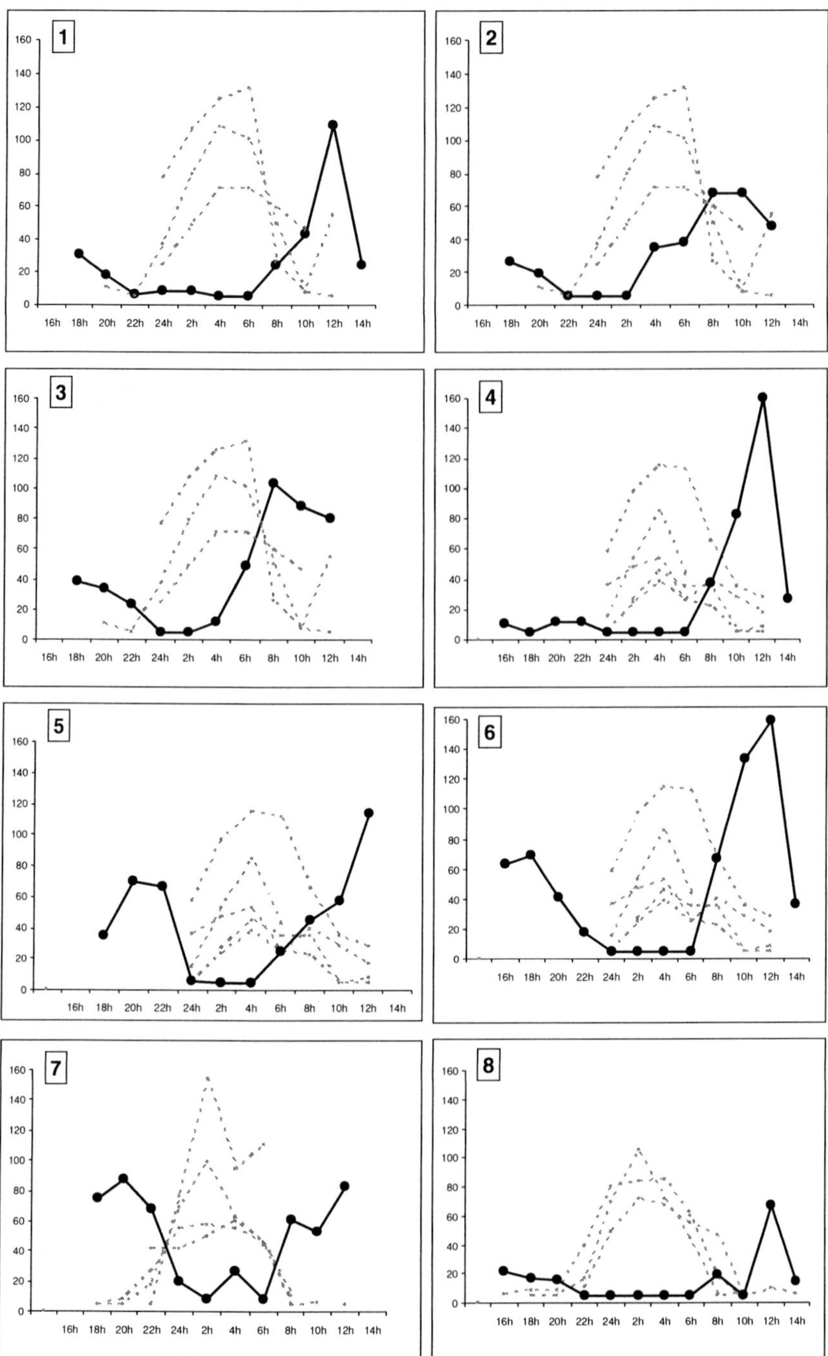

FIGURE 1. Circadian variation of plasma melatonin in 8 Smith-Magenis children and controls. Solid lines = SMS children aged 5-6 years (panels 1-3), 7-8 yrs (panels 4-6), 12 yrs (panel 7) and 17 yrs (panel 8). Dotted lines = age-matched controls. Aged-matched controls were healthy children or adolescents hospitalized for small stature, with normal results of cortisol, GH and melatonin values.

in SMS, 15.5 hrs ± 3.5 (controls : 8 hrs ± 1). Similarly urinary melatonin and 6-sulfatoxymelatonin revealed an inverted night/day ratio.

Cortisol follows an usual circadian secretion and is in normal range. Growth hormone (GH) and prolactin follow an usual circadian secretion and are in normal range.

Patients with SMS have a phase shift of the circadian melatonin rhythm of 9.6 ± 0.9 hours[15] but not a full reversal of the circadian melatonin rhythm. This disorder does not affect entrainment and pacemaker of other endocrine functions.

Interestingly, the abnormal circadian rhythm of melatonin parallels sleep disturbances and abnormal day behavior in SMS (Figure 2). During the night, early sleep onset, frequent awakenings and early sleep offset are consistent features of the disease and are highly specific diagnostic criteria in SMS. The sleep attacks occurring at the end of the day may represent in fact the endogenous sleep onset of the patient that could be regarded therefore as equivalent to a sleep phase advance. According to this hypothesis, the endogenous sleep onset time would be masked by the imposed social activities.

During the day patients are tired in the morning and tantrums appear when melatonin rise. Naps and sleep attacks occur when melatonin peaks at midday and in the evening, respectively.

Considering that behavioral problems correlate with the abnormal circadian rhythm of melatonin in SMS, it is tempting to hypothesize that at least part of hyperactivity and attention deficit occur because the patients struggled against sleep, when melatonin rise during the day.

This is particularly relevant for child behavior. For ethic reasons, no experience of free-running sleep has been reported in Smith-Magenis syndrome

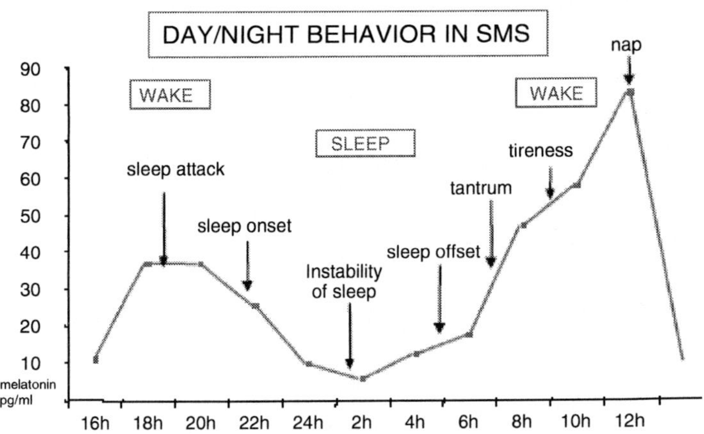

FIGURE 2. Sleep-wake patterns correlated with melatonin secretion in SMS.

today, but we can expect that if they were allow to do so, patients would sleep during the day and remain awaken during the night. Interestingly children travelling transmeridian clinically reset their clock and sleep well for a few nights.

4. Hypothesis Concerning Melatonin Dysfunction

Melatonin, the main hormone of the pineal gland, is synthesized from serotonin. Its synthesis and release are stimulated by darkness and inhibited by light. Light or dark entrainment proceed through the retino-hypothalamic tract (RHT) to reach the suprachiasmatic nuclei (SCN) of the anterior hypothalamus.[16,17] SCN contain biological clocks, which are endogenous pacemakers generating circadian rhythms entrained by environmental stimuli. A number of clock genes controlling circadian rhythms have been recently identified in higher eukaryotes.[18] Their expression shares common features across species,[19] it oscillates with a 24-hr rhythm and persists in the absence of environmental cues. It is reset by changes in light/dark cycle and undergoes negative feedback that down-regulates their activity. Considering that clock genes are expressed in a circadian pattern in SCN, one can hypothesized that haploinsufficiency for a clock gene could account for sleep disturbance in SMS. The Per1 gene, which maps to chromosome 17p12 is not deleted in SMS. Interestingly, subunit 3 of the COP9 signal transduction complex (*COPS3*) maps within the SMS critical region in 17p11.2.[20] COP9 is essential for the light control of gene expression during plant development and is conserved across species. It has been shown that the gene for subunit 3 of the COP9 signal transduction complex, *COPS3*, is expressed in transformed lymphoblastoid cell lines of SMS patients. However, haploinsufficiency in one gene probably does not play a significant role with respect to melatonin phase shift in SMS, there may be an age dependant penetrance or variability of the expression of the phenotype.

Circadian rhythmicity not only involves clock genes but also requires an input signaling pathway for detection of exogenous signals (zeitgeber) and their transmission to SCN via the RHT, and the output signaling pathway of postganglionic fibers ascending to the pineal gland is required to maintain melatonin secretion under the control of the SCN.[21] Consequently, the inversion of the circadian rhythm of melatonin in SMS may also result from an alteration of the input/output-signaling pathway (e.g. photic entrainment in the retina/RHT or β1-adrenergic signaling transduction to the pineal gland). Yet, how melatonin acts on sleep is presently unknown. It may modify brain levels of monoamine transmitters, thereby initiating a cascade of events culminating in the activation of sleep. Actually, the mechanism of this quantitatively normal but rhythmically abnormal melatonin secretion in SMS is not known.

5. Treatment of the Inverted Rhythm of Melatonin in Smith-Magenis Syndrome

Melatonin, the main hormone of the pineal gland, entrained by light-dark cycles, is normally secreted during the night. Interestingly, all SMS patients have a phase shift of their circadian rhythm of melatonin, with a diurnal secretion of the hormone. Tantrums and tiredness occur when melatonin rises and patients have naps and sleep attacks when melatonin peaks at midday and in the evening, respectively. Considering that behavioral problems and night sleep insufficiency in SMS may correlate with the inverted circadian rhythm of melatonin, we hypothesized that at least part of hyperactivity and attention deficit might occur because the patients struggle against sleep, when melatonin rises during the day. Sleep disturbances are extremely severe and difficult to manage. These observations are particularly relevant to therapeutic approaches in SMS. Indeed, melatonin administration alone is not necessarily warranted,[22] as the amount of secreted hormone is largely normal but its kinetic is erratic.

An original therapeutic approach including blockade of endogenous melatonin signaling pathways combined with on-time exogenous melatonin administration was studied in patients aged 6–18 years.[23,24] Because the circadian rhythm of melatonin is controlled by the sympathetic nervous system, SMS patients were given acebutolol, as β_1-adrenergic antagonists reduce the production of melatonin.[25] A cardiac and pneumologic examination was performed before trial. After a morning β_1-adrenergic antagonists administration, plasma melatonin levels rapidly decreased in all SMS patients. Mean melatonin levels fell from 68 pg/ml to 8 pg/ml after drug administration. Individual melatonin levels decreased 3-20 folds, remained low from 8 am to 6 am the next day and rised again from 6 to 8 am prior to drug administration (Figure 3).

With this treatment, day behavior markedly improved. While untreated patients had 1-3 naps/day and frequent sleep attacks at the end of the day, beta-blocker administration resulted in the disappearance of naps and sleep attacks. The explosive tantrums (1-2 each day) were less frequent (1 or 2 per week) and could be easily managed. Parents, teachers, friends and uninformed neighbors noted a more appropriate behavior.

Before treatment SMS patients had a poor concentration ability (less than 10 minutes, even for the eldest), while they could concentrate 30-60 min or more for parlor game, computer activity, looking TV, gardening or little jobs at home when given beta-blockers. For the children, teachers acknowledged a better concentration during school time and they were reported to be quieter and less hyperactive. Home and social behavior improved but remained problematic. No significant increase of cognitive performance was observed.

The combination of morning β_1-adrenergic antagonist and evening melatonin administration restored plasmatic circadian melatonin rhythmicity, improved behavioral disturbances and enhanced sleep in SMS. Studies were conducted with a control release (CR) melatonin.[24] After a single dose of exogenous melatonin, plasmatic melatonin levels rapidly peaked and slowly

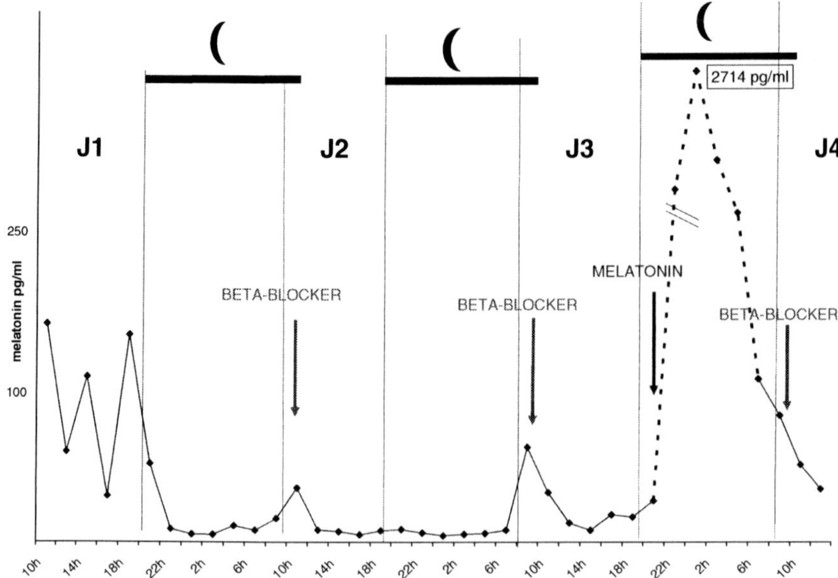

FIGURE 3. Circadian variation of plasma melatonin before (J1) and after morning beta adrenergic antagonist administration (J2-J3) and controled release melatonin administration (J3).

decreased thus mimicking the effects of endogenous melatonin on circadian rhythm. Mean melatonin levels raised from 12.7 ± 10,6 pg/ml to 2189 ± 1800 pg/ml two hours after drug administration. Individual melatonin levels increased 170 folds, remained high from 10 pm to 2 am, and slowly decreased till 6 am (Figure 3). Mean sleep onset was delayed by 30 minutes, sleep offset by 60 minutes and the mean gain of sleep was 30 min. Sleep awakenings disappeared in most cases and wake-up time was delayed.

Patients no more wake up during the night and EEG recordings confirmed a more regular sleep stage organization and a rapid access to sleep stage 3-4. Sleep was deep and quiet and day/night life was dramatically improved. No desensitisation was observed over a 3 years period of drug administration. Finally, no side effects of β_1-adrenergic antagonists or melatonin were noted and all parents and care giving are convince for continuation of the treatment.

6. Conclusions

Smith-Magenis syndrome is a rare genetic disorder ascribed to interstitial deletion of chromosome 17 (17p11.2). SMS is also a circadian disorder with extreme phase shift of melatonin secretion. This is the first biologic model of sleep and behavioral disorders in a genetic disease.

Elucidating pathophysiological mechanisms of behavioral phenotypes is particularly relevant to therapeutic approach in genetic diseases. Indeed, in SMS where the circadian rhythm of melatonin is shifted, β_1-adrenergic antagonists combined with evening melatonin administration restore a circadian rhythm of melatonin, suppress inappropriate diurnal melatonin secretion and improve sleep and behavioral disorders.

7. Acknowledgments. We are grateful to the individuals, parents and the ASM 17 family support group of France for participating in the studies. We thank Arnold Munnich, Bruno Claustrat, Marie Christine de Blois, Jean Louis Bresson, Agnes Mogenet, Daniel Sidi, Jean Claude Souberbielle for their assistance with the studies in the Necker-Enfants Malades Hospital, Paris, France; Ann Smith and Wallace Duncan in the NIH SMS Research Unit for helpfull cooperation. The studies were supported in part by Institut de Recherche Internationales Servier, Paris, France and Neurim, Pharmaceuticals, Tel Aviv, Israel.

8. *References*

1. Smith, A. C. M.; McGavran, L.; Robinson, J.; Waldstein, G.; Macfarlane, J.; Zonana, J.; Reiss, J.; Lahr, M.; Allen, L.; Magenis, E. : Interstitial deletion of (17)(p11.2p11.2) in nine patients. *Am. J. Med. Genet.* 24: 393-414, (1986).
2. Greenberg, F.; Guzzetta, V.; Montes de Oca-Luna, R.; Magenis, R. E.; Smith, A. C. M.; Richter, S. F.; Kondo, I.; Dobyns, W. B.; Patel, P. I.; Lupski, J. R. : Molecular analysis of the Smith-Magenis syndrome: a possible contiguous-gene syndrome associated with del(17)(p11.2). *Am. J. Hum. Genet.* 49: 1207-1218, (1991).
3. Moncla, A.; Piras, L.; Arbex, O. F.; Muscatelli, F.; Mattei, M.-G.; Mattei, J.-F.; Fontes, M. : Physical mapping of microdeletions of the chromosome 17 short arm associated with Smith-Magenis syndrome. *Hum. Genet.* 90: 657-660, (1993).
4. Juyal R.C., Figuera L.E., Hauge X., Elsea S.H., Lupski J.R., Greenberg F., Baldini A., Patel P.I., Molecular analysis of 17p11.2 deletion in 62 Smith-Magenin syndrome patients. *Am. J. Hum. Genet.* 58 :998-1007. (1996)
5. Greenberg F, Lewis RA, Potocki L, Glaze L, Parke J, Killian J, et al. Multi disciplinary clinical study of Smith-Magenis syndrome (deletion 17p11.2). *Am J Med Genet* 62:247-54. (1996)
6. Smith ACM, Dykens E, Greenberg F. Behavioral phenotype of Smith-Magenis syndrome (del 17p11.2) *Am J Med Genet*;81:179-185 (1998)
7. Boddaert N., De Leersnyder H., Bourgeois M., Munnich A., Brunelle F., Zilbovivius M. Anatomical and functional brain imaging evidence of lenticulo-insular anomalies in Smith Magenis syndrome. *Neuroimage 21 1021-1025.* (2004)
8. Smith ACM, Dykens E, Greenberg F. Sleep disturbance in Smith-Magenis syndrome (del 17p11.2). *Am J Med Genet*;81:186-91. (1998)
9. Lucas, R. E.; Vlangos, C. N.; Das, P.; Patel, P. I.; Elsea, S. H. : Genomic organisation of the ~1.5 Mb Smith-Magenis syndrome critical interval: transcription map, genomic contig, and candidate gene analysis. *Europ. J. Hum. Genet.* 9: 892-902 (2001).

10. Potocki, L.; Chen, K.-S.; Park, S.-S.; Osterholm, D. E.; Withers, M. A.; Kimonis, V.; Summers, A. M.; Meschino, W. S.; Anyane-Yeboa, K.; Kashork, C. D.; Shaffer, L. G.; Lupski, J. R. : Molecular mechanism for duplication 17p11.2—the homologous recombination reciprocal of the Smith-Magenis microdeletion. *Nature Genet.* 24: 84-87 (2000).

11. Slager, R. E.; Newton, T. L.; Vlangos, C. N.; Finucane, B.; Elsea, S. H.: Mutations in RAI1 associated with Smith-Magenis syndrome. *Nature Genet.* 33: 466-468, (2003).

12. Potocki, L.; Glaze, D.; Tan, D.-X.; Park, S.-S.; Kashork, C. D.; Shaffer, L. G.; Reiter, R. J.; Lupski, J. R. : Circadian rhythm abnormalities of melatonin in Smith-Magenis syndrome. *J. Med. Genet.* 37: 428-433,(2000).

13. De Leersnyder, H.; de Blois, M.-C.; Claustrat, B.; Romana, S.; Albrecht, U.; von Kleist-Retzow, J.-C.; Delobel, B.; Viot, G.; Lyonnet, S.; Vekemans, M.; Munnich, A. : Inversion of the circadian rhythm of melatonin in the Smith-Magenis syndrome. *J. Pediat.* 139: 111-116, (2001).

14. Cavallo A. Plasma melatonin rhythm in normal puberty : interactions of age and pubertal stages. *Neuroendocrinology*;55:372-9.(1992)

15. Cornelissen G., Halberg F., Tarquini R., Perfetto F., Salti R., Laffi G., Otsuka K. Point and interval estimations of circadian melatonin ecphasia in Smith-Magenis syndrome. *Biomed.Pharmacother.* 57.1: 39-44.(2003)

16. Moore RY. Circadian rhythms: basic neurobiology and clinical applications. *Annu. Rev. Med.*; 48: 253-266. (1997)

17. Brzezinski A. Melatonin in humans. *N Engl J Med*;336:186-95.(1997)

18. Van Esseveldt L.E. Lehman M.N. Boer G.J. The suprachiasmatic nucleus and the circadian time-keeping system revisited. *Brain Research Reviews* 33: 34-77. (2000)

19. Wager-Smith K. Kay S.A. Circadian rhythms genetics : from flies to mice to humans. *Nature genetics*. 26: 23-27. (2000)

20. Potocki L, Chen KS, Lupski J. Subunit 3 of the COP9 signal transduction complex is conserved from plants to humans and maps within the Smith-Magenis syndrome critical region in 17p11.2. *Genomics*; 57:180-2.(1999)

21. Arendt J. Mamalian pineal rhythms. *Pineal Res Rev.* 3: 161-213. (1985)

22. Jan J.E., Freeman R.D., Fast D.K. Melatonin treatment of sleep-wake cycle disorders in children and adolescents. *Dev. Med. Child.Neurol.* 41, 491-500 . (1999)

23. De Leersnyder, H.; de Blois, M.-C.; Vekemans, M.; Sidi, D.; Villain, E.; Kindermans, C.; Munnich, A.: Beta-1-adrenergic antagonists improve sleep and behavioural disturbances in a circadian disorder, Smith-Magenis syndrome. *J. Med. Genet.* 38: 586-590, (2001).

24. De Leersnyder, H.; Bresson J.L. de Blois, M.-C.; Souberbielle J.C., Mogenet A., Delhotal-Landes B., Salefranque F., Munnich, A.: Beta-1-adrenergic antagonists and melatonin reset the clock and restore sleep in a circadian disorder, Smith-Magenis syndrome. *J. Med. Genet.* 40: 74-78, (2003).

25. Stoschitzky K., Sakotnik A., Lercher P., Sweiker R.,Maier R., Liebmann P., Lindner W. Infdluence of beta blockers on melatonin release. *Eur. J. Clin. Pharmacol.* 55: 111-1. (1999).

Melatonin in Circadian Rhythm Sleep Disorders

V. Srinivasan, M. G. Smits, L. Kayumov, S. R. Pandi-Perumal, Daniel P. Cardinali and M. J. Thorpy

1. Abstract

Normal circadian rhythms are synchronized to a regular 24 hr environmental light/dark (L/D) cycle. Both suprachiasmatic nucleus (SCN) and melatonin are essential for this adaptation. Desynchronization of circadian rhythms as occurs in chronobiological disorders result in severe disturbances of sleep. The Circadian rhythm sleep Disorders (CRSDs) include delayed sleep phase syndrome (DSPS), Non 24 hr sleep/wake rhythm disorder, jet lag and shift-work. Depression also shows circadian rhythm disturbances. Disturbances in the phase position of plasma melatonin levels have been documented in all these disorders. Whether this melatonin disruption is a cause or a consequence of the disorders is not known. Further research of a larger number of patients with CRSDs can help to determine the association. At present there appears to be a role of *endogenous* melatonin in the pathophysiology of these circadian rhythm sleep disorders. *Exogenous* melatonin is useful in treating the disturbed sleep-wake rhythms seen in DSPS, Non 24 hr sleep/wake rhythm, Shift–work sleep disorder, jet lag and in depression. The magnitude and direction of the shift of the sleep-wake rhythm depend upon the time of melatonin onset. In most of these conditions an abnormal phase position of melatonin rhythm has been documented.

2. Introduction

Circadian Rhythm Sleep disorders (CSRD) have become the major focus of attention in recent years. Major industrial, air and train accidents have been generally attributed to inefficient handling of situations by individuals suffering from malfunctioning the circadian time keeping system.[1] Decrease in nighttime alertness and performance coupled with poor day time sleep as seen in night shift workers is the main cause for increased number of accidents seen during night shift. Sleep-related vehicle accidents are twenty times higher at 0600 h than at 10.00 h.[2] Working at the time of circadian trough associated with loss of sleep exerts a negative impact on work performance.[3]

Synchronization of the sleep/wake rhythm and the rest/activity cycles with L/D cycle of the external environment is essential to maintain man's normal mental and physical health.[4] The hormone melatonin formed mainly in the pineal gland, is essential for this physiological adaptation.[5] Indeed drastic alterations in the secretion of melatonin with disturbance of its rhythmicity have been shown to underlie circadian rhythm disorders.[6,7] Shifting circadian rhythms back to normal in these disorders is associated with the correct timing of melatonin rhythm.[8]

According to Arendt[9] melatonin serves as the window to view the functioning of the *"internal clock"*, the circadian time keeping system of the body. In this chapter the role of melatonin in circadian sleep rhythm disorders, such as the non-24 h sleep/wake rhythm disorder seen in blind human subjects, DSPS, shift work sleep disorder, jet lag, and depression is discussed. An understanding of the role of melatonin in circadian rhythm disorders will demonstrate how melatonin can be used to resynchronize the desynchronized sleep/rhythm disorders.[10-12]

3. Circadian Rhythms

Circadian rhythms (*circa* = about; *dies* = day) are rhythms that are close to 24 h in period length and are the most thoroughly explored biological rhythms. The medical implications of these rhythms have been well studied recently. De Mairan provided the first scientific evidence that self-sustained oscillators endogenously generate these rhythms.[13] He showed that leaves of *mimosa* plant open during daytime and close at night even when the leaves were placed in complete darkness. Self-sustained oscillators have become recognized as *"Biological clocks"*. By using ablation, transplantation or electrophysiological studies it has been shown that the suprachiasmatic nucleus (SCN) of the hypothalamus acts as the main biological clock in all animals,[14] as well as in human beings,[15] for the generation of circadian rhythms. Circadian rhythms differ from other daily rhythms in having a free running period that can be entrained by a *Zeitgeber (time cue)*. In the absence of temporal cues like such as the light/dark (L/D) cycle circadian rhythms free run and can be readjusted to 24 h by light acting through the monosynaptic retinohypothalamic (RHT) pathway.[14,15] Human beings have an endogenous free running period slightly longer than 24 h.

If the circadian pacemaker were not adjusted the timing of endogenous rhythms would lose up to an hour a day with respect to clock time of the day.[16] Abnormal phase positions and severe disturbances in sleep/wake circadian rhythms have become prominent features of the circadian rhythm disorders[17-22] (Figure 1).

These disorders do not respond well to conventional methods of treatments such as the use of hypnotics but respond to therapeutic manipulations

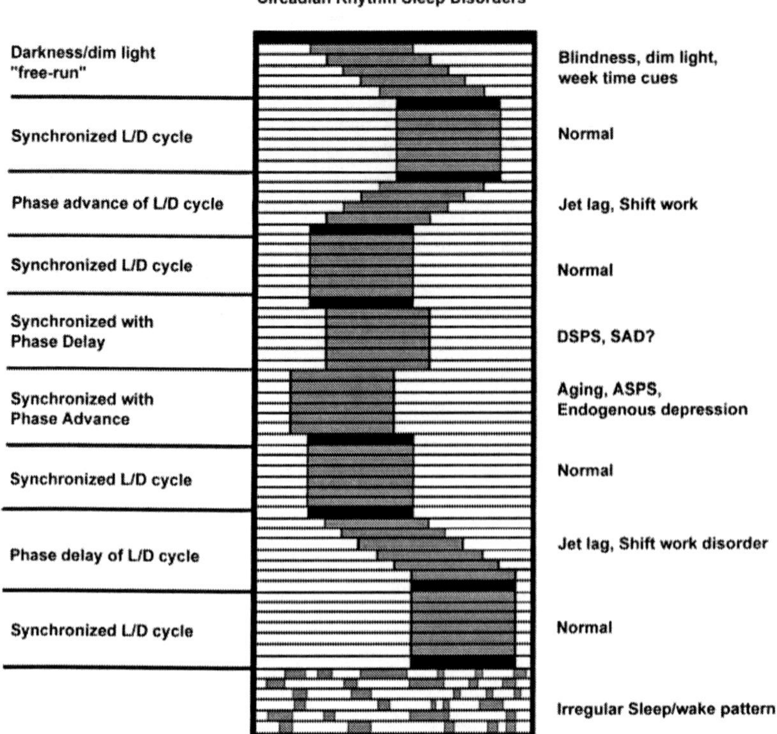

FIGURE 1. Schematic representation of the human circadian rhythm sleep disorders. L/D = light/dark. Modified from Dr. J. Redman, Monash University, Australia.

based upon chronobiologic principles such as the use of phototherapy or melatonin.[22-25] Melatonin can alter the timing of mammalian circadian rhythms and can reset disturbed circadian rhythms to their correct phase position.[26] Hence melatonin is termed as a *"chronobiotic"* or an *"internal Zeitgeber"*.[27]

4. Melatonin

Melatonin is synthesized primarily in the pineal gland of all animals and man. Synthesis also occurs in the retina, enterochromaffin cells of the gut, and erythrocytes in humans.[28] Most of the biosynthesis occurs only at night. The rate-limiting enzyme *N-acetyl transferase* (NAT) correlates closely with melatonin production.

The circadian rhythm of melatonin production is regulated by SCN,[29] but its duration, phase, and amplitude are influenced by changes in L/D cycles. The retinohypothalamic tract conveys information about the L/D cycle to the SCN (Figure 2). Ganglionic photoreceptor cells containing melanopsin in the retina are involved in the perception of light that regulates melatonin synthesis. The most effective wavelength that suppresses melatonin production is in the blue light range of 460-470 nm.[30]

The parenchyma of the mammalian pineal gland is predominately composed of a group of cells called pinealocytes. Pinealocytes take up the essential amino acid *L*-Tryptophan (Trp) from the circulation (Figure 3). Although the transport system for Trp into the pinealocytes has not been made clear yet; the incorporated Trp is hydroxylated to 5-hydroxytryptophan (HTrp) by *Tryptophan Hydroxylase* (TH or TPH; EC 1.14.16.4), that requires tetrahydrobiopterin (BH4) as a co-factor to hydroxylate its amino acid substrate. Subsequently, HTrp is decarboxylated to 5-Hydroxytryptamine (5HT; Serotonin) through the action of *aromatic-L-amino acid decarboxylase* (AAAD; EC 4.1.1.28). *Serotonin-N-acetyltransferase* (SNAT or NAT or AA-NAT; EC 2.3.1.87), which regulates the rate of melatonin biosynthesis in the pineal gland, catalyzes the acetylation of 5HT to N-acetylserotonin (NAS).

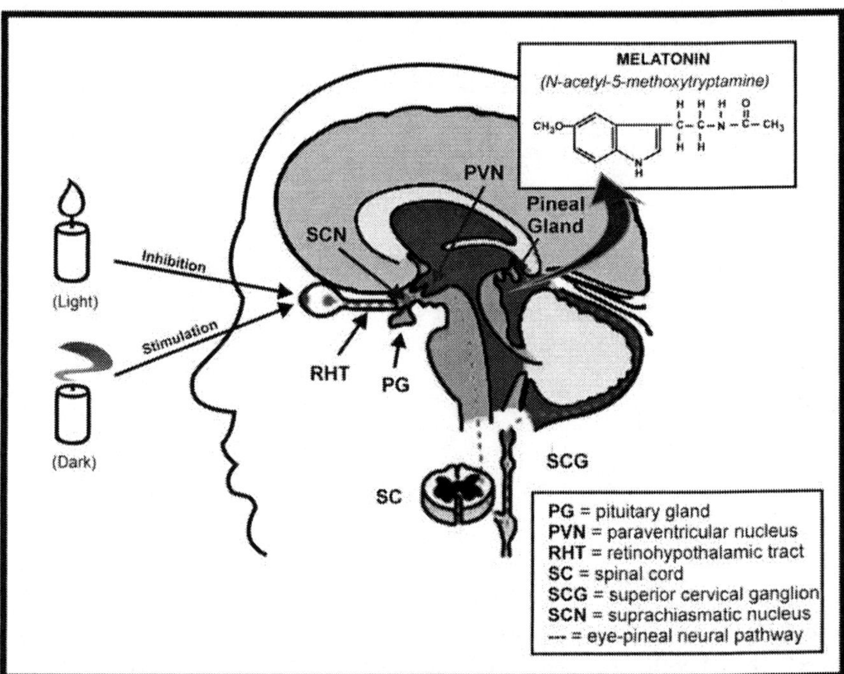

FIGURE 2. Photic regulation of melatonin.

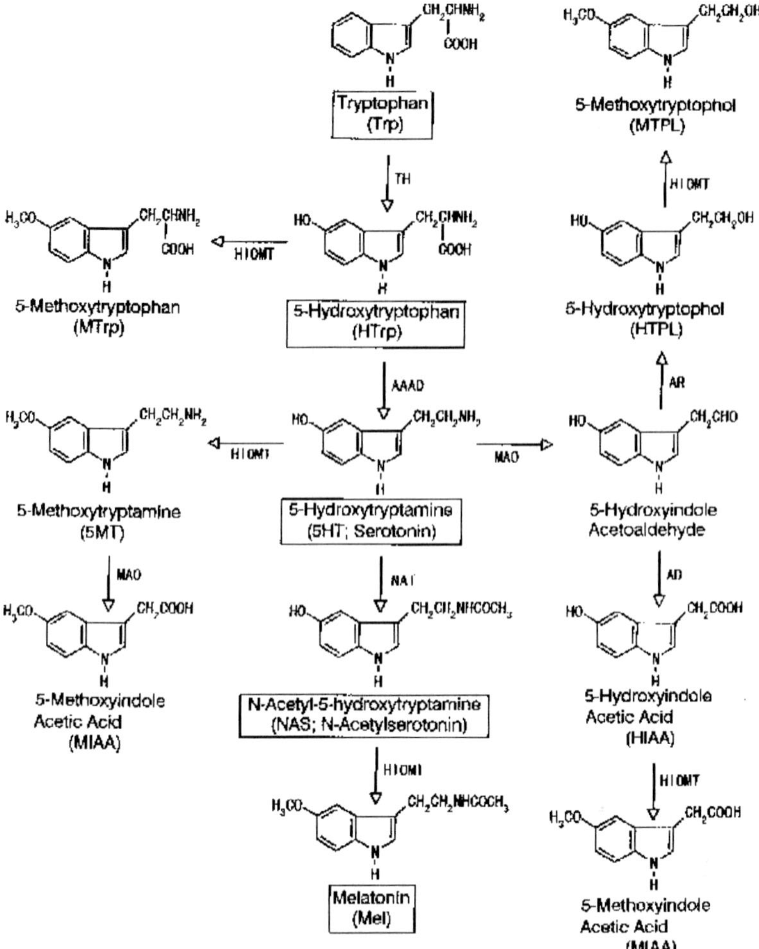

FIGURE 3. Schematic diagram of the biosynthetic and regulatory pathways of melatonin and indole metabolism in the mammalian pineal gland. Melatonin is synthesized from *L*-tryptophan by the consecutive action of four (TH, AAAD, NAT, HIOMT) separate enzymes: TH = *Tryptophan hydroxylase*; AAAD = *L-aromatic amino acid decarboxylase*; NAT = *N-acetyl transferase*; HIOMT = *hydroxyindole-O-methyltransferase*.

Finally, a methyl group from S-adenosylmethionine is transferred to NAS by *hydroxyindole-O-methyltransferase* (HIOMT; EC 2.1.1.4), and NAS is converted to 5-methoxy-N-acetyltryptamine (melatonin). The resulting melatonin is highly nonpolar because the charged 5-hydroxy and amine groups in 5HT are obliterated by the action of SNAT and HIOMT. One of the alternative metabolic pathways of 5HT is its oxidative deamination to 5-hydroxyindole acetaldehyde catalyzed by monoamine oxidase (MAO; EC 1.4.3.4) and is followed by the rapid conversion to 5-hydroxyindole acetic acid (HIAA) and 5-Hydroxytryptophol (HTPL). Another pathway is the methylation of 5HT to 5-methoxytryptamine (5MT) catalyzed by HIOMT. HIOMT also catalyzes the methylation of HTrp, HIAA, and HTPL to 5-methoxytryptophan (MTrp), 5-methoxyindole acetic acid (MIAA) and 5-methoxytryptophol (MTPL), respectively (for review see[31,32])

In most mammalian species studied, the content of melatonin and its precursor, NAS, in the pineal gland, show clear circadian changes with the highest level occurring during the dark period. This elevation of melatonin and NAS amounts observed in the dark period is due to the elevation of SNAT activity and the expression of the SNAT gene.[33] TPH activity also displays a diurnal rhythmicity, with distinct nocturnal elevations, regulated by circadian clock.[31] On the other hand, the 5HT amount in the pineal gland shows circadian changes opposite to the melatonin and NAS rhythm. These circadian changes of 5HT amount in the pineal gland may relate to its cyclic production from Trp, metabolism to NAS and release from the pinealocytes. The pineal levels of HIAA, MIAA, HTPL, and MTPL also show circadian changes parallel with the 5HT rhythm.[34,35]

Once formed, melatonin is not stored in the pineal gland but is immediately secreted into the bloodstream. Melatonin binds mostly to plasma albumin.[36] Melatonin then passes through the choroid plexus to the cerebrospinal fluid (CSF). Melatonin in plasma exhibits a circadian rhythm and is reproducible from day to day and from week to week and is almost like a *"hormonal finger print'* that can be reflected both in saliva and in urine.[37] In human beings plasma melatonin begins to increase steadily after 7.00 PM to 11.00 PM and reaches the peak value between 2.00 AM to 4.00 AM. The level then declines reaching a low level during daytime. The rhythm is well preserved from childhood to adulthood but in old age the night level reduces, which may be a contributing factor to insomnia seen in old age.[38]

Melatonin exerts its physiological actions such as the regulation of circadian rhythms or sleep induction, by acting through specific receptors present on cell membranes. Subtypes of melatonin receptors (MT1, MT2) have been identified in the plasma membrane of neural and peripheral tissues. Both receptors have been cloned and share general features with other G protein linked receptors.[39] MT 3 receptors homologous to quinone reductase 2[40] and nuclear[41] receptors for melatonin have also been described. Melatonin also exerts direct effects on intracellular proteins like calmodulin[42] and have strong free radical scavenger properties,[43] exerted in a non-receptor mediated way.

Within the SCN, melatonin reduces neuronal activity in a time-dependent manner. In rodents, the effects of melatonin on SCN activity are mediated by at least two different receptors. They are insensitive during the day, but sensitive at dusk and dawn (MT2; causes phase shifts) and during early night period (MT1; decreases neuronal firing rate).[39,44] Melatonin secreted during nighttime provides enough inertia to resist minor perturbations of the circadian timing system.

5. Circadian and Homeostatic Regulation of Sleep/Wakefulness

Sleep stages vary in a cyclic fashion that repeats approximately every 90 minutes. It consists of two forms namely slow wave sleep (SWS), and rapid eye movement (REM) sleep. A normal sleep cycle begins with light non-REM sleep going through four stages to slow wave sleep and then passes through the same stages in a reverse order approaching stage REM sleep in about 90 minutes. The daily sleep/wake cycle is strongly influenced by two separate processes. (1) an endogenous biological clock that drives the circadian rhythm of sleep/wake cycle (known as Process-C) and (2) a homeostatic component (Process S) that influences the sleep propensity depending upon the prior sleep/wakefulness and quality and duration of previous sleep episodes.[45] These two processes constantly interact with each other and it is not easy to separate the contribution of these two processes in sleep regulatory mechanisms.[46] However by using a *forced desynchrony model*, the contribution of these two processes on the regulation of sleep/wake have been studied. From these observations it has become evident that the homeostatic process drives slow wave and REM sleep is driven by circadian the component.[46] It is interesting to note that the pineal hormone melatonin plays a definite role both in sleep induction mechanism and regulation of the circadian component of sleep/wakefulness by acting on MT1 and MT2 receptors present on the SCN.[44] The temporal pattern of melatonin production by the pineal gland also correlates well with timing of human sleep. Melatonin secretion is initiated approximately 2 h before an individual's habitual bedtime and correlates with the onset of evening sleepiness.[47] According to Zhdanova and Tucci[48] melatonin's direct sleep promoting effect and a circadian regulating component of the sleep may occur jointly or separately. Melatonin induces sleep even when administered in physiologic doses (3-5 microgram) and is helpful in sleep maintenance as well.[49,50] In normal adults maximal sleep occurs at the time of low point in temperature rhythm and as melatonin is a major regulator of the circadian rhythm of core body temperature (cBT) in humans[51] there is every reason to designate melatonin as a physiological regulator of sleep mechanism.[52]

6. Melatonin and Sleep/Wake Rhythm

The relationship between melatonin production and occurrence of sleep at night prompted investigators to suggest that melatonin is involved in the physiological regulation of sleep.[53] That the evening rise of circulating melatonin is associated with an evening increase in sleep propensity has been well demonstrated.[54]

According to Wehr and co-workers[54] the switch process *"dusk"* and *"dawn"* is associated with melatonin onset and offset and also corresponds to the period of increase and fall of sleep propensity and the rise and fall of cBT. Exogenous melatonin administration can induce sleepiness at night even at very low doses.[49] Unlike some other hypnotic drugs, melatonin does not cause hangover effects the next morning. Melatonin may consolidate sleep by promoting interaction between sleep homeostasis and circadian rhythms[55] in elderly insomniacs.[56]

The peak of melatonin secretion occurs at the time of maximal sleepiness, whereas the nadir occurs at the peak of core body temperature, alertness and performance.[57] In the genetic disorder, Smith-Magenis syndrome, increased melatonin production occurs during daytime and is associated with inadequate nighttime sleep.

Administration of a beta adrenergic blocking agent suppresses both daytime melatonin and daytime sleepiness thereby supporting the involvement of melatonin in sleep mechanisms.[58]

7. REM Sleep and Melatonin

Rapid eye movement sleep (REM), also known as paradoxical sleep, occurs every 90 minutes and was first reported by Kleitman and coworkers in 1953.[59] Each REM sleep episode lasts around 20 to 30 minutes. The amount of REM sleep, its continuity, polarity, are all under strong circadian control by the SCN.[55,60] At nighttime melatonin has been shown to act by transducing and amplifying the circadian drive of the SCN. Melatonin intake restored the REM sleep percentage, as well as REM sleep quality accompanied by improvements in clinical symptoms, in patients with REM sleep behavior disorder.[61] It also increased the amplitude of temperature decline during nighttime. From these studies it is evident that melatonin helps to increase the output amplitude signal from SCN.[62] Thus if exogenous melatonin is administered at specific time, it can be beneficial in normalizing nighttime REM sleep. As the greatest increase in endogenous secreted melatonin occurs during the sleep time, melatonin administration at that time can supplement the sleep promoting effect of endogenous melatonin.[62]

8. Chronotyping of Individuals

Two different types of people with specific circadian sleep characteristics have been described, namely Morning types (*Larks*) and Evening types (*Owls*). These are called chronotypes and have been identified based upon the timing of their circadian phase. The self-report of the circadian phase can be evaluated by the administration of Horne-Ostberg Morningness-Eveningness Questionnaire (MEQ).[63] The Owl people go to sleep mainly past midnight while Lark people go to sleep around several hours earlier.

Morningness–Eveningness (M-E) score is a significant predictor of the circadian phase of cBT rhythm.[64] A significant association between M-E type and time of peak melatonin levels also has been found.[19] Evening type individuals have been found well suited for night shift work.[65] While some studies have indicated a correlation between MEQ scores and physiological measures of circadian phase, the reliability of MEQ score in assessing the phase of the circadian oscillator awaits further investigation.[66] While 10% MT, 10% ET and 80% of either type were found in European population, 75% MT, 16% ET and 9% of either type has been reported in human population living in tropical climatic conditions.[65]

9. Delayed Sleep Phase Síndrome

Weitzman and his co-workers[67] first identified delayed sleep phase syndrome (DSPS), which is mainly encountered in young individuals.[68,69] A common sleep/wake disorder that accounts for 10% of cases of chronic sleep disorders, DSPS is due altered physiological timing of our biological clock.[70] The sleep onset time and wake time are delayed.[71] The sleep onset is delayed in some cases to between 02.00 h and 06.00 h in the morning. Neither sleep architecture nor the maintenance of sleep is affected.[72] However, persons suffering from this disorder experience chronic sleep onset insomnia and forced early awakening results in daytime sleepiness. It has been shown that the peak melatonin secretion occurred between 08.00 h and 15.00 h in some DSPS patients demonstrating the abnormal phase position of melatonin in this sleep disorder.[72]

9.1. Clinical Correlates

The characteristic combination of symptoms of DSPS includes (a) chronic inability to fall asleep at desired clock time (b) a normal sleep pattern with a sleep of normal length and the ability to awaken spontaneously refreshed; and (c) a history of unsuccessful attempts to treat the problem. The minimal criteria for the diagnosis of DSPS includes delayed sleep onset for at least a month or excessive sleepiness with sleep log evidence of delayed sleep. When the patient is allowed to sleep undisturbed, the delayed sleep is of normal length and quality. In addition, 24-hour polysomnography or temperature monitoring should be consistent with this history, and the patient should have no other insomnia disorder.[73] The usual criteria of being unable able to go to sleep before 2.00 AM and difficulty in waking up before 10.00 AM for at least a year should be present.

A high sensitivity to light has been reported in 47% of patients suffering from DSPS and this super sensitivity to light could be involved in the pathophysiology of this disorder.[71,74] Patients with DSPS also have disturbances of other circadian rhythms such as body temperature, growth hormone and cortisol secretion, as well as hunger.[75] The subjects can have their "breakfast" at noon, "lunch" in the evening and "dinner" in the middle of the night.[69]

9.2. Causes and Prevalence

DSPS is the most frequently occurring circadian rhythm sleep disorders (CRSD).[76] Dagan and Eisenstein[69]found that 83.5% of 322 CRSD patients were of the DSPS type. The prevalence of DSPS in adolescence is more than 7%.[77-80] The onset of CRSD occurred in early childhood in 64.3%, the beginning of puberty in 25.3%, and during adulthood in 10.4% among DSPS.[69] Even a minor brain injury or a head trauma can act as a trigger for the development of DSPS.[81] DSPS can also follow whiplash injury.[82,83] Frequently occurring jet lag or frequently occurring shift-work are risk factors for developing DSPS.[71,84] Regenstein reported that 75% of their patients had severe symptoms or a prior history of depression.[72] DSPS persists even after remission of the depression suggesting that DSPS may be a cause rather a consequence of depression. Some patients developed a chronic fatigue syndrome-like clinical picture with late melatonin onset following viral infection suggesting that viral encephalitis could be a cause of DSPS.[85]

9.3. Pathophysiology

The endogenous circadian period length that is the internal cycle in the absence external timing information, may be increased. The sleep/wake cycle depends on a number of endogenous physiological parameters such as cBT, melatonin, cortisol, and other hormonal and metabolic profiles.[75] Therefore, a change in the timing of the intrinsic body clock can result in the development of a delayed sleep-wake rhythm as is seen in DSPS.[20] Longer endogenous circadian periods may underlie the preference of youth for late bedtimes.[70] An insufficient exposure to outdoor sunlight can also underlie the cause of DSPS.[8] A delayed melatonin secretory pattern with peak melatonin secretion, sometimes delayed until daytime could also underlie DSPS.[7,86]

Archer et al.investigated the link between delayed sleep phase syndrome and a length polymorphism in the PER3 clock gene.[45] They found that the length of the PER3 repeat region identifies a potential genetic marker for DSPS. This finding has been supported by number of other studies.[87-89]

9.4. Treatment Modalities

A procedure known as *chronotherapy* involves systematic delay of bedtimes and wake times over a period of days until the desired bedtime is achieved.

Once achieved, strict adherence to the new sleep/wake cycle is critical for maintaining a positive response.[67,90] Although effective, chronotherapy is demanding and compliance is low.

Following the successful application of bright light for resetting the endogenous pacemaker, bright light treatment was used (>2500 lux) in the treatment of DSPS.[91] The time of administration is important as phase advances are induced when light exposure is scheduled after the minimum of body temperature, and phase delays are induced when light exposure is scheduled before the minimum of cBT. Accurate assessment of the minimum body temperature can be difficult. As the maximum of the 24-hour melatonin secretion corresponds with the minimum of the 24-hour cBT, salivary melatonin measurement can be taken as a good index for determining the best time for exposure to bright light.[92] The duration of exposure to bright light in the treatment of DSPS differs from that used to treat other chronobiological disorders.

Rosenthal and co-workers used bright light of 2500 lux to treat DSPS.[93] Exposure to bright light daily for 2 h from 6.00 AM to 9.00 AM and then shielding the subject's eyes from sunlight by asking them to wear dark goggles for a period of one week advanced the sleep onset time and also improved daytime alertness. Pharmacological methods of treatment have been reported.[74,94] Vitamin B12 has been shown in case reports to benefit patients with DSPS.[66,95] As melatonin has proven to be an effective chronotherapeutic drug with few side effects, melatonin has become the drug of choice for treating DSPS.

9.5. Melatonin in the Treatment of DSPS

As exogenous administration of the pineal hormone melatonin has been shown to reset the biological clock, several investigators have used melatonin to treat DSPS patients. Dahlitz and co-workers were the first to report a placebo-controlled study that demonstrated the efficacy of melatonin in the treatment of DSPS patients.[96,97] Melatonin was administered orally in the dose of 5 mg at 22.00 h to a group of 8 patients suffering from DSPS for a period of 4 weeks. In this study, it was noted that melatonin significantly advanced the sleep onset time by an average of 82 minutes with a range from 19 to 124 minutes. The mean wakefulness time also advanced by 117 minutes.

Though the total duration of sleep remained unaltered (mean about 8 h) after melatonin treatment, there was significant improvement in sleep quality. Just as the effect of bright light depends on the time of its application, melatonin also needs to be administered at an appropriate time. Lewy et al[98] found that when melatonin was administered 5 h before endogenous melatonin onset, it advanced circadian time maximally. Therefore Nagtegaal et al.[99] administered melatonin 5 h before the onset of evening rise of endogenous melatonin (DLMO; *dim-light melatonin onset*) for a period of 4 weeks and found phase advancement of sleep/wake rhythm. The onset of the nocturnal melatonin profile was found to be phase advanced by 1.5 h. Following this report, Kayumov et al., administered melatonin to a group of 22 patients with DSPS who had their sleep time restricted to 24.00 h to 08.00 h.[7] Melatonin in the dose of 5 mg/day was administered 3 to 4 h before sleep onset for a period of 4 weeks. Melatonin significantly 'phase advanced' the sleep period, and decreased sleep onset latency as compared to placebo. No adverse effects of melatonin were noted in this study. In addition it was also found that exogenous melatonin normalized the circadian pattern of melatonin excretion in three of the five patients who had abnormal melatonin production, with peak melatonin excretion occurring between 08.00 h and 15.00 h.

Melatonin also was shown in placebo-controlled studies to be effective in children with idiopathic chronic sleep onset insomnia, which is related to child onset DSPS.[65,100] Many studies demonstrated the efficacy of melatonin in the treatment of DSPS.[71] The time of melatonin administration is the most crucial factor in correcting the phase abnormality of sleep/wake rhythm seen in patients suffering from DSPS. Melatonin may be the best treatment for DSPS, especially when combined with exposure to bright light in the morning.[92] Melatonin was found to be useful in treating sleep disorders following head injury.[101]

10. Advanced Sleep Phase Syndrome

As age-related advances, changes in sleep patterns can, in part, be attributed to changes in the functioning of the circadian oscillator.[20,102,103] The characteristic pattern of Advanced Sleep Phase Syndrome (ASPS) includes complaints of persistent early evening sleep onset and early morning awakenings.[104] Typically in ASPS, sleep onset occurs at around 08.00 PM and wakefulness occurs at around 03.00 AM.[105] The quality of sleep is associated with increased awakenings occurring during night.[106] The ability to maintain correct phase relationships among circadian rhythms is disturbed possibly due to attenuation of the melatonin rhythm.[107]

Recently Leger et al.[108] in studies undertaken in 517 human subjects aged 55 years and above noted a significant decline in 6-sulfatoxymelatonin excretion in subjects suffering from insomnia.9.0 microgram/night compared to 18.0 microgram/night seen in other subjects of the same age group. Melatonin replacement therapy in the dosage of 2 mg as controlled release tablets, Circadin, improved sleep quality significantly, in patients suffering from insomnia. Improvement in alertness and behavioral integrity was also noted in these subjects. From this study the authors concluded that decline in melatonin production associated with age impairs the sleep ability.[108] They also concluded that melatonin promotes sleep perhaps through circadian entraining effects rather than by sleep-regulating effect (Figure 4).[22]

The genetic testing of ASPS families by a number of investigators has shown that ASPS is also a familial disorder. So far in the world five families have been studied and they all provided evidence to support that ASPS is also an inherited

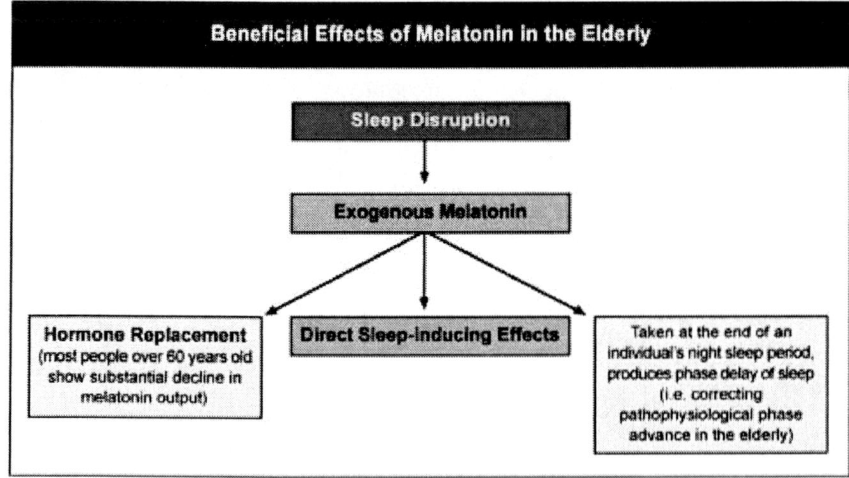

FIGURE 4. Schematic diagram of the mode of action of melatonin in elderly insomniacs.

sleep-wake rhythm disorder.[89,103,104,109-111] An autosomal dominant mode of inheritance of ASPS has been documented in these studies. An interesting finding in some of the studies is that there is not only an advancement of sleep-wake rhythm but also an advancement of melatonin rhythm as well.[104,109]

The familial ASPS gene hPer 2 has been localized near the telomere of the chromosome 2q.[89] Human Per 2 comprises 23 exons and study of the sequences of exon 17 in a large family with advanced sleep phase syndrome revealed that in the exon 17, a substitution of serine at amino acid 662 with glycine occurred.[89]

The genetic testing undertaken by Reid et al. demonstrated that mutation of other genes may also lead to the alteration of sleep-wake rhythm such as ASPS and DSPS.[104] In a recent study undertaken by Zucconi,[109] on a large family with 32 members, 8 were found to have clinical history of ASPS with four members exhibiting definite ASPS. These subjects also showed advancement of DLMO (18.30 h) compared to one healthy control that had DLMO at 22.00 h. Thus alteration in the function of clock genes has been documented as one of the major causes of CRSDs.

Administration of exogenous melatonin was effective in not only improving the sleep quality but also in improving the sleep/wake rhythm in both elderly human subjects and elderly patients suffering from Alzheimer Dementia.[112-115] Asayama and co-workers found melatonin to be effective in not only prolonging sleep time but also in reducing activity at night.[115] A severe disturbance of melatonin secretion has been reported in patients suffering from senile dementia of Alzheimer's type.[116] Exogenous melatonin administration has been found useful in improving the sleep/wake rhythm in these patients with out causing any side effects.[112-115]

11. Jet Lag

Rapid transmeridian flight across several time zones results in a temporary mismatch between the endogenous circadian rhythms and the new environmental L/D cycle.[117] As a result endogenous rhythms shift in the direction of the flight; an eastbound flight will result in a phase advance of rhythms while westbound flight will result in phase delay.[12] As different rhythms take different times to re-establish their normal phase relationship, a transient desynchronization of circadian rhythms occurs giving rise to symptoms such as altered (and transient) sleep pattern (e.g., disturbed night time sleep, impaired daytime alertness and performance), mood and cognitive performance (e.g. irritability and distress), appetite (e.g. anorexia), along with other physical symptoms such as disorientation, fatigue, gastrointestinal disturbances and light-headedness that are collectively termed as Jet lag.[12,118,119]

A number of reports have been published proving the efficacy of exogenous melatonin in alleviating the symptoms of jet lag, especially the sleep disturbances, both in placebo controlled field studies, as well as in laboratory studies.[120,121] Melatonin has been shown to act both as a hypnotic and a chronobiotic (for review, see[122]). Recommended melatonin ingestion times have been proposed for both for east and westward flights. Melatonin capsules taken in the evening at by local time in the new time zone have been found beneficial for traveling in both directions, irrespective of the timing of flight departure.[123,124]

Recent studies have shown that jet lag has a negative impact on athletic performance.[3] To accelerate resynchronization after transmeridian flight, athletes have resorted to take of various pharmacological agents. In a study on the effects of exogenous melatonin in soccer athletes[125] intake of melatonin (3 mg/day) at expected bedtime in Tokyo immediately after leaving Buenos Aires (12 time zone flight), combined with adequate exposure to sunlight with performance of physical exercise after reaching Tokyo, helped the athletes to overcome the consequences of jet lag. A positive correlation between the preflight melatonin production rate and sleep quality and morning alertness after flight was also noted.[125] These studies on the use of melatonin in athletes show that it is a potentially useful agent for improving athlete's performance after transmeridian flight.

12. Shift-Work Sleep Disorder

In our modern industrialized society, a large number of workers are engaged in work schedules that involve either daytime or night work. It has been estimated that at least one fifth of total global work force operates in rotating shift work. These individuals are forced to forego their nocturnal sleep while they are on a nightshift, and sleep during the day. This inversion of the sleep/wake rhythm with work at night at the low phase of the circadian temperature rhythm and sleep at the time of peak body temperature has given rise to insomnia like sleep disturbance. Sleep loss impairs the individual's alertness and performance that affects not only work productivity but also has been found to be a major cause for industrial and sleep related motor vehicle crashes.[2,126] Sleep related crashes occur most commonly in the early morning hours (0200-0600 h).[2] The loss of sleep is reported in night shift workers, even when engaged in day work soon after the onset of night shift.[127] Sleep deprivation and the associated desynchronization of circadian rhythms are common in Shift-work sleep disorder.

Many treatment procedures have been advocated. Czeisler and co-workers administered bright light for improving the physiological adaptation of the circadian rhythms of night-shift workers to their inverted sleep/wake schedules.[128] In this study bright light was found effective in resynchronization of alertness, cognition, performance, and body temperature to the new work schedules. Following the successful application of bright light, melatonin has been used in shift workers to accelerate adaptation of their circadian rhythms and sleep/wake rhythms to the new work schedules. Folkhard et al. published a report in which melatonin administration at the morning bedtime following a night shift was shown to be effective in not only improving the duration and quality of day time sleep, but also in improving alertness while on work during the night shift.[129]

A phase delay in plasma melatonin was noted in shift workers when melatonin was administered at the morning bedtime following the night shift.[130] The shift in melatonin secretion has been associated with increase in work performance as well.[131] Correctly timed administration of melatonin is advocated for hastening adaptation of circadian rhythms in shift-workers.[57] As melatonin administration in the evening (16.00 h) does not affect daytime sleepiness and mood, Rajaratnam et al. recommend the use of melatonin in situations where there is misalignment of the circadian clock to with external time cues such as with shift work.[132] Melatonin (1.5 mg at 16.00 h) was able to advance the timing of both endogenous melatonin and cortisol rhythms without causing any deleterious effects on endocrine function or daytime mood and sleepiness. Combinations of both bright light and melatonin could also be an effective and reliable strategy for treating shift work disorder.[57]

13. Non 24-hour Sleep/Wake Disorder

Non 24-hour Sleep/wake disorder is seen mostly in blind human subjects since their sleep/wake cycle is not synchronized to the 24-hour L/D cycle. These subjects suffer from recurrent insomnia and daytime sleepiness. The circadian rhythm of sleepiness has shifted out of phase with the desired time for sleeping.[133] Melatonin has been shown to phase shift the human circadian system, by an advance or delay, according to phase response curve changes both at physiological and pharmacological doses.[120] Therefore, melatonin has been employed to correct right abnormal sleep/wake rhythms in blind human subjects.

Two groups of investigators have reported that melatonin treatment completely synchronizes sleep/wake cycle to a 24-hour cycle in blind human subjects.[134,135] Lockley and his co-workers[134] phase advanced the rhythms to the correct position in three totally bind persons by using 5 mg of melatonin. Sack and his co-workers[135] administered 10 mg of melatonin for 3 to 9 weeks to 7 totally blind persons and found that melatonin was effective in inducing phase-advances of the sleep/wake rhythms by 0.6 h/day. On reaching complete entrainment, the dose was gradually reduced and synchronization of sleep/wake rhythms to the normal 24 hr day schedule was maintained with a low dose of 0.5 mg that resulted in plasma melatonin concentrations close to the physiological range. In this study the authors were able to show that the beneficial effects of melatonin could be attributed not only to its entrainment properties but also to its direct soporific effects. According to Arendt, the maintenance of synchronization of circadian sleep/wake rhythms with melatonin at physiologic concentrations supports the concept that melatonin forms an important part of human circadian system.[136]

The prevalence of Non 24-hour Sleep/wake disorder among sighted patients is unknown, but believed to be rare. Fewer than 50 cases have been reported in the world literature, the vast majority from Japanese publications.[66,137,138] Only 9 cases have been documented outside of Japan and these have been predominantly male and associated with avoidant or schizoid personalities.[139-142] Non 24-hour Sleep/wake disorder is a rare sleep disorder among sighted patients in Western populations. Only 3 patients were seen over a span of about 20 years in a sleep clinic that services a yearly average of over 500 patients, a small percentage of which are circadian rhythm disorders (Dr. Kayumov, personal communication). In the Japanese population it has been estimated that Non 24-hour Sleep/wake disorder comprises 23% of all circadian rhythm sleep disorders. The prevalence of circadian rhythm disorders in this population $(0.13-0.4\%)$[76,143] is consistent with that observed in other populations.[80]

It is likely that this sleep disorder is rare in Western populations because it is under-diagnosed. Diagnosis is complicated by the fact that at times Non 24-hour Sleep/wake disorder can resemble both ASPS and DSPS and in fact exhibit the same polysomnographic features. A recently completed polysomnographic study was performed on 22 untreated DSPS patients.[144] During imposed sleep periods (from 24:00 to 8:00 h) the main findings were delayed sleep onset latency (averaging one hour), abnormal distribution of SWS across the night with the highest amount of deep sleep in the early morning hours, short sleep duration (less than 6 hours) and increased amount of intervening wakefulness. The patients with non-24-h sleep/wake disorder displayed almost identical polysomnographic features since the baseline recordings were performed during delayed phases of their cycles. The fact that exogenous melatonin entrained the sleep cycles of the patients strongly suggests that the primary defect is a failure of the circadian clock to entrain normally to the environmental L/D cycle.Patients with non-24-h sleep/wake disorder demonstrate a psychiatric co-morbidity, such as depression, which may result from years of living out of synchrony with the rest of society.

14. Sleep/Wake Rhythm and Mood Disorders

Circadian rhythm disturbances, including sleep/wake rhythm disturbances, occur in endogenous depression, manic-depressive psychosis and seasonal affective disorder (SAD). Insomnia commonly occurs in patients suffering from endogenous depression, and REM sleep has been shown to occur early in the sleep episode.[145] In addition there is a phase advance of other physiological parameters such as prolactin and growth hormone secretion, cBT, and the nadir of cortisol secretion. Many rhythms are phase advanced by two to three hours.

Early work of Wetterberg[146] revealed that depression could be linked biochemically to disturbances in melatonin production and secretion. Low nocturnal melatonin levels have been reported in patients with depression.[147,148] Manic–depressive disorder may be due to disturbances in the timing of the phase position of circadian rhythms.[145] Melatonin secretion in these patients has been found to be abnormal and detection of an abnormal melatonin rhythm may give a biochemical clue to the presence of this disorder even before the clinical manifestations.[149] Patients with manic-depressive illness can exhibit a circabidian sleep/wake rhythm with sleep loss completely for one night, between two nights of normal sleep.[145]

In SAD or winter depression, recurrent episodes of depression occur during the winter months, and euthymia or hypomania occurs during the summer.[150] Most circadian rhythms are phase delayed in this disorder. Delayed onset of melatonin rhythms may be one of the key factors for the onset of SAD.[149] Patients suffering from SAD are more sensitive to dawn and disturbances in circadian rhythms may underlie this disorder.[151]

Bright light treatment has been used successfully in treating SAD. Morning exposure to bright light (2500 Lux) phase advanced all circadian rhythms. In addition, Exposure of SAD patients to natural sunlight in the morning hours for a period of one week caused complete remission of depressive symptoms.[151] It has been suggested that retinal-SCN-pineal gland disturbances can account for SAD.[152]

The hypnotic property of melatonin was evaluated in a group of depressed patients, in which administrations of slow-release melatonin was found to be effective in improving sleep but not improving the symptoms of depression.[153] Similar findings were made in an open study using a fast release melatonin preparation.[154] In a recent study of patients suffering from major depression,[155] it was found that both serum melatonin and urinary 6-sulfoxymelatonin were phase-shifted as compared to normal controls. Depressed patients exhibited 77 minutes peak time delay of serum melatonin secretion as compared to controls matched to age, gender, and season. The study confirms a shift in melatonin secretion in major depression.[155]

15. Conclusions

Normal circadian rhythms are synchronized to a regular 24 hr environmental L/D cycle. Both SCN and melatonin are essential for this adaptation. Desynchronization of circadian rhythms as occurs in chronobiological disorders result in severe disturbances of sleep. Common CRSDs are DSPS, ASPS, Non 24 hr sleep/wake rhythm disorder, Jet lag, and shift-work. Depression exhibited also circadian rhythm disturbances. Disturbances in the phase position of plasma melatonin levels have been documented in all these disorders. Whether this melatonin disruption is a cause or a consequence of these disorders is not known. Further research with of a large number of patients with circadian rhythm sleep disorders can help determine the association. However, at present there appears to be a role of melatonin in the pathophysiology of these circadian rhythm sleep disorders.

Melatonin has been found useful in treating the disturbed sleep/wake rhythms seen in DSPS, Non 24 hr sleep-wake rhythm, Shift-work sleep disorder, jet lag and depression. In most of these conditions an abnormal phase position of melatonin rhythm has clearly been documented. Recently ASPS has been found to be a familial disorder, the familial ASPS gene being localized near the telomere of the chromosome of 2q. In these disorders, both the sleep/wake rhythm and melatonin secretion have been found to be phase advanced.

16. Acknowledgments. We acknowledge Dr. Jenny Redman for her permission to modify use the figure that she originally published (Figure 1). We thank Puan Rosnida Said, Department of Physiology, School of Medical Sciences, University Sains Malaysia for excellent secretarial assistance. The authors would also like to thank Christiane Kochi for the preparation of the figures and graphics.

17. References

1. J. Arendt, S. Deacon, J. English, S. Hampton, and L. Morgan, Melatonin and adjustment to phase shift, work hours, sleepiness and accidents, *J Sleep Res*.4, 74 (1995).
2. J. Horne and L. Reyner, Vehicle accidents related to sleep: a review, *Occup Environ Med*.56, 289-294 (1999).
3. G. Atkinson, B. Drust, T. Reilly, and J. Waterhouse, The relevance of melatonin to Sports Medicine and Science, *Sports Med*.33, 809-831 (2003).
4. L. Wetterberg, J Beck-Friis, and B F Kjellman. Melatonin acts as a marker for a sub group of depression in: *Adults Biological rhythms, Mood disorders, Light therapy and Pineal gland* (American Psychiatric Press Inc; Washington DC, 1990).
5. V. Srinivasan, Melatonin biological rhythm disorders and phototherapy, *Ind J Physiol Pharmacol*.41, 309-328 (1997).
6. S. W. Lockley, D. J. Skene, H. Tabandeh, A. C. Bird, R. Defrance, and J. Arendt, Relationship between napping and melatonin in blind, *J Biol Rhythms*.12, 16-25 (1997).

7. L. Kayumov, G. Brown, R. Jindal, K. Butto, and C. M. Shapiro, A randomized double blind, placebo controlled cross over study of the effects of exogenous melatonin in delayed sleep phase syndrome, *Psychosomatic Med.*63, 40-48 (2001).

8. A. J. Lewy, V. K. Bauer, S. Ahmed et al, The human phase response curve (PRC) to melatonin is about 12 hours out of phase with the PRC to light, *Chronobiol Int.*15, 71-83 (1998).

9. J. Arendt, Importance and relevance of melatonin to biological rhythms, *J Neuroendocrinol.*15, 1-5 (2003).

10. L. Kayumov, V. Zhdanova, and C. M. Shapiro, Melatonin, Sleep, and circadian rhythms disorders, *Seminars in Clinical Neuropsychiatry.*5, 44-55 (2000).

11. T. Ninomiya, N. Iwatani, A. Tomoda, and T. Miike, Effects of exogenous melatonin on pituitary hormones in humans, *Clin Physiol.*21(3), 292-299 (2001).

12. J. Arendt and J. Deacon, Treatment of circadian rhythm disorders-Melatonin, *Chronobiol Int.*14, 185-204 (1997).

13. J. De Mairan, Observation botanique, *Histoire de L'Academie Royale des Sciences.*, 35-36 (1729).

14. R. Y. Moore and D. C. Klein, Visual pathways and central neural control of circadian rhythms in pineal N acetyltransferase activity, *Brain Res.*71, 17-33 (1974).

15. A. A. Sadun, J. D. Schaechter, and L. E. Smith, A retinohypothalamic pathway in man; light mediation of circadian Rhythms, *Brain Res.*302, 371-377 (1984).

16. W. J. Hrushesky and G. A. Bjarnason, Circadian cancer therapy, *J Clin Oncol.*11, 1403-1417 (1993).

17. A. Ayalon, H. Hermesh, and Y. Dagan, Case study of circadian rhythm sleep disorder following haloperidol treatment: reversal by risperidone and melatonin, *Chronobiol Int.*19, 947-959 (2002).

18. M. Morris, L. Lack, and D. Dawson, Sleep onset insomniacs have delayed temperature rhythms, *Sleep.*13, 1-14 (1990).

19. M. Gibertini, C. Graham, and M. R. Cook, Self report of circadian type reflects the phase of melatonin rhythm, *Biol Psychol.*50, 19-33 (1999).

20. D. R. Wagner, Disorders of the circadian sleep-wake cycle, *Neurol Clin.*14, 651-670 (1996).

21. R. L. Sack and A. J. Lewy, Circadian rhythm sleep disorders: lessons from the blind, *Sleep Med Rev.*5, 189-206 (2001).

22. V. Srinivasan, Melatonin sleep-wakefulness rhythm and its disorders, *Medical Nutritional Res Commun.*2, 1-8 (1994).

23. N. Zisapel, Circadian rhythim sleep disorders: pathophysiology and potential approaches to management, *CNS Drugs.*15, 311-328 (2001).

24. D. Dawson and S. M. Armstrong, Chronobiotics - drugs that shift rhythms, *Pharmacol Ther.*69, 15-36 (1996).

25. Y. Dagan, I. Yovel, D. Hallis, M. Eisenstein, and I. Raichick, Evaluating the role of melatonin in the long-term treatment of delayed sleep phase syndrome DSPS, *Chronobiol Int.*15, 181-190 (1998).

26. P. Pevet, B. Bothorel, B. Slotten, and M. Saboureau, The chronobiotic properties of melatonin, *Cell Tissue Res.*309, 183-191 (2002).

27. S. M. Armstrong, Melatonin the Internal zeitgeber of mammals, *Pineal Res Rev.*7, 157-202 (1989).

28. D. P. Cardinali and P. Pevet, Basic aspects of melatonin action, *Sleep Med Rev.*2, 175-190 (1998).

29. D. C. Klein and R. Y. Moore, Pineal N acetyltransferase and hydroxy indole-O-methyl transferase. Control by retino hypothalamic tract and suprachiasmatic nucleus, *Brain Res.*14, 245-254 (1979).
30. K. Thapan, J. Arendt, and D. Skene, An action spectrum for melatonin suppression; evidence for a novel non-rod,non-cone photoreceptor system in humans, *J Physiol.*535, 261-267 (2001).
31. J. Z. Nowak and J. B. Zawliska, Melatonin and its physiological and therapeutic properties, *Pharm World Sci.*20, 18-27 (1998).
32. T. Harumiand S. R. Pandi-Perumal. Pineal Gland and Melatonin Science, in: *Application of chromatographic technique to the analysis of Melatonin and its related indoleamines,* edited by C. Halder, M. Singaravel, and S. K. Maitra. (Publishers Inc Enfield, NH, USA, 2002) pp. 349-367.
33. D. C. Klein, S. L. Coon, P. H. Roseboom et al, The Melatonin rhythm generating enzyme molecular regulation of serotonin N-acetyltransferase in the pineal gland, *Recent Progr Horm Res.*52, 307-357 (1997).
34. J. N. Mefford, P. Chang, D. C. Klein, M. A. Namboodiri, d. Sugden, and J. Barchas, Reciprocal day/ night relationship between serotonin Oxidation and N-acetylation products in the rat pineal gland, *Endocrinology.*113, 1582-1586
35. R. J. Reiter, Pineal Melatonin: Cell biology of its synthesis and of its physiological significance, *Endocrine Rev.*12, 151-180.
36. D. P. Cardinali, H. J. Lynch, and R. J. Wurtman, Binding of melatonin to human and rat plasma proteins, *Endocrinology.*91, 1213-1218 (1972).
37. R. Nowak, I. C. McMillen, J. Redman, and R. V. Short, The correlation between serum and salivary melatonin concentration and urinary 6-hydroxy melatonin sulphate excretion rates; two non-invasive techniques for monitoring human circadian rhythmicity, *Cin Endocrinology.*27, 445-452 (1987).
38. V. Srinivasan, Melatonin Oxidative stress and aging, *Curr Sci.*76, 46-54 (1999).
39. M. L. Dubocovich, M. A. Rivera-Bermudez, M. J. Gerdin, and M. I. Masana, Molecular pharmacology, regulation and function of Mammalian melatonin receptors, *Front Biosci.*8, D1093-D1108 (2003).
40. P. A. Witt-Enderby, J. Bennett, M. J. Jarzynka, S. Firestine, and M. A. Melan, Melatonin receptors and their regulation: biochemical and structural mechanisms, *Life Sci.*72, 2183-2198 (2003).
41. L. Naji, A. Carrillo-Vico, J. M. Guerrero, and J. R. Calvo, Expression of membrane and nuclear melatonin receptors in mouse peripheral organs, *Life Sci.*74, 2227-2236 (2004).
42. G. Benitez-King, A. Rios, A. Martinez, and F. Anton-Tay, In vitro inhibition of Ca2+/calmodulin-dependent kinase II activity by melatonin, *Biochim Biophys Acta.*1290, 191-196 (1996).
43. R. J. Reiter, D. X. Tan, E. Gitto et al, Pharmacological utility of melatonin in reducing oxidative cellular and molecular damage, *Pol J Pharmacol.*56, 159-170 (2004).
44. Q. Wan, H. Q. Wan, F. Man et al, Differential modulation of GABA receptor function by Mel 1a and Mel 1b receptors, *Nat Neurosci.*X, 401-403 (1999).
45. S. N. Archer, D. L. Robiliard, D. J. Skene et al, A length polymorphism in the circadian clock gene Per 3 is linked to delayed sleep phase syndrome and extreme diurnal preference, *Sleep.*264, 413-415 (2003).
46. C. Gillin, E. Seifritz, R. K. Zoltoski, and R. J. Salin-Pascual. Basic science of Sleep, in: *Comprehensive Text Book of Psychiatry,* edited by B. J. Sadock and V. A.

Sadock. 1. ed. (Lippincott Williams & Wilkins., Philadelphia, USA, 2000) pp. 199-209.

47. O. Tzischinsky, A. Shlitner, and P. Lavie, The association between the nocturnal sleep gate and the nocturnal onset of urinary 6-sulphatoxymelatonin, *J Biol Rhythms*.8, 199-209 (1993).

48. I. V. Zhdanova and V. Tucci, Melatonin Circadian rhythms and sleep, *Curr Treat Opt Neurol*.5, 225-229 (2003).

49. B. M. Stone, C. Turner, S. L. Mills, and A. N. Nicholson, Hypnotic activity of melatonin, *Sleep*.23, 663-669 (2000).

50. V. Zhdanova, R. J. Wurtman, C. Morabito, V. R. Piotrovska, and H. J. Lynch, Effects of low doses of melatonin, given 2-4 hours before habitual bedtime, on sleep in normal young humans, *Sleep*.19 5, 423-431 (1997).

51. A. Cagnacci, J. Elliott, and S. Yen, Melatonin: a major regulator of the circadian rhythm of core body temperature in humans, *J Clin Endocrinol Metab*.75, 447-452 (1992).

52. T. Shochat, R. Luboshitzky, and P. Lavie, Nocturnal melatonin onset is phase locked to primary sleep gate, *Am J Physiol*.273 (1 pt 2), R364-R370 (1997).

53. V. Srinivasan, The Pineal gland: its physiological and pharmacological role, *Ind J Physiol Pharmacol*.33, 263-272 (1989).

54. T. A. Wehr, D. Aeschbach, and W. C. J. Duncan, Evidence for a biological dawn and dusk in human circadian timing system, *J Physiol*.535, 937-951 (2001).

55. D. J. Dijk and C. A. Czeisler, Contribution of the circadian pacemaker and sleep homeostasis to sleep propensity, sleep structure, electroencephalographic slow waves, and spindle activity in humans, *J Neurosci*.15, 3526-3538 (1995).

56. N. Zisapel, Development of a melatonin based formulation for the treatment of insomnia in the elderly, *Drug Dev Res*.50, 226-234 (2000).

57. S. M. W. Rajaratnam and J. Arendt, Health in 24 hr society, *Lancet*.358, 999-1005 (2001).

58. H. de Leersnyder, M. C. de Blois, M. Vekemas et al, Beta 1-adrenergic antagonists improve sleep and behavioral disturbances in a circadian disorder, Smith-Magenis syndrome, *J Med Genet*.38 (9), 586-590 (2001).

59. E. Aserinsky and N. Kleitman, Regularly occurring periods of eye motility, and concomitant phenomena during sleep, *Science*.118(3062), 273-274 (1953).

60. S. W. Wurts, W. F. Seidel, W. C. Dement, and D. M. Edger, Effects of suprachiasmatic nucleus lesions on REM sleep continuity, *Sleep Res*.26, 758 (1997).

61. D. Kunz and F. Bes, Melatonin as a therapy in REM sleep behaviour disorder patients: an open labeled pilot study on the possible influence of Melatonin on REM sleep regulation, *Mov Disord*.14, 507-511 (1999).

62. D. Kunz, R. Mahlberg, C. Muller, A. Tilmann, and F. Bes, Melatonin in patients with reduced REM sleep duration: Two randomized controlled trials, *J Clin Endocrinol Metab*.89, 128-134 (2004).

63. J. A. Horne and O. Ostberg, A self-assessment questionnaire to determine morningness-eveningness in human circadian rhythm, *Int J Chronobiol*.4 2, 97-110 (1976).

64. A. K. Pati, A. Chandrawanshi, and A. Reinberg, Shift Work: Consequences and management, *Curr Sci*.81, 32-50 (2001).

65. M. G. Smits, H. F. van Stel, H. K. van der, A. M. Coenen, and G. A. Kerkhof, Melatonin improves health status and sleep in children with idiopathic chronic

sleep-onset insomnia: a randomized placebo-controlled trial, *J Am Acad Child Adolesc Psychiatry*.42 11, 1286-1293 (2003).

66. H. Yamadera, K. Takahashi, and M. Okawa, A multicenter study of sleep-wake rhythm disorders: therapeutic effects of vitamin B12, bright light therapy, chronotherapy and hypnotics, *Psychiatry Clin Neurosci*.50, 203-209 (1996).

67. E. D. Weitzman, C. A. Czeisler, R. M. Coleman et al, Delayed sleep phase syndrome. A chronobiologic disorder with sleep onset insomnia, *Arch Gen Psychiat*.38(7), 737-746 (1981).

68. Y. Dagan, D. Stein, M. Steinbock, I. Yovel, and D. Hallis, Frequency of delayed sleep phase syndrome among hospitalized adolescent psychiatric patients, *J Psychosom Res*.45, 15-20 (1998).

69. Y. Dagan and M. Eisenstein, Circadian rhythm sleep disorders; towards more precise definition and diagnosis, *Chronobiol Int*.16, 213-222 (1999).

70. M. Hori and T. Asada, Life-style regularity and core body temperature phase in delayed sleep phase syndrome, *Sleep and Biological Rhythms*.1, 251-552 (2003).

71. T. Watanabe, N. Kajimura, M. Kato et al, Sleep and circadian rhythm disturbances in patients with delayed sleep phase syndrome, *Sleep*.26, 657-661 (2003).

72. Q. R. Regestein and T. H. Monk, Delayed sleep phase syndrome: a review of its clinical aspects, *Amer J psychiat*.152, 602-628 (1995).

73. M. J. Thorpy. *Diagnostic classification of sleep disorders Diagnostic and coding manual* 1990).

74. R. Rosenberg. Case studies in Insomnia, in: *Assessment and treatment of delayed sleep phase syndrome*, edited by P. J. Hauri . (Plenum, New York, 1991) pp. 193-205.

75. K. Shibui, M. Uchiyama, K. Kim et al, Melatonin, cortisol and thyroid-stimulating hormone rhythms are delayed in patients with delayed sleep phase syndrome, *Sleep and Biological Rhythms*.1, 209-214 (2003).

76. M. Yazaki, S. Shirakawa, M. Okawa, and K. Takahashi, Demography of sleep disturbances associated with circadian rhythm disorders in Japan, *Psychiatry Clin Neurosci*.53, 267-268 (1999).

77. R. P. Pelayo, M. J. Thorpy, and P. Glovinsky, Prevalence of delayed sleep phase syndrome among adolescents, *J Sleep Res*.17, 392 (1988).

78. Q. R. Regestein and M. Pavlova, Treatment of delayed sleep phase syndrome, *Gen Hosp Psychiat*.17, 335-345 (1995).

79. M. J. Thorpy, E. Korman, A. J. Speilman, and P. B. Glovinsky, Delayed sleep phase syndrome in adolescents, *J Adolesc Health Care*.9, 22-27 (1988).

80. H. Schrader, G. Bovim, and T. Sand, The prevalence of delayed and advanced sleep phase syndromes, *J Sleep Res*.2, 51-55 (1993).

81. C. Quinto, C. Gellido, S. Chokroverty, and J. Masdeu, Posttraumatic delayed sleep phase syndrome, *Neurology*.54, 250-252 (2000).

82. J. E. Nagtegaal, G. A. Kerkhof, M. G. Smits, A. C. Swart, and Y. G. van der Meer, Traumatic brain injury associated sleep phase syndrome, *Func Neurol*.126, 345-348 (1997).

83. M. G. Smits and J. E. Nagtegaal, Posttraumatic delayed sleep phase syndrome, *Neurology*.55, 902-903 (2000).

84. M. Shirayama, Y. Shirayama, H. Iida et al, The psychological aspects of patients with delayed sleep phase syndrome DSPS, *Sleep Medicine*.4, 427-433 (2003).

85. M. G. Smits, R. V. Rooij, and J. E. Nagtegaal, Influence of melatonin on Quality of life in patients with chronic fatigue and late melatonin onset, *JCFS*.48, 45-50 (2002).

86. M. Shibui, M. Uchiyama, and M. Okawa, Melatonin rhythms in delayed sleep phase syndrome, *J Biol Rhythms*.14, 72-76 (1999).

87. N. Cermakian and D. B. Boivin, A molecular perspectives of human circadian rhythms disorders, *Brain Res Rev*.42, 204-220 (2003).

88. T. Ebisawa, M. Uchiyama, N. Kajimura et al, Association of structural polymorphisms in the human period3 gene with delayed sleep phase syndrome, *EMBO Reports*.21, 342-346 (2001).

89. K. L. Toh, C. R. Jones, Y. He et al, An hPer2 phosphorylation site mutation in familial advanced sleep phase syndrome, *Science*.291, 1040-1043 (2001).

90. C. A. Czeisler, G. S. Richarson, R. M. Coleman et al, Chronotherapy: resetting the circadian clocks of patients with delayed sleep phase insomnia, *Sleep*.4(1), 1-21 (1981).

91. A. J. Lewy, R. L. Sack, L. S. Miller et al, The use of plasma melatonin levels and light in the assessment and treatment of chronobiologic sleep and mood disorders, *J Neural Trans Suppl*.21, 311-322 (1986).

92. M. G. Smitsand S. R. Pandi-Perumal. Delayed sleep phase syndrome: a melatonin onset disorder, in: *Melatonin: Biological basis of its function in health and disease,* edited by S. R. Pandi-Perumal and D. P. Cardinali. (2004) .

93. R. E. Rosenthal, J. R. Joseph-Vanderpool, A. A. Levendosky et al, Phase-shifting effects of morning light as treatment for delayed sleep phase syndrome, *Sleep*.134, 354-361 (1990).

94. S. S. Campbell, P. J. Murphy, C. J. van den Heuvel, M. L. Roberts, and T. N. Stauble, Etiology and treatment of intrinsic circadian rhythm sleep disorders, *Sleep Med Rev*.3, 179-200 (1999).

95. M. Okawa, K. Mishima, T. Nanami et al, Vitamin B 12 treatment for sleep-wake rhythm disorders, *Sleep*.131, 15-23 (1990).

96. D. Dahlitz, B. Alvarez, J. Vignau, J. English, J. Arendt, and J. D. Parks, Delayed sleep phase syndrome, response to melatonin, *Lancet*.337 8750, 1121-1124 (1991).

97. B. Alvarez, M. J. Dahlitz, J. Vignou, and J. D. Parkes, The delayed sleep phase syndrome: clinical and investigative findings in 14 subjects, *J Neurol Neurosurg Psychiat*.55, 665-670 (1992).

98. A. J. Lewy, S. Ahmed, J. M. Jackson, and R. L. Sack, Melatonin shifts human circadian rhythms according to a phase response curve, *Chronobiol Int*.9, 380-392 (1992).

99. J. E. Nagtegaal, G. A. Kerkhof, M. G. Smits, A. C. Swart, and Y. G. Deer Meer, Delayed sleep phase syndrome; A placebo controlled crossover study on the effects of melatonin administered five hours before the individualÆs dim light Melatonin onset, *J Sleep Res*.7, 135-143 (1998).

100. M. G. Smits, J. E. Nagtegaal, J. Van der Heijiden, A. M. Coenen, and G. A. Kerkhof, Melatonin for chronic sleep onset insomnia in Children: a randomized placebo controlled trial, *J Child Neurol*.162, 86-92 (2001).

101. S. Kemp, R. Biswas, V. Neumann, and A. Coughlan, The value of Melatonin for sleep disorders occurring post head injury a pilot RCT, *Brain Inj*.189, 911-919

102. S. M. Armstrong and J. R. Redman, Melatonin: a chronobiotic with anti-aging properties, *Medical hypothesis*.34, 300-309 (1199).

103. C. R. Jones, S. S. Campbell, S. E. Zone et al, Familial advanced sleep-phase syndrome: a short-period circadian rhythms variant in humans, *Nature Med*.5, 1062-1065 (1999).

292   V. Srinivasan, et al.

104. K. J. Reid, A. M. Chang, M. L. Dubocovich, F. W. Turek, J. S. Takahashi, and P. C. Zee, Familial advanced sleep-phase syndrome, *Arch Neurol*.587, 1089-1094 (2001).
105. K. Ando, D. F. Kripke, and S. Ancoli-Israel, Delayed and advanced sleep phase symptoms, *Isr J Psychiat Relat Sci*.39, 11-18 (2002).
106. C. A. Czeisler, R. E. Kronauer, J. S. Mooney, J. L. Anderson, and J. S. Allain. Biological rhythm disorders, depression and phototherapy. A new hypothesis Psychiatric clinics of North America,edited by M. K. Erman. (W B Saunders company, Phildelphia USA, 1987) pp. 687-709.
107. R. J. Reiter. Pineal rhythmicity, Neural, behavioral and endocrine consequences, in: *Biological rhythms, Mood disorders and Light therapy,* edited by M. Shafii and S. Shaffi. (American Psychiatric Press Inc, Washington DC, 1990) pp. 41-66.
108. D. Leger, M. Lauden, and N. Zisapel, Nocturnal 6-sulfatoxy melatonin excretion in insomnia and its relation to the response to melatonin replacement therapy, *Am J Med*.116, 91-95 (2004).
109. M. Zucconi, Familial advanced sleep-phase syndrome, *Sleep Medicine*.3, 177-178 (2002).
110. K. Satoh, K. K. Mishima, Y. Inoue, T. Ebisawa, and T. Shimuzu, Two pedigrees of familial advanced sleep phase syndrome in Japan, *Sleep*.264, 416-417 (2003).
111. K. J. Reid, A. M. Chang, and P. C. Zee, Circadian rhythm sleep disorders, *Med Clin N Am*.88, 631-651 (2004).
112. L. I. Brusco, M. Marquez, and D. P. Cardinali, Monozygotic twins with Alzheimer's disease treated with melatonin: Case report, *Journal of Pineal Research*. (1998).
113. L. I. Brusco, M. Marquez, and D. P. Cardinali, Melatonin treatment stabilizes chronobiologic and cognitive symptoms in Alzheimer's disease, *Neuroendocrinology Letters*.19, 111-115 (1998).
114. D. P. Cardinali, L. I. Brusco, C. Liberczuk, and A. M. Furio, The use of melatonin in Alzheimer's disease, *Neuroendocrinology Letters*.23 (suppl. 1), 26-29 (2002).
115. K. Asayama, H. Yamadera, T. Ito, H. Suzuki, Y. Kudo, and S. Endo, Double blind study of melatonin effects on the sleep-wake rhythm cognitive and Non-cognitive functions in AlzheimerÆs Type dementia, *J Nippon Med Sch*.70, 334-341 (2003).
116. K. Mishima, T. Tozawa, K. Satoh, Y. Matsumoto, Y. Hishikawa, and M. Okawa, Melatonin secretion rhythm disorders in senile Dementia of AlzheimerÆs Type with disturbed sleep-waking, *Biol Psychiatry*.45, 417-421 (1999).
117. J. Arendt and V. Marks, Physiological changes underlying Jet lag, *Brit Med J*.284, 141-146 (1982).
118. A. O. Jamieson, G. K. Zammit, R. S. Rosenberg, J. R. Davis, and J. K. Walsh, Zolpidem reduces the sleep disturbance of Jetlag, *Sleep Med*.2(5), 423-430 (2001).
119. B. L. Parry, Jetlag; minimizing its effects with critically timed bright light and melatonin administration, *J Mol Microbiol Biotechnol*.4(5), 463-466 (2002).
120. S. Deacon and J. Arendt, Melatonin induced temperature suppression and its acute phase shifting effects correlate in a dose dependent manner in humans, *Brain Res*.688, 77-85 (1995).
121. J. Arendt, J. Aldhous, and V. Marks, Alleviation of jet lag by melatonin. Preliminary results of double blind trial, *Brit Med J*.292, 1170 (1986).

122. G. F. Oxenkrug and P. J. Requintina, Melatonin and jet lag syndrome: Experimental model and clinical implications, *CNS Spect.*82, 139-148 (2003).

123. D. P. Cardinali, L. I. Brusco, S. Pérez Lloret, and A. M. Furio, Melatonin in sleep disorders and jet-lag, *Neuroendocrinology Letters.*23 (suppl. 1), 13-17 (2002).

124. T. Takahashi , M. Sasaki, H. Itoh et al, Melatonin alleviates jet lag symptoms caused by an 11-hour eastward flight, *Psychiatry Clin Neurosci.*56, 301-302 (2002).

125. D. P. Cardinali, G. P. Bortman, G. Liotta et al, A multifactorial approach employing melatonin to accelerate resynchronization of sleep-wake cycle after a 12 time-zone westerly transmeridian flight in elite soccer athletes, *J Pineal Res.*32, 41-46 (2002).

126. A. E. Reinbergand M. H. Smolensky. Biological rhythms in clinical and laboratory medicine, in: *Night shift work, transmeridian and space flights,* edited by Y. Touitou and E. Haus. (Springer-Verlag, Germany, 1992) pp. 243-255.

127. T. Akerstedt, Work hours, sleepiness and underlying mechanism, *Sleep Res.*4, 15-22 (1995).

128. C. A. Czeisler, M. P. Johnson, J. E. Duffy, E. N. Brown, J. M. Ronda, and R. E. Kronauer, Exposure to bright light and darkness to treat physiologic maladaptation to night shift work, *N Eng J Med.*322, 1253-1259 (1990).

129. S. Folkard, J. Arendt, and M. Clark, Can melatonin improve shift-workerÆs tolerance to night shift? Some preliminary findings, *Chronobiol Int.*10, 315-320 (1993).

130. R. Sack and A. Lewy, Melatonin as a chronobiotic: treatment of circadian desynchrony in night workers and the blind, *J Biol Rhythms.*12, 595-603 (1997).

131. M. A. Quera-Salva, C. Guilleminault, B. Claustrat et al, Rapid shift in peak melatonin secretion associated with improved performance in short shift work schedule, *Sleep.*2012, 1145-1150 (1997).

132. S. M. W. Rajaratnam, D. J. Dijk, B. Middleton, and B. M. Stone, Melatonin phase shifts Human circadian Rhythms with no evidence of changes in the duration of endogenous melatonin secretion or the 24-hour production of reproductive hormones, *J Clin Endocrinol Metab.*88, 4303-4309 (2003).

133. H. Nakagawa, R. L. Sack, and A. J. Lewy, Sleep propensity free-runs with the temperature, melatonin and cortisol rhythms in a totally blind person, *Sleep.*15, 330-336 (1992).

134. S. W. Lockley, D. J. Skene, K. James, K. Thapen, J. Wright, and J. Arendt, Melatonin administration can entrain the free running circadian system of blind subjects, *J Endocrinol.*164 1, R1-R6 (2000).

135. R. L. Sack, R. W. Brandes, A. R. Kendall, and A. J. Lewy, Entrainment of free running circadian rhythms by melatonin in blind people, *N Eng J Med.*343(15), 1070-1077 (2000).

136. J. Arendt, Melatonin, circadian rhythms and sleep, *N Eng J Med.*34, 1114-1116 (2000).

137. K. Shibui, M. Okawa, M. Uchiyama et al, Continuous measurement of temperature in non-24 hour sleep-wake syndrome, *Psychiatry Clin Neurosci.*52(2), 236-237 (1998).

138. Y. Kamei, J. Urata, M. Uchiyaya et al, Clinical characteristics of circadian rhythm sleep disorders, *Psychiatry Clin Neurosci.*52(2), 234-235 (1998).

139. A. L. Eliott, J. N. Mills, and J. M. Waterhouse, A man with too long a day, *J Physiol London.*212, 30-31 (1970).

140. T. M. Hoban, R. L. Sack, A. J. Lewy, L. S. Miller, and C. M. Singer, Entrainment of a free-running human with bright light?, *Chronobiol Int*.6, 347-353 (1989).
141. A. J. McArthur, A. J. Lewy, and R. L. Sack, Non-24-hour sleep-wake syndrome in a sighted man: circadian rhythm studies and efficacy of melatonin treatment, *Sleep*.19, 544-553 (1996).
142. D. A. Oren, H. A. Giesen, and T. A. Wehr, Restoration of detectable melatonin after entrainment to a 24-hour schedule in a "free-running" man, *Psychoneuroendocrinology*.22, 39-52 (1997).
143. Y. T. O. Kayukawa, Epidemiology of circadian rhythm sleep disorders, *Study on the diagnosis , treatment and epidemiology of sleep disorders*.Annual Report of the National Project Team 1995, 131-135 (1995).
144. L. Kayumov, G. Brown, R. Jindal, K. Buttoo, and C. M. Shapiro, A randomized, double-blind, placebo-controlled crossover study of the effects of exogenous melatonin on delayed sleep phase syndrome, *Psychosom Med*.63, 40-48 (2001).
145. T. A. Wehr, R. Sack, N. Rosenthal, W. Duncan, and J. C. Gillin, Circadian rhythm disturbances in manic-depressive illness, *Fed Proceed*.42, 2809-2814 (1983).
146. L. Wetterberg, The pineal hormone melatonin as marker for a subgroup of depression, *Medicographica*.7, 4-7 (1986).
147. L. Wetterberg, J. Beck-Friis, B. Aperia, and U. Petterson, Melatonin Cortisol ratio in depression, *Lancet*.2, 1361 (1979).
148. A. Venkobarao, S. Parvathi devi, and V. Srinivasan, Urinary melatonin in depression, *Ind J Psychiat*.25, 167-172 (1983).
149. A. Mayedaand J. Nurnberger. Biological rhythms, mood disorders, Light therapy and Pineal gland, in: *Melatonin and circadian rhythms in Bipolar disorder*, edited by M. Shafii and S. Shafii. (American Psychiatric Press Inc., Washington DC USA, 1990) pp. 119-137.
150. N. E. Rosenthal, R. A. Sack, J. C. Gillin et al, Seasonal affective disorder; A description of the syndrome and preliminary findings with light therapy, *Arch Gen Psychiat*.41, 72-80 (1984).
151. A. Wirz-Justice, P. Graw, K. Krauchi et al, Natural light treatment of seasonal affective disorder, *J Affective Dis*.37, 109-120 (1996).
152. D. A. Oren, Retinal melatonin and dopamine in seasonal affective disorder, *J neural transmission*.83, 85-95 (1991).
153. O. T. Dolberg, S. Hirschmann, and L. Grunhaus, Melatonin for the treatment of sleep disturbances in major depressive disorder, *Am J Psychiat*.155, 1119-1121 (1998).
154. I. Fainstein, A. Bonetto, L. I. Brusco, and D. P. Cardinali, Effects of melatonin in elderly patients with sleep disturbance. A pilot study., *Curr Ther Res*.58, 990-1000 (1997).
155. M. Crasson, S. Kjiri, A. Colin et al, Serum melatonin and urinary 6-sulfatoxy melatonin in major depression, *Psychoneuroendocrinology*.291, 1-12 (2004).

The Role of Melatonin in Sleep Regulation

NAVA ZISAPEL

1. Abstract

Sleep, a state marked by lessened consciousness, lessened movement of the skeletal muscles, and slowed-down metabolism, has an essential restorative function and an important role in memory consolidation. Sleep is an orchestrated neurochemical process involving sleep promoting and arousal centers in the brain. Sleep propensity depends on the amount of sleep deprivation and on the circadian clock phase. Typically, humans sleep during the dark phase of the 24 h cycle. Insomnia is a symptom, resulting from insufficient sleep or sleep of poor quality. Insomnia is the most prevalent sleep complaint in the general population affecting about 27% of adults, and 10% in the general population reporting the symptoms as serious. The most commonly prescribed insomnia medications are the benzodiazepines and non-benzodiazepines GABA-A receptor modulators. These drugs typically shorten the latency to sleep, but are associated with sedation, anterograde amnesia, rebound, abuse potential, dependence, and tolerance, are indicated for short-term use and do not treat non-restorative sleep. Melatonin, the hormone secreted at night for the pineal gland, is a physiological sleep regulator in humans. Administration of melatonin to blind people who free run in the absence of the light cue and to subjects with delayed sleep phase syndrome is useful to attain synchronization of the sleep/wake cycle with the external night period. Aging is associated with increase in the prevalence of insomnia and decrease in melatonin production. Insomnia patients aged 55 years and older produce significantly less melatonin than younger adults and many of them produce lesser amounts than normal for their age. Administration of melatonin, in a prolonged release formulation to mimic the physiological release of the hormone, has proven effective in the improvement of sleep quality and daytime alertness in insomnia patients aged 55 years and over. Melatonin therapy may be less effective in patients producing timely and sufficient amounts of the hormone endogenously. Melatonin is not associated with the side effects of GABA-A modulators and thus represents a new paradigm is sleep medicine.

2. The Neurophysiology and Neurobiology of Sleep

Sleep is a ubiquitous phenomenon in the animal kingdom. The sleep need is remarkably standardized in both quality and quantity and, if disturbed, results in problems during wakefulness. A prominent hypothesis implicates sleep in the plastic cerebral changes that underlie learning and consolidation of memory.[1] Sleep is associated with characteristic changes in central nervous activity as measured by polysomnography (electro-encephalogram, EEG; electromyogram, EMG; electrooculogram, EOG). Based on these signals sleep is divided into two distinct states known as non-rapid eye movement sleep (NREM) stages 1, 2, 3, 4 and rapid eye movement sleep (REM). Due to the EEG signals recorded during the NREM stages 3 and 4 (slow, high-amplitude waves in the frequency range below 4 Hz), these stages are also known as slow-wave sleep (SWS). NREM and REM sleep occur in a roughly 90 minute cycle, which is repeated 5 to 6 times a night. Each sleep cycle contains a non-REM period (or slow-wave sleep) and the REM period. However, the balance between the states shifts during the course of the night from NREM sleep predominance during the first ninety-minute cycle, to predominance of REM sleep in the final cycle of the night. Recent findings link NREM sleep to restoration of performance and REM sleep for consolidation of memory.[2,3]

Humans normally sleep at night. The timing and propensity of sleep are thought to reflect two interacting processes:[4] 1) an accumulated sleep need (S), that is manifested by increase in sleep propensity after sleep deprivation and decrease during sleep. 2) A circadian process (C) controlled by an endogenous pacemaker, which is basically independent of sleep and waking. Process C determines the times of onset and termination of sleep, respectively by changing the threshold of sleep need that will increase sleep propensity. Arousal levels also play a role in sleep initiation and whenever arousal level is reduced sleep propensity increases.

The arousal system comprises neurons of the reticular formation within the brainstem, which excite neurons of the nonspecific thalamo-cortical projection system, and neurons of the ventral extra-thalamic relay systems located in the posterior hypothalamus and basal forebrain, all of which predominantly utilize glutamate as a neurotransmitter. In addition, cholinergic neurons of the ponto-mesencephalic tegmentum and basal forebrain promote cortical activation during waking and REM sleep. Noradrenergic locus coeruleus neurons and histaminergic neurons of the posterior hypothalamus promote an arousal and prevent REM as well as SWS sleep.[5] A small set of orexin/hypocretin producing neurons in the lateral hypothalamic and perifornical areas interact with feeding centers in the hypothalamus, arousal and sleep-wakefulness centers in the brainstem, sympathetic and parasympathetic nuclei and the limbic system. These neurons may be an important cellular link in the integration of adaptive behavior associated with arousal and energy homeostasis and their improper functioning is responsible for narcolepsy, or the inability to maintain wakefulness.[6]

The arousal systems are redundant in the sense that no one system is absolutely necessary for the occurrence of waking but nevertheless, they have distinct roles in waking and sleep. During SWS, the arousal systems are inhibited in part by GABAergic neurons co-distributed with many neurons of the arousal systems. The ventrolateral preoptic nucleus (VLPO) contains a group of sleep-active, galanin producing neurons that appears to be a critical component of sleep circuitry across multiple species. The VLPO presumably inhibits the major ascending monoaminergic arousal systems during sleep; lesions of the VLPO cause insomnia.[7] Noradrenaline, acetylcholine and serotonin, all of which are transmitters of wakefulness, inhibit the electrical activity of these neurons. Hence, the upper and lower thresholds for sleep and waking presumably reflect relative inactivity levels of the sleep promoting neurons.

Recent finding suggest that adenosine, acting via the A1 receptor, is a key factor in the homeostatic control of sleep. In brain areas that regulate cortical vigilance, particularly in the basal forebrain, high extracellular adenosine concentrations, induced by prolonged wakefulness, decrease the activity of cholinergic cells and via this mechanism promote sleep. Local energy depletion in the basal forebrain of rats induced elevations in extracellular concentrations of lactate, pyruvate and adenosine, as well as increases in non-REM sleep during the following night. In addition microdialysis perfusion of antisense oligonucleotides against the mRNA of the A1 receptor in the magnocellular cholinergic region of the basal forebrain of freely behaving rats resulted in reduction NREM sleep with an increase in wakefulness. These changes may at least partially mediate the long-term effects of prolonged wakefulness (i.e. process S).[8-11]

Process C, among other physiological and behavioral conditions that fluctuate between states of high and low activity during the 24 hr. day/night cycle, is presumably regulated by the intrinsic body clock residing in the brain's suprachiasmatic nucleus (SCN). The endogenous clock cycle is somewhat slower or faster than the solar 24-h day/night cycle (in humans it is usually >24 h) and is normally entrained by the 24-h light dark cycle to match the environmental rhythm.[12] The SCN appears to function similarly in nocturnal and diurnal species, with maximal electrical activity during daytime. The "interpretation" of the sleep centers to SCN signals must therefore be downstream to the SCN. The projections from the SCN top the VLPO are scarce. Rather, the VLPO feeds-back into the SCN. The sleep/wake regulating activity of the SCN may possibly be mediated by direct connections to the dorsomedial hypothalamic nucleus (DMH) that innervates the VLPO.[13,14] In addition, the SCN-regulated synthesis of the pineal hormone melatonin (N-acetyl-5-methoxytryptamine) at night provides a humoral sleep/wake regulation signal.

3. Insomnia - the Disease and Current Approaches to Treatment

In Western industrial countries about one third of the adult population reports at least occasional difficulties with sleeping. Similarly, a World Health Organization study in 15 different countries revealed that an average of 27% of primary healthcare attenders were suffering from sleep problems. The prevalence of insomnia increases steeply with age during the fifth decade of life. About half of the people in the age group 50-59, 60-69 and 70+ stated that they had insomnia, and in half of them the insomnia was severe. Complaints of insomnia show consistency, persisting for months. Recent surveys, including the World Health Organization study, have clearly shown that at least half of the sleep-disordered population has had their sleep complaints for years. With respect to severity, epidemiological studies throughout the world (USA, Canada, Europe, Australia, Asia, and Russia) have found that at least 10% of the total population in these countries rated their sleep problem as serious. People with insomnia generally have more medical complaints, and seek medical care more often than people without insomnia.[15]

Primary Insomnia is defined in the fourth revision of the Diagnostic and Statistical Manual of the American Psychiatric Association (DSM-IV) as well as in the tenth revision of the International Classification of Diseases (ICD-10). Key elements for the definition of Primary Insomnia that are described in DSM-IV as well as in ICD-10 include insufficient sleep quantity and/or inadequate sleep quality over a certain period as well as negative effects on subsequent day-time functioning. People with insomnia generally have more medical complaints, and seek medical care more often than people without insomnia.[15,16] Recent surveys suggest that chronic insomnia is major risk factor in traffic accidents.[17-19]

The non-benzodiazepines hypnotics, zolpidem and zaleplon, with benzodiazepines and antidepressants -are the most commonly prescribed medications for insomnia, but they are associated with sedation, anterograde amnesia, rebound, abuse potential, dependence, and tolerance, and labeling for short-term use (up to 14 nights). Several new agents are in development to treat insomnia, including Es-zopiclone, indiplon and a sustained-release zolpidem-all of which act on the GABA-A receptor as well. On the basis of DSM-IV and ICD-10 criteria currently marketed hypnotics were developed and approved to address quantitative sleep problems, mainly time to sleep onset (TSO) and total sleep time (TST). A brief course of hypnotic treatment is deemed appropriate in transient or short-term insomnia. However, about a quarter of severe insomniacs use hypnotic drugs regularly and usually long-term usage increases markedly with age, with 5% of severe insomniacs in the age group 18 to 25, but 33% aged 56 to 65. Patients treated with hypnotic drugs, even those that are short acting, may experience daytime sleepiness and there is the potential for impaired performance and increased risk for accidents.[20,21] All benzodiazepines adversely affect cognition by disrupting both short and long term memory.[22,23] Because the new agents in development to treat insomnia are acting on the GABA-A receptor, experts suspect that they will also share the class side-effects profile.

In recent years, the focus of clinically oriented sleep medicine has shifted from sleep quantity to sleep quality. Several studies have shown that insomnia related to quality rather than quantity of sleep is associated with impaired daytime functioning (e.g. negative effects on memory, vigilance and psychomotor skills) and quality of life. It has been shown that low sleep quality, rather than quantity, is associated with physical health problems, anxiety, depression and fatigue.[24] Bonnet and Arand[25] demonstrated that experimentally-induced sleep disturbance, matched to the degree of sleep disturbance found in objective insomniacs (i.e. sleep complaint disturbance clinically demonstrable in the sleep laboratory) failed to produce significant performance decrements suggesting that sleep deprivation itself is not the cause for the performance decrements. Moreover, sleep quality does not appear to correlate well with next day performance within individuals[26] whereas those patients with subjective insomnia (i.e. no objectively demonstrable sleep disturbance-likely to have non restorative sleep) seem to perform no better at daytime, and in at least one case worse, than those with objective insomnia.[27,28] A recent study in Germany has demonstrated that patients with Global Sleep Dissatisfaction (GSD) were two times more likely to report excessive daytime sleepiness compared to insomnia patients without GSD.[29] Current hypnotics do not improve nonrestorative sleep. Moreover, they do not improve, and in many cases even impair daytime functioning. For the development of novel sleep medication it will therefore be crucial to address the restorative value of sleep. The understanding of restorative sleep as a key function of good sleep quality and unimpaired daytime functioning sets a new corridor for the clinical decision making in sleep medicine.

4. Melatonin as Sleep Regulator in Humans

Melatonin is a hormone produced and secreted by the pineal gland at night. Its production is regulated by the biological clock localized in the suprachiasmatic nuclei (SCN) of the hypothalamus. Melatonin is first released into the third ventricle and from there into the cerebrospinal fluid and circulating blood. The time and duration of the melatonin peak encode time-of-day and length-of-day information to the brain including the SCN and peripheral organs. Melatonin is rapidly metabolized by the liver and its half-life in plasma is only 40-50 min. Over 85% are eliminated in the urine as 6-sulphatoxymelatonin (6-SMT). Peak plasma concentrations are reached within 20-30 minutes of oral administration of melatonin and then decay rapidly. The mean urinary excretion profile for 6-SMT reflects blood levels and the amount of melatonin produced endogenously may be reliably based upon cumulative urinary excretion of 6-SMT over the 24 hours. In contrast to GABA-A CNS depressant hypnotics, melatonin acts via its own receptors (MT1, MT2), which are members of the G protein-linked receptor familyand perhaps other receptors that have not been cloned. In preparation for sleep, melatonin acts to induce heat loss and reduce arousal.[30]

A major difficulty in elucidating the role of melatonin in sleep regulation is the lack of appropriate animal models. Melatonin is a signal of darkness in nocturnal as well as diurnally active species. Yet, in nocturnal species melatonin is associated with waking whereas in diurnal species it is associated with sleep. Accordingly, administration of melatonin promotes sleep in humans[31] but not in rats and mice.[32,33] Because melatonin is not inducing sleep in mice, it is not possible to use the melatonin receptors MT1 and MT2 knockout mice, or even pinealectomy experiments, to explore the role of melatonin and its receptors in sleep. In addition, MT1 and MT2 knockout mice have not added much to our understanding of the role of these melatonin receptor subtypes in SCN regulation, perhaps because most mouse strains do not produce melatonin[34,35] and may not be sensitive to the absence of this signal.

Because the SCN has a similar function in nocturnal and diurnally active animals, the "interpretation" of the melatonin signal must be downstream to the SCN, and possibly involves a counter-balance between melatonin's effects on brain regions that are involved in arousal and those involved in suppression of arousal. This would call for the presence of melatonin receptors outside the SCN, e.g., the DMH, VLPO, hypothalamic orexin neurons, or arousal systems in the brainstem. The MT1 and MT2 melatonin receptors have a very narrow distribution in the brain, and there is no proof for functional activities of these receptors in brain regions other than SCN. Functional and ligand binding studies have however pointed out the presence of low affinity functional melatonin receptors in the preoptic area of the hypothalamus and in the medulla-pons in rats[36,37] but the proteins involved in this activity have not been identified.

Another difficulty in elucidating the role of melatonin in human sleep is the presence of the endogenous hormone, which interferes with the effects of exogenous melatonin. Nevertheless, there is a growing body of evidence for a role for endogenous melatonin in human sleep.[38] First of all, the period of increased melatonin secretion from the pineal gland is concurrent with the habitual hours of sleep in humans, and the onset of melatonin secretion highly correlates with the onset of the evening sleepiness.[39-41] Secondly, aging, presence of certain diseases (e.g. primary degeneration of the autonomic nervous system and diabetic neuropathy, some types of neoplasms, Alzheimer's disease), and certain drugs (e.g. b-blockers, clonidine, naloxone and non-steroidal anti-inflammatory drugs), abolish the nocturnal production of melatonin and are associated with impaired sleep (for review see ref.[30]). In totally blind people who cannot perceive the light-dark cycle (the major synchronizer of the circadian pacemaker) the circadian rhythms often "free run" on a cycle slightly longer than 24 h. Blind subjects with normally entrained rhythms (as evidenced from the excretion of its metabolite 6-SMT) had fewer naps of a shorter duration than abnormally entrained or free-running subjects. The timing of these naps was not random; significantly more naps occurred within a five-hour range before and after the 6-SMT peaks than outside this range.[42] Furthermore, in delayed sleep phase syndrome (DSPS), a circadian rhythm sleep disorder characterized by sleep-onset insomnia and difficulty in awakening at the desired time, the melatonin rhythms are delayed compared with those in normal individuals.[43] Significant association of delayed sleep phase syndrome with a mutation in arylalkylamine-N-acetyl transferase, the rate-limiting enzyme in melatonin synthesis has recently been reported.[44]

If indeed melatonin plays a role in sleep regulation, either through its chronobiotic, soporific activities or both, we would expect that exogenous melatonin administration will be able to induce sleep when the homeostatic drive to sleep is insufficient and to improve sleep in cases of deficiency in endogenous melatonin. There is much evidence in support of such notions. Sleep-inducing effects of melatonin (0.3-240 mg) were reported in humans following intravenous, intranasal and oral administration of melatonin at daytime (i.e. when it is not present in the blood and when the homeostatic drive to sleep is insufficient to induce sleep spontaneously). Unlike benzodiazepines hypnotics, melatonin-promoted sleep has normal or enhanced REM electroencephalographic patterns and normal sleep architecture (for review see ref.[30]).

Exogenous melatonin administration synchronized sleep to the day-night cycles in free-running totally blind subjects.[45,46] In some of these subjects, entrainment of other bodily rhythm was also achieved. However, in many of the patients the sleep-wake cycles were synchronized, while other rhythms were free-running. This apparent internal desynchronization suggests the melatonin directly predominantly acts at the sleep regulating centers rather than by synchronizing the biological clock to the 24 h cycle.

Similarly, exogenous melatonin administration advanced sleep in DSPS patients.[47] It is yet to be examined whether this effect is due to phase shifting of the biological clock or by direct interaction with sleep-promoting and inhibiting centers. In infants with Smith-Magenis syndrome, who have high melatonin production during daytime, and suffer from somnolence during the daytime and insomnia at night, administration of beta adrenergic receptors blockers during daytime, to suppress melatonin production, resulted in improved behavior and reduced somnolence, and concurrent administration of melatonin at night greatly improved their sleep.[48] In disabled brain damaged infants melatonin greatly improves night sleep.[49]

In blind and DSPS patients, as well in subjects given melatonin during daytime, the sleep promoting effects of melatonin are demonstrated at times when the endogenous production of the hormone is low or delayed. In insomnia patients, melatonin must be given at night, while the production of the endogenous melatonin is ongoing and might provide sufficient amounts of hormone internally. The efficacy of exogenous melatonin may therefore be best demonstrated in patients who produce insufficient amounts of melatonin.

5. Efficacy of Melatonin in Patients with Insomnia: A Question of Melatonin Deficiency?

With aging the timing of sleep is advanced whereas that of the melatonin peak is delayed.[50,51] There has been controversy in scientific and medical publications on whether melatonin production declines as part of normal aging or whether the decline is due to insomnia and other age related morbidities.[50,52,53] To investigate this controversial issue, we evaluated melatonin production at night, as measured by total nocturnal 6-SMT secretion, in a large study with a representative number of over 500 subjects including healthy elderly, elderly insomniacs and healthy young subjects (unpublished data). It is interesting to compare 6-SMT levels reported in the literature and results obtained by us (Figure 1). As can be seen in Figure 1, 6-SMT levels for healthy volunteers aged 20-35 years are remarkably consistent among published data.[53-55] Moreover, 6-SMT levels found for the insomnia patients aged 55 and older is also highly

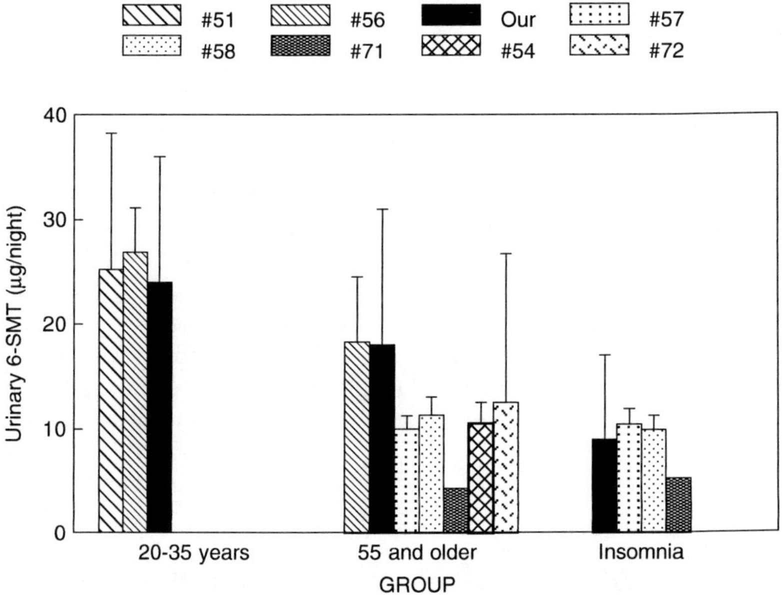

FIGURE 1. Nocturnal 6-SMT excretion by subjects without sleep complaints aged 20-35 years and 55 and older and patients with insomnia aged 55 and older. Results are from unpublished studies from our laboratory (black bars) and published studies (after ref.[51,54,56-58,71,72]).

consistent among different laboratories.[56-58] From all published data it can be concluded that the mean amount excreted by the insomnia patients aged 55 and older is approximately 40-60% of that in the healthy controls aged 20-35.

The inconsistencies in the literature are predominantly related to whether melatonin production is declining to the same degree in elderly subjects who do not have sleep problems. Despite differences in absolute 6-SMT values reported for healthy elderly controls, most studies, as well as a meta-analysis of literature data has concluded that 6-SMT decreases by approximately 20-37% between the age of 20-35 and 55 years and older (Figure 1). From the 6-SMT values reported for subjects aged 55 and older without insomnia in most published studies and found in our studies (unpublished) it may be possible to assume that production of 4 µg 6-SMT/night and over should be considered normal for this age group.

If the low production of melatonin is causally related to insomnia in the elderly, it should presumably be associated with a significantly higher likelihood of response to exogenous melatonin therapy. Early studies have shown Improvement of sleep in elderly insomnia patients by very high doses of melatonin.[59] The first attempts to study the efficacy of low doses of melatonin (acute administration; 1 ad 5 mg regular release at night) failed to show significant effects on the polysomnography-recorded sleep in insomnia patients.[60] Later on, the majority of studies showing efficacy of lower doses (0.3-5 mg per os.) of melatonin in insomnia were performed with elderly insomniacs or patients in whom melatonin production was suppressed. In most of these studies melatonin therapy resulted in shortening of sleep latency compared to placebo.[30,61,62]

Because of its fast clearance, regular melatonin formulations can produce physiological levels for only 2 hr, but in order to exert its effect, it needs to be present throughout the entire night. To circumvent this problem, a prolonged-release formulation of melatonin, (PR-melatonin) was developed (Circadin™, Neurim Pharmaceuticals Ltd, Israel), which provides a melatonin profile mimicking its normal physiological release (Figure 2). The effects of PR-melatonin on sleep architecture in elderly patients with insomnia were evaluated in a sleep laboratory study. The statistical analysis of the all night EEG spectral measures showed no significant modifications by treatment.

Attenuation of low frequency activity and enhancement of spindle frequency activity, characteristic of GABA-A receptor modulators did not occur with PR-melatonin. The clinical development of PR-melatonin was based on the observation that it improves sleep in elderly insomnia patients, as shown in exploratory placebo-controlled clinical studies.[63,64]

To date, three multi-centre Phase III clinical trials were conducted to evaluate the clinical efficacy of PR-melatonin on the improvement of sleep quality and showed significant treatment effects compared to placebo. In a pooled analysis of adverse event experience and a clinical laboratory summary carried out in all pivotal trials, no significant melatonin-induced adverse effects,

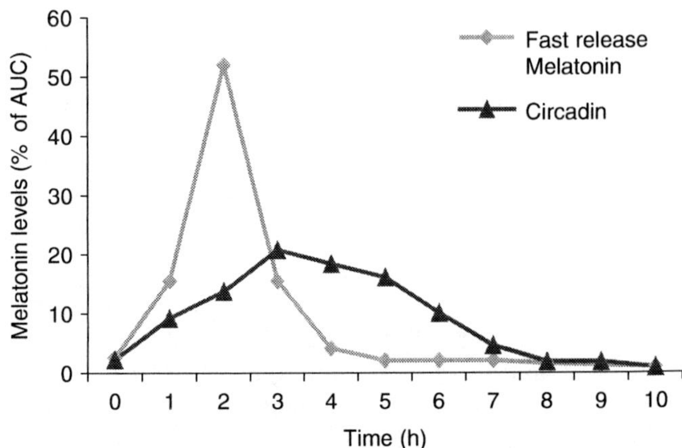

FIGURE 2. Blood melatonin levels after oral administration of PR-melatonin (Circadin™) compared to regular release melatonin formulations.

changes in laboratory parameters or vital signs were found neither in short-term (3 weeks) nor long-term studies (up to 18 months).

6. Pharmacodynamic Interactions between Melatonin and GABA-A Receptor Modulators

Of specific interest is the pharmacodynamic interaction between GABA-A receptor modulating drugs and melatonin. Benzodiazepines which are the most commonly used drugs for the treatment of insomnia in the elderly, have the potential to create tolerance and addiction and are therefore not recommended for chronic use. However, at least 10% of the population in France are thought to consume benzodiazepines regularly and similar figures have been described in other European countries and the US. At least 50% of benzodiazepines users have been reported to be willing to stop its use, but development of physical and/or psychological dependencies and rebound insomnia seem to interrupt most attempts at benzodiazepines discontinuation. Benzodiazepines paradoxically suppress the nocturnal rise in plasma melatonin and shift its day-night rhythmicity[65,66] thereby worsening the condition for the patient even more.

Previous reports have shown a potentiating effect of benzodiazepines with concomitant melatonin intake.[67,68] Recent reports show that melatonin similarly augments the hypnotic effects of non-benzodiazepines as well allowing the use of low doses of these hypnotic drugs to improve sleep quantity as well as quality.[69,70] This finding prompted the attempt to use PR-melatonin for benzodiazepines discontinuation in 34 volunteers suffering from insomnia who had been long-term benzodiazepines users.64 In a double blind placebo controlled trial, patients were randomised to PR-melatonin or placebo and were asked to taper off their benzodiazepines dose during 4 weeks and discontinue benzodiazepines completely on the 5th and 6th weeks while continuing administration of PR-melatonin or placebo. The results indicated that PR-melatonin effectively facilitated discontinuation of benzodiazepines while maintaining good sleep quality. A follow-up survey revealed that after 18 months 70% of the patients who discontinued benzodiazepines were still using PR-melatonin intermittently and none reverted to hypnotics.

7. Considerations on the Legal Status of Melatonin

Heavy media attention, including CNN, CBS and major newspaper coverage, was focused on melatonin in 1994 and 1995 over-claiming the effects of this compound even though these health claims have never been evaluated for efficacy and safety in large-scale clinical trails compliant with regulatory requirements. With passage of the "Dietary Supplement Health and Education Act" in 1994, melatonin became widely available in the United States under variable dosage forms of unknown purities. In Argentina and Poland, regular release melatonin products are registered without specific indications. In the UK melatonin was introduced shortly after the US market in 1993. However, at the end of 1995 the Medical Control Agencies ordered all the suppliers of melatonin to stop selling the hormone because it was believed that it requires a license obtained by regulatory approval. In the EEC, Japan Canada and Israel, melatonin has been classified as new chemical entity requiring a full application dossier with health authorities. PR-melatonin represents a new therapeutic principle in sleep medicine, in particular for elderly, improving quantity and quality of sleep. The clinical use of PR-melatonin awaits marketing approval by regulatory health agencies.

11. References

1. Maquet P. The role of sleep in learning and memory. Science 2001;294:1048-52.
2. Mednick SC, Nakayama K, Cantero JL, et al. The restorative effect of naps on perceptual deterioration. Nat Neurosci 2002;5:677-81.
3. Mednick SC NK, Stickgold R. Perceptual Learning after a Nap: The Mini-Me of Sleep. Sleep 2003;26(Suppl.):1082U.
4. Achermann P, Borbely AA. Mathematical models of sleep regulation. Front Biosci 2003;8:S683-93.
5. Jones BE. Arousal systems. Front Biosci 2003;8:S438-51.
6. Yamanaka A, Muraki Y, Tsujino N, Goto K, Sakurai T. Regulation of orexin neurons by the monoaminergic and cholinergic systems. Biochem Biophys Res Commun 2003;303:120-9.
7. Gallopin T, Fort P, Eggermann E, et al. Identification of sleep-promoting neurons in vitro. Nature 2000;404:992-5.
8. Porkka-Heiskanen T, Alanko L, Kalinchuk A, Stenberg D. Adenosine and sleep. Sleep Med Rev 2002;6:321-32.

9. Kalinchuk AV, Urrila AS, Alanko L, et al. Local energy depletion in the basal forebrain increases sleep. Eur J Neurosci 2003;17:863-9.

10. Thakkar MM, Winston S, McCarley RW. A1 receptor and adenosinergic homeostatic regulation of sleep-wakefulness: effects of antisense to the A1 receptor in the cholinergic basal forebrain. J Neurosci 2003;23:4278-87.

11. Arrigoni E CN, Saper CB, McCarley RW. The Effects of Adenosine on the Membrane Properties of Basal Forebrain Cholinergic Neurons. Sleep 2003;26 (Suppl.):108A.

12. Young MW. Life's 24-hour clock: molecular control of circadian rhythms in animal cells. Trends Biochem Sci 2000;25:601-6.

13. Deurveilher S, Burns J, Semba K. Indirect projections from the suprachiasmatic nucleus to the ventrolateral preoptic nucleus: a dual tract-tracing study in rat. Eur J Neurosci 2002;16:1195-213.

14. Chou TC, Bjorkum AA, Gaus SE, Lu J, Scammell TE, Saper CB. Afferents to the ventrolateral preoptic nucleus. J Neurosci 2002;22:977-90.

15. Hajak G. Epidemiology of severe insomnia and its consequences in Germany. Eur Arch Psychiatry Clin Neurosci 2001;251:49-56.

16. Roth T, Hajak G, Ustun TB. Consensus for the pharmacological management of insomnia in the new millennium. Int J Clin Pract 2001;55:42-52.

17. Roth T, Ancoli-Israel S. Daytime consequences and correlates of insomnia in the United States: results of the 1991 National Sleep Foundation Survey. II. Sleep 1999;22 (Suppl. 2):S354-8.

18. Ancoli-Israel S, Roth T. Characteristics of insomnia in the United States: results of the 1991 National Sleep Foundation Survey. I. Sleep 1999;22 (Suppl 2):S347-53.

19. Leger D, Levy E, Paillard M. The direct costs of insomnia in France. Sleep 1999;22 (Suppl 2):S394-401.

20. Hohagen F, Kappler C, Schramm E, et al. Prevalence of insomnia in elderly general practice attenders and the current treatment modalities. Acta Psychiatr Scand 1994;90:102-8.

21. Hyyppa MT, Kronholm E, Alanen E. Quality of sleep during economic recession in Finland: a longitudinal cohort study. Soc Sci Med 1997;45:731-8.

22. Everitt DE, Avorn J, Baker MW. Clinical decision-making in the evaluation and treatment of insomnia. Am J Med 1990;89:357-62.

23. Sateia MJ, Doghramji K, Hauri PJ, Morin CM. Evaluation of chronic insomnia. An American Academy of Sleep Medicine review. Sleep 2000;23:243-308.

24. Zeitlhofer J, Schmeiser-Rieder A, Tribl G, et al. Sleep and quality of life in the Austrian population. Acta Neurol Scand 2000;102:249-57.

25. Bonnet MH, Arand DL. Sleepiness as measured by modified multiple sleep latency testing varies as a function of preceding activity. Sleep 1998;21(5):477-83.

26. Hauri PJ. Insomnia. Clin Chest Med 1998;19:157-68.

27. Dorsey CM, Bootzin RR. Subjective and psychophysiologic insomnia: an examination of sleep tendency and personality. Biol Psychiatry 1997;41:209-16.

28. Sugerman JL, Stern JA, Walsh JK. Daytime alertness in subjective and objective insomnia: some preliminary findings. Biol Psychiatry 1985;20:741-50.

29. Ohayon MM, Zulley J. Correlates of global sleep dissatisfaction in the German population. Sleep 2001;24:780-7.

30. Zisapel N. The use of melatonin for the treatment of insomnia. Biol Signals Recept 1999;8:84-9.

31. Dollins AB, Zhdanova IV, Wurtman RJ, Lynch HJ, Deng MH. Effect of inducing nocturnal serum melatonin concentrations in daytime on sleep, mood, body temperature, and performance. Proc Natl Acad Sci U S A 1994;91:1824-8.

32. Huber R, Deboer T, Schwierin B, Tobler I. Effect of melatonin on sleep and brain temperature in the Djungarian hamster and the rat. Physiol Behav 1998;65:77-82.

33. Mailliet F, Galloux P, Poisson D. Comparative effects of melatonin, zolpidem and diazepam on sleep, body temperature, blood pressure and heart rate measured by radiotelemetry in Wistar rats. Psychopharmacology (Berl) 2001;156:417-26.

34. Jin X, von Gall C, Pieschl RL, et al. Targeted disruption of the mouse Mel(1b) melatonin receptor. Mol Cell Biol 2003;23:1054-60.

35. Liu C, Weaver DR, Jin X, et al. Molecular dissection of two distinct actions of melatonin on the suprachiasmatic circadian clock. Neuron 1997;19:91-102.

36. Laudon M, Nir I, Zisapel N. Melatonin receptors in discrete brain areas of the male rat. Impact of aging on density and on circadian rhythmicity. Neuroendocrinology 1988;48:577-83.

37. Zisapel N, Egozi Y, Laudon M. Inhibition of dopamine release by melatonin: regional distribution in the rat brain. Brain Res 1982;246:161-3.

38. Cajochen C, Krauchi K, Wirz-Justice A. Role of melatonin in the regulation of human circadian rhythms and sleep. J Neuroendocrinol 2003;15:432-7.

39. Akerstedt T, Gillberg M, Wetterberg L. The circadian covariation of fatigue and urinary melatonin. Biol Psychiatry 1982;17:547-54.

40. Lockley SW, Skene DJ, Tabandeh H, Bird AC, Defrance R, Arendt J. Relationship between napping and melatonin in the blind. J Biol Rhythms 1997;12:16-25.

41. Shochat T, Luboshitzky R, Lavie P. Nocturnal melatonin onset is phase locked to the primary sleep gate. Am J Physiol 1997;273:R364-70.

42. Lockley SW, Skene DJ, Butler LJ, Arendt J. Sleep and activity rhythms are related to circadian phase in the blind. Sleep 1999;22:616-23.

43. Shibui K, Uchiyama M, Okawa M. Melatonin rhythms in delayed sleep phase syndrome. J Biol Rhythms 1999;14:72-6.

44. Hohjoh H, Takasu M, Shishikura K, Takahashi Y, Honda Y, Tokunaga K. Significant association of the arylalkylamine N-acetyltransferase (AA-NAT) gene with delayed sleep phase syndrome. Neurogenetics 2003;4:151-3.

45. Sack RL, Lewy AJ. Circadian rhythm sleep disorders: lessons from the blind. Sleep Med Rev 2001;5:189-206.

46. Skene DJ. Optimization of light and melatonin to phase-shift human circadian rhythms. J Neuroendocrinol 2003;15:438-41.

47. Dahlitz M, Alvarez B, Vignau J, English J, Arendt J, Parkes JD. Delayed sleep phase syndrome response to melatonin. Lancet 1991;337:1121-4.

48. De Leersnyder H, Bresson JL, de Blois MC, et al. Beta 1-adrenergic antagonists and melatonin reset the clock and restore sleep in a circadian disorder, Smith-Magenis syndrome. J Med Genet 2003;40:74-8.

49. Jan MM. Melatonin for the treatment of handicapped children with severe sleep disorders. Pediatr Neurol 2000;23:229-32.

50. Haimov I, Laudon M, Zisapel N, et al. Sleep disorders and melatonin rhythms in elderly people. Bmj 1994;309:167.

51. Duffy JF, Zeitzer JM, Rimmer DW, Klerman EB, Dijk DJ, Czeisler CA. Peak of circadian melatonin rhythm occurs later within the sleep of older subjects. Am J Physiol Endocrinol Metab 2002;282:E297-303.

52. Zeitzer JM, Daniels JE, Duffy JF, et al. Do plasma melatonin concentrations decline with age? Am J Med 1999;107:432-6.
53. Kennaway DJ, Lushington K, Dawson D, Lack L, van den Heuvel C, Rogers N. Urinary 6-sulfatoxymelatonin excretion and aging: new results and a critical review of the literature. J Pineal Res 1999;27:210-20.
54. Luboshitzky R, Shen-Orr Z, Tzischichinsky O, Maldonado M, Herer P, Lavie P. Actigraphic sleep-wake patterns and urinary 6-sulfatoxymelatonin excretion in patients with Alzheimer's disease. Chronobiol Int 2001;18:513-24.
55. Cugini P, Touitou Y, Bogdan A, et al. Is melatonin circadian rhythm a physiological feature associated with healthy longevity? A study of long-living subjects and their progeny. Chronobiol Int 2001;18:99-107.
56. Baskett JJ, Wood PC, Broad JB, Duncan JR, English J, Arendt J. Melatonin in older people with age-related sleep maintenance problems: a comparison with age matched normal sleepers. Sleep 2001;24:418-24.
57. Lushington K, Lack L, Kennaway DJ, Rogers N, van den Heuvel C, Dawson D. 6-Sulfatoxymelatonin excretion and self-reported sleep in good sleeping controls and 55-80-year-old insomniacs. J Sleep Res 1998;7:75-83.
58. Lushington K, Dawson D, Kennaway DJ, Lack L. The relationship between 6-sulphatoxymelatonin rhythm phase and age in self-reported good sleeping controls and sleep maintenance insomniacs aged 55-80 years. Psychopharmacology (Berl) 1999;147:111-2.
59. MacFarlane JG, Cleghorn JM, Brown GM, Streiner DL. The effects of exogenous melatonin on the total sleep time and daytime alertness of chronic insomniacs: a preliminary study. Biol Psychiatry 1991;30:371-6.
60. James SP, Sack DA, Rosenthal NE, Mendelson WB. Melatonin administration in insomnia. Neuropsychopharmacology 1990;3:19-23.
61. Hughes RJ, Sack RL, Lewy AJ. The role of melatonin and circadian phase in age-related sleep-maintenance insomnia: assessment in a clinical trial of melatonin replacement. Sleep 1998;21:52-68.
62. Zisapel N. Circadian rhythm sleep disorders: pathophysiology and potential approaches to management. CNS Drugs 2001;15:311-28.
63. Haimov I, Lavie P, Laudon M, Herer P, Vigder C, Zisapel N. Melatonin replacement therapy of elderly insomniacs. Sleep 1995;18:598-603.
64. Garfinkel D, Laudon M, Nof D, Zisapel N. Improvement of sleep quality in elderly people by controlled-release melatonin. Lancet 1995;346:541-4.
65. McIntyre IM, Burrows GD, Norman TR. Suppression of plasma melatonin by a single dose of the benzodiazepine alprazolam in humans. Biol Psychiatry 1988;24(1):108-12.
66. Monteleone P, Forziati D, Orazzo C, Maj M. Preliminary observations on the suppression of nocturnal plasma melatonin levels by short-term administration of diazepam in humans. J Pineal Res 1989;6:253-8.
67. Ferini-Strambi L, Zucconi M, Biella G, et al. Effect of melatonin on sleep microstructure: preliminary results in healthy subjects. Sleep 1993;16:744-7.
68. Garfinkel D, Laudon, M. Zisapel, N. Improvement of sleep quality by controlled-release melatonin in benzodiazepine treated elderly insomniacs. Arch. Gerontol. Geriat 1997;24:223-231.
69. Reichardt RM WN, Kautz M, Belenky G, Balkin T. Daytime Melatonin-Zolpidem Cocktail: I. Effects on Daytime Sleep. Sleep 2003;26(Suppl.):0500I.

70. Wesensten NJ RR, Kautz M, Belenky G, Balkin T. Daytime Melatonin-Zolpidem Cocktail: II. Effects on Memory and Vigilance. Sleep 2003;26(Suppl.):0503I.
71. Sakotnik A, Liebmann PM, Stoschitzky K, et al. Decreased melatonin synthesis in patients with coronary artery disease. Eur Heart J 1999;20:1314-7.
72. Girotti L, Lago M, Ianovsky O, et al. Low urinary 6-sulphatoxymelatonin levels in patients with coronary artery disease. J Pineal Res 2000;29:138-42.

Sleep and Circadian Rhythms in the Blind

JONATHAN S. EMENS AND ALFRED J. LEWY

1. Introduction

In the majority of totally blind individuals, the biological clock is no longer synchronized, or entrained, by the light/dark cycle. Despite exposure to regular social cues, meal times and sleep/wake schedules, the circadian phase (timing) of biological events in these individuals continues to drift to a progressively later (or, rarely, earlier) hour. As a result, these individuals suffer from periodic bouts of nighttime insomnia and daytime hypersomnolence, as the circadian sleep propensity rhythm moves in and out of synchrony with the 24-hour day. Recently, oral melatonin has been shown to be highly effective in resetting the biological clock in these individuals.[1] Although several treatment parameters continue to be determined, melatonin is likely to prove a safe and effective treatment for this type of sleep disorder.

2. The Circadian Pacemaker and its Regulation by Light

The circadian pacemaker (or biological clock) regulates the timing of a host of physiologic parameters across the biological day.[2-4] In addition to sleep propensity and alertness, circadian variation can be seen in clinical phenomena such as myocardial infarction[5] and fasting glucose levels.[6] The anatomical locus for the biological clock is the hypothalamic suprachiasmatic nuclei (SCN),[7] and it is the molecular "clock-work" within individual SCN neurons that form the basis for pacemaker rhythmicity.[8,9]

The period (tau) of the pacemaker is not precisely 24-hours,[10,11] and therefore the pacemaker needs to be synchronized, or entrained, to the external 24-hour day. Light through the eyes has been shown to be the primary synchronizer.[12-17] Photic information conveyed from the retina via the retinohypothalamic tract to the SCN provides the daily phase shifts necessary to maintain entrainment.[18] Timing,[19] intensity,[20,21] and wavelength[22,23] have all been shown to be important factors in the resetting effects of light. The resetting effects of any time cue, such as light, can be described using a phase response curve (PRC), in which the timing of the stimulus is plotted against the resulting shift in circadian phase: shifts in the rhythm to an earlier hour are termed phase advances, whereas shifts to a later hour are phase delays.

Of all the parameters with endogenous circadian variation, the rhythm of melatonin secretion has proved to be the most reliable marker of circadian phase.[24] Melatonin is secreted by the pineal gland, and levels are high during the night and low during the day.[25] This pattern of secretion is the same in both nocturnal (night active) and diurnal (day active) species, and so melatonin secretion can be thought of as a marker for the biological night. Melatonin levels are suppressed by light[26] but are relatively unperturbed by other types of stimuli. Our laboratory uses the nightly onset of melatonin secretion gathered under dim light conditions (the dim light melatonin onset, or DLMO) as our marker of circadian phase.[27] There are many operational definitions of the DLMO. Recently, we have used a threshold of 2 pg/ml.[27] Under normal entrained conditions, the 2 pg/ml DLMO occurs about 3 hours before bedtime (that is, 13 hours after waketime). In totally blind individuals, melatonin does not need to be sampled under dim light conditions and so the acronym is simplified to MO.

3. Regulation of Sleep by the Circadian Pacemaker

The timing of the sleep/wake cycle in particular has been shown to be regulated by the circadian pacemaker.[28-33] The circadian drive for sleep and wakefulness is yoked to other endogenous rhythms, such as core body temperature and plasma melatonin, and experiments over the past decade have highlighted how the interaction between circadian sleep propensity and homeostatic sleep drive underlie normal patterns of human sleep and wakefulness.[34] The sleep dependent process is perhaps the more intuitive of the two: during the waking day, the homeostatic drive for sleep increases while during sleep it is dissipated. In contrast, the peak in the circadian rhythm of sleep propensity occurs around the time of the body temperature minimum (several hours before waketime under entrained conditions), while the circadian drive for wakefulness typically crests at the time of peak body temperature (just before the onset of sleep). The somewhat "paradoxical timing" of circadian sleep/wake propensity becomes clear when it is noted that the interaction between the two processes allow for consolidated episodes of sleep and wakefulness: in the latter half of the waking period, the increasing homeostatic drive for sleep is balanced by the rise in circadian drive for wakefulness.[35] The opposite is true during episodes of sleep: in the later half of the sleep period when the homeostatic drive for sleep has been dissipated the circadian drive for sleep is greatest.[35]

4. Circadian Rhythms in the Blind

In the majority of completely blind individuals (those without any light perception), the lack of photic input to the pacemaker results in circadian rhythms that are no longer synchronized to the 24-hour day and are said to be free-running.[11,36-41] Despite exposure to regular sleep/wake, rest/activity and light/dark schedules, as well as regular social cues, the rhythms of most such blind free runners (BFRs) continue to drift to a progressively later time, in accordance with the greater than 24 h tau of the pacemaker. Initially, there were only sporadic case reports of circadian abnormalities in blind individuals. Miles and colleagues were among the first to demonstrate that, despite rigorous adherence to a 24-hour schedule and the use of hypnotic and stimulant medications, the circadian rhythms of one blind individual continued to drift later each day.[41] Another blind individual was documented to have a free-running rhythm of plasma cortisol when she was studied over a period of 50 days.[40] Lewy and colleagues demonstrated abnormal timing of melatonin secretion in six of ten blind subjects and studied them longitudinally: based on these data they proposed that the circadian rhythms of the blind could be classified as either normally entrained, entrained at an abnormal phase or free-running.[39] This classification scheme has been borne out by continued research. Sack and colleagues found that roughly half (11/20) of a cohort of totally blind individuals had free-running rhythms of plasma melatonin and cortisol.[11] This may be an underestimate of the incidence of free-running among the totally blind, since free-running individuals with a tau very close to 24 hours might appear to be entrained. Nevertheless Sack and colleagues found that the timing of melatonin and cortisol rhythms drifted approximately half an hour later every day: the average tau of the circadian pacemaker ($\pm$ SD) among these BFRs was 24.55 $\pm$ 0.31 h. Subsequent studies have confirmed that among blind individuals lacking light perception, the majority have free-running circadian rhythms.[37] In keeping with the classification system proposed by Lewy et al, a minority of totally blind individuals maintain entrainment, although some do so at an abnormal phase; in the two studies noted above, approximately half of the totally blind individuals who maintained entrainment did so at an abnormal phase.[11,37]

It remains unclear what zeitgebers (time cues) mediate entrainment among those blind persons who are without light perception and yet are able to maintain synchronization to the 24-hour day. Light may have a role: the photoreceptive system that mediates the circadian resetting effects of light is likely distinct from the rods and cones that mediate vision.[42-50] Furthermore, light has been shown capable of constricting pupils,[11] suppressing plasma melatonin levels[51] and resetting circadian phase[52] among some individuals lacking subjective light perception. Non-photic cues, including activity or sleep itself, may also play a role.[39,53] It appears that orally administered melatonin is able to override these weaker zeitgebers in resetting circadian phase (see section 6 below).[54]

5. Sleep Disorders in the Blind

Given the prominent role of the circadian pacemaker in the regulation of sleep and wakefulness, it is not surprising that BFRs have subjective and objective decrements in sleep quality and daytime alertness. Several studies have documented higher rates of both daytime somnolence, as well as nighttime sleep complaints, among the blind.[55-59] A survey of 1,073 blind individuals demonstrated a significantly increased prevalence of both symptoms of insomnia, as well as daytime sleepiness, compared to controls.[57] Associated with these complaints was a higher use of hypnotics as well as napping during the day. A study of blind children by the same group documented similar symptoms of sleep disruption and daytime somnolence.[58]

A survey of 388 blind individuals using the Pittsburgh Sleep Quality Index (PSQI) found that nearly half (48.7%) had at least a mild degree of sleep disturbance (PSQI score > 5).[56] Those blind individuals without any light perception had an even higher prevalence of sleep disturbance (65.5%).

Nakagawa and colleagues underscored the importance of the circadian system in sleep disturbances found in the blind when they demonstrated that sleep propensity free-ran with the endogenous rhythms of core body temperature, cortisol and melatonin in a BFR: during four consecutive weeks, circadian phase was assessed in a BFR by measuring the endogenous rhythms of core body temperature, cortisol and plasma melatonin.[30] The subject was then given the opportunity to nap for up to 7 minutes out of every 20 for the next 24-hours, while sleep was measured with standard polysomnography.

Using this ultra-short sleep-wake schedule, it was shown that the sleep propensity rhythm moved 0.59 hours later each day in synchrony with the melatonin tau of 24.65 h.

A later study of 59 blind subjects, including 17 BFRs, found that periodic bouts of daytime somnolence and nighttime insomnia were predicted by circadian phase.[60] Such subjective decrements in sleep quality have been confirmed by polysomnography.

Sack and colleagues demonstrated that when BFRs attempted to sleep at a time that was 12 hours out of phase with the 24-hour day sleep efficiency was significantly lower and time-awake-after-sleep onset was higher, compared to sleep when their endogenous rhythms were in phase.[1] When 26 totally blind individuals with demonstrated free-running rhythms of salivary and urinary 6-sulfatoxymelatonin were compared to normal controls with polysomnography, the BFRs were shown to have lower total sleep times, sleep efficiencies and total REM sleep, as well as increased sleep latencies.[61]

6. Treatment of Free-Running Circadian Rhythms with Melatonin

6.1. Demonstration of Phase Shifts and Entrainment

Experiments in rodents were the first to demonstrate that the mammalian circadian pacemaker could be entrained by exogenous melatonin.[62] Soon afterwards, attempts were made to entrain BFRs with oral melatonin administration.

Early experiments in our laboratory involved administration of 5 mg melatonin at bedtime for three weeks.[63] Although this regime induced phase advances in the MO, entrainment did not occur.

Longer durations of administration were not possible at that time due to FDA limitations; however, entrainment was demonstrated in a BFR who had been taking about 7.5 mg of melatonin at bedtime on his own for a year.[64]

Our laboratory[1] and the University of Surrey research group[65] subsequently went on to conclusively demonstrate entrainment of BFRs with exogenous melatonin. We administered 10 mg melatonin one hour before preferred bedtime to seven BFRs for three to nine weeks: six of the seven entrained, while the seventh subject (who had the longest tau) had a decrease in his tau from 24.9 to 24.3 h.[1] In three of the subjects entrainment was maintained even as the dose was decreased to 0.5 mg.

In those subjects who entrained, significant improvements in sleep parameters, such as sleep efficiency and time-awake-after-sleep onset, resulted.[1] The Surrey group found that 5 mg melatonin administered at 21:00 for 35-71 days was able to entrain three of seven BFRs and shorten tau in a fourth.[65]

6.2. Timing of Melatonin Administration

The melatonin PRC is important for understanding entrainment of BFRs with melatonin: it describes how the timing of melatonin administration, relative to the biological clock, determines the direction of the resulting phase shifts.[66] Although the melatonin PRC was constructed using data from normally entrained sighted subjects,[67-70] it is applicable to the blind. Our melatonin PRC utilized 0.5 mg melatonin administered for four days.[70] We found that melatonin administered in the afternoon and evening caused phase advances, while melatonin administered late in the night and in the morning caused phase delays. Using the endogenous melatonin onset as a marker of circadian phase, we can delineate the boundaries of the phase-advance and phase-delay zones of the PRC. By convention, waketime is used as a reference point and is defined as circadian time (CT) 0. Because the 2 pg/ml MO occurs 13 hours after waketime, it is defined as CT 13. Using the MO as our reference point, the phase-advance zone extends from CT 6 to CT 18 and the phase-delay zone extends from CT 18 to CT 6.[70] The advance zone therefore extends from seven hours before the MO until five hours after the MO (Figure 1).

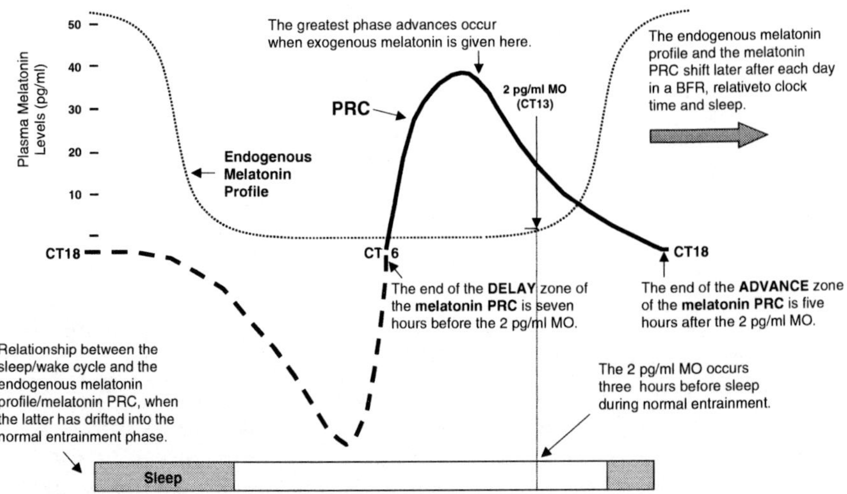

FIGURE 1. The relationship between the endogenous melatonin profile, the melatonin PRC, and the sleep/wake cycle. Adapted with permission from: Lewy AJ, Bauer VK, Hasler BP, Kendall AR, Pires LN, Sack RL. Brain Res. 2001;918:96-100. © 2001 Elsevier Inc.

The melatonin PRC explains entrainment to melatonin in BFRs. In most BFRs, who are administered melatonin at a set clock hour, circadian phase drifts to a progressively later time each day until the exogenous melatonin falls upon the advance zone. Once melatonin causes a phase advance equal in magnitude to the daily phase delay, the individual entrains.

The timing of melatonin administration must also take into account the normal phase relationship between the endogenous melatonin rhythm and the sleep/wake cycle (recall that the endogenous MO normally occurs several hours *before* bedtime under entrained conditions, Figure 1). We have found that entrainment with melatonin may result in an endogenous MO that occurs several hours *after* administration time, so that bedtime dosing results in entrainment at a delayed phase.[66,71] Therefore, the time of melatonin administration must be well before the desired bedtime to promote entrainment at a normal phase.

The other variable known to influence the interval between melatonin administration time and the MO (the phase angle of entrainment or PAE) is tau.

We were the first to demonstrate that the PAE to melatonin in BFRs is positively correlated with tau.[71,72] Wright and colleagues similarly demonstrated that the PAE to light in the sighted is also correlated with tau.[73]

6.3. Dose of Melatonin Administration

Following the initial successes entraining BFRs, we went on to treat them using the lower 0.5 mg dose de novo.[74] Exogenous melatonin was again administered 1-2 hours before bedtime in three BFRs with taus of 24.2 to

24.4 h. All three subjects were entrained. The Surrey group has also found that lower doses of melatonin are capable of entraining BFRs: four of ten subjects entrained to a nightly dose of 0.5 mg.[75]

Although low-dose melatonin proved effective, we first speculated that the subject from our initial study who had failed to entrain would require a longer duration of treatment and/or higher doses of melatonin. He had the longest tau of the group and therefore might require a proportionately stronger resetting signal [because he tended to delay almost an hour each day (due to his long tau) he should require a daily phase advance greater than individuals who only drift 0.2 to 0.6 hours later each day]. He was first treated with 9-10 mg of melatonin for 83 days and his tau was again shortened to 24.36 h without entrainment.

When the dose was increased to 20 mg and treatment was continued for 60 days, he again failed to entrain and his tau was shortened to only 24.58 h. However, when the dose was decreased to 0.5 mg entrainment quickly ensued.[76]

To explain this result we have hypothesized that lower doses of melatonin may result in greater phase shifts than higher doses of melatonin. High doses could result in elevated levels of melatonin that stimulate both the advance and delay regions of the melatonin PRC, resulting in smaller net phase shifts.[66,76] These "spillover" effects may not be as relevant in subjects with shorter taus. It is even possible that the administration of low-dose melatonin could have significant spillover effects in a person with a long melatonin half-life.

Low-dose melatonin has the advantage of not inducing significant sleepiness and therefore can be administered at any time of day, depending on the direction and magnitude of the phase shifts that are required. In other types of circadian sleep disorders (e.g., advanced sleep phase syndrome, delayed sleep phase syndrome, jet lag, and shift work sleep disorder), this means that melatonin can be administered in the morning to achieve phase delays or in the afternoon to achieve maximal phase advances without causing sleepiness, which is more of a problem with higher doses. In BFRs, as discussed above, this means that melatonin can be administered well before bedtime, so that the endogenous MO will occur at a more normal clock hour relative to sleep.[66]

6.4. Timing of Treatment Initiation

It has been hypothesized that the initial timing of melatonin administration may also impact treatment success or failure in BFRs.[65] In the study by the University of Surrey group, it was found that about half of seven BFRs entrained to a daily dose of 5 mg melatonin when it was initiated on the advance zone while the remaining subjects, in whom melatonin was initiated on the delay zone, failed to entrain.[65] They speculated that initiation of treatment on the delay zone might lead to either changes in the melatonin PRC or melatonin receptor sensitivity that could decrease the probability of future entrainment. These researchers noted that this explanation was inconsistent

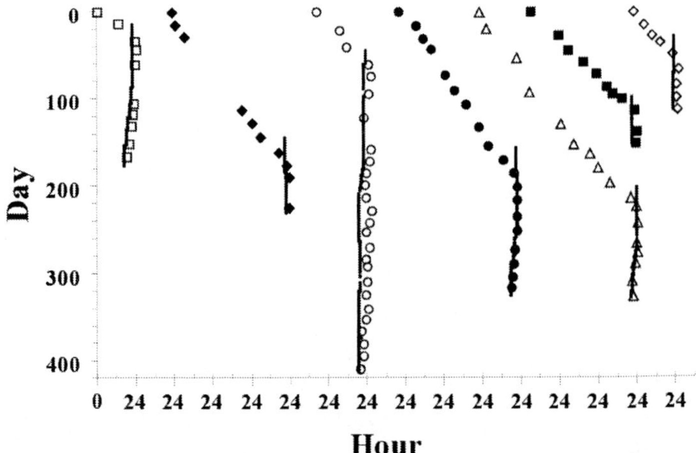

FIGURE 2. Entrainment of seven BFRs using melatonin [0.5 mg; in one case (◇) 0.05 mg] administered approximately one hour before bedtime. Symbols represent an assessment of circadian phase as determined by the time that endogenous plasma or saliva melatonin concentrations continuously rose above the 2-pg/ml or 0.7-pg/ml threshold respectively. Vertical lines represent the timing and duration of exogenous melatonin treatment. Melatonin was initially administered on the delay zone of the melatonin PRC. Used with permission from: Lewy AJ, Emens JS, Bernert RA, Lefler BJ. J. Biol. Rhythms. 2004;19(1):68-75. © 2004 Sage Publications.

with experiments in animals which demonstrated that entrainment eventually occurs once the time of administration falls on the advance zone.[62, 65] The clinical implication of this hypothesis is quite significant: circadian phase would have to be accurately assessed, most likely on more than one occasion, before BFRs could be successfully treated with precisely timed melatonin. We recently tested this hypothesis. Seven subjects were administered 0.5 mg (in one case, 0.05 mg) melatonin approximately one hour before bedtime (Figure 2). Treatment was always initiated when this coincided with the delay zone of the melatonin PRC. In all seven cases, the subjects eventually entrained as circadian phase free-ran and the time of melatonin administration came to fall upon the advance zone of the melatonin PRC.[66] We think that circadian time of treatment initiation using low-dose melatonin is not a factor in treatment success.

7. Conclusions

The majority of totally blind individuals have circadian rhythms that are no longer synchronized to the 24-hour day. The resulting symptoms of periodic daytime somnolence and nighttime insomnia rate only second to loss of vision in their impact on quality of life. Oral administration of low-dose melatonin can entrain almost all BFRs. We are currently trying to determine the lowest possible dose capable of entraining a BFR.

8. Acknowledgements. Supported by grants from the Public Health Service (R01 MH56874, R01 MH55703, R01 AG21826 and R01 HD42125 to Dr. Lewy; and 5 MO1 RR000334, to the General Clinical Research Center of OHSU) and the National Alliance for Research on Schizophrenia and Depression (2000 NARSAD Distinguished Investigator Award to Dr. Lewy). Dr. Emens is supported by a grant from the Public Health Service (K23RR017636-01). We are indebted to the nursing staff of the General Clinical Research Center and to Nancy Stahl, Rick Boney, Bryan Lefler, Angie Koenig and Krista Yuhas.

9. *References*

1. Sack RL, Brandes RW, Kendall AR, et al. Entrainment of free-running circadian rhythms by melatonin in blind people. The New England Journal of Medicine. 2000;343:1070-1077.
2. Waterhouse JM, DeCoursey PJ. Human Circadian Organization. In: Dunlap JC, Loros JJ, DeCoursey PJ, eds. Chronobiology: Biological Timekeeping. Sunderland: Sinauer Associates; 2004:291-323.
3. Van Cauter E, Copinschi G, Turek FW. Endocrine and Other Biologic Rhythms. In: DeGroot LJ, Jameson JL, Burger HG, et al., eds. Endocrinology. Philadelphia: W.B. Saunders Company; 2001:235-256.
4. Moore-Ede MC, Sulzman FM, Fuller CA. Circadian Timing of Physiological Systems. The Clocks That Time Us. Cambridge, MA: Harvard University Press; 1982:201-294.
5. Muller JE. Circadian Variation in Cardiovascular Events. American Journal of Hypertension. 1999;12:35S-42S.
6. Troisi RJ, Cowie CC, Harris MI. Diurnal variation in fasting plasma glucose: implications for diagnosis of diabetes in patients examined in the afternoon. Jama. 2000;284(24):3157-3159.
7. Card JP, Moore RY. The Organization of Visual Circuits Influencing the Circadian Activity of the Suprachiasmatic Nucleus. In: Klein DC, Moore RY, Reppert SM, eds. Suprachiasmatic Nucleus. New York, NY: Oxford University Press; 1991:51-76.
8. Dunlap JC. Molecular bases for circadian clocks. Cell. 1999;96:271-290.
9. Reppert SM, Weaver DR. Coordination of circadian timing in mammals. Nature. 2002;418:935-941.
10. Czeisler C, Duffy J, Shanahan T, et al. Stability, precision and near-24-hour period of the human circadian pacemaker. Science. 1999;284:2177-2181.
11. Sack RL, Lewy AJ, Blood ML, et al. Circadian rhythm abnormalities in totally blind people: incidence and clinical significance. J. Clin. Endocr. Metab. 1992;75(1):127-134.
12. Honma K, Honma S. A human phase response curve for bright light pulses. Jap. J. Psychiatry. 1988;42:167-168.

13. Minors DS, Waterhouse JM, Wirz-Justice A. A human phase-response curve to light. Neurosci. Lett. 1991;133:36-40.
14. Wever RA. Light effects on human circadian rhythms. A review of recent Andechs experiments. J. Biol. Rhythms. 1989;4(2):161-186.
15. Czeisler CA, Kronauer RE, Allan JS, et al. Bright light induction of strong (Type O) resetting of the human circadian pacemaker. Science. 1989;244:1328-1333.
16. Lewy AJ, Sack RL, Miller S, et al. Antidepressant and circadian phase-shifting effects of light. Science. 1987;235:352-354.
17. Czeisler CA. The effect of light on the human circadian pacemaker. In: Chadwick DJ, Ackrill K, eds. Circadian Clocks and Their Adjustment. Chichester: John Wiley & Sons Ltd; 1995:254-302.
18. Klein DC, Moore RY, Reppert SM. Suprachiasmatic Nucleus. New York: Oxford; 1991.
19. Khalsa SB, Jewett ME, Cajochen C, et al. A phase response curve to single bright light pulses in human subjects. J Physiol. Jun 15 2003;549(Pt 3):945-952.
20. Zeitzer JM, Dijk DJ, Kronauer RE, et al. Sensitivity of the human circadian pacemaker to nocturnal light: melatonin phase resetting and suppression. J Physiol. 2000;526(3):695-702.
21. Boivin DB, Duffey JF, Kronauer RE, et al. Dose-response relationships for resetting of human circadian clock by light. Nature. 1996;379:540-541.
22. Brainard GC, Hanifin JP, Greeson JM, et al. Action spectrum for melatonin regulation in humans: evidence for a novel circadian photoreceptor. J. Neurosci. Aug 15 2001;21(16):6405-6412.
23. Lockley SW, Brainard GC, Czeiler CA. High sensitivity of the human circadian melatonin rhythm to resetting by short wavelenght light. The Journal of Clinical Endocrinology and Matabolism. 2003;88(9):4502-4505.
24. Lewy AJ, Sack RL. The dim light melatonin onset (DLMO) as a marker for circadian phase position. Chronobiol. Int. 1989;6(1):93-102.
25. Lewy AJ. The Pineal Gland. In: Goldman L, Bennett JC, eds. Cecil Textbook of Medicine. 21 ed. Philadelphia: W.B. Saunders Company; 2000:1207-1208.
26. Lewy AJ, Wehr TA, Goodwin FK, et al. Light suppresses melatonin secretion in humans. Science. 1980;210:1267-1269.
27. Lewy AJ, Cutler NL, Sack RL. The endogenous melatonin profile as a marker for circadian phase position. J. Biol. Rhythms. 1999;14(3):227-236.
28. Czeisler CA, Weitzman ED, Moore-Ede MC, et al. Human sleep: its duration and organization depend on its circadian phase. Science. 1980;210:1264-1269.
29. Zulley J, Wever R, Aschoff J. The dependence of onset and duration of sleep on the circadian rhythm of rectal temperature. Eur. J. Physiol. 1981;391:314-318.
30. Nakagawa H, Sack RL, Lewy AJ. Sleep propensity free-runs with the temperature, cortisol and melatonin rhythms in a totally blind person. Sleep. 1992;15:330-336.
31. Dijk D, Czeisler C. Paradoxical timing of the circadian rhythm of sleep propensity serves to consolidate sleep and wakefulness in humans. Neurosci. Lett. 1994 1994;166:63-68.
32. Dijk DJ, Czeisler CA. Contribution of the circadian pacemaker and the sleep homeostat to sleep propensity, sleep structure, electroencephalographic slow waves, and sleep spindle activity in humans. J. Neurosci. 1995;15(5):3526-3538.
33. Dijk D, Duffy JF, Riel E, et al. Ageing and the circadian and homeostatic regulation of human sleep during forced desynchrony of rest, melatonin and temperature rhythms. Journal of Physiology. 1999;516(2):611-627.

34. Dijk D, Lockley SW. Invited Review: Integration of human sleep-wake regulation and circadian rhythmicity. J Appl Physiol. 2002 2002;92:852-862.
35. Dijk DJ, Czeisler CA. Paradoxical timing of the circadian rhythm of sleep propensity serves to consolidate sleep and wakefulness in humans. Neurosci. Lett. 1994 1994;166(1):63-68.
36. Klerman EB, Rimmer DW, Dijk D, et al. Nonphotic entrainment of the human circadian pacemaker. American Journal of Physiology-Regulatory Integrative and Comparative Physiology. 1998;43(4):R991-996.
37. Lockley SW, Skene DJ, Arendt J, et al. Relationship between melatonin rhythms and visual loss in the blind. J. Clin. Endocr. Metab. 1997;82(11):3763-3770.
38. Klein T, Martens H, Dijk DJ, et al. Circadian sleep regulation in the absence of light perception: chronic non-24-hour circadian rhythm sleep disorder in a blind man with a regular 24-hour sleep-wake schedule. Sleep. 1993;16(4):333-343.
39. Lewy AJ, Newsome DA. Different types of melatonin circadian secretory rhythms in some blind subjects. J Clin Endocrinol Metab. Jun 1983;56(6):1103-1107.
40. Orth DN, Besser GM, King PH, et al. Free-running circadian plasma cortisol rhythm in a blind human subject. Clin. Endocrinol. (Oxf.). Jun 1979;10(6):603-617.
41. Miles LE, Raynal DM, Wilson MA. Blind man living in normal society has circadian rhythms of 24.9 hours. Science. Oct 28 1977;198(4315):421-423.
42. Freedman MS, Lucas RJ, Soni B, et al. Regulation of mammalian circadian behavior by non-rod, non-cone, ocular photoreceptors. Science. 1999;284:502-504.
43. Provencio I, Rodriguez IR, Jiang G, et al. A novel human opsin in the inner retina. J of Neuroscience. 15 Jan, 2000 2000;20(2):600-605.
44. Provencio I, Rollag MD, Castrucci AM. Photoreceptive net in the mammalian retina. Nature. 2002;415:493.
45. Berson DM, Dunn FA, Takao M. Phototransduction by retinal ganglion cells that set circadian clock. Science. 2002;295:1070-1073.
46. Hattar S, Liao HW, Takao M, et al. Melanopsin-containing retinal ganglion cells: architecture, projections, and intrinsic photosensitivity. Science. 2002;295:1065-1070.
47. Ruby NF, Brennan TJ, Xie X, et al. Role of melanopsin in circadian responses to light. Science. 2002;298:2211-2213.
48. Panda S, Sato TK, Castrucci AM, et al. Melanopsin (*Opn4*) requirement for normal light-induced circadian phase-shifting. Science. 13 December 2002;298:2213-2216.
49. Hannibal J, Hindersson P, Knudson SM, et al. The photopigment melanopsin is exclusively present in pituitary adenylate cyclase-activating polypeptide-containing retinal ganglion cells of the retinohypothalamic tract. The Journal of Neuroscience. 2002;22:RC191.
50. Lucas RJ, Hattar S, Takao M, et al. Diminished pupillary light reflex at high irradiances in melanopsin-knockout mice. Science. 10 Jan, 2003 2003;299:245-247.
51. Czeisler CA, Shanahan TL, Klerman EB, et al. Suppression of melatonin secretion in some blind patients by exposure to bright light. N. Eng. J. Med. 1995; 332:6-11.
52. Klerman EB, Shanahan TL, Brotman DJ, et al. Photic resetting of the human circadian pacemaker in the absence of conscious vision. J. Biol. Rhythms. Dec 2002;17(6):548-555.

53. Klerman EB. Non-photic effects on the circadian system: results from experiments in blind and sighted individuals. In: Honma K, Honma S, eds. Zeitgebers, Entrainment and Masking of the Circadian System. Sapporo: Hokkaido University Press; 2001:155-169.

54. Lewy AJ, Emens JS, Sack RL, et al. Zeitgeber heirarchy in humans: resetting the circadian phase positions of blind people using melatonin. Chronobiol. Int. 2003;20(5):837-852.

55. Miles LEM, Wilson MA. High incidence of cyclic sleep/wake disorders in the blind. Sleep Res. 1977;6:192.

56. Tabandeh H, Lockley SW, Robert Buttery F, et al. Disturbance of sleep in blindness. Am. J. Ophthalmology. 1998;126:707-712.

57. Leger D, Guilleminault C, Defrance R, et al. Prevalence of sleep/wake disorders in persons with blindness. Clinical Science. 1999;97:193-199.

58. Leger D, Prevot E, Philip P, et al. Sleep disorders in children with blindness. Ann. Neurol. 1999;46:648-651.

59. Leger D, Guilleminault C, Defrance R, et al. Blindness and sleep patterns. The Lancet. 1996;348:830-831.

60. Lockley SW, Skene DJ, Butler LJ, et al. Sleep and the activity rhythms are related to circadian phase in the blind. Sleep. 1999;22(5):616-623.

61. Leger D, Guilleminault C, Santos C, et al. Sleep/wake cycles in the dark: sleep recorded by polysomnography in 26 totally blind subjects compared to controls. Clinical Neurophysiology. 2002;113:1607-1614.

62. Redman J, Armstrong S, Ng KT. Free-running activity rhythms in the rat: entrainment by melatonin. Science. 1983;219:1089-1091.

63. Sack RL, Lewy AJ, Blood ML, et al. Melatonin administration to blind people: phase advances and entrainment. J. Biol. Rhythms. 1991;6(3):249-261.

64. Sack RL, Stevenson J, Lewy AJ. Entrainment of a previously free-running blind human with melatonin administration. Sleep Res. 1990;19:404.

65. Lockley SW, Skene DJ, James K, et al. Melatonin administration can entrain the free-running circadian system of blind subjects. J. Endocrinol. 2000;164:R1-R6.

66. Lewy AJ, Emens JS, Bernert RA, et al. Eventual entrainment of the human circadian pacemaker by melatonin is independent of the circadian phase of treatment initiation: clinical implications. J. Biol. Rhythms. Feb 2004;19(1):68-75.

67. Middleton B, Arendt J, Stone BM. Complex effects of melatonin on human circadian rhythms in constant dim light. J. Biol. Rhythms. 1997;12(5):467-477.

68. Zaidan R, Geoffriau M, Brun J, et al. Melatonin is able to influence its secretion in humans: description of a phase-response curve. Neuroendocrinol. 1994;60:105-112.

69. Lewy AJ, Ahmed S, Jackson JML, et al. Melatonin shifts circadian rhythms according to a phase-response curve. Chronobiol. Int. 1992;9(5):380-392.

70. Lewy AJ, Bauer VK, Ahmed S, et al. The human phase response curve (PRC) to melatonin is about 12 hours out of phase with the PRC to light. Chronobiol. Int. 1998;15(1):71-83.

71. Lewy AJ, Hasler BP, Emens JS, et al. Pretreatment circadian period in free-running blind people may predict the phase angle of entrainment to melatonin. Neurosci. Lett. 2001;313:158-160.

72. Sack RL, Brandes RL, Lewy AJ. Totally blind people with free-running circadian rhythms can be normally entrained with melatonin. Sleep Research Online. 1999;2 (Supplement 1):624.

73. Wright KP, Jr., Hughes RJ, Kronauer RE, et al. Intrinsic near-24-h pacemaker period determines limits of circadian entrainment to a weak synchronizer in humans. Proc Natl Acad Sci U S A. Nov 20 2001;98(24):14027-14032.
74. Lewy AJ, Bauer VK, Hasler BP, et al. Capturing the circadian rhythms of free-running blind people with 0.5 mg melatonin. Brain Res. 2001;918:96-100.
75. Hack LM, Lockley SW, Arendt J, et al. The effects of low-dose 0.5-mg melatonin on the free-running circadian rhythms of blind subjects. J. Biol. Rhythms. Oct 2003;18(5):420-429.
76. Lewy AJ, Emens JS, Sack RL, et al. Low, but not high, doses of melatonin entrained a free-running blind person with a long circadian period. Chronobiol. Int. 2002;19(3):649-658.

Disturbances of Hormonal Circadian Rhythms in Shift Workers

DIANE B. BOIVIN

1. Disturbed Sleep and Circadian Rhythms in Shift Workers

In normal individuals living on a day-oriented schedule, it is hypothesized that a harmonious relationship between homeostatic and circadian processes serves to promote uninterrupted bouts of 8 hours of sleep and 16 hours of wakefulness per day.[1,2]

The structure of sleep is dependent on a complex interaction of circadian phase and sleep/wake dependent processes. When sleep is displaced, as is the case with transmeridian travelers and shift workers,[3,4] the normal phase relationship between the sleep/wake cycle and the endogenous circadian oscillator is perturbed, a situation which can lead to substantial deterioration in sleep quality.[5] Together, these observations indicate that the temporal alignment between the sleep/wake cycle and the endogenous circadian system is a determining factor in the quality of the subsequent sleep and waking episodes.

Sleep recording during ultrashort sleep/wake paradigms,[6,7] free-running conditions,[8,9] or forced desynchrony experiments[1] are consistent with these observations and revealed that the peak propensity for REM sleep is observed near the nadir of the core body temperature cycle, thus at the end of a normal night. In comparison, the occurrence of SWS ("slow wave sleep") is much more influenced by the duration of prior waking than by circadian phase[1] (Figure 1).

Some cross-influence still exists since REM sleep can also be influenced by sleep deprivation[10] just as slow wave activity in non-REM sleep shows a small but significant circadian modulation.[1]

Day sleep in shift workers is often measured as 1-3 hours shorter than night sleep on days off, or on day and evening work schedules.[11-16] Following night shifts, most workers report no difficulties falling asleep and latencies to sleep onset are short.[15]

The short sleep latencies are accompanied by short latencies to SWS, which may represent the response to a physiological sleep debt. Shift workers, who typically show great variation in their degrees of circadian entrainment to particular sleep/wake cycles,[17-24] also demonstrate a discrepancy in the amount and distribution of REM sleep[11,14,25,26] Increases in sleep stage transitions further substantiate the fragility of diurnal sleep[14,26] (Figure 2).

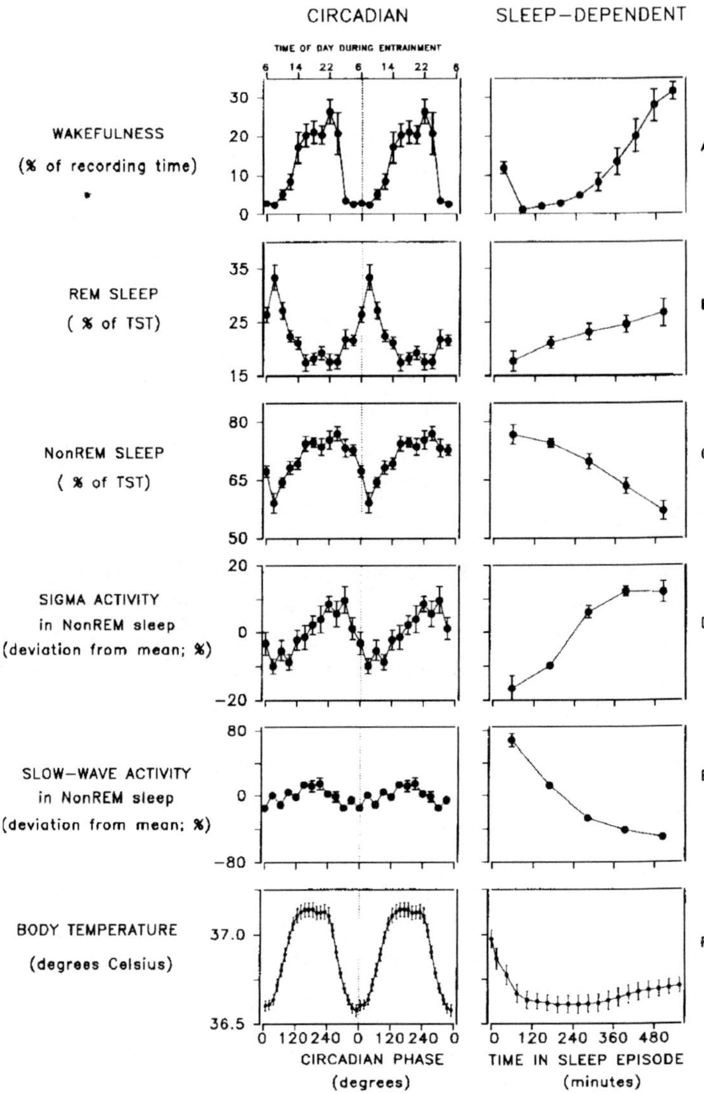

FIGURE 1. As part of a 'forced desynchrony' protocol, eight men lived under a 28-hour rest-activity cycle for 30-36 days. Under these conditions, the imposed day length is much longer than the length of the internal day and beyond the range of entrainment of the circadian pacemaker. Sleep episodes are therefore recorded at all circadian phases, as separated by periods of scheduled awakening of equal length. The sleep-dependent and circadian components of sleep could therefore be distinguished. In this illustration, the core body temperature is used as a physiological marker for circadian phase (described along the x-axis as degrees). The fitted minimum of temperature is attributed a circadian phase of 0 degrees, where a full cycle of the oscillation is 360 degrees. The modulation of the concentration of REM in sleep episodes, for example, is strongly influenced by circadian phase, while that of slow-wave activity is strongly influenced by the prior sleep duration. Reproduced from[1] with permission. © 1995 The Society for Neuroscience.

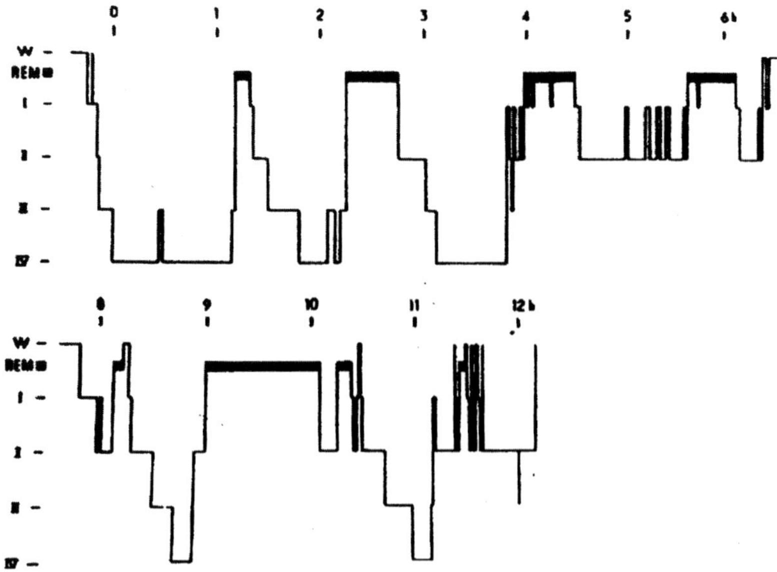

FIGURE 2. Hypnogram of nocturnal and diurnal sleep in shiftworkers. Example of two "hyponograms": above a reference night. Below a day sleep of the same subject. The latency of the first REM sleep period decreased and its duration increased. Reproduced from[16] with permission. © 1995 The Society for Neuroscience.

2. Two Process-Regulation of Hormones

The duration of prior sleep and waking influences rhythms is sensitive to homeostatic processes whereas the endogenous circadian system affects rhythms sensitive to circadian phase. To understand the disturbances of hormonal rhythms associated with atypical schedules, one must factor in the time of falling asleep, the time of prior waking, the degree of sleep disruption, and the phase relationship between the endogenous circadian system and the sleep-wake cycle.

Some hormones are more sensitive to sleep-wake dependent processes whereas others are more sensitive to circadian processes. Overall, it is appropriate to consider that both processes affect the diurnal variation of most hormones, although to various relative degrees.

Hormones such as melatonin and cortisol have a strong circadian component. They are considered as reliable circadian markers since their timing of secretion provides a useful evaluation of the state of the endogenous circadian system. When the timing of sleep is abruptly shifted, the endogenous circadian pacemaker remains closer to its initial position, at least for several days. Thus the levels of melatonin and cortisol at given clock times remain unchanged and closer to their initial positions for several days.

Their diurnal rhythm may however be distorted by light exposure and stressful situations. Other hormones such as GH (growth hormone) and PRL (prolactin) are more affected by sleep-wake dependent processes. Their secretion shifts rapidly when sleep is displaced and remains closely linked to the appearance of sleep, regardless of time-of-day. Hormones such as TSH are markedly affected by both processes. Some components rapidly shift with the new sleep schedule whereas others are influenced by the endogenous circadian system and continue to occur at their original time-of-day (Figure 3).[27] A good knowledge of the variation of specific hormones with the sleep-wake cycle and with circadian phase is important to understand how their secretion is disrupted in conditions of shifted sleep-wake schedules.

3. Circadian Variation of Hormones

Cortisol levels reach their minimal values early in the night and their maximum values around the regular time of awakening.[28-31] More specifically cortisol levels peak at the end of darkness, thus the start of the light period. Interestingly, cortisol levels rise by 50% in the morning when light levels change from dim to bright intensities, an effect that has been attributed to reduced melatonin levels.[32] The diurnal rhythm of cortisol secretion persists even in sleep deprivation experiments, namely during constant routine procedures, and depends on a robust and significant circadian component.[33-35] This rhythm persists when the sleep-wake cycle is desynchronized from the endogenous circadian system, as may occur in ultra-rapid sleep-wake paradigms.[28] Cortisol is secreted by pulses and its circadian variation is associated with a change in pulses amplitude rather than pulses frequency.[31,36] A stable phase relationship was noticed between cortisol and melatonin secretion. Namely cortisol levels tend to increase when melatonin levels go down such that both hormones are negatively correlated.[32,37] Melatonin also exhibits a robust and significant circadian variation.[38-41] Its levels peak at night during the middle of the sleep episode and are undetectable during the day.[40,42-44]

The endogenous circadian component of cortisol and melatonin rhythms is further demonstrated by their sensitivity to light-induced phase shifts. Figure 4 summarizes the results of a phase-resetting experiment, using light of indoor intensities.[35] Eight healthy young men, aged 18-30 years, underwent an initial assessment of their endogenous circadian system by a 30-40 hour constant routine. The constant routine is a procedure specially designed to unmask the endogenous circadian variation of physiological parameters.[45,46] Subjects were then exposed to a 3-cycle 5-hour light stimulus of an intensity of ~180 lux. The stimulus was centered 1.5 hour after the endogenous circadian temperature minimum, in order to produce a phase advance shift.

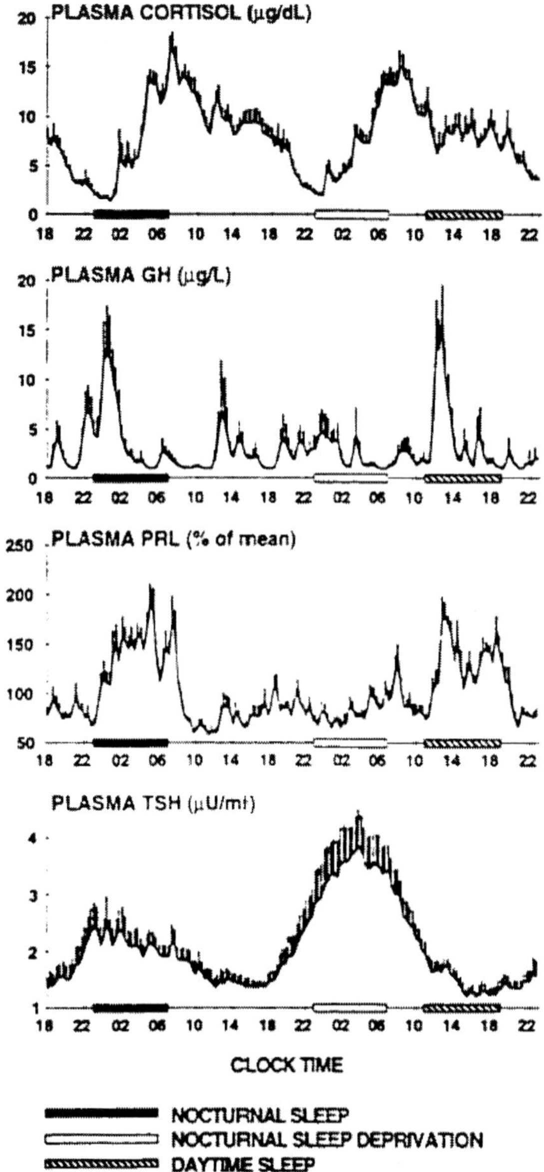

FIGURE 3. Mean 24-hr profiles of plasma cortisol. GH, PRL, and TSH in a group of eight normal young men (ages 20-27 years) studied during a 53-hr period including 8 hr of nocturnal sleep, 28 hr of sleep deprivation, and 8 hr of daytime sleep. The vertical bars at each time point represent the SEM. The black bars represent the sleep periods. The open bars represent the period of nocturnal sleep deprivation. The dashed bars represent the period of daytime sleep. Data were sampled at 20-min intervals. Reproduced from[27] with permission

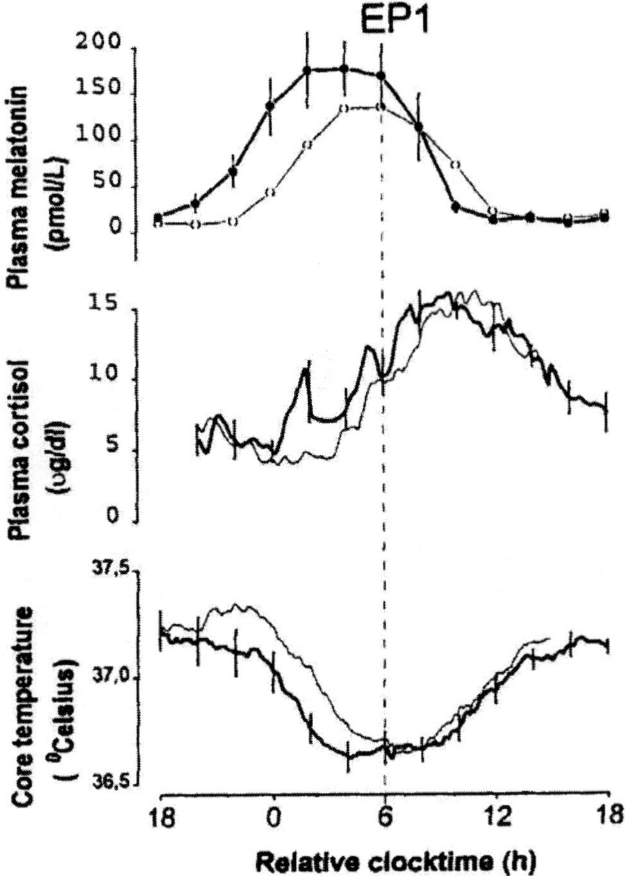

FIGURE 4. Resetting of melatonin and cortisol rhythms in humans by ordinary room light. Endogenous circadian curves of plasma melatonin, plasma cortisol, and core body temperature observed during the final constant routine in two groups of subjects. Treatment subjects (bold lines; ± s.e.m.) were exposed to ~180 lux during their 3 cycles of 5-h light stimulus; control subjects (thin lines, mean) were exposed to ~0.03 lux during their stimulus. Results are illustrated as a function of time-of-day; clock time is expressed in hours with respect to the initial endogenous circadian temperature nadir (vertical line) which was assigned a reference value of 06:00. Reproduced from[35] with permission

Circadian phase was reassessed by a final 40-50 hour constant routine. Blood was withdrawn every 20 and 60 min during constant routines in order to draw the endogenous circadian rhythms of plasma cortisol and plasma melatonin, respectively. A mean phase advance shift of +1.61 hour and +1.26 hour was observed in the circadian rhythms of plasma cortisol and melatonin, respectively. These results contrasted with the −0.21 hour and −1.51

hour phase delay of these rhythms in a group of aged-matched subjects who underwent the same experimental manipulations of their sleep-wake cycle but remained in darkness during the stimulus sessions.

In addition to its resetting effect, light exposure inhibits melatonin secretion at night in a dose-dependent way.[47-53] The evaluation of circadian phase based on ambulatory melatonin measurements must thus carefully control for light and darkness exposure.

GH demonstrates a small but significant variation with circadian phase with a small pulse at the beginning of the night.[54-58] This pulse can still be observed in sleep-deprivation studies and is presumed to be of circadian origin.[59] The increase starts in the late afternoon and continues after sleep onset.[60,61] Under normally entrained conditions, the primary pulse is observed prior to bedtime in men. In women this variation is less clear since pulses are observed throughout the day.[62,63] The predominant surge of GH secretion early in the night does not persist in ultra-rapid sleep-wake paradigms indicating that its circadian control is weak.[28]

PRL also shows a small but significant circadian variation that is maintained during sleep deprivation.[64] Peak levels are observed at night, 1-2 hours after peak melatonin concentrations whereas low levels occur during the day.[64,65] A secondary peak of PRL secretion, half the height of the main nocturnal peak, can be observed in the afternoon.[66]

TSH levels start rising late in the evening prior to the regular bedtime and reach peak levels at night.[67,69] TSH levels progressively decline during sleep and reach minimal values at the end of the night. A small rise of TSH can be observed near the regular time of awakening.[70] Under sleep deprivation conditions, TSH exhibits a robust circadian rhythm and its levels continue to rise throughout the night to attain very high values.[39,57] It was suggested that, under normally entrained condition, sleep would blunt the nocturnal rise of TSH since it exerts a strong inhibitory effect on TSH secretion.[59] The circadian rhythm of TSH secretion is presumed to be the result of TRH stimulation.[71]

A significant diurnal variation of glucose and insulin secretion was observed in experiments of sleep deprivation, with peak levels around the habitual bedtime.[72,73] Namely, glucose and insulin levels rise by 16% and 49%, respectively at the time-of-day corresponding to the habitual bedtime.[73] An interaction of cortisol and food metabolism was observed.[74] Namely, the diurnal rhythms of glucose and insulin are negatively correlated with that of cortisol.[73] Pharmacological increase of cortisol levels results in an inhibition of insulin secretion. A marked increase of cortisol levels leads to an apparent delayed resistance to insulin that can last up to 16 hours.[75] Cortisol administration in the evening, combined with pharmacological suppression of endogenous levels, results in a rapid drop of insulin followed by a delayed increase of glucose and insulin. The increase in glucose levels is more prolonged and delayed than that observed with morning cortisol administration.[76] The meal-induced increase in glucose and insulin is also greater in the evening compared to morning ingestion. This response has been correlated with the reduction of endogenous cortisol levels at night.[77,78]

A small but significant circadian rhythm of leptin secretion, a hormone produced by adipose tissues, was reported in experiments of sleep deprivation, with peak levels at night.[79,80] Leptin is secreted in pulses and its levels have been shown to be inversely correlated to those of ACTH and cortisol in humans.[81] This relationship might be the result of the inhibitory effects of glucocorticoids on leptin secretion. It was suggested that the increase of leptin levels at night could play a role in suppressing appetite.[82] The amplitude of this rhythm is correlated with the insulin response to food,[83] the body mass index,[84] and negatively correlated with the body fat mass.[84]

4. Sleep-Wake Dependent Variation of Hormones

GH, leptin, and PRL are primarily regulated by homeostatic processes whereas TSH secretion is affected significantly both by circadian and sleep-wake-dependent processes. The major peak of GH secretion is observed around sleep onset in young men,[54,56-58,85-87] even when the sleep episode is delayed.[86,88] This nocturnal increase of GH secretion appears related to a concomitant increase in pulsatile GHRH secretion.[89] The nocturnal peak of GH and GHRH secretion has been positively correlated with SWS and delta wave activity,[57,87,90-93] although controversial results have been presented.[60,94,95] GHRH-induced GH release is greater during SWS than during REM sleep.[96] This response is inhibited during awakening from sleep,[97] although no change in GH secretion were reported with sleep deprivation during the second half of the night.[98] Interestingly, gamma-hydroxybutyrate, known to be a potent promoter of SWS, was found to also increase the amplitude of the major peak of GH, without affecting melatonin or TSH secretion.[99]

Sleep exerts a strong stimulant effect on PRL secretion[64,100,101] regardless of the time-of-day.[103] Peak PRL levels are observed during the later part of the sleep episode such that their distribution is quite different from that of GH.[66,100,103] Nighttime PRL levels decrease with sleep deprivation during the second half of the night.[98] PRL levels increase during recovery night following sleep deprivation.[104]

TSH secretion reaches a plateau in the middle of the nighttime sleep episode and is inhibited by sleep.[30,68,105] TSH has been negatively correlated with SWS and delta wave as well as positively correlated with alpha wave, an index of awakening.[105,106] This could explain the reduced TSH pulse during recovery sleep following sleep deprivation, which is characterized by rebound SWS.[98,107] Sleep deprivation is associated with a significant increase in TSH as well as T3 and T4 concentrations.[30,69,104,107-109]

Leptin secretion increases during sleep and depends more on homeostatic than circadian processes.[79] Slow pulses of leptin are noticed throughout its diurnal rhythm and are negatively correlated with glucose levels.[79] Insulin and glucose also increase during sleep regardless of the time-of-day.[72,73] An association between the increase of glucose during sleep and GH secretion has been proposed.[73] Namely insulin and glucose levels were reported higher at the beginning of NREM sleep and continued to cycle at about every 100 minutes in association with REM/NREM cycles.[110] Plasma renin is another hormone which levels mirror the REM/NREM cycles and cycles at about 100 minutes.[111,112] Renin increases and decreases concomitantly with slow wave activity.[113]

As previously described, the diurnal variation of cortisol and melatonin secretion is mainly dependent on the endogenous circadian system. Nevertheless, a small inhibition of cortisol secretion at the start of the sleep episode and a small increase at the start of the waking episode are observable.[56,114] The inhibitory effect of sleep on cortisol secretion is observed at a variety of clock times when subjects are studied under an ultra-rapid sleep-wake paradigm.[28] A negative association between cortisol levels and SWS was reported.[36] Reduced ACTH and cortisol response to CRH were observed during sleep and especially during SWS, whereas increased stimulation were noticed during sleep deprivation.[104,115,116]

Some studies have shown increased cortisol secretion with sleep deprivation[31,104,114] whereas others have not.[117,118] Partial or total sleep deprivation was shown to delay the early night reduction in cortisol secretion during the recovery sleep episode.[119] In a prior study, awakening during the second REM sleep episode was associated with a rapid rise of cortisol levels followed by a temporary inhibition such that mean cortisol levels remained unchanged.[120] Melatonin does not appear to be much affected by sleep[59,104] Some studies reported increase melatonin levels with sleep deprivation,[117,121] although postural changes rather than sleep deprivation could be involved.[59,122]

5. Hormonal Patterns in Shifted Sleep Schedules

Working on irregular schedules results in acute and persistent misalignment between the endogenous circadian system and the shifted sleep-wake cycle. It also involves extended waking episodes as well as acute and chronic sleep disruption. These are the presumed mechanisms by which hormonal rhythms are affected.

The resulting rhythms can be interpreted as been the result of the relative influence of circadian and sleep-wake dependent processes. For these reasons, hormones such as GH and PRL are more influenced by sleep and waking disturbances whereas hormones such as cortisol and melatonin are more sensitive to circadian misalignment associated with atypical work schedules.

A complex interaction between these two processes is also to be expected since hormones are influenced by both processes simultaneously and because the quality of sleep and waking themselves is significantly influenced by circadian phase.

Spontaneous adjustment of circadian rhythms of core body temperature, melatonin, cortisol or TSH to shifts in the rest-activity cycle are reported in a minority of shift workers.[17,19-21,23,123] Appropriately timed nocturnal bright light has proven an effective intervention to promote the realignment of endogenous rhythms of core body temperature, melatonin, and cortisol to a nocturnal reorientation of the work schedule in laboratory simulations[124-128] and in field studies[129-131] (Figure 5 and 6).

Laboratory and field studies of atypical sleep schedules confirmed that the secretory pulses of GH are rapidly shifted with a rapid shift of the sleep episode as occurs in night shift workers.[132] The main GH pulse seems to be more dependent on the occurrence of SWS than of sleep onset.[29] This primary GH pulse is observed during the first half of daytime sleep in shift workers.[54]

However, the temporal distribution of GH secretion is a bit blunted compared to that of night sleepers, which points to an interaction between circadian and homeostatic processes. This interpretation is supported by the observation of reduced REM sleep latency in research volunteers sleeping during the day[54] as well as lower GH pulses in free-running subjects.[55] Normal daytime sleep structure was reported in regular middle-aged night workers in association with a nearly complete GH adaptation. In these workers, GH pulses were more randomly distributed.[54]

The amount of GH secreted per 24 hours was comparable between night workers, and control subjects sleeping either at night or during the day. GH was significantly higher during the first half of sleep episodes in controls on a day schedule and in night workers, but not in controls on a night schedule (Figure 7).[54] This indicates possible adaptation of GH secretion in permanent night workers, as compared to the acute effect of an abrupt sleep displacement observed in control day sleepers.

A field study of jet travel across 7 time zones was conducted in healthy young volunteers who flew westward from Brussels-Chicago, adapted to Chicago for 30 days, then returned to Brussels.[133] The westward travel to Chicago resulted in a marked increase in GH secretion, associated with larger secretory pulses.

GH secretion during sleep took up to 11 days to adjust to the new time zone. The main pulse of GH secretion occurred later in the first sleep episode following the return to Brussels. A GH pulse was also observed at sleep onset.

Changes of GH might be influenced by the disturbed REM/NREM cycles and nocturnal awakenings. Circadian misalignment and extended waking episodes can markedly influence the temporal distribution of REM sleep and SWS, respectively.

PRL rises at sleep onset and peaks during sleep episodes regardless of time-of-day (64). PRL secretion was found to be similar during daytime sleep episodes in shift workers but greater during their waking episodes.[19] PRL secretion was measured in healthy young volunteers participating in a field

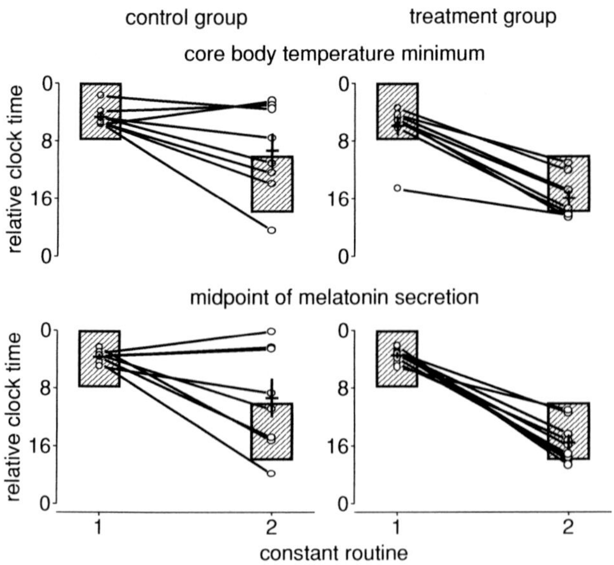

FIGURE 5. Adaptation of the temperature and melatonin rhythm in night workers. Fifteen nurses working permanent night schedules (≥ 8 shifts/15 days) were recruited from area hospitals. Following a vacation period of ≥ 10 days on a regular daytime schedule, workers were admitted to the laboratory for the assessment of circadian phase via a 36-hour constant routine. They returned to work ~12 night shifts on their regular schedules under one of two conditions. Treatment group workers (n = 10, mean age ± SD 41.7± 8.8 years) received an intervention including 6 hours of intermittent bright light exposure in the workplace (~2,000 lux) and shielding from bright morning outdoor light with tinted goggles (15% visual light transmission). Control group workers (n = 9, mean age ± SD 42.0 ± 7.2 years) were observed in their habitual work environments. On work days, participants maintained regular sleep/wake schedules including a single 8-hour sleep/darkness episode beginning 2 hours after the end of the night shift. A second 36-hour constant routine was performed following the series of night shifts. The figure illustrates the progression of circadian phase from a day to a night-oriented schedule. The time of circadian phase, determined via the fitted minimum of the core body temperature cycle (upper panels) and the midpoint of salivary melatonin concentration (lower panels) are plotted for each worker and are expressed relative to in-bed times (hatched rectangles). To facilitate visualization, workers were assigned a relative bedtime of 00:00 for night time sleep during vacation period and 10:00 for daytime sleep following night shifts. Sleep episodes are represented from bedtime (top) to waketime (bottom).Mean circadian phase, plotted for the groups, is expressed as a horizontal bar (± s.e.m.). Reproduced from[131] with permission.

Brussels-Chicago-Brussels jet travel.[134] Baseline PRL secretion was bimodal with peaks during midsleep and late afternoon.

Mean PRL secretion and relative secretion during sleep and waking episodes were not affected by jet lag. However, a small rise in PRL occurred at its original sleep time. The main PRL pulse during sleep was shifted earlier

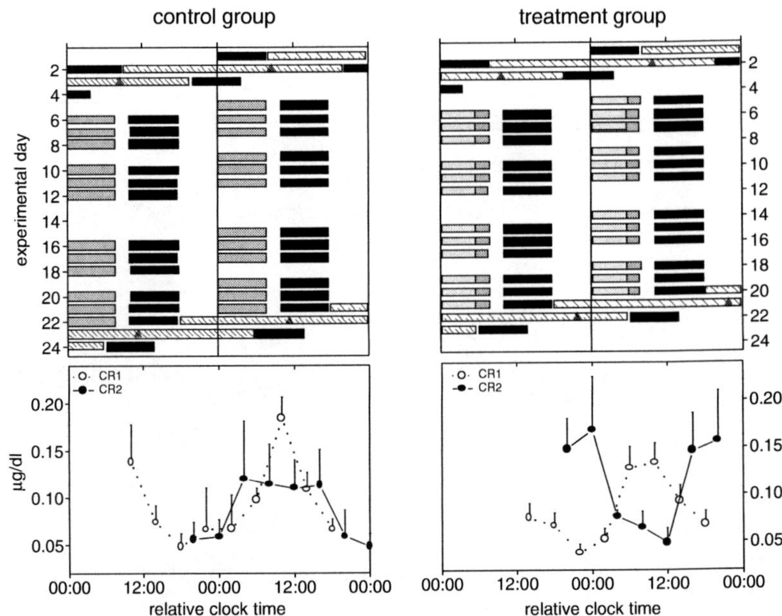

FIGURE 6. Cortisol rhythm in night workers. Results from a subgroup of 6 nurses from the treatment group and 5 nurses from the control group, as described in the legend of Figure 4. Upper panels show the protocol for the two experimental groups of shift workers. Successive days are shown side-by-side and along the vertical axis. Sleep periods are indicated as black bars. Hatched bars indicate constant routine procedures. Dark grey bars indicate periods of night shift work performed in habitual lighting. In the treatment group, intermittent exposure to bright light during the first 6 hours of night shifts is shown as light grey bars. In lower panels, mean salivary cortisol concentration per 4 hours is shown (± s.e.m.) in both groups of workers for initial and final constant routines. This experiment revealed that treatment group workers displayed significant shifts in the time of peak cortisol expression and realignment of the rhythm with the night-oriented schedule. Smaller phase shifts suggesting an incomplete adaptation to the shift work schedule were observed in the control group. Reproduced from[146] with permission.

in Chicago compared to its baseline position in Brussels. It also returned slowly to its original timing following the return flight to Brussels. The PRL decline during waking was slow to return to its usual pattern. These changes are not totally explained by sleep disturbances and suggest an interaction between circadian and sleep-wake dependent processes.

Prior studies demonstrated increased cortisol and blunted ACTH response to CRH during night shifts compared to day shifts.[135] Compared to day shifts, basal cortisol and ACTH levels were significantly lower and higher during night shifts, respectively. It was hypothesized that reduced cortisol

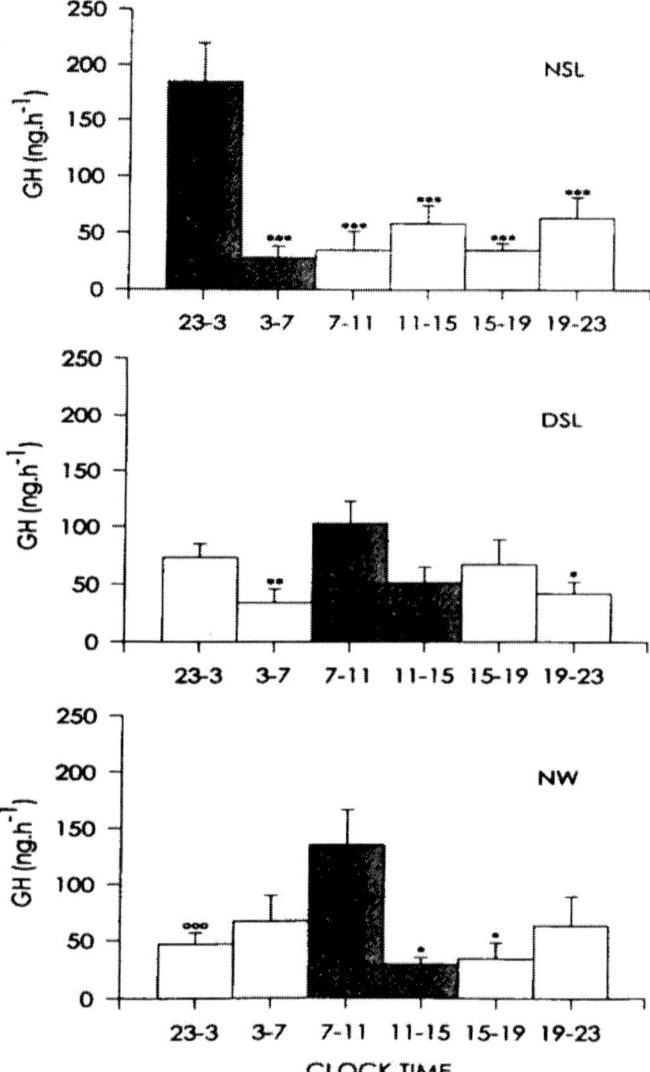

FIGURE 7. GH secretion in night workers. Mean ± s.e.m. amount of GH secreted per hour during successive 4-hour periods for 3 groups of subjects (NSL: night sleepers; DSL: day sleepers; NW: night workers). In each group, the amount of GH secreted per hour during the sleep period is represented by shaded areas. Intragroup comparisons: ***$p < 0.001$; **$p < 0.01$; *$p < 0.05$: significant differences from the first 4 h of sleep (23-3 for NSL and 7-11 for DSL and NW). Intergroup comparisons [000] $p < 0.001$: significant differences for the period (23-3) in NW versus NSL, but not in DSL versus NSL. Reproduced from[54] with permission

levels during night shifts could reduce the negative feedback and enhance CRH and ACTH secretion.

In shift workers, cortisol levels are significantly higher during daytime sleep episodes than during nocturnal sleep in day active workers[19] (Figure 8). The change of cortisol secretion during daytime sleep is smaller than that observed during nighttime sleep.[58] Cortisol levels are lower during nighttime waking episodes in shift workers than during daytime waking episodes in day workers.[19] These observations are consistent with persistent misalignment of the endogenous circadian rhythm of cortisol secretion still very much adjusted to a day-oriented schedule. Nevertheless, partial adaptation of cor-

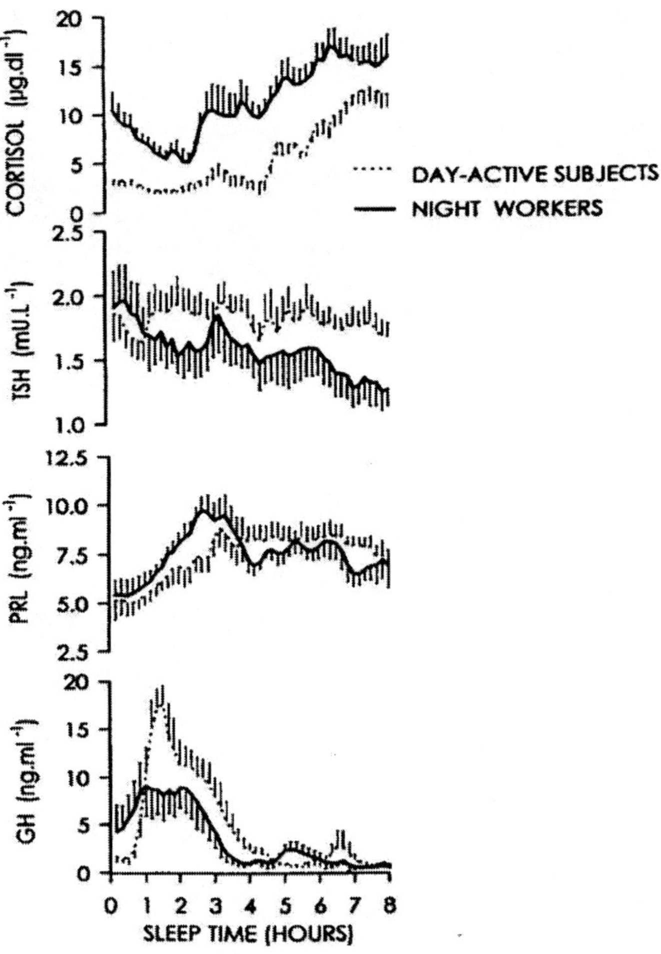

FIGURE 8. Disturbances in hormonal rhythms of night workers. Mean ± s.e.m. hormonal rhythms during the usual sleep time of day-active subjects (23h00-07h00) and of night workers (07h00-15h00)

tisol secretion has been reported in shift workers, with a large interworkers variability.[136]

A study of 24 nurses in cardiac emergency room showed that during the first five days on night shifts, the circadian rhythm of cortisol secretion is still adjusted to a day-oriented schedule, with increased concentrations in the morning and decreasing levels at the end of the evening.[137] The pattern was reversed after the fifth night in workers who adapted to the night schedule (Figure 9). Morning cortisol levels decreased across the shift cycle in 22 middle-aged men on rapidly counterclockwise rotation.[138] Morning cortisol levels were significantly reduced after night shifts in 6 male residents in obstetrics whereas afternoon levels were normal or high.[139] Progressive reduction of morning cortisol levels is consistent with some degree of circadian adaptation to a night oriented schedule.[137,140]

Some studies reported greater cortisol levels and distorted rhythms in shift workers[137,141,142] whereas others did not.[117,118] There is considerable individual variability in the group of shift workers studied and work-related conditions may mask the expression of cortisol rhythm in the field and account for differences across studies.[23,127]

Adaptation eventually occurs following rapid travel across time zones since environmental synchronizers are also shifted. Adaptation of ACTH and cortisol rhythms took about 10 days to adapt to a jet travel from Brussels-Chicago-Brussels.[143] The quiescent period at the start of the shifted sleep episode was slower to adapt compared to the peak secretory period following waketime. No

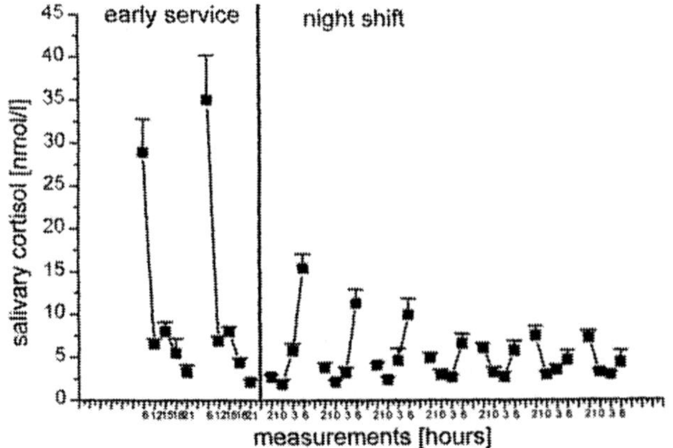

FIGURE 9. Changes in cortisol secretion during shiftwork. A total of 24 nurses working on rotating shifts were studied. After two day shifts (left), nurses worked 7 consecutive nights (right). Salivary cortisol was measured every 3 hours. Mean ± s.e.m. of the means of cortisol concentration are illustrated. ANOVA for repeated measures indicates significant effects for night, hour and the interaction night × hour (all $p < 0.001$). Reproduced from[137] with permission

change was noticed in mean levels and amplitude of the 24-hour rhythm of cortisol secretion in these travelers. In shift workers, the time lag between melatonin peak levels and the start of the quiescent period of cortisol secretion was found to be stable and comparable to that of day subjects.[21,144] These rhythms were still mainly adjusted to a day-oriented schedule.[21]

Adaptive phase delays of cortisol and melatonin rhythms to an atypical work schedule is possible[145] and is influenced by the pattern of light and darkness exposure.[62,123,127,129,140,146-149] The ability to phase shift has been negatively correlated to the amplitude of the circadian rhythm of cortisol secretion,[150] although this issue remains a matter of controversy.

In a previous investigation, we determined that a comprehensive light/darkness intervention including phototherapy in the workplace, shielding from bright morning light, and the maintenance of a stable and diurnal sleep/darkness period could promote circadian adaptation of plasma melatonin[131] and cortisol rhythms[140] in nurses working permanent night shifts. Following a 10-day vacation period on a day-oriented schedule, peak cortisol secretion occurred as expected in the hours following the time of habitual awakening, and lowest concentrations were observed close to the onset of night time sleep (Figure 6). Baseline melatonin peaked in the middle of the scheduled nocturnal sleep episode and levels were undetectable in the afternoon. Following the period of night shift work, workers who remained in their habitual light environments displayed a misalignment between the endogenous circadian cortisol and melatonin rhythms and their sleep-wake schedule. This resulted in higher cortisol concentrations near diurnal bedtimes whereas minimal levels were detected near the start of the night shift. In these workers, melatonin was still secreted at night, namely at the end of their night shifts. Conversely, workers who received the intervention displayed a complete adaptation of cortisol and melatonin rhythms to the inverted sleep-wake cycle.

TSH secretion is inhibited by daytime sleep.[105] During daytime sleep, a faster descent of TSH levels is observed whereas during nighttime waking TSH levels are greatly enhanced[19,108]

In night workers with an 8-hour delay shift of their sleep episode, the TSH acrophase can occur around sleep onset[151] (Figure 10). Some degree of adaptation of TSH secretion, consistent with that of the endogenous circadian temperature rhythm, can occur in night shift work.[151]

A simulated jet lag experiment with an 8-hour advance of the sleep episode revealed a marked increase in TSH during shift days.[70] A TSH pulse was still observable near the habitual onset of baseline sleep at night, although this circadian rise was smaller than baseline. TSH did not show its progressive decline during sleep episodes. In this experiment, TSH levels were reported to be 2-3 fold higher during the first two post-shift days.

Sleep was disrupted during these post-shift days and TSH increased during awakening from sleep. Exposure to bright light from 23h00 to 15h00 limited the TSH increase at the end of the second shifted sleep episode. This sleep

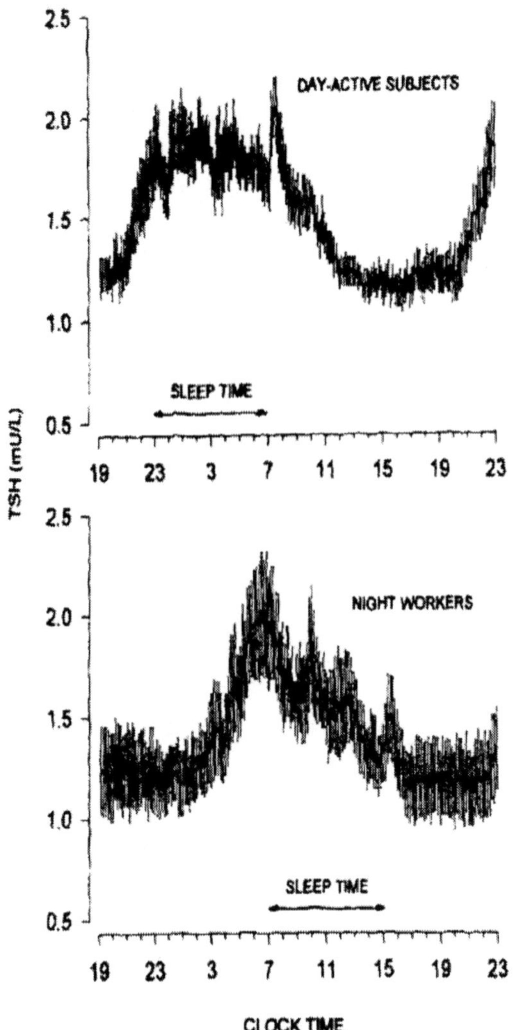

FIGURE 10. Delay of the TSH rhythm in night workers. Mean ± s.e.m. plasma TSH rhythm obtained from blood sampled every 10 min in 8 day-active subjects (upper panel) and in 7 regular night workers (lower panel). The acrophase of the TSH rhythm was located in both groups at about the time of sleep onset. Reproduced from[151] with permission.

episode had a better sleep efficiency and was associated with a normal TSH decline throughout the episode.[70]

A significant increase in glucose and insulin secretion rate occurs during the early nocturnal sleep and returns to baseline during late sleep. During nocturnal sleep deprivation, glucose and insulin secretion remain stable.[152]

Glucose levels peak earlier during daytime recovery sleep following a night of sleep deprivation than during nocturnal sleep. The reduction of glucose and insulin secretion was smaller late during daytime sleep than during night-time sleep.[152]

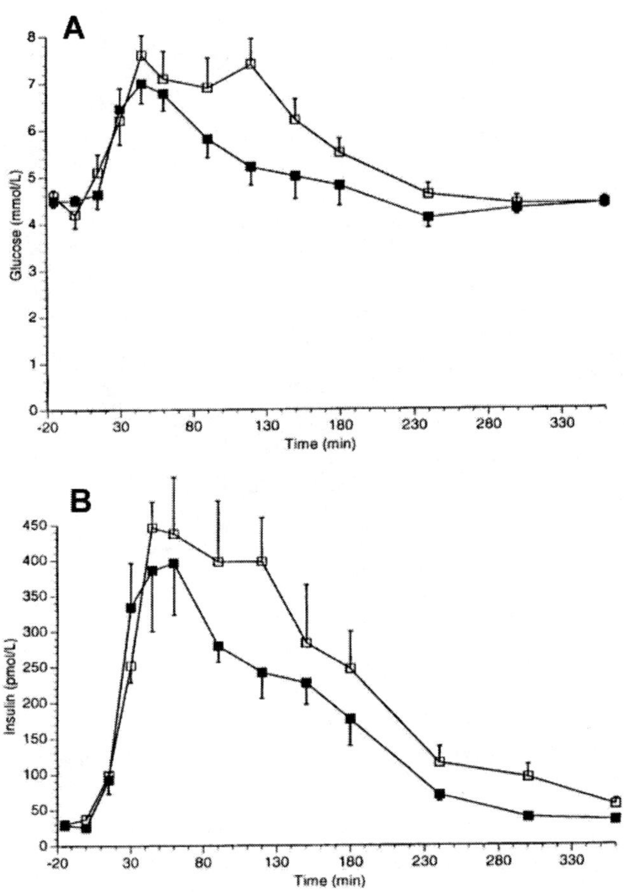

FIGURE 11. Postprandial hormone and metabolic responses in simulated shift work. Nine healthy young subjects participated to a simulated shift work experiment. A +9-hour advance shift of the endogenous circadian pacemaker was achieved by exposure to artificial bright light. Insulin and glucose response to a high carbohydrate test meal was investigated. The test meal was given at 13:30 during baseline and shifted days. In the shifted condition, it was thus administered at a circadian phase that was delayed by 9 hour compared to the baseline condition. Postprandial glucose (A) and insulin (B) response to the test meal before (filled squares) and after (open squares) the phase shift. Results are expressed as means ± s.e.m. Glucose and insulin response was significantly increased following the shift. Reproduced from[156] with permission.

The adaptation of glucose and insulin secretion is only partial in shift workers.[24] A 15% and 55% increase in glucose and insulin secretion was observed in response to meal when the sleep episode was displaced during the day.[73] A study was conducted in Antarctica on 2 women and 10 men, aged 24-34 years, working on a rotating schedule.[153] During this study, glucose tolerance deteriorated at the start of night shift but returned to normal after return to a day-oriented schedule. Volunteers studied during free-running conditions presented a relative glucose and lipid resistance.[154] Similar results were found after a simulated 9-hour advance shift using bright light exposure[155,136] (Figure 11). The increase in glucose, insulin, and triacylglycerol was dependent on meal composition.[155]

Simon and colleagues studied the 24-hour leptin secretion in seven young men during a regular day-oriented schedule and during an acute 8-hour delay shift of the sleep schedule.[79] In the baseline condition, leptin levels were increased during nocturnal sleep and were lower during the daytime waking episode. In the shifted condition, two periods of increased leptin levels were observed, namely during the night and during daytime sleep.

These results support a complex interaction between circadian and sleep-wake dependent processes. Changes in leptin levels paralleled those of plasma glucose and insulin.

A simulated jet lag study with a 12-hour shift in the light/dark cycle, sleep/wake cycle, and meal timing revealed that the leptin rhythm was rapidly shifted to the new schedule, much faster than the shift of the endogenous circadian pacemaker.[157] This suggests that leptin secretion is much more influenced by the timing of meals than by circadian phase.

Altogether, these studies suggest that disturbed metabolism of carbohydrates and lipids can increase the risk of cardiovascular diseases in shift workers. Indeed, greater serum triglycerides, cholesterol, and impaired glucose metabolism were found in this population.[158-160]

6. Conclusions

Shift work has been associated with a number of health problems including cardiovascular disease, impaired glucose and lipid metabolism, gastrointestinal discomfort, reproductive difficulties, and breast cancer.[158-168]

Despite years of night work, only a minority of workers show an appropriate reorientation of physiological rhythms such as those of 6-sulphatoxymelatonin, plasma melatonin, cortisol, and TSH. This can relate to the observable troughs in vigilance[169-172] and the fragility of sleep[12,16,173,174] particular to night work. Disturbed hormonal rhythms in shift work and jet travelers can be linked to a complex interaction of homeostatic and circadian processes. The specific contribution of disturbed hormonal rhythms, circadian misalignment, and sleep debt to the various medical problems encountered by shift workers remains to be clarified.

7. Acknowledgments. D. B. Boivin is supported by the Canadian Institutes of Health Research and the "Fonds de la Recherche en Santé du Québec". The author is grateful to Annny Casademont, Natacha Castilloux, Francine O. James, and Martha Popielarz, for their editorial support.

8. References

1. Dijk DJ, Czeisler CA. Contribution of the circadian pacemaker and the sleep homeostat to sleep propensity, sleep structure, electroencephalographic slow waves, and sleep spindle activity in humans. J Neurosci 1995; 15(5):3526-3538.
2. Daan S, Beersma DGM, Borbély AA. Timing of human sleep: recovery process gated by a circadian pacemaker. Am J Physiol 1984; 246(2 pt 2):R161-R183.
3. Akerstedt T, Gillberg M. Sleep disturbances and shift work. In: Reinberg A, Vieux N, Andlauer P, editors. Night and shift work : biological and social aspects : proceedings of the Fifth International Symposium on Night and Shift Work, Scientific Committee on Shift Work of the Permanent Commission and International Association on Occupational Health (PCIAOH) Rouen, 12-16 May 1980. Oxford: Pergamon Press, 1981: 127-137.
4. Vokac Z, Lund L. Patterns and duration of sleep in permanent security night guards. J Hum Ergol (Tokyo) 1982; 11(Suppl.):311-316.
5. Fookson JE, Kronauer RE, Weitzman ED, Monk TH, Moline ML, Hoey E. Induction of insomnia on a non-24 hour sleep-wake schedule. Sleep Res 1984; 13:220.
6. Carskadon MA, Dement WC. Distribution of REM sleep on a 90 minute sleep-wake schedule. Sleep 1980; 2(3):309-317.
7. Lavie P. Ultrashort sleep-wake cycle: timing of REM sleep. Evidence for sleep-dependent and sleep-independent components of the REM cycle. Sleep 1987; 10(1):62-68.
8. Czeisler CA, Weitzman ED, Moore-Ede MC, Zimmerman JC, Kronauer RE. Human sleep: its duration and organization depend on its circadian phase. Science 1980; 210(4475):1264-1267.
9. Weitzman ED, Czeisler CA, Zimmerman JC, Ronda JM. Timing of REM and stages 3 and 4 sleep during temporal isolation in man. Sleep 1980; 2(4): 391-407.
10. Brunner DP, Dijk DJ, Tobler I, Borbely AA. Effect of partial sleep deprivation on sleep stages and EEG power spectra: evidence for non-REM and REM sleep homeostasis. Electroencephalogr Clin Neurophysiol 1990; 75(6):492-499.
11. Forêt J, Benoît O. Structure du sommeil chez des travailleurs à horaires alternants. Electroencephalogr Clin Neurophysiol 1974; 37(4):337-344.
12. Frese M, Harwich C. Shiftwork and the length and quality of sleep. J Occup Med 1984; 26(8):561-566.
13. Rosekind MR, Hurd S, Buccino KR. Relationship of day versus night sleep to physician performance and mood. Ann Emerg Med 1994; 24(5):928-934.
14. Bryden G, Holdstock TL. Effects of night duty on sleep patterns of nurses. Psychophysiology 1973; 10(1):36-42.
15. Tepas DI, Walsh JK, Moss PD, Armstrong D. Polysomnographic correlates of shift worker performance in the laboratory. In: Reinberg A, Vieux N, Andlauer P, editors. Night and shift work : biological and social aspects : proceedings of the Fifth International Symposium on Night and Shift Work, Scientific Committee

on Shift Work of the Permanent Commission and International Association on Occupational Health (PCIAOH) Rouen, 12-16 May 1980. Toronto: Pergamon Press, 1980: 179-186.

16. Forêt J, Benoît O. Predictable effects on individual sleep patterns during a rapidly rotating shift system. Int Arch Occup Environ Health 1980; 45:49-56.

17. Weibel L, Spiegel K, Gronfier C, Follenius M, Brandenberger G. Twenty-four-hour melatonin and core body temperature rhythms: their adaptation in night workers. Am J Physiol 1997; 272(3 Pt 2):R948-R954.

18. Weibel L, Spiegel K, Follenius M, Ehrhart J, Brandenberger G. Internal dissociation of the circadian markers of the cortisol rhythm in night workers. Am J Physiol 1996; 270(4 Pt 1):E608-E613.

19. Weibel L, Brandenberger G. Disturbances in hormonal profiles of night workers during their usual sleep and work times. J Biol Rhythms 1998; 13(3):202-208.

20. Sack RL, Blood ML, Lewy AJ. Melatonin rhythms in night shift workers. Sleep 1992; 15(5):434-441.

21. Roden M, Koller M, Pirich K, Vierhapper H, Waldhauser F. The circadian melatonin and cortisol secretion pattern in permanent night shift workers. Am J Physiol 1993; 265(1 Pt 2):R261-R267.

22. Costa G, Bertoldi A, Kovacic M, et al. Hormonal secretion of nurses engaged in fast-rotating shift systems. Int Arch Occup Environ Health 1997; 3(Suppl. 2): S35-S39.

23. Goh VH, Tong TY, Lim CL, Low EC, Lee LK. Circadian disturbances after night-shift work onboard a naval ship. Mil Med 2000; 165(2):101-105.

24. Simon C, Weibel L, Brandenberger G. Twenty-four-hour rhythms of plasma glucose and insulin secretion rate in regular night workers. Am J Physiol Endocrinol Metab 2000; 278(3):E413-E420.

25. Akerstedt T, Kecklund G, Knutsson A. Spectral analysis of sleep electroencephalography in rotating three-shift work. Scand J Work Environ Health 1991; 17(5):330-336.

26. Kripke DF, Cook B, Lewis OF. Sleep of night workers: EEG recordings. Psychophysiology 1971; 7(3):377-384.

27. Van Cauter E, Spiegel K. Circadian and sleep control of hormonal secretions. In: Turek FW, Zee PC, editors. Regulation of sleep and circadian rhythms. New York: Marcel Dekker, Inc., 1999: 397-425.

28. Weitzman ED, Nogeire C, Perlow M et al. Effects of a prolonged 3-hour sleep-wake cycle on sleep stages, plasma cortisol, growth hormone and body temperature in man. J Clin Endocrinol Metab 1974; 38(6):1018-1030.

29. Weitzman ED, Fukushima D, Nogeire C, Roffwarg H, Gallagher TF, Hellman L. Twenty-four hour pattern of the episodic secretion of cortisol in normal subjects. J Clin Endocrinol Metab 1971; 33(1):14-22.

30. Orth DN, Island DP, Liddle GW. Experimental alteration of the circadian rhythm in plasma cortisol (17-OHCS) concentration in man. J Clin Endocrinol Metab 1967; 27(4):549-555.

31. Weibel L, Follenius M, Spiegel K, Ehrhart J, Brandenberger G. Comparative effect of night and daytime sleep on the 24-hour cortisol secretory profile. Sleep 1995; 18(7):549-556.

32. Leproult R, Colecchia EF, L'Hermite-Baleriaux M, Van Cauter E. Transition from dim to bright light in the morning induces an immediate elevation of cortisol levels. J Clin Endocrinol Metab 2001; 86(1):151-157.

33. Czeisler CA, Kronauer RE, Allan JS et al. Bright light induction of strong (Type 0) resetting of the human circadian pacemaker. Science 1989; 244(4910):1328-1333.
34. Van Cauter E, Sturis J, Byrne MM et al. Demonstration of rapid light-induced advances and delays of the human circadian clock using hormonal phase markers. Am J Physiol 1994; 266(6 Pt. 1):E953-E963.
35. Boivin DB, Czeisler CA. Resetting of circadian melatonin and cortisol rhythms in humans by ordinary room light. Neuroreport 1998; 9(5):779-782.
36. Follenius M, Brandenberger G, Bandesapt JJ, Libert JP, Ehrhart J. Nocturnal cortisol release in relation to sleep structure. Sleep 1992; 15(1):21-27.
37. Rivest RW, Schulz P, Lustenberger S, Sizonenko PC. Differences between circadian and ultradian organization of cortisol and melatonin rhythms during activity and rest. J Clin Endocrinol Metab 1989; 68(4):721-729.
38. Van Coevorden A, Mockel J, Laurent E et al. Neuroendocrine rhythms and sleep in aging men. Am J Physiol 1991; 260(4 pt 1):E651-E661.
39. Allan JS, Czeisler CA. Persistence of the circadian thyrotropin rhythm under constant conditions and after light-induced shifts of circadian phase. J Clin Endocrinol Metab 1994; 79(2):508-512.
40. Trinchard-Lugan I, Waldhauser F. The short term secretion pattern of human serum melatonin indicates a pulsatile hormone release. J Clin Endocrinol Metab 1989; 69(3):663-669.
41. Weitzman ED, Weinberg U, D'Eletto R et al. Studies of the 24 hour rhythm of melatonin in man. J Neural Transm Suppl 1978; 13:325-337.
42. Provencio I, Foster RG. Circadian rhythms in mice can be regulated by photoreceptors with cone-like characteristics. Brain Res 1995; 694(1-2):183-190.
43. Shanahan TL, Czeisler CA. Light exposure induces equivalent phase shifts of the endogenous circadian rhythms of circulating plasma melatonin and core body temperature in men. J Clin Endocrinol Metab 1991; 73(2):227-235.
44. Lewy AJ, Wehr TA, Rosenthal NE et al. Melatonin secretion as a neurobiological "marker" and effects of light in humans. Psychopharmacol Bull 1982; 18(4):127-129.
45. Czeisler CA, Allan JS, Strogatz SH et al. Bright light resets the human circadian pacemaker independent of the timing of the sleep-wake cycle. Science 1986; 233(4764):667-671.
46. Mills JN, Minors DS, Waterhouse JM. Adaptation to abrupt time shifts of the oscillator[s] controlling human circadian rhythms. J Physiol 1978; 285:455-470.
47. Wetterberg L. Melatonin in humans -physiological and clinical studies. J Neural Transm 1978; 13:289-310.
48. Lewy AJ, Wehr TA, Goodwin FK, Newsome DA, Markey SP. Light suppresses melatonin secretion in humans. Science 1980; 210(4475):1267-1269.
49. Brainard GC, Lewy AJ, Menaker M et al. Dose-response relationship between light irradiance and the suppression of plasma melatonin in human volunteers. Brain Res 1988; 454(1-2):212-218.
50. Brainard GC, Rollag MD, Hanifin JP. Photic regulation of melatonin in humans: ocular and neural signal transduction. J Biol Rhythms 1997; 12(6):537-546.
51. Bojkowski CJ, Aldhous ME, English J et al. Suppression of nocturnal plasma melatonin and 6-sulphatoxymelatonin by bright and dim light in man. Horm Metabol Res 1987; 19(9):437-440.
52. McIntyre IM, Norman TR, Burrows GD, Armstrong SM. Quantal melatonin suppression by exposure to low intensity light in man. Life Sci 1989; 45(4):327-332.

53. Trinder J, Armstrong SM, O'Brien C, Luke D, Martin MJ. Inhibition of melatonin secretion onset by low levels of illumination. J Sleep Res 1996; 5(2):77-82.
54. Weibel L, Follenius M, Spiegel K, Gronfier C, Brandenberger G. Growth hormone secretion in night workers. Chronobiol Int 1997; 14(1):49-60.
55. Moline ML, Monk TH, Wagner DR et al. Human growth hormone release is decreased during sleep in temporal isolation (free-running). Chronobiologia 1986; 13(1):13-19.
56. Weitzman ED, Czeisler CA, Zimmerman JC, Moore-Ede MC. Biological rhythms in man: relationship of sleep-wake, cortisol, growth hormone and temperature during temporal isolation. In: Martin JB, Reichlin S, Bick K, editors. Adv Biochem Psychopharmacol: Neurosecretion and Brain Peptides. New York: Raven Press, 1981: 475-499.
57. Alford FP, Baker HW, Burger HG et al. Temporal patterns of integrated plasma hormone levels during sleep and wakefulness. I. Thyroid-stimulating hormone, growth hormone and cortisol. J Clin Endocrinol Metab 1973; 37(6):841-847.
58. Pietrowsky R, Meyrer R, Kern W, Born J, Fehm HL. Effects of diurnal sleep on secretion of cortisol, luteinizing hormone, and growth hormone in man. J Clin Endocrinol Metab 1994; 78(3):683-687.
59. Czeisler CA, Klerman EB. Circadian and sleep-dependent regulation of hormone release in humans. Recent Prog Horm Res 1999; 54:97-132.
60. Weitzman ED, Pollak CP. Effects of flurazepam on sleep and growth hormone release during sleep in healthy subjects. Sleep 1982; 5(4):343-349.
61. Plotnick LP, Thompson RG, Kowarski A, de LL, Migeon CJ, Blizzard RM. Circadian variation of integrated concentration of growth hormone in children and adults. J Clin Endocrinol Metab 1975; 40(2):240-247.
62. Suvanto S, Harma M, Laitinen JT. The prediction of the adaptation of circadian rhythms to rapid time zone changes. Ergonomics 1993; 36(1-3):111-116.
63. Mullington J, Hermann D, Holsboer F, Pollmacher T. Age-dependent suppression of nocturnal growth hormone levels during sleep deprivation. Neuroendocrinology 1996; 64(3):233-241.
64. Waldstreicher J, Duffy JF, Brown EN, Rogacz S, Allan JS, Czeisler CA. Gender differences in the temporal organization of prolactin (PRL) secretion: Evidence for a sleep-independent circadian rhythm of circulating PRL levels - A clinical research center study. J Clin Endocrinol Metab 1996; 81(4):1483-1487.
65. Nokin J, Vekemans M, L'Hermite M, Robyn C. Circadian periodicity of serum prolactin concentration in man. Br Med J 1972; 3(826):561-562.
66. Van Cauter E, L'Hermite M, Copinschi G, Refetoff S, Desir D, Robyn C. Quantitative analysis of spontaneous variations of plasma prolactin in normal man. Am J Physiol 1981; 241(5):E355-E363.
67. Parker DC, Pekary AE, Hershman JM. Effect of normal and reversed sleep-wake cycles upon nyctohemeral rhythmicity of plasma thyrotropin: Evidence suggestive of an inhibitory influence in sleep. J Clin Endocrinol Metab 1976; 43:318-329.
68. Brabant G, Prank K, Ranft U et al. Physiological regulation of circadian and pulsatile thyrotropin secretion in normal man and woman. J Clin Endocrinol Metab 1990; 70(2):403-409.
69. Parker DC, Rossman LG, Pekary AE, Hershman JM. Effect of 64-hour sleep deprivation on the circadian waveform of thyrotropin (TSH): further evidence of sleep-related inhibition of TSH release. J Clin Endocrinol Metab 1987; 64(1):157-161.

70. Hirschfeld U, Moreno-Reyes R, Akseki E et al. Progressive elevation of plasma thyrotropin during adaptation to simulated jet lag: effects of treatment with bright light or zolpidem. J Clin Endocrinol Metab 1996; 81(9):3270-3277.
71. Brabant G, Prank K, Hoang-Vu C, Hesch RD, von zur MA. Hypothalamic regulation of pulsatile thyrotopin secretion. J Clin Endocrinol Metab 1991; 72(1):145-150.
72. Frank SA, Roland DC, Sturis J et al. Effects of aging on glucose regulation during wakefulness and sleep. Am J Physiol 1995; 269(6 Pt 1):E1006-E1016.
73. Van Cauter E, Blackman JD, Roland D, Spire JP, Refetoff S, Polonsky KS. Modulation of glucose regulation and insulin secretion by circadian rhythmicity and sleep. J Clin Invest 1991; 88(3):934-942.
74. Dallman MF, Strack AM, Akana SF et al. Feast and famine: critical role of glucocorticoids with insulin in daily energy flow. Front Neuroendocrinol 1993; 14(4):303-347.
75. Plat L, Byrne MM, Sturis J et al. Effects of morning cortisol elevation on insulin secretion and glucose regulation in humans. Am J Physiol 1996; 270(1 Pt 1):E36-E42.
76. Plat L, Leproult R, L'Hermite-Baleriaux M et al. Metabolic effects of short-term elevations of plasma cortisol are more pronounced in the evening than in the morning. J Clin Endocrinol Metab 1999; 84(9):3082-3092.
77. Van Cauter E, Shapiro ET, Tillil H, Polonsky KS. Circadian modulation of glucose and insulin responses to meals: relationship to cortisol rhythm. Am J Physiol 1992; 262(4 Pt 1):E467-E475.
78. Van Cauter E, Polonsky KS, Scheen AJ. Roles of circadian rhythmicity and sleep in human glucose regulation. Endocr Rev 1997; 18(5):716-738.
79. Simon C, Gronfier C, Schlienger JL, Brandenberger G. Circadian and ultradian variations of leptin in normal man under continuous enteral nutrition: relationship to sleep and body temperature. J Clin Endocrinol Metab 1998; 83(6): 1893-1899.
80. Bornstein SR, Licinio J, Tauchnitz R et al. Plasma leptin levels are increased in survivors of acute sepsis: associated loss of diurnal rhythm, in cortisol and leptin secretion. J Clin Endocrinol Metab 1998; 83(1):280-283.
81. Licinio J, Mantzoros C, Negrao AB et al. Human leptin levels are pulsatile and inversely related to pituitary-adrenal function. Nat Med 1997; 3(5):575-579.
82. Sinha MK, Ohannesian JP, Heiman ML et al. Nocturnal rise of leptin in lean, obese, and non-insulin-dependent diabetes mellitus subjects. J Clin Invest 1996; 97(5):1344-1347.
83. Laughlin GA, Yen SS. Hypoleptinemia in women athletes: absence of a diurnal rhythm with amenorrhea. J Clin Endocrinol Metab 1997; 82(1):318-321.
84. Saad MF, Riad-Gabriel MG, Khan A et al. Diurnal and ultradian rhythmicity of plasma leptin: effects of gender and adiposity. J Clin Endocrinol Metab 1998; 83(2):453-459.
85. Born J, Muth S, Fehm HL. The significance of sleep onset and slow wave sleep for nocturnal release of growth hormone (GH) and cortisol. Psychoneuroendocrinology 1988; 13(3):233-243.
86. Honda Y, Takahashi K, Takahashi S et al. Growth hormone secretion during nocturnal sleep in normal subjects. J Clin Endocrinol Metab 1969; 29(1):20-29.
87. Parker DC, Sassin JF, Mace JW, Gotlin RW, Rossman LG. Human growth hormone release during sleep: electroencephalographic correlation. J Clin Endocrinol Metab 1969; 29(6):871-874.

88. Sassin JF, Parker DC, Mace JW, Gotlin RW, Johnson LC, Rossman LG. Human growth hormone release: relation to slow-wave sleep and sleep-walking cycles. Science 1969; 165(892):513-515.

89. Jaffe CA, Turgeon DK, Friberg RD, Watkins PB, Barkan AL. Nocturnal augmentation of growth hormone (GH) secretion is preserved during repetitive bolus administration of GH-releasing hormone: potential involvement of endogenous somatostatin-a clinical research center study. J Clin Endocrinol Metab 1995; 80(11):3321-3326.

90. Gronfier C, Luthringer R, Follenius M et al. A quantitative evaluation of the relationships between growth hormone secretion and delta wave electroencephalographic activity during normal sleep and after enrichment in delta waves. Sleep 1996; 19(10):817-824.

91. Vaughan GM, Allen JP, Tullis W, Siler-Khodr TM, de la Pena A, Sackman JW. Overnight plasma profiles of melatonin and certain adenohypophyseal hormones in men. J Clin Endocrinol Metab 1978; 47(3):566-571.

92. Beck U, Brezinova V, Hunter WM, Oswald I. Plasma growth hormone and slow wave sleep increase after interruption of sheep. J Clin Endocrinol Metab 1975; 40(5):812-815.

93. Holl RW, Hartman ML, Veldhuis JD, Taylor WM, Thorner MO. Thirty-second sampling of plasma growth hormone in man: correlation with sleep stages. J Clin Endocrinol Metab 1991; 72(4):854-861.

94. Steiger A, Herth T, Holsboer F. Sleep-electroencephalography and the secretion of cortisol and growth hormone in normal controls. Acta Endocrinol (Copenh) 1987; 116(1):36-42.

95. Jarrett DB, Greenhouse JB, Miewald JM, Fedorka IB, Kupfer DJ. A reexamination of the relationship between growth hormone secretion and slow wave sleep using delta wave analysis. Biol Psychiatry 1990; 27(5):497-509.

96. Van Cauter E, Caufriez A, Kerkhofs M, Van OA, Thorner MO, Copinschi G. Sleep, awakenings, and insulin-like growth factor-I modulate the growth hormone (GH) secretory response to GH-releasing hormone. J Clin Endocrinol Metab 1992; 74(6):1451-1459.

97. Spath-Schwalbe E, Hundenborn C, Kern W, Fehm HL, Born J. Nocturnal wakefulness inhibits growth hormone (GH)-releasing hormone-induced GH secretion. J Clin Endocrinol Metab 1995; 80(1):214-219.

98. Baumgartner A, Dietzel M, Saletu B et al. Influence of partial sleep deprivation on the secretion of thyrotropin, thyroid hormones, growth hormone, prolactin, luteinizing hormone, follicle stimulating hormone, and estradiol in healthy young women. Psychiatry Res 1993; 48(2):153-178.

99. Van Cauter E, Plat L, Scharf MB et al. Simultaneous stimulation of slow-wave sleep and growth hormone secretion by gamma-hydroxybutyrate in normal young Men. J Clin Invest 1997; 100(3):745-753.

100. Sassin JF, Frantz AG, Weitzman ED, Kapen S. Human prolactin: 24-hour pattern with increased release during sleep. Science 1972; 177(55):1205-1207.

101. Spiegel K, Weibel L, Gronfier C, Brandenberger G, Follenius M. Twenty-four-hour prolactin profiles in night workers. Chronobiol Int 1996; 13(4): 283-293.

102. Sassin JF, Frantz AG, Kapen S, Weitzman ED. The nocturnal rise of human prolactin is dependent on sleep. J Clin Endocrinol Metab 1973; 37(3):436-440.

103. Frantz AG. Prolactin. N Engl J Med 1978; 298(4):201-207.

104. von Treuer K, Norman TR, Armstrong SM. Overnight human plasma melatonin, cortisol, prolactin, TSH, under conditions of normal sleep, sleep deprivation, and sleep recovery. J Pineal Res 1996; 20(1):7-14.
105. Goichot B, Brandenberger G, Saini J, Wittersheim G. Nocturnal plasma thyrotropin variations are related to slow-wave sleep. J Sleep Res 1992; 1(3):186-190.
106. Gronfier C, Luthringer R, Follenius M et al. Temporal link between plasma thyrotropin levels and electroencephalographic activity in man. Neurosci Lett 1995; 200(2):97-100.
107. Gary KA, Winokur A, Douglas SD, Kapoor S, Zaugg L, Dinges DF. Total sleep deprivation and the thyroid axis: effects of sleep and waking activity. Aviat Space Environ Med 1996; 67(6):513-519.
108. Goichot B, Weibel L, Chapotot F, Gronfier C, Piquard F, Brandenberger G. Effect of the shift of the sleep-wake cycle on three robust endocrine markers of the circadian clock. Am J Physiol 1998; 275(2 Pt 1):E243-E248.
109. Palmblad J, Akerstedt T, Froberg J, Melander A, von SH. Thyroid and adrenomedullary reactions during sleep deprivation. Acta Endocrinol (Copenh) 1979; 90(2):233-239.
110. Kern W, Offenheuser S, Born J, Fehm HL. Entrainment of ultradian oscillations in the secretion of insulin and glucagon to the nonrapid eye movement/rapid eye movement sleep rhythm in humans. J Clin Endocrinol Metab 1996; 81(4):1541-1547.
111. Schulz H, Brandenberger G, Gudewill S et al. Plasma renin activity and sleep-wake structure of narcoleptic patients and control subjects under continuous bedrest. Sleep 1992; 15(5):423-429.
112. Brandenberger G, Follenius M, Simon C, Ehrhart J, Libert JP. Nocturnal oscillations in plasma renin activity and REM-NREM sleep cycles in humans: a common regulatory mechanism? Sleep 1988; 11(3):242-250.
113. Luthringer R, Brandenberger G, Schaltenbrand N et al. Slow wave electroencephalic activity parallels renin oscillations during sleep in humans. Electroencephalogr Clin Neurophysiol 1995; 95(5):318-322.
114. Weitzman ED, Zimmerman JC, Czeisler CA, Ronda J. Cortisol secretion is inhibited during sleep in normal man. J Clin Endocrinol Metab 1983; 56(2):352-358.
115. Spath-Schwalbe E, Uthgenannt D, Voget G, Kern W, Born J, Fehm HL. Corticotropin-releasing hormone-induced adrenocorticotropin and cortisol secretion depends on sleep and wakefulness. J Clin Endocrinol Metab 1993; 77(5):1170-1173.
116. Bierwolf C, Struve K, Marshall L, Born J, Fehm HL. Slow wave sleep drives inhibition of pituitary-adrenal secretion in humans. J Neuroendocrinol 1997; 9(6):479-484.
117. Salin-Pascual RJ, Ortega-Soto H, Huerto-Delgadillo L, Camacho-Arroyo I, Roldan-Roldan G, Tamarkin L. The effect of total sleep deprivation on plasma melatonin and cortisol in healthy human volunteers. Sleep 1988; 11(4):362-369.
118. Davidson JR, Moldofsky H, Lue FA. Growth hormone and cortisol secretion in relation to sleep and wakefulness. J Psychiatry Neurosci 1991; 16(2):96-102.
119. Leproult R, Copinschi G, Buxton O, Van Cauter E. Sleep loss results in an elevation of cortisol levels the next evening. Sleep 1997; 20(10):865-870.
120. Spath-Schwalbe E, Gofferje M, Kern W, Born J, Fehm HL. Sleep disruption alters nocturnal ACTH and cortisol secretory patterns. Biol Psychiatry 1991; 29(6):575-584.

121. Akerstedt T, Froberg JE, Friberg Y, Wetterberg L. Melatonin excretion, body temperature and subjective arousal during 64 hours of sleep deprivation. Psychoneuroendocrinology 1979; 4(3):219-225.
122. Deacon SJ, Arendt J, English J. Posture: a possible masking factor of the melatonin circadian rhythm. In: Touitou Y, Arendt J, Pévet P, editors. Melatonin and the pineal gland-from basic science to clinical application. Elsevier Science, 1993: 387-390.
123. Koller M, Harma M, Laitinen JT, Kundi M, Piegler B, Haider M. Different patterns of light exposure in relation to melatonin and cortisol rhythms and sleep of night shift workers. J Pineal Res 1994; 16:127-135.
124. Czeisler CA, Johnson MP, Duffy JF, Brown EN, Ronda JM, Kronauer RE. Exposure to bright light and darkness to treat physiologic maladaptation to night work. N Engl J Med 1990; 322(18):1253-1259.
125. Martin SK, Eastman CI. Medium-intensity light produces circadian rhythm adaptation to simulated night-shift work. Sleep 1998; 21(2):154-165.
126. Dawson D, Lack L, Morris M. Phase resetting of the human circadian pacemaker with use of a single pulse of bright light. Chronobiol Int 1993; 10(2): 94-102.
127. Horowitz TS, Cade BE, Wolfe JM, Czeisler CA. Efficacy of bright light and sleep/darkness scheduling in alleviating circadian maladaptation to night work. Am J Physiol Endocrinol Metab 2001; 281(2):E384-E391.
128. Crowley SJ, Lee C, Tseng CY, Fogg LF, Eastman CI. Combinations of bright light, scheduled dark, sunglasses, and melatonin to facilitate circadian entrainment to night shift work. J Biol Rhythms 2003; 18(6):513-523.
129. Gibbs M, Hampton S, Morgan L, Arendt J. Adaptation of the circadian rhythm of 6-sulphatoxymelatonin to a shift schedule of seven nights followed by seven days in offshore oil installation workers. Neurosci Lett 2002; 325(2):91-94.
130. Stewart KT, Hayes BC, Eastman CI. Light treatment for NASA shiftworkers. Chronobiol Int 1995; 12(2):141-151.
131. Boivin DB, James FO. Circadian adaptation to night shift work by judicious light and darkness exposure. J Biol Rhythms 2002; 17(6):556-567.
132. Takahashi Y, Kipnis DM, Daughaday WH. Growth hormone secretion during sleep. J Clin Invest 1968; 47(9):2079-2090.
133. Golstein J, Van Cauter E., Désir D et al. Effects of "jet lag" on hormonal patterns. IV. Time shifts increase growth hormone release. J Clin Endocrinol Metab 1983; 56(3):433-440.
134. Désir D, Van Cauter E., L'Hermite M et al. Effects of "jet lag" on hormonal patterns. III. Demonstration of an intrinsic circadian rhythmicity in plasma prolactin. J Clin Endocrinol Metab 1982; 55(5):849-857.
135. Leese G, Chattington P, Fraser W, Vora J, Edwards R, Williams G. Short-term night-shift working mimics the pituitary-adrenocortical dysfunction in chronic fatigue syndrome. J Clin Endocrinol Metab 1996; 81(5):1867-1870.
136. Quera-Salva MA, Defrance R, Claustrat B, De Lattre J, Guilleminault C. Rapid shift in sleep time and acrophase of melatonin secretion in short shift work schedule. Sleep 1996; 19(7):539-543.
137. Hennig J, Kieferdorf P, Moritz C, Huwe S, Netter P. Changes in cortisol secretion during shiftwork: implications for tolerance to shiftwork? Ergonomics 1998; 41(5):610-621.

138. Axelsson J, Akerstedt T, Kecklund G, Lindqvist A, Attefors R. Hormonal changes in satisfied and dissatisfied shift workers across a shift cycle. J Appl Physiol 2003; 95(5):2099-2105.

139. Chatterton RT, Jr., Dooley SL. Reversal of diurnal cortisol rhythm and suppression of plasma testosterone in obstetric residents on call. J Soc Gynecol Investig 1999; 6(1):50-54.

140. James FO, Walker CD, Boivin DB. Controlled exposure to light and darkness realigns the salivary cortisol rhythm in night shift workers. Chronobiol Int 2004; *in press.*

141. Touitou Y, Motohashi Y, Reinberg A et al. Effect of shift work on the night-time secretory patterns of melatonin, prolactin, cortisol and testosterone. Eur J Appl Physiol Occup Physiol 1990; 60(4):288-292.

142. Goh VH, Tong TY, Lim CL, Low EC, Lee LK. Effects of one night of sleep deprivation on hormone profiles and performance efficiency. Mil Med 2001; 166(5):427-431.

143. Désir D, Van Cauter E., Fang VS et al. Effects of "jet lag" on hormonal patterns. I. Procedures, variations in total plasma proteins, and disruption of adrenocorticotropin-cortisol periodicity. J Clin Endocrinol Metab 1981; 52(4): 628-641.

144. Weibel L, Brandenberger G. The start of the quiescent period of cortisol remains phase locked to the melatonin onset despite circadian phase alterations in humans working the night schedule. Neurosci Lett 2002; 318(2):89-92.

145. Whitson PA, Putcha L, Chen Y-M, Baker E. Melatonin and cortisol assessment of circadian shifts in astronauts before flight. J Pineal Res 1995; 18(3): 141-147.

146. Budnick LD, Lerman SE, Nicolich MJ. An evaluation of scheduled bright light and darkness on rotating shiftworkers: trial and limitations. Am J Int Med 1995; 27(6):771-782.

147. Barnes RG, Forbes MJ, Arendt J. Shift type and season affect adaptation of the 6-sulphatoxymelatonin rhythm in offshore oil rig workers. Neurosci Lett 1998; 252(3):179-182.

148. Dawson D, Encel N, Lushington K. Improving adaptation to simulated night shift: timed exposure to bright light versus daytime melatonin administration. Sleep 1995; 18(1):11-21.

149. Dumont M, Benhaberou-Brun D, Paquet J. Profile of 24-h light exposure and circadian phase of melatonin secretion in nightworkers. J Biol Rhythms 2001; 16(5):502-511.

150. Reinberg A, Vieux N, Ghata J, Chaumont AJ, LaPorte A. Is the rhythm amplitude related to the ability to phase-shift circadian rhythms of shift-workers? J Physiol (Paris) 1978; 74(4):405-409.

151. Weibel L, Brandenberger G, Goichot B, Spiegel K, Ehrhart J, Follenius M. The circadian thyrotropin rhythm is delayed in regular night workers. Neurosci Lett 1995; 187(2):83-86.

152. Scheen AJ, Byrne MM, Plat L, Leproult R, Van Cauter E. Relationships between sleep quality and glucose regulation in normal humans. Am J Physiol 1996; 271 (2 Pt 1):E261-E270.

153. Lund J, Arendt J, Hampton SM, English J, Morgan LM. Postprandial hormone and metabolic responses amongst shift workers in Antarctica. J Endocrinol 2001; 171(3):557-564.

154. Morgan L, Arendt J, Owens D et al. Effects of the endogenous clock and sleep time on melatonin, insulin, glucose and lipid metabolism. J Endocrinol 1998; 157(3):443-451.

155. Ribeiro DC, Hampton SM, Morgan L, Deacon S, Arendt J. Altered postprandial hormone and metabolic responses in a simulated shift work environment. J Endocrinol 1998; 158(3):305-310.
156. Hampton SM, Morgan LM, Lawrence N et al. Postprandial hormone and metabolic responses in simulated shift work. J Endocrinol 1996; 151(2):259-267.
157. Schoeller DA, Celle LK, Sinha MK, Caro JF. Entrainment of the diurnal rhythm of plasma leptin to meal timing. J Clin Invest 1997; 100(7):1882-1887.
158. Romon M, Nuttens MC, Fievet C et al. Increased triglyceride levels in shift workers. Am J Med 1992; 93(3):259-262.
159. Theorell T, Akerstedt T. Day and night work: changes in cholesterol, uric acid, glucose and potassium in serum and in circadian patterns of urinary catecholamine excretion. A longitudinal cross-over study of railway workers. Acta Med Scand 1976; 200(1-2):47-53.
160. Nagaya T, Yoshida H, Takahashi H, Kawai M. Markers of insulin resistance in day and shift workers aged 30-59 years. Int Arch Occup Environ Health 2002; 75(8):562-568.
161. Knutsson A, Hallquist J, Reuterwall C, Theorell T, Akerstedt T. Shiftwork and myocardial infarction: a case-control study. Occup Environ Med 1999; 56(1):46-50.
162. Nakamura K, Shimai S, Kikuchi S et al. Shift work and risk factors for coronary heart disease in Japanese blue-collar workers: serum lipids and anthropometric characteristics. Occup Med (Lond) 997; 47(3):142-146.
163. Harrington JM. Shift work and health - a critical review of the literature on working hours. Ann Acad Med Singapore 1994; 23(5):699-705.
164. Romon M, Bertin Lebrette C. Travail posté et alimentation. Cahiers de Nutrition et de Diétetique 1998; 33(6):390-394.
165. Nurminen T. Shift work and reproductive health. Scandinavian Journal of Work, Environment & Health 1998; 24(Suppl 3):28-34.
166. Hansen J. Increased breast cancer risk among women who work predominantly at night. Epidemiology 2001; 12(1):74-77.
167. Koller M, Kundi M, Cervinka R. Field studies of shift work at an Austrian oil refinery. I: health and psychosocial wellbeing of workers who drop out of shiftwork. Ergonomics 1978; 21(10):835-847.
168. Koller M. Health risks related to shift work. An example of time-contingent effects of long-term stress. Int Arch Occup Environ Health 1983; 53(1): 59-75.
169. Hanecke K, Tiedemann S, Nachreiner F, Grzech-Sukalo H. Accident risk as a function of hour at work and time of day as determined from accident data and exposure models for the German working population. Scand J Work Environ Health 1998; 24(Suppl 3):43-48.
170. Akerstedt T. Sleepiness as a consequence of shift work. Sleep 1988; 11(1): 17-34.
171. Fathallah FA, Brogmus GE. Hourly trends in workers' compensation claims. Ergonomics 1999; 42(1):196-207.
172. Hamelin P. Lorry driver's time habits in work and their involvement in traffic accidents. Ergonomics 1987; 30(9):1323-1333.
173. Dumont M, Montplaisir J, Infante-Rivard C. Sleep quality of former night-shift workers. Int J Occup Environ Health 1997; 3(Supplement 2):S10-S14.
174. Niedhammer I, Lert F, Marne MJ. Effects of shift work on sleep among French nurses. A longitudinal study. J Occup Med 1994; 36(6):667-674.

Melatonin, The Pineal Gland, and Headache Disorders

MARIO F. P. PERES

1. Introduction

There is increasing evidence that headache disorders have many connections with melatonin secretion and pineal function. I will overview in this chapter the putative role of melatonin in the pathophysiology and treatment of headache disorders.

The nervous system evolved over the millennia to meet the demands of environmental conditions, including the light-dark cycle, in order to assure survival and reproduction of living organisms.[1] It has been demonstrated in the past decades that the circadian biological rhythms are not only the response to the 24-hour day night environment but in fact is due to a system in the brain.[2] A synchronization system to adapt the internal to the external environment is one of the key elements of the central nervous system to maintain life. The pineal gland and its main secretory product, melatonin, are the main elements for the synchronization of the internal biological events to the environment. Melatonin is absent during the day in humans and its nocturnal secretion is the main biological event signaling the when is night to the organism.

In humans, the pineal gland lies in the center of the brain, behind the third ventricle. Because its pine shape format the organ was coined "pineal" gland. The gland is usually 8mm in diameter weighing about 1g.[3] It consists of two types of cells: pinealocytes, which predominate and produce both indoleamines (melatonin) and peptides (such as arginine vasotocin), and neuroglial cells.[4] The gland is highly vascular. Melatonin is a derivative of the essential amino acid tryptophan. The pinealocytes are the principal location for melatonin biosynthesis, after its uptake into cells. Tryptophan is first hydroxylated and then decarboxylated, resulting in the formation of 5-hydroxytryptamine (serotonin).[5] Serotonin is N-acetylated, with the resulting formation of N-acetylserotonin which is subsequently O-methylated to form melatonin.[5] Melatonin is present in the earliest life forms and is found in all organisms including bacteria, algae, fungi, plants, insects, and vertebrates including humans. In all species melatonin synthesis exhibits a circadian rhythm.[6]

Once melatonin is synthesized in the pineal, it is quickly released, generating a blood melatonin rhythm reminiscent of that seen in the gland. Being an amphiphilic molecule, melatonin is capable of entering every cell in the organism; additionally, it readily crosses all morphophysiological barriers, including, the blood-brain barrier and the placenta.[7] Melatonin is enzymatically degraded in the liver to 6-hydroxymelatonin[8] and finally excreted in the urine as 6-sulfatoxymelatonin. Urine analysis is widely used as a measure of melatonin secretion, since it is correlated with the nocturnal profile of plasma melatonin secretion.[9]

Melatonin was first identified in bovine pineal tissue,[10] and has been subsequently portrayed exclusively as a hormone. Recently accumulated evidence has challenged this concept. Several characteristics of melatonin distinguish it from a classic hormone such as its direct, non-receptor-mediated free radical scavenging activity.[11] As melatonin is also ingested in vegetables, fruits, rice, wheat, and herbal medicines, from the nutritional point of view, melatonin can also be classified as a vitamin. It seems likely that melatonin initially evolved as an antioxidant, becoming a vitamin in the food chain, and in multicellular organisms, where it is produced, it has acquired autacoid, paracoid, and hormonal properties.[12]

A family of receptors for melatonin on cell membranes has recently been cloned.[13] The distribution of the receptors seems to be broad, species-specific, and G-protein–coupled. Principally, there are three high-affinity melatonin receptors, MEL1a, MEL1b, and MEL1c, from which the first two are found in humans. The gene for MEL1a receptor is localized into human chromosome 4q35.1[14] and is present in both kidney and intestine. MEL1b receptor is mapped to human chromosome 11q21 to 22.[15]

At present, indications for melatonin therapeutic applications include sleeping disorders, circadian rhythm disorders, insomnia in blind people, insomnia in elderly patients, aging, Alzheimer disease, and as an adjuvant in cancer therapy.[16]

2. Relevance of Melatonin and Chronobiology in Neurology

Chronobiological disorders occurring in men can be divided in two types: 1) environmental or external, due to the life style or environment, as in shift workers, individuals crossing time zones in the jet lag syndrome, and in maladaptation to daylight savings; 2) the endogenous or internal, including the delayed and advanced sleep phase syndromes, and the non-24-hour sleep – wake disorder with free-running circadian rhythm[2]. It has been proposed that the endogenous type may underlie many conditions including depression, chronic fatigue, fibromyalgia, and migraine.[17]

Sleep is well known to play an important role as a restorative function. In human beings, it has a circadian rhythm, normally occurring at night, usually together with the nocturnal melatonin secretion.[18] This has led to the idea that melatonin is an internal sleeping facilitator in humans, and therefore useful in the treatment of insomnia and the readjustment of circadian rhythms. There is evidence that administration of melatonin is able to induce sleep when the drive to sleep is insufficient; to inhibit the drive for wakefulness from the SCN; and to induce phase shifts in the circadian clock such that the circadian phase of increased sleep propensity occurs at a new, desired time.[19]

Many neurological disorders occur with a marked rhythmicity, dependent either on the 24-hour or the seasonal cycle. Thus maladaptation is probably linked to pineal function and melatonin secretion, including stroke, multiple sclerosis, facial paralysis, and seasonal affective disorder.[2,20]

The pineal gland is a photoneuroendocrine organ, converting external luminous stimuli into a hormone secretion, being responsible for the synchronization between the internal homeostasis and the environment. Consequently, an altered synchronization system may interfere with all neurological diseases. Sleep and circadian rhythms are often disrupted in people with neurological disorders.[21] The symptoms associated with neurological diseases may be due in part to disruption of the sleep-wake cycle. Alternatively, various neurological disorders may themselves disrupt the sleep-wake cycle, resulting in a positive feedback loop whereby disrupted sleep and wake exacerbate the neurological disorders while the disease itself has a negative effect on the sleep-wake states.[22]

Symptoms associated to those disorders may fluctuate according to a specific rhythm (circaannual, circamenal, circadian) and are often related to either sleep or wake periods. Epilepsy, dementia, movement disorders, multiple sclerosis, cerebrovascular disorders, neuromuscular disorders, and brain tumors have all been linked to an altered chronobiology, melatonin dysfunction, or benefited from melatonin treatment.[23] Primary headaches also follow this rule. Migraines, cluster headaches, indomethacin responsive headaches, and hypnic headaches have all been related to melatonin dysfunction.

3. Melatonin and Migraine

Melatonin and migraine are linked in several ways. Clinical symptoms may fluctuate, some patients reporting their headaches predominantly or specifically at a certain period of the day. Both episodic (55%) and chronic (62,5%) migraineurs reported waking up in the morning with headaches or being woken up at night by the headache.[24] In addition, headaches occurring in the morning period have been attributed to sleep disorders.[25] A distribution of migraine attacks according to the estrous cycle is evident in migraine. True menstrual migraine occurs in 14% of migraineurs.[26] Menstrually associated migraine can occur in up to 55% of cases.[27] A circaannual variation can be observed in cyclic migraine[28] or in the cluster-migraine association.[29]

Peres et al.[30] studied the chronobiological features in 200 chronic or episodic migraine patients, 93 (46.5%) reported headaches after changing the sleep schedule. A significant delay was presented in 54% of patients, ranging from −02:30 to +05:00 hs. Most patients (69%) delayed the sleeping phase, as opposed to those (31%) who advanced it. Individual's shifts above 02:00 hs represented 12.5% of patients. There is enough evidence for the relevancy and impact of changes in biological rhythm in migraine patients, and also the possible opposite effect of migraine in sleeping schedule of patients.

Melatonin was first studied in migraine patients by Claustrat et al.[31] in 1989, showing lower plasma levels on samples from patients drawn at 11 p.m. compared to controls. Migraine patients without depression had lower levels than controls, but migraineurs with superimposed depression exhibited the greatest melatonin deficiency. Murialdo et al.[32] also found nocturnal urinary melatonin significantly decreased throughout the ovarian cycle of migraine patients without aura as compared to controls. During the luteal phase, when melatonin levels should normally increase, migraine patients showed a less pronounced change when compared to controls. Melatonin excretion was further decreased when patients suffered a migraine attack.

Brun et al.[33] studied urinary melatonin in women with migraine without aura attacks associated with menses and controls. Melatonin levels throughout the cycle were significantly lower in the migraine patients than in controls. In the control group, melatonin excretion increased significantly from the follicular to the luteal phase, whereas no difference was observed in the migraine group. Melatonin may be implicated in the pathogenesis of menstrual migraine.

Peres et al.[34] studied plasma melatonin nocturnal profile. Thirteen sets of samples collected hourly from 07:00 pm to 07:00 am in chronic migraine patients and controls. It was observed lowered melatonin levels in patients with insomnia compared to those without insomnia, and a phase delay in the melatonin peak in patients vs controls, suggesting a chronobiological dysfunction in chronic migraineurs.

Only small studies showed a benefit in migraine patients treated with melatonin. Claustrat et al,[35] looked at nocturnal plasma melatonin profile and melatonin kinetics during melatonin infusion in six patients with status migrainous. Individual plasma profiles were disturbed in three migraine patients; two had a phase-delay and one a phase-advance. Four of the six patients reported headache relief the next morning after melatonin infusion began and the remaining two patients did so after the third night of infusion. In addition, three patients described that during migraines there was a decrease in the pulsatility of pain.

Another study[36] investigated the effects of melatonin on varying headaches and their relation to delayed sleep phase syndrome (DSPS). Thirty DSPS patients were treated; in one patient with migraine, his attacks dramatically decreased after beginning melatonin treatment. One patient was successfully treated during a migraine attack by means of external (pico Tesla) magnetic fields.[37]

An open label trial[38] investigating the effectiveness of melatonin 3 mg for migraine prevention in 32 patients showed a significant improvement in the first month of treatment, sustained to the third month. Thirty-two of 34 patients completed the study; 78.1% of patients (25/32) who completed the study had at least 50% reduction. No patients reported increase in headaches. Complete (100%) response was achieved in 8 patients (25%), higher than 75% reduction was found in 7 patients (21.8%), and 50 to 75% reduction in 10 patients (31.3%) after 3 months of therapy. Melatonin improved significantly headache frequency, headache intensity, and duration. Menstrually associated migraines equally decreased in frequency. Two patients withdrew from the study, 1 because excessive sleepiness and the other because of alopecia. Three patients spontaneously reported increase in libido. No changes in body weight occurred.

Melatonin may be implicated in the pathogenesis of migraine, menstrual migraine, cyclic migraine, and chronic migraine. It might also play a role in migraine co morbid disorders, particularly depression and insomnia.

4. Melatonin and Cluster Headaches

It has been suspected that melatonin may be involved in cluster headache genesis, primarily because melatonin is a sensitive marker of endogenous rhythms, which are disrupted in cluster headache.[39] In 1984, Chazot et al.[40] identified a decrease in nocturnal melatonin secretion and abolished melatonin rhythm in cluster headache patients. Waldenlind et al.[41] also showed lowered nocturnal melatonin levels during cluster periods than remissions. Determining urinary levels of 6-sulphatoximelatonin throughout the year, Waldenlind et al.[42] found had higher melatonin levels in women than men. Swedes had higher melatonin levels than Italians, and smokers lower levels than non-smoking cluster headache patients. Leone et al.[43] observed melatonin and cortisol acrophases significantly correlated in controls but not in cluster headache patients, indicating a chronobiological disorder in these patients.

Blau and Engel[44] found that increases in body temperature after exercise, a hot bath or elevated environmental temperature triggered cluster headaches in 75 of 200 cluster headache patients. This finding can be explained by a decrease in melatonin secretion caused by temperature increase.[45] Melatonin for cluster headache prevention was then studied in a double-blind, placebo-controlled trial by Leone et al.[46] with significant decrease in cluster headache attacks in the melatonin-treated group compared with placebo.

Two patients with chronic cluster headache in the trial did not respond to melatonin therapy, but Peres and Rozen[47] described two chronic cluster headache patients who responded to melatonin 9 mg at bedtime. Melatonin not only prevented nocturnal cluster attacks, but daytime attacks as well. Nagtegaal et al.[48] studying melatonin treatment in delayed sleep phase syndrome, identified a patient with episodic cluster headache in whom both disorders improved after melatonin treatment. Melatonin plays an important role in the pathophysiology and treatment of cluster headaches.

5. Melatonin and Other Headaches Disorders

Hypnic headache is a benign, recurrent headache disorder that occurs only during diurnal and nocturnal sleep. Headaches are often frequent, usually occurring every night, with striking consistency the same time every night.[49]

Hypnic headache is typical in the elderly. Since melatonin secretion significantly declines with aging, one may speculate the melatonin deficiency as a possible cause of hypnic headache.[50]

Scarce reports have been shown improvement with melatonin treatment. Good response was detected in three patients, however no controlled trial has been conducted so far.[51] Martins and Gouveia[52] reported an interesting case of hypnic headache related to travel across time zones: a 68-year-old woman flew from Portugal to Brazil, shifting 3 hours in the sleep phase. Melatonin has not been given to the patient.

Peres et al.[53] described a patient with hemicrania continua with seasonal variation, proposing that the melatonin chemical structure similarity to that of indomethacin, could be one of the possible mechanisms of action involved in indomethacin responsive headaches. Rozen[54] reported the case of a hemicrania continua patient who responded to melatonin 9 mg, and described 3 idiopathic stabbing headache patients treated with melatonin with excellent clinical response and side effect profile.[55] Future studies will clarify the role of melatonin in indomethacin responsive headaches, and hypnic headaches.

6. Headaches, Pineal Cysts, and Other Chronobiological Disorders

Pineal cysts are common findings in neuroimaging studies, found in 1.3 to 2.6% of brain MRIs. Sawamura et al.[56] examined brain MRIs of 6023 consecutive patients finding pineal cysts (> 5 mm) in 1.3% of patients imaged. The cysts predominantly occurred in females; 29 cysts in 3008 males (0.96%) and 50 cysts in 3015 females (1.65%). Young women between the ages of 21 and 30 years (the decade of life migraine increases its prevalence to three-fold) had the highest frequency (5.82%). All the studies showing high prevalence of pineal cysts in brain MRI are biased by the recruitment criteria, usually not controlling for the reason (neurological signs or symptoms) the exam was ordered. Only one study[57] showed what should be the right incidence of pineal cysts analyzing 1000 asymptomatic volunteers. Incidental pineal cysts were found in brain MRI of only 2 patients (0.2%).

Pineal region lesions (benign, malignant tumors or cysts) can be clearly symptomatic, when the lesion size (usually bigger than 15mm) compromises other brain structures, causing hydrocephalus. Headache is the most common symptom. Unilateral headaches has been reported in pinealectomized subjects.[58]

Mandera et al[59] analyzed 24 pediatric patients (17 girls, mean age 9, and 7 boys, mean age 14) with pineal cysts and found an abnormal melatonin secretion profile. Wober-Bingol[60] et al found 2 pineal cysts (2,1%), 7 to 8 mm, in 96 headache patients with MRI incidental findings. One patient with migraine without aura, and another with chronic tension-type headache. Follow-up of 16 months in both patients revealed no change in headache frequency and pattern, suggesting a benign nature of the lesion.

Peres et al[61] reported 5 headache patients with pineal cysts, 4 women, 1 man, cysts ranging from 8 to 13,5 mm. Two patients had migraine without aura, 1 had migraine with aura, 1 chronic migraine, and 1 hemicrania continua. Three patients had strictly unilateral headaches. Pineal cysts may be not incidental in headache patients; abnormal melatonin secretion might hypothetically be involved in the mechanism of those patients, however this has yet to be determined.

Jet lag is a travel-induced circadian rhythm phenomenon that afflicts healthy individuals following long-distance flights through several time zones. Typical symptoms are daytime sleepiness, fatigue, impaired alertness, headaches, irritability, loss of appetite, decrease in physical performance and trouble initiating and maintaining sleep. It is attributed to transient desynchronization in the circadian rhythm until the internal biological clock is rephased to the new environmental conditions.[62] Headache features, diagnosis and impact have never been properly studied in patients with jet lag syndrome. Peres et al[30] studied migraine patients reporting frequent traveling across time zones; the majority of them (79%) had their headaches worse after it.

Shift work involves a substantial employed population worldwide. Pucci et al.[63] studied 157 employees, finding primary headaches in 26.7%, shift work being a major risk factor. Ho and Ong[64] studying 2096 individuals in a randomized survey in Singapore found that individuals who performed shift work had more frequent headaches. Peres et al.[30] found 86% of migraineurs had their headaches made worse after shift work.

Delayed and advanced sleep phase syndromes (DSPS, ASPS) are chronic, long-term circadian schedule disorders. Various primary headaches were found in DSPS patients:[48] 3 women had chronic tension-type headache, one patient migraine with aura, one cluster headache. Their headaches decreased dramatically after melatonin therapy with 5 mg.

7. Mechanisms

Melatonin may play a role in headache pathophysiology via several mechanisms (Table 1).

Melatonin has been shown to possess anti-inflammatory effects, among many others actions. By virtue of its ability to directly scavenge toxic free radicals,[5] it reduces macromolecular damage in all organs. The free radicals and reactive oxygen and nitrogen species known to be scavenged by melatonin include the highly toxic hydroxyl radical, peroxynitrite anion and hypochlorous acid, among others.

Melatonin also prevents the translocation of nuclear factor-kappa B to the nucleus and its binding to DNA, thereby reducing the up regulation of a variety of proinflammatory cytokines, interleukins and tumor necrosis factor-alpha.[65] Melatonin inhibits the production of adhesion molecules that promote the sticking of leukocytes to endothelial cells, attenuating transendothelial cell migration and edema.[66]

Melatonin inhibits the activity of nitric oxide synthase,[67] besides acting in membrane stabilization.[68]

Inhibition of dopamine release by melatonin has been demonstrated in specific areas of the mammalian central nervous system (hypothalamus, hippocampus, medulla-pons, and retina).[69] A growing body of biological, pharmacologic, and genetic data supports a role for dopamine in the pathophysiology of migraine.[70]

Melatonin has been related to GABA and glutamate neurotransmission, and both to headache pathophysiology.[71] It is thought that the hypnotic activity of melatonin is mediated by the GABA ergic system.[72] Melatonin rapidly and reversibly potentiated the GABA-A receptor-mediated response.[73]

TABLE 1. Melatonin mechanisms potentially related to headache disorders.

anti-inflammatory effect

toxic free radicals scavenging

nitric oxide synthase activity inhibition

dopamine release inhibition

membrane stabilization

similar chemical structure to indomethacin

GABA potentiation

Opioid analgesia

glutamate neurotoxicity protection

Neurovascular regulation

Serotonin modulation

Neuroprotection by melatonin from glutamate-induced excitotoxicity during development of the brain,[74] and the antagonistic effects of melatonin on glutamate release and neurotoxicity in cerebral cortex[75] have been reported.

Melatonin induces activated T lymphocytes to release opioid peptides with immunoenhancing and anti-stress properties. A melatonin-immuno-opioids network has been proposed. Cytokines named melatonin-induced-opioids (MIO) has been found to act on an opioid-binding site. Since melatonin may behave as a mixed opioid receptor agonist-antagonist, it is possible to potentiate the opioid analgesic efficacy.[76] Melatonin is also involved in cerebrovascular regulation,[77] and modulation of serotonin neurotransmission (spontaneous efflux and evoked release).[78]

Future studies are necessary for a better understanding of the role of melatonin in the pathophysiology and treatment of headache disorders. A more careful clinical examination of circadian and circaannual variation in primary and secondary headaches in different latitudes is needed, as the study of headache diagnosis, impact, and treatment response in chronobiological disorders. Animal models of cephalic pain and the genetics involved in chronobiological rhythms are important tools for further research.

Treatment of headache disorders with melatonin is promising, particularly in cluster headaches, hypnic headaches, indomethacin responsive headaches, and migraine. Melatonin may also be important in migraine co morbidity. Insomnia in headache patients is the most likely associated condition in migraine to respond to melatonin therapy. Lower melatonin levels may predict response to melatonin treatment as occur in insomnia associated to other diseases.[79] Other chronobiotic agents, such as melatonin receptors agonists, light therapy, and magnetic fields can also be tested.

Melatonin plays an important role in headache disorders, offering new avenues for studying its pathophysiology and treatment.

8. References

1. Arendt, J. Importance and relevance of melatonin to human biological rhythms. *J. Neuroendocrinol.* 15, 427-431 (2003).
2. Turek, F. W., Dugovic, C. & Zee, P. C. Current understanding of the circadian clock and the clinical implications for neurological disorders. *Arch. Neurol.* 58, 1781-1787 (2001).
3. Erlich, S. S. & Apuzzo, M. L. The pineal gland: anatomy, physiology, and clinical significance. *J. Neurosurg.* 63, 321-341 (1985).
4. Reiter, R. J. Pineal melatonin: cell biology of its synthesis and of its physiological interactions. *Endocr. Rev.* 12, 151-180 (1991).
5. Reiter, R. J., Calvo, J. R., Karbownik, M., Qi,W. & Tan, D. X. Melatonin and its relation to the immune system and inflammation. *Ann. N. Y. Acad. Sci.* 917, 376-386 (2000).
6. Dubbels, R. *et al.* Melatonin in edible plants identified by radioimmunoassay and by high performance liquid chromatography-mass spectrometry. *J. Pineal Res.* 18, 28-31 (1995).

7. Menendez-Pelaez, A. *et al.* Nuclear localization of melatonin in different mammalian tissues: immunocytochemical and radioimmunoassay evidence. *J. Cell Biochem.* 53, 373-382 (1993).

8. Reiter, R. J. Pineal melatonin: cell biology of its synthesis and of its physiological interactions. *Endocr. Rev.* 12, 151-180 (1991).

9. Markey, S. P., Higa, S., Shih, M., Danforth, D. N. & Tamarkin, L. The correlation between human plasma melatonin levels and urinary 6-hydroxymelatonin excretion. *Clin. Chim. Acta* 150, 221-225 (1985).

10. Orts, R. J. & Benson, B. Inhibition effects on serum and pituitary LH by a melatonin-free extract of bovine pineal glands. *Life Sci. II* 12, 513-519 (1973).

11. Zang, L. Y., Cosma, G., Gardner, H. & Vallyathan,V. Scavenging of reactive oxygen species by melatonin. *Biochim. Biophys. Acta* 1425, 469-477 (1998).

12. Tan, D. X. *et al.* Melatonin: a hormone, a tissue factor, an autocoid, a paracoid, and an antioxidant vitamin. *J. Pineal Res.* 34, 75-78 (2003).

13. Reppert, S. M. Melatonin receptors: molecular biology of a new family of G protein-coupled receptors. *J. Biol. Rhythms* 12, 528-531 (1997).

14. Godson, C. & Reppert, S. M. The Mel1a melatonin receptor is coupled to parallel signal transduction pathways. *Endocrinology* 138, 397-404 (1997).

15. Reppert, S. M. *et al.* Molecular characterization of a second melatonin receptor expressed in human retina and brain: the Mel1b melatonin receptor. *Proc. Natl. Acad. Sci. U. S. A* 92, 8734-8738 (1995).

16. Bubenik, G. A. *et al.* Prospects of the clinical utilization of melatonin. *Biol. Signals Recept.* 7, 195-219 (1998).

17. Gordijn, M. C., Beersma, D. G., Korte, H. J. & van den Hoofdakker,R.H. Testing the hypothesis of a circadian phase disturbance underlying depressive mood in nonseasonal depression. *J. Biol. Rhythms* 13, 132-147 (1998).

18. Rodenbeck, A., Huether, G., Ruther, E. & Hajak, G. Nocturnal melatonin secretion and its modification by treatment in patients with sleep disorders. *Adv. Exp. Med. Biol.* 467, 89-93 (1999).

19. Cajochen, C., Krauchi, K. & Wirz-Justice, A. Role of melatonin in the regulation of human circadian rhythms and sleep. *J. Neuroendocrinol.* 15, 432-437 (2003).

20. Checkley,S.A. *et al.* Melatonin rhythms in seasonal affective disorder. *Br. J. Psychiatry* 163, 332-337 (1993).

21. Turek, F. W., Dugovic, C. & Zee, P. C. Current understanding of the circadian clock and the clinical implications for neurological disorders. *Arch. Neurol.* 58, 1781-1787 (2001).

22. Zee PC, G. Z. Neurological disorders associated with disturbed sleep and circadian rhythms. Turek FW, Zee PC. Regulation of Sleep and Circadian Rhythms. 557-596. 1999. New York, NY, Marcel Dekker Inc.

23. Turek, F. W., Dugovic, C. & Zee, P. C. Current understanding of the circadian clock and the clinical implications for neurological disorders. *Arch. Neurol.* 58, 1781-1787 (2001).

24. Galego, J. C., Cipullo, J. P., Cordeiro, J. A. & Tognola, W. A. Clinical features of episodic migraine and transformed migraine: a comparative study. *Arq Neuropsiquiatr.* 60, 912-916 (2002).

25. Paiva, T., Batista, A., Martins, P. & Martins, A. The relationship between headaches and sleep disturbances. *Headache* 35, 590-596 (1995).

26. Silberstein, S. D. The role of sex hormones in headache. *Neurology* 42, 37-42 (1992).

27. Granella, F. *et al.* Migraine without aura and reproductive life events: a clinical epidemiological study in 1300 women. *Headache* 33, 385-389 (1993).
28. Medina, J. L. & Diamond, S. Cyclical migraine. *Arch. Neurol.* 38, 343-344 (1981).
29. Young, W. B., Peres, M. F. & Rozen, T. D. Modular headache theory. *Cephalalgia* 21, 842-849 (2001).
30. Peres, M. F. *et al.* Chronobiological features in episodic and chronic migraine. *Cephalalgia 23*, 590-591 (2003).
31. Claustrat,B. *et al.* Nocturnal plasma melatonin levels in migraine: a preliminary report. *Headache* 29, 242-245 (1989).
32. Murialdo, G. *et al.* Urinary melatonin excretion throughout the ovarian cycle in menstrually related migraine. *Cephalalgia* 14, 205-209 (1994).
33. Brun, J., Claustrat, B., Saddier, P. & Chazot, G. Nocturnal melatonin excretion is decreased in patients with migraine without aura attacks associated with menses. *Cephalalgia* 15, 136-139 (1995).
34. Peres, M. F. *et al.* Hypothalamic involvement in chronic migraine. *J. Neurol. Neurosurg. Psychiatry* 71, 747-751 (2001).
35. Claustrat, B. *et al.* Nocturnal plasma melatonin profile and melatonin kinetics during infusion in status migrainosus. *Cephalalgia* 17, 511-517 (1997).
36. Nagtegaal, J. E. *et al.* Effects of melatonin on the quality of life in patients with delayed sleep phase syndrome. *J. Psychosom. Res.* 48, 45-50 (2000).
37. Sandyk, R. The influence of the pineal gland on migraine and cluster headaches and effects of treatment with picoTesla magnetic fields. *Int. J. Neurosci.* 67, 145-171 (1992).
38. Peres, M. F., Zukerman, E., Moreira, F., Tanuri, F. C. & Cipolla-Neto, J. Melatonin 3 mg is effective for migraine prevention. Neurology in press. 2004.
39. May, A., Bahra, A., Buchel, C., Frackowiak, R. S. & Goadsby, P. J. Hypothalamic activation in cluster headache attacks. *Lancet* 352, 275-278 (1998).
40. Chazot, G. *et al.* A chronobiological study of melatonin, cortisol growth hormone and prolactin secretion in cluster headache. *Cephalalgia* 4, 213-220 (1984).
41. Waldenlind, E., Gustafsson, S. A., Ekbom, K. & Wetterberg, L. Circadian secretion of cortisol and melatonin in cluster headache during active cluster periods and remission. *J. Neurol. Neurosurg. Psychiatry* 50, 207-213 (1987).
42. Waldenlind, E. *et al.* Lowered circannual urinary melatonin concentrations in episodic cluster headache. *Cephalalgia* 14, 199-204 (1994).
43. Leone, M. *et al.* Twenty-four-hour melatonin and cortisol plasma levels in relation to timing of cluster headache. *Cephalalgia* 15, 224-229 (1995).
44. Blau, J. N. & Engel, H. O. A new cluster headache precipitant: increased body heat. *Lancet* 354, 1001-1002 (1999).
45. Peres, M. F., Seabra, M. L., Zukerman, E. & Tufik, S. Cluster headache and melatonin. *Lancet* 355, 147 (2000).
46. Leone, M., D'Amico, D., Moschiano, F., Fraschini, F. & Bussone, G. Melatonin versus placebo in the prophylaxis of cluster headache: a double-blind pilot study with parallel groups. *Cephalalgia* 16, 494-496 (1996).
47. Peres, M. F. & Rozen, T. D. Melatonin in the preventive treatment of chronic cluster headache. *Cephalalgia* 21, 993-995 (2001).
48. Nagtegaal, J. E., Smits, M. G., Swart, A. C., Kerkhof, G. A. & van der Meer, Y. G. Melatonin-responsive headache in delayed sleep phase syndrome: preliminary observations. *Headache* 38, 303-307 (1998).

49. Dodick, D. W., Jones, J. M. & Capobianco, D. J. Hypnic headache: another indomethacin-responsive headache syndrome? *Headache* 40, 830-835 (2000).

50. Dodick, D. W., Eross, E. J. & Parish, J. M. Clinical, anatomical, and physiologic relationship between sleep and headache. *Headache* 43, 282-292 (2003).

51. Capo, G. & Esposito, A. Hypnic headache: a new Italian case with a good response to pizotifene and melatonin. *Cephalalgia 21*, 505-506 (2001).

52. Martins, I. P. & Gouveia, R. G. Hypnic headache and travel across time zones: a case report. *Cephalalgia* 21, 928-931 (2001).

53. Peres, M. F., Stiles, M. A., Oshinsky,M. & Rozen, T. D. Remitting form of hemicrania continua with seasonal pattern. *Headache* 41, 592-594 (2001).

54. Rozen, T. D. Melatonin as treatment for indomethacin-responsive headache syndromes. *Cephalalgia* 23, 734-735 (2003).

55. Rozen, T. D. Melatonin as treatment for idiopathic stabbing headache. *Neurology* 61, 865-866 (2003).

56. Sawamura, Y. *et al.* Magnetic resonance images reveal a high incidence of asymptomatic pineal cysts in young women. *Neurosurgery* 37, 11-15 (1995).

57. Katzman, G. L., Dagher, A. P. & Patronas, N. J. Incidental findings on brain magnetic resonance imaging from 1000 asymptomatic volunteers. *JAMA* 282, 36-39 (1999).

58. Chazot G, Claustrat B, Broussolle E & Lapras C. Headache and depression. Nappi G (ed.), pp. 299-303 (Raven Press, New York,1991).

59. Mandera, M., Marcol, W., Bierzynska-Macyszyn, G. & Kluczewska, E. Pineal cysts in childhood. *Childs Nerv. Syst.* 19, 750-755 (2003).

60. Wober-Bingol, C. *et al.* Magnetic resonance imaging for recurrent headache in childhood and adolescence. *Headache* 36, 83-90 (1996).

61. Peres, M., Brandt, R., Porto, P. & Zukerman, E. Headaches and pineal cyst: a (more than) coincidental relationship? *Cephalalgia* 23, 676 (2003).

62. Atkinson, G., Drust, B., Reilly, T. & Waterhouse, J. The relevance of melatonin to sports medicine and science. *Sports Med.* 33, 809-831 (2003).

63. Pucci, E. *et al.* Shift work and headache: observations in a group of hospital workers. *Cephalalgia* 23, 605 (2003).

64. Ho, K. H. & Ong, B. K. A community-based study of headache diagnosis and prevalence in Singapore. *Cephalalgia* 23, 6-13 (2003).

65. Chuang, J. I., Mohan, N., Meltz, M. L. & Reiter, R. J. Effect of melatonin on NF-kappa-B DNA-binding activity in the rat spleen. *Cell Biol. Int.* 20, 687-692 (1996).

66. Shaikh, A. Y., Xu, J., Wu, Y., He, L. & Hsu, C. Y. Melatonin protects bovine cerebral endothelial cells from hyperoxia-induced DNA damage and death. *Neurosci. Lett.* 229, 193-197 (1997).

67. Bettahi, I. *et al.* Melatonin reduces nitric oxide synthase activity in rat hypothalamus. *J. Pineal Res.* 20, 205-210 (1996).

68. Garcia, J. J. *et al.* Melatonin prevents changes in microsomal membrane fluidity during induced lipid peroxidation. *FEBS Lett.* 408, 297-300 (1997).

69. Zisapel, N. Melatonin-dopamine interactions: from basic neurochemistry to a clinical setting. *Cell Mol. Neurobiol.* 21, 605-616 (2001).

70. Peroutka, S. J. Dopamine and migraine. *Neurology* 49, 650-656 (1997).

71. Ramadan, N. M. The link between glutamate and migraine. *CNS. Spectr.* 8, 446-449 (2003).

72. Wang, F. *et al.* Hypnotic activity of melatonin: involvement of semicarbazide hydrochloride, blocker of synthetic enzyme for GABA. *Acta Pharmacol. Sin.* 23, 860-864 (2002).
73. Wu, F. S., Yang, Y. C. & Tsai, J. J. Melatonin potentiates the GABA(A) receptor-mediated current in cultured chick spinal cord neurons. *Neurosci. Lett.* 260, 177-180 (1999).
74. Espinar, A. *et al.* Neuroprotection by melatonin from glutamate-induced excitotoxicity during development of the cerebellum in the chick embryo. *J. Pineal Res.* 28, 81-88 (2000).
75. Zhang, Q. Z., Gong, Y. S. & Zhang, J. T. Antagonistic effects of melatonin on glutamate release and neurotoxicity in cerebral cortex. *Zhongguo Yao Li Xue. Bao.* 20, 829-834 (1999).
76. Ebadi,M., Govitrapong,P., Phansuwan-Pujito,P., Nelson,F. & Reiter, R. J. Pineal opioid receptors and analgesic action of melatonin. *J. Pineal Res.* 24, 193-200 (1998).
77. Savaskan, E. *et al.* Cerebrovascular melatonin MT1-receptor alterations in patients with Alzheimer's disease. *Neurosci. Lett.* 308, 9-12 (2001).
78. Monnet, F. P. Melatonin modulates [3h]serotonin release in the rat hippocampus: effects of circadian rhythm. *J. Neuroendocrinol.* 14, 194-199 (2002).
79. deVries, M. W. & Peeters, F. P. Melatonin as a therapeutic agent in the treatment of sleep disturbance in depression. *J. Nerv. Ment. Dis.* 185, 201-202 (1997).

The Role of Peptides in Disturbed Sleep in Depression

Axel Steiger

1. Introduction

Most patients with depression report disturbed sleep.[1] Correspondingly sleep-EEG changes are frequent symptoms of depression. Furthermore a set of neuroendocrine aberrances, particularly overactivity of the hypothalamo-pituitary-adrenocortical (HPA) system and changes of the hypothalamo-pituitary-somatotrophic (HPS) system is characteristically in affective disorders. Since sleep is a time of considerable activity in endocrine systems it is an attractive approach to study sleep EEG and, by blood sampling during sleep via long catheter, nocturnal hormone secretion simultaneously. In young normal human subjects characteristical patterns of sleep-endocrine activity are found. During the first half of the night the major amounts of slow-wave sleep (SWS), slow-wave activity (SWA) and the growth hormone (GH) surge preponderate, whereas cortisol secretion reaches its nadir. During the second half of the night cortisol levels rise, the major portion of rapid-eye-movement sleep (REMS) occurs, and the amounts of SWS and GH are low[2,3] (see Figure 1). Already this pattern points to a reciprocal interaction between the HPS and the HPA systems with GH and cortisol, respectively, as their peripheral endpoints. Furthermore it is likely that common factors are links between the electrophysiological and the endocrine activity during sleep. This hypothesis is supported, since during depression (and similarly during normal aging) SWS and GH decrease, whereas HPA activity is enhanced. Clinical and preclinical studies showed that neuropeptides are these factors. Ehlers and Kupfer[4] submitted first that a reciprocal interaction of the peptides corticotropin-releasing hormone (CRH) and GH-releasing hormone (GHRH) plays a key role in the pathophysiology of changes of sleep-endocrine activity in depression. Since then this view was corroborated and amplified. This chapter intends to present the state of the art in this field.

2. Sleep-Endocrine Activity in Patients with Depression

Characteristically sleep-EEG findings in patients with depression include disturbed sleep continuity (prolonged sleep latency, enhanced intermittent wakefulness, early morning awakenings), a decrease of SWS (in younger

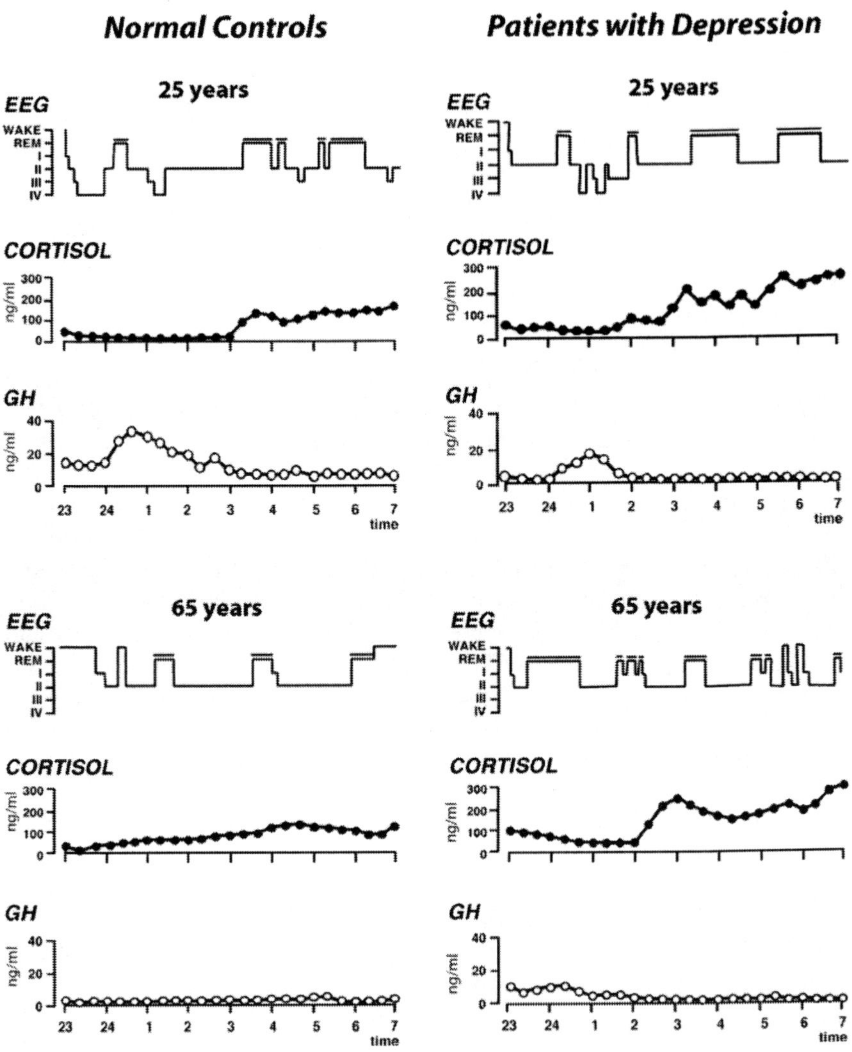

FIGURE 1. Individual hypnograms and patterns of cortisol and growth hormone (GH) secretion in 4 male subjects (young and old patients with depression and normal controls).

patients a shift of the major portion of SWS from the first to the second sleep cycle) and REMS desinhibition (shortened REMS latency, prolonged first REMS period and elevated REMS density, a measure for the amount of rapid-eye-movements during REMS (reviewed in[5,6]). The endocrine aberrances in affective disorders are reviewed elsewhere, (HPA-system overactivity reviewed in[7] and HPS-system dysfunction reviewed in[8]). Elevated cortisol and adrenocorticotropic hormone (ACTH) levels were reported in most sleep-endocrine studies in patients with depression throughout the night or 24 h respectively in comparison to healthy volunteers[9;10]. The circadian pattern of cortisol secretion is preserved during depression. Particularly in females a positive correlation between age and cortisol levels was reported[10]. Enhanced cortisol plasma and norepinephrine levels but normal ACTH plasma and CRH cerebrospinal fluid (CSF) levels throughout 30 h were found in one study[11]. GH secretion was blunted in most[8,12-14] but not in all[9] studies. As mentioned before, these findings suggest a crucial relationship between shallow sleep, blunted GH and HPA overdrive in depression. Interestingly normal aging goes ahead with changes of sleep EEG resembling those in depression (see below). Therefore it is likely that there are common mechanisms underlying the changes of sleep-endocrine activity during depression and aging.

A sharp decline in the sigma frequency range in women during menopause was found, whereas in men these changes occurred more gradually.[15] After menopause sleep-endocrine alterations associated with depression are accentuated. This view is supported by a study on sleep-endocrine activity in pre- and postmenopausal women with depression and in matched controls. Cortisol levels were elevated in postmenopausal patients, while a decrease was found in postmenopausal controls. Decrease of SWS and an increase of REMS density were prominent in post-but not in premenopausal patients. An inverse correlation was found between the decline in SWS and sleep continuity and follicle stimulating hormone (FSH) secretion in the patients. These observations suggest a role of menopause for these sleep-EEG changes. In premenopausal patients a shift in SWS and SWA from the first to the second NREMS period was found, which was not related to age or hormone secretion.[16]

Depression is frequently associated with loss of appetite, reduced food intake and weight loss. Therefore the secretion of leptin in patients with depression is of interest since leptin is involved in the regulation of food intake. In drugfree depressed patients nocturnal leptin levels were significantly higher than in age and sex matched controls.[17] The sexual dimorphism with higher leptin concentrations in women than in men as described before was confirmed in patients and in healthy subjects. Serum leptin was correlated with body mass index in normal controls but not in patients. This observation points to an altered regulation of leptin release in depression. Neither in the patients with depression nor in the male controls an increase in leptin between 0000 and 0400 h (as reported by[18] or at any other timepoint during the night was found. However in the group of young healthy women (younger than 35 years) an increase in leptin concentrations between 0000 and 0400 h occurred and showed a trend to be greater than in young female patients. Hence in young female patients with depression the nocturnal leptin surge appears to be blunted. As expected cortisol levels were enhanced in the patients. As glucocorticoids are capable to prevent the fasting induced decline of serum leptin, we suggest that hypercortisolism in depression might counteract the reduction in leptin release due to decreased food intake and weight loss. Elevated leptin levels in depression might in turn further promote the release of CRH as shown in animals[19;20] and contribute to HPA system overactivity in depression.

Simultaneous investigation of nocturnal thyroid stimulating hormone (TSH) and ACTH levels in drugfree male patients with depression and matched controls showed a blunted TSH and elevated ACTH secretion in the patients.

ACTH was negatively correlated to TSH in the first half of the night. These findings suggest that both pituitary hormones reflect a common dysregulation of the HPA and the hypothalamic-pituitary thyroid systems in depression.[21]

Sleep-endocrine activity was compared longitudinally between acute depression and recovery in 2 studies in patients with depression. A decrease of ACTH and cortisol throughout 24 h and a normalization of REMS was reported after remission.[9] Since not all of the patients in this study were drugfree at the retest and most antidepressants suppress REMS[22] it is difficult to differentiate between the effects of remission and of drugs.

Intraindividual comparison of male drugfree patients[8] confirmed a decrease of cortisol after remission. In this study prolactin levels of the patients did not differ from those of younger healthy subjects and remained unchanged after recovery.[23] Similarly nocturnal prolactin levels did not differ between depressed patients and matched controls.[24] Furthermore testosterone levels increased after recovery in male patients.[25]

The aberrant sleep EEG and the blunted GH levels, however, did not differ between acute depression and recovery. Sleep EEG showed even a further deterioration as the number of awakenings increased and stage 4 sleep decreased[8]. Both studies[8;9] confirm that hypersecretion of HPA hormones is a symptom of acute depression.

The decline of cortisol in remitted patients[8;9] is similar to the normalization of results of challenge test of the HPA system and of CRH CSF levels after recovery (reviewed in[7]). The persistence of most sleep-EEG changes[26] and of blunted GH levels[14] after remission has been confirmed over a period of 3 years.

Obviously HPA activity normalizes independently from sleep architecture. Therefore hypercortisolism in depression is not a consequence of shallow sleep. On the other hand blunted testosterone levels in male patients appear to be secondary to HPA overdrive. Finally prolactin is not affected either by depression or by aging. The persistence of changes of sleep EEG and GH release after recovery may be explained as a biological scar due to the metabolic aberrances during acute depression. An alternative explanation would be that these changes represent a trait in depressed patients. This issue remains unclear as long as no data are available comparing intraindividually the premorbid and depressed state in patients.

In young male patients who survived severe brain injury the sleep-endocrine pattern resembled that of remitted patients with a history of depression. Several months after the injury the cortisol concentrations of these patients did not differ from normal controls. The GH levels and sleep stage 2 time of the patients however were lower than in the controls.[27] It appears likely that in the survivors of brain injury either HPA overactivity due to stress under the intensive care situation or treatment with glucocorticoids in a subgroup contributed to the changes of NREMS and GH.

Interestingly, in patients with primary insomnia increased nocturnal cortisol levels and a shorter quiescent period of cortisol was found when compared to controls.[28] Similarly increases of ACTH and cortisol throughout 24 h were reported in these patients in comparison to controls.[29] These observations suggest similarities in the pathophysiology of primary insomnia and of depression. This is of particular interest since epidemiological studies showed an elevated risk for depression in patients with persisting insomnia.[30]

3. Changes of Sleep-Endocrine Activity During Aging

Sleep EEG and nocturnal hormone secretion change throughout the life span. These changes resemble those during depression. There is a controversy whether deterioration of sleep due to aging starts during puberty or during early adolescence.[31] It is well established however that already during the third decade of the life span distinct parallel decreases of SWS, SWA and GH secretion start. Near to the onset of the fifth decade the GH pause occurs. In females the menopause is a major turning point towards impaired sleep,[32] whereas in men the sleep quality declines continuously during aging. Correspondingly in the elderly the amounts of SWS and SWA are low or SWS is even totally absent (Figure 1). Furthermore sleep continuity is disturbed as the sleep onset latency, and the number and duration of nocturnal awakenings increase and the sleep efficiency and the sleep period time decrease. Finally REMS time and REMS latency decrease in senescence. Controversial data were reported on the effects of age on HPA hormones. Elevated and unchanged cortisol levels have been found in the elderly. Most studies agree that the amplitude of the cortisol rhythm is blunted. Accordingly cortisol concentration at the nadir is elevated. Similar results were found in the largest sample of normal male control subjects (16-83 years old) investigated to far.[33] Only a modest effect of aging on 24 h mean cortisol levels was observed. Aging was associated with an elevation of the cortisol nadir, but morning maximum cortisol values remained stable across all ages. Increases in evening cortisol concentrations occurred after the age of 50 years, when sleep became more fragmented and REMS decreased. Also melatonin and, in males, testosterone levels decline during aging. In contrast to most other hormones prolactin levels remain widely unaffected by aging.[34]

In adult depressed patients aging and depression exert synergistic effects on sleep-endocrine activity.[10] As a result of this synergism sleep-endocrine changes are most distinct in elderly patients with depression (Figure 1).

Many preclinical and human studies demonstrated specific roles of various neuropeptides in the regulation of sleep-endocrine activity.

4. The Role of Peptides in Sleep Regulation

4.1. Growth Hormone-Releasing Hormone (GHRH)

4.1.1. Preclinical Studies

Beside its role as major endogenous stimulus of GH release GH-releasing hormone (GHRH) is an important endogenous sleep-promoting substance. Hypothalamic GHRH mRNA displays a circadian rhythm. In the rat the highest levels occur when sleep propensity reaches its maximum in these night active animals at the beginning of the light period.[35] Furthermore hypothalamic GHRH contents show sleep-related variations with a nadir in the morning, increases in the afternoon and decline at night.[36]

Very big supermice slept more than normal mice.[37] In giant transgenic mice (MT-rGH mice) plasma GH is permanently elevated and secretion pulses are absent. During the light period NREMS was modestly increased and REMS was almost doubled in these mice in comparison to normal mice, whereas sleep did not differ between groups during the dark period. Also after sleep deprivation the MT-rGH mice slept more than control mice.[38] Dwarf rats with deficits in the central GHRHergic transmission and reduced hypothalamic GHRH contents showed less NonREMS (NREMS) than normal rats.[39] In dwarf homozygous *(lit/lit)* mice the GHRH receptor is non-functional. Their amounts of NREMS and REMS were lower than in normal mice. Infusion of GH by Alzet minipumps in the dwarf mice prompted normalization of REMS, but not of NREMS within 9 days. GHRH, ghrelin and octreotide did not influence sleep EEG in the dwarf mice. These data suggest that (i) GHRH deficiency is associated with decreases in REMS, (ii) decreases in GH lead to decreases in REMS, (iii) the actions of GHRH, ghrelin and octreotide on sleep EEG require intact GHRH receptor signaling.[40]

Intracerebroventriuclar (icv) administration of GHRH results in increases of SWS in rats and rabbits.[41,42] The same effect occurred when GHRH was injected into the medial preoptic area in rats[43] or systemically to rats.[44] Vice versa in the rat NREMS decreased when GHRH was inhibited by receptor antagonists.[45] A major role of GHRH in the sleep promotion by sleep deprivation is suggested by the observation that GHRH antibodies antagonized this effect in rats,[46] and hypothalamic GHRH mRNA increased after sleep deprivation in rats.[43,47] The sleep rebound after sleep deprivation was inhibited in rats by microinjections of a GHRH antagonist into the area preoptica.[43] Calcium levels in GABAergic neurons cultured from rat fetal hypothalamus increased when perfused with GHRH.[48] It is thought that many hypothalamic GHRH responsive neurons are GABAergic.

4.1.2. Human Studies

Similarly to the findings in animals[41-44] repetitive intravenous (iv) administration of GHRH during the first few hours of the night (2200 to 0100) increased SWS and GH release and decreased cortisol in young normal men.[49] Mimicking the pulsatile endogenous release appears to be an important methodological issue since sleep remained unchanged after GHRH infusion.[50] Sleep promotion in young men by GHRH was confirmed after iv[50;51] and intranasal[52] administration. The effects of repetitive iv GHRH on sleep-endocrine activity were investigated in states with a change of the GHRH/CRH ratio in favor of CRH – (i) the second half of the night in young normal men, and (ii) in elderly normal men and women. (i) After repetitive iv GHRH during the early morning hours (0400 to 0700) no major changes of sleep EEG were found. GH increased whereas ACTH and cortisol remained unchanged.[53] (ii) The response of GH to GHRH is blunted in older men at daytime.[54] Similarly only a weak sleep-promoting effect of GHRH was found in a sample of elderly healthy women and men. The first NREMS period increased and the number of awakenings decreased.[55] In a similar vein intranasal GHRH in elderly healthy volunteers had only a weak sleep-promoting effect.[56]

In contrast to the findings in male subjects sleep was impaired and HPA hormones were elevated in normal and depressed women of a wide age range after repetitive iv GHRH (see 5.1.). Similarly we found decreases of sleep stage 4 and of REMS, but no change of HPA hormone secretion in young normal women after repetitive iv GHRH.[57]

4.2. Somatostatin

4.2.1. Preclinical Studies

Somatostatin acts as major antagonist to GHRH in the regulation of GH release. After icv somatostatin selective increases of REMS were reported in rats.[58] Systemic[59] and icv[60] administration of the somatostatin analogue octreotide decreased NREMS in rats in a dose-dependent fashion. This change was followed by distinct increases of SWA 2-3 hours after administration.

4.2.2. Humans Studies

Similarly to the findings in rats[59,60] SWS decreased in young normal men after subcutaneous octreotide.[61] Intermittent wakefulness increased during the second half of the night, and EEG activity in the sigma range decreased.

Octreotide is known to be more potent than exogenous somatostatin. This explains that repetitive iv somatostatin impaired sleep in normal elderly subjects (total sleep time and REMS decreased and wakefulness increased during the first sleep cycle),[62] whereas it exerted no effect in young normal men.[49,63,64] In all these data point to a reciprocal interaction of GHRH and somatostatin in sleep regulation similarly to their relationship in GH regulation. The same dose of somatostatin which was not effective in young men impaired sleep in the elderly probably due to a decline of endogenous GHRH.

4.3. Ghrelin and GH Secretagogues

4.3.1. Preclinical Findings

Ghrelin was identified in the stomach and hypothalamus of humans and rats as endogenous ligand of the GH secretagogue (GHS) receptor.[65] Beside of GHRH and somatostatin ghrelin is thought to be another player in the regulation of GH release. Furthermore ghrelin stimulates HPA hormones and has a key role in the energy balance by stimulation of food intake and weight gain (review[66]). Already prior to the cloning of the GHS receptor, which preceded the detection of ghrelin, synthetic GHSs were known.[67] In normal mice systemic administration of ghrelin promoted NREMS.[40] In mice with non-functional GHRH receptors, however, ghrelin did not affect sleep EEG.

4.3.2. Human Studies

Similar to the effects of GHRH repetitive iv ghrelin enhanced SWS and GH in young normal men.[68] In contrast to the effects of GHRH, which blunted cortisol in young men.[49] ACTH and cortisol increased, particularly during the first half of the night after ghrelin.[68] The response of GH to ghrelin was most potent after the first injection and lowest after the fourth injection. Vice versa cortisol secretion showed an inverse pattern of response. We suggest that ghrelin acts as an interface between the HPA and HPS systems. The pattern of hormone changes after ghrelin resembles the effects of repetitive iv administration of the synthetic GHSs GH-releasing peptide-6 (GHRP-6)[69] and hexarelin.[70] The sleep EEG-effects of these compounds however differed from the influence of ghrelin. After GHRP-6 sleep stage 2 increased.[69] whereas after hexarelin SWS and SWA decreased, probably due to a change of the GHRH/CRH ratio in favor of CRH.[70] Oral administration of the GHS, MK-677, for one week had a distinct sleep-promoting effect in young men and only a weak effect in elderly controls.[71]

4.4. Corticotropin-Releasing Hormone (CRH)

4.4.1. Preclinical Studies

The HPA system mediates the reaction to acute physical and psychological stress. The stress reaction starts with the release of CRH from the parvocellular portion of the paraventricular nucleus of the hypothalamus. This results in the secretion of ACTH from the anterior pituitary and finally in the secretion of cortisol (in humans) or corticosterone (in rats) from the adrenocortex. Various co-factors contribute to this cascade (review:[7]).

Icv administration of CRH decreased SWS in rats[41] and rabbits.[72] Furthermore in rats even after 72 h of sleep deprivation CRH reduced SWS, prolonged sleep latency and increased REMS.[73]

In the Lewis rat the synthesis and release of CRH is reduced due to a hypothalamic gene defect in comparison to the related Fisher 344 and Sprague-Dawley rat strains. In Lewis rats wakefulness was reduced and SWS was elevated in comparison to intact strains. REMS did not differ between strains. After icv administration of CRH waking was enhanced similarly in Lewis and Sprague-Dawley rats. Obviously the mechanisms mediating the response to exogenous CRH are intact in the Lewis rats.[74] Studies on the effects of CRH antagonists in rats reported conflicting results. The CRH antagonists, α-helical CRH and astressin reduced wakefulness in a dose-related manner when given before the dark period.[75] These findings suggest that CRH contributes to the regulation of physiological waking periods. In another study[76] α-helical CRH was effective only in stressed rats. In these rats REMS was enhanced. It decreased to values of the non-stressed condition after the substance. Astressin, however, did not influence REMS in stressed rats, whereas it lead to normalization of increased wakefulness.[77] Since after sleep deprivation α-helical CRH diminished selectively the REMS rebound in rats, stress acting via CRH is thought to be the major factor inducing this rebound.[78] Some of this preclinical work[76,78] suggests, that CRH promotes REMS. According to their work Chang and Opp,[77] however, suggested, that CRH is not directly involved in the regulation of REMS. They hypothesized that the increase of REMS in stressed rats[76] is mediated by prolactin. The CRH receptor antagonist NBI 30775 (R121919) attenuated stress-induced sleep disturbances in rats, particularly in those animals with high innate anxiety.[79]

4.4.2. Human Studies

Similarly to the preclinical findings[41,72] pulsatile iv administration of human CRH (2200 to 0700) in young normal male subjects decreased SWS and REMS. In addition the GH surge was blunted and cortisol increased during the first half of the night.[80] Hourly iv injections of 10 μg CRH (0800 – 1800) did not affect sleep EEG during the following night, whereas melatonin decreased.[81] This finding suggests a reciprocal interaction between HPA activity and melatonin secretion. In young healthy male subjects EEG activity in the sigma frequency range increased during the first 3 sleep cycles both after a single iv bolus of CRH during the first SWS period and during wakefulness.[82] The responsiveness of the sleep EEG to CRH appears to increase during aging. This hypothesis is supported by a study comparing the influence of a single dose of ovine CRH given 10 min after sleep onset in young and middle-aged normal male subjects.[83] In the young men sleep

EEG remained unchanged, whereas SWS decreased and wakefulness increased in the middle-aged subjects.

From the human study the influence of endogenous CRH on REMS is uncertain since CRH suppressed REMS.[80] Studies on the sleep-EEG effects of ACTH and cortisol help to differentiate the central and peripherally mediated sleep-EEG changes after CRH in humans. Nocturnal infusions of ACTH decreased REMS in normal controls,[84] whereas cortisol and GH increased.[85] Repetitive iv administration of the synthetic ACTH (4-9) analogue ebiratide prompted a set of sleep-EEG changes corresponding to a general central nervous system (CNS) activation, whereas REMS, GH and cortisol remained unchanged.[86]

Since the pioneering work of Gillin[87] it is known that certain steroids modulate sleep.[88] Also continuous infusion (2300 – 0700)[89] and pulsatile iv administration (1700 – 0700) of cortisol, increased SWS[90] and SWA[91] and reduced REMS in young normal male subjects. GH increased after cortisol.[90,91] Similarly SWS and GH increased and REMS decreased after pulsatile iv boluses of cortisol in normal elderly males.[92] Since CRH[80] and cortisol exerted opposite influences on SWS[89,90] and GH[90,92] it appears unlikely that these effects are mediated by the changes of peripheral cortisol levels. It is more likely that negative feedback inhibition of endogenous CRH induces these changes. Because CRH,[80] ACTH[85] and cortisol[89,90] diminish REMS in contrast to ebiratide, REMS suppression may be mediated by cortisol after each of these hormones. Similarly the inhibition of cortisol synthesis by metyrapone reduced SWS and cortisol in normal male controls whereas REMS was not affected.[93] In this study endogenous CRH was probably enhanced since ACTH was distinctly elevated. In the rat subcutaneous administration of corticosterone increased wakefulness.[94]

Sleep-EEG changes after 9 days treatment in patients with multiple sclerosis with the glucocorticoid receptor (GR) agonist methylprednisolone differed from acute effects of cortisol. REMS latency was shortened, REMS density increased and a major portion of SWS shifted from the 1st to the 2nd sleep cycle. These changes are similar to the sleep EEG in patients with depression.[95]

4.5. Neuropeptide Y (NPY)

4.5.1. Preclinical Studies

Opposite effects of CRH and neuropeptide (NPY) were found in animal models of anxiety (review:[96]). Similarly in sleep regulation NPY, besides GHRH appears to be a physiological antagonist of CRH. (i) After icv administration of NPY to rats EEG spectral activity changed similarly to the effects of benzodiazepines.[97] (ii) The prolongation of sleep latency by CRH was antagonized dose-dependently by NPY in rats.[98]

4.5.2. Human Studies

In young normal men repetitive iv administration of NPY prompted decreases of sleep latency, the first REMS period, and increases of stage 2 sleep and sleep period time and blunted cortisol and ACTH secretion.[99] These data suggest that NPY participates in sleep regulation, particularly in the timing of sleep onset as an antagonist of CRH acting via the $GABA_A$ receptor.

4.6. Galanin

4.6.1. Preclinical Studies

Galanin is a peptide that is widely located in the mammalian brain and coexists in neurons with various peptides and classical neurotransmitters participating in sleep regulation. It is also known to stimulate GH via GHRH in man.[100] A cluster of GABAergic and galaninergic neurons was identified in the ventrolateral preoptic area, which is thought to stimulate NREMS.[101] Sleep in the rat remained unchanged after icv galanin, whereas REMS deprivation induced galanin gene expression.[102]

4.6.2. Human Study

The effects of two dosages of galanin (4×50 μg and 4×150 μg respectively) administered hourly between 2200 and 0100 h on sleep endocrine activity in healthy young men were tested. Galanin increased REMS during the 3rd sleep cycle with no difference between the doses. EEG spectral analysis revealed an increase in the delta and theta frequency range during the total night after the lower dose of galanin, but not after the higher dose. The secretion of GH, cortisol and prolactin remained unchanged after each of the doses. These data suggest that galanin is another sleep-promoting peptide. Since GH levels remained unchanged it is unlikely that GHRH is involved in the increase of SWA after galanin. An inhibitory effect of galanin at the locus coeruleus may contribute to the increases in REMS and in SWA.[103]

5. Effects of Peptides in Patients with Depression

5.1. GHRH

The influence of pulsatile iv administration of GHRH during the first few hours of the night was tested in drugfree patients of both sexes with depression (age range 19-76 years) and in matched controls. A sexual dimorphism in the response to GHRH was found. In male patients and controls GHRH inhibited ACTH during the first half of the night and cortisol during the second half of the night. In contrast these hormones were enhanced in females, regardless whether they were healthy or depressed. Similarly NREMS and particularly stage 2 sleep increased and wakefulness decreased in male patients and controls. Opposite sleep-impairing effects were found in women. These data confirm a reciprocal antagonism of GHRH and CRH in males, whereas a synergism of GHRH and CRH is suggested in females. The latter issue may contribute to the increased prevalence of mood disorders in women.[10,104]

5.2 Galanin

4 × 50 µg galanin or placebo were given hourly between 0900 and 1200 to patients with depression who were on a stable dose of trimipramine.[105] In the night following administration of galanin the REMS latency increased. The Hamilton Depression Rating Scale Score was determined 30 min before the first, and 30 min after the last injection. The improvement in this rating was significantly greater after galanin than after placebo. These data suggest an acute antidepressive effect of galanin by a mechanism related to that of therapeutic sleep deprivation.

5.3. NPY

In patients with depression of both sexes with a wide age range and age-matched controls the sleep latency was shortened after NPY whereas cortisol and ACTH levels and other sleep-EEG variables including the first REMS period remained unchanged.[106]

Obviously the influence of NPY on the timing of sleep onset is a robust effect since it was found in in the rat model;[41] in young normal controls;[99] in normal controls of both sexes with a higher age and in patients with depression as well. On the other hand HPA hormones were suppressed in the young men, but not in patients with depression and matched controls.

5.4. CRH Antagonism

5.4.1. Cortisol

Patients with depression received hourly injections of cortisol (1 mg per kg bodyweight) or placebo from 1900 to 0700. After cortisol GH, NREMS and particularly SWS increased.[107]

These effects resemble the findings after cortisol administration in young and elderly normal controls.[89-92] The changes of sleep-endocrine activity may be explained by feedback inhibition of CRH.

5.4.2. The CRH-1 Receptor Antagonist NBI 30775

In a first clinical trial the CRH-1 receptor antagonist NBI 30775 (R121919) showed antidepressive effects and was well tolerated.[108] A random subgroup of these patients underwent three sleep-EEG recordings (before treatment, at the end of the first week and at the end of the fourth week of active treatment). SWS increased compared with baseline after one week and after four weeks. The number of awakenings and REMS density showed a trend to decline during the same time period.

Separate evaluation of these changes with panels of two dose ranges showed no significant effect at lower doses, whereas in the higher dose range after the substance REMS density decreased and SWS increased between baseline and week 4. Furthermore positive associations between Hamilton Depression Rating Scale Scores and SWS at the end of active treatment were found. Taken together CRH-1 receptor antagonism appears to induce normalization of sleep-EEG changes. This study suggests that (i) CRH overdrive contributes to shallow sleep and REMS desinhibition in depression as well, and (ii) CRH-1 receptor antagonism is a way to counteract these changes.

6. Conclusions

A bidirectional interaction between sleep EEG and endocrine activity is well established. Various peptides exert specific effects on the sleep EEG. In Figure 2 a model of peptidergic sleep regulation in humans is displayed.

It is well documented that a reciprocal interaction of the neuropeptides GHRH and CRH plays a key role in sleep regulation. GHRH promotes sleep, at least in males, whereas CRH enhances vigilance and impairs sleep. Furthermore GHRH and CRH exert opposing effects on GH and the HPA hormones. Changes in the CRH:GHRH ratio in favor of CRH contribute to shallow sleep, elevated cortisol secretion and blunted GH during depression and aging. Due to synergism of these processes aberrances of sleep-endocrine activities are most distinct in elderly depressed patients. Interestingly recent data suggest CRH-like effects of GHRH in women. Several studies, particularly the administration of the CRH-1 receptor agonist NBI-30775 in patients with depression and some, but not all studies on the effects of CRH antagonists in the rat model suggest that CRH promotes REMS. NPY is another antagonist of CRH, particularly in the timing of sleep onset in humans and rats. Besides of CRH somatostatin appears to be another sleep-impairing peptide. Acute administration of cortisol was shown to enhance SWS and GH in young and elderly normal human controls and in patients with depression, probably due to feedback inhibition of CRH. Furthermore REMS was suppressed in healthy volunteers after cortisol. In contrast subchronic administration of the GR agonist methylprednisolone in patients with multiple sclerosis prompted some sleep-EEG changes resembling those in patients with depression. These data suggest that synergism of elevated CRH and of its peripheral consequence, elevated glucocorticoid levels contributes to REMS desinhibition during depression. Besides of GHRH galanin and ghrelin were shown to promote SWS. Studies in dwarf mice suggest that intact GHRH receptors are the prerequisite for the effect of ghrelin

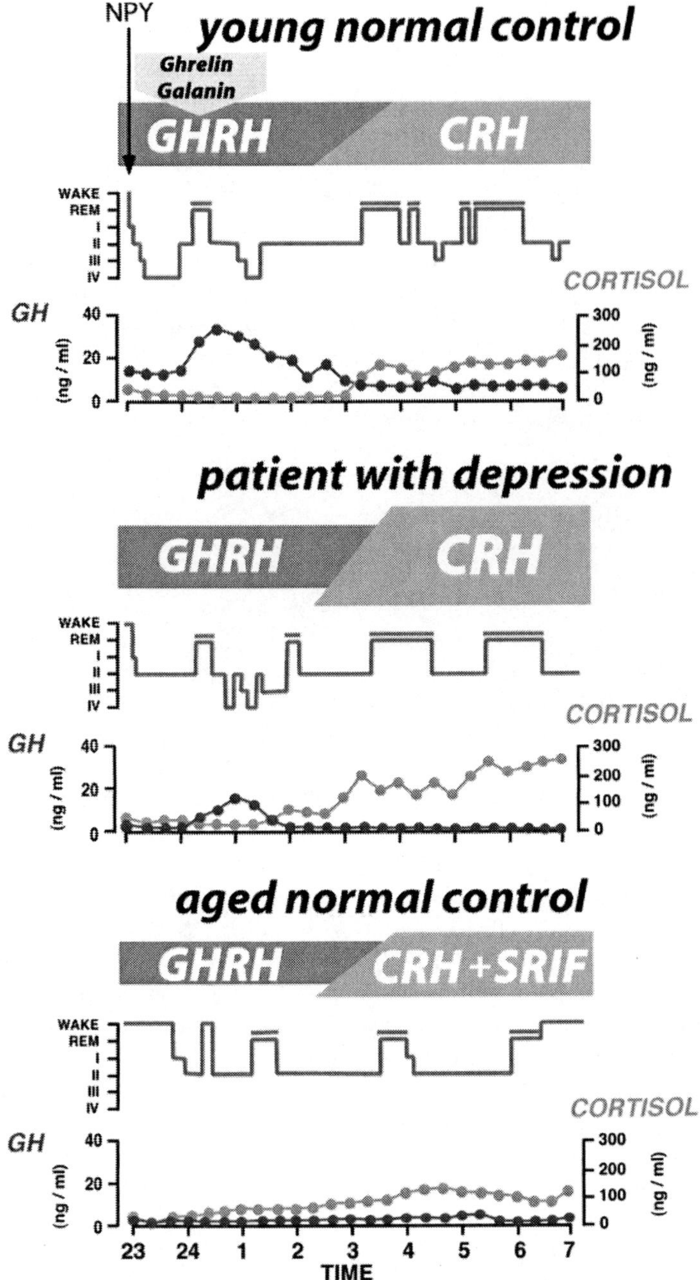

FIGURE 2. Model of peptidergic sleep regulation.

on sleep. Galanin is colocalized with GABA in the ventrolateral preoptic nucleus. Many hypothalamic GHRH responsive neurons are GABAergic. Galanin, ghrelin and GHRH may either act in a synergistic fashion or these peptides may be part of a cascade resulting in the promotion of NREMS. It appears likely that GABAergic neurons mediate the effects of these peptides. Treatment of patients with depression with the CRH receptor-1 antagonist NBI 30775 resulted in a normalization of the characteristical sleep-EEG changes. This is the first encouraging application of research on peptides in sleep regulation.

7. Acknowledgement. Studies from the author's laboratory were supported by grants from the Deutsche Forschungsgemeinschaft (Ste 486/1-2, 5-1 and 5-2 and 5-3).

8. *References*

1. E. Holsboer-Trachsler and E. Seifritz, Sleep in depression and sleep deprivation: a brief conceptual review, *World J. Biol. Psychiatry* 1, 180-186 (2000).
2. E. D. Weitzman, Circadian rhythms and episodic hormone secretion in man., Annu. Rev. Med. 27, 225-243 (1976).
3. A. Steiger, Sleep and endocrine regulation, Frontiers in Bioscience 8, 358-376 (2003).
4. C. L. Ehlers and D. J. Kupfer, Hypothalamic peptide modulation of EEG sleep in depression: a further application of the S-process hypothesis, Biol. Psychiatry 22, 513-517 (1987).
5. R. M. Benca, W. H. Obermeyer, R. A. Thisted and J. C. Gillin, Sleep and psychiatric disorders. A meta-analysis, *Arch. Gen. Psychiatry* 49, 651-668 (1992).
6. C. F. 3rd. Reynolds and D. J. Kupfer, Sleep research in affective illness: state of the art circa 1987., *Sleep* 10, 199-215 (1987).
7. F. Holsboer, The rationale for corticotropin-releasing hormone receptor (CRH-R) antagonists to treat depression and anxiety, *J. Psychiatr. Res.* 33, 181-214 (1999).
8. A. Steiger, U. von Bardeleben, T. Herth and F. Holsboer, Sleep EEG and nocturnal secretion of cortisol and growth hormone in male patients with endogenous depression before treatment and after recovery, *J. Affect. Disord.* 16, 189-195 (1989).
9. P. Linkowski, J. Mendlewicz, M. Kerkhofs, R. Leclercq, J. Golstein, M. Brasseur, G. Copinschi and E. Van Cauter, 24-hour profiles of adrenocorticotropin, cortisol, and growth hormone in major depressive illness: effect of antidepressant treatment, *J. Clin. Endocrinol. Metab.* 65, 141-152 (1987).
10. I. A. Antonijevic, H. Murck, R. M. Frieboes and A. Steiger, Sexually dimorphic effects of GHRH on sleep-endocrine activity in patients with depression and normal controls -part II: hormone secretion, *Sleep Research Online* 3, 15-21 (2000).
11. M. L. Wong, M. A. Kling, P. J. Munson, S. Listwak, J. Licinio, P. Prolo, B. Karp, I. E. McCutcheon, T. D. Jr. Geracioti, M. D. De Bellis, K. C. Rice, D. S.

Goldstein, J. D. Veldhuis, G. P. Chrousos, E. H. Oldfield, S. M. McCann and P. W. Gold, Pronounced and sustained central hypernoradrenergic function in major depression with melancholic features: relation to hypercortisolism and corticotropin-releasing hormone, *Proc. Natl. Acad. Sci. USA* 97, 325-330 (2000).

12. P. N. Sakkas, C. R. Soldatos, J. D. Bergiannaki and C. N. Stefanis, Growth hormone secretion during sleep in male depressed patients, *Prog. Neuropsychopharmacol. Biol. Psychiatry* 22, 467-483 (1998).

13. U. Voderholzer, G. Laakmann, R. Wittmann, C. Daffner-Bujia, A. Hinz, C. Haag and T. Baghai, Profiles of spontaneous 24-hour and stimulated growth hormone secretion in male patients with endogenous depression, *Psychiatry Res.* 47, 215-227 (1993).

14. D. B. Jarrett, J. M. Miewald and D. J. Kupfer, Recurrent depression is associated with a persistent reduction in sleep-related growth hormone secretion, *Arch. Gen. Psychiatry* 47, 113-118 (1990).

15. C. L. Ehlers and D. J. Kupfer, Slow-wave sleep: do young adult men and women age differently?, *J. Sleep Res.* 6, 211-215 (1997).

16. I. A. Antonijevic, H. Murck, R. M. Frieboes, M. Uhr and A. Steiger, On the role of menopause for sleep-endocrine alterations associated with major depression, *Psychoneuroendocrinology* 28, 401-418 (2003).

17. I. A. Antonijevic, H. Murck, R. M. Frieboes, R. Horn, G. Brabant and A. Steiger, Elevated nocturnal profiles of serum leptin in patients with depression, *J. Psychiatr. Res.* 32, 403-410 (1998).

18. J. Licinio, C. Mantzoros, A. B. Negrao, G. Cizza, M. L. Wong, P. B. Bongiorno, G. P. Chrousos, B. Karp, C. Allen, J. S. Flier and P. W. Gold, Human leptin levels are pulsatile and inversely related to pituitary-adrenal function, *Nat. Med.* 3, 575-579 (1997).

19. M. W. Schwartz, R. J. Seeley, L. A. Campfield, P. Burn and D. G. Baskin, Identification of targets of leptin action in rat hypothalamus, *J. Clin. Invest.* 98, 1101-1106 (1996).

20. J. Raber, S. Chen, L. Mucke and L. Feng, Corticotropin-releasing factor and adrenocorticotrophic hormone as potential central mediators of OB effects, *J. Biol. Chem.* 272, 15057-15060 (1997).

21. C. Peteranderl, I. A. Antonijevic, A. Steiger, H. Murck, R. M. Frieboes, M. Uhr and L. Schaaf, Nocturnal secretion of TSH and ACTH in male patients with depression and healthy controls, *J. Psychiatr. Res.* 36 (3), 189-196 (2002).

22. C. N. Chen, Sleep, depression and antidepressants., *Br. J. Psychiatry* 135, 385-402 (1979).

23. A. Steiger and F. Holsboer, Nocturnal secretion of prolactin and cortisol and the sleep EEG in patients with major endogenous depression during an acute episode and after full remission, *Psychiatry Res.* 72, 81-88 (1997).

24. D. B. Jarrett, J. M. Miewald, I. B. Fedorka, P. Coble, D. J. Kupfer and J. B. Greenhouse, Prolactin secretion during sleep: a comparison between depressed patients and healthy control subjects, *Biol. Psychiatry* 22, 1216-1226 (1987).

25. A. Steiger, U. von Bardeleben, K. Wiedemann and F. Holsboer, Sleep EEG and nocturnal secretion of testosterone and cortisol in patients with major endogenous depression during acute phase and after remission, *J. Psychiatr. Res.* 25, 169-177 (1991).

26. D. J. Kupfer, C. L. Ehlers, E. Frank, V. J. Grochocinski, A. B. McEachran and A. Buhari, Electroencephalographic sleep studies in depressed patients during long-term recovery, *Psychiatry Res.* 49, 121-138 (1993).

27. R. M. Frieboes, U. Müller, H. Murck, D. Y. von Cramon, F. Holsboer and A. Steiger, Nocturnal hormone secretion and the sleep EEG in patients several months after traumatic brain injury, *Journal of Neuropsychiatry and Clinical Neurosciences* 11, 354-360 (1999).

28. A. Rodenbeck, G. Huether, E. Rüther and G. Hajak, Interactions between evening and nocturnal cortisol secretion and sleep parameters in patients with severe chronic primary insomnia, *Neurosci. Lett.* 324, 159-163 (2002).

29. A. N. Vgontzas, E. O. Bixler, H. M. Lin, P. Prolo, G. Mastorakos, A. Vela-Bueno, A. Kales and G. P. Chrousos, Chronic insomnia is associated with nyctohemeral activation of the hypothalamic-pituitary-adrenal axis: Clinical implications, *J. Clin. Endocrinol. Metab.* 86, 3787-3794 (2001).

30. M. Vollrath, W. Wicki and J. Angst, The Zurich Study. VIII. Insomnia: Association with depression, anxiety, somatic syndromes, and course of insomnia, *Eur. Arch. Psychiatry Neurol. Sci.* 239, 113-124 (1989).

31. D. L. Bliwise, Sleep in normal aging and dementia., *Sleep* 16, 40-81 (1993).

32. C. L. Ehlers, W. M. Kaneko, M. J. Owens and C. B. Nemeroff, Effects of gender and social isolation on electroencephalogram and neuroendocrine parameters in rats, *Biol. Psychiatry* 33, 358-366 (1993).

33. E. Van Cauter, R. Leproult and L. Plat, Age-related changes in slow wave sleep and REM sleep and relationship with growth hormone and cortisol levels in healthy men, *JAMA* 284, 861-868 (2000).

34. A. Van Coevorden, J. Mockel, E. Laurent, M. Kerkhofs, M. L'Hermite-Balériaux, C. Decoster, P. Neve and E. Van Cauter, Neuroendocrine rhythms and sleep in aging men, *Am. J. Physiol.* 260, E651-E661 (1991).

35. S. Bredow, P. Taishi, F. Jr. Obál, N. Guha-Thakurta and J. M. Krueger, Hypothalamic growth hormone-releasing hormone mRNA varies across the day in rat., *NeuroReport* 7, 2501-2505 (1996).

36. J. Gardi, F. Jr. Obál, J. Fang, J. Zhang and J. M. Krueger, Diurnal variations and sleep deprivation-induced changes in rat hypothalamic GHRH and somatostatin contents, *Am. J. Physiol.* 277, R1339-R1344 (1999).

37. E. Lachmansingh and C. D. Rollo, Evidence for a trade-off between growth and behavioural activity in giant "Supermice" genetically engineered with extra growth hormone genes., *Can. J. Zool.* 72, 2158-2168 (1994).

38. I. Hajdu, F. Jr. Obál, J. Fang, J. M. Krueger and C. D. Rollo, Sleep of transgenic mice producing excess rat growth hormone, *Am. J. Physiol. Regulatory Integrative Comp. Physiol.* 282, R70-R76 (2002).

39. F. Jr. Obál, J. Fang, P. Taishi, B. Kacsóh, J. Gardi and J. M. Krueger, Deficiency of growth hormone-releasing hormone signaling is associated with sleep alterations in the dwarf rat, *The Journal of Neuroscience* 21, 2912-2918 (2001).

40. F. Jr. Obál, J. Alt, P. Taishi, J. Gardi and J. M. Krueger, Sleep in mice with non-functional growth hormone-releasing hormone receptors, *Am. J. Physiol. Regulatory Integrative Comp. Physiol.* 284, R131-R139 (2003).

41. C. L. Ehlers, T. K. Reed and S. J. Henriksen, Effects of corticotropin-releasing factor and growth hormone-releasing factor on sleep and activity in rats, *Neuroendocrinology* 42, 467-474 (1986).

42. F. Jr. Obál, P. Alföldi, A. B. Cady, L. Johannsen, G. Sary and J. M. Krueger, Growth hormone-releasing factor enhances sleep in rats and rabbits, *Am. J. Physiol.* 255, R310-R316 (1988).
43. J. Zhang, F. Jr. Obál, T. Zheng, J. Fang, P. Taishi and J. M. Krueger, Intrapreoptic microinjection of GHRH or its antagonist alters sleep in rats, *The Journal of Neuroscience* 19, 2187-2194 (1999).
44. F. Jr. Obál, R. Floyd, L. Kapás, B. Bodosi and J. M. Krueger, Effects of systemic GHRH on sleep in intact and in hypophysectomized rats., *Am. J. Physiol.* 270, E230-E237 (1996).
45. F. Jr. Obál, L. Payne, L. Kapás, M. Opp and J. M. Krueger, Inhibition of growth hormone-releasing factor suppresses both sleep and growth hormone secretion in the rat, *Brain Res.* 557, 149-153 (1991).
46. F. Jr. Obál, L. Payne, M. Opp, P. Alföldi, L. Kapás and J. M. Krueger, Growth hormone-releasing hormone antibodies suppress sleep and prevent enhancement of sleep after sleep deprivation, *Am. J. Physiol.* 263, R1078-R1085 (1992).
47. J. Toppila, L. Alanko, M. Asikainen, I. Tobler, D. Stenberg and T. Porkka-Heiskanen, Sleep deprivation increases somatostatin and growth hormone-releasing hormone messenger RNA in the rat hypothalamus, *J. Sleep Res.* 6, 171-178 (1997).
48. A. De, L. Churchill, F. Jr. Obál, S. M. Simasko and J. M. Krueger, GHRH and IL1 beta increase cytoplasmic Ca2+ levels in cultured hypothalamic GABAergic neurons, *Brain Res.* 949, 209-212 (2002).
49. A. Steiger, J. Guldner, U. Hemmeter, B. Rothe, K. Wiedemann and F. Holsboer, Effects of growth hormone-releasing hormone and somatostatin on sleep EEG and nocturnal hormone secretion in male controls, *Neuroendocrinology* 56, 566-573 (1992).
50. L. Marshall, L. Derad, C. J. Starsburger, H. L. Fehm and J. Born, A determinant factor in the efficacy of GHRH administration in the efficacy of GHRH administration in promoting sleep: high peak concentration versus recurrent increasing slopes, *Psychoneuroendocrinology* 24, 363-370 (1999).
51. M. Kerkhofs, E. Van Cauter, A. Van Onderbergen, A. Caufriez, M. O. Thorner and G. Copinschi, Sleep-promoting effects of growth hormone-releasing hormone in normal men, *Am. J. Physiol.* 264, E594-E598 (1993).
52. B. Perras, L. Marshall, G. Köhler, J. Born and H. L. Fehm, Sleep and endocrine changes after intranasal administration of growth hormone-releasing hormone in young and aged humans, *Psychoneuroendocrinology* 24, 743-757 (1999).
53. T. Schier, J. Guldner, M. Colla, F. Holsboer and A. Steiger, Changes in sleep-endocrine activity after growth hormone-releasing hormone depend on time of administration., *J. Neuroendocrinol.* 9, 201-205 (1997).
54. M. Iovino, P. Monteleone and L. Steardo, Repetitive growth hormone-releasing hormone administration restores the attenuated growth hormone (GH) response to GH-releasing hormone testing in normal aging, *J. Clin. Endocrinol. Metab.* 69, 910-913 (1989).
55. J. Guldner, T. Schier, E. Friess, M. Colla, F. Holsboer and A. Steiger, Reduced efficacy of growth hormone-releasing hormone in modulating sleep endocrine activity in the elderly, *Neurobiol. Aging* 18, 491-495 (1997).
56. B. Perras, H. Pannenborg, L. Marshall, R. Pietrowsky, J. Born and H. L. Fehm, Beneficial treatment of age-related sleep disturbances with prolonged intranasal vasopressin, *J. Clin. Psychopharmacol.* 19, 28-36 (1999).

57. A. Steiger, K. Held, S. Mathias, P. Schüssler, and J. Weikel, GHRH impairs sleep in youn normal women, *Archives of Pharmacology* 369 (Suppl. 1), R148 (2004).
58. J. Danguir, Intracerebroventricular infusion of somatostatin selectively increases paradoxical sleep in rats, *Brain Res.* 367, 26-30 (1986).
59. L. Beranek, F. Jr. Obál, P. Taishi, B. Bodosi, F. Laczi and J. M. Krueger, Changes in rat sleep after single and repeated injections of the long-acting somatostatin analog octreotide, *Am. J. Physiol.* 273, R1484-R1491 (1997).
60. L. Beranek, I. Hajdu, J. Gardi, P. Taishi, F. Jr. Obál and J. M. Krueger, Central administration of the somatostatin analog octreotide induces captopril-insensitive sleep responses, *Am. J. Physiol.* 277, R1297-R1304 (1999).
61. M. Ziegenbein, K. Held, H. Künzel, H. Murck, I. A. Antonijevic and A. Steiger, The somatostatin analogue octreotide impairs sleep and decreases EEG sigma power in young male subjects, *Neuropsychopharmacology* 29, 146-151 (2004).
62. R. M. Frieboes, H. Murck, T. Schier, F. Holsboer and A. Steiger, Somatostatin impairs sleep in elderly human subjects., *Neuropsychopharmacology* 16, 339-345 (1997).
63. D. C. Parker, L. G. Rossman, T. M. Siler, J. Rivier, S. S. Yen and R. Guillemin, Inhibition of the sleep-related peak in physiologic human growth hormone release by somatostatin, *J. Clin. Endocrinol. Metab.* 38, 496-499 (1974).
64. D. J. Kupfer, D. B. Jarrett and C. L. Ehlers, The effect of SRIF on the EEG sleep of normal men, *Psychoneuroendocrinology* 17, 37-43 (1992).
65. M. Kojima, H. Hosoda, Y. Date, M. Nakazato, H. Matsuo and K. Kangawa, Ghrelin is a growth hormone-releasing acylated peptide from stomach, *Nature* 402, 656-660 (1999).
66. T. L. Horvath, S. Diano, P. Sotonyi, M. Heiman and M. Tschöp, Minireview: ghrelin and the regulation of energy balance–a hypothalamic perspective, *Endocrinology* 142, 4163-4169 (2001).
67. C. Y. Bowers, GH releasing peptides-structure and kinetics, *J. Pediatr. Endocrinol.* 6, 21-31 (1993).
68. J. C. Weikel, A. Wichniak, M. Ising, H. Brunner, E. Friess, K. Held, S. Mathias, D. A. Schmid, M. Uhr and A. Steiger, Ghrelin promotes slow-wave sleep in humans, *Am. J. Physiol. -Endoc. M.* 284, E407-E415 (2003).
69. R. M. Frieboes, H. Murck, P. Maier, T. Schier, F. Holsboer and A. Steiger, Growth hormone-releasing peptide-6 stimulates sleep, growth hormone, ACTH and cortisol release in normal man, *Neuroendocrinology* 61, 584-589 (1995).
70. R. M. Frieboes, I. A. Antonijevic, K. Held, H. Murck, T. Pollmächer, M. Uhr, and A. Steiger, Hexarelin decreases slow-wave sleep and stimulates the sleep-related secretion of GH, ACTH, cortisol and prolactin in healthy volunteers, *Psychoneuroendocrinology*, in press.
71. G. Copinschi, R. Leproult, A. Van Onderbergen, A. Caufriez, K. Y. Cole, L. M. Schilling, C. M. Mendel, I. De Lepeleire, J. A. Bolognese and E. Van Cauter, Prolonged oral treatment with MK-677, a novel growth hormone secretagogue, improves sleep quality in man, *Neuroendocrinology* 66, 278-286 (1997).
72. M. Opp, F. Jr. Obál and J. M. Krueger, Corticotropin-releasing factor attenuates interleukin 1-induced sleep and fever in rabbits, *Am. J. Physiol.* 257, R528-R535 (1989).
73. F. Marrosu, G. L. Gessa, M. Giagheddu and W. Fratta, Corticotropin-releasing factor (CRF) increases paradoxical sleep (PS) rebound in PS-deprived rats, *Brain Res.* 515, 315-318 (1990).

74. M. R. Opp, Rat strain differences suggest a role for corticotropin-releasing hormone in modulating sleep, *Physiol. Behav.* 63, 67-74 (1997).

75. F. C. Chang and M. R. Opp, Blockade of corticotropin-releasing hormone receptors reduces spontaneous waking in the rat, *Am. J. Physiol.* 275, R793-R802 (1998).

76. M. M. C. Gonzalez and J. L. Valatx, Effect of intracerebroventricular administration of alpha-helical CRH (9-41) on the sleep/waking cycle in rats under normal conditions or after subjection to an acute stressful stimulus, *J. Sleep Res.* 6, 164-170 (1997).

77. F. C. Chang and M. R. Opp, Role of corticotropin-releasing hormone in stressor-induced alterations of lseep in rat, *Am. J. Physiol. Regulatory Integrative Comp. Physiol.* 283, R400-R4007 (2002).

78. M. M. C. Gonzalez and J. L. Valatx, Involvement of stress in the sleep rebound mechanism induced by sleep deprivation in the rat: use of alpha-helical CRH (9-41), *Behav. Pharmacol.* 9, 655-662 (1998).

79. M. Lancel, T. C. Wetter, A. Steiger and S. Mathias, Effect of the GABAA agonist gaboxadol on nocturnal sleep and hormone secretion in healthy elderly subjects, *Am. J. Physiol. -Endoc. M.* 281, E130-E137 (2001).

80. F. Holsboer, U. von Bardeleben and A. Steiger, Effects of intravenous corticotropin-releasing hormone upon sleep-related growth hormone surge and sleep EEG in man, *Neuroendocrinology* 48, 32-38 (1988).

81. M. Kellner, A. Yassouridis, B. Manz, A. Steiger, F. Holsboer and K. Wiedemann, Corticotropin-releasing hormone inhibits melatonin secretion in healthy volunteers -a potential link to low-melatonin syndrome in depression?, *Neuroendocrinology* 16, 339-345 (1997).

82. I. A. Antonijevic, H. Murck, R. M. Frieboes, F. Holsboer and A. Steiger, Hyporesponsiveness of the pituitary to CRH during slow wave sleep is not mimicked by systemic GHRH, *Neuroendocrinology* 69, 88-96 (1999).

83. A. N. Vgontzas, E. O. Bixler, A. M. Wittman, K. Zachman, H. M. Lin, A. Vela-Bueno, A. Kales and G. P. Chrousos, Middle-aged men show higher sensitivity of sleep to the arousing effects of corticotropin-releasing hormone than young men: clinical implications, *J. Clin. Endocrinol. Metab.* 86, 1489-1495 (2001).

84. H. L. Fehm, E. Späth-Schwalbe, R. Pietrowsky, W. Kern and J. Born, Entrainment of nocturnal pituitary-adrenocortical activity to sleep processes in man-a hypothesis., *Exp. Clin. Endocrinol.* 101, 267-276 (1993).

85. J. Born, E. Späth-Schwalbe, H. Schwakenhofer, W. Kern and H. L. Fehm, Influences of corticotropin-releasing hormone, adrenocorticotropin, and cortisol on sleep in normal man, *J. Clin. Endocrinol. Metab.* 68, 904-911 (1989).

86. A. Steiger, J. Guldner, H. Knisatschek, B. Rothe, C. Lauer and F. Holsboer, Effects of an ACTH/MSH(4-9) analog (HOE 427) on the sleep EEG and nocturnal hormonal secretion in humans, *Peptides* 12, 1007-1010 (1991).

87. J. C. Gillin, L. S. Jacobs, D. H. Fram and F. Snyder, Acute effect of a glucocorticoid on normal human sleep, *Nature* 237, 398-399 (1972).

88. E. Friess, K. Wiedemann, A. Steiger and F. Holsboer, The hypothalamic-pituitary-adrenocortical system and sleep in man., *Adv. Neuroimmunol.* 5, 111-125 (1995).

89. J. Born, E. R. De Kloet, H. Wenz, W. Kern and H. L. Fehm, Gluco- and antimineralocorticoid effects on human sleep: a role of central corticosteroid receptors, *Am. J. Physiol.* 260, E183-E188 (1991).

90. E. Friess, U. von Bardeleben, K. Wiedemann, C. Lauer and F. Holsboer, Effects of pulsatile cortisol infusion on sleep-EEG and nocturnal growth hormone release in healthy men, *J. Sleep Res.* 3, 73-79 (1994).

91. E. Friess, H. Tagaya, C. Grethe, L. Trachsel and F. Holsboer, Acute cortisol administration promotes sleep intensity in man, *Neuropsychopharmacology* 29, 598-604 (2004).

92. S. Bohlhalter, H. Murck, F. Holsboer and A. Steiger, Cortisol enhances non-REM sleep and growth hormone secretion in elderly subjects, *Neurobiol. Aging* 18, 423-429 (1997).

93. H. Jahn, F. Kiefer, M. Schick, A. Yassouridis, A. Steiger, M. Kellner and K. Wiedemann, Sleep-endocrine effects of the 11-b-Hydroxasteroiddehydrogenase inhibitor metyrapone, *Sleep* 26, 823-829 (2003).

94. G. Vázquez-Palacios, S. Retana-Márquez, H. Bonilla-Jaime and J. Velázquez-Moctezuma, Further definition of the effect of corticosterone on the sleep-wake pattern in the male rat, *Pharmacol. Biochem. Behav.* 70, 305-310 (2001).

95. I. A. Antonijevic and A. Steiger, Depression-like changes of the sleep-EEG during high dose corticosteroid treatment in patients with multiple sclerosis, *Psychoneuroendocrinology* 28, 401-418 (2003).

96. A. Steiger and F. Holsboer, Neuropeptides and human sleep, *Sleep* 20, 1038-1052 (1997).

97. C. L. Ehlers, C. Somes, A. Lopez, D. Kirby and J. E. Rivier, Electrophysiological actions of neuropeptide Y and its analogs: new measures for anxiolytic therapy?, *Neuropsychopharmacology* 17, 34-43 (1997).

98. C. L. Ehlers, C. Somes, E. Seifritz and J. E. Rivier, CRF/NPY interactions: a potential role in sleep dysregulation in depression and anxiety, *Depression & Anxiety* 6, 1-9 (1997).

99. I. A. Antonijevic, H. Murck, S. Bohlhalter, R. M. Frieboes, F. Holsboer and A. Steiger, NPY promotes sleep and inhibits ACTH and cortisol release in young men, *Neuropharmacology* 39, 1474-1481 (2000).

100. T. M. Davis, J. M. Burrin and S. R. Bloom, Growth hormone (GH) release in response to GH-releasing hormone in man is 3-fold enhanced by galanin, *J. Clin. Endocrinol. Metab.* 65, 1248-1252 (1987).

101. C. B. Saper, T. C. Chou and T. E. Scammell, The sleep switch: hypothalamic control of sleep and wakefulness, *Trends Neurosci.* 24, 726-731 (2001).

102. J. Toppila, D. Stenberg, L. Alanko, M. Asikainen, J. H. Urban, F. W. Turek and T. Porkka-Heiskanen, REM sleep deprivation induces galanin gene expression in the rat brain, *Neurosci. Lett.* 183, 171-174 (1995).

103. H. Murck, I. A. Antonijevic, R. M. Frieboes, P. Maier, T. Schier and A. Steiger, Galanin has REM-sleep deprivation-like effects on the sleep EEG in healthy young men, *J. Psychiatr. Res.* 33, 225-232 (1999).

104. I. A. Antonijevic, H. Murck, R. M. Frieboes, J. Barthelmes and A. Steiger, Sexually dimorphic effects of GHRH on sleep-endocrine activity in patients with depression and normal controls - part I: the sleep EEG, *Sleep Research Online* 3, 5-13 (2000).

105. H. Murck, K. Held, M. Ziegenbein, H. Künzel, F. Holsboer, and A. Steiger, Intravenous administration of the neuropeptide galanin has fast antidepressant efficacy and affects the sleep EEG, *Psychoneuroendocrinology*, in press.

106. K. Held, H. Murck, I. A. Antonijevic, H. Künzel, M. Ziegenbein, and A. Steiger, Neuropeptide Y (NPY) does not differentially affect sleep-endocrine regulation in depressed patients and controls, *Pharmacopsychiatry* 32, 184 (1999).
107. D. A. Schmid, H. Brunner, F. Holsboer, and E. Friess, Cortisol promotes nonREM sleep in patients with major depression, *International Journal of Neuropsychopharmacology* 3 (Suppl. 1), S302 (2000).
108. A. W. Zobel, H. Künzel, A. Sonntag and F. Holsboer, Effects of the high-affinity corticotropin-releasing hormone receptor 1 antagonist R121919 in major depression: the first 20 patients treated., *J. Psychiatr. Res.* 34, 171-181 (2000).

Neuroendocrine Correlates of Sleep in Premenstrual, Pregnant, Postpartum and Menopausal Depression

CHARLES J. MELISKA, EVA L. MAURER, L. FERNANDO MARTINEZ, DIANE L. SORENSON, GINA G. ZIRPOLI, ANA M. LOPEZ, NEAL BASAVARAJ, AND BARBARA L. PARRY

1. Abstract

1.1. Objective

To compare groups of normal control (NC) women with depressed patients (DP) diagnosed with Premenstrual Dysphoric Disorder (PMDD) or major depressive episode (MDE) during pregnancy, postpartum, peri- or post-menopause, with respect to neuroendocrine correlates of sleep architecture. We hypothesized that NC and DP would differ in the pattern of relationships between neuroendocrine and sleep measures across reproductively related depressions.

1.2. Methods

We measured sleep by polysomnography (PSG), and levels of plasma melatonin and serum prolactin, cortisol, and thyroid stimulating hormone (TSH) at 30-minute intervals, from 18:00-11:00 h in NC and DP. We sampled estradiol, progesterone, luteinizing hormone and follicle stimulating hormone at 18:00 and 06:00 h.

1.3. Results

Correlations between sleep and neuroendocrine measures varied widely, as a function of diagnosis. Cortisol was related to sleep variables in PMDD and in pregnant and postpartum DP. In menopausal DP, FSH was related to Stage 4 and Delta sleep. Women with PMDD had more Stage 2 and less Stage 3 sleep than pregnant and postpartum patients.

1.4. Conclusion

Levels of circulating cortisol, melatonin, prolactin, and the cortisol-to-melatonin ratio may be associated with differences in sleep architecture in a diagnosis-specific fashion.

2. Sleep Disturbances in Mood Disorders

Major depression is about twice as common in women as in men,[1-3] and women's depressive episodes often occur during times of reproductive hormonal change – at puberty, during oral contraceptive use, the premenstrual or ovulatory phase of the menstrual cycle, during pregnancy or the postpartum period, or during the transition to menopause (see reviews of Parry[4,5]). A growing body of objective polysomnographic (PSG) data also implicates sleep disturbances in depression (e.g., see review of Benca et al.[6]). Compared with insomniacs and normal controls (NC), the sleep of depressed patients (DP) is characterized by reduced total sleep time (TST), lower sleep efficiency, early awake time, shorter rapid eye movement (REM) latency, increased REM density, and increased number of awakenings.[7] Similarly, Taub[8] observed reduced TST, REM latency, and REM duration in DP as compared with NC.

Studies have examined the association of neuroendocrine measures of melatonin, cortisol, prolactin, thyroid stimulating hormone (TSH), and reproductive hormones – estrogen (E2), progesterone (P4), luteinizing hormone (LH), and follicle stimulating hormone (FSH) – on sleep during the menstrual cycle, in premenstrual dysphoric disorder (PMDD), pregnancy, the postpartum period, and menopause. For a review of sleep studies, see Parry et al.[9]

2.1. Neuroendocrine Studies in Premenstrual Dysphoric Disorder

As defined in the Diagnostic and Statistical Manual of Mental Disorders, Fourth Edition, (DSM-IV),[10] Premenstrual Dysphoric Disorder (PMDD), formerly Premenstrual Syndrome (PMS), may be linked to mood disorders. Neuroendocrine disturbances have been implicated in PMDD, as described below.

Melatonin: In healthy women, Nair et al.[11] found a phase delay of the nocturnal peak of melatonin secretion during the mid-menstrual period (midcycle), although melatonin was sampled only every 2 to 4 hours at two points in the cycle and without documentation of estradiol or progesterone levels. Brzezinski et al.[12] and Berga and Yen[13] found that in NC women, melatonin circadian rhythms are relatively stable and resistant to hormonal influences during the menstrual cycle. Shibui et al.[14] reported that serum melatonin secretion was significantly decreased in the luteal compared with the follicular menstrual cycle phase in 8 healthy women undergoing an ultrashort sleep-wake cycle schedule.

Parry et al.[15] observed that although melatonin rhythms in eight NC women did not change significantly during four phases of the menstrual cycle, the eight women with PMDD had an earlier (phase-advanced) offset of melatonin secretion, which contributed to a shorter secretion duration and a decreased area under the curve (AUC). In a larger follow-up study[16] of 21 PMDD and 11 NC subjects, in PMDD subjects, during the luteal compared with the follicular menstrual cycle phase, melatonin onset time was delayed, duration was compressed, and AUC, amplitude and mean levels were decreased. In NC subjects, melatonin rhythms did not change significantly with the menstrual cycle.

Cortisol: In 8 healthy women undergoing ultrashort sleep-wake cycles, Shibui et al.[14] observed that the amplitude of cortisol rhythms was significantly decreased in the luteal compared with the follicular menstrual cycle phase.

In a pilot study of eight women with prospectively documented PMS who underwent multiple samplings for cortisol during midcycle (late follicular) and the late luteal menstrual cycle phase, Parry et al.[17] reported increased serum cortisol levels during the midcycle phase. In a study of 20 women with late luteal phase dysphoric disorder (LLPDD)[18] and 11 NC subjects, in whom cortisol levels were measured every 30 minutes from 18:00-09:00 h during the midfollicular (MF) and late luteal (LL) menstrual cycle phase, Parry et al.[19] found that in NC, but not in LLPDD, subjects, the cortisol peak was significantly delayed in the LL compared with the MF menstrual cycle phase. In a separate study of 15 women with PMDD and 15 NC subjects, Parry et al.[20] again observed altered timing but not quantitative measures of cortisol secretion in PMDD: in the LL versus MF phase, the cortisol acrophase occurred one hour earlier, on average, in NC, but not in PMDD subjects.

TSH: In a study of 8 healthy women undergoing ultra-short sleep-wake cycle schedules, Shibui et al.[14] found that the amplitude of the TSH rhythm was significantly decreased in the luteal compared with the follicular menstrual cycle phase.

In a study of the circadian rhythms of TSH in 23 PMDD and 18 NC subjects in which samples for TSH were measured every 30 minutes from 18:00-09:00 h during MF and LL menstrual cycle phases, Parry et al.[21] observed that TSH rhythms occurred earlier in PMDD than in NC subjects.

Prolactin: In a study of 20 women with LLPDD and 11 NC subjects, Parry et al.[19] measured prolactin every 30 minutes from 18:00-09:00 h during MF and LL menstrual cycle phases. In LLPDD patients, prolactin peak and amplitude were higher, and acrophase earlier, than in NC subjects. In a separate study of 23 PMDD and 18 NC subjects, Parry et al.[21] measured prolactin every 30 minutes from 18:00-09:00 h during MF and LL menstrual cycle phases and found that PMDD patients had higher prolactin concentrations, consistent with previous findings.

2.2. Neuroendocrine Studies of Depression During Pregnancy

Melatonin, cortisol, thyroid, prolactin and reproductive hormones have been linked to mood changes occurring during pregnancy and postpartum (see review, Parry et. al.,[22,23] Wisner and Stowe[24]).

Melatonin: Parry et al.[25] studied 7 women with a MDE during pregnancy and 2 normal control subjects matched for age and month pregnant in which plasma melatonin was measured every 30 minutes from 18:00-11:00 h in dim (<50 lux)/dark conditions. Compared with NC subjects, DP tended to have lower mean melatonin peak and AUC, shorter duration, and earlier offset (but not onset) time. These differences, however, were not significant in this relatively small sample. Suzuki et al.[26] reported significant differences in the cortisol/melatonin ratio between poor sleepers (lower values) and good sleepers (higher values) during late pregnancy, as assessed by subjective sleep logs. Other investigators have observed increased melatonin levels or delayed offset in MDE.[9,27-31]

Cortisol: Mean cortisol between 18:00-24:00 h was significantly lower (p < .05) in DP compared with NC subjects.

TSH: Parry et al.[25] observed a non-significant trend toward lower TSH in DP compared with NC subjects.

Prolactin: Parry et al.[25] found no significant differences in mean prolactin levels between DP and NC subjects.

2.3. Neuroendocrine Studies of Postpartum Depression

Postpartum depression is classified as a MDE in DSM-IV (1994).

Melatonin: Parry et al.[23] measured plasma melatonin every 30 minutes from 18:00-11:00 h in dim (< 50 lux)/dark light in 11 women with a MDE postpartum and 4 NC women matched for age and postpartum month. DP had a trend towards higher mean melatonin peak and AUC compared with NC subjects. Timing measures (onset, offset, duration) were not significantly different between groups.

Although not systematically studied in postpartum women, melatonin levels were reported lower in a majority of studies comparing DP with NC subjects,[11,32-40] and we found and replicated lower melatonin circadian rhythms in patients with PMDD and in *combined* pregnant and postpartum DP compared with NC.[15,16,25,41,42]

Cortisol: Harris et al.[43] in 120 primiparous women measured saliva twice daily for cortisol and progesterone from 2 weeks before delivery to day 35 postpartum. Seven women developed major depression postpartum. Lower levels of evening cortisol, but not progesterone, in the immediate peripartum period were associated with postnatal depression. In a study of 11 postpartum DP and 4 NC subjects, in which serum samples for cortisol were obtained every 30 minutes from 18:00-11:00 h, Parry et al.[44] observed no significant differences in cortisol between DP and NC subjects.

TSH: In a study of 11 postpartum DP and 4 NC subjects, in which serum samples for TSH were obtained every 30 minutes from 18:00-11:00 h, Parry et al. (unpublished observations) found a non-significant trend toward lower TSH in DP compared with NC subjects.

Prolactin: Hyperprolactinemia has been associated with depressive symptoms.[45-47] Harris et al.[48] observed that in 147 postpartum mothers, 14.9% of whom were depressed, plasma prolactin levels were inappropriately low in depressed women who breast-fed. Abou-Saleh et al.[49] in a single morning sample found significantly lower plasma prolactin levels in women with postpartum depression. In a study of 11 postpartum DP and 4 NC subjects, in which serum samples for prolactin were obtained every 30 minutes from 18:00-11:00 h (Parry et al., unpublished observations), we found that in breastfeeding mothers, PRL was elevated in DP relative to NC [mean ($\pm$SD) = 104.9 ($\pm$19.2) vs. 25.6 ($\pm$24.8) ng/ml, p = .045].

2.4. Neuroendocrine Studies of Menopause

Melatonin: Blaicher et al.[28] evaluated overnight urinary 6-sulfatoxymelatonin (6-SMT) excretion in 60 postmenopausal women. Compared with controls, 6-SMT values were significantly higher in depressed females. Patients with hyperprolactinemia showed a trend towards an elevated average nocturnal melatonin concentration. Melatonin levels were significantly lower in patients with insomnia and obesity. In a study of the effects of HRT in menopausal DP versus NC, Parry et al.[9], reported that DP did not differ from NC subjects on melatonin variables of onset, offset, duration, or AUC at baseline or after E2 treatment.

Cortisol: Prinz et al.[50] in a study of 42 women (mean age 69.6 years), of whom 20 were on estrogen replacement therapy (ERT), found that elevated 24-hour urinary free cortisol was associated with impaired sleep and earlier awakening in older women not on ERT, but not in women on ERT. Antonijevic et al.[51] reported that nocturnal cortisol secretion, sampled every 30 minutes between 20:00-22:00 h and every 20 minutes between 22:00 and 07:00 h, was increased in 9 postmenopausal patients with depression, while a decrease was noted in 9 postmenopausal controls.

Parry et al.[9] studied menopausal DP versus NC after hormone replacement therapy (HRT): Mean levels of serum cortisol measured at baseline did not differ between NC and DP (n = 12, 7 respectively). In the subgroup that received estradiol treatment (oral 17-β estradiol, 1-2mg; 6 NC, 4 DP), however, total serum cortisol across time was lower in DP than in NC subjects.

E2 or other treatments (progesterone: medroxyprogesterone acetate, 2.5-5.0 mg in NC; antidepressant (AD): fluoxetine, 10-40 mg in DP) did not significantly alter mean total cortisol levels in either group. Cortisol amplitude and mesor also were lower in DP than in NC. Amplitude was lower in DP than in NC after E2 treatment. After E2 plus P4 treatment in NC subjects, cortisol acrophase occurred earlier compared with baseline.

TSH: Ballinger et al.[52] reported that clinically depressed late premenopausal women had significantly higher levels of TSH. In a study of the effects of HRT in menopausal DP versus NC by Parry et al.[9], mean TSH levels did not differ between NC and DP at baseline. After E2, but not other treatments, TSH values increased relative to baseline in NC, but not in DP. For TSH nadir, there was a significant treatment x diagnosis interaction. TSH nadir and mesor increased with E2 treatment in NC but not in DP.

Prolactin: Among non-depressed women, Schiff et al.[53] studied the effects of conjugated estrogens (0.625 mg) and placebo in 10 postmenopausal women during sleep and found that estrogen compared with placebo administration blunted LH and prolactin changes, but not the rise in cortisol. Fernandez et al.[54] found that morning levels of serum prolactin were lower during menopause compared with the pre- or peri-menopause. In a study of the effects of HRT in menopausal DP versus NC,[9] mean prolactin levels did not differ in NC and DP at baseline. E2 treatment produced a treatment x time x diagnosis interaction: DP, but not NC subjects, had elevated mean prolactin levels from 18:00 to 23:00 h and from 05:00 to 10:00 h, but not from 23:30 to 04:30 h. E2 plus AD treatment also raised mean prolactin levels in DP. Prolactin amplitude increased after E2 treatment in both NC and DP (p = .046). Prolactin mesor was higher in DP than in NC.

3. Methods

3.1. Subjects: General Characteristics

DP met DSM-IV criteria for a MDE, without psychotic features; NC subjects were without DSM-IV Axis I or II disorders in themselves or first-degree relatives. Prior to admission, all subjects had a history and physical examination and laboratory tests for a chemistry panel, thyroid indices, and complete blood count, urinalysis and urine toxicology screen. All were non-smokers, without significant cardiovascular, pulmonary or metabolic disease, were off psychoactive or other medication, including oral contraceptives (for four weeks), alcoholic beverages or other over-the-counter or recreational substances (for 72 hours prior to entering the study) that would interfere with neuroendocrine measures. Patients with bipolar illness, primary anxiety disorders (e.g., panic or obsessive-compulsive disorder) or substance abuse or dependence were excluded. In DP and NC subjects, there could be no history of alcohol abuse within the last year. Patients with primary sleep disorders, habitual sleep durations of < 6 or > 10 hours per night, or sleep onset times not entrained to between 21:30-00:30 h, sleep/wake cycles not synchronized with the 24 h environmental day/night cycle, and shift workers were excluded.

3.1.1. Characteristics of Subjects in Individual Studies

PMDD: Aged 18-45 years, 26 NC (mean age: 36.5 ± 5.6 years) plus 30 DP (mean age: 36.5 ± 4.6 years) who met DSM-IV criteria for PMDD were studied. Subjects were tested during the follicular and luteal phases of the menstrual cycle; we report only results of the luteal phase, when mood differences between NC and DP are the greatest.

Pregnant: Aged 19–45 years, 2 NC (mean age: 28.0 ± 12.7 years) plus 7 DP (mean age: 30.0 ± 8.1 years) who developed a non-psychotic MDE within their first 34 weeks of pregnancy participated.

Postpartum: Aged 18-45 years, 4 NC (mean age: 28.8 ± 5.1 years) plus 11 DP (mean age: 31.3 ± 6.6 years) who developed a DSM-IV non-psychotic MDE with onset within three months postpartum participated. Both breast-feeding and non-breast-feeding women were included.

Menopause: Aged 45-72 years, 10 DP (mean age: 55.7 ± 7.4 years) and 14 NC (mean age: 53.1 ± 3.6 years) participated. DP were peri- or post-menopausal women who met DSM-IV criteria for a MDE. NC were postmenopausal, without menses for at least one year. FSH was greater than 40 mIU/ml.[55] Participants were required to provide documentation of a normal PAP smear and mammogram within the last year. Both groups had no HRT treatment for 3 months prior to the study. DP needed to be off antidepressant medication at least 2 weeks (for fluoxetine, 1 month) prior to entering the study.

3.2. Screening Measures

Subjects were screened by an initial telephone interview and underwent a Structured Clinical Interview for DSM-IV (SCID)[56] to confirm a diagnosis. To document mood symptoms, NC (for one month; 2 months for PMDD) and DP (for at least 2 weeks; 2 months for asymptomatic/control patients in the menstrual study) completed daily ratings for symptoms of depression, anxiety, fatigue, irritability, withdrawal, appetite, physical discomfort and alertness on a 100 mm line visual analogue scale. Each week trained clinicians (inter-rater reliability kappa coefficient > .85), blinded to diagnosis and treatment condition, evaluated mood with a 24-item Hamilton Depression Rating Scale (HDRS)[57] which included an atypical depressive symptom inventory[58] and a mania rating scale.[59] Subjects also completed the Beck Depression Inventory (BDI).[60] Postpartum women completed the Edinburgh Post-Natal Depression Scale (EPDS)[61] weekly. For inclusion, DP needed to have mean HDRS ratings ≥ 14; BDI ratings ≥ 10, and ≥ 10 on the EPDS for a minimum of 2 weeks. NC subjects had no clinically significant mood changes during the month, with mean HDRS ratings ≤8, BDI ratings ≤ 5, and EPDS ratings ≤ 5.

3.3. Hospital Admissions

Following screening, evaluation and pre-admission baseline measures, we admitted subjects meeting criteria to the General Clinical Research Center (GCRC) of the University of California, San Diego (UCSD) Medical Center. The 1st night of admission was an adaptation to the sleep laboratory in which oximetry and periodic leg movements were recorded to rule out sleep disorders, after which subjects were allowed to go out on pass for the day. Following the adaptation night, subjects returned to the CRC at 16:30 h and were placed in relatively dim light (<30 lux). CRC nurses checked their hemoglobin levels and blood pressures. During Night 2, C3 or C4 electroencephalogram (EEG), bilateral extraorbital electro-oculogram (EOG), and sub-mental electromyogram (EMG) recordings were obtained, and blood samples (3cc) were collected every 30 minutes from 18:00-11:00 h. Serum samples for E2, P4, FSH and LH were obtained at 18:00 and 06:00 h. In a preliminary subset of subjects, we found less than 30 minutes of difference in wake time between nights of sleep with and without the intravenous catheter.

3.4. Analyses

Assay Methodology: E2, P4, FSH, LH, plasma melatonin, serum cortisol, TSH and prolactin were assayed by previously described methods.[16,19]

Plasma Melatonin: For pregnant, postpartum, and menopause studies, onset time, offset time, duration, peak, and area under the curve (AUC) were determined as follows: Onset time was defined at the first value exceeding the mean of 18:00-20:00 h values plus 2 standard deviations (SD) (threshold) followed by 2 consecutive points of equal or greater magnitude; Offset time was the first value below the threshold followed by 2 consecutive points of equal or lesser magnitude; Duration (h) was the time between onset and offset; Peak was the single highest concentration; AUC (pg/ml/h) was the integrated value for the concentration between onset and offset time. For PMDD studies, onset, offset, duration, peak and AUC were calculated by previously described methods.[15]

Sleep measures: All-night PSG recordings (EEG, EOG, sub-mental EMG) were digitized, stored on optical discs and scored visually in 30 second epochs without knowledge of conditions for sleep stages according to Rechtschaffen and Kales[62] criteria by sleep technicians (interrater reliability κ coefficient 0.85). The technicians were trained in the J. Christian Gillin Laboratory of Sleep and Chronobiology to produce standard sleep variables [total sleep time, sleep efficiency, sleep latency, stage 1-4 (minutes and percent), delta (minutes and percent), Rapid Eye Movement (REM) latency, density, and time (minutes and percent), wake time after sleep onset].

Statistical Methods: PSG, neuroendocrine, and reproductive hormone differences were evaluated using Group X Diagnosis Multivariate Analyses of Variance (MANOVA), univariate ANOVAs, and unadjusted t-tests for follow-up comparisons. Two-tailed (non-directional) hypotheses were assumed for all comparisons. Relationships between PSG indices and neuroendocrine measures were assessed with Pearson correlation coefficients, with the exception of some of the evaluations of the relationship of PRL with sleep, which were carried out using partial correlations, controlling for breastfeeding status and BMI. To reduce bias (Type 1 error) associated with multiple comparisons, only correlations whose two-tailed significance exceeded p = .01 were reported.

4. Results

4.1. *Relationship of Sleep Variables to Neuroendocrine Measures*

4.1.1. Menstrual Cycle (MC) Studies: PMDD (DP) and NC

Sleep with Melatonin: In NC, plasma melatonin from 04:00-11:00 h was positively correlated with sleep end time (r = .563, p = .008). Mean melatonin throughout the collection period (18:00-11:00 h) was positively correlated with Stage 3 minutes (r = .767, p = .000) and Stage 3% sleep (r = .748, p = .000). In DP, mean melatonin was not significantly correlated with sleep variables.

Sleep with Cortisol: In NC, mean serum cortisol from 18:00-11:00 h was correlated negatively with total sleep time (r = − .577, p = .008), sleep efficiency (r = − .658, p = .002), and Stage 2 minutes (r = − .570, p = .009), and positively with wakenings after sleep onset (r = .666, p = .001). In DP, cortisol from 18:00-01:00 h was negatively correlated with REM minutes (r = − .446, p = .017), and REM% (r = − .484, p = .009). Cortisol from 18:00-01:00 h was positively correlated with Stage 2% sleep (r = .546, p = .003).

Sleep with Cortisol/Melatonin Ratio. In NC, the log total cortisol/total melatonin ratio was negatively correlated with Stage 3 sleep minutes (r = − .708, p = .000) and Stage 3% sleep (r = − .684, p = .001). (In DP, the log total cortisol/total melatonin ratio was negatively correlated with Stage 3 sleep minutes (r = − .414, p = .029) when the conventional (p < .05) criterion was applied).

Sleep with Prolactin: In NC and DP mean serum prolactin was not significantly correlated with sleep variables.

Sleep with TSH: In NC and DP, TSH was not significantly correlated with sleep variables.

Sleep with Reproductive Hormones:

Estrogen: In NC, mean serum E2 concentration was positively correlated with Stage 2 minutes (r = .623, p = .003). In DP, mean serum E2 concentration was not significantly correlated with sleep variables.

Progesterone: In NC and DP, P4 was not significantly correlated with sleep variables.

LH: In NC and DP, LH was not significantly correlated with sleep variables.
FSH: In NC and DP, FSH was not significantly correlated with sleep variables.

4.1.2. Pregnancy

Sleep with Melatonin: In NC and DP, melatonin was not significantly correlated with sleep variables.

Sleep with Cortisol: In pregnant DP, cortisol from 01:30 – 08:30 h was positively correlated with REM minutes (r = .939, p = .005).

Sleep with Cortisol/Melatonin Ratio. In pregnant DP, the cortisol/melatonin ratio from 01:30–08:30 h was negatively correlated with Stage 2 sleep minutes (r = −.934, p = .006).

Sleep with Prolactin: In NC and DP, prolactin was not significantly correlated with sleep variables.

Sleep with TSH: In Pregnant DP, mean TSH from 04:00-11:00 h was positively correlated with REM density (r = .939, p = .006).

Sleep with Reproductive Hormones:

Estrogen: In pregnant NC and DP, E2 was not significantly correlated with sleep variables.

Progesterone: In pregnant DP, mean serum P4 concentration was not significantly correlated with sleep variables.

LH: In pregnant NC and DP, correlations between LH and sleep variables were not statistically significant.

FSH: In pregnant DP, FSH was positively correlated with Stage 2 minutes (r = .894, p = .007); in pregnant NC, correlations between FSH and sleep variables were not statistically significant.

4.1.3. Postpartum

Sleep with Melatonin: In postpartum DP, mean total melatonin (18:00-11:00 h) was positively correlated with Stage 3% sleep (r = .805, p = .005).

Sleep with Cortisol: In postpartum DP, mean serum cortisol from 01:30-08:30 h was negatively correlated with total sleep time (r = − .780, p = .008) and REM minutes (r = − .794, p = .006). In postpartum NC, correlations of cortisol with sleep variables were not statistically significant.

Sleep with Cortisol/Melatonin Ratio. In postpartum DP, the log total cortisol/total melatonin ratio was negatively correlated with Stage 3 sleep minutes (r = − .767, p = .010). In NC, the cortisol/melatonin ratio was not significantly correlated with sleep variables.

Sleep with Prolactin: In postpartum NC and DP, zero-order correlations between prolactin and sleep variables were not statistically significant. However, partial correlations – controlling for breastfeeding and body mass index (BMI) – indicated that in DP (but not NC), mean total PRL from 01:30–08:30 h was positively correlated with Delta minutes (r = .904, p = .005), and negatively correlated with wake after sleep onset (r = − .894, r = .007) and Stage 2% sleep (r = −.898, p = .006). PRL from 18:00-01:00 h was positively correlated with Stage 3 minutes (r = .924, p = .003) and Stage 3% sleep (r = .946, p = .001).

Sleep with TSH: In postpartum NC and DP, TSH was not significantly correlated with sleep variables

Sleep with Reproductive Hormones:

Estrogen: In postpartum NC and DP, E2 was not significantly correlated with sleep variables.

Progesterone: In postpartum NC and DP, P4 was not significantly correlated with sleep variables.

LH: In postpartum NC and DP, LH was not significantly correlated with sleep variables.

FSH: In postpartum NC and DP, FSH was not significantly correlated with sleep variables.

4.1.4. Menopause

Sleep with Melatonin: In NC, mean plasma melatonin from 04:00-11:00 h was positively correlated with sleep onset time (r = .819, p = .000). In DP, melatonin was not significantly correlated with sleep variables.

Sleep with Cortisol: In NC, mean serum cortisol from 09:00-11:00 h was negatively correlated with Stage 2% sleep (r = − .787, p = .002). In DP, cortisol was not significantly correlated with sleep variables.

Sleep with Cortisol/Melatonin Ratio. In menopausal DP, the log total cortisol/total melatonin ratio was negatively correlated with sleep latency (r = − .887, p = .010). In NC, the cortisol/melatonin ratio was not significantly correlated with sleep variables.

Sleep with Prolactin: In menopausal NC and DP, serum prolactin was not significantly correlated with sleep variables.

Sleep with TSH: In menopausal NC and DP, TSH was not significantly correlated with sleep variables.

Sleep with Reproductive Hormones:

Estrogen: In menopausal NC and DP, E2 was not significantly correlated with sleep variables.

Progesterone: In menopausal NC and DP, P4 was not significantly correlated with sleep variables.

LH: In menopausal NC, serum LH was positively correlated with REM latency (r = .752, p = .005). In menopausal DP, correlations between LH and sleep variables were not significant.

FSH: In menopausal NC, serum FSH was not significantly correlated with sleep variables. In menopausal DP, FSH was positively correlated with Stage 4 minutes (r = .907, p = .002), Delta minutes (r = .957, p = .000), and Delta % sleep (r = .901, p = .002).

4.2. PSG Variables: Across the Lifecycle in Diagnostic Groups

Between-group differences in sleep characteristics were evaluated separately for NC and DP. Salient group differences (p < .05) are highlighted below.

4.2.1. Normal Controls

In the menstrual cycle studies, NC had more Stage 2% sleep than postpartum and menopausal NC; less Stage 3% sleep than pregnant, postpartum and menopausal NC; and less Delta % sleep than postpartum NC.

Menopausal NC had more REM% sleep and more REM density than postpartum NC (see Table 1).

TABLE 1. Group Differences in Sleep Characteristics in Normal Control vs. Depressed Patients (p-values in parentheses). (Analyses adjusted for Age as a covariate.) Some non-significant contrasts are presented [in brackets] for comparison purposes.

	Normal Control (Mean ± SEM)	Depressed (Mean ± SEM)
Total Sleep Time		Pregnant > Postpartum (.022)
Sleep Efficiency		
		Postpartum > Pregnant (.002)
Sleep Latency		Postpartum > PMDD (.000)
		[Postpartum < Menopause (.118)]
Wake After Sleep Onset		
Stage 1%		Pregnant > PMDD (.020)
	[PMDD > Pregnant (.296)]	PMDD > Pregnant (.023)
Stage 2%	PMDD > Postpartum (.045)	[PMDD > Postpartum (.066)]
	PMDD > Menopause (.003)	Menopause > Pregnant (.020)
		Menopause > Postpartum (.040)
	PMDD < Pregnant (.028)	PMDD < Pregnant (.000)
Stage 3%	PMDD < Postpartum (.001)	PMDD < Postpartum (.000)
	PMDD < Menopause (.002)	Menopause < Pregnant (.014)
		Menopause < Postpartum (.036)
Stage 4%		Menopause < PMDD (.028)
		Menopause < Pregnant (.033)
Delta %	Postpartum > PMDD (.002)	Menopause < Postpartum (.018)
		[Menopause < PMDD (.072)]
REM %	Menopause > Postpartum (.044)	
REM Latency		
		PMDD < Pregnant (.004)
REM Density	Menopause > Postpartum (.043)	[PMDD < Postpartum (.068)]
		PMDD < Menopause (.003)
Sleep Onset Time		Postpartum > Pregnant (.013)
		Postpartum > PMDD (.011)
Sleep End Time		Pregnant > PMDD (.043)

4.2.2. Depressed Patients

Women with PMDD had more Stage 2% sleep than pregnant DP, and more Stage 4% sleep than menopausal DP. They had a shorter sleep latency than postpartum DP; less Stage 3% sleep than pregnant and postpartum DP; lower REM density than pregnant and menopausal DP; earlier sleep end time; and less Delta % sleep than postpartum DP.

Unlike NC, pregnant DP had increased total sleep time (Figure 1), reduced Stage 1 and Stage 3 minutes, reduced REM density, and earlier sleep end time when compared with the other groups (combined).

Pregnant DP also had (1) more Stage 1% sleep, Stage 3% sleep, Delta % sleep, REM density, and later sleep end time than PMDD women; and (2) more Stage 3% sleep and Delta % sleep than menopausal DP. Pregnant DP had less Stage 2% sleep than PMDD and menopausal DP.

Menopausal DP had greater REM density than postpartum DP.

4.3. PSG Variables: Differences between NC and DP across Reproductively-Related Groups

4.3.1. Menstrual Cycle

Differences between NC and DP were not significant.

4.3.2. Pregnancy

Mean plasma melatonin was lower in pregnant DP than in NC from 04:00–11:00 h (p = .003; see Figure 2). Mean serum cortisol was elevated in both NC and DP pregnant subjects, relative to the other groups (p = .000). However, mean serum cortisol was lower in pregnant DP than in NC from 18:00-01:00 h (p = .006). Pregnant DP (but not NC) had elevated mean LH levels relative to PMDD, postpartum, and menopausal DP (all p-values < .000).

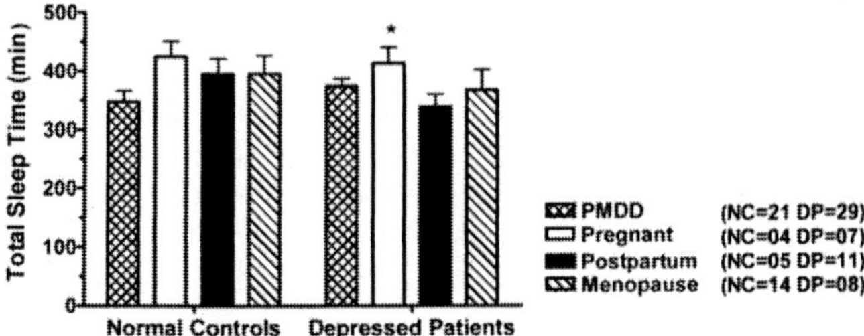

FIGURE 1. Mean total sleep time per group in NC vs. DP. Numerals in parentheses indicate N/group. Asterisk denotes significant differences between Pregnant DP and other DP groups, combined: * = p < .05.

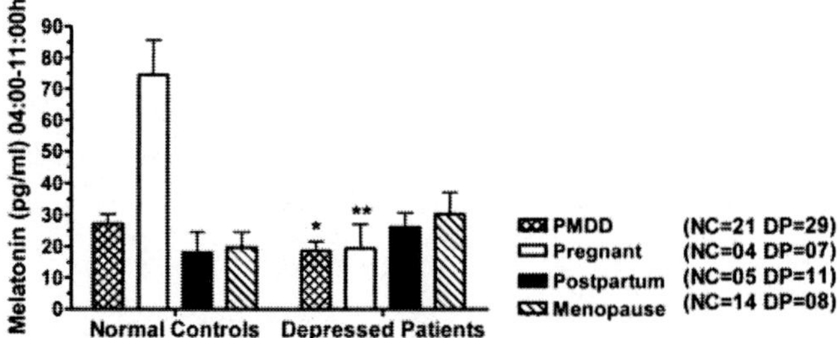

FIGURE 2. Mean morning plasma melatonin levels in NC vs. DP. Numerals in parentheses indicate N/group. Asterisks denote significant differences between DP and the corresponding NC group: * = p < .05; ** = p<.01.

4.3.3. Postpartum

DP had more REM% sleep than NC (22.1% vs. 16.4%, respectively; p = .029); no other DP vs. NC differences were statistically significant

4.3.4. Menopause

NC and DP did not differ significantly in any neuroendocrine or sleep characteristics.

5. Discussion

While few reliable correlations between neuroendocrine and sleep variables were noted and differences in patterns between NC and DP were small, reliable correlations between neuroendocrine measures and some sleep parameters were found.

5.1. Sleep and Melatonin

We found lower melatonin in DP relative to NC in PMDD and pregnant subjects, as reported previously for other MDEs.[11,32-40] In contrast to Blaicher,[28] however, we found no difference in melatonin levels in menopausal NC and DP.

Increased melatonin is associated with behavioral inhibition, increased total sleep time, and deeper stages of sleep, such as delta sleep. We found plasma melatonin to be correlated with sleep end time, stage 3 minutes and Stage 3% sleep in NC, in menstrual cycle studies, and with Stage 3% sleep in postpartum DP.

5.2. Sleep and Cortisol

Consistent with the present findings, Eriksson et al.[63] reported increased cortisol concentrations in pregnant women. Our results suggest that, more than other neuroendocrine measures, serum cortisol may be associated with changes in sleep architecture in PMDD, as suggested by significant correlations with Stage 2 sleep % and REM sleep %. In studies of NC during the menstrual cycle, cortisol was correlated with total sleep time, sleep efficiency, wake after sleep onset and Stage 2 sleep. Cortisol also was related to Stage 2 sleep in menopausal subjects, and to REM sleep in pregnant and postpartum subjects; however the direction of the observed correlations (both positive and negative) was highly dependent on diagnosis. Because of the relatively small n's involved in these studies, these results must be regarded as tentative.

Some of our results support the notion that, in contrast to melatonin, cortisol elevation is associated with increased alertness and arousal (lighter stages of sleep, reduced deeper stages of sleep, less restorative sleep, reduced total sleep time, lower sleep efficiency, increased wake after sleep onset, less delta, stage 3 and stage 4 sleep). Among menstrual cycle NC and postpartum DP, higher cortisol was associated with reduced total sleep time; in menstrual cycle NC, higher cortisol was associated with lighter, rather than deeper sleep (e.g., negative correlation between serum cortisol and sleep efficiency, positive correlation with wake after sleep onset). As suggested by Suzuki,[26] the cortisol/melatonin ratio also was related to some sleep variables. In support of this hypothesis, we found that log total cortisol/melatonin ratio was negatively correlated with Stage 3 sleep minutes in menstrual cycle NC and postpartum DP, and with sleep latency in menopausal DP. Also, in pregnant DP (but not NC), higher cortisol/melatonin ratio was associated with an increase in lighter (Stage 2) sleep.

5.3. Sleep and Prolactin

Hyperprolactinemia may be associated with depressive symptoms.[45-47] In the present studies, higher prolactin in postpartum DP was associated with elevated Stage 3 and delta sleep, fewer awakenings, and reduced "light" (Stage 2) sleep, when breastfeeding status and BMI were controlled for statistically. Thus, prolactin, which we found to be elevated in breastfeeding postpartum DP, relative to breastfeeding NC,[25] might promote deeper, more restorative sleep in postpartum DP.

5.4. Sleep and Reproductive Hormones

Touzet et al.[64] reported a positive correlation between serum FSH levels and sleep duration in normally cycling women aged 19-44 years. In the present study, we observed a strong correlation between FSH and measures of Stage 4 and Delta sleep in older, menopausal DP; few sleep measures were related to reproductive hormone measures in the other diagnostic groups.

Pregnant DP, who had longer sleep times, also had higher serum LH levels than PMDD, postpartum and menopausal DP. This finding contrasts with earlier reports[65,66] indicating that elevated LH secretion was associated with sleep disruption. The fact that sleep was systematically interrupted in the earlier studies (as a treatment for depression) may account for the discrepancy, since subjects were allowed to sleep through the night in the present study.

5.5. *Reproductive Condition and Restorative Sleep*

Pregnant subjects slept more and had deeper stages of sleep compared with other diagnostic groups (see Figure 1). We also observed alterations in the relative proportions of Stage 2 and Stage 3 sleep in some of the groups studied (see Table 1). For example, pregnant, postpartum, and menopausal NC had less Stage 2 and more Stage 3 sleep than menstrual cycle NC, reflecting lighter stages of sleep and more deep, restorative sleep, respectively. This trend was evident, but less pronounced among DP. Thus, a shift from lighter toward deeper, more restorative sleep may characterize pregnant and postpartum DP (which may reflect altered progesterone levels[67]), relative to PMDD. In contrast, that menopausal DP participants had more Stage 2%, and less Stage 3% sleep and Delta% sleep than pregnant and postpartum subjects suggests that reduced Stage 3 and Delta sleep may characterize menopausal depression, consistent with findings in other MDE and in aging individuals.[68] More enduring depressive symptoms over time may become more manifest on PSG measures.[69]

6. Summary

Based on these initial observations, further work specifying neuroendocrine correlates with sleep during different reproductive-related events in well-defined diagnostic groups is warranted. Elucidating the relationship of the cortisol/melatonin ratio to sleep measures especially would be valuable.

7. Acknowledgments. Portions of this research were supported by National Institutes of Mental Health Grants MH-59919 and MH-063462, and the National Institute of Health Research Center Grant # M01-RR-00827. We appreciate the work of Alan Turken and Richard Hauger, M.D. in performing the biochemical assays, and the advice and comments of Daniel F. Kripke, M.D.

8. *References*

1. M. M. Weissman, P. J. Leaf, C. E. Holzer, 3rd, J. K. Myers, and G. L. Tischler, The epidemiology of depression. An update on sex differences in rates, J Affect Disord 7(3-4), 179-188 (1984).
2. M. W. Weissman, Epidemiology of major depression in women. Women and the contoversies in hormonal replacement therapy (American Psychiatric Association, New York, 1996).

3. R. C. Kessler, K. A. McGonagle, M. Swartz, D. G. Blazer, and C. B. Nelson, Sex and depression in the National Comorbidity Survey. I: Lifetime prevalence, chronicity and recurrence, J Affect Disord 29(2-3), 85-96 (1993).

4. B. L. Parry, in: Psychopharmacology: The Fourth Generation of Progress, edited by F. E. Bloom and D. J. Kupfer (Raven Press, Ltd, New York, 1995), pp. 1029-1042.

5. B. L. Parry, in: Postpartum Mood Disorders, edited by L. J. Miller (American Psychiatric Press, Washington D.C., 1999), pp. 21-46.

6. R. M. Benca, W. H. Obermeyer, R. A. Thisted, and J. C. Gillin, Sleep and psychiatric disorders. A meta-analysis, Arch Gen Psychiatry 49(8), 651-668 (1992).

7. J. C. Gillin, W. Duncan, K. D. Pettig0rew, B. L. Frankel, and F. Snyder, Successful separation of depressed, normal, and insomniac subjects by EEG sleep data, Arch Gen Psychiatry 36(1), 85-90 (1979).

8. J. M. Taub, Individual variations in the sleep of depression, Int J Neurosci 23(4), 269-280 (1984).

9. B. L. Parry, C. J. Meliska, L. F. Martinez, N. Basavaraj, G. G. Zirpoli, D. Sorenson, E. L. Maurer, A. Lopez, K. Markova, A. Gamst, T. Wolfson, R. Hauger, and D. F. Kripke, Menopause: neuroendocrine changes and hormone replacement therapy, J Am Med Womens Assoc 59(2), 135-145 (2004).

10. American Psychiatric Association. Diagnostic and Statistical Manual of Mental Disorders (American Psychiatric Association, Washington DC, 1994).

11. N. P. Nair, N. Hariharasubramanian, and C. Pilapil, Circadian rhythm of plasma melatonin in endogenous depression, Prog Neuropsychopharmacol Biol Psychiatry 8(4-6), 715-718 (1984).

12. A. Brzezinski, H. J. Lynch, M. M. Seibel, M. H. Deng, T. M. Nader, and R. J. Wurtman, The circadian rhythm of plasma melatonin during the normal menstrual cycle and in amenorrheic women, J Clin Endocrinol Metab 66(5), 891-895 (1988).

13. S. L. Berga, and S. S. Yen, Circadian pattern of plasma melatonin concentrations during four phases of the human menstrual cycle, Neuroendocrinology 51(5), 606-612 (1990).

14. K. Shibui, M. Uchiyama, M. Okawa, Y. Kudo, K. Kim, X. Liu, Y. Kamei, T. Hayakawa, T. Akamatsu, K. Ohta, and K. Ishibashi, Diurnal fluctuation of sleep propensity and hormonal secretion across the menstrual cycle, Biol Psychiatry 48(11), 1062-1068 (2000).

15. B. L. Parry, S. L. Berga, D. F. Kripke, M. R. Klauber, G. A. Laughlin, S. S. Yen, and J. C. Gillin, Altered waveform of plasma nocturnal melatonin secretion in premenstrual depression, Arch Gen Psychiatry 47(12), 1139-1146 (1990).

16. B. L. Parry, S. L. Berga, N. Mostofi, M. R. Klauber, and A. Resnick, Plasma melatonin circadian rhythms during the menstrual cycle and after light therapy in premenstrual dysphoric disorder and normal control subjects, J Biol Rhythms 12(1), 47-64 (1997).

17. B. L. Parry, R. H. Gerner, J. N. Wilkins, A. E. Halaris, H. E. Carlson, J. M. Hershman, M. Linnoila, J. Merrill, P. W. Gold, R. Gracely, and et al., CSF and endocrine studies of premenstrual syndrome, Neuropsychopharmacology 5(2), 127-137 (1991).

18. American Psychiatric Association. DSM-III-R: Diagnostic and Statistical Manual of Mental Disorders. (APA, Washington, D. C., 1987).

19. B. L. Parry, R. Hauger, E. Lin, B. Le Veau, N. Mostofi, P. L. Clopton, and J. C. Gillin, Neuroendocrine effects of light therapy in late luteal phase dysphoric disorder, Biol Psychiatry 36(6), 356-364 (1994).
20. B. L. Parry, S. Javeed, G. A. Laughlin, R. Hauger, and P. Clopton, Cortisol circadian rhythms during the menstrual cycle and with sleep deprivation in premenstrual dysphoric disorder and normal control subjects, Biol Psychiatry 48(9), 920-931 (2000).
21. B. L. Parry, R. Hauger, B. LeVeau, N. Mostofi, H. Cover, P. Clopton, and J. C. Gillin, Circadian rhythms of prolactin and thyroid-stimulating hormone during the menstrual cycle and early versus late sleep deprivation in premenstrual dysphoric disorder, Psychiatry Res 62(2), 147-160 (1996).
22. B. L. Parry, and J. A. Hamilton, in: The Art of Psychopharmacology, edited by S. C. Risch and D. S. Janowsky (Guilford Press, New York, 1992), pp.
23. B. L. Parry, C. J. Meliska, N. Basavaraj, D. L. Sorenson, G. G. Zirpoli, E. L. Maurer, L. F. Martinez, A. M. Lopez, D. F. Kripke, J. C. Gillin, J. A. Elliot, R. Hauger, A. Gamst, and T. Wolfson, Biological rhythms in postpartum depression, 1st World Congress of Chronobiology (Sapporo, Japan, Sep, 2003).
24. K. L. Wisner, and Z. N. Stowe, Psychobiology of postpartum mood disorders, Semin Reprod Endocrinol 15(1), 77-89 (1997).
25. B. L. Parry, D. L. Sorenson, C. J. Meliska, N. Basavaraj, G. G. Zirpoli, A. Gamst, and R. Hauger, Hormonal basis of mood and postpartum disorders, Curr Womens Health Rep 3(3), 230-235 (2003).
26. S. Suzuki, L. Dennerstein, K. M. Greenwood, S. M. Armstrong, T. Sano, and E. Satohisa, Melatonin and hormonal changes in disturbed sleep during late pregnancy, J Pineal Res 15(4), 191-198 (1993).
27. L. K. Sekula, J. F. Lucke, E. K. Heist, R. K. Czambel, and R. T. Rubin, Neuroendocrine aspects of primary endogenous depression. XV: Mathematical modeling of nocturnal melatonin secretion in major depressives and normal controls, Psychiatry Res 69(2-3), 143-153 (1997).
28. W. Blaicher, E. Speck, M. H. Imhof, D. M. Gruber, C. Schneeberger, M. O. Sator, and J. C. Huber, Melatonin in postmenopausal females, Arch Gynecol Obstet 263(3), 116-118 (2000).
29. T. A. Wehr, W. C. Duncan, Jr., L. Sher, D. Aeschbach, P. J. Schwartz, E. H. Turner, T. T. Postolache, and N. E. Rosenthal, A circadian signal of change of season in patients with seasonal affective disorder, Arch Gen Psychiatry 58(12), 1108-1114 (2001).
30. A. Tuunainen, D. F. Kripke, J. A. Elliott, J. D. Assmus, K. M. Rex, M. R. Klauber, and R. D. Langer, Depression and endogenous melatonin in postmenopausal women, J Affect Disord 69(1-3), 149-158 (2002).
31. D. F. Kripke, S. D. Youngstedt, K. M. Rex, M. R. Klauber, and J. A. Elliott, Melatonin excretion with affect disorders over age 60, Psychiatry Res 118(1), 47-54 (2003).
32. J. Beck-Friis, D. von Rosen, B. F. Kjellman, J. G. Ljunggren, and L. Wetterberg, Melatonin in relation to body measures, sex, age, season and the use of drugs in patients with major affective disorders and healthy subjects, Psychoneuroendocrinology 9(3), 261-277 (1984).
33. J. Beck-Friis, B. F. Kjellman, B. Aperia, F. Unden, D. von Rosen, J. G. Ljunggren, and L. Wetterberg, Serum melatonin in relation to clinical variables in patients with major depressive disorder and a hypothesis of a low melatonin syndrome, Acta Psychiatr Scand 71(4), 319-330 (1985).

34. R. Brown, J. H. Kocsis, S. Caroff, J. Amsterdam, A. Winokur, P. E. Stokes, and A. Frazer, Differences in nocturnal melatonin secretion between melancholic depressed patients and control subjects, Am J Psychiatry 142(7), 811-816 (1985).

35. A. Cavallo, K. G. Holt, M. S. Hejazi, G. E. Richards, and W. J. Meyer, 3rd, Melatonin circadian rhythm in childhood depression, J Am Acad Child Adolesc Psychiatry 26(3), 395-399 (1987).

36. B. Claustrat, G. Chazot, J. Brun, D. Jordan, and G. Sassolas, A chronobiological study of melatonin and cortisol secretion in depressed subjects: plasma melatonin, a biochemical marker in major depression, Biol Psychiatry 19(8), 1215-1228 (1984).

37. S. H. Kennedy, P. E. Garfinkel, V. Parienti, D. Costa, and G. M. Brown, Changes in melatonin levels but not cortisol levels are associated with depression in patients with eating disorders, Arch Gen Psychiatry 46(1), 73-78 (1989).

38. A. J. Lewy, T. A. Wehr, P. W. Gold, and F. K. Goodwin, in: Catecholamines: Basic and Clinical Frontiers, edited by E. Usdin, I. J. Kopin and J. Barchas (Pergaman Press, New York, 1979), pp. 1173 -1175.

39. J. Mendlewicz, P. Linkowski, L. Branchey, U. Weinberg, E. D. Weitzman, and M. Branchey, Abnormal 24 hour pattern of melatonin secretion in depression, Lancet 2(8156-8157), 1362 (1979).

40. L. Wetterberg, B. Aperia, J. Beck-Friis, B. F. Kjellman, J. G. Ljunggren, A. Nilsonne, U. Petterson, A. Tham, and F. Unden, Melatonin and cortisol levels in psychiatric illness, Lancet 2(8289), 100 (1982).

41. B. L. Parry, B. LeVeau, N. Mostofi, H. C. Naham, R. Loving, P. Clopton, and J. C. Gillin, Temperature circadian rhythms during the menstrual cycle and sleep deprivation in premenstrual dysphoric disorder and normal comparison subjects, J Biol Rhythms 12(1), 34-46 (1997).

42. B. L. Parry, C. Udell, J. A. Elliott, S. L. Berga, M. R. Klauber, N. Mostofi, B. LeVeau, and J. C. Gillin, Blunted phase-shift responses to morning bright light in premenstrual dysphoric disorder, J Biol Rhythms 12(5), 443-456 (1997).

43. B. Harris, L. Lovett, J. Smith, G. Read, R. Walker, and R. Newcombe, Cardiff puerperal mood and hormone study. III. Postnatal depression at 5 to 6 weeks postpartum, and its hormonal correlates across the peripartum period, Br J Psychiatry 168(6), 739-744 (1996).

44. B. L. Parry, C. J. Meliska, N. Basavaraj, D. L. Sorenson, G. G. Zirpoli, E. L. Maurer, L. F. Martinez, A. M. Lopez, A. Gamst, T. Wolfson, J. A. Elliott, R. Hauger, J. C. Gillin, and D. F. Kripke, Neuroendocrine rhythms in portpartum depression: preliminary results, Chronobiol Int 20(6), 1173-1174. (Abstract) (2003).

45. M. C. Koppelman, B. L. Parry, J. A. Hamilton, S. W. Alagna, and D. L. Loriaux, Effect of bromocriptine on affect and libido in hyperprolactinemia, Am J Psychiatry 144(8), 1037-1041 (1987).

46. A. Reavley, A. D. Fisher, D. Owen, F. H. Creed, and J. R. Davis, Psychological distress in patients with hyperprolactinaemia, Clin Endocrinol (Oxf) 47(3), 343-348 (1997).

47. M. C. Oliveira, C. B. Pizarro, L. Golbert, and C. Micheletto, Hyperprolactinemia and psychological disturbance, Arq Neuropsiquiatr 58(3A), 671-676 (2000).

48. B. Harris, S. Johns, H. Fung, R. Thomas, R. Walker, G. Read, and D. Riad-Fahmy, The hormonal environment of post-natal depression, Br J Psychiatry 154(660-667 (1989)).

49. M. T. Abou-Saleh, R. Ghubash, L. Karim, M. Krymski, and I. Bhai, Hormonal aspects of postpartum depression, Psychoneuroendocrinology 23(5), 465-475 (1998).

50. P. Prinz, S. Bailey, K. Moe, C. Wilkinson, and J. Scanlan, Urinary free cortisol and sleep under baseline and stressed conditions in healthy senior women: effects of estrogen replacement therapy, J Sleep Res 10(1), 19-26 (2001).
51. I. A. Antonijevic, H. Murck, R. M. Frieboes, M. Uhr, and A. Steiger, On the role of menopause for sleep-endocrine alterations associated with major depression, Psychoneuroendocrinology 28(3), 401-418 (2003).
52. C. B. Ballinger, M. C. Browning, and A. H. Smith, Hormone profiles and psychological symptoms in peri-menopausal women, Maturitas 9(3), 235-251 (1987).
53. I. Schiff, Q. Regestein, J. Schinfeld, and K. J. Ryan, Interactions of oestrogens and hours of sleep on cortisol, FSH, LH, and prolactin in hypogonadal women, Maturitas 2(3), 179-183 (1980).
54. B. Fernandez, J. L. Malde, A. Montero, and D. Acuna, Relationship between adenohypophyseal and steroid hormones and variations in serum and urinary melatonin levels during the ovarian cycle, perimenopause and menopause in healthy women, J Steroid Biochem 35(2), 257-262 (1990).
55. L. W. Tam, V. Stucky, R. E. Hanson, and B. L. Parry, Prevalence of depression in menopause: a pilot study, Arch Women's Mental Health 2(4), 175-181 (1999).
56. M. B. First, R. L. Spitzer, M. Gibbon, and J. B. Williams, Structured Clinical Inerview for DSM-IV Axis I Disorders-Patient Edition (SCID-I/Pm, version 2.0) Biometerics Research Dept., New York State Psychiatric Institute. New York, 1995).
57. M. Hamilton, Development of a rating scale for primary depressive illness, Br J Soc Clin Psychol 6(4), 278-296 (1967).
58. J. B. Williams, M. J. Link, N. E. Rosenthal, L. Amira, and M. Terman, Structured Interview Guide for the Hamilton Depression Rating Scale, Seasonal Affective Disorders Version (SIGH-SAD), revised edition (New York Psychiatric Institute, New York, 1994).
59. N. E. Rosenthal, and M. M. Heffernan, in: Nutrition and the Brain, edited by R. J. Wurtman and J. J. Wurtman (Raven Press, New York, 1986), pp.
60. A. T. Beck, C. H. Ward, M. Mendelson, J. Mock, and J. Erbaugh, An inventory for measuring depression, Arch Gen Psychiatry 4(561-571 (1961)).
61. J. L. Cox, J. M. Holden, and R. Sagovsky, Detection of postnatal depression. Development of the 10-item Edinburgh Postnatal Depression Scale, Br J Psychiatry 150(782-786 (1987)).
62. A. Rechtschaffen, and A. A. Kales, A Manual of Standardized Terminology, Techniques and Scoring System for Sleep Stages of Human Subjects (Superintendent of Documents, US Government Printing Office, Washington, D. C., 1968).
63. L. Eriksson, S. Eden, J. Holst, G. Lindstedt, and B. von Schoultz, Diurnal variations in thyrotropin, prolactin and cortisol during human pregnancy, Gynecol Obstet Invest 27(2), 78-83 (1989).
64. S. Touzet, M. Rabilloud, H. Boehringer, E. Barranco, and R. Ecochard, Relationship between sleep and secretion of gonadotropin and ovarian hormones in women with normal cycles, Fertil Steril 77(4), 738-744 (2002).
65. A. Baumgartner, M. Dietzel, B. Saletu, R. Wolf, A. Campos-Barros, K. J. Graf, I. Kurten, and U. Mannsmann, Influence of partial sleep deprivation on the secretion of thyrotropin, thyroid hormones, growth hormone, prolactin, luteinizing hormone, follicle stimulating hormone, and estradiol in healthy young women, Psychiatry Res 48(2), 153-178 (1993).

66. B. Saletu, M. Dietzel, O. M. Lesch, M. Musalek, H. Walter, and J. Grunberger, Effect of biologically active light and partial sleep deprivation on sleep, awakening and circadian rhythms in normals, Eur Neurol 25(Suppl 2), 82-92 (1986).

67. S. S. Yen, R. B. Jaffe, and R. L. Barbieri, Reproductive Endocrinology: Physiology, Pathophysiology, and Clinical Management (Saunders, Philadelphia, 1999).

68. V. Crasset, S. Mezzetti, M. Antoine, P. Linkowski, J. P. Degaute, and P. van de Borne, Effects of aging and cardiac denervation on heart rate variability during sleep, Circulation 103(1), 84-88 (2001).

69. J. Puig-Antich, R. Goetz, C. Hanlon, M. Davies, J. Thompson, W. J. Chambers, M. A. Tabrizi, and E. D. Weitzman, Sleep architecture and REM sleep measures in prepubertal children with major depression: a controlled study, Arch Gen Psychiatry 39(8), 932-939 (1982).

Neuroendocrine and Behavioral Correlates of Sleep Deprivation: A Synthesis of Neurobiological and Psychological Mechanisms

MEHMET YUCEL AGARGUN AND LUTFULLAH BESIROGLU

1. Introduction

There is a burgeoning literature related to an alteration in mood and behavioral variables after total or partial sleep deprivation. For example, acute sleep manipulations usually are followed by a temporary improvement in mood in major depression; however, there is a regression to the previous status of depression after any subsequent sleep.[1] It is may also be asserted that morning mood improves when REM sleep is intact but worsens after a night of sleep deprivation in normal volunteers.[2] Given that sleep deprivation results in different treatment responses in normal healthy or depressed subjects, emotional or behavioral concomitants of sleep deprivation correlate with its neuroendocrine bases. In this chapter, firstly, we briefly review neuroendocrine aspects of sleep deprivation in healthy and depressed subjects. Secondly, we look at emotional and behavioral results in response to sleep deprivation in clinical samples. Another aim of this chapter is to try to figure out how these correlates concomitantly interact with each other. Finally, we discuss the therapeutic consequences of sleep deprivation in a within-sleep mood regulation process.

2. Neurohormones and Sleep Deprivation

We will briefly review the basic data in the literature on hormonal response to sleep deprivation including Growth Hormone, Ghrelin, Somatostatin, Hypothalamic-Pituitary-Adrenal Axis (HPA), Hypothalamic-Pituitary-Thyroid Axis (HPT), Prolactin, Neuroactive steroids, Vasoactive intestinal peptide, Insulin, and Sex Hormones in this chapter.

GHRH is one of the best documented sleep-promoting substances. Its administration enhances non-rapid-eye-movement (NREM) sleep whereas its inhibition reduces spontaneous sleep and sleep rebound. Growth hormone's major peak occurs shortly after sleep onset in association with the first slow wave sleep (SWS) episode is blunted during sleep deprivation (SD).[3] However, the blunting of the normal sleep-related GH pulse is compensated during the day. The amount of GH secreted during a 24 h period is similar whether or not a person has slept during the night.[4] The sleep deprivation-induced changes in GH secretion might depend on the age of the subject.[5]

Ghrelin, an endogenous ligand of the growth hormone secretagogue receptor, has been shown to promote slow-wave sleep.[6] During the first hours of sleep, ghrelin might promote sleep-associated GH secretion and contribute to the promotion of SWS.[7] During sleep, ghrelin levels increases during the early part of the night and decreased in the morning. This nocturnal increase is blunted during SD.

There is a reciprocal interaction of GHRF and somatostatin in sleep regulation. Inhibition of endogenous GHRF by somatostatin suppresses both NREM and GH secretion.[8] SD suppress somatostatin (SRIH) mRNA levels significantly in rats, and this suppression is significantly higher after dark onset than in the morning.[9]

HPA has been generally proposed as the integrator of the behavioral and autonomic responses to stress. Sleep, in particular deep sleep, has an inhibitory influence on the HPA. While none of the adrenal hormones rise during SD were found in early reports,[10,11] more recently, it has been reported that SD like a stressor can lead to a mild activation of the HPA axis and elevated plasma concentrations of glucocorticoids in both humans and rats.[12-14] However, RSD may not be sufficient to induce the activation of the HPA axis.[15]

Corticotropin-releasing factor (CRF) administration results in EEG activation associated with increased behavioral activity and decreased sleep time. CRF decreases SWS sleep time both in rats and humans and modifies the sleep pattern in sleep deprived rats.[16,17] R121919, a CRH-receptor-antagonist decreases REM and sleep time and increases SWS density significantly.[18] In the rats, after the SD period, CRF levels significantly increases in the striatum, limbic areas, pituitary and decreases in the hypothalamus.[19] In depressed subjects, baseline nonsupression of cortisol with dexametasone in depressed patients has been shown to normalize after one night of SD.[20,21] Increased levels of cortisol have been found after SD by Bouhuys et al[22] though most studies do not demonstrate between HPA response and SD.[23,24] These results do not support enough a role for the HPA axis in the antidepressant effects of SD. Furthermore, 83% of those who experience a beneficial response after a single night of SD relapse within 2 days.[25]

There is a close link between sleep, wakefulness and HPT axis. Sleep suppresses TSH releases while the normal circadian pattern of TSH levels reaches a peak in the early morning hours and a nadir during the day. SD stimulates TSH secretion in animals, healthy subjects and depressed patients.[26,27] This increased activity is perhaps secondary to the increased energy requirements of continuous wakefulness.

SD activates the tubuloinfundibular dopaminergic system and enhances dopamine release. Increasing dopamine activity is the final inhibitory pathway of prolactin regulation.[28]

Some neuroactive steroids are potent modulators of an array of ligand-gated ion channels and of distinct G-protein coupled receptors, and they can influence sleep and memory.[29] Whereas the concentrations of neuroactive steroids are altered in depression and normalize after antidepressant pharmacotherapy, partial SD does not affect the concentrations of neuroactive steroids either in responders.[30]

Vasoactive intestinal peptide (VIP) has been shown to increase rapid eye movement (REM) sleep in normal and insomniac animals. REM SD produces an increase in the density of VIP receptors in several brainstem and forebrain structures in rats.[31]

SD results in decreased insulin sensitivity at peripheral receptor sites which can eventually lead to insulin exhaustion at pancreatic sites.[32,33]

A decreased levels of androstanedione, dihydrotestosterone, estradiol were produced by sleep deprivation period in early stage (24-h restless period) after SD.[34,35] These alterations are not observed in the levels of FSH and LH. All alterations tend to return base line levels after 24-h recovery period.[36]

3. Emotional Consequences of Sleep Deprivation

It is well-known that therapeutic SD's effects on brain metabolic rates, especially in limbic areas, may correlate with a therapeutic response to a night of sleep loss and to antidepressant medication.[38] According to a meta-analysis of sleep deprivation conducted by Wu and Bunney,[25] 50-60% of depressed patients display a transient improvement of mood after TSD. However, almost about 80% of unmedicated TSD responders relapse into depressed mood after the next night of sleep. Thus, the clinical usefulness of SD is limited. Furthermore, there is specific question: Is there a variation in response to SD? Several factors may affect treatment response to SD. First of all, a shortened REM latency predicted a positive response to sleep deprivation. Several authors[39,40] reported that TSD responders, in contrast to non-responders, exhibited a prolongation of REM latency after TSD. In accordance with this finding, some evidence shows that those with an "endogenous" pattern to their depression have a better response rate to sleep deprivation than "neurotic" depressions.[41] Thus, subtypes of depressive disorders are also related with treatment response to SD. Wu and Bunney[25] reported a response rate of 67% vs. 48% for "endogenous" vs. "neurotic" depressive subtypes. Another clinical variable affecting the efficacy of SD is diurnaL variation. Diurnal mood variation has been positively correlated with response to sleep deprivation.[42]. Unipolar vs. bipolar diagnosis also appears to affect treatment response. Bipolar depressed patients respond to sleep deprivation significantly more often than unipolar depressed patients.[43] Interestingly, patients who show more "activation," who are less tired, show more behavioral activation, increased vigilance.[44] Finally there is a difference in response variation in terms of neuroimaging. A relative increase in cingulate activity has been seen before sleep deprivation in responders vs. nonresponders and in responders vs. control populations. Responders clearly demonstrate a decrease in this cingulate activity after response to sleep deprivation.[45]

Selective REM sleep deprivation has recently been found to be associated with a positive response in depressed mood. In a recent study Cartwright et al[46] examining the immediate effect on morning mood of a reduction of REM sleep by interruptions to collect reports of ongoing mental activity, and the long-term effect of this manipulation when it is repeated over several months on mood regulation and remission from major depression shown that sixty-four percent of the variance in remission could be accounted for by four variables: the initial level of self-reported symptoms, the reported diurnal variability in mood, the degree of overnight reduction in depressed mood following interruptions of REM sleep and the quality of dream reports from these awakenings.

Recently, we conducted a study examining whether a relationship exists between nightmares and terminal insomnia in unipolar depressed patients with melancholic features or diurnal mood symptoms. The subjects of the study were 82 depressed patients with melancholic features and 75 depressed patients without melancholic features. Using HDRS item 6, item 7, item 8, we classified the patients as having initial insomnia, as having middle insomnia, and terminal insomnia. We also assessed the patients in terms of the presence of frequent nightmares (at least twice a week during the depressive episode). The rates of terminal insomnia and nightmares were higher both in males and in females with melancholic features than the patients without melancholic features. It may be suggested that REM sleep deprivation closer to morning have a therapeutic effect on mood regulation and improve negative dream affect and content in depressed subjects with melancholic features or diurnal mood symptoms. In melancholia, an adaptive function may be suggested for spontaneous awaking early in the morning to prevent mood from the effects of negative dream affect. On the other hand, this reflects a within-sleep mood regulation process taking place. Nightmares reflect a negative dream affect and terminal insomnia plays a role to prevent morning mood although a failure in the completion of this process takes place during a depressive episode, in particular with melancholic features.

4. Neurobiological Basis of Mood Regulation in Response to Sleep Deprivation

A number of neurotransmitter systems and neuroendocrine systems have been implicated in the antidepressant effects of sleep deprivation.

4.1. Dopamine

Dopamine involved in the therapeutic effects of SD. Almost for three decades it has been known that patients who respond to sleep deprivation treatment have lower levels of cerebral spinal fluid (CSF) homovanillic acid, a dopamine metabolite, at baseline and higher after SD than non-responders.[47] More recently, Benedetti et al.[48] showed that antidepressant response is blocked by a dopamine reuptake inhibitor, amineptine, via down-regulation of postsynaptic dopamine binding. Interestingly, an amelioration of the tremor and rigidity in patients with Parkinson's disease was reported after SD.[49] Moreover, increased eye blinks, associated with dopaminergic activity, have been found following sleep deprivation in depressed patients, with increase in eye blink rate being related to degree of clinical improvement.[50]

4.2. Serotonin

There is relatively limited data on role of serotonin in the antidepressant effects of sleep deprivation. A recent study suggests a better response to sleep deprivation regarding a functional polymorphism in the 5-HT promoter gene.[51]

4.3. Norepinephrine

SD responders have been reported to be higher plasma norepinephrine metabolites.[52] It seems to be, on the other hand, increase in thyroid hormones, which may itself mediate antidepressant effects of sleep deprivation, have also been found in sleep deprivation responders, possibly attributable to increased norepinephrine function.

4.4. Acetylcholine

Acetylcholine (Ach) is associated with REM sleep.[53] Release of ACh in the cortex is highest during waking and REM sleep, and lowest during delta sleep. It appears that REM sleep initiation begins in the ACh neurons located in the peribrachial area of pons. This area of the brain initiates the effects of all of components of REM sleep including cortical desynchrony, rapid eye movement, and skeletal paralysis. This area controls cortical desynchrony directly via pathways that pass through areas of the thalamus, and, indirectly, through ACh neurons in the basal forebrain. Rapid eye movement is initiated and maintained via ACh pathways that go to the tectum at the back of the brain stem.

Although there is the limited data on monoamines' possible roles in response to sleep deprivation, it is possible to draw conclusions about serotonin interactions of the brain in terms of consequences of SD via discussing the reciprocal-interaction model of REM sleep regulation by McCarley and Hobson.[54]

McCarley and Hobson described a reciprocal interaction between the cholinergic REM-on and aminergic REM-off neurons in 1975. According to this model, the REM-off neurons in the LC were inhibitory to the REM-on neuronal population, while the REM-on neurons exerted an excitatory effect on the LC REM-off neurons. It was proposed that the LC-neurons were active throughout, except during REM sleep, and their continuous activation inhibited the activity of cholinergic REM-on neurons. The LC-neurons cease firing and this results in withdrawal of the tonic inhibition from the REM-on neurons in REM sleep. Cessation of noradrenergic neuronal activity allows an increased number of REM-on neurons to escape from inhibition resulting in generation of REM sleep. The activation of REM-on neurons exerts an excitatory effect on the LC-neurons resulting in inhibition of REM-on neurons and termination of REM sleep episodes. Thus, it is reasonable to suggest that REM sleep deprivation causes an important increase in serotonergic and noradrenergic activity of the brain. This seems to be explanatory for mood regulation in response to sleep deprivation in depressed subjects.

5. Clinical Implications and Future Directions

Sleep deprivation is a therapeutic manipulation of sleep. Although the onset of its effect is rapid, clinical research suggests a high rate of relapse in depression. Thus, its therapeutic effects can be maintained for weeks to months with adjunctive treatments. It seems to be combination of sleep deprivation with either a medication or some other biological manipulations causes a sustained effect. For example, a positive response to SD may be facilitated by heightened activity or sensitivity of dopaminergic and hypothalamic-pituitary-thyroid systems.[55] It is inevitable to suggest that not only biological mechanisms but also psychological approaches involved in therapeutic consequences of sleep deprivation. Future research needs to demonstrate and focus on the efficacy of selective REM sleep deprivation in mood regulation in healthy and depressive subjects. For example, masochistic character of REM sleep dreams in depressive subjects may be related to melancholic features. As a pathognomic feature of melancholia, being worse in affective state in the morning or diurnality of mood symptoms related to the intervening dream content and its affect. REM sleep deprivation closer to morning by dream collection method may have a therapeutic effect on mood regulation and improve negative dream affect and content in depressed subjects with melancholic features or diurnal mood symptoms.

6. *References*

1. A. Wirz-Justice, R. Van den Hoofdakker, Sleep deprivation in depression: what do we know, where do we go?, Bio.l Psychiatry 46, 445–453 (1999).
2. E. Colecchia, A. Berardi, U.Hirschfeld, S. Kosslyn, R. Leprout, T. Levin, A. Riffle, E, Van Cauter. Temporal variations in cognitive function measured by the Harvard

Cognitive Performance Battery (HCPB): sleepiness and mood across 40 h of continuous wakefulness, Sleep Res. 26, 611 (2004).

3. G. Brandenberger, The ultradian rhythm of sleep: diverse relations with pituitary and adrenal hormones, Rev. Neurol. 159, 6S5-10 (2003).

4. G. Brandenberger, C. Gronfier, F, Chapotot, C. Simon, F. Piquard, Effect of sleep deprivation on overall 24 h growth-hormone secretion, Lancet 356, 1408 (2000).

5. J. Mullington, D. Herman, F. Holsboer, T. Pollmacher, Age-dependent suppression of nocturnal growth hormone levels during sleep deprivation, Neuroendocrinology 64, 233–241 (1996).

6. M. Kojima, H. Hosoda, Y. Date, M. Nakazato, H. Matsuo, K. Kangawa, Ghrelin is a growth-hormone-releasing acylated peptide from stomach. Ghrelin is a growth-hormone-releasing acylated peptide from stomach, Nature 402, 656-660 (1999).

7. A, Dzaja, M. A. Dalal, H. Himmerich, M. Uhr, T. Pollmacher, A. Schuld, Sleep enhances nocturnal plasma ghrelin levels in healthy subjects, Am. J. Physiol. Endocrinol. Metab. 286(6), E963-7 (2004).

8. F. Jr. Obal, J. M. Krueger, The somatotropic axis and sleep, Rev. Neurol (Paris). 157(11 Pt 2), S12-5 (2001).

9. J. Zhang, Z. Chen, P. Taishi, F. Jr. Obal, J. Fang, J. M. Krueger, Sleep deprivation increases rat hypothalamic growth hormone-releasing hormone mRNA, Am J. Physiol. 275(6 Pt 2), R1755-61 (1998).

10. T. Akerstedt, J. Palmblad, B. de la Torre, R. Marana, M. Gillberg, Adrenocortical and gonadal steroids during sleep deprivation, Sleep 3, 23-30 (1980).

11. J. Palmblad, T. Akerstedt, J. Froberg, A. Melander, H. von Schenck H, Thyroid and adrenomedullary reactions during sleep deprivation, Acta Endocrinol. (Copenh). 90, 233-239 (1979).

12. R. Leproult, G. Copinschi, O. Buxton, E. Van Cauter, Sleep loss results in an elevation of cortisol levels the next evening, Sleep 20, 865–870 (1997).

13. M. Meerlo, M. Koehl, K. van der Borght, F. W. Turek, Sleep Restriction Alters the Hypothalamic-Pituitary-Adrenal Response to Stress, J. Neuroendocrinology 14, 397–402 (2002).

14. F. Chapotot, A. Buguet, C, Gronfier, G. Brandenberger, Hypothalamo-pituitary-adrenal axis activity is related to the level of central arousal: effect of sleep deprivation on the association of high-frequency waking electroencephalogram with cortisol release, Neuroendocrinology 73, 312-21 (2001).

15. H. Fujihara, R. Serino, Y. Ueta, H. Sei, Y. Morita, Six-hour selective REM sleep deprivation increases the expression of the galanin gene in the hypothalamus of rats, Brain. Res. Mol. Brain Res. 119, 152-159 (2003).

16. C. L. Ehlers, T. K. Reed, S. J. Henriksen, Effects of corticotropin-releasing factor and growth hormone-releasing factor on sleep and activity in rats, Neuroendocrinology 42, 467-474 (1986).

17. F. Holsboer, U. von Bardeleben, A. Steiger, Effects of intravenous corticotropin-releasing hormone upon sleep-related growth hormone surge and sleep EEG in man, Neuroendocrinology 48, 32-38 (1988).

18. K. Held, H. Kunzel, M. Ising, D. A. Schmid, A. Zobel, H. Murck, F. Holsboer, A. Steiger, Treatment with the CRH1-receptor-antagonist R121919 improves sleep-EEG in patients with depression, J. Psychiatr. Res. 38, 129-136 (2004).

19. P. Fadda, W. Fratta, Stress-induced sleep deprivation modifies corticotropin releasing factor (CRF) levels and CRF binding in rat brain and pituitary, Pharmacol. Res. 35, 443-446 (1997).

20. E. Halsboer-Trachsler, K. Wiedemann, F. Halsboer, Serial partial sleep deprivation in depression: clinical effects and dexamethasone supression test results, Neuropsychobiology 19, 73-78 (1988).

21. H. A. Nasrallah, S. Kuperman, W. Coryell, Reversal of dexamethasone nonsuppression with sleep deprivation in primary depression, Am. J. Psychiatry 137, 1463-1464 (1980).

22. A. L. Bouhuys, F. Flentge, R.H. Van den Hoofdakker, Effects of total sleep deprivation on urinary cortisol, self-rated arousal, and mood in depressed patients, Psychiatry Res. 34, 149-162 (1990).

23. D. Ebert, H. Feistel, W. Kaschka, A. Barocka, A. Pirner, Single photon emission computerized tomography assessment of cerebral dopamine D2 receptor blockade in depression before and after sleep deprivation–preliminary results, Biol. Psychiatry 35, 880-885 (1994).

24. P. Heiser, B. Dickhaus, W. Schreiber, H.W. Clement, C. Hasse, J. Hennig, H. Remschmidt, J. C. Krieg, W. Wesemann, C. Opper, White blood cells and cortisol after sleep deprivation and recovery sleep in humans, Eur. Arch. Psychiatry Clin. Neurosci. 250, 16-23 (2000).

25. J. C. Wu, W. E. Bunney, The biological basis of an antidepressant response to sleep deprivation and relapse: review and hypothesis, Am. J. Psychiatry 147, 14-21 (1990).

26. K. A. Gary, A. Winokur, S. D. Douglas, S. Kapoor, L. Zaugg, D.F. Dinges, Total sleep deprivation and the thyroid axis: effects of sleep and waking activity. Aviat. Space Environ. Med. 67, 513-519 (1996).

27. P.I. Parekh, T.A. Ketter, L. Altshuler, M.A. Frye, A. Callahan, L. Marangell, R. M. Post, Relationships between thyroid hormone and antidepressant responses to total sleep deprivation in mood disorder patients. Biol. Psychiatry 43, 392-394 (1998).

28. S. Kasper, D. A. Sack, T. A. Wehr, H. Kick, G. Voll, A. Vieira, Nocturnal TSH and prolactin secretion during sleep deprivation and prediction of antidepressant response in patients with major depression, Biol. Psychiatry, 24, 631-641 (1988).

29. M. Darnaudery, M. Pallares, J. J. Bouyer, M. Le Moal, W. Mayo, Infusion of neurosteroids into the rat nucleus basalis affects paradoxical sleep in accordance with their memory modulating properties. Neuroscience; 92, 583-588 (1999).

30. C. Schule, F. di Michele, T. Baghai, E. Romeo, G. Bernardi, P. Zwanzger, F. Padberg, A. Pasini, R. Rupprecht, Influence of sleep deprivation on neuroactive steroids in major depression. Neuropsychopharmacology 28, 577-81 (2003).

31. A. Jimenez-Anguiano, F. Garcia-Garcia, J. L. Mendoza-Ramirez, A. Duran-Vazquez, R. Drucker-Colin, Brain distribution of vasoactive intestinal peptide receptors following REM sleep deprivation, Brain Res. 728, 37-46 (1996).

32. M. Gonzalez-Ortiz, E. Martinez-Abundis, B. R. Balcazar-Munoz, S. Pascoe-Gonzalez, Effect of sleep deprivation on insulin sensitivity and cortisol concentration in healthy subjects, Diabetes Nutr. Metab. 13, 80-83 (2000).

33. T. VanHelder, J. D. Symons, M. W. Radomski, Effects of sleep deprivation and exercise on glucose tolerance. Aviat Space Environ. Med. 64, 487-92 (1993).

34. M. R. Gonzalez-Santos, O. V. Gaja-Rodriguez, R. Alonso-Uriarte, I. Sojo-Aranda, V. Cortes-Gallegos, Sleep deprivation and adaptive hormonal responses of healthy men, Arch. Androl. 22, 203-207 (1989).

35. K. Remes, K. Kuoppasalmi, H. Adlercreutz, Effect of physical exercise and sleep deprivation on plasma androgen levels: modifying effect of physical fitness. Int. J. Sports Med. 6, 131-135 (1985).

36. T. Akerstedt, J. Palmblad, B. de la Torre, R. Marana, M. Gillberg, Adrenocortical and gonadal steroids during sleep deprivation, Sleep 3, 23-30 (1980).

37. L. Lustberg, C. F. Reynolds, Depression and insomnia: questions of cause and effect. Sleep Med. Rev. 4, 253-262 (2000).

38. W. C. Duncan, J. C. Gillin, R. M. Post, R. H. Gerner, T. A. Wehr. Relationship between EEG sleep patterns and clinical improvement in depressed patients treated with sleep deprivation. Biol. Psychiatry 15, 879-889 (1980).

39. D. Riemann, M. Berger, The effects of total sleep deprivation and subsequent treatment with clomipramine on depressive symptoms and sleep electroencephalography in patients with a major depressive disorder. Acta Psychiatr. Scand. 81, 24-31 (1990).

41. R. H. Van den Hoofdakker, Total sleep deprivation: clinical and theoretical aspects. in: Depression: neurobiological, psychopathological and therapeutic advances, edited by A. V. Honig , H. M. Praag, (Chichester: John Wiley and Sons. 1997), pp. 564-589.

42. H. J. Haug, Prediction of sleep deprivation outcome by diurnal variation in mood. Biol. Psychiatry 31, 271-278 (1992).

43. B. Barbini, C. Colombo, F. Benedetti, E. Campori, L. Bellodi, E. Smeraldi, The unipolar-bipolar dichotomy and the response to sleep deprivation, Psych. Res. 79, 43-50 (1998).

44. C. Clark, R. Dupont, J. Golshan, J. C. Gillin, HMPAO SPECT, polysomnography, and antidepressant response to partial sleep deprivation, Sleep Res. 26, 247 (1997).

45. J. C. Gillin, M. Buchsbaum, J. Wu, C. Clark, W. Bunney, Sleep deprivation as a model experimental antidepressant treatment: findings from functional brain imaging, Depression Anxiety 14, 37-39 (2001).

46. R. Cartwright, E. Baehr, J. Kirkby, S. R. Pandi-Perumal, J. Kabat, REM sleep reduction, mood regulation and remission in untreated depression, Psychiatr Res. 121, 159-167 (2003).

47. R. H. Gerner, R. M. Post, J. C. Gillin, W. E. Bunney Jr, Biological and behavioral effects of one nights sleep deprivation in depressed patients and normals, J. Psychiatry Res. 28, 47-61 (1979).

48. F. Benedetti, B. Barbini, E. Campori, C. Colombo, E. Smeraldi, Dopamine agonist amineptine prevents the antidepressant effect of sleep deprivation, Psych. Res. 65, 179-1841996.

49. E. M. Demet, A. Chicz-Demet, J. H. Fallon, K. N. Sokolski, Sleep deprivation therapy in depressive illness and Parkinsons disease, Prog. Neuropsychopharmacol. Biol. Psychiatry. 23, 753-784 (1999).

50. D. Ebert, R. Albert, G. Hammon, B. Strasser, A. May, A. Merz, Eye-blink rates and depression. Is the antidepressant effect of sleep deprivation mediated by the dopamine system?, Neuropsychopharmacology 15, 332-339 (1996).

51. F. Benedetti, A. Serretti, C. Colombo, E. Campori, B. Barbini, D. di Bella, E. Smeraldi, Influence of functional polymorphism within the promoter of serotonin transporter gene on the effects of total sleep deprivation in bipolar pts, J. Psychiatry 156, 1450-1452 (1999).

52. D, Ebert, K. Ebmeier, The role of the cingulate gyrus in depression: from functional anatomy to neurochemistry, Biol. Psychiatry 39, 1044-1050 (1996).

53. M. Y. Agargun, H. Ozbek, Drug effects on dreaming. in: Sleep and Sleep Disorders: A Neuropsychopharmacological Approach, edited by M. Lader, D. P. Cardinali, S. R. Pandi-Perumal (Eurokah.com, 2003).
54. J. A. Hobson, R. W. McCarley, P. W. Wyzinski, Sleep cycle oscillation: Reciprocal discharge by two brainstem neuronal groups, Science 189, 55–58 (1975).
55. B. L. Ringel, M. P. Szuba, Potential mechanisms of the sleep therapies for depression. Depress and Anxiety 14, 29-36 (2001).

Endocrine Changes in Male Patients with Obstructive Sleep Apnea

RAFAEL LUBOSHITZKY

1. Introduction

1.1. Endocrine Changes During Sleep

Sleep is a major modulator of endocrine function, particularly of pituitary-dependent hormonal release (GH, prolactin and TSH). Some of the 24-h hormonal rhythms depend on the circadian clock (melatonin, ACTH and cortisol). Growth hormone secretion is stimulated during sleep and, in young men; about 70% of daily GH secretion occurs during the first third of the night. As a result of the consolidation of the sleep period in humans, the wake-sleep transition is associated with physiological changes with the endocrine system being part of the adaptive mechanism to reduce physical activity during sleep.[1,2] The timing and consolidation of nocturnal sleep change remarkably with aging. Sleep becomes brief and fragmented. Deep sleep is decreased associated with specific hormonal alterations.[3]

1.2. Endocrine Changes with Aging

Aging-associated alterations in body composition include increase in fat mass, loss of muscle size and strength, loss of bone and a change in blood levels of several hormones. Among these, leptin, DHEA, DHEAS and testosterone decline while gonadotropins and sex hormone-binding globulin (SHBG) gradually increase in aging males.[4-6] The decline in deep sleep from early adulthood to midlife is paralleled by a major decline in GH secretion. This suggests that age-related alterations in GH secretion may partially reflect decreased sleep quality.[7]

1.3. Endocrine Aspects of Obesity

Several endocrine abnormalities are reported in obesity. Some of these abnormalities are considered as causative factors for the development of obesity, whereas others are considered to be secondary effects of obesity and usually are restored after weight loss. GH is low and GH response to stimuli is blunted. IGF-I levels are normal or elevated. Cortisol, ACTH, and urine free cortisol levels are usually normal. Total testosterone and SHBG levels are low, but free testosterone levels are usually normal in obese men. LH and

FSH levels usually are normal and estrogens are elevated.[8] Patients who are deficient in either testosterone or growth hormone, show a reduction in visceral adiposity when their hormone levels are normalized. Stress has been shown to activate the HPA axis and may cause the hormonal changes associated with obesity.[9]

1.4. Hormones and Breathing

The current body of evidence suggests that hormones have an important role in the regulation of breathing. Hormones may stimulate breathing at the CNS, at peripheral chemoreceptors (progesterone), or indirectly by adjustment for the changes in body temperature, acid-base balance, fat and muscle mass. Leptin and CRH have been suggested to act as respiratory stimulants while somatostatin and neuropeptide Y have a depressing effect on breathing. Hormones may reduce upper airway collapsibility during sleep or may have bronchodilatory effects.[10]

The prevalence of sleep-disordered breathing is increased in several endocrine diseases such as acromegaly, hypothyoidism, diabetes mellitus, Cushing disease and polycystic ovary syndrome.[11] On the other hand sleep-disordered breathing such as obstructive sleep apnea (OSA) affects blood hormone levels, which may be recuperated with nasal continuous positive airway pressure (CPAP) treatment.[10]

Obstructive sleep apnea (OSA) is a common disorder affecting approximately 4-9% of middle-aged men. The disorder is characterized by repetitive collapse of the pharyngeal airway during sleep yielding hypoxia and hypercapnia. Apnea-induced hypoxia leads to arousal and termination of the obstructive events.

The repetitive nocturnal hypoxia and sleep fragmentation are associated with activation of a number of neural, humoral, thrombotic, metabolic and inflammatory disease mechanisms, all of which have also been implicated in the pathophysiology of cardiovascular disease.

Systemic hypertension, pulmonary hypertension or cor pulmonale is common in severely affected patients.[12,13] Obesity is a major risk factor for OSA, occurring in up to 50% of obese men.

About two thirds of middle-aged men with OSA suffer from obesity, mainly central type. Increased insulin levels in OSA patients and the association of insulin resistance and diabetes mellitus may suggest that sleep apnea is a manifestation of the metabolic syndrome.[14-16] We have investigated the nocturnal secretion of hormones in OSA patients and tested whether hormone levels are correlated with severity of OSA.

We determined serum LH, GH, leptin, cortisol and testosterone in middle-aged obese men with OSA and in control subjects with similar body weight with simultaneous sleep recordings. Also, we determined these variables after nine months of CPAP treatment.

2. Materials and Methods

We measured the concentrations of LH, GH, leptin, cortisol and testosterone in serum samples from 1900 to 0700 h with simultaneous sleep recordings in 10 men with obstructive sleep apnea (OSA) and in five control men. All participants underwent whole night sleep recording intended for the diagnosis of OSA and for habituation to the sleep laboratory. Diagnostic sleep recordings included electroencephalography, electromyography, electrooculography, respiration and oxygen saturation. Respiration was recorded using nasal and oral thermistor and thoracic strain gauge. The oxygen saturation was determined by a finger oximeter. The diagnosis of OSA was based on a combination of characteristic symptoms (loud and disturbing snoring, excessive daytime sleepiness, morning fatigue) and a finding of respiratory disturbance index (RDI) > 10 and a decrease in oxygen saturation of 4%. Hypoxia was evaluated by measuring oxygen saturation below 90% (PaO2 < 90%). During the experimental night, an IV catheter was inserted into an antecubital vein and kept patent with a slow infusion of 0.9% NACL. Blood samples were collected every 20 min from 1900–0700 h. From 2200 to 0700h, lights were off and subjects retired to sleep. Sleep quality and architecture was determined according to conventional criteria.[17] We determined the following parameters: total recording time, real sleep time [total recording time minus sleep latency and minus waking periods], sleep latency [time from lights off until 3 consecutive min. of sleep stage 2], sleep efficiency [total sleep time/total recording time] and REM latency [time to first REM]. In a second set of experiments, patients with OSA were fitted with a continuous positive airway pressure (CPAP) device (Respironics, Murrysville, PA). A second experimental night was performed after nine months of CPAP treatment for the determination of post treatment nocturnal serum hormone levels and sleep recordings.

LH and testosterone levels were determined in serum samples obtained every 20 min from 1900–0700 h. Leptin, GH and cortisol were determined every 60 min from 1900–0700 h. Serum leptin levels were determined by a two-site immunoradiometric assay (Diagnostic System Labs, Webster, Texas, USA) with intra-assay coefficients of variation (CV) of 3.7% at 2.9 ng/ml and 2.6% at 14.2 ng/ml; the interassay CVs were 6.6%, and 3.7%, respectively. Serum LH and testosterone levels were determined by immunoradiometric technique (Biodata Diagnostics, Rome, Italy). The LH intraassay CVs were 2.1% at 3.3 IU/L and 3.2% at 41 IU/L; the interassay CVs were 3.7% and 0.8%, respectively. The testosterone intraassay CVs were 6.0% at 4.0 nmol/L and 3.0% at 29.4 nmol/L; the interassay CVs were 1.9% and 1.6%, respectively. Serum GH levels were determined by immunometric assay (DPC, Los Angeles, CA). The GH intraassay CVs were 5.3% at 1.7 ng/ml and 6.2% at 15 ng/ml; the interassay CVs were 5.7% and 5.6%, respectively. Serum cortisol levels were determined by a competitive immunoassay (DPC, Los Angeles, CA) with intra-assay CVs of 9.0% at 110 nmol/L and

6.8% at 660 nmol/L; the interassay CVs were 10.3% and 9.9%, respectively. Mean and integrated [area under the curve (AUC)] serum leptin, LH, testosterone, GH and cortisol levels from 1900-0700 h were determined in all participants. Age, BMI and the sleep data of OSA patients and control men were compared by Wilcoxon 2 sample tests. The mean and AUC values of leptin, LH, testosterone, GH and cortisol were analyzed by Wilcoxon 2 sample tests to assess the overall difference between the two groups. Spearman correlations were used to evaluate the relationship between mean and AUC values of each hormone and BMI and the parameters of sleep apnea severity (RDI and Pa02 < 90%). Also, age, BMI and sleep data of the control group and the five patients with OSA at base line and after nCPAP treatment were compared by Wilcoxon 2 sample tests. Paired t tests were used to compare the OSA group at base line and during nCPAP treatment. Mean and AUC hormone values were analyzed by Wilcoxon 2 sample tests to assess the overall difference between the control and OSA and control and during nCPAP groups. Values were considered to be significantly different when p was less than or equal to 0.05.

3. Results

Patients with OSA had significantly higher RDI, PaO2 < 90% and BMI values compared with the control groups (Table 1). Sleep quality and architecture is shown in Table 2.

TABLE 1. Clinical characteristics of sleep apneic and control men studied (Values are given as Mean ± SD).

Parameter	OSA	Controls	P value
Age (years)	46.1 ± 7.3	42.0 ± 13.3	NS
BMI(Kg/m^2)	30.6 ± 4.9	26.3 ± 2.4	0.1
RDI (events/hour)	52.6 ± 17.4	7.0 ± 0.8	0.01
Apnea (no.)	252 ± 202	1.7 ± 1.7	0.006
Hypopnea (no.)	120 ± 120	38 ± 30	0.006
SaO$_2$ < 90% (% time)	6.2 ± 8.1	0.14 ± 0.25	0.05

TABLE 2. Characteristics of sleep quality in OSA and control subjects (Values are given as mean ± SD).

Parameter	OSA	Controls	P value
Sleep duration (h; min)	6:19 ± 1:06	5:37 ± 1:15	NS
Sleep efficiency (%)	75.3 ± 9.2	67.4 ± 11.9	NS
REM latency (h; min)	2:42 ± 1:01	2:19 ± 0:49	NS
Stage 1 (%)	2.4 ± 1.3	3.5 ± 2.6	NS
Stage 2 (%)	57.7 ± 9.8	43.7 ± 9.5	0.02
Stage 3/4 (%)	5.8 ± 7.2	11.0 ± 5.7	0.05
Stage REM (%)	12.7 ± 3.8	13.3 ± 6.4	NS
First REM (clock hour)	22:55 ± 0:56	22:55 ± 0:56	NS

Sleep efficiency in OSA patients and controls was about 70%, suggesting that the sampling procedure from the same room equally influenced the two groups. The OSA group spent a larger percent of their sleep in stage 2 (p < 0.02) and had a smaller percent of time in deep sleep stage 3/4(p < 0.05). All participants had REM sleep episodes. The nocturnal hormones secretory patterns at baseline are shown in figure 1A.

3.1. LH-Testosterone

The control group had significantly higher means and integrated (area under the curve, AUC) values of both LH and testosterone compared with OSA patients. In 4 of the 10 patients, the morning (0700 h) testosterone levels were below the lower adult male limit of 10.0 nmol/L.

After adjustment for BMI, there was a borderline statistically significantly difference in the 2 groups in mean LH and testosterone levels [LH-OSA, 2.2 ± 0.27 IU/L; control, 3.2 ± 0.4 ($P < 0.076$); testosterone-OSA, 9.4 ± 1.1 nmol/L; control, 13.7 ± 1.6 ($P < 0.05$)] and a significant difference in AUC [LH-OSA, 13.3 ± 6.3 IU/Lx h; control, 41.3 ± 10.0 ($P < 0.008$); testosterone-OSA, 67.2 ± 11.6 nmol/Lx h; control, 113.3 ± 26.8 ($P < 0.003$)].

There was a statistically significant negative correlation between LH-AUC and RDI ($r < 0.58$; $P < 0.03$) and between testosterone AUC and RDI ($r = 0.71$; $P < 0.0006$), but not with PaO2 < 90% (LH AUC: $r = 0.28$; $P = 0.34$; testosterone AUC: $r = 0.51$; $P = 0.07$).

3.2. Leptin

OSA patients had significantly higher mean and AUC leptin values (20.7 ± 11.2 ng/ml; 269.6 ± 145.6 ng/ml x h, respectively) compared with controls (10.4 ± 1.5 ng/ml; 135.2 ± 19.6 ng/ml xh, respectively).

Morning (07:00 h) leptin values were significantly higher in OSA compared with control men (21.6 ± 9.4 ng/ml vs. 9.2 ± 1.4 ng/ml, respectively; p<0.02). Mean and AUC leptin values positively correlated with BMI (p<0.006) but not with severity of OSA (RDI and PaO2 < 90%).

3.3. Growth Hormone

The nocturnal GH curve showed a significant major secretory burst at 24:00 in controls and a lesser peak at midnight in OSA patients (p < 0.0006). The control group had statistically significantly higher mean and AUC values compared with OSA patients (p < 0.004). A statistically significant difference between the two groups in GH levels was observed between 23:00-02:00h (p < 0.006). Mean GH levels and AUC were negatively correlated with RDI, Pa O2 < 90%, sleep efficiency and sleep stage 2.

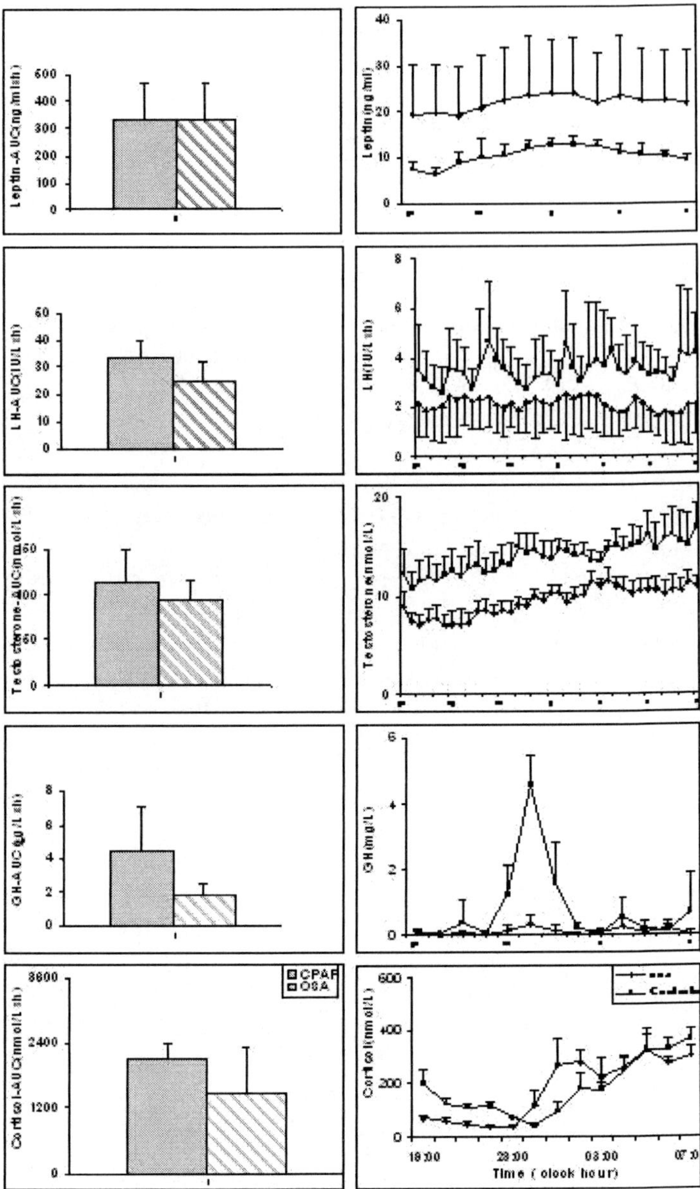

FIGURE 1. Effect of sleep apnea (A), on nocturnal LH, testosterone, GH, leptin and cortisol secretion. An arrow indicates the first REM sleep episode. The effect of nCPAP treatment on hormones secretion is shown in panel B.

3.4. Cortisol

Mean cortisol level, AUC, onset time or onset levels were similar in OSA and control groups. Analysis of the cortisol 13-hour time series data revealed that there was a statistically significant time effect (p < .0001) and a time by group interaction (p < 0.001). Analysis of each hour (Wilcoxon 2 sample test) revealed statistically significant difference in cortisol between 20:00 and 23:00 (p < .003) suggesting a delayed onset during sleep in OSA patients.

Cortisol onset time and onset level were negatively correlated with sleep efficiency, sleep stage 3 and positively correlated with percent awake. Cortisol values did not correlate with RDI or PaO2 < 90%.

3.5. Correlations Between Hormones

Mean leptin negatively correlated with LH(r = −0.58, p < 0.02) and testosterone (r = −0.47, p < 0.04). After controlling for BMI, mean leptin negatively correlated with mean testosterone (r = −0.52, p < 0.05) but not with mean LH. Mean and AUC GH values were positively correlated with mean and AUC testosterone values. A significant positive correlation was observed at lag time −4 h with GH leading testosterone by approximately 4 hours. GH was not correlated with LH. Cortisol onset level tended to be positively correlated with mean GH (r = .48, p < .07). Testosterone onset time tended to be negatively correlated with cortisol AUC (r = −.50, p < .06) and positively correlated with cortisol onset time (r = .70, p < .004).

3.6. CPAP Treatment

During nCPAP treatment parameters of OSA severity (RDI and PaO2 < 90%) returned to normal (table 3). Patients with OSA had similar BMI values compared with pretreatment values. They had significantly less sleep stage 2 and

TABLE 3. Sleep parameters at base line and during nCPAP treatment (Values are given as mean ± SD).

Parameter	OSA	Controls	CPAP
Age (years)	51.0 ± 5.5	42.8 ± 11.8	51.0 ± 6.5
BMI (Kg/m^2)[a]	31.2 ± 4.3	26.3 ± 2.4	31.4 ± 4.2
RDI (events/hour)	55.0 ± 22.5	6.8 ± 0.8	3.2 ± 1.9
SaO2 < 90% (% time)[b]	10.8 ± 0.6	0.16 ± 0.25	0.6 ± 0.6
Sleep time	6:26 ± 0: 58	5:37 ± 1: 15	6:29 ± 0: 45
Stage 1 (%)	76.5 ± 8.5	67.4 ± 11.9	76.1 ± 6.0
Stage 2 (%)	2.52 ± 1.43	3.46 ± 2.6	1.28 ± 0.96
Stage 3/4 (%)	57.0 ± 10.3	43.70 ± 9.5	35.76 ± 6.8
% REM[c]	15.88 ± 6.8	12.70 ± 3.8	21.51 ± 5.2

[a] p < 0.009 (OSA vs. CPAP); [b] p < 0.01(controls vs. OSA), p < 0.0001 (OSA vs. CPAP); [c] p < 0.01 (CPAP vs. controls).

more deep sleep stage 3/4 and REM sleep (table 3). The nocturnal hormones secretory patterns during CPAP treatment are shown in figure 1B.

Mean testosterone levels during nCPAP treatment were significantly higher than pre-treatment values ($p < 0.05$) with no change in LH concentrations. However, testosterone levels were still lower than levels observed in controls.

Mean leptin and AUC-leptin values in OSA patients both pre- and post-CPAP treatments were statistically significantly higher than in controls ($p < 0.02$). It is clear that CPAP treatment was not associated with any change in leptin levels

Mean and AUC leptin values during nCPAP treatment did not correlated with RDI or with $PaO2 < 90\%$.

Mean and AUC GH values were higher than pretreatment values ($p < 0.04$) and the peak nocturnal GH level was observed at 24:00h. When compared with controls, treated patients had lower mean and AUC GH values between 23:00 and 02:00h ($p < 0.02$). During CPAP treatment mean and AUC GH values were negatively correlated with $PaO2 < 90\%$. During CPAP treatment mean cortisol, AUC, onset time and onset level did not change compared with pre-treatment values. There was a significant difference in cortisol levels between basal and during CPAP treatment at 03:00-04:00h ($p < 0.04$). Cortisol onset during CPAP treatment was still significantly delayed when compared with control values ($p < 0.02$)

4. Discussion

In this study we evaluated the nocturnal secretion of LH, testosterone, GH, leptin and cortisol secretion in middle-aged obese men with obstructive sleep apnea. Hormone measurements were conducted at base line and following nine months of nCPAP treatment. Patients spent a larger percent of their sleep in stage 2, had a smaller percent of time in deep sleep but demonstrated well-defined REM sleep episodes.

OSA patients maintained a normally oriented diurnal rhythm of testosterone but had a significantly suppressed testosterone rise. Morning testosterone levels were in the hypogonadal range in 4 of the 10 patients (40%). The amounts of LH and testosterone secreted at night were significantly lower in OSA compared with controls independent of age and degree of obesity. Patients had also suppressed GH secretion at night. LH, testosterone and GH negatively correlated with RDI and $Pa\,O2 < 90\%$. Patients had increased leptin levels, which were positively correlated with BMI but not with severity of OSA. Although patients and controls had similar amounts of cortisol secreted during sleep, the nocturnal cortisol rise was delayed in OSA. Cortisol onset time and onset level were negatively correlated with deep sleep. After nine months of nCPAP treatment parameters of OSA severity were normalized, BMI, leptin and LH values remained unchanged whereas testosterone and

GH concentrations increased, yet to subnormal values. Cortisol onset during nCPAP treatment was partially corrected during nCPAP treatment.

Several factors may be responsible for the decreased LH-testosterone secretion observed in OSA including: hypoxia, sleep fragmentation, obesity and advanced age. Previous studies have demonstrated decreased fasting serum testosterone levels with LH levels not different from values in aged-matched controls.[18,19] Total and free testosterone levels were reduced in obese OSA men compared with age and body weight matched controls. A negative correlation was demonstrated between severity of sleep apnea and testosterone levels which persisted after adjusting for BMI and waist values.[20] These data have suggested that severity of sleep apnea is the major factor responsible for the decrease in testosterone secretion in OSA.[19] This hypothesis was examined in few studies by treating OSA patients with CPAP and revealed no conclusive findings. Oxygen treatment given to OSA patients for eight months was not associated with a change in testosterone or gonadotropin levels.[21], while CPAP therapy for three months was associated with a significant increase in testosterone but no change in free testosterone and LH levels. A concomitant decrease in body weight was also observed during CPAP therapy.[19] Similar findings were seen during uvolopalatopharyngoplasty therapy for OSA.[18] Others have found that decreased serum LH levels did not normalize during CPAP therapy.[21,22] These data and the present study suggest that correction of hypoxia and sleep fragmentation during CPAP therapy is not associated with complete recovery of the pituitary-gonadal axis function in OSA. Our finding that obesity is a major contributing factor to the reduced pituitary-gonadal function in OSA patients is further supported by previous observations that the increase in testosterone levels three months after uvolopalatopharyngoplasty therapy for OSA was associated with a significant weight reduction.[18]

Several studies have shown increased blood leptin levels in patients with OSA, which were correlated with severity of disease.[15,23] In all of these studies, hyperleptinemia was established based on one blood sample obtained in the morning or 3 hours after the last meal.[24] Patel et al.[25] measured evening and morning leptin levels in OSA and found that an increased ratio of evening to morning levels was independently associated with both an increased amount of total body fat and an increased RDI. The authors have suggested that sleep apnea suppress secretion of leptin in the morning or that the relative elevation in evening leptin may influence apnea pathogenesis. CPAP treatments for 3 days were associated with a significant reduction in morning hyperleptinemia. While others have demonstrated that leptin levels did not change from baseline levels after 2 days of CPAP treatment.[27] We did not find any change in serum leptin concentrations after nine months of nCPAP treatment. These findings are in contrast with Ip et al.[28] who demonstrated that treatment with CPAP for six months normalized fasting hyperleptinemia without a change in BMI or skin fold thickness. The discrepancy between this study and our observations may be explained by differences in timing and frequency of blood sampling, inclusion of male and female patients, severity of sleep apnea, degree of obesity and absolute leptin levels, as our patients had higher RDI, BMI and absolute leptin levels than patients studied by Ip et al.[28]

Since hyperleptinemia in OSA patients is primarily related to obesity, it is possible that at a certain degree of obesity in OSA patients, a much longer treatment period with CPAP is necessary to induce an effect on circulating leptin levels. On the other hand, it is also likely that the metabolic and hormonal changes observed in these patients already started at a very early stage after sleep apnea onset. By the time patients are diagnosed with the syndrome, usually at the age of 47-50 years when symptoms become apparent, the metabolic changes may be irreversible.[29]

OSA patients show a more marked impairment of the maximal secretory capacity of somatotroph cells ,together with reduced IGF-I sensitivity to rhGH stimulation.[30] In children with OSA, reduced morning IGF-I concentrations were observed with a significant increase in IGF-I and IGF-BP3 following adenotonsillectomy.[31] In adults, CPAP treatment resulted in a significant increase in nocturnal GH secretion.[19] Studied in obese OSA patients at baseline and after one night of CPAP treatment revealed that patients spent more time in light sleep at base line and more time in both SWS and REM sleep during CPAP treatment. GH levels were suppressed on the non-treatment night. With CPAP treatment, the normally observed peak in GH levels after 2-3 hours of sleep was restored.[32] In agreement with these observations, we have demonstrated that chronic nCPAP treatment partially restored GH secretion during sleep.

Cortisol secretion in OSA was investigated in only few studies. In one study nocturnal Cortisol secretion was normal in OSA and did not change after one night of CPAP treatment.[32] Others had demonstrated that OSA is associated with elevated serum Cortisol levels which were not altered after seven months of CPAP therapy.[22] CPAP withdrawal in OSA patients was associated with an immediate recurrence of sleep apnea with increase in apnea index, arousal index and oxygen desaturation but no change in levels of Cortisol or ACTH.[33] These data are in agreement with our findings of normal mean and AUC Cortisol values secreted at night in OSA at base line with no change after chronic CPAP therapy. We also found that OSA patients had a delayed onset during sleep. These observations in OSA patients are comparable to data observed in lean young men showing that the circadian Cortisol rhythm is little affected by changes in sleep schedule or sleep deprivation.[34] Also, human obesity is not characterized by dysregulation of the hypothalamic-pituitary-adrenal axis and Cortisol production rate was not associated with leptin levels in obese men.[35]

Our results are in agreement with the hypothesis that OSA may represent a leptin resistant state.[15,36] but point at obesity as the major determinant of hyperleptinemia in OSA patients. The observation that nCPAP treatment changed testosterone, GH and cortisol concentrations but not leptin levels may suggest that gonadotrophs, corticotrophs and somatotrophs are more sensitive to the insulting effects of hypoxia and sleep fragmentation than adipocytes. It may also imply that leptin resistance in OSA may be due to decreased cerebrospinal transport of leptin.

We conclude that obese middle-aged men with OSA have hyperleptinemia, decreased androgen and GH secretion, normal cortisol secretion and leptin resistance. These changes in hormones secretion are partially corrected during chronic CPAP treatment. Our findings support the hypothesis that obstructive sleep apnea is a manifestation of the metabolic syndrome, syndrome X, promoting atherosclerosis and cardiovascular disease in patients with OSA.

5. *References*

1. Van Cauter E, Leproult R, and L.Plat L. Age-related changes in slow wave sleep and REM sleep and relationship with growth hormone and cortisol levels in healthy men. JAMA 2000; 284(7): 861-868.
2. Tozawa T, Mishima K. Satoh M et al. Stability of sleep timing against the melatonin secretion rhythm with advancing age: Clinical implications. J Clin Endocrinol Metab 2003; 88:4689-4695.
3. Benca RM, Obermeyer WH, Thirsted RA, and Collin JC. Sleep and psychiatric disorders, Arch Psychiatry. 1992; 49(8): 651-668.
4. Lamberts SWJ, van den Beld AW, and van der Lely AJ. The endocrinology of aging. Science 1997; 278(10): 414-424.
5. Isidori AM, Strollo F, More M et al. Leptin and aging: Correlation with endocrine changes in male and female healthy adult populations of different body weights. J Clin Endocrinol Metab 2000; 85 (5): 1954-1962.
6. Feldman HA, Longcope C, Derby CA et al. Age trends in the level of serum testosterone and other hormones in middle-aged men: Longitudinal results from the Massachusetts Male Aging Study. J Clin Endocrinol Metab 2002; 87 (2): 589-598.
7. Herman-Bonert VS and Melmed S. In: The Pituitary, edited by Melmed S. (Blackwell Publishing, London, 2002, p.82.
8. Kokkoris P and Pi-Sunyer FX. Obesity and endocrine disease. Endocrinol Metab Clin North Am 2003; 32(4), 895-914.
9. Bjorntorp P. Endocrine abnormalities of obesity. Metabolism1995; 44(Suppl 3): 21-3.
10. Saaresranta T, and Polo O. Hormones and breathing. Chest 2002; 122(6): 2165-2182.
11. Grunstein R, Kian KY, and Sullivan CE. Sleep apnea in acromegaly. Ann Intern Med 1991; 115 (7): 527-532.
12. Lavie P, Herer P and Hoffstein V. Obstructive sleep apnea syndrome as a risk factor for hypertension: Population study. Br Med J 2000; 320(6):479-482.
13. Shamsuzzaman ASM, Gersh BJ, and Somers VK. Obstructive sleep apnea: Implications for cardiac and vascular disease. JAMA 2003; 290(10): 1906-1914.
14. Lavie P, Herer P, Peled R et al. Mortality in sleep apnea patients: a multivariate analysis of risk factors. Sleep 1995; 18 (2):149–157.

15. Vgontzas AN, Papanicolaou DA, Bixler EO et al. Sleep apnea and daytime sleepiness and fatigue: Relation to visceral obesity, insulin resistance, and hypercytokinemia. J Clin Endocrinol Metab 2000; 85 (7): 1151-1158.

16. Vgontzas AN, Bixler EO and Chrousos GP. Metabolic disturbances in obesity versus sleep apnea: the importance of visceral obesity and insulin resistance. J Intern Med 2003; 254 (1); 32-44.

17. Rechtschaffen A and Kales B. A manual standardized terminology, techniques and scoring system for sleep stages of human subjects, Washington DC: US Government Printing Office, 1968.

18. Santamaria JD, Prior JC and Fleetham JA. Reversible reproductive dysfunction in men with obstructive sleep apnea, Clin Endocrinol 1988; 28:461-470.

19. Grunstein RR, Handelsman J, Lawrence J et al. Neuroendocrine dysfunction in sleep apnea: reversal by continuous positive airway pressure therapy. J Clin Endocrinol Metab 1989;68 (3):352-358.

20. Gambineri A, Pelusi C and.Pasquali R. Testosterone levels in obese males with obstructive sleep apnea syndrome: relation to oxygen desaturation, body weight, fat distribution and the metabolic parameters. J Endocrinol Invest 2003; 26: 493-498.

21. Bratel T, Wennlund A and Carlstrom K. Impact of hypoxaemia on neuroendocrine function and catecholamines secretion in chronic obstructive pulmonary disease (COPD). Effects of long-term oxygen treatment. Respir Med 2000; 94: 1221-1228.

22. Bratel T, Wennlund A and Carlstrom K. Pituitary reactivity, androgens and catechlamines in obstructive sleep apnea. Effects of continuous positive airway pressure treatment (CPAP). Respir Med 1999; 93 (1): 1-7.

23. Schafer H, Pauleti D, Sudhop T et al. Body fat distribution, serum leptin, and cardiovascular risk factors in men with obstructive sleep apnea. Chest 2002; 122:829-839.

24. Phillips BG, Kato M, Markiewicz K et al. Increases in leptin levels, sympathetic rise, and weight gain in obstructive sleep apnea. Am J Physiol (Heart Circ. Physiol.) 2000; 279: H234-H 27.

25. Patel SR, Palmer E.K. Larkin EK et al. Relationship between obstructive sleep apnea and diurnal leptin rhythms. Sleep (in press 2004).

26. Chin K, Shimizu K, Nakamura T et al. Changes in intra-abdominal visceral fat and serum leptin levels in patients with obstructive sleep apnea syndrome following nasal continuous positive airway pressure therapy. Circul 1999; 100:706-712.

27. Harsch LA, Konturek PC, Koebnick C et al. Leptin and ghrelin levels in patients with obstructive sleep apnoea: effect of CPAP treatment. Eur Respir J 2003; 22:251-257.

28. Ip MSM, Lam KSL, Ho CM et al. Serum leptin and vascular risk factors in obstructive sleep apnea. Chest 2000; 118:580-586.

29. Lavie P. Pro: Sleep apnea causes cardiovascular disease.Am J Respir Crit Care Med 2004: 169:147-150.

30. Giannoti L, Pivetti F, Lanfranco F et al. Concomitant impairment of growth hormone secretion and peripheral sensitivity in obese patients with obstructive sleep apnea syndrome. J Clin Endocrinol Metab 2002; 87(11): 5052-5057.

31. Nieminen P, Lopponen T, Tolonen U et al. Growth and biochemical markers of growth in children with snoring and obstructive sleep apnea. Pediatr 2002; 109(4): 1-6.

32. Cooper BG, White JES, Ashworth LA et al. Hormonal and metabolic profiles in subjects with obstructive sleep apnea syndrome and the acute effects of nasal continuous positive airway pressure (CPAP) treatment. Sleep 1995; 18(3): 172-179.

33. Grunstein RG, Stewart DA, Lloyd H et al. Acute withdrawal of nasal CPAP in obstructive sleep apnea does not cause a rise in stress hormones. Sleep 1996; 19(10): 774-482.

34. Weibel L, Follenius M, Spiegel K et al. Comparative effects of night and daytime sleep on the 24-hour Cortisol secretory profile. Sleep 1995; 18(7): 549-556.

35. Purnell JQ, Brandon DD, Isabelle LM et al. Association of 24-hour Cortisol production rates, Cortisol-binding globulin, and plasma-free Cortisol levels with body composition, leptin levels, and aging in adult men and women. J Clin Endocrinol Metab 2004; 89 (1): 281-287.

36. Marik PE. Leptin, obesity, and obstructive sleep apnea. Chest 2000; 118:569-570.

Sleep-Disordered Breathing and Hormones

TARJA SAARESRANTA AND OLLI POLO

1. Introduction

Falling asleep removes postural muscle tone, voluntary respiratory control and the wakefulness stimulus for breathing. These changes predispose to respiratory abnormalities during sleep, including periodic breathing, repetitive episodes of obstructive, mixed or central apnea or hypopnea, or prolonged episodes of partial upper airway obstruction with increased respiratory resistance. In short term, the performance of the respiratory muscles is readily controlled through neural mechanisms, which continuously monitor the metabolic needs of the body by reading the output of chemoreceptors and adjusting alveolar ventilation accordingly. Acute failure of the respiratory control during sleep rapidly results in arousal and activation of the sympathetic division of the autonomic nervous system with increased catecholamine secretion.

The most important role of hormones in control of breathing is in determining the level of metabolism, oxygen consumption, carbon dioxide production and the cellular acid-base balance. These factors set the overall ventilatory needs and are therefore carefully monitored by chemoreceptors. The manifestation of the upper airway obstruction during sleep is critically linked with the underlying level of metabolism setting the drive for breathing.

Both low respiratory drive (typical for postmenopausal women with low progesterone) or high drive (eg. acromegalics with high insulin-like growth factor-I) predispose to episodes of upper airway obstruction during sleep. Sleep apnoea seems to be an epidemic, which spreads rapidly with obesity, another major health problem in Western societies. In the U.S., 24% of male and 9% of female government workers present with 5 or more episodes of sleep apnoea or hypopnoea per hour.[1]

Daytime symptoms of sleep apnoea appear in 4% of men and 2% of women.[1] Similar prevalence rates of symptomatic sleep apnoea in adults from 20 to 100 years of age are reported in a community-based study: 3.9% in men and 1.2% in women.[2] A number of hormones interact with sleep[3] and breathing.[4] Sleep-disordered breathing (SDB) affects hormones via a number of mechanisms.

On the other hand, hormones and endocrine states induce, aggravate or alleviate SDB. Finally, nasal continuous positive airway pressure (CPAP) therapy influences hormone secretion.

SDB and sleep disturbances may interact with hormones in several ways. Episodes of apnoea or hypopnoea cause sleep fragmentation and disturb sleep cycles and stages. Arousals may induce a stress response resulting in increased levels of stress hormones.[5] Hypoxia may also have direct effects on central neurotransmitters,[6] which result in alterations in the hypothalamo-pituitary axis and in stimulation of peripheral endocrine glands.[7] Hypercapnia alone or combined with hypoxia may increase levels of renin, adrenocorticotrophic hormone (ACTH), corticosteroids, aldosterone, and vasopressin.[8,9] Finally, disorganization of sleep, sleep loss and naps disturb sleep-controlled endocrine rhythms resulting in endocrine and metabolic abnormalities.

The direct and indirect effects of hormones and endocrine disorders on sleep and breathing are mediated via several pathways. The male gender and the postmenopausal state are risk factors[10,11] that link the sex-hormones to the pathophysiology of SDB. Sleep apnoea is common in acromegaly,[12-17] hypothyroidism[18-21] or Cushing's syndrome (Table 1).[22,23] Recent studies suggest that SDB may not only complete the clinical picture but play a central role in the pathophysiology of obesity,[24] leptin resistance[25-27] and the metabolic syndrome.[24,28,29] The prevalence estimates of sleep apnoea among various endocrine states and disorders are shown in Table 1. Unfortunately, due to the lack of well-documented epidemiological studies most prevalence estimates are based on small study populations. Prevalence estimates are also limited because of different definitions of SDB.

Much of our current knowledge on the interactions between hormones and obstructive sleep apnoea syndrome (OSAS) is based on intervention studies with nasal CPAP. Nasal CPAP is the most efficient method for maintaining upper airway patency during sleep. Its efficacy in controlling sleep apnoea and hypopnoea starts from the very first night of therapy.[30] Changes in the levels of several hormones (Table 2) have been interpreted to be related to SDB or associated sleep disturbance, if they consistently respond to on-off nasal CPAP interventions. Hormonal changes are potential mediators to link SDB with various comorbidities. Many studies investigating the effects of SDB or nasal CPAP therapy on hormone levels have only assessed single morning levels of hormones, and therefore the effects on the 24-h secretory profile are poorly known.

2. Glucose and Energy Metabolism

2.1. Insulin Resistance and Diabetes Mellitus

Insulin release is a complex oscillatory process with rapid pulses (10 min) superimposed on slower fluctuations (50 -100 min).[31] Sleep stimulates insulin secretion by increasing the oscillation amplitude, which may be partly mediated by growth hormone (GH).[31,32] In rats, the intracerebroventricular administration of insulin results in a 25% increase of slow wave sleep (SWS).[33] In humans,

TABLE 1. Prevalence of sleep-disordered breathing in some endocrine disorders and states.[*] = The prevalence in women is without hormone replacement therapy. AN = autonomic neuropathy.

Endocrine disorder	Prevalence of sleep apnoea	Authors	Sample size
Diabetes, type 1	31%	Rees et al. 1981[268]	16
	42%	Mondini et al. 1985[38]	12
Diabetes, type 2	1.9% (vs. 0.3% in non-diabetics)	Katsumata et al. 1991[41]	579
		Elmasry et al. 2001[43]	25
	36% (vs. 14.5% in non-diabetics)	Meslier et al. 2003[45]	494
	30.1% (vs. 13.9% of non-apnoeic snorers)		
Diabetes with autonomic neuropathy	37% (vs. 0% in those without AN)	Rees et al. 1981[268]	8 and 8 without AN)
		Catterall et al. 1984[39]	8 and 8 without AN)
	0% (vs. 6% in those without AN)	Ficker et al. 1998[44]	23 and 25 without AN)
	26% (vs. 0% in those without AN)		
Hypothyroidism	82%	Rajagopal et al. 1984[18]	11
	100%	Grunstein & Sullivan 1988[19]	10
	25%		20
	7.7% (vs. 1.9% in controls)	Lin et al. 1992[20]	26
		Pelttari et al. 1994[21]	
Acromegaly	40% with active, 0% with inactive disease	Hart et al. 1985[12]	10 with active, 11 with inactive disease
	45%	Pekkarinen et al. 1987[13]	11
	81%	Grunstein et al. 1991[14]	53
	91%	Pelttari et al. 1995[15]	11
	39%	Rosenow et al. 1996[16]	54
	75%	Weiss et al. 2000[17]	55
Cushing disease/ syndrome	45%	Shipley et al.1992[22,23]	22
Polycystic ovary syndrome	17%	Vgontzas et al. 2001[37]	53
	44%	Fogel et al. 2001[223]	18
Postmenopause	2.7%[*] (vs. 0.6% in premenopausal women)	Bixler et al. 2001[2]	314

no systematic relationship appears to exist between insulin secretion rate oscillations and NREM or REM sleep.[31] Sleep fragmentation resulting in sleep deprivation is likely to have an impact on the hormones, which regulate glucose tolerance.

Partial sleep restriction with 4 hours sleep per night for 6 days resulted in increased cortisol levels and impaired glucose tolerance even in healthy nonobese young men.[34] These metabolic and endocrine alterations were recuperated during recovery sleep. There is an accumulating body of evidence

TABLE 2. Various hormones in obstructive sleep apnoea syndrome and the effect of nasal CPAP therapy on hormones. ↑ = increased, ↓ = decreased, ↔ = no change, ANP = atrial natriuretic peptide, IGF-I = insulin-like growth factor-I, LH = luteinizing hormone, TSH = thyroid stimulating hormone (thyrotrophin).

Hormone	OSAS	Effect of nasal CPAP	Authors
Growth hormone	↓	↑	Grunstein et al. 1989,[124] Saini et al. 1993,[67] Cooper et al. 1995[68]
IGF-I	↓	↑	Grunstein et al. 1989[124]
TSH	↓ or ↔	↓	Bratel et al. 1999,[79] Meston et al. 2003[160]
Free T4	↓	↔	Meston et al. 2003[160]
Leptin	↑	↓	Saarelainen et al. 1997,[25] Chin et al. 1999,[26] Ip et al. 2000[27]
Noradrenaline	↑ or ↔	↓ or ↔	Grunstein et al. 1996,[269] Cooper et al. 1995,[68] Bratel et al. 1999[79]
Cortisol	↑	↔	Grunstein et al. 1989[125] &1996,[269] Bratel et al. 1999[79]
Aldosterone	↑ or ↔	↑ or ↓	Follenius et al. 1991,[270] Saarelainen et al. 1996,[271] Moller et al. 2003,[272] Meston et al. 2003[160]
Renin	↔	↑ or ↓ or ↔	Follenius et al. 1991,[270] Moller et al. 2003,[272] Meston et al. 2003[160]
ANP	↑	↓	Krieger et al. 1991,[273] Ichioka et al. 1992[274]
LH	↓ or ↔	?	Grunstein et al. 1989,[124] Bratel et al. 1999,[79] Luboshitzky et al. 2002[128]
Testosterone	↓ or ↔	↑ or ↔	Santamaria et al. 1988,[127] Grunstein et al. 1989,[124] Bratel et al. 1999,[79] Luboshitzky et al. 2002,[128] Meston et al. 2003[160]
Prolactin	↔	↓ or ↔	Grunstein et al. 1989,[124] Spiegel et al. 1995,[275] Bratel et al. 1999,[79] Meston et al. 2003[160]
Substance P	↑	?	Gislason et al. 1992[276]

that SDB is linked with insulin resistance and metabolic syndrome independently of body mass index (BMI) and other known risk factors (Figure 1).[24,28,29,35] Oxyhaemoglobin desaturation index (drops of oxygen saturation of 4% or more per hour, ODI_4) is a better predictor of insulin resistance than BMI.[36)] In women with polycystic ovary syndrome, insulin resistance is a stronger risk factor for sleep apnoea than BMI or serum testosterone levels.[37]

The prevalence of SDB in type 1 diabetes remains to be confirmed. Some authors have reported a prevalence rate of sleep apnoea as high as 42%,[38] whereas others have not observed a difference from the general population.[39] Small sample sizes and different diagnostic criteria for sleep apnoea may explain some of the discrepancies. Diabetic children n = 25) have more episodes of apnoea during sleep and the duration of apnoeic events is longer than in healthy controls.[40] Furthermore, the degree of severity of sleep apnoea correlates with the glucose control and the duration of diabetes.

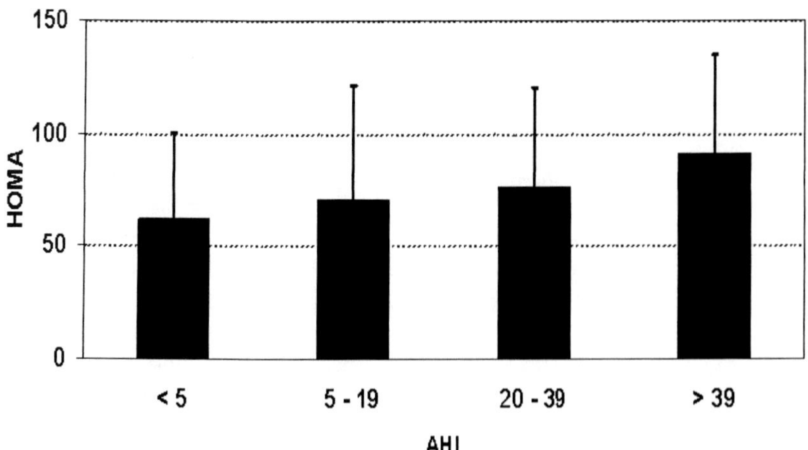

FIGURE 1. Index of insulin sensitivity calculated with a homeostasis model assessment (HOMA = $G_0 x I_0 / 22.5$, where G_0 and I_0 represent fasting serum glucose and insulin, respectively) in different apnoea-hypopnoea index (AHI) categories (total n = 150 men). P-value is significant for trend across AHI categories (p < 0.05). Subjects with increasing AHI are increasingly resistant to insulin. Based on data by Punjabi et al. (2002)[29]

Among nearly 13 000 japanese hospital inpatients, the prevalence of sleep apnoea was 0.3%.[41] In the same study, in a subgroup of nearly 600 male type 2 diabetics, the prevalence of sleep apnoea was higher than in nondiabetics (1.9% vs. 0.3%). In a Swedish 10-year follow-up study, snoring was a risk factor for diabetes independently of other risk factors.[42] Among hypertensive diabetics, the prevalence of sleep apnoea, defined as apnoea-hypopnoea index (AHI) ≥ 20, was 36% compared to 14.5% in nondiabetics.[43] Autonomic diabetic neuropathy may be[a] associated with sleep apnoea. Among 23 diabetics with autonomic neuropathy (one had type 1 diabetes, the others type 2 diabetes), six had sleep apnoea whereas none of the diabetics without autonomic neuropathy were affected.[44]

In male patients with OSAS n = 494), type 2 diabetes was present in 30.1% and in non-apneic snorers (n = 101) in 13.9%.[45] Impaired glucose tolerance was observed in 20.0% of OSAS patients and 13.9% of nonapnoeic snorers. Fasting and postload blood glucose increased and insulin sensitivity decreased with increasing severity of OSAS defined by AHI.[45]

2.1.1. Effect of Nasal CPAP

Investigations addressing the question of whether OSAS itself is an independent risk factor of insulin resistance in OSAS have led to conflicting results.[25,28,29,46,47] Among morbidly obese (average BMI 42.7 kg/m^2) patients with sleep apnoea and type 2 diabetes, nasal CPAP treatment for four months improved insulin responsiveness.[46] The study population was highly selected, and therefore these results cannot be extrapolated to all type 2 diabetics. Two other studies found no effect of a 2-month or a 3-month nasal CPAP therapy on glucose and insulin metabolism.[25,47] Because the duration of nasal CPAP therapy was for only two months, it cannot be excluded that a longer treatment period would improve glucose tolerance. Also the lack of statistical power due to the small number of patients could affect the results. A recent study in patients with moderate to severe OSAS (n = 40) demonstrated that insulin resistance measured by hyperinsulinemic euglycemic clamp improved within two days of effective CPAP use and remained stable after three months of treatment.[48] The improvement in insulin sensitivity was greater in patients with BMI less than 30 kg/m^2 than in more obese patients. The improved insulin sensitivity may reflect a decreasing sympathetic activity, indicating that OSAS is an independent risk factor for increased insulin resistance.

2.2. Leptin

Sleep influences the nocturnal leptin profile. The diurnal amplitude of leptin is reduced during total sleep deprivation and returns toward normal during recovery sleep.[49] Besides its best known function as a satiety hormone, leptin is also a powerful respiratory stimulant.[50] Leptin receptors are abundant in the nucleus of the solitary tract and other centres in the medulla involved in the respiratory responses to CO_2 and pH.[51,52] It is possible that leptin may act directly through leptin receptors on respiratory neurons in the medulla, in addition to acting through hypothalamic pathways. A recent study suggests that leptin acts through melanocortin 4 pathways to stimulate the respiratory responses to hypercapnia in an analogous manner to the hypothalamic control of appetite.[53] There is emerging evidence that the melanocortin pathway is linked both to obesity[54,55] and OSAS.[56]

Plasma leptin levels are higher in patients with sleep apnoea than in controls matched for BMI.[57] Further, hypercapnic patients with OSAS have higher leptin levels than eucapnic BMI-matched controls with sleep apnoea.[58] Fasting leptin levels correlate positively with the degree of sleep-disordered breathing as defined by AHI or the percentage of sleep time spent with an oxygen saturation below 90%[59,60] but are not dependent on AHI.[59] Leptin secretion could provide an adaptive mechanism to enhance ventilation in patients with severe respiratory impairment.

On the other hand, high circulating leptin levels suggest leptin resistance at the level of the central nervous system. Elevated leptin levels are likely to contribute to comorbidity of OSAS because high leptin levels are associated with coronary heart disease,[61] insulin resistance,[62] impaired fibrinolysis,[63] development of obesity,[64] or type 2 diabetes,[65] which are all highly prevalent in patients with OSAS.

2.2.1. Effect of Nasal CPAP

Serum leptin levels decrease with nasal CPAP therapy (Figure 2)[25-27] without weight loss.[26,27] A decrease in leptin levels is found already after the first night on nasal CPAP.[66] Nasal CPAP does not affect the secretory profile of leptin. A nocturnal increase in serum leptin levels is observed both on and off nasal CPAP.[66] The CPAP-induced reduction in leptin level is likely due to both improved sleep and breathing. Nasal CPAP therapy increases SWS and increases growth hormone (GH) secretion,[67-69] which in turn inhibits leptin secretion.[70,71]

While normalizing nocturnal breathing, hypoxic and hypercapnic stimuli may not any longer increase leptin secretion.[58,72-75] These findings suggest that nasal CPAP therapy either reduces the leptin resistance in obese patients or as ventilation improves less leptin is needed to stimulate breathing. It is also possible that nasal CPAP therapy normalizes sleep structure and growth hormone secretion thereby normalizing leptin.

The magnitude of the decrease in leptin levels with CPAP therapy correlates with cardiac sympathetic function measured before CPAP treatment by iodine-123-meta-iodobenzylguanidine imaging.[66] By reducing leptin levels, nasal CPAP therapy is likely to decrease the comorbidity related to OSAS.

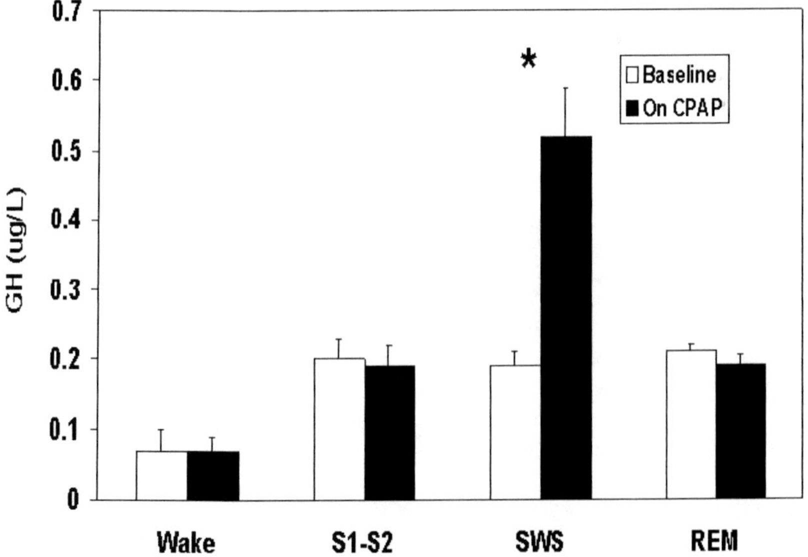

FIGURE 2. Serum leptin levels in 30 patients with OSAS and in 30 matched controls. A) Comparison with controls. B) Effect of a 6-month nasal CPAP treatment. From Ip et al. (2000),[27] with permission.

3. Catecholamines

Sleep onset is associated with a decrease of circulating concentrations of noradrenaline and adrenaline, with a nocturnal nadir occurring about one hour after sleep onset.[76] In contrast, 24-hour, daytime and nocturnal urinary noradrenaline levels as well as both daytime[77] and nocturnal[78,79] plasma noradrenaline concentrations are increased or within normal limits in patients with OSAS. When awake, no correlation between the severity of OSAS and plasma noradrenaline concentrations has been found,[77] whereas nocturnal noradrenaline levels correlate with OSAS severity and oxygen saturation.[78,79] Noradrenaline has been identified in the carotid bodies, suggesting that it is involved in the control of breathing.[80]

In blood and urine, high levels of catecholamines and their metabolites reflect increased sympathetic activity. Muscle sympathetic nerve activity is greater in obese than in normal-weight subjects,[81] and greater in sleep apnoeics than in age and BMI matched controls.[82,83] Hypoxia and hypercapnia induce sympathetic nervous system overactivity.[84] The sympathetic responses to hypoxia and hypercapnia are further potentiated during apnoea, when the inhibitory influence of the thoracic afferent nerves is eliminated.[81,83] Sleep fragmentation leading to chronic partial sleep loss is also likely to contribute to the increased sympathoadrenal activity and increased circulating catecholamine levels encountered in OSAS. This assumption is supported by observations in healthy male volunteers.[76] One night of partial sleep deprivation resulted in increases in circulating noradrenaline and adrenaline levels.[76] Most studies report a positive relationship between episodes of obstructive apnoea and noradrenaline levels, whereas a minority of studies have found a relationship between adrenaline and episodes of obstructive apnoea.[85,86]

3.1. Effect of Nasal CPAP

In most studies, nasal CPAP treatment decreases plasma or urinary noradrenaline levels, whereas adrenaline levels in most cases remain unchanged.[79,87,88] In contrast, in noninsulin-dependent diabetics with OSAS, neither fasting adrenaline nor noradrenaline levels changed on CPAP.[46]

4. Thyrotrophic Axis Hormones

4.1. Thyrotrophic Axis Hormones, Sleep and Breathing

Neuroanatomical findings support interactions between the thryrotropic axis hormones and sleep. Thyrotropin-releasing hormone (TRH) neurons are located in the wake-promoting histaminergic neurons in the posterior hypothalamus. However, histamine H1-receptors are also found in the paraventricular nucleus, anatomically enabling interactions with TRH within the

some authors have suggested a decrease after sleep onset.[113] The role of GH and IGF-I in the control of breathing is unclear, but some evidence suggests their direct or indirect involvement in the regulation of breathing.[114,115]

Ghrelin, the main endogenous ligand for growth hormone secretagogue (GHS) receptors, stimulates GH release, appetite, and weight gain in humans and rodents.[116] Circulating ghrelin acts via GHS-receptors located mainly on GHRH and NPY neurons.[116] Ghrelin enhances SWS in humans but no data is available about its possible role in the control of breathing. In patients with OSAS fasting ghrelin levels are higher than in BMI-matched controls.[117]

Somatostatin is the counterpart of GHRH in GH release. In young persons, administration of somatostatin either in a continuous[118] or a pulsatile[119] fashion does not alter sleep. However, in elderly persons, repetitive administration of somatostatin decreased total sleep time and REM time and induced increased intermittent wakefulness.[120] The age-dependent response of GHRH can be explained by the age-related decrease in the activity of GHRH, but not somatostatin.[121] Somatostatin appears to be a central respiratory inhibitor.[122,123]

Both in obesity and in OSAS, spontaneous GH secretion is reduced,[67,68,124] whereas IGF-I levels are normal or slightly decreased in obesity and decreased in OSAS.[124,125] These alterations in OSAS could simply reflect overweight or hypoxia-induced endocrine changes. However, a recent study showed that in OSAS the maximal secretory capacity of somatotroph cells and the IGF-I sensitivity to recombinant human growth hormone stimulation are impaired compared to weight-matched patients with simple obesity.[126] The alteration in GH/IGF-I axis activity in OSAS may reflect the reduced testosterone concentrations in OSAS[124,127,128] since testosterone increases IGF-I synthesis and release.[129,130]

5.2. Acromegaly

The association of snoring, daytime sleepiness and acromegaly was reported over a century ago.[131] Macroglossia and pharyngeal swelling are the most probable reasons for the high incidence of SDB in acromegaly (Table 1).[132-135] Accordingly, sleep apnoea is alleviated when tissue hypertrophy decreases with somatostatin analogue treatment.[136-138] GH or IGF-I may also have a direct role in the pathogenesis of sleep apnoea but observations are controversial.[12-14,16,139] Some investigators report an association between the presence of sleep apnoea and high growth hormone and IGF-I levels,[12,16,139] whereas others fail to show any association between obstructive sleep apnoea and the biochemical activity of acromegaly.[13,14] One study found an association between the biochemical activity of acromegaly and central sleep apnoea.[14] The high IGF-I levels in acromegaly may drive breathing and result in the increased hypercapnic ventilatory response measured during wakefulness[136] and increased frequency of central apnoea[136] or periodic breathing with symmetric waxing and waning respiratory efforts[15] during sleep.

Treatment of acromegaly with adenomectomy[16] or with somatostatin analogues[136,138] may cure acromegaly-related OSAS. The operative team should be aware of the risks of performing transsphenoidal adenoma resection in acromegalic patients with sleep apnoea in whom upper airway oedema could potentially further aggravate gas exchange postoperatively.[132] Because octreotide treatment may promptly alleviate OSAS, its preoperative administration is recommended.[137,140-142] Preoperative nasal CPAP therapy could also reduce perioperative risks.[142] Sedatives have to be avoided and monitoring of breathing should be extended beyond the immediate postoperative period. Perioperative tracheostomy is the safest and sometimes the only alternative to secure breathing after surgery.

After adenomectomy, sleep apnoea persists in every fifth patient, in particular in those, whose GH levels remain high.[16] In addition to endocrine factors, the high prevalence of residual SDB after adenomectomy could be related to soft tissue hypertrophy, which remains unaltered. However, uvulopalatopharyngoplasty is not feasible in the treatment of acromegaly-related OSAS.[143] Nasal CPAP with new pressure titration is often needed after surgery.[142]

5.3. Growth Hormone Deficiency

Not only excessive GH production but also GH deficiency could be linked with sleep apnoea. Syndromes with hereditary GH deficiency are often associated with obesity as well as craniofacial and pharyngeal abnormalities predisposing to SDB. Comprehensive studies are not available, so only anecdotal case-reports about Laron dwarfism[144] and Turner syndrome[145] support this. In Prader-Willi syndrome severe GH deficiency occurs in 38% of adults.[146] The pathological somnolence in Prader-Willi patients could be due to nonapnoeic breathing abnormalities rather than episodes of obstructive sleep apnoea.[147]

Sleep apnoea patients have low GH levels without any specific causes of GH deficiency.[124] GH secretion occurs mostly during sleep, and 70% of nocturnal GH pulses are associated with SWS.[110,148] In OSAS, GH secretion is decreased not only due to obesity[126,149,150] but because of sleep fragmentation resulting in decreased amount of SWS.[151] Repetitive hypoxemia may also affect GH secretion. In animals, hypoxia inhibits GH release or biosynthesis.[152] GH deficiency in adults is associated with impaired psychological well-being, insulin resistance, endothelial dysfunction, increased visceral fat, increased cardiovascular mortality, and accelerated ageing.[111,153] Similar features are typical in OSAS, which raises the question of a possible link between OSAS-related GH deficiency and the comorbidity seen in OSAS. Indeed, patients with severe OSAS have similar levels of IGF-I to adult patients with GH deficiency.[124] Low IGF-I may contribute to an increased risk for cardiovascular diseases among patients with sleep apnoea. The vascular effects of IGF-I are endothelium-dependent,[154] and endothelial cells have IGF-I receptors.[155] IGF-I increases endothelial cell nitric oxide production.[156] Nitric oxide is an important paracrine mediator of vasodilatation and inhibition of vascular smooth muscle cell growth.[157]

Two recent reports suggest that GH replacement therapy may also affect sleep and breathing.[158,159] Among 145 children on GH replacement therapy, four developed sleep apnoea, which in three cases was associated with tonsillar and adenoidal hypertrophy.[158] Sleep apnoea improved in one patient after cessation of GH therapy, and in all patients following tonsillectomy and adenoidectomy. In five male middle-aged patients with postoperative pituitary insufficiency, cessation of GH replacement for six months resulted in a decrease of obstructive apnoeic events but in an increase of central apnoeic events.[159] Following cessation of GH replacement, SWS decreased markedly.[159]

At least in theory, an unfortunate coexistence of GH deficiency and SDB would result in a potentially vicious interaction between two altered physiological functions, resulting in severe anatomical abnormalities. A primary GH deficiency could predispose to SDB through short stature, craniofacial growth retardation and low respiratory drive. SDB would further aggravate GH deficiency through sleep disturbance. A primary SDB could aggravate itself by affecting craniofacial and upper airway soft tissue growth through induction of secondary GH deficiency.

5.4. *Effect of Nasal CPAP*

Nasal CPAP therapy increases SWS and normalizes GH secretion without changes in body weight (Figure 3).[67-69] Fasting plasma ghrelin levels decrease with nasal CPAP therapy.[117] Increases in IGF-I concentrations with

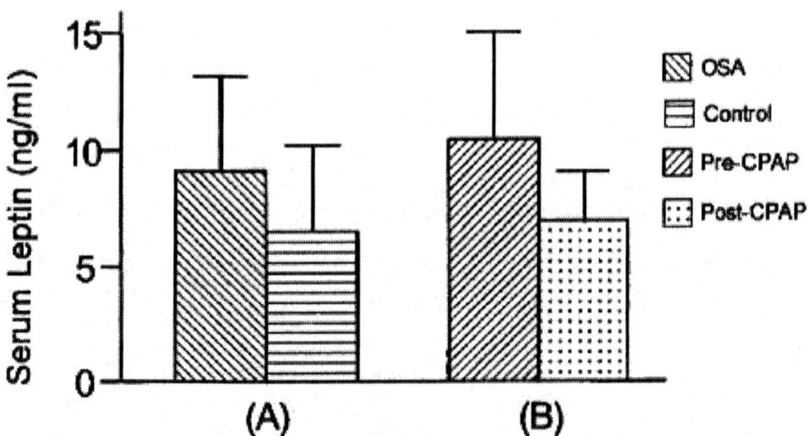

FIGURE 3. Plasma growth hormone (GH) levels measured with frequent sampling during different sleep stages in eight patients with severe OSAS before and after nasal CPAP treatment. The relationship between SWS and GH concentrations becomes significant on CPAP treatment. Modified from Grunstein (1996).[69]

CPAP[124,160] are most probably mediated via increased GH secretion. However, the possible effects of improved nocturnal breathing on GH release cannot be excluded.[152] Increased production of IGF-I[124] and circulating nitric oxide[161] are plausible mediators of the beneficial effect of nasal CPAP on cardiovascular disorders.

In patients with GH deficiency and with predisposing anatomical abnormalities for SDB, systematic screening for SDB is recommended. Nasal CPAP treatment and maxillomandibular surgery are feasible therapeutical approaches in these patients.

Treatment of SDB may result in normalization of GH secretion and normal growth in children.[162-164] On the other hand, symptoms of SDB should be monitored during GH replacement therapy because of the increased risk of SDB.

6. Adrenocorticotrophic Axis Hormones

6.1. Adrenocorticotrophic Axis, Sleep and Breathing

Corticotrophin-releasing hormone (CRH) has the opposite effect to sleep-promoting GHRH. It decreases SWS, inhibits GH release but stimulates cortisol secretion.[165] Cortisol secretion enhances SWS,[166,167] probably by feedback inhibition of CRH.[109]

CRH receptors are widely distributed in the brain areas involved in the control of breathing.[168] CRH acts as a central respiratory stimulant.[169,170] Cortisol is not known to be involved in control of breathing.

Early reports suggested a stimulatory stress-effect of OSAS on the pituitary-glucocorticoid axis. Later studies have found neither alterations in cortisol levels in OSAS nor any effect of CPAP treatment on this axis.[124,160]

6.2. Cushing's Syndrome and Cushing's Disease

Shipley and coworkers found sleep apnoea in 45% of their 22 patients with Cushing's disease or Cushing's syndrome.[22,23] Mechanisms could be related to uncovered endocrine interactions or alterations in morphology of the upper airways. Also long-term high-dose corticosteroid therapy may contribute to SDB.[142] This is of importance especially in patients with juvenile rheumatoid arthritis, whose craniofacial abnormalities (micrognathia) also predispose to SDB.

7. Sex Hormones

7.1. Sex Hormones, Sleep and Breathing

The brain is a major target organ of steroid hormones that are synthesized primarily in the gonads and the adrenal glands. Due to their high lipid solubility they easily cross the blood-brain barrier. The brain also contains steroid biosynthetic enzymes.[171,172] There are two major pathways for steroid metabolism in the brain: cytochrome P450 aromatase converts both testosterone and androstenedione into oestradiol and the second pathway uses 5-alpha-reductase to convert a number of steroids into their reduced forms.[173] Sex hormone receptors are widely distributed throughout the brain, progesterone receptor mRNA being identified e.g in the hypoglossal nucleus and the nucleus of the solitary tract (NTS)[174] and oestrogen alpha and beta receptors in NTS neurons.[175] Androgen receptors have been found in the hypoglossal nucleus.[176] Androgen binding is lower in female than in male hypoglossal nucleus.[177] High levels of oestradiol downregulate oestrogen receptors, whereas progesterone receptors are upregulated by oestradiol and downregulated by androgen withdrawal.[178] Androgen receptors are upregulated by androgens in both male and female rats and downregulated by androgen withdrawal.[179-182] Oestrogen can also act via non-genomic effects.[178] Sex steroids influence neuromodulatory neurons including serotonergic, noradrenergic, dopaminergic and cholinergic neurons[178] i.e. neurons involved in the control of the sleep/wake cycle and breathing.[178]

In humans, oestrogen increases the amount of REM sleep[183,184] and decreases the latency to REM sleep.[183] Oestrogen therapy decreases the latency to sleep onset,[183,185-187] decreases awakenings after sleep onset,[184,188] increases total sleep time,[48,183,184] increases the amount of SWS,[48] and decreases the rate of cyclic alternating patterns.[189] Some studies have found oestrogen to have no effect on objective sleep quality.[190] However, there is a discrepancy between objective and subjective sleep quality in postmenopausal women.[48,191]

Progesterone has sedative, benzodiazepine-like agonistic effects on $GABA_A$ receptors.[109] The effects on sleep appear to be mediated via the conversion of progesterone to its major metabolite allopregnanolone.[192]

Testosterone secretion follows a near-cosine waveform with its maximal value observed towards the end of the sleep period[193] or at about the time of the first REM sleep period.[194] In young healthy men, sleep deprivation decreases serum testosterone levels.[195] Age-dependent positive associations exist between sleep efficiency, shortened latency to REM sleep, and number of REM episodes, and circulating testosterone levels.[196]

Sex steroids have a profound effect on breathing.[4,197-199] Sex steroids might influence peripheral or central chemoreceptors or their sensory nuclei, or cortical, cerebellar or hypothalamic nuclei that project to the brainstem and spinal cord respiratory-related neurons.

Progesterone is the sex steroid most closely linked with the control of breathing, acting both centrally and via peripheral chemoreceptors.[200-203] The mechanisms of action of oestrogen on breathing are mostly unknown. The serotonergic system is involved in the neural control of breathing. In rats, 5HT levels in the hypoglossal nucleus closely reflect circulating oestradiol levels whereas in the phrenic nucleus 5HT levels correspond to circulating progesterone levels.[178]

The role of testosterone in control of breathing is not thoroughly investigated. In some hypogonadal men, testosterone supplementation increased ventilation and metabolic rate without change in arterial or end-tidal PCO_2.[204] Similar ventilatory increases has been reported in neutered cats.[204] Testosterone also suppressed hypercapnic ventilatory drive in infant monkeys during sleep.[206]

7.2. Pregnancy

Pregnancy has a marked impact on breathing, which is largely mediated through hormones. The levels of the female sex hormones progesterone and oestrogen increase markedly. Progesterone increases ventilation[207] and may cause hypocapnia and respiratory alkalosis, and result in respiratory instability and episodes of central apnoea during NREM sleep.[208] Pharyngeal dimensions decrease during pregnancy,[209] nasal congestion and rhinitis are frequent,[210] and the enlarging uterus compromises the performance of the diaphragm. Increased oestrogen levels may cause oedema in the upper airway mucosa, and thereby be responsible for upper airway symptoms.[211] On the other hand, increased female hormone levels may protect upper airway patency, assuming that upper airway dilators are able to respond appropriately.[205] Despite marked "central obesity", neither normal[212,213] nor multiple pregnancy[214] seems to predispose to SDB. However, in obese women pre-existing SDB may deteriorate during pregnancy.[213] In pre-eclampsia, partial upper airway obstruction during sleep is common.[215,216] The long periods of partial upper airway obstruction are associated with increased systemic arterial blood pressure, which can be lowered with nasal CPAP therapy.[215]

Snoring is frequent among pregnant women (12-23% vs. 4% in non-pregnant women).[217-219] Snoring or OSAS during pregnancy have been suggested to cause intrauterine foetal growth restriction and lower APGAR scores at birth.[217,218,220] Nasal CPAP therapy also seems to be both safe and effective during pregnancy,[215,221,222] and early intervention may improve the outcome of the mother and her baby.

7.3. Polycystic Ovary Syndrome

Recent studies show a high prevalence rate of SDB in women with polycystic ovary syndrome.[37,223] Previously, SDB in polycystic ovary syndrome was attributed entirely to obesity, but Vgontzas and collaborators[37] showed that insulin resistance was a stronger determinant of SDB than BMI or serum testosterone levels. The AHI correlates with waist-hip-ratio and serum total and free testosterone concentrations.[223] SDB should be suspected and verification should be sought when women with polycystic ovary syndrome present with compatible symptoms.

7.4. Menopause

In clinical studies, the male to female ratio of OSAS is about 10:1.[224-226] In community based populations, the prevalence of OSAS is higher, and the male to female ratio ranges from 2:1 and 4:1.[227-230]

Female hormones are thought to protect women from SDB until menopause.[231] Among women referred to the sleep clinic, 47% of the post-menopausal and 21% of the premenopausal women had sleep apnoea.[232] The observations from community-based studies on the impact of menopause on the prevalence of SDB are not consistent, although most studies show increased prevalence estimates of sleep apnoea after menopause.[2,228,229,233,234] Much of the discrepancy could be attributed to variations in the definition of SDB. Episodes of sleep apnoea seem to grossly underestimate SDB in women since partial upper airway obstruction is far more common. Out of 62 healthy postmenopausal women, 17% had significant amount of partial upper airway obstruction during sleep.[11] In a large community-based study, 1.9% of postmenopausal women and 0.6% of premenopausal women had OSAS, defined as an AHI of ≥ 10 and occurrence of daytime symptoms.[2]

Postmenopausal hormone therapy (HT) may prevent SDB. In a cross-sectional study, the prevalence estimates of sleep apnoea in postmenopausal women without HT were almost similar to those in men, whereas in HT users they were compatible with those in premenopausal women (Table 3).[2] In another study, hormone users had lower AHI than nonusers.[235] No significant

TABLE 3. Effect of gender, menopause and hormone replacement therapy on prevalence of sleep-disordered breathing. In postmenopausal women without hormone replacement therapy, prevalence of sleep-disordered breathing does not differ from that of men, whereas in postmenopausal women on hormone replacement therapy it is at the same level than in premenopausal women. Modified from the text and table of Bixler and co-workers.[2] 1) = Difference non-significant between the various modes of hormone replacement therapy.

		Sleep-disordered breathing			
Group	Sample size	Symptoms +AHI$\geq$10	AHI$\geq$15	Snoring+ 0<AHI<15	Snoring + AHI = 0
Men	741	3.9%	7.2%	17.3%	17.4%
Women	1000	1.2%	2.2%	5.4%	10.4%
Effect of menopause					
Premenopause	503	0.6%	0.6%	3.2%	7.9%
Postmenopause	497	1.9%	3.9%	7.5%	13.0%
Effect of hormone replacement					
Without	314	2.7%	5.5%	9.7%	14.8%
With	183	0.5%	1.1%	3.8%	9.8%
Mode of hormone replacement[1]					
Oestrogen	?		1.5%		
Oestrogen + progestin	?		0.3%		

difference was found between the effect of oestrogen alone or oestrogen plus progestin users.

Short-term administration of progestin alone (Figure 4)[199,236] or in combination with oestrogen[237,238] has shown only slight, if any, improvement in SDB in postmenopausal women. A recent preliminary study, however, showed a marked improvement in SDB with short-term oestrogen therapy, whereas addition of micronized progesterone attenuated the oestrogen-induced beneficial effects.[239]

However, it has not been excluded that a long-term HT might be beneficial in terms of improving SDB. Increasing evidence suggests that menopause is an independent risk factor for SDB. Nasal CPAP therapy is the treatment of choice in postmenopausal women with SDB. Indeed, SDB, partial obstruction in particular, should be considered in the differential diagnosis of depression, insomnia or restless legs syndrome to explain excessive sleepiness, and fatigue among postmenopausal women.

7.5. *Androgens*

The male predominance of OSAS has been attributed to testosterone-mediated aggravation of SDB or to the lack of protective effect of female hormones.

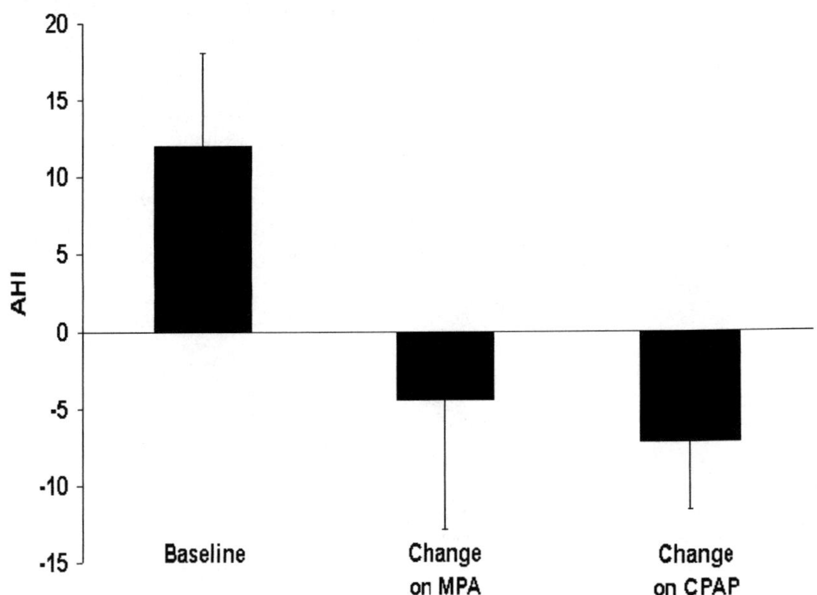

FIGURE 4. Decrease in apnoea-hypopnoea index with a two-week medroxyprogesterone acetate (60 mg daily) or with one-night nasal CPAP treatment in six postmenopausal women. Values are expressed as median (interquartile range). Based on data by Saaresranta et al. (2001)[199]

Androgens do not affect oestradiol or progesterone levels but may reduce their effect by down-regulating oestrogen and progesterone receptors.[240] Oestrone levels increase with androgens.[240] Among seven obese men, all except the one with hypogonadism presented with sleep apnoea.[241] In men, exogenous testosterone may suppress[242] or augment[243,244] hypoxic respiratory responses and lead to periodic breathing and sleep apnoea.[241,243,244] Exogenous testosterone does not affect the upper airway dimensions in men.[244]

Few studies have systemically evaluated the effects of exogenous androgen replacement therapy on SDB. Testosterone replacement therapy induced OSAS in one out of five men and aggravated pre-existing SDB in another.[242] In 11 hypogonadal men, testosterone replacement increased apnoeic events but only in three subjects the increase was considered clinically significant.[244] In a placebo-controlled randomised cross-over study of 17 elderly men with partial androgen deficiency, testosterone replacement therapy decreased total sleep time and sleep efficiency, and aggravated sleep apnoea.[245]

The few available data from women with SDB support the link between androgens and SDB. Irrespective of the menopausal state, obese women have higher androgen levels than nonobese women.[246,247] The prevalence of snoring plateaus or decreases in men after the age of 60 years.[248,249] Contrary to the observations in men, women continue to increase their snoring even beyond the age of 60 years.[248] Those observations suggest that decline of androgens in ageing males alleviates snoring, whereas menopause-induced oestrogen and progesterone deficiency and increased androgenicity continue to aggravate snoring in postmenopausal women. In a lean 70-year-old woman, testosterone-producing tumour caused sleep apnoea, which disappeared after removal of the tumour.[250] Exogenous testosterone induces sleep apnoea and even alters the upper airway dimensions in women.[251]

However, it is still somewhat controversial whether testosterone contributes to the development or aggravation of SDB. In men with OSAS, androgen blockade with flutamide did not affect ventilatory responses or SDB.[252] However, basal testosterone levels may be decreased in such patients and thus the therapeutic response to blockade might be decreased.

After discontinuing testosterone therapy for two months in haemodialysis patients, no change in AHI occurred.[253] On the other hand, 75% of the patients with a clinical history of OSAS were on testosterone therapy, compared to only 35% of those without history of SDB. It is not clear, whether these observations reflect the effect of testosterone per se, or possibly the severity of underlying renal disease and disturbances in the acid-base balance leading in respiratory changes.

In OSAS, both morning and nocturnal testosterone concentrations may be decreased[124,127,128] but increase after uvulopalatal resection.[127] OSAS may impair testosterone levels via several mechanisms. First, in obese individuals, total testosterone is decreased, and in massively obese patients also the free testosterone levels may decrease.[254,255] Second, patients with OSAS are sleep-deprived. Testosterone concentrations fall with prolonged physical stress, sleep deprivation and sleep fragmentation in normal young men.[256-258] Third, repetitive episodes of hypoxaemia is typical for OSAS. Hypoxia decreases luteinizing hormone (LH) and testosterone levels, and alters circadian rhythm of testosterone secretion.[7,259,260] Depression of testosterone levels correlates with the severity of hypoxemia in patients with sleep apnoea.[7,259,260] Testosterone levels rise with oxygen therapy in COPD[261] and with weight reduction in obesity hypoventilation syndrome.[262] Fourth, decreased testosterone levels may be part of an adaptive homeostatic mechanism to reduce SDB assuming that testosterone aggravates SDB.

Both LH and testosterone secretion increase during sleep.[263,264] Since CPAP therapy improves sleep quality, it is logical that also the decreased testosterone levels[124,127,128] are normalized on nasal CPAP therapy.[124,127]

Androgen replacement therapy is likely to become more common in the treatment of andropausal symptoms in ageing men. With the availability of preparations developed specifically for women, androgen replacement therapy is likely to become more widespread also as a treatment of fatigue, decreased libido or osteoporosis in postmenopausal women.[265] The appearance of symptoms suggesting SDB should be monitored in both men and in women during androgen replacement therapy.

8. Conclusions

SDB is still underdiagnosed,[266] also when appearing in connection with endocrine disorders. There are complex interactions between hormones and SDB. Nasal CPAP is currently the treatment of choice in mild to severe OSAS,[30,267] also partly restoring the endocrine balance. A better understanding of hormones and breathing may open new perspectives in developing strategies to prevent, alleviate or cure SDB and its systemic consequences. Sleep apnoea is a system disorder challenging both clinicians and scientists to combat the sleep apnoea epidemic.

9. *References*

1. Young T, Palta M, Dempsey J, et al. The occurrence of sleep-disordered breathing among middle-aged adults. N Engl J Med 1993;328:1230-1235.
2. Bixler EO, Vgontzas AN, Lin H-M, et al. Prevalence of sleep-disordered breathing in women: effects of gender. Am J Respir Crit Care Med 2001;163:608-613.
3. Van Cauter E. Endocrine physiology. In: Kryger MH, Roth T, Dement WC, eds. Principles and practise of sleep medicine. Third ed. New York: W.B. Saunders Company, 2000:266-278.
4. Saaresranta T, Polo O. Hormones and breathing. Chest 2002;122:2165-2182.
5. Späth-Schwalbe E, Gofferje M, Kern W, et al. Sleep disruption alters nocturnal ACTH and cortisol secretory patterns. Biol Psychiatry 1991;29:575-584.
6. Pastuszko A, Wilson DF, Ericinsk M. Neurotransmitter metabolism in rat brain synaptosomes: effect of anoxia and pH. J Neurochem 1982;38:1657-1667.

7. Semple PD, Beastall GH, Watson WS, et al. Hypothalamic-pituitary dysfunction in respiratory hypoxia. Thorax 1981;36:605-609.

8. Raff H, Roarty TP. Renin, ACTH, and aldosterone during acute hypercapnia and hypoxia in conscious rats. Am J Physiol 1988;254:R431-R435.

9. Raff H, Shinsako J, Keil LC, et al. Vasopressin, ACTH, and corticosteroids during hypercapnia and graded hypoxia in dogs. Am J Physiol 1983;244:E453-E458.

10. Young T, Peppard PE, Gottlieb DJ. Epidemiology of obstructive sleep apnea: a population health perspective. Am J Respir Crit Care Med 2002;165:1217-1239.

11. Polo-Kantola P, Rauhala E, Helenius H, et al. Breathing during sleep in menopause: a randomized, controlled, cross-over trial with estrogen therapy. Obstet Gynecol 2003;102:68-75.

12. Hart TB, Radow SK, Blackard WG, et al. Sleep apnea in acromegaly. Arch Intern Med 1985;145:865-866.

13. Pekkarinen T, Partinen M, Pelkonen R, et al. Sleep apnoea and daytime sleepiness in acromegaly: relationship to endocrinological factors. Clin Endocrinol Oxf 1987;27:649-654.

14. Grunstein R, Kian KY, Sullivan CE. Sleep apnea in acromegaly. Ann Intern Med 1991;115:527-532.

15. Pelttari L, Polo O, Rauhala E, et al. Nocturnal breathing abnormalities in acromegaly after adenomectomy. Clin Endocrinol 1995;43:175-182.

16. Rosenow F, Reuter S, Deuss U, et al. Sleep apnoea in treated acromegaly: relative frequency and predisposing factors. Clin Endocrinol 1996;45:563-569.

17. Weiss V, Sonka K, Pretl M, et al. Prevalence of the sleep apnea syndrome in acromegaly population. J Endocrinol Invest 2000;23:515-519.

18. Rajagopal KR, Abbrecht PH, Derderian SS, et al. Obstructive sleep apnea in hypothyroidism. Ann Intern Med 1984;101:491-494.

19. Grunstein RR, Sullivan CE. Sleep apnea and hypothyroidism: mechanisms and management. Am J Med 1988;85:775-779.

20. Lin C-C, Tsan K-W, Chen P-J. The relationship between sleep apnea syndrome and hypothyroidism. Chest 1992;102:1663-1667.

21. Pelttari L, Rauhala E, Polo O, et al. Upper airway obstruction in hypothyroidism. J Intern Med 1994;236:177-181.

22. Shipley JE, Schteingart DE, Tandon R, et al. Sleep architecture and sleep apnea in patients with Cushing's disease. Sleep 1992;15:514-518.

23. Shipley JE, Schteingart DE, Tandon R, et al. EEG sleep in Cushing disease and Cushing syndrome: comparison with patients with major depressive disorder. Biol Psychiatry 1992;32:146-155.

24. Vgontzas AN, Papanicolaou DA, Bixler EO, et al. Sleep apnea and daytime sleepiness and fatigue: relation to visceral obesity, insulin resistance, and hypercytokinemia. J Clin Endocrinol Metab 2000;85:1151-1158.

25. Saarelainen S, Lahtela J, Kallonen E. Effect of nasal CPAP treatment on insulin sensitivity and plasma leptin. J Sleep Res 1997;6:146-147.

26. Chin K, Shimizu K, Nakamura T, et al. Changes in intra-abdominal visceral fat and serum leptin levels in patients with obstructive sleep apnea syndrome following nasal continuous positive airway pressure therapy. Circulation 1999;100:706-712.

27. Ip MS, Lam KSL, Ho C, et al. Serum leptin and vascular risk factors in obstructive sleep apnea. Chest 2000;118:580-586.

28. Ip MSM, Lam B, Ng MMT, et al. Obstructive sleep apnea is independently associated with insulin resistance. Am J Respir Crit Care Med 2002;165:670-676.

29. Punjabi NM, Sorkin JD, Katzel LI, et al. Sleep-disordered breathing and insulin resistance in middle-aged and overweight men. Am J Respir Crit Care Med 2002;165:677-682.
30. Polo O, Berthon-Jones M, Douglas NJ, et al. Management of obstructive sleep apnoea/hypopnoea syndrome. Lancet 1994;344:656-660.
31. Simon C. Ultradian pulsatility of plasma glucose and insulin secretion rate: circadian and sleep modulation. Horm Res 1998;49:185-190.
32. Boden G, Ruiz J, Urbain JL, et al. Evidence for a circadian rhytm of insulin secretion. Am J Physiol 1996;271 2 Pt 1):E246-E252.
33. Danguir J, Nicolaidis S. Chronic intracerebroventricular infusion of insulin causes selective increase of slow wave sleep in rats. Brain Res 1984;306:97-103.
34. Spiegel K, Leproult R, Van Cauter E. Impact of sleep debt on metabolic and endocrine function. Lancet 1999;354:1435-1439.
35. Manzella D, Parillo M, Razzino T, et al. Soluble leptin receptor and insulin resistance as determinant of sleep apnea. Int J Obes 2002;26:370-375.
36. Tiihonen M, Partinen M, Närvänen S. The severity of obstructive sleep apnoea is associated with insulin resistance. J Sleep Res 1993;2:56-61.
37. Vgontzas AN, Legro RS, Bixler EO, et al. Polycystic ovary syndrome is associated with obstructive sleep apnea and daytime sleepiness: role of insulin resistance. J Clin Endocrinol Metab 2001;86:517-520.
38. Mondini S, Guilleminault C. Abnormal breathing patterns during sleep in diabetes. Ann Neurol 1985;17:391-395.
39. Catterall JR, Calverley PMA, Ewing DJ, et al. Breathing, sleep, and diabetic autonomic neuropathy. Diabetes 1984;33:1025-7.
40. Villa P, Multari G, Montesano M, et al. Sleep apnoea in children with diabetes mellitus: effect of glycaemic control. Diabetologia 2000;43:696-702.
41. Katsumata K, Okada T, Miyao M, et al. High incidence of sleep apnea syndrome in a male diabetic population. Diabetes Res Clin Pract 1991;131-2):45-51.
42. Elmasry A, Janson C, Lindberg E, et al. The role of habitual snoring and obesity in the development of diabetes: a 10-year follow-up study in a male population. J Intern Med 2000;248:13-20.
43. Elmasry A, Lindberg E, Berne C, et al. Sleep-disordered breathing and glucose metabolism in hypertensive men: a population-based study. J Intern Med 2001;249:153-161.
44. Ficker JH, Dertinger SH, Siegfried W, et al. Obstructive sleep apnoea and diabetes mellitus: the role of cardiovascular autonomic neuropathy. Eur Respir J 1998; 11:14-19.
45. Meslier N, Gagnadoux F, Giraud P, et al. Impaired glucose-insulin metabolism in males with obstructive sleep apnoea syndrome. Eur Respir J 2003;22:156-160.
46. Brooks B, Cistulli PA, Borkman M, et al. Obstructive sleep apnea in obese non-insulin-dependent diabetic patients: effect of continuous positive airway pressure treatment on insulin responsiveness. J Clin Endocrinol Metab 1994;79:1681-1685.
47. Smurra M, Philip P, Taillard J, et al. CPAP treatment does not affect glucose-insulin metabolism in sleep apneic patients. Sleep Med 2001;2:207-213.
48. Harsch IA, Schahin SP, Radespiel-Tröger M, et al. Continuous positive airway pressure treatment rapidly improves insulin sensitivity in patients with obstructive sleep apnea syndrome. Am J Respir Crit Care Med 2004;169:156-162.
49. Mullington JM, Chan JL, Van Dongen HP, et al. Sleep loss reduces diurnal rhythm amplitude of leptin in healthy men. J Neuroendocrinol 2003;15:851-854.

50. O'Donnell CP, Schaub CD, Haines AS, et al. Leptin prevents respiratory depression in obesity. Am J Respir Crit Care Med 1999;159:1477-1484.

51. Mercer JG, Moar KM, Hoggard N. Localization of leptin receptor Ob-R) messenger ribonucleic acid in the rodent hindbrain. Endocrinology 1998;139:29-34.

52. Shioda S, Funahashi H, Nakajo S, et al. Immunohistochemical localization of leptin receptor in the rat brain. Neurosci Lett 1998;243:41-44.

53. Polotsky VY, Smaldone MC, Scharf MT, et al. Impact of interrupted leptin pathways on ventilatory control. J Appl Physiol 2004;96:991-998.

54. Branson R, Protoczna N, Kral JG, et al. Binge eating as a major phenotype of melanocortin 4 receptor gene mutations. N Engl J Med 2003;348:1096-1103.

55. Farooqi IS, Keogh JM, Yeo GS, et al. Clinical spectrum of obesity and mutations in the melanocortin 4 receptor gene. N Engl J Med 2003;348:1085-95.

56. Palmer LJ, Buxbaum SG, Larkin E, et al. A whole-genome scan for obstructive sleep apnea and obesity. Am J Hum Genet 2003;72:340-350.

57. Phillips BG, Kato M, Narkiewicz K, et al. Increases in leptin levels, sympathetic drive, and weight gain in obstructive sleep apnea. Am J Physiol Heart Circ Physiol 2000;279:H234-H237.

58. Phipps PR, Starritt E, Caterson I, et al. Association of serum leptin with hypoventilation in human obesity. Thorax 2002;57:75-76.

59. Schafer H, Pauleit D, Sudhop T, et al. Body fat distribution, serum leptin, and cardiovascular risk factors in men with obstructive sleep apnea. Chest 2002;122:774-778.

60. Öztürk L, Ünal M, Tamer L, et al. The association of the severity of obstructive sleep apnea with plasma leptin levels. Arch Otolaryngol Head Neck Surg 2003;129:538-540.

61. Wallace AM, McMahon AD, Packard CJ, et al. Plasma leptin and risk of cardiovascular disease in the West of Scotland Coronary Prevention Study WOSCOPS). Circulation 2001;104:3052-3056.

62. Segal KR, Landt M, Klein S. Relationship between insulin sensitivity and plasma leptin concentration in lean and obese men. Diabetes 1996;45:988-91.

63. Söderberg S, Olsson T, Eliasson M, et al. Plasma leptin levels are associated with abnormal fibrinolysis in men and postmenopausal women. J Intern Med 1999;245:533-543.

64. Chessler SD, Fujimoto WY, Shofer JB, et al. Increased plasma leptin levels are associated with fat accumulation in Japanese Americans. Diabetes 1998;47:239-243.

65. McNeely MJ, Boyko EJ, Weigle DS, et al. Association between baseline plasma leptin levels and subsequent development of diabetes in Japanese Americans. Diabetes Care 1999;22:65-70.

66. Shimizu K, Chin K, Nakamura T, et al. Plasma leptin levels and cardiac sympathetic function in patients with obstructive sleep apnoea-hypopnoea syndrome. Thorax 2002;57:429-434.

67. Saini J, Krieger J, Brandenberger G, et al. Continuous positive airway pressure treatment. Effects on growth hormone, insulin and glucose profiles in obstructive sleep apnea patients. Horm Metab Res 1993;25:375-381.

68. Cooper BG, White JE, Ashworth LA, et al. Hormonal and metabolic profiles in subjects with obstructive sleep apnea syndrome and the effects of nasal continuous positive airway pressure CPAP) treatment. Sleep 1995;18:172-179.

69. Grunstein RR. Metabolic aspects of sleep apnea. Sleep 1996;19:S218-S220.

70. Isozaki O, Tsushima T, Miyakawa M, et al. Interaction between leptin and growth hormone GH)/IGF-I axis. Endocr J 1999;46 Suppl):S17-S24.

71. Randeva HS, Murray RD, Lewandowski KC, et al. Differential effects of GH replacement on the components of the leptin system in GH-deficient individuals. J Clin Endocrinol Metab 2002;87:798-804.
72. Grosfeld A, Zilberfarb V, Turban S, et al. Hypoxia increases leptin expression in human PAZ6 adipose cells. Diabetologia 2002;45:527-530.
73. Ambrosini G, Nath AK, Sierra-Honigmann MR, et al. Transcriptional activation of the human leptin gene in response to hypoxia: involvement of hypoxia-inducible factor 1. J Biol Chem 2002;277:34601-34609.
74. Tschöp M, Strasburger CJ, Hartmann G, et al. Raised leptin concentrations at high altitude associated with loss of appetite. Lancet 1998;352:1119-1120.
75. Saaresranta T, Irjala K, Polo O. Effect of medroxyprogesterone acetate on arterial blood gases, serum leptin levels, and neuropeptide Y. Eur Respir J 2002;20:1413-1418.
76. Irwin M, Thompson J, Miller C, et al. Effects of sleep and sleep deprivation on catecholamine and interleukin-2 levels in humans: clinical implications. J Clin Endocrinol Metab 1999;84:1979-1985.
77. Carlson JT, Hedner J, Elam M, et al. Augmented resting sympathetic activity in awake patients with obstructive sleep apnea. Chest 1993;103:1763-1768.
78. Eisenberg E, Zimlichman R, Lavie P. Plasma norepinephrine levels in patients with sleep apnea syndrome. N Engl J Med 1990;322:932-933.
79. Bratel T, Wennlund A, Carlström K. Pituitary reactivity, androgens and catecholamines in obstructive sleep apnoea. Effects of continuous positive airway pressure treatment CPAP). Respir Med 1999;93:1-7.
80. Fidone SJ, Gonzalez C, Yoshizaki K. Putative neurotransmitters in the carotid body: the case for dopamine. Federation Proc 1980;39:2636-2640.
81. Narkiewicz K, Kato M, Pesek CA, et al. Human obesity is characterized by a selective potentiation of central chemoreflex sensitivity. Hypertension 1999;33:1153-1158.
82. Somers VK, Dyken ME, Clary MP, et al. Sympathetic neural mechanisms in obstructive sleep apnea. J Clin Invest 1995;96:1897-1904.
83. Narkiewicz K, van de Borne PJH, Pesek CA, et al. Selective potentiation of peripheral chemoreflex sensitivity in obstructive sleep apnea. Circulation 1999;99:1183-1189.
84. Somers VK, Mark AL, Zalava DC, et al. Contrasting effects of hypoxia and hypercapnia on ventilatory and sympathetic activity in humans. J Appl Physiol 1989;67:2101-2106.
85. Hedner J, Darpo B, Ejnell H, et al. Reduction in sympathetic activity after long-term CPAP treatment in sleep apnoea: cardiovascular implications. Eur Respir J 1995;8:222-229.
86. Coy TV, Dimsdale JE, Ancoli-Israel S, et al. Sleep apnoea and sympathetic nervous system activity: a review. J Sleep Res 1996;5:42-50.
87. Rodenstein DO, D'Odemont JP, Pieters T, et al. Diurnal and nocturnal diuresis and natriuresis in obstructive sleep apnea. Effects of nasal continuous positive airway pressure therapy. Am Rev Respir Dis 1992;145:1367-1371.
88. Minemura H, Akashiba T, Yamamoto H, et al. Acute effects of nasal continuous positive airway pressure on 24-hour blood pressure and catecholamines in patients with obstructive sleep apnea. Intern Med 1998;37:1009-1013.
89. Ookuma K, Yoshimatsu H, Sakata T, et al. Hypothalamic sites of neuronal histamine action on food intake by rats. Brain Res 1989;490:268-275.

128. Luboshitzky R, Aviv A, Hefetz A, et al. Decreased pituitary-gonadal secretion in men with obstructive sleep apnea. J Clin Endocrinol Metab 2002;87:3394-3398.

129. Ho KK, Weissberger AJ. Secretory patterns of growth hormone according to sex and age. Horm Res 1990;33:7-11.

130. Wehrenberg WB, Giustina A. Mechanisms and pathways of gonadal steroid modulation of growth hormone secretion. Endocr Rev 1992;13:299-308.

131. Roxburgh F, Collis A. Notes on a case of acromegaly. BMJ 1896:63-65.

132. Guilleminault C, Cummiskey J, Dement WC. Sleep apnea syndrome: recent advances. Adv Intern Med 1980;26:347-372.

133. Guilleminault C, Simmons FB, Motta J, et al. Obstructive sleep apnea syndrome and tracheostomy. Long-term follow-up experience. Arch Intern Med 1981;141:985-988.

134. Cadieux RJ, Kales A, Santen RJ, et al. Endoscopic findings in sleep apnea associated with acromegaly. J Clin Endocrinol Metab 1982;55:18-22.

135. Mezon BJ, West P, MacClean J, et al. Sleep apnea in acromegaly. Am J Med 1980;69:615-618.

136. Grunstein RR, Ho KY, Berthon-Jones M, et al. Central sleep apnea is associated with increased ventilatory response to carbon dioxide and hypersecretion of growth hormone in patients with acromegaly. Am J Respir Crit Care Med 1994;150:496-502.

137. Rosenow F, Reuter S, Szelies B, et al. Sleep apnoea in acromegaly -prevalence, pathogenesis and therapy. Presse Med 1994;23:1203-1208.

138. Ip MS, Tan KC, Peh WC, et al. Effect of Sandostatin LAR on sleep apnoea in acromegaly: correlation with computerized tomographic cephalometry and hormonal activity. Clin Endocrinol 2001;55:477-483.

139. Perks WH, Horrocks PM, Cooper RA, et al. Sleep apnea in acromegaly. BMJ 1980;280:894-897.

140. Chanson P, Timsit J, Benoit O, et al. Rapid improvement in sleep apnoea of acromegaly after short-term treatment with somatostatin analogue SMS 201-995. Lancet 1986;2:1270-1271.

141. Leibowitz G, Shapiro MS, Salameh M, et al. Improvement of sleep apnoea due to acromegaly during short-term treatment with octreotide. J Intern Med 1994;236:231-235.

142. Rosenow F, McCarthy V, Caruso AC. Sleep apnoea in endocrine diseases. J Sleep Res 1998;7:3-11.

143. Mickelson SA, Rosenthal LD, Rock JP, et al. Obstructive sleep apnea syndrome and acromegaly. Otolaryngol Head Neck Surg 1994;111:25-30.

144. Dagan Y, Abadi J, Lifschitz A, et al. Severe obstructive sleep apnoea syndrome in an adult patient with Laron syndrome. Growth Hormone & IGF Research 2001;11:247-249.

145. Orliaguet O, Pepin JL, Bettega G, et al. Sleep apnoea and Turner's syndrome. Eur Respir J 2001;17:153-155.

146. Partsch CJ, Lammer C, Gillessen-Kaesbach G, et al. Adult patients with Prader-Willi Syndrome: clinical characteristics, life circumstances and growth hormone secretion. Growth Hormone & IGF Research 2000;10 Suppl B):S81-85.

147. Hertz G, Cataletto M, Feinsilver SH, et al. Sleep and breathing patterns in patients with Prader Willi Syndrome (PWS): effects of age and gender. Sleep 1993;16:366-371.

148. Van-Cauter E, Caufriez A, Kerkhofs M, et al. Sleep, awakenings, and insulin-like growth factor-I modulate the growth hormone GH) secretory response to GH-releasing hormone. J Clin Endocrinol Metab 1992;74:1451-1459.
149. Veldhuis JD, Iranmanesh A, Ho KK, et al. Dual effects in pulsatile growth hormone secretion and clearance subserve the hyposomatotropism of obesity in man. J Clin Endocrinol Metab 1991;72:51-59.
150. Scacchi M, Pincelli AI, Cavagnini F. Growth hormone in obesity. Int J Obes Relat Metab Disord 1999;23:260-271.
151. Issa FG, Sullivan CE. The immediate effects of nasal continuous positive airway pressure treatment on sleep pattern in patients with obstructive sleep apnea syndrome. Electroencephalogr Clin Neurophysiol 1986;63:10-17.
152. Zhang Y-S, Du J-Z. The response of growth hormone and prolactin of rats to hypoxia. Neuroscience Letters 2000;279:137-140.
153. Conceicao FL, Bojensen A, Jorgensen JOL, et al. Growth hormone therapy in adults. Frontiers in Neuroendocrinology 2001;22:213-246.
154. Wu H, Jeng YY, Yue C, et al. Endothelial-dependent vascular effects of insulin and insulin-like growth factor I in the perfused rat mesenteric artery and aortic ring. Diabetes 1994;43:1027-1032.
155. Bar RS, Boes M, Dake BL, et al. Insulin, insulin-like growth factors, and vascular endothelium. Am J Med 1988;85Suppl 5a):59-70.
156. Tsukahara H, Gordienko DV, Tonshoff B, et al. Direct demonstration of insulin-like growth factor-I-induced nitric oxide production by endothelial cells. Kidney Int 1994;45:598-604.
157. Radomski MW, Salas E. Nitric-oxide biological mediator, modulator and factor of injury: its role in the pathogenesis of atherosclerosis. Atherosclerosis 1995;118Suppl):S69-S80.
158. Gerard JM, Garibaldi L, Myers SE, et al. Sleep apnea in patients receiving growth hormone. Clin Ped 1997;1997:321-326.
159. Nolte W, Radisch C, Rodenbeck A, et al. Polysomnographic findings in five adult patients with pituitary insufficiency before and after cessation of human growth hormone replacement therapy. Clin Endocrinol 2002;56:805-810.
160. Meston N, Davies RJ, Mullins R, et al. Endocrine effects of nasal continuous positive airway pressure in male patients with obstructive sleep apnoea. J Intern Med 2003;254:447-454.
161. Ip MSM, Lam B, Chan L-Y, et al. Circulating nitric oxide is suppressed in obstructive sleep apnea and is reversed by nasal continuous positive airway pressure. Am J Respir Crit Care Med 2000;162:2166-2171.
162. Waters KA, Kirjavainen T, Jimenez M, et al. Overnight growth hormone secretion in achondroplacia: deconvolution analysis, correlation with sleep state, and changes after treatment of obstructive sleep apnea. Pediatr Res 1996;39:547-553.
163. Goldstein SJ, Shprintzen RJ, Wu RH, et al. Achondroplasia and obstructive sleep apnea: correction of apnea and abnormal sleep-entrained growth hormone release by tracheostomy. Birth Defects Orig Artic Ser 1985;21:93-101.
164. Nieminen P, Löppönen T, Tolonen U, et al. Growth and biochemical markers of growth in children with snoring and obstructive sleep apnea. Pediatrics 2002;109:e55.
165. Holsboer F, von Bardeleben U, Steiger A. Effects of intravenous corticotropin-releasing hormone upon sleep-related growth hormone surge and sleep EEG in man. Neuroendocrinology 1988;48:32-38.

166. Friess E, von Bardeleben U, Wiedemann K, et al. Effects of pulsatile cortisol infusion on sleep-EEG and nocturnal growth hormone release in healthy men. J Sleep Res 1994;3:73-79.

167. Bohlhalter S, Murck H, Holsboer F, et al. Cortisol enhances non-REM sleep and growth hormone secretion in elderly subjects. Neurobiol Aging 1997;18: 423-429.

168. Huber I, Krause U, Nink M, et al. Dexamethasone does not suppress the respiratory analeptic effect of corticotropin-releasing hormone. J Clin Endocrinol Metab 1989;69:440-442.

169. Mann K, Röschke J, Benkert O, et al. Effects of corticotropin-releasing hormone on respiratory parameters during sleep in normal men. Exp Clin Endocrinol 1995;103:233-240.

170. Nink M, Beyer J, Krause U, et al. Effects of corticotropin-releasing hormone on the postoperative course of elderly patients under long-term artificial respiration. Acta Endocrinol Copenh) 1992;127:200-204.

171. Mensah-Nyagan AG, Do-Rego JL, Beaujean D, et al. Neurosteroids: expression of steroidogenic enzymes and regulation of steroid biosythesis in the central nervous system. Pharmacological Reviews 1999;51:63-81.

172. Stoffel-Wagner B. Neurosteroid metabolism in the human brain. Eur J Endocrinol 2001;145:669-679.

173. Lephart ED, Lund TD, Horvath TL. Brain androgen and progesterone metabolizing enzymes: biosynthesis, distribution and function. Brain ResRev 2001;37:25-37.

174. Kastrup Y, Hallbeck M, Amandusson A, et al. Progesterone receptor expression in the brainstem of the female rat. Neurosci Lett 1999;275:85-88.

175. Shugrue PJ, Lane MV, Merchemthaler I. Comparative distributiojn of estrogen receptor-alpha and -beta mRNA in the rat central nervous system. J Comp Neurol 1997;388:507-525.

176. Simerly RB, Chang C, Muramatsu M, et al. Distribution of androgen and estrogen receptor mRNA-containing cells in the rat brain: an in situ hybridization study. J Comp Neurol 1990;294:76-95.

177. Yu WH, McGinnis MY. Androgen receptors in cranial nerve motor nuclei of male and female rats. J Neurobiol 2001;46:1-10.

178. Behan M, Zabka AG, Thomas CF, et al. Sex steroid hormones and the neural control of breathing. Respir Phys Neurobiol 2003;136:249-263.

179. Menard CS, Harlan RE. Up-regulation of androgen receptor immunoreactivity in the rat brain by androgenic-anabolic steroids. Brain Res 1993;622:226-236.

180. Lu SF, McKenna SE, Cologer-Clifford A, et al. Androgen receptor in mouse brain: sex differences and similarities in autoregulation. Endocrinology 1998;139:1594-1601.

181. Lynch CS, Story AJ. Dihydrotestosterone and estrogen regulation of rat brain androgen-receptor immunoreactivity. Physiol Behav 2000;69:445-453.

182. Blanco CE, Davenport T, Wachi S, et al. Androgen receptor immunoreactivity of male rat cervical motor neurons is increased by chronic pharmacologic testosterone treatment. Acta Physiol Pharmacol Bulg 2001;26:7-10.

183. Schiff I, Regestein Q, Tulchinsky D, et al. Effects of estrogens on sleep and psychological state of hypogonadal women. Jama 1979;242:2405-2414.

184. Thomson J, Oswald I. Effect of estrogen on the sleep, mood, and anxiety of menopausal women. BMJ 1977;2:1317-1319.

185. Brugge KL, Kripke DF, Ancoli-Israel S. The association of menopausal status and age with sleep disorders. Sleep Research 1989;18:208 Abstract).
186. Polo-Kantola P, Erkkola R, Helenius H, et al. When does estrogen replacement therapy improve sleep quality? Am J Obstet Gynecol 1998;178:1002-1009.
187. Regestein QR, Schiff I, Tulchinsky D, et al. Relationship among estrogen-induced psychophysiological changes in hypogonadal women. Psychosom Med 1981;43:147-155.
188. Polo-Kantola P, Erkkola R, Irjala K, et al. Effect of short-term transdermal estrogen replacement therapy on sleep: a randomized, double-blind crossover trial in postmenopausal women. Fertil Steril 1999;71:873-880.
189. Scharf MB, McDannold MD, Stover R, et al. Effects of estrogen replacement therapy on rates of cyclic alternating patterns and hot-flush events during sleep in postmenopausal women: a pilot study. Clin Ther 1997;19:304-311.
190. Purdie DW, Empson JAC, Crichton C, et al. Hormone replacement therapy, sleep quality and psychological wellbeing. Br J Obstet Gynaecol 1995;102:735-739.
191. Polo O. Sleep in postmenopausal women: better sleep for less satisfaction? Sleep 2003;26:652-653.
192. Friess E, Tagaya H, Trachsel L, et al. Progesterone-induced changes in sleep in male subjects. Am J Physiol 1997;272:E885-E891.
193. Empson JAC, Purdie DW. Effects of sex steroids on sleep. Ann Med 1999;31:141-145.
194. Luboshitzky R, Herer P, Levi M, et al. Relationship between rapid eye movement sleep and testosterone secretion in normal men. J Androl 1999;20:731-737.
195. González-Santos MR, Gajá-Rodríguez OV, Alonso-Uriarte R, et al. Sleep deprivation and adaptive hormonal responses of healthy men. Arch Androl 1989;22:203-207.
196. Schiavi RC, White D, Mandeli J. Pituitary-gonadal function during sleep in healthy aging men. Psychoneuroendocrinology 1992;17:599-609.
197. Saaresranta T, Polo-Kantola P, Irjala K, et al. Respiratory insufficiency in postmenopausal women: sustained improvement of gas exchange with short-term medroxyprogesterone acetate. Chest 1999;115:1581-1587.
198. Saaresranta T, Polo-Kantola P, Rauhala E, et al. Medroxyprogesterone in postmenopausal females with partial upper airway obstruction during sleep. Eur Respir J 2001;18:989-995.
199. Saaresranta T, Polo O. Sleep-disordered breathing and hormones. Eur Respir J 2003;22:161-172.
200. Skatrud JB, Dempsey JA, Kaiser DG. Ventilatory response to medroxyprogesterone acetate in normal subjects: time course and mechanism. J Appl Physiol 1978;446):939-944.
201. Zwillich CW, Natalino MR, Sutton FD, et al. Effects of progesterone on chemosensitivity in normal men. J Lab Clin Med 1978;92:262-269.
202. Hannhart B, Pickett CK, Moore LG. Effects of estrogen and progesterone on carotid body neural output responsiveness to hypoxia. J Appl Physiol 1990;68:1909-1916.
203. Saaresranta T, Aittokallio T, Polo-Kantola P, et al. Effect of medroxyprogesterone on inspiratory flow shapes during sleep in postmenopausal women. Respir Physiol Neurobiol 2003;134:131-143.
204. White DP, Schneider BK, Santen RJ, et al. Influence of testosterone on ventilation and chemosensitivity in male subjects. J Appl Physiol 1985;59:1452-1457.

205. Tatsumi K, Hannhart B, Pickett CK, et al. Effects of testosterone on hypoxic ventilatory and carotid body neural responsiveness. Am J Respir Crit Care Med 1994;149:1248-1253.
206. Emery MJ, Hlastala MP, Matsumoto AM. Depression of hypercapnic ventilatory drive by testosterone in the sleeping infant primate. J Appl Physiol 1994;76:1786-1793.
207. White DP, Douglas NJ, Pickett CK, et al. Sexual influence on the control of breathing. J Appl Physiol 1983;54:874-879.
208. Skatrud JB, Dempsey JA. Interaction of sleep state and chemical stimuli in sustaining rhythmic ventilation. J Appl Physiol 1983;55:813-822.
209. Pilkington S, Carli F, Dakin MJ, et al. Increase in Mallampati score during pregnancy. Br J Anaesth 1995;74:638-642.
210. Bende M, Gredmark T. Nasal stuffiness during pregnancy. Laryngoscope 1999;109:1108-1110.
211. Mabry RL. Rhinitis of pregnancy. South Med J 1986;79:965-971.
212. Brownell LG, West P, Kryger MH. Breathing during sleep in normal pregnant women. Am Rev Respir Dis 1986;133:38-41.
213. Maasilta P, Bachour A, Teramo K, et al. Sleep-related disordered breathing during pregnancy in obese women. Chest 2001;120:1448-1454.
214. Nikkola E, Ekblad U, Ekholm E, et al. Sleep in multiple pregnancy: breathing patterns, oxygenation, and periodic leg movements. Am J Obstet Gynecol 1996;174:1622-1625.
215. Edwards N, Blyton DM, Kirjavainen TT, et al. Hemodynamic responses to obstructive respiratory events during sleep are augmented in women with preeclampsia. Am J Hypertens 2001;14:1090-1095.
216. Connolly G, Razak AR, Hayanga A, et al. Inspiratory flow limitation during sleep in pre-eclampsia: comparison with normal pregnant women. Eur Respir J 2001;18:672-673.
217. Loube MDI, Poceta JS, Morales MC, et al. Self-reported snoring in pregnancy. Association with fetal outcome. Chest 1996;109:885-889.
218. Franklin KA, Holmgren PÅ, Jönsson F, et al. Snoring, pregnancy-induced hypertension, and growth retardation of the fetus. Chest 2000;117:137-141.
219. Guilleminault C, Querra-Salva M-A, Chwdhuri S, et al. Normal pregnancy, daytime sleeping, snoring and blood pressure. Sleep Med 2000;1:289-297.
220. Lefcourt LA, Rodis JF. Obstructive sleep apnea in pregnancy. Obstet Gynecol Survey 1996;51:503-506.
221. Polo O, Ekholm E. Nocturnal hyperventilation in pregnancy -reversal with nasal continuous positive airway pressure. Am J Obstet Gynecol 1995;173:238-239.
222. Guilleminault C, Kreutzer M, Chang JL. Pregnancy, sleep disordered breathing and treatment with nasal continuous positive airway pressure. Sleep Med 2004;5:43-51.
223. Fogel RB, Malhotra A, Pillar G, et al. Increased prevalence of obstructive sleep apnea syndrome in obese women with polycystic ovary syndrome. J Clin Endocrinol Metab 2001;86:1175-1180.
224. Block AJ, Boysen PG, Wynne JW, et al. Sleep apnea, hypopnea and oxygen desaturation in normal subjects. A strong male predominance. N Engl J Med 1979;300:513-517.
225. Guilleminault C, Quera-Salva MA, Partinen M, et al. Women and the obstructive sleep apnea syndrome. Chest 1988;93:104-109.

226. Vgontzas AN, Tan TL, Bixler EO, et al. Sleep apnea and sleep disruption in obese patients. Arch Intern Med 1994;154:1705-1711.
227. Young T, Palta M, Dempsey J, et al. The occurrence of sleep-disordered breathing among middle-aged adults. N Engl J Med 1993;328:1230-1235.
228. Redline S, Kump K, Tishler PV, et al. Gender differences in sleep disordered breathing in a community-based sample. Am J Respir Crit Care Med 1994;149:722-726.
229. Kripke DF, Ancoli-Israel S, Klauber MR, et al. Prevalence of sleep-disordered breathing in ages 40-64 yr: a population-based survey. Sleep 1997;20:65-76.
230. Olson LG, King MT, Hensley MJ, et al. A community study of snoring and sleep-disordered breathing: prevalence. Am J Respir Crit Care Med 1995;152:711-716.
231. Block AJ, Wynne JW, Boysen PG. Sleep-disordered breathing and nocturnal oxygen desaturation in postmenopausal women. Am J Med 1980;69:75-79.
232. Dancey DR, Hanly PJ, Soong C, et al. Impact of menopause on the prevalence and severity of sleep apnea. Chest 2001;120:151-155.
233. Gislason T, Benediktsdottir B, Björnsson JK, et al. Snoring, hypertension, and the sleep apnea syndrome. An epidemiologic study of middle-aged women. Chest 1993;103:1147-1151.
234. Young T. . Am J Respir Crit Care Med 2003;167:1181-1185.
235. Shahar E, Redline S, Young T, et al. Hormone-Replacement Therapy and Sleep-Disordered Breathing. Am J Respir Crit Care Med 2003;167:1186-1192.
236. Collop NA. Medroxyprogesterone acetate and ethanol-induced exacerbation of obstructive sleep apnea. Chest 1994;106:792-799.
237. Cistulli PA, Barnes DJ, Grunstein RR, et al. Effect of short term hormone replacement in the treatment of obstructive sleep apnoea in postmenopausal women. Thorax 1994;49:699-702.
238. Keefe DL, Watson R, Naftolin F. Hormone replacement therapy may alleviate sleep apnea in menopausal women: a pilot study. Menopause 1999;6:196-200.
239. Manber R, Kuo TF, Cataldo N, et al. The effects of hormone replacement therapy on sleep-disordred breathing in postmenopausal women: a pilot study. Sleep 2003;26:163-168.
240. Poulin R, Simard J, Labrie C, et al. Down-regulation of estrogen receptors by androgens in the ZR-75-1 human breast cancer cell line. Endocrinology 1989;125:392-399.
241. Harman E, Wynne JW, Block AJ, et al. Sleep-disordered breathing and oxygen desaturation in obese patients. Chest 1981;79:256-260.
242. Matsumoto AM, Sandblom RE, Schoene RB, et al. Testosterone replacement in hypogonadal men: effects on obstructive sleep apnoea, respiratory drives, and sleep. Clin Endocrinol 1985;22:713-721.
243. Sandblom R, Matsumoto A, Schoene R, et al. Obstructive sleep apnea syndrome induced by testosterone administration. N Engl J Med 1983;308:508-510.
244. Schneider BK, Pickett CK, Zwillich CW, et al. Influence of testosterone on breathing during sleep. J Appl Physiol 1986;61:618-623.
245. Liu PY, Yee B, Wishart SM, et al. The short-term effects of high-dose testosterone on sleep, breathing, and function in older men. J Clin Endocrinol Metab 2003;84:3605-3613.
246. Mantzoros CS, Georgiadis EI, Evangelopoulou K, et al. Dehydroepiandrosterone sulfate and testosterone are independently associated with body fat distribution in premenopausal women. Epidemiology 1996;7:513-516.

247. Goodman-Gruen D, Barrett-Connor E. Total but not bioavailable testosterone is a predictor of central adiposity in postmenopausal women. Int J Obes Relat Metab Disord 1995;19:293-298.
248. Koskenvuo M, Partinen M, Kaprio J. Snoring and disease. Ann Clin Res 1985;175):247-251.
249. Lindberg E, Taube A, Janson C, et al. A 10-year follow-up of snoring in men. Chest 1998;114:1048-1055.
250. Dexter DD, Dovre EJ. Obstructive sleep apnea due to endogenous testosterone production in a woman. Mayo Clin Proc 1998;73:246-248.
251. Johnson MW, Anch AM, Remmers JE. Induction of the obstructive sleep apnea syndrome in woman by exogenous androgen administration. Am Rev Respir Dis 1984;129:1023-1025.
252. Stewart DA, Grunstein RR, Berthon-Jones M, et al. Androgen blockade does not affect sleep-disordered breathing or chemosensitivity in men with obstructive sleep apnea. Am Rev Respir Dis 1992;146:1389-1393.
253. Millman RP, Kimmel PL, Shore ET, et al. Sleep apnea in hemodialysis patients: the lack of testosterone effect on its pathogenesis. Nephron 1985;40:407-410.
254. Glass AR. Endocrine aspects of obesity. Med Clin North Am 1989;73:139-160.
255. Lima N, Cavaliere H, Knobel M, et al. Decreased androgen levels in massively obese men may be associated with impaired function of the gonadostat. Int J Obes Relat Metab Disord 2000;24:1433-1437.
256. Elman I, Breier A. Effects of acute metabolic stress on plasma progesterone and testosterone in male subjects: relationship to pituitary-adrenocortical axis activation. Life Sci 1997;61:1705-1712.
257. Luboshitzky R, Zabari Z, Shen-Orr Z, et al. Disruption of the nocturnal testosterone rhythm by sleep fragmentation in normal men. J Clin Endocrinol Metab 2001;86:1134-1139.
258. Singer F, Zumoff B. Subnormal serum testosterone levels in male internal medicine residents. Steroids 1992;57:86-89.
259. Kouchiyama S, Masuyama S, Shinozaki T, et al. [Prediction of the degree of nocturnal oxygen desaturation in sleep apnea syndrome by estimating the testosterone level]. Nippon Kyobu Shikkan Gakkai Zasshi 1989;27:941-945.
260. Kouchiyama S, Honda Y, Kuriyama T. Influence of nocturnal oxygen desaturation on circadian rhythm of testosterone secretion. Respiration 1990;57:359-363.
261. Aasebo U, Gyltnes A, Bremnes RM, et al. Reversal of sexual impotence in male patients with chronic obstructive pulmonary disease and hypoxemia with long-term oxygen therapy. J Steroid Biochem Molec Biol 1993;46:799-803.
262. Semple PA, Graham A, Malcolm Y, et al. Hypoxia, depression of testosterone, and impotence in pickwickian syndrome reversed by weight reduction. Br Med J Clin Res Ed 1984;289:801-802.
263. Pietrowsky R, Meyrer R, Kern W, et al. Effects of diurnal sleep on secretion of cortisol, luteinizing hormone, and growth hormone in man. J Clin Endocrinol Metab 1994;783):683-687.
264. Pincus SM, Mulligan T, Iranamanesh A, et al. Older males secrete luteinizing hormone and testosterone more irregularly, and jointly more asynchronously, than younger males. Proc Natl Acad Sci 1996;93:14100-14005.
265. Davis S. Androgen replacement in women: a commentary. J Clin Endocrinol Metab 1999;84:1886-1891.

266. Young T, Evans L, Finn L, et al. Estimation of the clinically diagnosed proportion of sleep apnea syndrome in middle-aged men and women. Sleep 1997;20:705-706.
267. Polo O. Continuous positive airway pressure for treatment of sleep apnoea. Lancet 1999;353:2086-2087.
268. Rees PJ, Prior JG, Cochrane GM, et al. Sleep apnoea in diabetic patients with autonomic neuropathy. J R Soc Med 1981;74:192-195.
269. Grunstein R, Stewart DA, Lloyd H, et al. Acute withdrawal of nasal CPAP in obstructive sleep apnea does not cause a rise in stress hormones. Sleep 1996;19:774-782.
270. Follenius M, Krieger J, Krauth MO, et al. Obstructive sleep apnoea treatment: peripheral and central effects on plasma renin activity and aldosterone. Sleep 1991;14:211-217.
271. Saarelainen S, Hasan J, Siitonen S, et al. Effect of nasal CPAP treatment on plasma volume, aldosterone and 24-h blood pressure in obstructive sleep apnoea. J Sleep Res 1996;5:181-185.
272. Moller DS, Lind P, Strunge B, et al. Abnormal vasoactive hormones and 24-h blood pressure in obstructive sleep apnea. Am J Hypertens 2003;16:274-280.
273. Krieger J, Follenius M, Sforza E, et al. Effects of treatment with nasal continuous positive airway pressure on atrial natriuretic peptide and arginine vasopressin release during sleep in patients with obstructive sleep apnoea. Clin Sci 1991;80:443-449.
274. Ichioka M, Hirata Y, Inase N, et al. Changes of circulating atrial natriuretic peptide and antidiuretic hormone in obstructive sleep apnea syndrome. Respiration 1992;59:164-168.
275. Spiegel K, Follenius M, Krieger J, et al. Prolactin secretion during sleep in obstructive sleep apnoea patients. J Sleep Res 1995;4:56-62.
276. Gislason T, Hedner J, Terenius L, et al. Substance P, thyrotropin-releasing hormone, and monoamine metabolites in cerebrospinal fluid in sleep apnea patients. Am Rev Respir Dis 1992;146:784-786.

Autonomic Nervous System Activity during Sleep in Humans

Heart Rate Variability during Sleep

GABRIELLE BRANDENBERGER AND ANTOINE U. VIOLA

1. Introduction

Heart rate (HR) is determined by the rate of depolarization of the cardiac pacemaker which is found in the sinoatrial node, the atrioventricular node, and the Purkinje tissue. The intrinsic HR in absence of any neurohumoral influence is about 100 to 120 bpm. In the intact individual, HR at any time represents the net effect of the parasympathetic nerves which slow it and the sympathetic nerves which accelerate it. In most physiological conditions, sympathetic and parasympathetic (vagal) activities modulating HR undergo a reciprocal regulation, leading to the classic notion of sympathovagal balance.

Sympathetic excitation and simultaneous vagal inhibition, or vice versa, are presumed to contribute to the beat-to-beat fluctuations of R-R intervals, also known as heart rate variability (HRV) which occur even at rest (Figure 1). There has been increasing interest in the study of HRV because of the predictive association between reduced HRV and higher risk for coronary disease, mortality after myocardial infarction, and all-cause mortality.[1-8] High sympathetic activity is a potent descriptor of poor survival,[9] whereas high vagal tone provides some degree of cardioprotection.[10]

A thorough understanding of the cardiac and vascular effects of sleep may clarify why some cardiovascular events occur less often and some others more often during sleep than during wakefulness. Reports from Somers et al.[11] and Murali et al.[12] indicate that in humans, slow wave sleep (SWS) lowers arterial blood pressure and causes bradycardia, and that both changes become more pronounced as sleep progresses from its minimal to its maximal slow wave activity. During rapid eye movement (REM) sleep large transient increases in blood pressure reverse the low pressure that characterizes the preceding SWS.

More importantly, by directly recording sympathetic nerve traffic, these investigators have provided evidence that sympathetic nerve activity is reduced by more than half from wakefulness to sleep stage 4 but increased to levels above waking values during REM sleep. These striking changes strongly suggest a primary role for sympathetic nervous system in the production of both the inhibitory and the excitatory cardiovascular phenomena of sleep.[13]

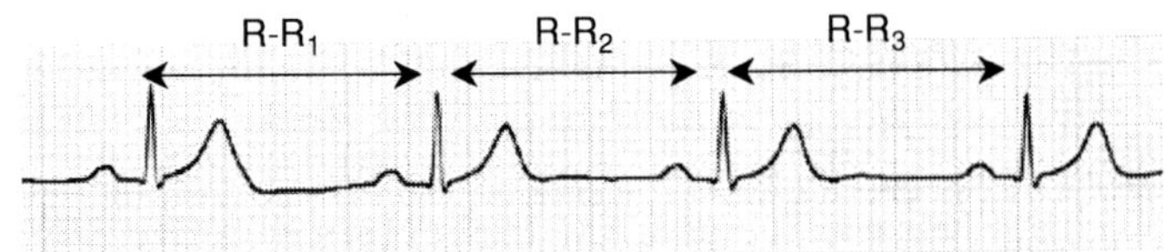

FIGURE 1. Successive R-R intervals. Heart rate variability is defined as the time fluctuations between R-R intervals.

Here we provide data on the relationship between the time and frequency domain HRV indexes and sleep stages in young (20-30 years) and older persons (55-66 years), throughout the night. We use a method for averaging HRV data derived from Achermann et al.[23] which normalizes the individual differences in the duration of sleep episodes and reconciles sleep stage patterning and spectral analysis of EEG power.

Finally, we propose an original approach for the evaluation of the autonomic nervous system activity based on measures of HR and HRV during the first complete NREM-REM sleep cycle with the idea that it represents a spontaneous autonomic maneuver free of any external influence.

2. HRV Indexes: Methodological Aspects

To quantify HRV, the analog electrocardiogram (ECG) signal is recorded using chest electrodes to obtain a QRS complex of sufficient amplitude and a stable baseline. After analog-to-digital conversion, a computer stores the time intervals between successive R waves.

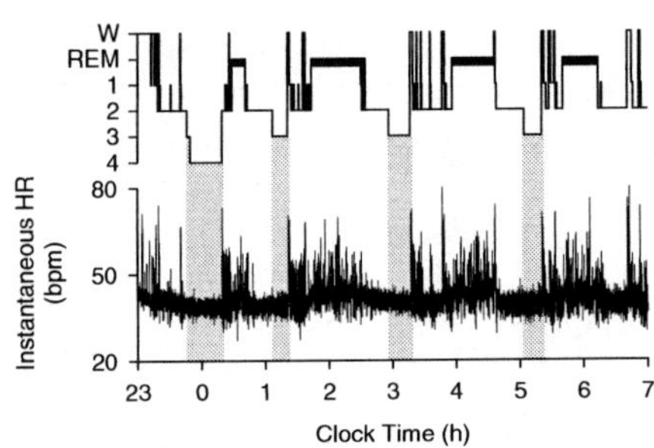

FIGURE 2. Overnight pattern of instantaneous HR (beat/min) in a young subject together with the hypnogram. Quiet periods (i.e. SWS and the preceding sleep stage 2) alternate throughout the night with active periods (i.e. REM sleep and the preceding sleep stage 2).

Simple descriptive statistics are then computed providing mean HR and time domain estimation of HRV, while various algorithms have been used to extract from the tackogram the characteristics of the rhythmic components embedded in its variability.

2.1. *Time Domain Measures*

Some of the more common time domain measures of HRV are a) the standard deviation of normal-to-normal intervals (SDNN) which expresses the overall variability and b) the root-mean-square difference among successive normal R-R intervals (RMSSD) which is considered an index of parasympathetic activity.[24]

The series of R-R intervals can be visualized in a geometrical pattern, such as the Poincaré plot, which is a scatter plot of cardiac R-R intervals (R-Rn+1) against the previous intervals (R-Rn) (Figure 3).

This simple construction provides summary information about overall and instantaneous beat-to-beat variability. Each sleep stage has a characteristic and distinctive Poincaré plot. During SWS when the basic pattern of R-R intervals is very regular with little interbeat variability, the Poincaré plot is characterized by a tight cluster of points, reflecting low overall variability (Figure 3).

During REM sleep, the R-R profiles display larger and slower oscillations. Data on Poincaré plots are then spread along the diagonal showing their high overall variability (Figure 4).

SDNN relates to the variance of the data distributed along the diagonal line of the Poincaré plot projected on to the x-axis. RMSSD relates to the

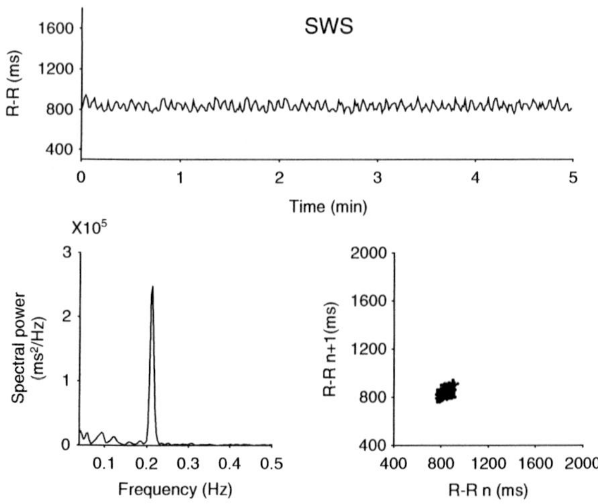

FIGURE 3. Representative pattern of R-R intervals during SWS together with the 5-min Poincaré plot and power spectrum.

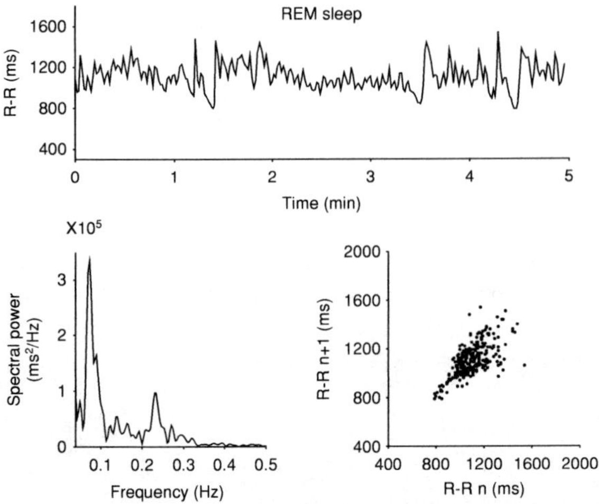

FIGURE 4. Representative pattern of R-R intervals during REM sleep together with the 5-min Poincaré plot and power spectrum.

variance of the distribution of data points perpendicular to the diagonal line. This procedure has been used to identify abnormalities of cardiac dynamic rhythms in numerous pathological conditions, including sudden infant death syndrome,[25,26] heart failure,[27-29] and obstructive sleep apnea.[30] Recently, we proposed a new marker of sympathovagal balance during sleep, the interbeat autocorrelation coefficient (rRR) derived from Poincaré plot, which is closely related to variations in EEG activity.[31,32]

2.2. Frequency Domain Indexes

Power spectrum analysis provides the basic information of how power is distributed as a function of frequency.[3,33-36] Three main spectral components are distinguished in a spectrum calculated from short-term recordings of 5 min: the respiratory rhythm of the heart period variation (whose frequency is synchronous with the respiratory rate), defined as the high-frequency (HF) spectral component (0.15-0.50 Hz; 6.6-2.0 s), is generally considered a marker of vagal modulation. The rhythm corresponding to vasomotor waves and present in heart period and arterial pressure variabilities (also referred to as Mayer waves) is defined as the low-frequency (LF) component (0.04-0.15 Hz; 25.0-6.6 s). It is considered an indicator of sympathetic activation with a parasympathetic component. The very low frequency (VLF) component (< 0.04 Hz) is assumed to be due to long-term regulatory mechanisms such as humoral factors, temperature, and other slow components.

When absolute indexes are used, the changes in total power are usually seen to influence SDNN, RMSSD, LF and HF components in the same direction.

Therefore, some further indexes focusing mainly on the fractional distribution of power and independent of the absolute values of variance are also necessary. With the use of appropriate autonomic blocking agents and experimental strategies, several studies have demonstrated that the LF/HF ratio, or the normalized LF/(LF+HF) and HF/(LF+HF) ratios, are markers of the dynamic balance between sympathetic and parasympathetic activities.[33,35-43] In addition, a high coherence was found between the LF and HF components in R-R variability and similar components present in the discharge variability of cardiac[40,44,45] or muscle sympathetic nerve activity.[46,47] As largely reported in the literature,[14-17] SWS is characterized by low power in the LF band, high power in the HF band (Figure 3), and an increased HF/(LF+HF) ratio, reflecting parasympathetic dominance. In contrast, REM sleep is associated with high LF/(LF+HF), reflecting sympathetic dominance (Figure 4). It should be noted that only Zemaitytë et al.[14] used appropriate drugs (propranol, atropine) during sleep to validate HRV indexes. Thus, the results obtained by microneurography, which revealed increased bursts of sympathetic nerve activity during REM sleep, are of the highest importance.[11]

Spectral analysis that is used to detect possible rhythmicities hidden in the signal necessitates stationary conditions that, in strict terms, are unknown to biology. A practical compromise has to be found between the length of the series and theoretical mathematical requirements. Shorter data segments are more likely to be stationary, and segments with sudden HR variations (arousals) have to be excluded.

Since the initial studies, the interpretation of the various spectral components has become more debatable with the demonstration that LF fluctuations in the R-R interval variability do not accurately reflect changes in sympathetic activity.[48-50] Eckberg[51] argues that the physiology behind the parameters associated with the HRV spectrum may be in danger of being obscured by the language and mathematical manipulations used to describe them. Therefore, care should be taken when evaluating HRV, so much the more as the data appears to depend on the experimental means used to modulate the sympathovagal balance. New methods, based on the wavelet transform[52,53] or adapted from chaos theory,[54-56] will give further insight into the complex changes in beat-to-beat intervals during sleep.

3. Ultradian Oscillations in EEG Activity and HRV Indexes in Young Men: The Duality in Sleep Stage 2

In man, sleep displays one of the most prominent ultradian rhythms characterized by the regular occurrence of two basic sleep states : NREM sleep and REM sleep which alternate with an 80-120 min period. Using EEG spectral analysis which provides a dynamic description of sleep processes, certain authors have reported that slow wave activity (0.5-3.5 Hz) characterizing the depth of sleep oscillates with a similar period. NREM sleep is not uniform, but is composed of successive stages: stages 1 and 2, SWS and stage 2 preceding REM sleep, all of which have to be taken into account, as each of these sleep stages shows widely differing EEG properties. Normalizing the individual hypnograms allowed pure stage 2, REM sleep and SWS to be characterized and averaged among subjects.[57] An ultradian rhythm in SDNN, rRR and the LF/(LF+HF) ratio emerges coupled in a mirror image of the overnight oscillations in slow wave activity (Figure 5). During SWS, SDNN, rRR, and LF/(LF+HF) are the lowest, and during REM sleep, the highest. There is a duality in sleep stage 2, which becomes

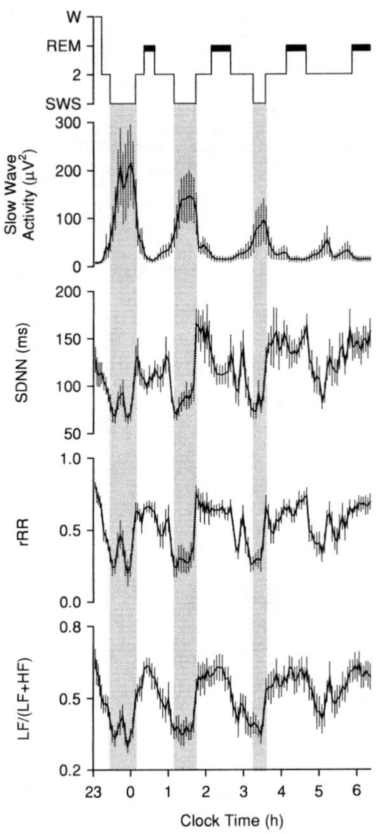

FIGURE 5. Mean (± SE) normalized time courses of slow wave activity, standard deviation of normal R-R intervals (SDNN), interbeat autocorrelation coefficient of R-R intervals (rRR), and the ratio of low frequency power to high-plus low frequency power (LF/(LF+HF) during the NREM-REM sleep cycles in nine good sleepers with normal respiratory patterns. An ultradian rhythm in HRV indexes, inversely coupled to the overnight oscillations in slow wave activity, emerges in good sleepers.

progressively "quieter" with a decrease in SDNN, rRR, and LF/(LF+HF) from REM sleep to SWS, and abruptly "active" during the transition from SWS to REM sleep.

Overnight cross-correlation between the profiles of consecutive 5-min segments in slow wave activity and HRV indexes indicates that cardiac changes precede EEG changes by about 5 min. It is then tempting to propose that an oscillatory process in sympathovagal balance adjusts itself in anticipation of EEG activity, and that sleep disorders might result from an alteration of the autonomic nervous system activity as previously reported[58] or from inadequate coupling between autonomic and EEG ultradian rhythms.

4. Age-Related Changes in HRV

Aging is commonly associated with decreased sleep quality which can influence HRV.[59] The changes with age include reduced lower total sleep time and SWS duration with a disappearance in sleep stage 4, whereas the duration of sleep stage 2 and REM sleep remains quite stable.[60-62] The sudden HR increase associated with the emergence from SWS in the first sleep cycle is lower in elderly than in young subjects (Figure 6). Elderly subjects are also prone to periodic breathing characterized by rises and falls in ventilation without apnea.[63-65] Periodic breathing is found in about 20% of the sleep time and is frequent during REM sleep and the preceding sleep stage 2. It strongly influences HRV which can prevent evaluation of autonomic tone.[66-68]

Two distinct features that depend on respiratory pattern characterize HRV in the elderly:[69] 1) During periods of normal breathing, there is a large drop

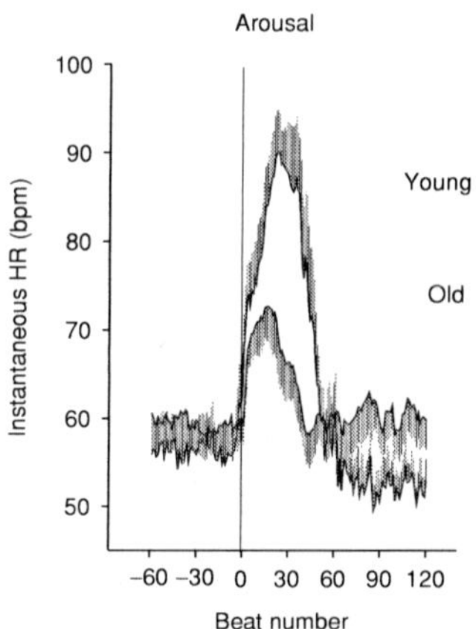

FIGURE 6. Increase in instantaneous HR (means ± SE) during arousal at the transition from SWS to active sleep stage 2 in twelve young and twelve older subjects.

in all absolute HRV indexes compared to young subjects, without any significant sleep-stage dependent variations (Figure 7), and a predominant loss of parasympathetic activity (Figure 8) which may be related to decreased sleep quality; 2) Periodic breathing induces substantial modifications in HRV by triggering important oscillations in the VLF range (i.e. $\approx$ 40 s), which simply reflect the respiratory pattern, with additional LF and HF oscillations.

Periodic breathing represents a powerful modulator of HRV which possibly acts through a direct mechanical effect and implies substantial autonomic drive to the heart, both of sympathetic and parasympathetic nature. Such a major influence of respiration is generally ignored in studies concerned with age-related changes, despite the large number of studies that have emphasized the influence of ventilatory changes on HRV in various experimental conditions, even in young subjects.[70-75] Controlled breathing has been recommended to maintain the breathing frequency in the HF band. However, this

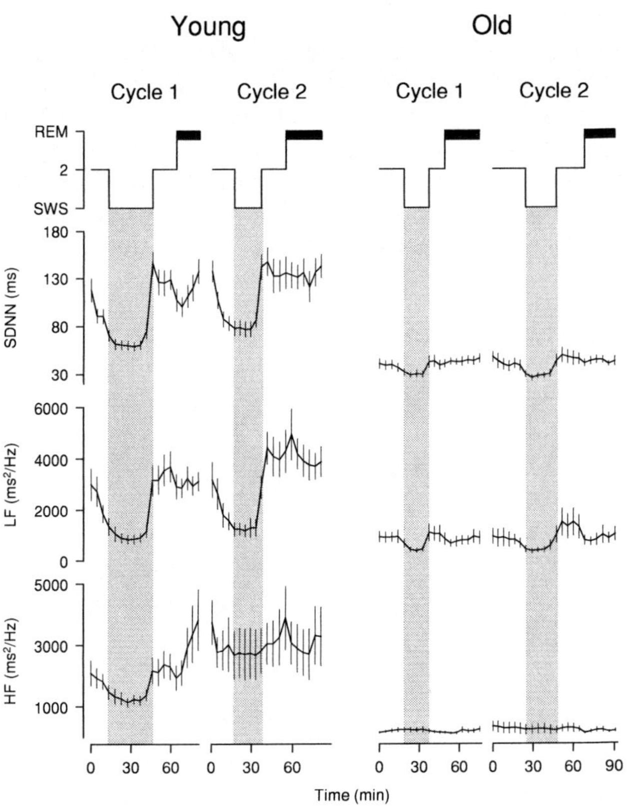

FIGURE 7. Mean ($\pm$ SE) time courses of SDNN, LF and HF power during the two first normalized NREM-REM sleep cycles with normal respiratory patterns in twelve young and twelve older subjects (abbreviations as in Figure 5). Note the large fall in absolute HRV indexes in the elderly.

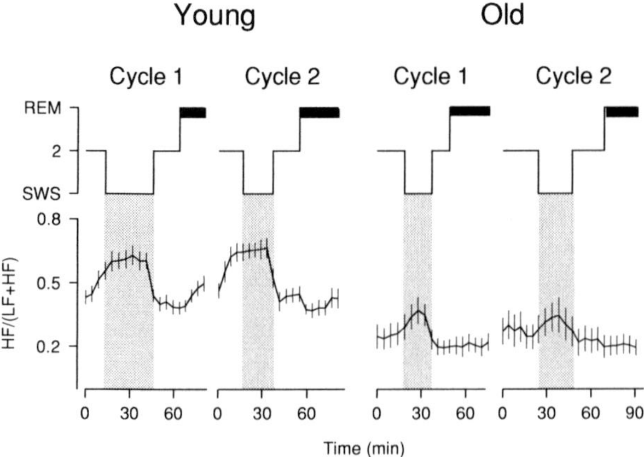

FIGURE 8. Mean (± SE) time courses of HF/(LF+HF) during the two first NREM-REM sleep cycles with normal respiratory patterns in twelve young and twelve elderly subjects. Aging is associated with a predominant loss in parasympathetic activity, as inferred from low HF/(LF+HF). (abbreviations as in Figure 5).

is not a physiological breathing and the mental engagement in following the metronome may produce sympathetic excitation.

5. Sleep Stage Alternation as an Autonomic Maneuver

We propose then an original approach for the evaluation of the activity of the autonomic nervous system, by using the spontaneous switch from a pronounced vagal tone during SWS to a sympathetic dominance during REM sleep. SWS is characterized by a regular respiratory pattern[76] and high ECG stationarity required for spectral HRV analysis. The emergence from SWS is associated with an arousal characterized by a sudden HR increase, well-documented in the literature.[77-79] Thus, sleep stage alternation represents an autonomic maneuver free of any external influence with three distinct phases: two steady states (SWS, and the subsequent "active" sleep stage 2) which permit the use of HRV analysis in time and frequency domains, and a transient phase, the arousal associated with the emergence from SWS that allows determination of HR responsiveness to a spontaneous stimulus (Figure 9).

With this approach, we have demonstrated that patients who underwent cardiac transplantation 4-14 years earlier showed two distinct profiles during arousal from SWS.[80] Some patients had no HR variation, while others presented an HR increase, but smaller than in control subjects (Figure 4).

Patients with a HR increase also display sleep stage dependent changes in sympathetic HRV indexes, whereas the non-reactive group was characterized by a simultaneous inability of HR to respond to arousal and of HRV to

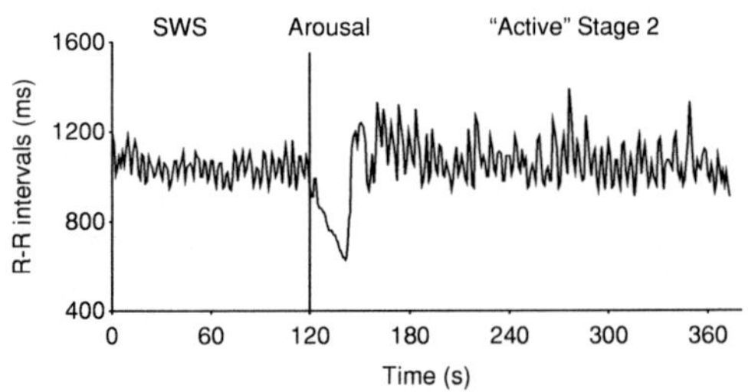

FIGURE 9. HR and HRV measurements during the first NREM-REM sleep cycle as a tool for the investigation of autonomic activity. This spontaneous autonomic maneuver comprises three distinct phases: two steady states, i.e. SWS and the subsequent active sleep stage 2 for measurement of HRV, and a transient phase, the arousal associated with the emergence from SWS, for determining HR responsiveness.

change with sleep stage alternation. The increased HR reactivity corroborated by surrounding HRV variations supports the concept of a partial cardiac sympathetic reinnervation in some patients with no indication of enhanced parasympathetic activity, but other signs of reinnervation have to be identified to validate this hypothesis. Thus, sleep stage alternation provides discrimination between parasympathetic and sympathetic influences, as has been previously recognized by Vanoli et al[81] who studied patients after myocardial infarction and found a loss of the capability of the vagus to physiologically activate during sleep. The R-R profile during the first NREM-REM sleep cycle may be used to detect cardiovascular dysfunction and to assess the extent of a recovery process following treatment.

With this procedure, the state of sympathovagal balance modulating the sinus node pacemaker activity can be quantified in a variety of physiological and pathophysiological conditions, and may be particularly useful for diagnosis of syndrome characterized by specific sleep-dependent changes in HRV, such as sleep apnea,[30] congestive heart failure,[28] myocardial infarction,[81] atrioventricular blocks[82] and sudden cardiac death syndrome.[25]

6. Summary

Numerous studies have demonstrated that global HRV varies according to sleep stages and that there is a change in the distribution of power spectra all along the NREM-REM sleep cycles. Aging causes a large drop in HRV and a predominant loss of parasympathetic activity which may be related to decreased sleep quality. Completing the conceptual background of Malliani and Montano[83] who emphasized power spectral analysis of R-R intervals as a clinical tool for the investigation of autonomic activity, we propose the first NREM-REM sleep cycle as an appropriate moment of observation, with three distinct phases : two steady states for the measurement of HRV, i.e. the parasympathetic-dominated SWS and the subsequent active stage 2 under sympathetic dominance, with the arousal at the emergence from SWS as a transient excitatory event for the measure of HR responsiveness. This simple non-invasive method can reveal impairment in autonomic cardiovascular control or altered target function that may otherwise remain undetected and could furnish interesting prognostic markers.

7. Acknowledgments. The results presented in the chapter are based on computer programs from Chantal Simon. The editorial assistance of Michèle Siméoni and Fabienne Goupilleau is gratefully acknowledged. We are indebted to Jean Ehrhart and Daniel Joly for sleep recordings and analysis.

8. *References*

1. Bigger JT, Fleiss JL, Rolnitzky LM et al. The ability of several short-term measures of RR variability to predict mortality after myocardial infarction. Circulation 1993; 88(3):927-934.
2. Tsuji H, Venditti FJ, Manders Jr ES et al. Reduced heart rate variability and mortality risk in an elderly cohort. The Framingham Heart Study. Circulation 1994; 90(2):878-883.
3. Stein KM, Lippman N, Lerman BB. Heart rate variability and cardiovascular risk assessment. In: Laragh JH, Brenner BM, eds. Hypertension: Pathophysiology, Diagnosis and Management. New York: Raven, 1995: 889-903.
4. Tsuji H, Larson MG, Venditti FJ et al. Impact of reduced heart rate variability on risk for cardiac events. The Framingham Heart Study. Circulation 1996; 94(11):2850-2855.
5. Barron HV, Viskin S. Autonomic markers and prediction of cardiac death after myocardial infarction. Lancet 1998; 351(9101):461-462.
6. La Rovere MT, Bigger Jr JT, Marcus F et al. Baroreflex sensitivity and heart rate variability in prediction of total cardiac mortality after myocardial infarction. Lancet 1998; 351(9101):478-484.
7. Galinier M, Pathak A, Fourcade J et al. Depressed low frequency power of heart rate variability as an independent predictor of sudden death in chronic heart failure. Eur Heart J 2000; 21(6):475-482.
8. Lombardi F, Mäkikallio TH, Myerburg RJ et al. Sudden cardiac death: role of heart rate variability to identify patients at risk. Cardiovasc Res 2001; 50(2):210-217.
9. Cohn JN, Rector TS. Prognosis of congestive heart failure and predictors of mortality. Am J Cardiol 1988; 62(2):25A-30A.
10. Singer DH, Martin GJ, Magid N et al. Low heart rate variability and sudden cardiac death. J Electrocardiol 1988; 21:s46-s55.
11. Somers VK, Phil D, Dyken ME et al. Sympathetic-nerve activity during sleep in normal subjects. N Engl J Med 1993; 328(5):303-307.
12. Murali NS, Svatikoua A, Somers VK. Cardiovascular physiology and sleep. Front Biosci 2003; 8:s636-s652.
13. Mancia G. Autonomic modulation of the cardiovascular system during sleep. N Engl J Med 1993; 328(5):347-349.

14. Zemaitytë D, Varoneckas G, Sokolov E. Heart rhythm control during sleep. Psychophysiology 1984; 21(3):279-289.

15. Berlad I, Shlitner A, Ben-Haim S et al. Power spectrum analysis and heart rate variability in stage 4 and REM sleep: evidence for state-specific changes in autonomic dominance. J Sleep Res 1993; 2(2):88-90.

16. Baharav A, Kotogal S, Gibbons V et al. Fluctuations in autonomic nervous activity during sleep displayed by power spectrum analysis of heart rate variability. Neurology 1995; 45(6):1183-1187.

17. Vaughn BV, Quint SR, Messenheimer JA et al. Heart period variability in sleep. Electroencephalogr Clin Neurophysiol 1995; 94(3):155-162.

18. Toscani L, Gangemi PF, Parigi A et al. Human heart rate variability and sleep stages. Ital J Neurol Sci 1996; 17(6):437-439.

19. Bonnet MH, Arand DL. Heart rate variability: sleep stage, time of night and arousal influences. Electroencephalogr Clin Neurophysiol 1997; 102(5):390-396.

20. Trinder J, Kleiman J, Carrington M et al. Autonomic activity during human sleep as a function of time and sleep stage. J Sleep Res 2001; 10(4):253-264.

21. Burgess HJ, Penev PD, Schneider R et al. Estimating cardiac autonomic activity during sleep: impedance cardiography, spectral analysis and Poincaré plots. Clin Neurophy 2003; 115(1):19-28.

22. Jurysta F, Van de Borne P, Migeotte PF et al. A study of the dynamic interactions between sleep EEG and heart rate variability in healthy young men. Clin Neurophysiol 2003; 114(11):2146-2155.

23. Achermann P, Dijk DJ, Brunner DP et al. A model of human sleep homeostasis based on EEG slow-wave activity: quantitative comparison of data and simulations. Brain Res Bull 1993; 31(1-2):97-113.

24. Kamen PW, Krum H, Tonkin AM. Poincaré plot of heart rate variability allows quantitative display of parasympathetic nervous activity in humans. Clin Sci 1996; 91(2):201-208.

25. Schechtman VL, Raetz SL, Harper RK et al. Dynamic analysis of cardiac R-R intervals in normal infants and in infants who subsequently succumbed to the sudden infant death syndrome. Pediatr Res 1992; 31(6):606-612.

26. Schechtman VL, Harper RK, Harper RM. Development of heart rate dynamics during sleep-waking states in normal infants. Pediatr Res 1993; 34(5):618-623.

27. Woo MA, Stevenson WG, Moser DK et al. Patterns of beat-to-beat heart rate variability in advanced heart failure. Am Heart J 1992; 123(3):704-710.

28. Woo MA, Stevenson WG, Moser DK et al. Complex heart rate variability and serum norepinephrine levels in patients with advanced heart failure. J Am Coll Cardiol 1994; 23(3):565-569.

29. Kamen PW, Tonkin AM. Application of the Poincaré plot to heart rate variability: a new measure of functional status in heart failure. Aust NZ J Med 1995; 25(1):18-26.

30. Aljadeff G, Gozal D, Schechtman VL et al. Heart rate variability in children with obstructive sleep apnea. Sleep 1997; 20(2):151-157.

31. Otzenberger H, Simon C, Gronfier C et al. Temporal relationship between dynamic heart rate variability and electroencephalographic activity during sleep in man. Neurosci Lett 1997; 229(3):173-176.

32. Otzenberger H, Gronfier C, Simon C et al. Dynamic heart rate variability: a tool for exploring sympathovagal balance continuously during sleep in men. Am J Physiol 1998; 275(3 Pt 2):H946-H950.

33. Akselrod S, Gordon D, Ubel FA et al. Power spectrum analysis of heart rate fluctuation: a quantitative probe of beat-to-beat cardiovascular control. Science 1981; 213(4504):220-222.

34. Pagani M, Lombardi F, Guzzetti S et al. Power spectral analysis of heart rate and arterial pressure variabilities as a marker of sympatho-vagal interaction in man and conscious dog. Circ Res 1986; 59(2):178-193.

35. Bootsma M, Swenne CA, Van Bolhuis HH et al. Heart rate and heart rate variability as indexes of sympathovagal balance. Am J Physiol (Heart Circ Physiol 35) 1994; 266(4 Pt 2):H1565-H1571.

36. Task Force of the European Society of Cardiology and the North American Society of Pacing and Electrophysiology. Heart rate variability. Standards of measurement, physiological interpretation, and clinical use. Circulation 1996; 93(5):1043-1065.

37. Pomeranz B, Macaulay RJB, Caudill MA et al. Assessment of autonomic function in humans by heart rate spectral analysis. Am J Physiol (Heart Circ Physiol 17) 1985; 248(1 Pt 2):H151-H153.

38. Malliani A, Lombardi F, Pagani M. Power spectrum analysis of heart rate variability: a tool to explore neural regulatory mechanisms. Br Heart J 1994; 71(1):1-2.

39. Montano N, Gnecchi Ruscone T, Porta A et al. Power spectrum analysis of heart rate variability to assess the changes in sympathovagal balance during graded orthostatic tilt. Circulation 1994; 90(4):1826-1831.

40. Malliani A, Pagani M, Furlan R et al. Individual recognition by heart rate variability of two different autonomic profiles related to posture. Circulation 1997; 96(12): 4143-4145.

41. Lazzeri C, La Villa G, Mannelli M et al. Effects of acute clonidine administration on power spectral analysis of heart rate variability in healthy humans. J Auton Pharmacol 1998; 18(5):307-312.

42. Montano N, Cogliati C, Porta A et al. Central vagotonic effects of atropine modulate spectral oscillations of sympathetic nerve activity. Circulation 1998; 98(14): 1394-1399.

43. Malliani A. The pattern of sympathovagal balance explored in the frequency domain. News Physiol Sci 1999; 14:111-117.

44. Montano N, Lombardi F, Gnecchi Ruscone T et al. Spectral analysis of sympathetic discharge, R-R interval and systolic arterial pressure in decerebrate cats. J Auton Nerv Syst 1992; 40(1):21-31.

45. Montano N, Cogliati C, Dias da Silva VJ et al. Effects of spinal section and of positive-feedback excitatory reflex on sympathetic and heart rate variability. Hypertension 2000; 36(6):1029-1034.

46. Pagani M, Montano N, Porta A et al. Relationship between spectral components of cardiovascular variabilities and direct measures of muscle sympathetic nerve activity in humans. Circulation 1997; 95(6):1441-1448.

47. Furlan R, Porta A, Costa F et al. Oscillatory patterns in sympathetic neural discharge and cardiovascular variables during orthostatic stimulus. Circulation 2000; 101(8):886-892.

48. Skyschally A, Breuer HW, Heusch G. The analysis of heart rate variability does not provide a reliable measurement of cardiac sympathetic activity. Clin Sci (Lond) 1996; 91 Suppl:102-104.

49. Houle MS, Billman GE. Low-frequency component of the heart rate variability spectrum: a poor marker of sympathetic activity. Am J Physiol 1999; 276(1 Pt 2): H215-H223.

50. Hammer JW, Morin RJ, Rudolph JL et al. Inconsistent link between low-frequency oscillations: R-R interval responses to augmented Mayer waves. J Appl Physiol 2001; 90(4):1559-1564.
51. Eckberg D. Sympathovagal balance: a critical appraisal. Circulation 1997; 96(9): 3224-3232.
52. Ivanov PC, Rosenblum MG, Peng CK et al. Scaling behaviour of heartbeat intervals obtained by wavelet-based time-series analysis. Nature 1996; 383(6598):323-327.
53. Pichot V, Gaspoz JM, Molliex S et al. Wavelet transform to quantify heart rate variability and to assess its instantaneous changes. J Appl Physiol 1999; 86(3):1081-1091.
54. Mayer-Kress G. Monitoring chaos of cardiac rhythms. In: Biotech USA, Proceedings of the 6th annual industry conference and exhibition. San Francisco, October 2-4, 1989:435-443.
55. Braun C, Kowallik P, Freking A, et al. Demonstration of nonlinear components in heart rate variability of healthy persons. Am J Physiol (Heart Circ Physiol 44) 1998; 275:H1577-H1584.
56. Wagner CD, Persson PB. Chaos in the cardiovascular system: an update. Cardiovasc Res 1998; 40(2):257-264.
57. Brandenberger G, Ehrhart J, Piquard F et al. Inverse coupling between ultradian oscillations in delta wave activity and heart rate variability during sleep. Clin Neurophysiol 2001; 112(6):992-996.
58. Bonnet MH, Arand DL. Heart rate variability in insomniacs and matched normal sleepers. Psychosom Med 1998; 60(5):610-615.
59. Crasset V, Mezzetti S, Antoine M et al. Effects of aging and cardiac denervation on heart rate variability during sleep. Circulation 2001; 103(1):84-88.
60. Dijk DJ, Beersma DG, Van den Hoofdakker RH. All night spectral analysis of EEG sleep in young adult and middle-aged male subjects. Neurobiol Aging 1989; 10(6):677-682.
61. Ehlers CL, Kupfer DJ. Effects of age on delta and REM sleep parameters. Electroencephalogr Clin Neurophysiol 1989; 72(2):118-125.
62. Landolt HP, Dijk DJ, Achermann P et al. Effect of age on the sleep EEG: slow-wave activity and spindle frequency activity in young and middle-aged men. Brain Res 1996; 738(2):205-212.
63. Aserinsky E. Periodic respiratory pattern occurring in conjunction with eye movements during sleep. Science 1965; 150(697):763-766.
64. Webb P. Periodic breathing during sleep. J Appl Physiol 1974; 37(6):899-903.
65. Chapman KR, Bruce EN, Gothe B et al. Possible mechanisms of periodic breathing during sleep. J Appl Physiol 1988; 64(3):1000-1008.
66. Sanderson JE, Yeung LY, Yeung DT et al. Impact of changes in respiratory frequency and posture on power spectral analysis of heart rate and systolic blood pressure variability in normal subjects and patients with heart failure. Clin Sci (Lond) 1996; 91(1):35-43.
67. Mortara A, Sleight P, Pinna GD et al. Abnormal awake respiratory patterns are common in chronic heart failure and may prevent evaluation of autonomic tone by measures of heart rate variability. Circulation 1997; 96(1):246-252.
68. Lorenzi-Filho G, Dajani HR, Leung RS et al. Entrainment of blood pressure and heart rate oscillations by periodic breathing. Am J Respir Crit Care Med 1999; 159(4 Pt 1):1147-1154.
69. Brandenberger G, Viola AU, Ehrhart J et al. Age-related changes in cardiac autonomic control during sleep. J Sleep Res 2003; 12(3):173-180.

70. Elghozi JL, Laude D, Girard A. Effects of respiration on blood pressure and heart rate variability in humans. Clin Exp Pharmacol Physiol 1991; 18(11):735-742.
71. Brown TE, Beightol AL, Koh J et al. Important influence of respiration on human R-R interval power spectra is largely ignored. J Appl Physiol 1993; 75(5):2310-2317.
72. Novak V, Novak P, De Champlain J et al. Influence of respiration on heart rate and blood pressure fluctuations. J Appl Physiol 1993; 74(2):617-626.
73. Hayano J, Mukai S, Sakakibara M et al. Effects of respiratory interval on vagal modulation of heart rate. Am J Physiol (Heart Circ Physiol 36) 1994; 267(1 Pt 2):H33-H40.
74. Bernardi L, Wdowczyk-Szulc J, Valenti C et al. Effects of controlled breathing, mental activity and mental stress with or without verbalization on heart rate variability. J Am Coll Cardiol 2000; 35(6):1462-1469.
75. Bernardi L, Porta C, Gabutti A et al. Modulatory effects of respiration. Auton Neurosci 2001; 90(1-2):47-56.
76. Krieger J. Breathing during sleep in normal subjects. In: Kryger MH, Roth TR, Dement WC, eds. Principles and Practice of Sleep Medicine. Philadelphia: Saunders, 1994:212-223.
77. Sforza E, Jouny C, Ibanez V. Cardiac activation during arousal in humans: further evidence for hierarchy in the arousal response. Clin Neurophysiol 2000; 111(9): 1611-1619.
78. Gosselin N, Michaud M, Carrier J et al. Age difference in heart rate changes associated with micro-arousals in humans. Clin Neurophysiol 2002; 113(9):1517-1521.
79. Trinder J, Allen N, Kleiman J et al. On the nature of cardiovascular activation at an arousal from sleep. Sleep 2003; 26(5):543-551.
80. Viola AU, Brandenberger G, Buchheit M et al. Sleep as a tool for evaluating autonomic drive to the heart in cardiac transplant patients. Sleep 2004; in press.
81. Vanoli E, Adamson PB, Ba-Lin MS et al. Heart rate variability during specific sleep stages. A comparison of healthy subjects with patients after myocardial infarction. Circulation 1995; 91(7):1918-1922.
82. Viola AU, Simon C, Doutreleau S et al. Abnormal heart rate variability in a subject with second degree atrioventicular blocks during sleep. Clin Neurophysiol 2004; 115(4):946-950.
83. Malliani A, Montano N. Heart rate variability as a clinical tool. Ital Heart J 2002; 3(8):439-445.

Neuroendocrine Correlates of Infectious Disease: Implications for Sleep

LINDA A. TOTH AND JENNIFER M. ARNOLD

1. Introduction

The optimal outcome of an infection is elimination of the pathogen, recuperation, and survival. Achieving this outcome requires a host response that is both effective and controlled. For example, an inadequate immune response can permit excessive microbial proliferation and damage and perhaps result in death, whereas an excessive or unregulated host immune response can result in unnecessary inflammatory tissue damage, as well as potential disability or death.

An effective immune response and its appropriate homeostatic regulation are both critically important in mediating recovery from infectious challenge. However, alterations in sleep during infectious disease may also confer less-recognized contributions to recovery. Sleep alterations may reflect both immune-enhancing physiologic responses to infection and homeostatic immune-regulatory mechanisms.

Alterations in sleep during infection may in turn have a direct impact on the magnitude and efficacy of immune and stress-related responses to infectious challenge. Infectious challenge triggers mutual interactions among host defense systems, immune regulatory systems, and perhaps sleep that ideally generate a coordinated and optimally recuperative response to the perturbations imposed by infectious disease.

2. Overview: Stress, Host Defense, and Sleep During Infectious Disease

The immune system is the primary means of generating a host defense response to infectious challenge.

Like the immune system, the hypothalamic-pituitary-adrenal (HPA)$^{\Gamma}$ axis also becomes activated in response to internal stressors such as a state of infectious or inflammatory disease. The HPA system serves the important role of modulating the immune response elicited by microbial challenge.

Although an effective immune response and its appropriate homeostatic regulation are both critically important in mediating recovery from infectious challenge, alterations in sleep during infectious disease may also contribute to recovery. One likely advantage of increased sleep is reduced energy utilization due to inactivity and a lower metabolic rate.[1,2] This type of energy conservation could help the host to maintain metabolic homeostasis despite the generation of fever during periods of infection-related anorexia. Furthermore, because animals typically seek a protected location for sleep, increased somnolence may promote survival by causing the animal to remain in a relatively safe location during periods of disease-related debilitation. Finally, sleep is postulated to promote immune responsiveness and could perhaps in that way also help to facilitate recovery.[3]

The precise mechanisms that underlie changes in sleep during infectious disease have not yet been elucidated, but likely mediators are components of both the immune and the HPA responses. Many of the same substances that regulate the immune response also influence sleep in consistent ways. For example, pro-inflammatory cytokines, including interleukin 1 (IL1), tumor necrosis factor α (TNFa), and interferon α (IFNa), promote both immune responsiveness and slow-wave sleep (SWS).[3] In contrast, anti-inflammatory mediators, including corticotropin releasing hormone (CRH), adrenocorticotropic hormone (ACTH), and glucocorticoids (GC), tend to limit inflammation and the immune response and also promote arousal.[4,5] Alterations in sleep during infection thus reflect and may in turn impact the magnitude and efficacy of immune and stress-related responses to infectious challenge.

3. HPA Activation During Infectious Disease

3.1. Immunomodulation

Pro-inflammatory cytokines induced during the host response to infectious challenge (e.g., IL1, TNFa, interleukin 6) activate the HPA axis.[6,7] This communication between peripheral cytokines and central nervous system (CNS) circuitry occurs via a variety of mechanisms: direct activation of peripheral sensory nerves, active translocation of cytokines across the blood-brain barrier, diffusion of cytokines into the brain at circumventricular organs, and central penetration of immunocytes.[8-13] Microbial components such as bacterial lipopolysaccharide can also signal the CNS from the periphery via hepatic mechanisms[14] and by stimulating the production of inflammatory cytokines.[15]

Various cells of the CNS, including neurons, cells of the choroid plexus, astrocytes, and microglia, can also produce cytokines.[16,17] The detection of mRNA for cytokines such as IL1 in these cell types indicates that they can produce these mediators *de novo*, rather than simply respond to cytokines produced elsewhere.[16] IL1, TNFa, and other cytokines, produced in the CNS or in the periphery, induce fever and other signs of disease via interactions with cytokine receptors in the hypothalamus and elsewhere in the brain. In addition, these cytokines induce so-called "sickness behaviors" that include lethargy, decreased social interaction, decreased appetite, and increased sleep.[9,15,18] Finally, they also impact the hypothalamus, eliciting the release of CRH, and the pituitary, stimulating cells that secrete ACTH.[5]

Just as some neurons and other CNS cells can produce both cytokine mediators and neurotransmitters, some peripheral immunocytes can produce neurotransmitters, as well as cytokines.[19] By acting on receptors in lymphatic or inflamed tissue, immunocyte-derived neurotransmitters can directly influence transmitter release from peripheral nerves.[15] These neurotransmitters also act on the immune cells directly, generally either promoting or attenuating cytokine-induced effects on immunocyte function.[19] These complex interactions suggest that classification of individual effector substances as neurotransmitters or cytokines may, at least in some situations, be a semantic or historical, rather than a functional, distinction.[19]

Macrophages and other immunocytes have also receptors for CRH, ACTH, and GC (20). GC are generally anti-inflammatory in nature.[6] For example, GC inhibit the production of interleukin-2 and interferon-gamma by lymphocytes and the expression of IL1 and TNFa by macrophages. GC also inhibit lymphocyte adhesion and T-cell binding to endothelial cells.[20] In general, HPA axis activation, via the release of GC, allows the CNS to regulate the immune response, reducing the likelihood of severe inflammatory damage.

The sympathetic nervous system also influences the immune response. Norepinephrine and epinephrine influence the numbers and distribution of leukocytes, typically increasing the numbers of circulating neutrophils and natural killer (NK) cells, decreasing numbers of circulating T and B lymphocytes, and increasing lymphocyte numbers in spleen and lymph nodes.[21] Elevated norepinephrine and epinephrine have been associated with increased risk for illness.[16]

3.2. Stress

The CNS reaction to stress also involves activation of the HPA axis and the release of GC from the adrenal cortex. Stress has been associated with both immune-enhancing and immunosuppressive effects. During acute stress (e.g., exercise or escape), the sympathetic nervous system reacts by releasing catecholamines. At the onset of stress, catecholamines mobilize leukocytes, triggering them to enter bloodstream and lymphatic system. Thus, epinephrine and norepinephrine increase numbers of neutrophils and NK cells in the blood and numbers of T and B lymphocytes in the spleen and lymph nodes. Basically, the body responds to an acute stressor as a potential antigen and prepares the immune system to respond appropriately.[22] As the stress response continues and the HPA axis is activated, GC cause leukocytes to leave circulation and enter the skin, lymph nodes, gastrointestinal tract, and other sites in preparation for immune challenges potentially associated with the stressor, thereby causing a reduction in circulating numbers of leukocytes.[22]

In contrast to acute stress, chronic stress is generally immunosuppressive. Chronic activation of the HPA axis reduces the numbers of helper and cytotoxic T lymphocytes, B lymphocytes, NK cells, and monocytes in the blood. Chronic stress is often associated with blunted immune function.[15] For example, persons undergoing high levels of chronic stress in their normal lives have a greater likelihood of developing a cold,[16] or may have poor antibody responses to immune challenge.[23,24] In contrast, people with a strong social support structure characterized by numerous social relationships (e.g., family, friends, work, etc) tend to live longer.[16] Individual differences in the magnitude of stress-related influences on immune function are prevalent, stable across time, and may have clinical significance.[25]

Stress and infection induce similar activation of the HPA axis.[26] For example, in mice, infection with influenza virus is associated with significant elevations in serum corticosterone levels.[27,28] HPA activation is not dependent solely on disease-induced damage, as it occurs even in response to challenges with minimal associated pathology.[6,29]

4. HPA Activation During Infectious Disease: Implications for Sleep

4.1. HPA Mediators and Sleep

The HPA axis is influenced by sleep and also participates in the regulation of sleep.[30-33] In humans, sleep onset is usually associated with the short-term inhibition of cortisol secretion, whereas awakening is associated with a surge of secretion.[34] Cortisol secretion during sleep is inversely related to EEG slow-wave amplitude,[35] which is an index of the depth of sleep, and during waking is temporally coupled to EEG activity,[36,37] which is a measure of central alertness. Human subjects who are deprived of sleep lose the modulatory impact of sleep-wake transitions on cortisol regulation, which results in a reduced amplitude of the cortisol circadian rhythm, higher basal levels during quiescent periods, and elevated evening levels.[30,38]

A potential role for endogenous GC in influencing sleep during microbial disease, either directly via CNS effects or indirectly via immunosuppressive and anti-inflammatory actions, is supported by numerous observations that CRH and GC, as well as numerous other endogenous and exogenous immune-modulating substances, influence sleep patterns in normal animals.[4] In healthy rabbits, for example, the administration of an immunosuppressive dose of cortisone reduces both the time spent in SWS and delta-wave amplitude during SWS.[39]

In rabbits undergoing *Candida albicans* infections, administration of cortisone dramatically attenuates infection-related alterations in sleep, allowing the rabbits to maintain a fairly normal circadian organization of sleep. Consistent with its anti-inflammatory effects, cortisone also alleviates some of the clinical signs of disease, such as fever.[39]

Studies in mice also suggest a relationship between HPA responsiveness and sleep after infectious challenge. During influenza infection, C57BL/6J mice spend more time in SWS, whereas BALB/cByJ mice show fragmentation of sleep.[40] Data from a variety of models indicate that C57BL/6 mice generate a more modest HPA response to stress than do BALB/c mice, which show a robust response.[41-44]

Such observations are consistent with the hypothesis that GC may contribute to the suppression of sleep that develops in BALB/c mice after infectious challenge. Because CRH, ACTH, and GC elicit increased arousal,[4,5,45] greater adrenal responsiveness in BALB/c mice as compared to C57BL/6 mice is consistent with a response of reduced sleep during infection-associated stress.

4.2. Effects of Sleep and Sleep Loss on Host Defense

Immune challenge can alter sleep in animals and people. Some observations even suggest that sleep patterns reflect the progression of the disease process, the prognosis, or the clinical outcome. For example, in rabbits inoculated with *E. coli*, *S. aureus* or *C. albicans*, a prolonged phase of enhanced sleep after microbial challenge is associated with a more favorable prognosis and less severe clinical signs than is a short period of enhanced sleep that is followed by a prolonged reduction in sleep.[46,47] Similarly, HIV-infected humans who are seropositive but are otherwise healthy demonstrate excess Stage 4 sleep,[48] but sleep deteriorates and becomes disrupted as the disease progresses.[49] Disturbed sleep is highly prevalent among patients with HIV infections.[50-52] Path analysis has indicated that in HIV patients, psychological distress precipitates poor sleep that in turn reduces suppressor T cell numbers.[53]

Studies of sleep-deprived rodents undergoing antigenic challenge indicate that sleep loss may cause functionally significant immune perturbations. For example, secondary antibody responses to antigenic challenge are impaired in sleep-deprived mice and rats.[54,55] Sleep loss is also reported to retard both viral clearance and the development of a protective antibody response in influenza-infected mice,[54] although this result has not been replicated.[56,57] Rats subjected to chronic sleep deprivation did not show changes in splenocyte responses to mitogens, but numbers of circulating lymphocytes were reduced.[58]

The hypothesis that sleep loss impairs immune competence is most strongly supported by the observation that chronic sleep deprivation of rats results in intestinal bacterial proliferation, microbial penetration into lymph nodes, septicemia, and eventually death.[59-62] The penetration of bacteria into normally sterile tissues during prolonged sleep deprivation implies an abnormal host defense and the gradual development of immune insufficiency. Thus, sleep loss could render healthy individuals susceptible to disease, exacerbate existing disease, or complicate recovery in patient populations. Sleep deprivation may also modulate physiologic signs of immune or acute phase responses. For example, sleep deprivation exacerbates fever induced by inoculation of rabbits with *E. coli*,[63] administration of sheep red blood cells to rats,[64] and intracerebroventricular administration of saline or immunoglobulins to rats.[65] Sleep deprivation is also reported to exacerbate anticoagulant-induced anemia[66] and to retard tumor growth in rats.[67]

Sleepiness and sleep quality broadly influence measures of general health status, particularly impacting perceptions about energy and fatigue. In addition, chronic sleep fragmentation, non-restorative sleep, and inadequate sleep may have significant adverse immune consequences in human populations. Host defense responses such as antibody production or microbial clearance may be altered by sleep loss, just as they can be impacted by stress- or illness-induced activation of the GC system (e.g.,[68]). Conversely, sleep may promote recovery from some types of medical conditions by promoting an internal hormonal environment that favors cell proliferation.[69] Sleep disruption can be profound in hospitalized patients and nursing home residents.[70-73] For example, slow-wave sleep occupies less than 1% of the night during the 5 to 8 days after open-heart surgery.[74] In humans, chronic stress that is associated with sleep loss or experimentally imposed sleep restriction impairs immune responses to vaccination against influenza and hepatitis A viruses.[23,75] Greater knowledge about such interactions could have important health implications for hospitalized patients and nursing home residents, who commonly experience severe disruptions in normal patterns of sleep.

Some epidemiological studies of human populations support a relationship between insomnia or unusually short nighttime sleep durations and decreased life expectancy,[76-79] although other studies do not.[80-83] Associations between absent or diminished sleep, reduced EEG amplitude, and imminent death also occur in aged mice prior to spontaneous death,[84] in mice with fatal experimental rabies infections,[85,86] and in rats that die subsequent to chronic sleep deprivation and septicemia.[59,60,62] The total amount of sleep and the EEG amplitude during sleep also gradually decline in rabbits with trypanosome infections,[87] which are eventually fatal. Sleep deprivation also causes death in *cyc*[01] mutant *Drosophila melanogaster* in association with reduced expression of heat-shock genes.[88]

Despite these provocative correlations, the impact of sleep loss on immune competence is difficult to assess experimentally. One important concern is distinguishing between effects attributable to a lack of sleep *per se* and those associated with non-specific stress. Stress-related immune impairments are well documented and impact the response to microbial challenge in humans and animals.[23,89] Experimental models of sleep deprivation in animals generally attempt to minimize non-specific stress. Some methods of inducing sleep loss in animals elicit relatively few of the classical physiological signs of non-specific stress (e.g.,[90]), whereas others are characterized by increases in markers traditionally considered to reflect stress (e.g., elevated catecholamines or GC) (e.g.,[91,92]). Because humans can voluntarily choose to forego sleep, the immunologic impact of sleep loss in humans may be difficult to replicate in animals that undergo forced waking.[93] However, the sleep loss or insomnia that people endure as a "normal" facet of life is frequently associated with stressors or environmental factors that necessitate or otherwise contribute to loss of sleep (e.g., examinations, bereavement, shift work, depression).

The potential association between infection-related release of sleep-modulatory cytokines and hormones and alterations in sleep raises an important question. Is altered sleep merely a by-product of infectious disease and the immune response, or does sleep in some way facilitate recovery from microbial infections? Data that are accruing from both animal and human studies suggest that short-term sleep loss may be accompanied by enhancement of non-specific host defense mechanisms, whereas chronic or prolonged sleep loss may result in immune suppression.[94,95] Similar arguments have been posed for the relationship between immune function and acute or chronic stress, particularly as reflected by elevations in circulating GC levels.[6,96]

5. Conclusions and Perspectives

In general, a good host defense response to microbial or inflammatory challenge requires activation of an appropriate immune defense and also influences patterns of sleep. These sleep changes are likely to be mediated by the same mechanisms that influence sleep in healthy animals. Appropriate transitions between sleep and arousal during infectious disease may behaviorally facilitate recuperation and survival, and the processes of sleep and host defense may be mutually reinforcing. The nature of these relationships is a major unanswered question about the role of sleep in human health. Over the long term, identifying factors that influence these relationships will help us to delineate the mechanisms that cause fatigue, excessive sleepiness, or non-restorative or poor sleep during acute and chronic disease, and perhaps even under normal conditions. Identifying those mechanisms might contribute to the eventual development of interventions that can control or prevent these debilitating problems.

6. Acknowledgments. This work was supported by the Southern Illinois University School of Medicine and by NIH grants HL70522, NS40220, and RR16421. The authors thank Drs. Mark Opp and Deborah Suchecki for reviewing a preliminary version of this manuscript.

7. *References*

1. Zepelin H, Rechtschaffen A. Mammalian sleep, longevity, and energy metabolism. Brain Behav Evol 1974; 10:425-470.
2. Zepelin H. Mammalian sleep. In: Kryger MH, Roth T, Dement WC, editors. Principles and Practice of Sleep Medicine. New York: W.B. Saunders Co., 2000: 82-92.
3. Krueger JM, Fang J. Host defense. In: Kryger MH, Roth T, Dement WC, editors. Principles and Practice of Sleep Medicine. New York: W.B. Saunders Co., 2000: 255-265.
4. Chang F-C, Opp MR. Corticotropin-releasing hormone (CRH) as a regulator of waking. Neurosci Biobehav Rev 2001; 25:445-453.
5. Vgontzas AN, Chrousos GP. Sleep, the hypothalamic-pituitary-adrenal axis and cytokines: multiple interactions and disturbances ins sleep disorders. Endocrinol Metab Clin N Am 2002; 31:15-36.
6. Bailey M, Engler H, Hunzeker J, Sheridan JF. The hypothalamic-pituitary-adrenal axis and viral infection. Viral Immunol 2003; 16:141-157.
7. Beishuizen A, Thijs LG. Endotoxin and the hypothalamo-pituitary-adrenal axis. J Endotoxin Res 2003; 9:2-24.
8. Banks WA, Ortiz L, Plotkin SR, Kastin AJ. Human interleukin (IL) 1a, murine IL-1a and murine IL-1b are transported from blood to brain in the mouse by a shared saturable mechanism. J Pharmacol Exp Ther 1991; 259:988-996.
9. Dantzer R, Konsman J-P, Bluthe R-M, Kelley KW. Neural and humoral pathways of communication from the immune system to the brain: parallel or convergent? Autonomic Neuroscience: Basic and Clinical 2000; 85:60-65.
10. Goehler LE, Gaykema RPA, Hansen MK, Anderson K, Maier SF, Watkins LR. Vagal immune-to-brain communication: a visceral chemosensory pathway. Autonomic Neuroscience: Basic and Clinical 2000; 85:49-59.
11. Quan N, Stern EL, Whiteside MB, Herkenham M. Induction of pro-inflammatory cytokine mRNAs in the brain after peripheral injection of subseptic doses of lipopolysaccharide in the rat. J Neuroimmunol 1999; 93:72-80.
12. Quan N, Whiteside M, Herkenham M. Time course and localization patterns of interleukin-1b messenger RNA expression in brain and pituitary after peripheral administration of lipopolysaccharide. Neuroscience 1998; 83:281-293.
13. Saper CB, Breder CD. The neurologic basis of fever. N Engl J Med 1994; 330:1880-1886.
14. Blatteis CM, Li A, Li Z, Perlik V, Feleder C. Signaling the brain in systemic inflammation: the role of complement. Front Biosci 2004; 9:915-931.
15. Wilson CJ, Finch CE, Cohen HJ. Cytokines and cognition -the case for a head-to-toe inflammatory paradigm. J Am Geriatr Soc 2002; 50:2041-2056.
16. Cohen N, Kinney KS. Exploring the phylogenetic history of neural-immune interactions. In: Ader R, Felten DL, Cohen N, editors. Psychoneuroimmunology. New York: Academic Press, 2001: 21-54.
17. Reichlin S. Secretion of immunomodulatory mediators from the brain: an alternative pathway of neuroimmunomodulation. In: Ader R, Felten DL, Cohen N, editors. Psychoneuroimmunology. New York: Academic Press, 2001: 499-516.
18. Kent S, Bluthe R, Kelley KW, Dantzer R. Sickness behavior as a new target for drug development. Trends Pharmacol Sci 1992; 13:24-28.
19. Bellinger DL, Lorton D, Lubahn C, Felten DL. Innervation of lymphoid organs - association of nerve with cells of the immune system and their implications for

disease. In: Ader R, Felten DL, Cohen N, editors. Psychoneuroimmunology. New York: Academic Press, 2001: 55-112.

20. Bonneau RH, Padgett DA, Sheridan JF. Psychoneuroimmune interactions in infectious disease: studies in animals. In: Ader R, Felten DL, Cohen N, editors. Psychoneuroimmunology. New York: Academic Press, 2001: 483-498.

21. Dhabhar FS, McEwen BS. Bidirectional effects of stress and glucocorticoid hormones on immune function: possible explanations for paradoxical observations. In: Ader R, Felten DL, Cohen N, editors. Psychoneuroimmunology. New York: Academic Press, 2000: 301-338.

22. Dhabhar FS, Miller AH, McEwen BS, Spencer RL. Effects of stress on immune cell distribution - dynamics and hormonal mechanisms. J Immunol 1995; 154: 5511-5527.

23. Kiecolt-Glaser JK, Glaser R, Gravenstein S, Malarkey WB, Sheridan JF. Chronic stress alters the immune response to influenza virus vaccine in older adults. Proc Nat Acad Sci USA 1996; 93:3043-3047.

24. Smith A, Vollmer-Conna U, Bennett B, Wakefield D, Hickie I, Lloyd A. The relationship between distress and the development of a primary immune response to a novel antigen. Brain Behav Immun 2004; 18:65-75.

25. Marsland AL, Bachen EA, Cohen S, Rabin B, Manuck SB. Stress, immune reactivity and susceptibility to infectious disease. Physiol Behav 2002; 77:711-716.

26. Maier SF, Watkins LR, Nance DM. Multiple routes of action of interleukin-1 on the nervous system. In: Ader R, Felten DL, Cohen N, editors. Psychoneuroimmunology. New York: Academic Press, 2001: 563-584.

27. Dunn AJ, Powell ML, Meitin C, Small PA. Virus infection as a stressor: Influenza virus elevates plasma concentrations of corticosterone, and brain concentrations of MHPG and tryptophan. Physiol Behav 1989; 45:1-4.

28. Hermann G, Beck FM, Tovar CA, Malarkey WB, Allen C, Sheridan JF. Stress-induced changes attributable to the sympathetic nervous system during experimental influenza viral infection in DBA/2 inbred mouse strain. J Neuroimmunol 1994; 53:173-180.

29. Miller AH, Spencer RL, Pearce BD, Pisell TL, Leung JJ, Dhabhar FS et al. Effects of viral infection on corticosterone secretion and glucocorticoid receptor binding in immune tissues. Psychoneuroendocrinology 1997; 22:455-474.

30. Buxton OM, Spiegel K, Van Cauter E. Modulation of endocrine function and metabolism by sleep and sleep loss. In: Sleep Medicine (TL Lee-Chiong, MJ Sateia, and MA Carskadon, eds.), Hanley and Belfus, Philadelphia, PA. Pp 59-70, 2002.

31. Leese G, Chattington P, Fraser W, Vora J, Edwards R, Williams G. Short-term night-shift working monoms the pituitary-adrenocortical dysfunction of chronic fatigue syndrome. J Clin Endocrinol Metab 1996; 81:1867-1870.

32. Spath-Schwalbe E, Scholler T, Kern W, Fehm HL, Born J. Nocturnal adrenocorticotropin and cortisol secretion depends on sleep duration and decreases in association with spontaneous awakening in the morning. J Clin Endocrinol Metab 1992; 75:1431-1435.

33. Spiegel K, Leproult R, Van Cauter E. Impact of sleep debt on metabolic and endocrine function. Lancet 1999; 354:1435-1439.

34. Van Cauter E, Leproult R, Plat L. Age-related changes in slow wave sleep and REM sleep and relationship with growth hormone and cortisol levels in healthy men. J Am Med Assoc 2000; 284:861-868.

35. Kupfer DJ, Bulik CM, Jarrett DB. Nighttime plasma cortisol secretions and EEG sleep - are they associated? Psychiatry Res 1983; 10:191-199.
36. Chapotot F, Gronfier C, Jouny C, Muzet A, Brandenberger G. Cortisol secretion is related to electroencephalographic alertness in human subjects during daytime wakefulness. J Clin Endocrinol Metab 1998; 83:4263-4268.
37. Chapotot F, Buguet A, Gronfier C, Brandenberger G. Hypothalamo-pituitary-adrenal axis activity is related to the level of central arousal: effect of sleep deprivation on the association of high-frequency waking electroencephalogram with cortisol release. Neuroendocrinol 2000; 73:312-321.
38. Leproult R, Copinschi G, Buxton O, Van Cauter E. Sleep loss results in an elevation of cortisol levels in the next evening. Sleep 1997; 20865870.
39. Toth LA, Gardiner TW, Krueger JM. Modulation of sleep by cortisone in normal and bacterially infected rabbits. Am J Physiol 1992; 263:R1339-R1346.
40. Toth LA, Rehg JE, Webster RG. Strain differences in sleep and other pathophysiological sequelae of influenza virus infection in naive and immunized mice. J Neuroimmunol 1995; 58:89-99.
41. Eleftheriou BE, Bailey DW. Genetic analysis of plasma corticosterone levels in two inbred strains of mice. J Endocrinol 1972; 55:415-420.
42. Flint MS, Tinkle SS. C57BL/6 mice are resistant to acute restraint modulation of cutaneous hypersensitivity. Toxicol Sci 2001;62:250-256.
43. Shanks N, Griffiths J, Zalcman S, Zacharko RM, Anisman H. Mouse strain differences in plasma corticosterone following uncontrollable footshock. Pharmacol Biochem Behav 1990; 36:515-519.
44. Shanks N, Kusnecov AW. Differential immune reactivity to stress in BALB/cByJ and C57BL/6J mice: in vivo dependence on macrophages. Physiol Behav 1998; 65:95-103.
45. Steiger A. Sleep and the hypothalamo-pituitary-adrenocortical system. Sleep Med Rev 2002; 6:125-138.
46. Toth LA, Krueger JM. Alteration of sleep in rabbits by *Staphylococcus aureus* infection. Infect Immun 1988; 56:1785-1791.
47. Toth LA, Tolley EA, Krueger JM. Sleep as a prognostic indicator during infectious disease in rabbits. Proc Soc Exp Biol Med 1993; 203:179-192.
48. Norman SE, Chediak HD, Kiel M, Cohn MA. Sleep disturbances in HIV-infected homosexual men. AIDS 1990; 4:775-781.
49. Kubicki S, Henkes H, Terstegge K, Ruf B. AIDS related sleep disturbances -a preliminary report. In: Kubicki S, Henkes H, Bienzle U, Pohle HD, editors. HIV and the Nervous System. New York: Gustav Fischer, 1988: 97-105.
50. Rubinstein ML, Selwyn PA. High prevalence of insomnia in an outpatient population with HIV infection. J Acq Imm Defic Synd 1998; 19:260-265.
51. Robbins JL, Phillips KD, Dudgeon WD, Hand GA. Physiological and psychological correlates of sleep in HIV infections. Clin Nurs Res 2004; 13:33-52.
52. Hand GA, Phillips KD, Sowell RL, Rojas M, Becker J. Prevalence of poor sleep quality in a HIV+ population of Americans. J S C Med Assoc 2003; 99:183-187.
53. Cruess DG, Antoni MH, Gonzalez J, Fletcher MA, Klimas N, Duran R et al. Sleep disturbance medicates the association between psychological distress and immune status among HIV-positive men and women on combination retroviral therapy. J Psychosom Res 2003; 54:185-189.

54. Brown R, Pang G, Husband AJ, King MG. Suppression of immunity to influenza virus infection in the respiratory tract following sleep disturbance. Reg Immunol 1989; 2:321-325.

55. Brown R, Price RJ, King MG, Husband AJ. Interleukin-1 b and muramyl dipeptide can prevent decreased antibody response associated with sleep deprivation. Brain Behav Immun 1989; 3:320-330.

56. Renegar KB, Floyd R, Krueger JM. Effects of short-term sleep deprivation on murine immunity to influenza virus in young adult and senescent mice. Sleep 1998; 21:241-248.

57. Toth LA, Rehg JE. Effects of sleep deprivation and other stressors on the immune and inflammatory responses of influenza-infected mice. Life Sci 1998; 63:701-709.

58. Benca RM, Kushida CA, Everson CA, Kalski R, Bergmann BM, Rechtschaffen A. Sleep deprivation in the rat: VII. Immune function. Sleep 1989; 12:47-52.

59. Everson CA, Bergmann BM, Rechtschaffen A. Sleep deprivation in the rat: III. Total sleep deprivation. Sleep 1989; 12:13-21.

60. Everson CA. Sustained sleep deprivation impairs host defense. Am J Physiol 1993; 265:R1148-R1154.

61. Everson CA, Toth LA. Systemic bacterial invasion induced by sleep deprivation. Am J Physiol 2000; 278:R905-R916.

62. Rechtschaffen A, Gilliland MA, Bergmann BM, Winter JB. Physiological correlates of prolonged sleep deprivation in rats. Science 1983; 221:182-184.

63. Toth LA, Opp MR, Mao L. Somnogenic effects of sleep deprivation and *Escherichia coli* inoculation in rabbits. J Sleep Res 1995; 4:30-40.

64. Brown R, Pang G, Husband AJ, King MG, Bull DF. Sleep deprivation and the immune response to pathogenic and non-pathogenic antigens. In: Husband AJ, editor. Behaviour and Immunity. Ann Arbor: CRC Press, 1992: 127-133.

65. Opp MR, Krueger JM. Interleukin-1 is involved in responses to sleep deprivation in the rabbit. Brain Res 1994; 639:57-65.

66. Drucker-Colin RR, Jaques LB, Winocur G. Anemia in sleep-deprived rats receiving anticoagulants. Science 1971; 174:505-507.

67. Bergmann BM, Rechtschaffen A, Gilliland MA, Quintans J. Effect of extended sleep deprivation on tumor growth in rats. Am J Physiol 1996; 271:R1460-R1464.

68. Hermann DM, Mullington J, Hinze-Selch D, Schreiber W, Galanos C, Pollmächer T. Endotoxin-induced changes in sleep and sleepiness during the day. Psychoneuroendocrinology 1998; 23:427-437.

69. Yarrington A, Mehta P. Does sleep promote recovery after bone marrow transplantation? -a hypothesis. Pediatr Tansplantation 1998; 2:51-55.

70. Honkus VL. Sleep deprivation in critical care units. Crit Care Nurs 2003; 26:179-189.

71. Gabor JY, Cooper AB, Hanly PJ. Sleep disruption in the intensive care unit. Curr Opin Crit Care 2001; 7:21-27.

72. Parthasarathy S, Tobin MJ. Sleep in the intensive care unit. Intensive Care Med 2003; 30:197-206.

73. Freedman NS, Gazendam J, Levan L, Pack AI, Schwab RJ. Abnormal sleep/wake cycles and the effect of environmental noise on sleep disruption in the intensive care unit. Am J Respir Crit Care Med 2001; 163:451-457.

74. Orr WC, Stahl ML. Sleep disturbances after open heart surgery. Am J Cardiol 1977; 39:196-201.

75. Lange T, Perras B, Fehm HL, Born J. Sleep enhances the human antibody response to hepatitis A vaccination. Psychosom Med 2003; 65:831-835.

76. Kripke DF, Simons RN, Garfinkel L, Hammond C. Short and long sleep and sleeping pills: Is increased mortality associated? Arch Gen Psychiatry 1991; 36:103-116.

77. Newman AB, Spiekerman CF, Enright P, Lefkowitz D, Manolio T, Reynolds CF et al. Daytime sleepiness predicts mortality and cardiovascular disease in older adults. J Am Geriatr Soc 2000; 48:115-123.

78. Pollack CP, Perlick D, Linsner JP, Wenston J, Hsieh F. Sleep problems in the community elderly as predictors of death and nursing home placement. J Commun Health 1990; 15:123-135.

79. Wingard DL, Berkman LF. Mortality risk associated with sleeping patterns among adults. Sleep 1983; 6:102-107.

80. Bursztyn M, Ginsberg G, Stessman J. The siesta and mortality in the elderly: effect of rest without sleep and daytime sleep duration. Sleep 2002; 25:187-191.

81. Kripke DF, Garfinkel L, Wingard DL, Klauber MR, Marler MR. Mortality associated with sleep duration. Arch Gen Psychiatry 2002; 59:131-136.

82. Manabe K, Matsui T, Yamaya M, Sato-Nakagawa T, Okamura N, Arai H et al. Sleep patterns and mortality among elderly patients in a geriatric hospital. Gerontology 2000; 46:318-322.

83. Nilsson PM, Nilsson J-A, Hedblad B, Berglund G. Sleep disturbance in association with elevated pulse rate for prediction of mortality -consequences of mental strain? J Intern Med 2001; 250:521-529.

84. Welsh DK, Richardson GS, Dement WC. Effect of age on the circadian pattern of sleep and wakefulness in the mouse. J Gerontol 1986; 41:579-586.

85. Gourmelon P, Briet D, Court L, Tsiang H. Electrophysiological and sleep alterations in experimental mouse rabies. Brain Res 1986; 398:128-140.

86. Gourmelon P, Briet D, Clarencon D, Court L, Tsiang H. Sleep alterations in experimental street rabies virus infection occur in the absence of major EEG abnormalities. Brain Res 1991; 554:159-165.

87. Toth LA, Tolley EA, Broady R, Blakely B, Krueger JM. Sleep during experimental trypanosomiasis in rabbits. Proc Soc Exp Biol Med 1994; 205:174-181.

88. Shaw PJ, Tononi G, Greenspan RJ, Robinson DF. Stress response genes protect against lethal effects of sleep deprivation in *Drosophila*. Nature 2002; 417:287-291.

89. Peterson PK, Chao CC, Molitor T, Murtaugh M, Strgar F, Sharp BM. Stress and pathogenesis of infectious disease. Rev Infect Dis 1991; 13:710-720.

90. Everson CA. Clinical manifestations of sleep deprivation. In: Schwartz WJ, editor. Monographs in Clinical Neuroscience. Basil: Larger, 1997: 34-59.

91. Suchecki D, Lobo LL, Hipólide DC, Tufik S. Increased ACTH and corticosteroid secretion induced by different methods of paradoxical sleep deprivation. J Sleep Res 1998; 7:276-281.

92. Suchecki D. Paradoxical sleep deprivation and the hypothalamic-pituitary-adrenal axis. SRS Bulletin 2002; 8:9-14.

93. Dinges DF. An overview of sleepiness and accidents. J Sleep Res 1995; 4(S2):4-14.

94. Krueger JM, Fang J, Majde JA. Sleep in health and disease. In: Ader R, Felten DL, Cohen N, editors. Psychoneuroimmunology. New York: Academic Press, 2001: 667-685.

95. Rogers NL, Sziba MP, Staab JP, Evans DL, Dinges DF. Neuroimmunologic aspects of sleep and sleep loss. Sem Clin Neuropsychiatr 2001; 6:295-307.

96. Dhabhar FS, McEwen BS. Acute stress enhances while chronic stress suppresses cell-mediated immunity *in vivo*: a potential role for leukocyte trafficking. Brain Behav Immun 1997; 11:286-306.

Neuroendocrine Correlates
of Restless Legs Syndrome

THOMAS C. WETTER AND STEPHANY FULDA

1. Introduction

Restless legs syndrome (RLS) is a common but often underdiagnosed sensorimotor disorder of sleep/wake motor regulation with prevalence rates estimated from population surveys between 1 and 10%, increasing with age.[1] RLS is characterized by an imperative desire to move the extremities associated with paraesthesias, motor restlessness, worsening of symptoms at rest and in the evening or at night and, as a consequence, sleep disturbances. Additionally, most patients with RLS have periodic limb movements (PLM) during sleep and relaxed wakefulness, which can be assessed by recording a surface electromyogram (EMG) of both tibialis anterior muscles in addition to standard polysomnographic parameters.[2] When PLM occur as an isolated cause of insomnia or daytime sleepiness without symptoms of RLS, the condition is termed "periodic limb movement disorder" (PLMD).

Two forms with almost identical clinical features are distinguished, i.e. idiopathic and symptomatic RLS. A positive family history is often reported in the idiopathic form and large pedigrees with familial RLS suggest that the disorder follows a pattern of autosomal dominant inheritance with a high degree of penetrance.[3]

Apart from the idiopathic form, RLS can be secondary to other disorders or conditions. The main causes include end stage renal disease, iron deficiency and pregnancy. There have been reports of an association with a variety of disorders such as diabetes mellitus, hypothyroidism, rheumatoid arthritis, amyloidosis, and folic acid deficiency.[4] Drug-induced forms of RLS have occured after treatment with neuroleptics, selective serotonin reuptake inhibitors (SSRI), tricyclic antidepressants, mirtazapine, mianserin and lithium.

Today, most authors agree that RLS has its origin in the central nervous system, however, complex interactions between central and peripheral structures may contribute to the disorder. Based on the knowledge of the efficacy of dopaminergic and opioidergic drugs and the provocation or exacerbation of RLS symptoms following treatment with dopamine receptor blocking agents,[5] there is evidence of the involvement of the dopaminergic and opioid system in the pathogenesis of RLS. Recent PET and SPECT studies revealed some controversial results of the nigrostriatal dopaminergic neurotransmission probably reflecting a dysfunction of the central dopaminergic system.[6]

The etiology, however, remains unclear, despite what is known about the conditions that may induce the syndrome.[4] Although there is some evidence that hormones may play a role in the pathophysiology of the disorder, neuroendocrine studies in RLS patients are limited. In the following sections we review the relevant studies, including those primarily focussing on the involvement of neurotransmitters or neuropeptides.

2. Methodological Considerations and Research Strategies

Research concerning neuroendocrine correlates in RLS can be roughly classified into two categories.

First, there are some studies which transform the clinically characteristic features of RLS into a neuroendocrine hypothesis (see Table 1). A common feature of these studies is the small number of participants, which makes non-significant results hard to interpret. Within this approach there are some studies which do not test a neuroendocrine hypothesis explicitly but are of a more explorative, descriptive nature.

Secondly, indirect evidence can be found in large-scaled or epidemiologic studies that describe associations with comorbid disorders in RLS patients or explore the prevalence of these disorders in specific populations such as patients with diabetes mellitus. Most of these studies include more than a hundred subjects, which makes a detailed assessment of RLS symptoms often unfeasible. Due to the large sample sizes one has to keep in mind that very small effects will be statistically significant. Therefore, the results have to be interpreted cautiously.

This makes it noticeable that the studies are very heterogeneous. A common feature, however, is the fact that neuroendocrine measures were not the primary target parameter of the study but were taken as a marker for other physiological systems such as the dopaminergic neurotransmission. This made it quite difficult to group the available studies into meaningful categories without too much overlap between them.

3. Neuroendocrine Studies in Patients with RLS

3.1. Prolactin, Growth Hormone and Cortisol

Dopamine is the major inhibitory factor for prolactin release[7] and also influences human growth hormone secretion.[8] Measurements of both parameters under normal conditions or in response to pharmacological challenges have

TABLE 1. Delineation of hypothesis and neuroendocrine measurements from clinical findings in RLS.

Clinical finding	Specific hypothesis	Operationalization
Response to L-dopa	Alterations in dopamine synthesis	Measurement of CSF metabolites of dopamine synthesis HVA, neopterin, BH4[15,16]
	Primary altered dopaminergic tone	Challenge with dopamine antagonist and response of prolactin, growth hormone[9]
	Hypothalamus-pituitary axis influenced by alterations in extrastriatal dopaminergic systems	24-hour profile of prolactin and growth hormone as indirect markers of dopamine system[12]
	Circadian changes in dopaminergic function	Challenge with L-dopa at different times of the day and response of growth hormone and prolactin[13]
	Increased amplitude of circadian variation of dopaminergic function	Effect of dopaminergic medication on melatonin secretion[22]
Response to opioid agents	Primary altered opiodergic tone	Challenge with opioid antagonist naloxone and response of prolactin, growth hormone[9]
Circadian variation	Abnormalities of circadian system	24-hour profile of cortisol[12]
		Daytime and nightime cumulative excretion of aMLT[20]
		24-hour profile of salivary melatonin in a constant routine procedure[21]
	Decreased output amplitude of the circadian system	Treatment with melatonin[19]
Worsening with SSRI	Involvement of serotonin	Measurement of CSF serotonin metabolites[15,16]
Other:		
Hypocretin	Alterations in hypocretin levels	Measurement of CSF hypocretin-1 levels[33-35]
Secondary forms of RLS :		
Pregnancy	Increased levels of estrogens, progesterone and prolactin may provoke RLS	Measurement of oestradiol, progesterone and prolactin during pregnancy[47]
Diabetes mellitus	Diabetes-associated polyneuropathy as a peripheral trigger of RLS	Prevalence of diabetes mellitus in patients with RLS[24,28,29,40,49]
		Prevalence of RLS in patients with diabetes mellitus[52,54,55]

been taken as indirect tests to measure the activity of the dopaminergic and opiodergic system.

Winkelmann and coworkers[9] measured plasma levels of adrenocorticotropic hormone (ACTH), cortisol, prolactin, and growth hormone in eight subjects with RLS after administration of placebo, naloxone, and metoclopramide at 14:00 in a randomized cross-over design.

Prior studies have shown that administration of the opioid antagonist naloxone increased plasma concentrations of ACTH, cortisol, and growth hormone[10] and that the dopamine antagonist metoclopramide increased prolactin and growth hormone concentrations in normal subjects.[11] Winkelmann et al.[9] have found a similar pattern in RLS patients and concluded that the endogenous opiodergic or the dopaminergic system in general was not modified in RLS patients.

Wetter et al.[12] have explored the time course of prolactin, growth hormone, and cortisol in ten drug-naïve RLS patients and eight matched controls in a 25-hour bedrest procedure where sleep was recorded overnight between 23:00 and 07:00. Blood samples were drawn every 20 minutes. No differences in overall plasma levels, frequency or amplitude of the pulses of the respective hormones between patients and controls were found. This finding caused the authors to infer that a possible alteration of the dopaminergic system in RLS does not affect the release of prolactin and growth hormone from the pituitary gland.

Garcia-Borreguero and coworkers[13] have explored hormone levels in response to administration of L-dopa in twelve RLS patients and twelve matched controls. In a bedrest condition they administered 200 mg L-dopa at 11:00 in the morning and at 23:00 at night in a randomized order and one week apart. Blood samples were drawn five and 20 minutes before administration and every 15 minutes after L-dopa administration for two hours. Interestingly, there was no difference in prolactin, growth hormone or cortisol response to L-dopa after daytime administration, but in the night growth hormone levels were increased and prolactin was decreased after L-dopa intake in RLS patients when compared to controls. This finding might suggest an enhanced circadian variation in dopaminergic function with an increased sensitivity of dopamine receptors at night rather than an overall altered dopaminergic tone.

3.2. Dopamine and Biopterin Metabolites

The hypothesis of a decreased dopaminergic activity in RLS has lead to studies measuring factors involved in the synthesis of dopamine.

As early as 1985 high levels of dopamine and homovanillic acid (HVA) in the cerebrospinal fluid (CSF), but not serum, have been reported in a single patient with severe, familial RLS.[14]

Earley and coworkers[15] have measured HVA, biopterin (BH4), and neopterin in CSF taken in a lumbar punctuation in the morning in 16 patients with RLS and two weeks off medication. All values were compared to those of

14 control subjects whose values were provided by the laboratory with only information about gender and age available. Due to differences in age between RLS patients and control subjects, several analyses with subgroups matched for age were performed. When comparing both groups there was an increase in neopterin in RLS patients, whereas when comparing an age-matched subsample of eleven RLS patients and ten controls, RLS patients showed decreased 5-hydroxyindoleacetic acid (5-HIAA) levels (see 3.4). Finally, Earley et al.[15] computed an age-adjusted regression; here only increased BH4 concentrations were apparent in RLS patients. The finding that all three analyses yielded different non-overlapping results makes this study difficult to interpret.

None of the findings from the study of Earley et al.[15] have been confirmed in a subsequent study by Stiasny-Kolster and coworkers[16] who measured the same and additional CSF metabolites by means of an evening lumbar puncture in 22 RLS patients and eleven control subjects. They found no difference between patients and controls for HVA, other dopaminergic metabolites (3-ortho-methyl-dopa, L-dopa), pterin metabolites (BH4, BH2, neopterin), or 5-methyltetrahydrofolate, a key metabolite in cytosolic methyl group transfer.

Due to the low number of studies and their conflicting results no attempt is made to draw any conclusions.

3.3. Melatonin

A typical feature of RLS is an increase of symptoms in the evening or at night.[2] Although the presence of a circadian variation of symptoms could be biased by the fact that RLS symptoms become apparent during inactivity, which typically occurs in the evening and night, at least two studies have shown a "true" circadian variation of symptoms.[17,18]

In 2001 Kunz and Bes[19] applied melatonin daily for six weeks to nine patients with PLMD (without RLS) and found a significant reduction of PLM. However, the study was uncontrolled and open-labeled, therefore the results have to be interpreted cautiously.

Tribl and coworkers[20] have collected daytime (07:00–22:00) and nighttime (22:00–07:00) urinary 6-OH-melatonin-sulfate (aMLT) excretion in 15 untreated patients with RLS and eleven control subjects and found no difference in daytime or night-time urinary aMLT excretion. They concluded that cumulative melatonin production is not reduced in RLS.

A more detailed analysis of the time course of melatonin in RLS subjects was conducted by Michaud et al.[21] They compared hourly collected salivary melatonin levels in seven RLS patients and matched controls in a 28-hour modified constant routine procedure. There was a significantly shorter duration of melatonin secretion (time from dim light melatonin onset (DLMO) and melatonin synthesis offset) in RLS patients, a trend for earlier melatonin synthesis offset, and a shorter duration of the melatonin episode (time from DLMO to melatonin secretion offset). Michaud et al.[21] also found a strong cross-correlation between symptoms of RLS and melatonin levels with changes in melatonin concentrations preceding the increase in leg discomfort by two hours.

Garcia-Borreguero and coworkers[22] have measured salivary melatonin hourly from 17:00 to 03:00 under dim light conditions in eight patients with previously untreated RLS before and after treatment with L-dopa for an average of 21 days. DLMO was significantly earlier at the end of the treatment; this was mainly due to DLMO advances occurring in the four patients who showed signs of augmentation.

Taken together, these studies suggest a role of melatonin in RLS. To date it has been unclear if melatonin is merely a marker of an altered circadian system in RLS patients or more directly involved in the pathophysiology of RLS, possibly by exerting an inhibitory effect on dopamine secretion as has been shown for some areas of the mammalian central nervous system.[23]

3.4. Serotonin

It is well documented that SSRIs can provoke or aggravate symptoms of RLS.[24] The role of serotonin has been explored in the same few studies that investigated dopamine metabolites (see 3.2).

Long before studies measured parameters of the serotonin metabolism in 1987, Guilleminault and coworkers[25] reported that treatment with 500 mg 5-hydroxytryptophane (five patients) or with L-tryptophane (six patients) had no effect on the number of PLM or PLM-associated arousals in patients with PLMD without restless legs symptoms.

Earley et al.[15] have determined 5-HIAA levels, the major urinary metabolite of serotonin, in CSF taken in the morning. In a comparison between an age-matched subsample of eleven RLS patients and ten control subjects, RLS patients showed decreased 5-HIAA levels. This difference, however, was not apparent when comparing all 16 RLS patients to 14 control subjects, or in an age-adjusted regression analysis.

Stiasny-Kolster et al.[16] have not found any differences in the serotonin metabolites 5-HIAA and 5-hydroxytryptophan between 22 RLS patients and eleven control subjects in an evening CSF sample.

Given the low number of studies and diversity of their findings, it seems premature to draw any conclusions regarding the role of serotonin in RLS.

3.5. Thyroid and Parathyroid Hormones

A close temporal association between recurring hypothyroidism and the occurence of restless legs was described in a single patient in 1985.[26] In 1996 another case report linked RLS to pharmacologically induced hypothyroidism, which resolved with treatment of the thyroid disorder.[27]

Rothdach et al.[28] have found no difference in the use of thyroid medication between RLS positive elderly subjects and those without restless legs symptoms. On the other hand, Banno and coworkers[29] have reported a higher incidence of acquired hypothyroidism in females but not males suffering from RLS when compared to matched controls. More recently, Tan and coworkers[30] have specifically explored the prevalence of RLS in patients with thyroid disorders and found no difference in prevalence rates compared with a control sample.

In uremic patients with RLS, Collado-Seidel and coworkers[31] have found reduced levels of intact parathyroid hormone as compared to uremic patients without RLS. This finding has not been replicated by Goffredo Filho et al.[32] who observed intact parthyroid hormone levels comparable for dialysis patients with and without RLS.

So far, the available evidence in humans has not suggested a major role of thyroid and parathyroid hormones in RLS.

3.6. Orexin/Hypocretin

Hypocretins are neuropeptides involved in the pathophysiology of narcolepsy.[33] Hypocretins increase EEG arousal and locomotor activity and interact with the dopamine system,[33] important aspects of RLS.

To date, three studies[34-36] have measured hypocretin in patients with RLS. Allen et al.[34] have found increased evening CSF hypocretin-1 levels in 16 RLS patients as compared to eight control subjects. A subsequent subgroup analysis showed that this overall increase was due to patients with early onset of RLS who had significantly higher levels of hypocretin-1 whereas late onset RLS subjects did not differ from controls. This finding, however, has not been replicated neither by Mignot and coworkers[35] who have found normal daytime hypocretin levels in ten RLS patients, nor by Stiasny et al.[36] who have measured CSF hypocretin-1 levels at 18:00 in 13 RLS patients and nine controls. In the latter study,[36] no difference between early and late onset RLS patients was apparent.

Taken together, further studies are needed to clarify the role of hypocretins in the pathophysiology of RLS.

3.7. Steroid Hormones/Estrogen, Progesterone

Some[24,27,28,37,38] but not all[39,40] epidemiological studies have shown a higher prevalence of RLS in women than in men. The prevalence increases with age and elderly women seem to be particularly affected,[41] which points to the postmenopausal hormonal status as a possible cause. On the other hand, there is a markedly high incidence of RLS during pregnancy.[42]

3.7.1. RLS in Pregnancy

There is repeated and solid evidence that pregnant women have a markedly higher risk for RLS, and that those women with pre-existing RLS experience a worsening of RLS symptoms.[42]

Ekbom[39] and later his son[43] have favoured a hormonal hypothesis regarding the incidence of RLS in pregnancy. During pregnancy plasma levels of estrogen, progesterone and prolactin increase and then decrease after delivery.[44] This corresponds well with the increase in RLS symptoms during pregnancy and the distinct relief from RLS symptoms after delivery.

However, iron and folate requirements during pregnancy are very much increased and both may play a role in the etiology of RLS.[45] Interestingly, a recent study[46] that followed pregnant women with and without RLS from the 35th week of gestation to approximately 12 weeks postpartum has found markedly elevated oestradiol levels in pregnant women with RLS during late term pregnancy but not after delivery. Iron, ferritin and magnesium did not differ between subjects with and without RLS during pregnancy.

3.7.2. RLS and Hormone Replacement Therapy (HRT)

Given that estrogen, progesterone and prolactin have been implicated in the pathogenesis of RLS during pregnancy, hormone replacement therapy (HRT) would be expected to worsen RLS symptomatology.

Although Rothdach et al.[28] have reported that more elderly women with RLS took estrogens than women without RLS, this difference was not statistically significant. Polo-Kantola and coworkers[47] have examined the incidence of PLM during estrogen replacement therapy for up to seven months. They have found no effect of HRT on PLM and PLM-associated arousals. Corrobating this finding, Saletu-Zyhlarz et al.[48] have reported no effect of estrogen or an estrogen-progesteron combination on PLM.

Female hormones are candidate measures to explain prevalence differences between men and women and the high incidence of RLS during pregnancy. The available evidence, however, is sparse and only tentatively identifies estrogens as a possible parameter. It seems likely that only very high levels of steroid hormones, as they are found in pregnancy, may play a role in RLS.

3.8. Insulin

3.8.1. Prevalence of Diabetes Mellitus in RLS

Several large-scale epidemiological studies have assessed the prevalence of RLS and relevant comorbidites.[24,28,29,40,49] Although sample sizes ranged from 200[29] to nearly 20,000,[24] only the three largest studies with 1506,[49] 1803[40] and 18980[24] persons found a higher prevalence of diabetes in subjects with RLS,[24,40] RLS symptoms[49] or PLMD[24]. A low prevalence of RLS was a problem in some[28,29] but not all of the other studies.[29]

Four studies[31,32,50,51] have assessed comorbidities in patients with end-stage renal disease undergoing dialysis. None of the studies have found a higher prevalence in patients with renal disease and RLS in comparison to those without RLS concerning diabetes mellitus,[31,32] diabetes as a primary diagnosis[50] or use of antidiabetic medication.[51]

Finally, Machtey and Weitz[52] have compared 80 patients with rheumatism and RLS with 150 control subjects with rheumatism but without RLS and found that 23% of RLS subjects but only 4.7% of control patients had known diabetes. A closer investigation in a subsample of 58 RLS subjects revealed that 13 patients had a known diabetes mellitus. An oral glucose tolerance test (OGTT) identified a further five patients with diabetes mellitus and another 23 subjects had an abnormal OGTT result.

From these studies it can be concluded that diabetes mellitus seems to be unrelated to secondary, uremic RLS. In subjects with idiopathic forms, a higher prevalence of diabetes mellitus has been reliably shown in very large studies. However, the effect size is likely very small so that diabetes mellitus may play a role in only a minority of patients with RLS.

3.8.2. Prevalence of RLS in Patients with Diabetes Mellitus

Studies exploring the prevalence of RLS in patients with diabetes have yielded mixed results.

Banerji and Hurwitz[53] have assessed the prevalence of RLS in 53 patients with diabetes mellitus and 50 control subjects with a standardized clinical interview. They have found that 17% of patients with diabetes but only 2% of control subjects had RLS, a significantly higher number. In a similar study, Skomro et al.[54] have compared 58 consecutive type 2 diabetic patients to 48 matched non-diabetic control patients. Although twice as many patients with diabetes (24%) had RLS compared to 12.5% in control patients, this difference was not statistically significant.

Two larger-scaled studies have reported more converging evidence. Oboler et al.[55] have investigated RLS symptoms in 515 outpatients. In that sample patients with diabetes mellitus had no increased incidence of restless legs symptoms (odds ratio 1.11). However, the incidence of RLS was assessed with a single questionnaire item. O'Hare and coworkers[56] have determined prevalence of RLS in 800 consecutive patients with diabetes mellitus in comparison to 100 non-diabetic controls. The prevalence of RLS did not differ between diabetic patients (8.8%) and controls (7%), but type 2 diabetic patients had a significantly higher prevalence of RLS (10.8%) than type I diabetic patients (6%).

Taken together, these studies suggest that patients with diabetes mellitus may only have a slightly increased risk for RLS.

4. Concluding Remarks

Reviewing the studies pertaining to neuroendocrine correlates in RLS, some features are evident. On the one hand, studies are very diverse ranging from well designed large-scale epidemiological studies to small uncontrolled case series. On the other hand, only a small number of studies were designed to assess neuroendocrine mechanisms *per se* in patients with RLS. For example, some studies used neuroendocrine measures primarily to unmask a presumed dopaminergic dysfunction. So far, it has remained unclear whether or not neuroendocrine changes are causative or secondary phenomena. Nevertheless, many of the studies have been published only recently and reflect a growing interest in this question.

Summarizing the overview, the present studies have revealed some controversial results and do not suggest that there are consistent differences in neuroendocrine measures between RLS patients and appropriate controls. Furthermore, for most neuroendocrine parameters there is substantial inter-individual and intraindividual variability. Thus, it will be more favourable to utilize 'within' rather than 'between subjects' designs with repeated measurements where possible to increase sample sizes and include well-defined and homogenous groups of subjects. There is also some suggestion that neuroendocrine measures may only differ at a specific time of the day, e.g. in the evening or at night. However, what constitutes this specific time of day may vary from patient to patient so that in a repeated measures design overall group differences will be unspecific and small. Here, more detailed analyses and comparisons of individual time courses might be promising. Given that some evidence has accumulated so far, further research into neuroendocrine correlates of RLS is needed but also seems a promising avenue of research.

5. Acknowledgement. The study[46] was supported by a European Union grant QLK6-CT-2000-00499.

6. *References*

1. M. Zucconi and L. Ferini-Strambi, Epidemiology and clinical findings of restless legs syndrome, *Sleep Med.* 5, 293-299 (2004).
2. R.P. Allen, D. Picchietti, W.A. Hening, C. Trenkwalder, A.S. Walters, and J. Montplaisir, Restless legs syndrome: diagnostic criteria, special considerations, and epidemiology. A report from the restless legs syndrome diagnosis and epidemilogy workshop at the National Institutes of Health, *Sleep Med.* 4, 101-119 (2003).
3. L. Ferini-Strambi, M. T. Bonati, A. Oldani, P. Aridon, M. Zucconi, and G. Casari, Genetics in restless legs syndrome, *Sleep Med.* 5, 301-304 (2004).
4. R.P. Allen and C.J. Earley, Restless legs syndrome: A review of clinical and pathophysiological features, *J. Clin. Neurophysiol.* 18, 128-147 (2001).
5. T.C. Wetter, J. Winkelmann, and I. Eisensehr, Current treatment options for restless legs syndrome, *Expert. Opin. Pharmacother.* 4, 1727-1738 (2003).
6. T.C. Wetter, I. Eisensehr, and C. Trenkwalder, Functional neuroimaging studies in restless legs syndrome, *Sleep Med.* (in press).
7. D. A. Leong, L. S. Frawley, and J. D. Neill, Neuroendocrine control of prolactin secretion, *Annu. Rev. Physiol.* 45, 109-127 (1983).
8. G. Verde, G. Oppizzi, G. Colussi, G. Cremascoli, L. Botalla, E. E. Muller, F. Silvestrini, P. G. Chiodini, and A. Liuzzi, Effect of dopamine infusion on plasma

levels of growth hormone in normal subjects and in acromegalic patients, *Clin. Endocrinol.* 5, 419-423 (1976).

9. J. Winkelmann, J. Schadrack, T. C. Wetter, W. Zieglgänsberger, and C. Trenkwalder, Opoid and dopamine antagonist drug challenges in untreated restless legs syndrome patients, *Sleep Med.* 2, 57-61 (2001).

10. J. Volavka, J. Bauman, J. Pevnick, D. Reker, B. James and D. Cho, Short-term hormonal effects of naxolone in man, *Psychoneuroendocrinology* 5, 225-234 (1980).

11. G. P. Bernini, A. R. Lucarini, F. Franchi, and A. Salvetti, Humoral effects of metoclopramide and domperidone in normal subjects and in hypertensive patients, *J. Endocrinol. Invest.* 11, 711-716 (1988).

12. T. C. Wetter, V. Collado-Seidel, H. Oertel, M. Uhr, A. Yassouridis, and C. Trenkwalder, Endocrine rhythms in patients with restless legs syndrome, *J. Neurol.* 249, 146-151 (2002).

13. D. Garcia-Borreguero, O. Larossa, J. J. Granizo, Y. de la Llave, and W. Hening, Circadian variation in neuroendocrine response to L-dopa in restless legs syndrome, *Sleep* (in press).

14. J. Montplaisir, R. Godbout, D. Boghen, J. DeChamplain, S. N. Young, and G. Lapierre, Familial restless legs with periodic movements in sleep: electrophysiologic, biochemical, and pharmacological study, *Neurology* 35, 130-134 (1985).

15. C. J. Earley, K. Hyland, and R. P. Allen, CSF dopamine, serotonin, and biopterin metabolites in patients with restless legs syndrome, *Mov. Disord.* 16, 144-149 (2001).

16. K. Stiasny-Kolster, J. C. Möller, J. Zschocke, O. Bandmann, W. Cassel, W. H. Oertel, G. and F. Hoffmann, Normal dopaminergic and serotonergic metabolites in crebrospinal fluid and blood of restless legs syndrome patients, *Mov. Disord.* 19, 192-196 (2004).

17. W. A. Hening, A. S. Walters, M. Wagner, R. Rosen, V. Chen, S. Kim, M. Shah, and O. Thai, Circadian rhythm of motor restlessness and sensory symptoms in the idiopathic restless legs syndrome, *Sleep* 22, 901-912 (1999).

18. C. Trenkwalder, W. A. Hening, A. S. Walters, S. S. Campbell, K. Rahman, and S. Chokroverty, Circadian rhythm of periodic limb movements and sensory symptoms of restless legs syndrome, *Mov. Disord.* 14, 102-110 (1999).

19. D. Kunz and F. Bes, Exogenous melatonin in periodic limb movement disorder: an open clinical trial and a hypothesis, *Sleep* 24, 183-187 (2001).

20. G. G. Tribl, F. Waldhauser, T. Sycha, E. Auff, and J. Zeitlhofer, Urinary 6-hydroxymelatonin-sulfate excretion and circadian rhythm in patients with restless legs syndrome, *J. Pineal Res.* 35, 295-296 (2003).

21. M. Michaud, M. Dumont, B. Selmaoui, J. Paquet, M. L. Fantini, and J. Montplaisir, Circadian rhythm of restless legs syndrome: relationship with biological markers, *Ann. Neurol.* 55, 372-380 (2004).

22. D. Garcia-Borreguero, C. Serrano, O. Larossa, and J. J. Granizo, Circadian effects of dopaminergic treatment in restless legs sindrome, *Sleep Med.* (in press).

23. N. Zisapel, Melatonin-dopamine interactions: from basic neurochemistry to a clinical setting, *Cell. Mol. Neurobiol.* 21, 605-616 (2001).

24. M. M. Ohayon and T. Roth, Prevalence of restless legs syndrome and periodic limb movement disorder in the general population, *J. Psychosom. Res.* 53, 547-554 (2002).

25. C. Guilleminault, S. Mondini, J. Montplaisir, J. Mancuso, D. Cobaska, and W. C. Dement, Periodic leg movement, L-dopa, 5-hydroxytryptophan, and L-tryptophan, *Sleep* 10, 393-397 (1987).

26. J. L. Schlienger, Restless legs syndrome due to moderate hypothyroidism, *Presse Med.* 14, 791 (1985).
27. E. Bon, Y. Rolland, M. Laroche, A. Cantagrel, and B. Mazieres, Hypothyroidism on cochimax revealed by restless legs syndrome, *Rev. Rhum.* 63, 304 (1996).
28. A. J. Rothdach, C. Trenkwalder, J. Haberstock, U. Keil, and K. Berger, Prevalence and risk factors of RLS in an elderly population, *Neurology* 54, 1064-1068 (2000).
29. K. Banno, K. Delaive, R. Walld, and M. H. Kryger, Restless legs syndrome in 218 patients: associated disorders, *Sleep Med.* 1, 221-229 (2000).
30. E. K. Tan, S. C. Ho, P. Eng, L. M. Loh, L. Koh, S. Y. Lum, M. L. Teoh, Y. Yih, and D. Khoo, Restless legs symptoms in thyroid disorders, *Parkinsonism Relat. Disord.* 10, 149-151 (2004).
31. V. Collado-Seidel, R. Kohnen, W. Samtleben, G. F. Hillebrand, W. H. Oertel, and C. Trenkwalder, Clinical and biochemical findings in uremic patients with and without restless legs syndrome, *Am. J. Kidney Dis.* 31, 324-328 (1998).
32. G. S. Goffredo Filho, C. C. Gorini, A. S. Purysko, H. C. Silva, and I. E. F. Elias, Restless legs sindrome in patients on chronic hemodialyses in a Brazilian city, *Arq. Neuropsiquiatr.* 61, 3-8 (2003).
33. S. Taheri, J. M. Zeitzer, and E. Mignot, The role of hypocretins (orexins) in sleep regulation and narcolepsy, *Annu. Rev. Neurosci.* 25, 283-313 (2002).
34. R. P. Allen, E. Mignot, B. Ripley, S. Nishino, and C. J. Earley, Increased CSF hypocretin-1 (orexin-A) in restless legs syndrome, *Neurology* 59, 639-641 (2002).
35. E. Mignot, G. J. Lammers, B. Ripley, M. Okun, S. Nevsimalova, S. Overeem, J. Vankova, J. Black, J. Harsh, C. Basetti, H. Schrader, and S. Nishino, The role of cerebrospinal fluid hypocretin measurement in the diagnosis of narcolepsy and other hypersomnias, *Arch. Neurol.* 59, 1553-1562 (2002).
36. K. Stiasny-Kolster, E. Mignot, L. Ling, J. C. Möller, W. Cassel, and W. H. Oertel, CSF hypocretin-1 levels in restless legs syndrome, *Neurology* 61, 1426-1429 (2003).
37. G. J. Lavigne and J. Y. Montplaisir, Restless legs syndrome and sleep bruxism: prevalence and association among Canadians, *Sleep* 17, 739-743 (1994).
38. C. Purvis, B. Phillips, and K. Asher, Self report of RLS: 1996 Kentucky behavior risk factor surveillance survey, *Sleep Res.* 26, 474 (abstract) (1997).
39. K. A. Ekbom, Restless legs syndrome, *Acta Med. Scand.* 158, 122 (1945).
40. B. Phillips, T. Young, L. Finn, K. Asher, W. A. Henning, and C. Purvis, Epidemiology of restless legs syndrome in adults, *Arch. Int. Med.* 160, 2137-2141 (2000).
41. D. Garcia Borreguero, Restless legs syndrome. A clinical study in general Chilean population and in uremic patients (article reviewed), *Sleep Med.* 2, 557-559 (2001).
42. M. Manconi, V. Govoni, A. De Vito, N. T. Economou, E. Cesnik, G. Mollica, and E. Granieri, Pregnancy as a risk factor for restless legs, *Sleep Med.* 5, 305-308 (2004).
43. K. Ekbom, Akroparestesier och restless legs under graviditet, *Lakartidningen* 57, 2597-2603 (1960).
44. P. Coats, E. Walter, E. Youssefnejadian, and I. L. Craft, Variations in plasma steroid and prostaglandin concentrations during human pregnancy, *Acta Obstet. Gynecol. Scand.* 56, 453-457 (1977).
45. J. Krieger and C. Schroeder, Iron, brain and restless legs syndrome, *Sleep Med. Rev.* 5(4), 277-286 (2001).
46. S. Fulda, A. Dzaja, M. Lancel, and T. Pollmächer, Sleep and perioidic leg movements in late term pregnancy and postpartum in women with and without RLS, *J. Sleep. Res.* (abstract) (in press).

47. P. Polo-Kantola, E. Rauhala, R. Erkkola, K. Irjala, and O. Polo, Estrogen replacement therapy and nocturnal periodic limb movements: a randomized controlled trial, *Obstet. Gynecol.* 97, 548-554 (2001).

48. G. Saletu-Zyhlarz, P. Anderer, G. Gruber, M. Mandl, D. Gruber, M. Metka, J. Huber, M. Oettel, T. Gräser, M. H. Abu-Bakr, E. Grätzhofer, and B. Saletu, Insomnia related to postmenopausal syndrome and hormone replacement therapy: sleep laboratory studies on baseline differences between patients and controls and double-blind placebo-controlled investigation on the effects of a novel estrogen-progesteron combination (Climoden®, Lafamme®) versus estrogen alone, *J. Sleep Res.* 12, 239-254 (2003).

49. D. Foley, S. Ancoli-Israel, P. Britz, and J. Walsh, Sleep disturbances and chronic disease in older adults, results of the 2003 National Sleep Foundation Sleep in America survey, *J. Psychosom. Res.* 56, 497-502 (2004).

50. N. G. Kutner and D. L. Bliwise, Restless legs complaint in African-American and Caucasian hemodialysis patients, *Sleep Med.* 3, 497-500 (2002).

51. G. L. Gigli, M. Adorati, P. Dolso, A. Piani, M. Valente, S. Brodini, and R. Budai, Restless legs sindrome in end-stage renal disease, *Sleep Med.* 5, 309-315 (2004).

52. I. Machtey and Z. Weitz, in: *Progress in Rheumatology*, edited by I. Machtey (Rheomatology Service, Golda Medical Center, Petah-Tiqva, 1987), pp. 206-210.

53. N. K. Banerji and L. J. Hurwitz, Restless legs syndrome, with particular reference to its occurence after gastric surgery, *Br. Med. J.* 738, 774-775 (1970).

54. R. P. Skomro, S. Ludwig, E. Salamon, and M. H. Kryger, Sleep complaints and restless legs syndrome in adult type 2 diabetes, *Sleep Med.* 2, 417-422.

55. S. K. Oboler, A. V. Prochazka, and T. J. Meyer, Leg symptoms in outpatient veterans, *West. J. Med.* 155, 256-259 (1991).

56. J. A. O'Hare, F. Abuaisha, and M. Geoghegan, Prevalence and forms of neuropathic morbidity in 800 diabetics, *Ir. J. Med. Sci.* 163, 132-135 (1994).

Neuroendocrine and Neuroimmune Correlates of Narcolepsy

MICHELE L. OKUN AND MARY COUSSONS-READ

1. Introduction

The term "narcolepsy" originated in 1880 when it was first described as a pathological condition characterized by irresistible episodes of sleep of short duration.[1] In the 1930s, Daniels (as cited in Guilleminault)[1] introduced the clinical "tetrad" known today as cataplexy, sleep paralysis (SP), and hypnagogic hallucinations (HH) that are included in the clinical definition of narcolepsy. Cataplexy is a sudden loss of muscle tone, typically elicited by strong emotion: laughter, joking or anger. The muscles involved may be inconsequential, such as a fluttering of the cheeks, or functionally significant, as in the case of whole body muscle atonia resulting in complete body collapse. Sleep paralysis is the inability to move either upon falling asleep or upon awakening. Typically narcoleptics are frightened by this experience of not being able to move their limbs or to speak. Frequently, sleep paralysis is accompanied by hypnagogic hallucinations, which often serve to intensify and prolong the paralysis. Hypnagogic hallucinations are visual and/or auditory experiences occurring most often at sleep onset. These are thought to be a cause of immediate entry into REM sleep. Commonly, narcoleptics report these hallucinations as negative or frightening in nature, although the reason for the unpleasant content is unknown. These symptoms are often referred to as REM intrusion behaviors due to another clinical feature of narcolepsy: the rapid onset to REM sleep. Narcoleptics show a subsequent bypass of the initial stages of sleep on polysomnography tests by entering abnormally quickly into REM sleep. Where most adults enter the first REM sleep period after about 90 minutes of sleep, most narcoleptics enter REM sleep within 20 minutes.[1] Diagnosis of narcolepsy is concluded by a positive outcome on the Multiple Sleep Latency Test (MSLT). This test measures the propensity to sleep and is comprised of 4-5 20-minute naps scheduled every 2-hours beginning in the morning hours following a full night of PSG. If the subject falls asleep and enters REM sleep within 15 minutes, a sleep onset REM period (SOREM) is noted. A patient must have at least 2 SOREMs to have a positive MSLT.

2. Epidemiology of Narcolepsy

Narcolepsy occurs more often in the general population than most would anticipate. Studies of prevalence rates have been conducted on a variety of populations and have yielded variable results. In the early 1970s, William Dement and colleagues assessed prevalence rates in the San Francisco Bay and the Los Angeles areas where rates were calculated at 0.05 per cent and 0.067 per cent respectively.[2,3] Since then the prevalence in other populations has been studied, including the Israelis,[4] the Finnish,[5,6] various ethnic groups in the US,[7,8] the British, Germans, Italians, Portuguese, and Spanish,[9] the Koreans,[10] and the Chinese.[11] The prevalence rates reported for these studies differ dramatically. The reason lies in not only the methodology used to determine what constitutes narcolepsy, but possibly in a genetic predisposition that may vary by population and/or ethnic group. The Israeli study,[4] for instance, indicates that the prevalence for narcolepsy in Israel is quite low. An extensive assessment of over 1526 patients with EDS over a 9-year period indicates a prevalence rate of about 0.002 per 1000 people. On the other hand, epidemiological studies in Hong Kong conducted via structured phone interviews in the general population have determined the prevalence rate to be about .34 per 1000 people.[11] In Europe, the prevalence rate for five different countries was .47 per 1000 people;[9] in a small county in Minnesota, the prevalence rate was 0.01 per 1000 people.[8] The European study used the Sleep-EVAL expert system to interview people in the general population by telephone, while the Minnesota study looked at medical records for patients who had one of seven H-ICDA diagnostic codes. Because of these diverse data collection methods, it is difficult to establish an accurate prevalence rate for narcolepsy. Although no clear prevalence rates have been established for the US, it is thought that approximately 1 out of every 2000 people have or will develop narcolepsy.[12] While this number is larger than most would contemplate, only about 15-30% of narcoleptic individuals are ever diagnosed or treated.[13] Hence, narcolepsy is not a rare disease and can be quite debilitating.

3. Genetic Association

Several pivotal studies over the past two decades have advanced the notion that narcolepsy has a genetic foundation.[12,14-16] Initially, the major histocompatibility antigen complex DR2 was thought to be the culprit in the development of narcolepsy because Honda et al., reported that 100% of the Japanese narcoleptic patients studied were DR2+.[14] Further investigations confirmed this fact in Caucasian populations, including those in the United States;[12,17] however, advancing scientific techniques were able to delve deeper into the genome revealing many alleles and subtypes of each allele.[12] The genetic factor strongly associated with narcolepsy has been found in the human leukocyte antigen (HLA) class II region.[18-20] The haplotype most frequently associated with narcolepsy is DRB1*1501-DQB1*0602. Several reports have shown a 90-95% frequency rate for this haplotype for patients with clear-cut narcolepsy, while there is a 25% occurrence rate in the general population.[7,21-23] Despite the strong association, no mutations in the haplotype have been revealed that would indicate a causative relationship.

Until a few years ago, no other susceptibility genes had been identified besides DRB1*1501-DQB1*0602. Researchers are currently considering tumor necrosis factor (TNF)-alpha genes as a possible factor in the development of narcolepsy. TNF-α is a cytokine with varied functions, e.g. activation of macrophages, apoptosis, and sleep promotion.[18-20] An association has been identified between the promoter variants of the TNF-α gene and autoimmune and infectious diseases, such as cerebral malaria.[18-20] In addition studies have shown high levels of TNF-α in narcoleptics versus controls;[24-26] Hohjoh et al.[20] have been investigating the possibility that TNF-α is involved in the pathophysiology of human narcolepsy. Comparing Japanese narcoleptic (47) and control (111) subjects on HLA type and TNF-α gene frequency, the researchers concluded that there was a significant difference in the frequency of a specific TNF-α allele at position −857 and it could be another susceptibility gene for narcolepsy. A follow-up study by the same group assessed members of four Japanese narcoleptic families to determine the haplotypes of HLA and TNF-α (−857T). Results indicated that this haplotype DRB1*1501-TNF-α(−857T) is quite common in family members of narcoleptic patients and is quite rare in the general Japanese population.[19] Thus, this particular haplotype may lead to a strong predisposition to develop narcolepsy.

4. Autoimmune Hypothesis

A genetic predisposition is involved in autoimmune disease, specifically with particular HLA types, principally HLA class II antigens.[12,15,22,23,27-29] Most autoimmune diseases have onset in late adolescence, a non-progressive evolution with exacerbations and remissions, and a complex genetic susceptibility.[12,27,28] A low concordance rate in monozygotic twins hints at the involvement of environmental factors in the development of narcolepsy.[27,30] Narcolepsy has been described as a potential autoimmune disease as it meets these criteria.[12] The data regarding the association of the DQB1*0602-DRB1*1501 haplotype and the more recent discovery of absent or abnormally low levels of the neuropeptide hypocretin[31,32] with narcolepsy is very tight and supports the idea that narcolepsy may have an autoimmune origin that has yet to be revealed. Additional information that bolsters the autoimmune hypothesis stems from the analysis of patterns of birth data. In various diseases, such as multiple sclerosis[33] and schizophrenia,[34] a strong association with month of birth has been established. For narcoleptic patients, a significant excess of people were born in March compared to the general population.[30]

Although these correlative data are not completely convincing because they do not provide clear causal factors for this disorder, some researchers contend that scientific techniques are too immature to be able to detect the autoimmune nature of diseases such as narcolepsy. If the autoimmune process involved in narcolepsy occurs inside the central nervous system, it would be extremely difficult to identify disease onset. Imaging techniques are not sensitive enough to identify the inflammatory response in the hypothalamus, which is where the hypocretin neurons are found.[27] Additionally, inflammation will not occur with disease onset and it may be plausible that the hypocretin neurons are an "innocent bystander" in the process and not the intended target.[27] Hence, although no definitive data exist to indicate that narcolepsy is an autoimmune disease, there is much evidence that suggests it is.

5. Immune Relationship

Related to the possibility that narcolepsy may have an autoimmune component is recent work addressing a possible role of cytokines in the regulation of sleep and sleep disorders. Several inflammatory cytokines, including TNF-α and IL-1, have been shown to play a role in the regulation of sleep,[35-37] while IL-6 has been shown to augment fatigue and sleepiness.[38] Recently these cytokines have been implicated in the development and continuation of several sleep disorders including narcolepsy.[25,39,40]

5.1. IL-1

IL-1 is a cytokine of particular interest when discussing sleep and immunity because it is deeply involved in both the regulation of the immune system and in physiological sleep.[41,42] In the historic search for a sleep-promoting substance, IL-1 turned out to be the frontrunner for this title. It was shown that muramyl peptides induced production of lymphocyte-activating factor, also referred to as endogenous pyrogen, and now called interleukin 1 (IL-1).[35] At that point in time, the process of sleep was thought to be a function solely of the brain[35] and up until now, IL-I was presumed to be solely an immune substance. Years later, IL-1 was shown by Fontana et al.[43] to be a product of astrocytes; hence IL-1 was a brain product as well as an immune product.[35] Thus the current hypothesis is that IL-1 operates both within the CNS and the periphery, creating an interrelationship between the immune and endocrine systems and the sleep-wake cycle.[37] Although the complete function of IL-1 is still under investigation, the role it plays in sleep and sleep inhibition has initiated great interest in this cytokine.

Since this discovery, a plethora of research exploring the possibility that various cytokines may be sleep-promoting as well as sleep-inhibiting has occurred. Returning to the Moldofsky et al. work,[37] IL-1-like activity showed peaks at midday (1-3pm) and during the nighttime hours (11-midnight), which the authors maintain correspond to subjective times of sleepiness. In a study designed to evaluate the role of B-endorphins and IL-1β during sleep, Covelli et al.[44] discovered significant differences in IL-1β production in subjects who slept and those who could not. The two subjects who were unable to fall asleep had no IL-1β secretion during the study. Hence, the authors conclude that normal sleep is associated with nocturnal rises in IL-1β, while disturbed sleep inversely affects IL-1β secretion.[41] Corroborating data followed shortly, showing that IL-1β incubated in the presence of endotoxin was at a maximum around the time of sleep onset and in the first few hours of sleep, with a decline as the sleep cycle progressed.[42] There is also evidence of sleep-inhibiting cytokines. Anti-inflammatory cytokines that inhibit production of IL-1, such as IL-4, IL-10 and IL-13, inhibit spontaneous sleep.[45]

5.2. TNF-α

TNF-α, like IL-1, is a sleep regulatory substance[46] and evidence suggests that it is involved in NREMS (non-rapid eye movement sleep) regulation. It is produced by astrocytes, monocytes, and macrophages and is involved in most inflammatory processes, such as rheumatoid arthritis.[35] When central or systemic administration of exogenous TNF-α is given to rabbits or rodents, time spent in NREMS is increased.[47] Subsequently, serum levels of TNF-α have been shown to vary with the sleep-wake cycle. The data for TNF-α stem primarily from sleep deprivation studies. Increased levels of TNF-α are produced by circulating monocytes during sleep deprivation, but not when a person experiences other stressors.[47]

5.3. IL-6

IL-6 is a multifunctional cytokine secreted by macrophages in response to infectious challenge. It is involved in the regulation of the acute phase response, which produces fatigue and sleepiness.[48] Many of the biological responses initiated by IL-1 and TNF are mediated by IL-6.[49] There is a circadian rhythm to IL-6 indicated by low values during the daytime and maximum levels at night.[50] Although not considered a sleep-regulatory substance, IL-6 produces sleepiness in individuals when it is administered exogenously during the daytime.[48] Additionally, IL-6 levels are higher in individuals who have disorders of excessive daytime sleepiness.[25,51] The strong association with IL-6 and the sleep cycle has been confirmed with sleep deprivation studies: sleep onset is associated with an increase in serum levels of IL-6.[50]

5.4. How Sleep Deprivation Affects Cytokine Production

Cytokine production is different in normal sleep as compared to sleep deprivation, and there exists a circadian rhythm to the production not only of cytokines, but also of the various immune cells that produce the cytokines.[52]

This understanding guided Born and colleagues[52] to assess the role of nocturnal sleep on normal immune regulation in a design to assess acute sleep loss rather than excessive sleep loss. Each of the 10 men served as his own control in a two-part procedure. During the first 51-hour session, the subjects slept two consecutive regular sleep-wake cycles. In the second session, the subjects were kept awake for 24 hours then allowed to sleep normally during the next 24 hours. They found no alteration in the absolute production of IL-1β and TNF-α, and they note that when the increase in monocytes is taken into account, there appears to be no influence of sleep on cytokine production.[52] Other studies, however, have found sleep deprivation to be associated with a delayed nocturnal release of sleep-associated cytokines, IL-1, IL-6 and TNF-α, with subsequent recuperation of normal levels on recovery nights.[25,50,53] A study designed to compare the immune effects of both total sleep deprivation (TSD) and partial sleep deprivation (PSD) revealed dysregulation of sleep-regulatory cytokines and cytokine receptors; however, only in total sleep loss were there significant increases in TNF-αRI and IL-6. This suggests that the benefit of some sleep, in this case two 2-hour naps per day, could be prevention of the negative effects. It is important to note that the somnogenic actions of IL-1β and TNF-α are linked to each other;[47] that is, they induce each other's production. Additionally, their linkage and effects are tied to other substances comprising a biochemical network, which ultimately regulates sleep.[47]

5.5. *What Can These Immune Substances Tell Us About Narcolepsy?*

These and other studies linking cytokines to the regulation of sleep and sleepiness have led to speculation that these mediators may play a role in the pathogenesis of narcolepsy, and some studies have suggested that proinflammatory cytokines may be involved in sleep disorders. For example, Vgontzas and colleagues[25] evaluated plasma cytokine levels in patients with obstructive sleep apnea, idiopathic hypersomnia, and narcolepsy. The concentrations of TNF-α in plasma were elevated in apneics and narcoleptics compared to controls. Moreover, although neither IL-1β nor IL-6 were significantly elevated in narcoleptics, IL-6 was markedly elevated in the group of sleep apneics who exhibited obesity.[25] The relationship between IL-6 and obesity is interesting since narcoleptics tend to be heavier than controls.[7] Narcoleptics have been reported to have disturbances in metabolism and food intake as a result of the depleted hypocretin;[54] interestingly, hypocretin neurons closely interact with the leptin system.[55] Studies assessing leptin levels in narcoleptics reported that a reduction of more than 50% in leptin serum levels was seen in patients with narcolepsy compared to controls.[56] Leptin secretion, a hormone that informs the brain of the size of adipose tissue, could possibly be intensified to compensate for the lack of hypocretin, explaining the increased BMI seen in this population. Interestingly, increased BMI has been related to higher levels of circulating proinflammatory cytokines,[57] further suggesting that BMI and/or leptin levels may be related to proinflammatory cytokine levels and potentially, narcoleptic symptoms. Although not definitive, it leaves open the possibility that certain cytokines may be involved in the development of narcolepsy.

Other researchers have suggested that the observed changes in proinflammatory cytokines seen in sleep-disorder patients are not merely the result of impaired T-cell function, but may, in fact, represent a more subtle interaction between immune function and the regulation of sleep patterns. For example, Hinze-Selch et al.[39] assessed cytokine levels of IL-1β, IL-1ra, IL-2, IL-6 TNF-α and TNF-β in plasma and in mitogen-stimulated monocytes and lymphocytes in narcoleptics and in HLA-matched controls. They only found elevated secretion of IL-6 from lipopolysaccharide (LPS)-stimulated monocytes compared to controls. The authors conclude that there are no major T-cell functional abnormalities in narcoleptics, but the elevation in IL-6 may play a role in the REM sleep-associated symptoms of narcolepsy since IL-6 promotes the growth of cholinergic neurons.[39]

5.6. Recent Data

The idea that cytokines are involved in the regulation of sleep and sleep disorders has been recently re-evaluated with one study focusing on narcolepsy. Data from Vgontzas et al.[58] investigated the effect on inflammatory cytokines by inducing modest sleep restriction on subjects, mimicking more realistic conditions of sleep loss. Results indicate that both IL-6 and TNF-α showed increased secretion after 12 nights of modest sleep loss. Okun et al.[24] specifically addressed differences in proinflammatory cytokines levels in narcoleptics versus matched control subjects. Serum levels of TNF-α and IL-6 were assessed in 39 narcoleptics and 40 controls. Results indicate that both cytokines were significantly elevated in narcoleptic subjects (see Figure 1). Despite the methodological concern of only one blood draw and the unavailability of when the samples were drawn, the data is consistent with previous work.[25,58] These data further support the notion that cytokines may play a role in the development and/or exacerbation of narcolepsy symptoms.

The data have several interesting implications. First, the dysregulation of sleep observed in narcoleptics correlates with the immune and endocrine dysregulation seen in these subjects. Narcoleptics have a disrupted and fragmented 24-hour sleep cycle, which is thought to stem from the depletion of the neuropeptide, hypocretin.[59] Frequent daytime naps can intrude upon the narcoleptic, while frequent awakenings can plague the nighttime hours. This persistent disruption of sleep could potentially be a primary explanatory variable in the altered cytokine levels seen in this and other studies.[25,35,50,60] Redwine et al.[50] has reported that partial sleep deprivation staves off the release of nocturnal IL-6. This may subsequently disturb the homeostatic role that nocturnal IL-6 secretion plays in regulating sleep on subsequent

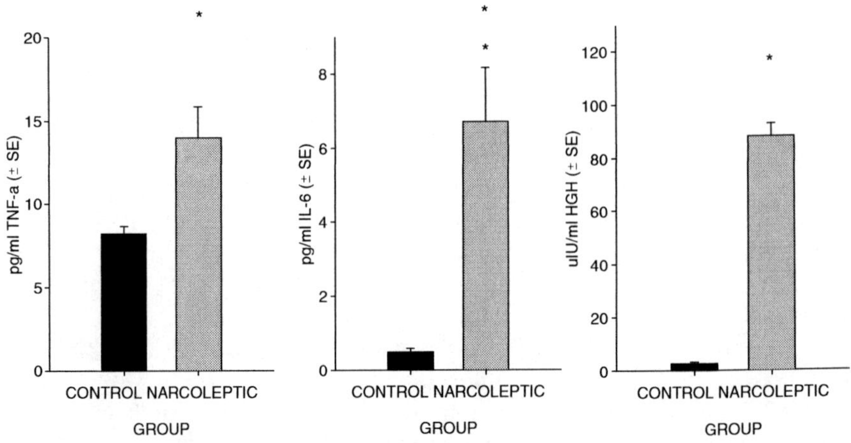

FIGURE 1. Mean (± SE) levels of serum tumor necrosis factor alpha (TNF-α), serum interleukin-6 (IL-6), and serum human growth hormone (hGH) obtained from control (N = 40, black bars) and narcoleptic (N = 39, gray bars) subjects. The asterisk indicates significance at p < .05. Figure reprinted from Okun et al (2004).

nights. Moreover, IL-6 is inversely associated with SWS.[50] So, in narcoleptics, the higher levels of IL-6 are consistent with the decreased SWS and the increased REM sleep[25] observed in this population. Moreover, the observed changes in cytokine levels may in fact contribute to the higher likelihood of disturbed sleep and/or increased incidence of infection. Irwin[60] has reported that even partial sleep deprivation can produce a significant reduction in cellular immunity. However, it must be noted that all the cited studies have assessed peripheral blood cytokine measurements. This does not clearly inform the researcher as to where the cytokines originated from, only that they are present in serum.

5.7. Endocrine Relationship

Endocrine substances are important to both the sleep-wake cycle and the immune system. Several endocrine substances have been shown to influence sleep. For example, melatonin, a hormone produced by the pineal gland, is known to play a significant role in the maintenance of the circadian rhythm, and is thought to stimulate various components of the immune system. Although the exact effects of melatonin are unclear, data suggest that melatonin exerts hypnotic effects rather than shaping the sleep-wake cycle by phase resetting.[61] Progesterone, one of the sex steroid hormones, acts as an immunosuppressive agent on lymphocytes and influences the onset of SWS.[37] During the menstrual cycle, women experience fluctuations in progesterone levels depending on which phase they are in luteal (high progesterone) or follicular (low progesterone). In an attempt to understand progesterone's contribution to sleep, Moldofsky et al.[37] evaluated immune and endocrine parameters in healthy nonpregnant women during high and low progesterone phases of the menstrual cycle. Results indicate that onset to Slow Wave Sleep (SWS) is delayed and the amount of SWS is reduced during the high progesterone phase.[37] In addition to the relationship between SWS and high progesterone levels, there were also differences in sleep-related Natural Killer (NK) cell activity and progesterone levels: high progesterone levels corresponded to a decline in nocturnal NK cell activity.[37]

The temporal relationship of Growth Hormone (GH) and Growth Hormone Releasing Hormone (GHRH) with SWS has led to increased attention to their overall role in the sleep-wake process. GH is thought to be completely sleep-dependent because whether sleep is delayed, interrupted, or impending, sleep onset will elicit a pulse in GH secretion.[62] The somatotropic axis is now considered a potent stimulator of sleep.[63] The function of the somatotropic axis is to regulate anabolism and tissue growth by producing various hormones, including GH. It also produces two neurohormones: GHRH, which stimulates GH synthesis and release, and somatostatin, which inhibits GH synthesis and release.[63] Of note is the sensitive feedback loop this system incorporates. Both somatostatin and GH, in addition to IGF-1, inhibit GHRH production. Hence, diminished production of GH, usually corresponding to a decrease in SWS, would signal an increase of GHRH to subsequently elevate production of GH.

The majority of the data supporting the role of GH and GHRH in sleep stems from disordered sleep observed in aging populations.[64] Studies assessing how GH and GHRH can affect sleep provide a clear indication of the importance of these hormones. In a study assessing an intranasal injection of GHRH on endocrine function and sleep in young and old men, SWS-associated GH elevation was reduced while REM and SWS were enhanced regardless of age.[64] It appears that exogenous introduction of GHRH mimicked endogenous GHRH via the negative feedback loop and subsequently exerted effects on sleep parameters.[64] REM sleep and SWS show increases as a result of this administration, particularly in the second half of the sleep time.[64]

Few studies have assessed GH levels in patients with narcolepsy, although the conclusions of these studies suggest that GH secretion is blunted in narcolepsy. It should be noted, however, that the methodology of these studies and the time period in which they were conducted warrants skeptical acceptance of the results. One study[65] showed GH concentrations remain stable after administration of L-DOPA. The other two studies[66,67] concentrated on the sleep onset secretion of GH as their determining factor rather than the entire 24-hour period. Besset et al.[67] found that narcoleptics had very low GH secretion with rare and small secretory peaks not clearly linked to sleep stages. Higuchi et al.[66] showed that the expected peak of GH during nocturnal sleep onset was absent or markedly decreased in the 4 narcoleptic subjects studied. Knowing that sleep is distributed throughout the 24-hour period for narcoleptics, rather than in a concise 8-hour period, GH levels should be altered to some degree from normal subjects.

6. Summary

The cytokine and GH alterations reported in the studies described above may be explanatory variables for the diverse variations seen in narcoleptic sleep patterns. It is well established that narcoleptics have fragmented sleep architecture with non-normal SWS patterns in the first third of the night, often accompanied by frequent daytime naps. The shortened REM latency may contribute to the altered levels of cytokines and GH observed since IL-6 is secreted during Stages 1 & 2 and during REM sleep, and GH secretion is associated with SWS. GH secretion is stimulated not only by GHRH, but also by IL-6.[50, 68] In addition, GH has been reported to promote proinflammatory and T cell function.[50] This helps to solidify evidence of the complex interaction between sleep and immunity. In attempting to explain the extreme levels of GH, a possible hypothesis may stem from the additional SWS acquired via the frequent daytime naps. GH is secreted in a pulsatile fashion, especially during SWS, so it is possible that narcoleptic episodes would be associated with increases in SWS secretions of GH as the attainment of sleep is spread throughout the 24-hour period rather than obtained in the consolidated 8-hour nighttime cycle.

In attempting to explain the extreme levels of HGH, Okun et al.[24] suggest a possible hypothesis stemming from the additional SWS acquired via the frequent daytime naps. HGH is secreted in a pulsatile fashion, especially during SWS, so it is possible that narcoleptic episodes would be associated with increased in SWS secretions of HGH as the attainment of sleep is spread throughout the 24-hour period rather than obtained in the consolidated 8-hour nighttime cycle. Additionally, data generated by rat experiments involving ICV injection of hypocretin suggest that hypocretin deficiency may exert a direct excitatory effect on HGH release in narcoleptics.[69] Because most classic narcoleptics are hypocretin deficient,[31] the increased levels of HGH observed in the present study are consistent with this observation, and may reflect direct modulation of HGH by hypocretin in this population. Hence, the distribution and timing of HGH secretion over the 24-hour period in narcoleptics may be the more important measure to assess, rather than overall levels.

The data described have several interesting implications. First, the dysregulation of sleep observed in narcoleptics correlates with the immune and endocrine dysregulation seen in these subjects. Narcoleptics have a disrupted and fragmented 24-hour sleep cycle, which is thought to stem from the depletion of the neuropeptide, hypocretin.[70] Frequent daytime naps can intrude upon the narcoleptic, while frequent awakenings can plague the nighttime hours. This persistent disruption of sleep could potentially be a primary explanatory variable in the altered cytokine levels seen in this and other studies.[25, 35, 36, 50, 60] Moreover, the observed changes in cytokine levels may in fact contribute to the higher likelihood of disturbed sleep and/or increased incidence of infection. Irwin[60] has reported that even partial sleep deprivation can produce a significant reduction in cellular immunity. However, it must be noted that all the cited studies have assessed peripheral blood cytokine measurements. This does not clearly inform the researcher as to where the cytokines originated from, only that they are present in serum.

The cumulative data indicates that proinflammatory cytokines and HGH may be involved in the development of or the exacerbation of narcolepsy. As the role of the immune system in sleep becomes clearer, a better interpretation will be available to assess why narcoleptics have altered cytokine and HGH levels. Future studies should be conducted to fully characterize the sources, patterns, and significance of alterations in patterns of proinflammatory cytokines and HGH in narcolepsy. In addition, future studies should include detailed assessment of sleep patterns and health status to account for potential cumulative deleterious effects of sleep deprivation upon the immune system. Even though the number of individuals who develop narcolepsy appears small in comparison to other sleep disorders such as sleep apnea, the present data provide further support for the use of narcolepsy as a model for understanding how chronic sleep disruption/fragmentation can impact the functioning of the immune system.

7. References

1. C. Guilleminault, Narcolepsy Syndrome, in: *Principles and Practice of Sleep Medicine*, edited by M.H.Kryger, T.Roth, and W.C.Dement, (W.B. Saunders Company, Philadelphia, 1994), pp. 549-561.
2. W. C. Dement, Zarcone V, Varner V, and et al, The Prevalence of Narcolepsy, *Sleep Research* 1, 148 (1972).
3. W. C. Dement, Carskadon MA, and Ley R, The Prevalence of Narcolepsy, *Sleep Research* 2, 148 (1973).
4. P. Lavie and R. Peled, Narcolepsy is a rare disease in Israel, *Sleep* 10(6), 608-609 (1987).
5. C. Hublin, M. Partinen, J. Kaprio, M. Koskenvuo, and C. Guilleminault, Epidemiology of narcolepsy, *Sleep* 17(8 Suppl), S7-12 (1994).
6. C. Hublin, J. Kaprio, M. Partinen, K. Heikkila, and M. Koskenvuo, Daytime sleepiness in an adult, Finnish population, *J Intern. Med.* 239(5), 417-423 (1996).
7. M. L. Okun, L. Lin, Z. Pelin, S. Hong, and E. Mignot, Clinical aspects of narcolepsy-cataplexy across ethnic groups, *Sleep* 25(1), 27-35 (2002).
8. M. H. Silber, L. E. Krahn, E. J. Olson, and V. S. Pankratz, The epidemiology of narcolepsy in Olmsted County, Minnesota: a population-based study, *Sleep* 25(2), 197-202 (2002).
9. M. M. Ohayon, R. G. Priest, J. Zulley, S. Smirne, and T. Paiva, Prevalence of narcolepsy symptomatology and diagnosis in the European general population, *Neurology* 58(12), 1826-1833 (2002).

10. S. C. Hong, K. Leen, S. A. Park, J. H. Han, S. P. Lee, L. Lin, M. Okun, S. Nishino, and E. Mignot, HLA and hypocretin studies in Korean patients with narcolepsy, *Sleep* 25(4), 440-444 (2002).

11. Y. K. Wing, R. H. Li, C. W. Lam, C. K. Ho, S. Y. Fong, and T. Leung, The prevalence of narcolepsy among Chinese in Hong Kong, *Ann.Neurol* 51(5), 578-584 (2002).

12. E. Mignot, M. Tafti, W. C. Dement, and F. C. Grumet, Narcolepsy and immunity, *Adv.Neuroimmunol.* 5(1), 23-37 (1995).

13. S. S. Chakravorty and D. B. Rye, Narcolepsy in the older adult: epidemiology, diagnosis and management, *Drugs Aging* 20(5), 361-376 (2003).

14. Y. Honda, A. Asaka, M. Fanimura, and T. Furosho, A genetic study of narcolepsy and excessive daytime sleepiness in 308 families with narcolepsy or hypersomnia proband, in: *Sleep/Wake Disorders: Natural History, Epidemiology and Long Term Evolution*, edited by C. Guilleminault and E. Lugaresi, (Raven Press, New York, 1983), pp. 187-199.

15. E. Mignot, L. Lin, W. Rogers, Y. Honda, X. Qiu, X. Lin, M. Okun, H. Hohjoh, T. Miki, S. Hsu, M. Leffell, F. Grumet, M. Fernandez-Vina, M. Honda, and N. Risch, Complex HLA-DR and -DQ interactions confer risk of narcolepsy-cataplexy in three ethnic groups, *Am J Hum.Genet.* 68(3), 686-699 (2001).

16. C. Peyron, J. Faraco, W. Rogers, B. Ripley, S. Overeem, Y. Charnay, S. Nevsimalova, M. Aldrich, D. Reynolds, R. Albin, R. Li, M. Hungs, M. Pedrazzoli, M. Padigaru, M. Kucherlapati, J. Fan, R. Maki, G. J. Lammers, C. Bouras, R. Kucherlapati, S. Nishino, and E. Mignot, A mutation in a case of early onset narcolepsy and a generalized absence of hypocretin peptides in human narcoleptic brains, *Nat.Med.* 6(9), 991-997 (2000).

17. N. Langdon, C. Lock, K. Welsh, D. Vergani, R. Dorow, H. Wachtel, D. Palenschat, and J. D. Parkes, Immune factors in narcolepsy, *Sleep* 9(1 Pt 2), 143-148 (1986).

18. H. Hohjoh, N. Terada, M. Kawashima, Y. Honda, and K. Tokunaga, Significant association of the tumor necrosis factor receptor 2 (TNFR2) gene with human narcolepsy, *Tissue Antigens* 56(5), 446-448 (2000).

19. H. Hohjoh, N. Terada, T. Miki, Y. Honda, and K. Tokunaga, Haplotype analyses with the human leucocyte antigen and tumour necrosis factor-alpha genes in narcolepsy families, *Psychiatry Clin Neurosci.* 55(1), 37-39 (2001).

20. H. Hohjoh, T. Nakayama, J. Ohashi, T. Miyagawa, H. Tanaka, T. Akaza, Y. Honda, T. Juji, and K. Tokunaga, Significant association of a single nucleotide polymorphism in the tumor necrosis factor-alpha (TNF-alpha) gene promoter with human narcolepsy, *Tissue Antigens* 54(2), 138-145 (1999).

21. L. Lin, M. Hungs, and E. Mignot, Narcolepsy and the HLA region, *J Neuro-immunol.* 117(1-2), 9-20 (2001).

22. E. Mignot, X. Lin, J. Arrigoni, C. Macaubas, F. Olive, J. Hallmayer, P. Underhill, C. Guilleminault, W. C. Dement, and F. C. Grumet, DQB1*0602 and DQA1*0102 (DQ1) are better markers than DR2 for narcolepsy in Caucasian and black Americans, *Sleep* 17(8 Suppl), S60-S67 (1994).

23. E. Mignot, R. Hayduk, J. Black, F. C. Grumet, and C. Guilleminault, HLA DQB1*0602 is associated with cataplexy in 509 narcoleptic patients, *Sleep* 20(11), 1012-1020 (1997).

24. M. L. Okun, S. Giese, L. Lin, M. Einen, E. Mignot, and M. E. Coussons-Read, Exploring the cytokine and endocrine involvement in narcolepsy, *Brain Behav.Immun.* 18(4), 326-332 (2004).

25. A. N. Vgontzas, D. A. Papanicolaou, E. O. Bixler, A. Kales, K. Tyson, and G. P. Chrousos, Elevation of plasma cytokines in disorders of excessive daytime sleepiness: role of sleep disturbance and obesity, *J Clin Endocrinol Metab* 82(5), 1313-1316 (1997).

26. A. N. Vgontzas, G. Mastorakos, E. O. Bixler, A. Kales, P. W. Gold, and G. P. Chrousos, Sleep deprivation effects on the activity of the hypothalamic-pituitary-adrenal and growth axes: potential clinical implications, *Clin Endocrinol (Oxf)* 51(2), 205-215 (1999).

27. D. Chabas, S. Taheri, C. Renier, and E. Mignot, The genetics of narcolepsy, *Annu. Rev Genomics Hum. Genet.* 4, 459-483 (2003).

28. E. Mignot, C. Guilleminault, F. Grumet, and W. C. Dement, Is narcolepsy an autoimmune disease?, in: *Sleep, Hormones and Immunological System*, edited by S. Smirne, Franceschi M, Ferini-Strambi L, and Zucconi M, (Masson, Milan, 1991), pp. 29-38.

29. E. Mignot, Genetic and familial aspects of narcolepsy, *Neurology* 50(2 Suppl 1), S16-S22 (1998).

30. Y. Dauvilliers, B. Carlander, N. Molinari, A. Desautels, M. Okun, M. Tafti, J. Montplaisir, E. Mignot, and M. Billiard, Month of birth as a risk factor for narcolepsy, *Sleep* 26(6), 663-665 (2003).

31. E. Mignot, G. J. Lammers, B. Ripley, M. Okun, S. Nevsimalova, S. Overeem, J. Vankova, J. Black, J. Harsh, C. Bassetti, H. Schrader, and S. Nishino, The role of cerebrospinal fluid hypocretin measurement in the diagnosis of narcolepsy and other hypersomnias, *Arch. Neurol.* 59(10), 1553-1562 (2002).

32. S. Nishino, B. Ripley, S. Overeem, S. Nevsimalova, G. J. Lammers, J. Vankova, M. Okun, W. Rogers, S. Brooks, and E. Mignot, Low cerebrospinal fluid hypocretin (Orexin) and altered energy homeostasis in human narcolepsy, *Ann. Neurol.* 50(3), 381-388 (2001).

33. D. I. Templer, N. H. Trent, D. A. Spencer, A. Trent, M. D. Corgiat, P. B. Mortensen, and M. Gorton, Season of birth in multiple sclerosis, *Acta Neurol Scand.* 85(2), 107-109 (1992).

34. P. B. Mortensen, C. B. Pedersen, T. Westergaard, J. Wohlfahrt, H. Ewald, O. Mors, P. K. Andersen, and M. Melbye, Effects of family history and place and season of birth on the risk of schizophrenia, *N. Engl. J Med.* 340(8), 603-608 (1999).

35. J. M. Krueger, S. Takahashi, L. Kapas, S. Bredow, R. Roky, J. Fang, R. Floyd, K. B. Renegar, N. Guha-Thakurta, and S. Novitsky, Cytokines in sleep regulation, *Adv. Neuroimmunol.* 5(2), 171-188 (1995).

36. J. M. Krueger and J. A. Majde, Cytokines and sleep, *Int. Arch. Allergy Immunol.* 106(2), 97-100 (1995).

37. H. Moldofsky, Sleep and the immune system, *Int. J. Immunopharmacol.* 17(8), 649-654 (1995).

38. M. R. Opp and L. Imeri, Sleep as a behavioral model of neuro-immune interactions, *Acta Neurobiol. Exp. (Warsz.)* 59(1), 45-53 (1999).

39. D. Hinze-Selch, T. C. Wetter, Y. Zhang, H. C. Lu, E. D. Albert, J. Mullington, H. Wekerle, F. Holsboer, and T. Pollmacher, In vivo and in vitro immune variables in patients with narcolepsy and HLA-DR2 matched controls, *Neurology* 50(4), 1149-1152 (1998).

40. A. N. Vgontzas, M. Zoumakis, D. A. Papanicolaou, E. O. Bixler, P. Prolo, H. M. Lin, A. Vela-Bueno, A. Kales, and G. P. Chrousos, Chronic insomnia is associated with a shift of interleukin-6 and tumor necrosis factor secretion from nighttime to daytime, *Metabolism* 51(7), 887-892 (2002).

41. V. Covelli, L. D'Andrea, S. Savastano, R. Valentino, A. P. Tommaselli, A. Selleri, F. Massari, and G. Lombardi, Interleukin-1 beta plasma secretion during diurnal spontaneous and induced sleeping in healthy volunteers, *Acta Neurol.(Napoli)* 16(3), 79-86 (1994).

42. F. Hohagen, J. Timmer, A. Weyerbrock, R. Fritsch-Montero, U. Ganter, S. Krieger, M. Berger, and J. Bauer, Cytokine production during sleep and wakefulness and its relationship to cortisol in healthy humans, *Neuropsychobiology* 28(1-2), 9-16 (1993).

43. A. Fontana, Astrocytes and lymphocytes: intercellular communication by growth factors, *J Neurosci Res.* 8(2-3), 443-451 (1982).

44. V. Covelli, F. Massari, C. Fallacara, I. Munno, E. Jirillo, S. Savastano, A. P. Tommaselli, and G. Lombardi, Interleukin-1 beta and beta-endorphin circadian rhythms are inversely related in normal and stress-altered sleep, *Int J Neurosci.* 63(3-4), 299-305 (1992).

45. J. M. Krueger and F. Obal, Jr., Sleep function, *Front Biosci.* 8, d511-d519 (2003).

46. J. M. Krueger, F. Obal, Jr., and J. Fang, Humoral regulation of physiological sleep: cytokines and GHRH, *J Sleep Res.* 8 Suppl 1, 53-59 (1999).

47. J. M. Krueger, J. Fang, P. Taishi, Z. Chen, T. Kushikata, and J. Gardi, Sleep. A physiologic role for IL-1 beta and TNF-alpha, *Ann.N.Y.Acad.Sci.* 856, 148-159 (1998).

48. E. Spath-Schwalbe, K. Hansen, F. Schmidt, H. Schrezenmeier, L. Marshall, K. Burger, H. L. Fehm, and J. Born, Acute effects of recombinant human interleukin-6 on endocrine and central nervous sleep functions in healthy men, *J Clin Endocrinol Metab* 83(5), 1573-1579 (1998).

49. M. R. Opp and J. D. Morrow. Lipopolusaccharide-Induced Alterations in Sleep and Body Temperature of Interleukin-6-Deficient Mice. 2003 (Unpublished Work).

50. L. Redwine, R. L. Hauger, J. C. Gillin, and M. Irwin, Effects of sleep and sleep deprivation on interleukin-6, growth hormone, cortisol, and melatonin levels in humans, *J.Clin.Endocrinol.Metab* 85(10), 3597-3603 (2000).

51. A. N. Vgontzas, D. A. Papanicolaou, E. O. Bixler, A. Lotsikas, K. Zachman, A. Kales, P. Prolo, M. L. Wong, J. Licinio, P. W. Gold, R. C. Hermida, G. Mastorakos, and G. P. Chrousos, Circadian interleukin-6 secretion and quantity and depth of sleep, *J Clin Endocrinol Metab* 84(8), 2603-2607 (1999).

52. J. Born, T. Lange, K. Hansen, M. Molle, and H. L. Fehm, Effects of sleep and circadian rhythm on human circulating immune cells, *The Journal of Immunology* 158(9), 4454-4464 (1997).

53. H. Moldofsky, F. A. Lue, J. R. Davidson, and R. Gorczynski, Effects of sleep deprivation on human immune functions, *FASEB J* 3(8), 1972-1977 (1989).

54. A. Schuld, W. F. Blum, M. Uhr, M. Haack, T. Kraus, F. Holsboer, and T. Pollmacher, Reduced leptin levels in human narcolepsy, *Neuroendocrinology* 72(4), 195-198 (2000).

55. S. Overeem, E. Mignot, J. G. van Dijk, and G. J. Lammers, Narcolepsy: clinical features, new pathophysiologic insights, and future perspectives, *J Clin Neurophysiol.* 18(2), 78-105 (2001).

56. A. Schuld, W. F. Blum, M. Uhr, M. Haack, T. Kraus, F. Holsboer, and T. Pollmacher, Reduced leptin levels in human narcolepsy, *Neuroendocrinology* 72(4), 195-198 (2000).

57. K. Esposito, A. Pontillo, C. Di Palo, G. Giugliano, M. Masella, R. Marfella, and D. Giugliano, Effect of weight loss and lifestyle changes on vascular inflammatory markers in obese women: a randomized trial, *JAMA* 289(14), 1799-1804 (2003).

58. A. N. Vgontzas, E. Zoumakis, E. O. Bixler, H. M. Lin, H. Follett, A. Kales, and G. P. Chrousos, Adverse effects of modest sleep restriction on sleepiness, performance, and inflammatory cytokines, *J Clin Endocrinol Metab* 89(5), 2119-2126 (2004).

59. E. Mignot, G. J. Lammers, B. Ripley, M. Okun, S. Nevsimalova, S. Overeem, J. Vankova, J. Black, J. Harsh, C. Bassetti, H. Schrader, and S. Nishino, The role of cerebrospinal fluid hypocretin measurement in the diagnosis of narcolepsy and other hypersomnias, *Arch.Neurol.* 59(10), 1553-1562 (2002).

60. M. Irwin, Effects of sleep and sleep loss on immunity and cytokines, *Brain Behav.Immun.* 16(5), 503-512 (2002).

61. I. Haimov, T. Shochat, and P. Lavie, Melatonin–a possible link between sleep and the immune system, *Isr.J Med.Sci.* 33(4), 246-250 (1997).

62. E. Van Cauter, Slow wave sleep and release of growth hormone, *JAMA* 284(21), 2717-2718 (2000).

63. F. Obal, Jr. and J. M. Krueger, The somatotropic axis and sleep, *Rev Neurol.(Paris)* 157(11 Pt 2), S12-S15 (2001).

64. B. Perras, L. Marshall, G. Kohler, J. Born, and H. L. Fehm, Sleep and endocrine changes after intranasal administration of growth hormone-releasing hormone in young and aged humans, *Psychoneuroendocrinology* 24(7), 743-757 (1999).

65. R. W. Clark, H. S. Schmidt, and W. B. Malarkey, Disordered growth hormone and prolactin secretion in primary disorders of sleep, *Neurology* 29(6), 855-861 (1979).

66. T. Higuchi, Y. Takahashi, K. Takahashi, Y. Niimi, and A. Miyasita, Twenty-four-hour secretory patterns of growth hormone, prolactin, and cortisol in narcolepsy, *J Clin Endocrinol Metab* 49(2), 197-204 (1979).

67. A. Besset, A. Bonardet, M. Billiard, B. Descomps, A. C. de Paulet, and P. Passouant, Circadian patterns of growth hormone and cortisol secretions in narcoleptic patients, *Chronobiologia* 6(1), 19-31 (1979).

68. Marshall L and Born J, Brain-Immune Interactions in Sleep, in: *Neurobiology of the Immune System*, edited by Clow A and Hucklebridge F, (Academic Press, San Diego, 2002), pp. 93-131.

69. S. Overeem, S. W. Kok, G. J. Lammers, A. A. Vein, M. Frolich, A. E. Meinders, F. Roelfsema, and H. Pijl, Somatotropic axis in hypocretin-deficient narcoleptic humans: altered circadian distribution of GH-secretory events, *Am.J Physiol Endocrinol Metab* 284(3), E641-E647 (2003).

70. E. Mignot, G. J. Lammers, B. Ripley, M. Okun, S. Nevsimalova, S. Overeem, J. Vankova, J. Black, J. Harsh, C. Bassetti, H. Schrader, and S. Nishino, The role of cerebrospinal fluid hypocretin measurement in the diagnosis of narcolepsy and other hypersomnias, *Arch.Neurol.* 59(10), 1553-1562 (2002).

Sleep Disorders and Neuroendocrine Investigations

An Arctic Perspective

TROND BRATLID

1. Introduction

There has been considerable interest in sleep-waking patterns of man in the polar and subpolar environment because of the special biogeographically and physiological factors of the environment. Extreme shifts in the ratio of light to darkness during the course of the year are prominent features in these areas, ranging from almost total darkness in the polar night to 24 h of light during the midnight sun period. There is rich evidence that these fluctuations affect plants and animals, but there is still meager information about the degree to which humans are affected. Numerous complaints are expressed among the general population in all age groups about disturbances in sleep pattern, especially during the dark period of the Arctic winter. Some persons also report low mood symptoms and lack of energy and drive.

2. Some Historical and Epidemiological Perspectives

North of the Polar Circle (above 66.5 degree latitude) there is a period around winter solstice when the sun does not rise above the horizon. In Tromsø, situated at about 69 degrees latitude (corresponding to the northern part of Alaska), this period lasts from about November 20 to about January 20 and is called the "darkness period". In the middle of this period there is total darkness most of the day, with only a few hours of dim daylight around noon. (For information about the maximal hours of sunshine in Tromsø see Figure 1 on the following page). Correspondingly we have Midnight sun with daylight from end of May to end of July.

During the last 50 years there have been some studies investigating how seasonal extremes with prolonged periods of months of continuous darkness and light affect the sleep routine. In a study carried out in Tromsø during the summer of 1951, Kleitman[1] reported that previous anecdotal

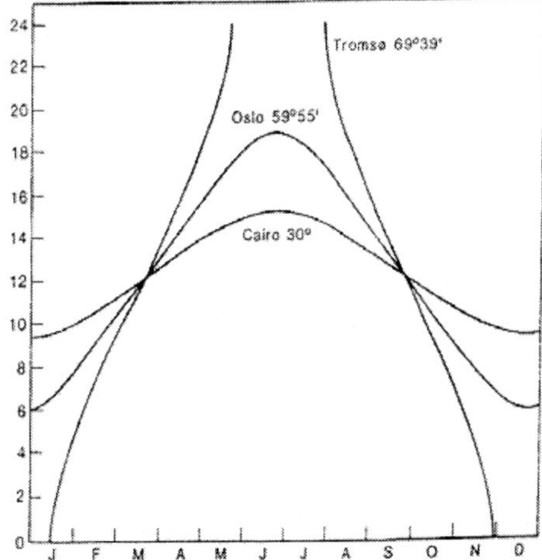

FIGURE 1. Theoretical maximum hours of sunlight for each month in Tromsø. Compared to Oslo and Cairo.

information indicating that the residents slept very little during the all light summer months, was highly inaccurate. He found in interviews that time spent in bed during the summer months averaged 7 h and 26 min, and 8 h and 25 min during winter months. Although there was considerably greater variability of the time of going to bed in the winter as compared to the summer, the time of getting up in the morning was the same for the 2 seasons.

The first regular study of sleep problems during the "darkness period" in Tromsø was performed in 1957 by Devold[2] among pupils (age range 15-22 years) from various schools in Tromsø. He reported that 38.7% of the pupils complained of sleep difficulties (mainly insomnia) from September to January, compared with 5.8% for a corresponding group living in Southern Norway.

In January 1959, Kjos[3] made another questionnaire study in schools in Tromsø. Of the students 28.2% reported that they had problems initiating sleep in the evening and about 60% had to be awakened by someone every morning. The investigation was repeated same school year at the end of April. At that time only 6.1% reported insomnia problems.

In a later questionnaire study among employees at the University Psychiatric Hospital in Tromsø, carried out in the darkness period, it was found that 23.6% of the respondents complained of sleep difficulties that were considered to be associated with the darkness period as such.[4]

As a part of a comprehensive population health survey in Tromsø,[5] questions about sleeplessness and its possible association with season were included. Of the 14,667 respondents, 41.7% of the women and 29.9% of the men said they were sometimes bothered by insomnia. Insomnia not associated with any specific time of the year was reported by 16% of women and 16.2% of men. Insomnia in the darkness period was reported by 17% of women and 9% of men. Insomnia in the midnight sun period or in spring-autumn was less common. Difficulty falling asleep was the most common type of insomnia both in winter and summer.

Overall the frequency of insomnia increased with increasing age, but with some notable differences with regard to type. Initial insomnia showed less relation to age whereas middle and late insomnia increased markedly with age. The seasonal type of insomnia in the midnight sun period decreased with age, whereas all other seasonal types increased with age.

3. Neuroendocrine Studies in Polar/Subpolar Environments

In 1970, Natani et al.[6] described long-term changes in sleep patterns in 4 men on an Antarctic base, some of which persisted up to 11 months after leaving the base. Stage 4 was virtually eliminated. These experiments, however, were performed on only 4 men under severely restricted physical and psychological conditions. Inspired by this, Weitzman et al.[7] studied a group of 7 healthy male subjects in Tromsø in regard to sleep stages and 24 h plasma cortisol and growth hormone patterns during the four seasons. No difference in total sleep or sleep stage per cents was found for any of the yearly seasons. A small but statistical significant increase in mean plasma cortisol concentration and amount secreted for 24 hour was found for the autumn–winter seasons, as compared with the spring and summer. However, no difference in the circadian curve of cortisol hormonal pattern was found. All subjects secreted growth hormone shortly after sleep onset at night and no difference was found as a function of season of the year.

Stokkan et al.[8] measured salivary melatonin in healthy volunteers in Tromsø at midwinter, midsummer and vernal and autumnal equinox. The hormonal patterns varied widely between individuals, but, in general, they were consistent within most individuals between the seasons. Highest peak values occurred in January when the mean level was also significantly higher than at any other time of the year. The lowest mean levels occurred in June. Although individual rhythms were not always apparent, the mean patterns showed significantly elevated melatonin concentrations during the night at all seasons. The June melatonin peak was similar to that in March and September, but appeared to be phase delayed with increased melatonin concentrations from midnight until morning. It was assumed that the delayed melatonin peak in June may be associated with a tendency among people to shift their rest/activity rhythm and that the pineal sensitivity to light is altered in sunny summer days.

4. Midwinter Insomnia (MI)

In its transient form most individuals have experienced insomnia at times of emotional crises or life stress. In some individuals, however, their sleep difficulties persist for months or even years. The use of different definitions, different phrasing of questionnaires, different prevalence periods, different season and latitude etc, makes it difficult to compare various epidemiological studies and evaluate the exact prevalence of insomnia. In medical practice insomnia is one of the most frequently encountered health complaints.

The clinical symptoms of Midwinter Insomnia (MI) gradually start at the beginning of the darkness period (ultimo November) and lasts until the sun returns over the horizon (ultimo January). Some individuals, mainly elderly, may have insomnia problems even several weeks before and/or after the darkness period. Cloudy and stormy weather may contribute to this.

MI is predominantly an initial type of insomnia, varying in degree from moderate difficulty in falling asleep after going to bed, to almost total inability to sleep during the whole night. Our clinical impression, not based on systematic epidemiological studies, is that MI is usually a mono-symptomatic disorder, seldom accompanied by manifest depression, anxiety or other psychiatric symptoms. Those who suffer complain about daytime sleepiness and a somewhat reduced working and social capacity. This is probably secondary to reduced sleep.

MI does not seem to be associated with any particular eating or living problems. On average, those suffering from MI show a lower level of physical activity during the darkness period than during the rest of the year. Whether this is a secondary phenomenon or related to the etiology should be further explored.

Based on clinical criteria, we consider it reasonable to hypothesize that MI is the expression of a disturbance-more particularly a phase-delay of the sleep-wake rhythm, caused by lack of entraining effect of normal daylight. According to the International Classification of Sleep Disorders (ICSD) such disturbances of the sleep-wake rhythm belongs to the chronobiological or circadian phase disorders.[9] Another type of chronobiologic sleep disorder is associated with hypersomnia continuing into the day, leading to excessive daytime somnolence. This disorder often accompanies winter depression or Seasonal Affective Disorder (SAD).[10]

In some respect, MI seems to be analogues with the Delayed Sleep Phase Syndrome (DSPS).[11] However, there are obvious differences between the two conditions. Whereas DSPS is a long-lasting disorder that does not seem to be associated with any special time of the year, MI appears exclusively during the darkness period. In previous controlled, double blind trials we found that MI shows lack of response to placebo, but a favorable response to moderate doses of benzodiazepine hypnotics.[12] It should be noted, however, that most people suffering from MI refrain from using hypnotics, partly because they consider their insomnia as an almost "normal" annual recurring phenomenon, partly because they wish to avoid the risk of being habituated to hypnotics.[13]

5. Neuroendocrine Studies of Midwinter-Insomnia

In a previous study it was found that subjects suffering from MI had, on the average, a higher plasma level of growth hormone, especially in the evening, than normal controls, whereas plasma cortisol did not differ.[14] In another study a group of individuals with initial insomnia during the dark period were compared to a group of healthy controls. They were followed with repeated measures of melatonin and questioned on ten different sleep variables from the beginning of January to vernal equinox in March.[15]

The distribution of melatonin over a 24 hour period (five times point) indicated an increase in the amplitude in both groups in the middle of January and a decrease in amplitude of melatonin at the time of year when sun first rise over the horizon (23rd-24th of January) (see Figure 2 and 3 on the following page). A positive correlation between morning tiredness and morning levels of melatonin was found among individuals with sleep disturbances, but not in controls.

The significant decrease in melatonin when the sunlight returns to Tromsø in January may be explained by a shift of timing of melatonin onset. Given that the endogenous period of the human circadian pacemaker is slightly longer than 24 h, the finding suggest that a naturalistic dawn signal is sufficient to prevent this natural delay drift. Zeitgeber transduction and circadian system

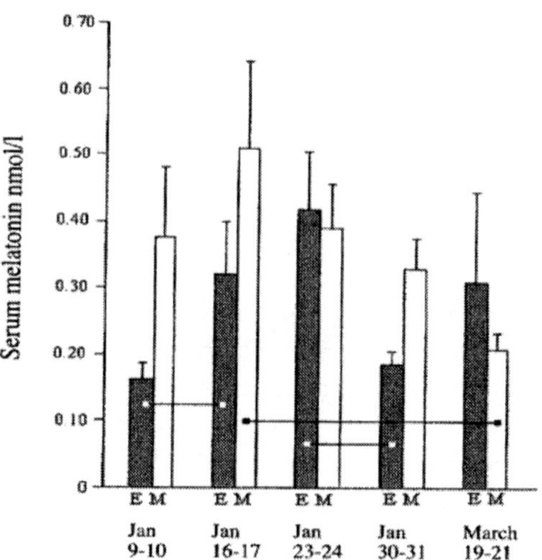

FIGURE 2. Evening (E) and morning (M) melatonin levels in insomnia patients during the dark period. Ordinate The ordinate shows the mean melatonin levels (mean ± SEM) of mean of melatonin between of the three times-points 22:00, 22:45 and 23:00, as well as the mean of melatonin levels between the two times time points 07:30 and 08:15. The abscissa shows, pair-wise, the evening melatonin values (E) and the morning values (M) melatonin values of the four dates in January and in March.

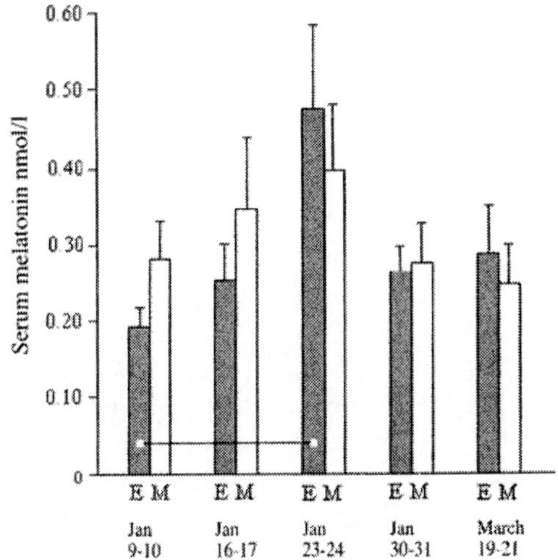

FIGURE 3. Evening (E) and morning (M) melatonin levels in controls during the dark period. For explanation of the bars and their significance, see Figure 2. A significant rise in the evening melatonin level was found in on the 23rd–24th of January as compared to the 9th–10th of January. No other significant differences were found. Melatonin is expressed as nmoll/l. The line on each bar indicates standard error of mean. (Printed with permission from International journal of Circumpolar Health)

response may be tuned to the time-rate-of-change of naturalistic twilight signals as reported to be the case in an experimental study by Danilenko et al.[16]

Both the 'sleep complain group' and the control group reported morning tiredness in January, while the control group reported a decrease in morning tiredness with increase of day length and hours of sunshine. The sleep complaining group still had some sleep problems in March and reported some morning tiredness. This may be due to an endogenous vulnerability in this group of persons.

One possible reason for the group difference may also be difference in exercise pattern in the two groups. Weydahl[17] has showed that melatonin in January will be influenced by hours of habitual exercise during fall among individuals living in a subarctic region in Norway. A further possible explanation may be different sensitivity in retinal light receptors between the two groups. Other possible underlying mechanisms which may also be of interest in the regulation of the sleep-wake cycle as has been reviewed by Lavie.[18]

6. Future Research

Inter-individual differences in sleep patterns, namely depth of sleep, have been attributed to individual diurnal type related features. Morning-types (M) generally retire and arise earlier and show less variable sleep length and awakening time than Evening-types (E).[19] Differences between M- and E-types with regard to circadian parameters have been observed in various behavioral, psychological and biological measures and are reviewed in detail by Kerkof.[20] M- and E-types appears to be a relatively stable individual characteristic that nevertheless can vary in degree across the seasons.

In an earlier review[21] about seasonal rhythms in healthy subjects, it was shown that changes in metabolic processes, increased sleep and circadian phase delay occurs somewhat physiologically in winter, and that winter impairment of mood, sleep-/wake- and vegetative functions is not restricted to midwinter insomnia and seasonal affective disorders, but occurs in an attenuated form in most normal healthy persons as well.

Prospective studies of the general population in Northern Norway[22] have found a decrease in depression and general sleeping problems over years, but failed to show variation in sleeping problems connected to the dark period. This support the theory that Midwinter Insomnia (MI) represents a more biologically, constitutionally-determined phenomenon.

An idea that is gaining favor is that the different sleep patterns in humans are probably regulated by the phasing of many cellular clocks interacting with recurring,meaningful events in the environment. One study supporting this view has recently been reported by Benedetti et al.[23] Their preliminary data show e g that the antidepressant effects of light therapy combined with sleep deprivation are influenced by a functional polymorphism within the promotor region of the serotonin transporter gene. Whether similar genetic mchinery may be involved in extreme phase-position such as morningness and eveningness, and predispose to seasonal variations in sleep disorders, e.g. Midwinter Insomnia, ought to be investigated in further studies.

7. *References*

1. N. Kleitman, H. Kleitman. The Sleep-Wakefullnes Pattern in the Artic, Scient. Monthly 76 , 349-356 (1953)
2. O. Devold, E. Barlindhaug, J. E. Backer. Søvnforstyrrelser i mørketiden, Tidsskr. Nor. Lægeforen. 77, 836-837 (1957)
3. K. Kjos, Søvnforstyrrelser som skolehygienisk problem i Nord-Norge, Tidsskr. Nor. Lægeforen, 79, 1241-1243 (1959)
4. O. Lingjaerde, T. Bratlid, Søvnvansker i mørketiden i Nord-Norge, ISM Skriftserie No. 6, 77-93, (1983)
5. R. Husby, O. Lingjærde, Prevalence of reported sleeplessness in northern Norway in relation to sex, age and season, Acta Psychiatr Scand 81, 542-547, (1990)
6. K. Natani et al, Long-therm changes in sleep paterns in men on the south polar plateau Arch Intern Med 125, 655-659 (1970)
7. E. Weitzman, A. DeGraaf, J. Sassin, T. Hansen, Seasonal patterns of sleep stages and secretion of cortisol and growth hormone during 24 hour periods in Northern Norway, Acta Endocrinologica 78, 65-76 (1975)

8. K. A. Stokkan, R. J. Reiter, Melatonin in Arctic urban residents, J. Pineal Res Jan, 16 (1) 33-36 (1994)

9. ICDS. International Classification of Sleep Disorders (1990)

10. N. E. Rosenthal et al, Diagnosis and treatment of seasonal affective disorder, JAMA 8, 270, 2717-2720 (1993)

11. E. D. Weitzman, C. A. Czeisler, R. M. Coleman, A. J. Speilman, I. C. Zimmerman, W. Dement, Delayed sleep phase syndrome. A chronobiological disorder with sleep-onset insomnia, Arc Gen Psychiatr 38, 737-746 (1981)

12. O. Lingjærde, T. Bratlid, O. C. Westby, J. O. Gordeladze, Effects of midazolam, flunitrazepam, and placebo against midwinter Insomnia in Northern Norway, Acta Psychiatr Scand, 67, 118-129 (1983)

13. O. Lingjærde, T. Bratlid, Triazolam versus flunitrazepam against midwinter insomnia in Northern Norway, Acta Psychiatr Scand, 64, 260-269 (1981)

14. T. Hansen, T. Bratlid, O. Lingjærde, T. Brenn, Midwinter insomnia in the subarctic region: evening levels of serum melatonin and Cortisol before and after treatment with bright artificial light, Acta Psychiatr Scand, 75, 428-434 (1987)

15. T. Bratlid, B. Wahlund, Alterations in serum melatonin and sleep in a subarctic region from winter to spring, International Journal of Circumpolar Health 62: 3 (2003)

16. K. V. Danilenko, A. Wirz-Justice, K. Krauchi, C. Cajochen, J. M. Weber, S. Fairhurst, M. Terman, Phase advance after one or three simulated dawns in humans, Chronobiol Int 17, 659-668 (2000)

17. A. Weydahl, Evening melatonin in January after changes in hours of habitual exercise during fall among youths living in subarctic, Arctic Med Res Jul; 53 (3), 146-151, (1994)

18. P. Lavie, Sleep-wake as a biological rhythm, Annu Rev Psychol 52, 277-303 (2001)

19. V. Lacoste, L. Wetterberg, Individual Variations of Rhythms in Morning and Evening Types with Special Emphasis on Seasonal Differences, in Light and Biological Rhythms in Man, edited by L. Wetterberg (Pergamon Press, Oxford, 1993) pp 287-304

20. G. A. Kerkhof, Interindividual differences in the human circadian system: A review, Biol Psychol, 20, 83-112

21. V. Lacoste, A. Wirz-Justice, Seasonal variation in normal subjects: an update of variables current in depression research, in Seasonal Affective Disorder and Phototherapy, eds N. E. Rosenthal and M. Blehar (Guilford Press, New York, 1989) pp 167-229

22. L. Skou Nilsen, V. Hansen, R. Olstad, Improvement in mental health over time in Northern Norway. A prospective study of a general population followed for 9 years, with special emphasis on the influence of darkness in winter, Soc Psychiatry Psychiatry Epidemiol, 39, 273-279 (2004)

23. F. Benedetti, C. Colombo, A. Seettti, C. Lorenzi, A. Pontiggia, B. Barbini, E. Smeraldi, Antidepressent effects of light therapy combinded with sleep deprivation arre influenced by a functional polymorphism within the promotor of the serotonin transporter, Biol Psychiatry, 54, 687-692 (2003)

Role of Pro-Inflammatory Cytokines in Sleep Disorders

PAOLO PROLO, FRANCISCO J. IRIBARREN, NEGOITA NEGAOS
AND FRANCESCO CHIAPPELLI

1. Summary

This chapter is designed to characterize how pro-inflammatory cytokines are involved in sleep regulation. Basic observations demonstrate that cytokines have a physiological role in sleep regulation with both somnogenic and inhibitory effects depending on the cytokine, dose and diurnal and/or circadian phase. These basic mechanisms have not yet been translated into clinical applications, although they have tremendous implications for answering why sleep is disturbed in so many individuals, and may eventually provide effective treatment. Yet, relationships between cytokine expression and sleep in healthy and insomniac populations at different ages are a current topic of investigation in several laboratories. Circadian expressions of interleukin (IL)-6, tumor necrosis factor α (TNF), adrenocorticotropic hormone (ACTH) and cortisol have been evaluated in various studies. Generally speaking, older individuals do not show increased daytime sleepiness when compared to controls, but display more daytime fatigue in all studies. Episodes of daytime sleepiness increase in the elderly following sleep deprivation. Older individuals show high levels of IL-6 and TNF prior to sleep onset. This increase is associated with over-production and release of these cytokines during recovery sleep as compared to younger controls. Daytime IL-6 hypersecretion and circadian shift of IL-6 secretion following sleep deprivation lead to daytime sleepiness and fatigue in younger subjects and deeper nighttime sleep in old subjects. Chronic insomnia is associated with a shift of IL-6 and TNF secretion from nighttime to daytime, which may explain the daytime fatigue and performance decrements associated with this disorder. The daytime shift of IL-6 and TNF secretion, combined with a 24-hour hypersecretion of cortisol, an arousal hormone, may explain the insomniacs' fatigue, which in contrast to disorders of excessive daytime sleepiness (EDS), is not associated with an increased sleep propensity at daytime or nighttime. These findings may lead to novel approaches in treating chronic insomnia.

2. Introduction

Since the early 1970s, quite a few studies have demonstrated a strong association between poor sleep and psychological factors, especially in relation to perceived stress.[1] Stress has been associated with activation of the hypothalamic-pituitary-adrenal (HPA) axis.[2] Corticotropin releasing hormone (CRH) and cortisol, respectively the hypothalamic and adrenal products of the HPA axis, both lead to arousal and sleeplessness in human subjects[3] and experimental animals.[4] Autonomic activation, and the elevation of hormones, including those produced by the HPA axis, play pivotal role in regulating immune surveillance mechanisms, including the production of cytokines that control the inflammatory process as well as events responsible for healing.[5-8] Conversely, sleep, particularly deep sleep, has an inhibitory influence on the stress system, of which the HPA axis is a major component.[9]

However conceptualized, psychoemotional stress is consistently associated with psychological manifestations, including anxiety, irritability and anger, sad and depressed moods, tension, fatigue, and with certain bodily manifestations, including perspiration, blushing or blanching of the face, increased heart beat, decreased blood pressure, and intestinal cramps and discomfort. Sleep duration and quality are also severely impaired by psychoemotional stress.

Three fundamental issues arise: 1) the consequences of stress upon normal subjects and patients are not uniform, and the psychopathological and physiopathological impact of stress may be significantly greater in certain people than others; 2) the impact of stress is dynamic and multi-faceted, such that the same person may exhibit a variety of manifestations of the psychoneuroendocrine-immune stress response with varying degrees of severity at different visits; and 3) the outcome of stress can be ambivalent in the sense that subjects and patients may position themselves along the spectrum of allostatic regulation, somewhere between the allostatic state (= toward regaining physiological balance), and the allostatic overload (= toward physiological collapse, with associated potential onset of varied pathologies). In brief, subjects with different levels of chronic stress (i.e., major life events or minor hassles) may be expected to respond dramatically differently.

Case in point is a recent study in which the participants were stratified in terms of their glucocorticoid response (i.e., plasma or salivary cortisol) following a mild cognitive challenge (i.e., Stroop color dissonance), which correlated inversely with the cytokine response. Cortisol low-responders reported less subjective stress and showed higher responses of pro-inflammatory cytokines (e.g., IL-1β, IL-6).[10] On the other hand, IL-6 responds with a delayed increase to acute stress. This stress may be tested by behavioral tasks[11] that may cause a bout of sleepiness in those subjects.

3. Role of Pro-Inflammatory Cytokines

The field of cytokine research has expanded remarkably over the past two decades. Cytokines were once believed to be solely products of the immune system that had definite immunological and hematological functions. It has become increasingly evident that in addition to those functions, cytokines play a critical role in the psychoneurological network, in sickness behavior, as well as in sleep deprivation. Due to the paucity of reliable data on other cytokines[13] we will focus our attention mostly on IL-6 and TNF.

Two factors are well established: IL-6 and TNF are somnogenic and fatigue-inducing proinflammatory cytokines;[9,12,13] the HPA axis stimulates arousal and suppresses sleepiness. Inflammatory cytokines, including IL-6 and TNF, are elevated in disorders of excessive daytime sleepiness (EDS), *i.e.* sleep apnea and narcolepsy.

Moreover, T helper (Th)1/Th2 cytokines play a fundamental role in human stress. Th1 cells primarily secrete IFN-γ, IL-2, and TNF, which activate cellular immunity, whereas Th2 cells secrete a different set of cytokines, primarily IL-4, IL-6, IL-10, and IL-13, which activate humoral immunity. Increased levels of IL-6 are associated with the presence of stress and depression.[14,15] The presence of IL-6 and IL-10 has been used to measure stress in humans as affected by behavioral interventions.[16] TNF has been utilized as a marker for Th1 cytokines when measuring psychological stress in humans,[17-19] including in the context of non-pharmacological interventions.[16] Certain kinds of cancer treatments (i.e., IL-2 therapy by itself or in combination with IFN-γ) are associated with depression symptomatology: Elevated IL-10 levels have been associated with depression among cancer patients.[20] Similarly, elevated IL-6 levels have been associated with stress and depression.[14,15] Apparently, this relationship is mediated by the activation of the cytokine network, more specifically, IL-6, soluble receptors of IL-2 (sIL-2R), IL-1 receptor antagonists and IL-10.[21]

The psycho-immunological framework offers exciting possibilities to understand sleep. There is increasing evidence suggesting a bi-directional integration between sleep and the immune system. Within the psychoimmunological framework, there is evidence that psychological distress is associated with immunological changes,[26] Such conceptual approach integrates the interaction between mind and biological processes, and by extension, the influence of non-pharmacological interventions on both. Meditation strategies have proven beneficial in both the psychological and immunological realms. In fact, meditation and other non-traditionally western techniques produce positive changes at the immunological level.[24,25] Mindfulness-based stress reduction (MBSR) is part of the meditative tradition; it is producing exciting evidence that it promotes healthy changes in quality of life, including in the improvement of depression, stress and sleep quality, which may be reflected at the cytokine level.[22]

MBSR has proven efficacious to decrease anxiety and depression among medical patients.[23] MBSR, derived from the contemplative traditions, teaches its practitioners meditative processes focusing on breath awareness, which is conducive to attain states of relaxation and observant detachment.

MBSR practice is rooted within non-judgment, patience and acceptance. In a study targeting early stage breast and prostate cancer patients, MBSR proved an effective strategy that promoted enhanced quality of life and a reduction of stress symptomatology.[22]

Among the benefits, improvement of sleep quality was one of the most noteworthy. In that study, over 40% of participants reported difficulty sleeping before participating in the 8-week MBSR regime. By the end of the 8-week treatment period, that figure was reduced to 20%. Moreover, and although this improvement did not reach statistical significance, the hours of night sleep increased by half hour. As significant, MSBR had an impact on stress biological markers (i.e., cytokines) associated with levels of stress: Increase of T cell production of IL-4, decrease of IFN-γ plus a decrease in natural killer (NK) cell production of IL-10.

Such data may point at a relationship between cytokines, sleeping and behavioral interventions. In that study, the improvement in sleep quality was coupled with a balance shift in the cytokine environment: from Th1 to Th2, which in turn, may be an indication of movement away from a depressive immune profile into a normalized one. Another non-pharmacological intervention, relaxation training, has shown in a 10-week relaxation-training regime, a positive effect among tinnitus sufferers.

It reduced the perception of stress, depression, and had an impact on biological markers associated with stress such as TNF.[16] These two behavioral interventions, MBSR, and relaxation training, may be exciting proof that non-pharmacological interventions can be useful tools to alleviate stress, and/or sleeping difficulties, with effects observable at the immunological level.

4. Circadian Oscillation of Pro-Inflammatory Cytokines

The mammalian circadian pacemaker resides just above the base of the brain, in a region of the hypothalamus called the suprachiasmatic nucleus (SCN).[27] Individual neurons within the SCN are capable of generating self-contained autonomous circadian timing. However, these thousands of cellular oscillators typically remain in synchrony and thus generate the coherent output necessary for a pacemaker (Figure 1).

In its simplest form, the circadian, or rhythmometric, system can be conceived as a central circadian clock, or pacemaker; a series of input pathways mediating the effects of several environmental synchronizers (such as light and darkness) on the pacemaker; and a set of output pathways conveying pacemaker signals to other regulatory systems of the brain and body.

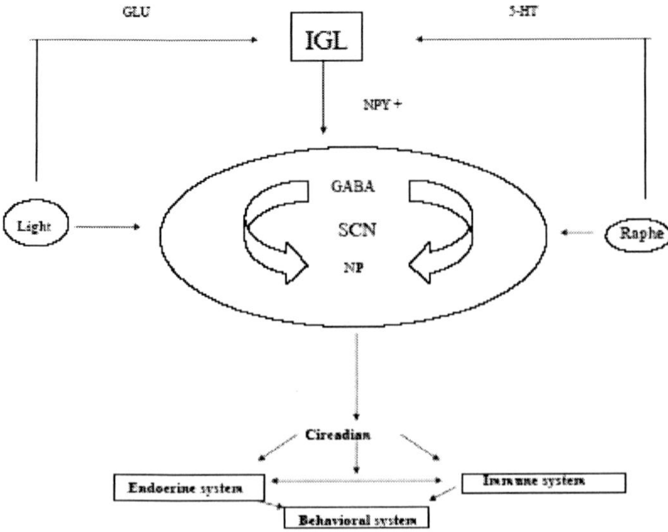

FIGURE 1. Schematic representation of the neural circuitry at the basis of circadian rhythms. The circadian pacemaker is positioned above the base of the brain in a distinct region of the hypothalamus called the suprachiasmatic nucleus (SCN). Cells of the SCN mutually interact and contain a number of chemicals important in cellular interaction, including the neurotransmitter gamma-aminobutyric acid (GABA) and one or more neuropeptides (NP). The SCN receives incoming information from three primary sources: (1) the retina of the eye, which sends signals about environmental lighting to the SCN along a direct pathway that uses glutamate (GLU) as its primary transmitter, together with at least two different NPs; (2) the intergeniculate leaflet (IGL) of the visual thalamus, which receives light signals from the retina and sends signals to the SCN via a pathway that employs both GABA and neuropeptide Y (NPY) as transmitters; and (3) the raphe cell groups of the midbrain, which employ the monoamine serotonin (5 HT) as primary neurotransmitter and signals to the IGL. Ending point is a complex interwining of the endocrine, immune and behavioral systems, which still needs to be fully clarified.

A stimulus or treatment that affects any one of these components may have an impact on the expression of circadian rhythms. On the other hand, stimuli that bypass the circadian system, and that instead act directly on physiological control systems, can modify the overall expression of circadian rhythms.[28]

Researchers on circadian rhythms are usually interested in distinguishing between "upstream" effects on the circadian pacemaker and/or its input pathways and those mediated "down-stream" on circadian output pathways and/or physiological control systems. Stimuli that alter circadian rhythm expression via downstream mechanisms can produce "masking" of circadian rhythms. That means, such effects are generally able to obscure the underlying circadian pacemaker.

Given the complexity of the pathways influencing the body's circadian rhythms, how can mechanisms underlying the chronobiological effects of any stimulus be determined? Assume, for example, that sleep deprivation alters the normal daily pattern of secretion of a particular hormone or cytokine. Several different types of alterations are possible, including changes in the overall level (i.e., amount) of substance secretion, the time of day at which the highest peak (or lowest trough) of secretion occurs, or the pattern of secretion over the course of the day and night.

Reported withdrawal-associated effects on circadian rhythms include phase-advances (i.e., earlier timing) of circadian rhythms in body temperatura,[29] rapid eye-movement (REM) sleep,[30] and levels of 5-hydroxyindoleacetic acid (5-HIAA, the primary metabolic by-product of serotonin, an important chemical involved in communication among nerve cells).[31,32] In contrast, phase delays (i.e., later timing) have been described for circadian rhythms in blood cortisol.[33]

Generally speaking, biological rhythms are variations of biological phenomena that are periodic and foreseeable in time. Temporal variations in cycles of rest-activity give temporal markers to the organism, and impose their period as a synchronizer of that activity. These rhythms can be characterized by different periods, leading to the division of circadian (a period of approximately 24 h), ultradian (a cycle that is shorter than 1 day), and infradian (a cycle longer than 1 day that may last weeks, months, or seasons). These clocks influence how the human body changes during the day, affecting temperature, blood pressure, hormone secretion and immune function.[34] This variation may be fit to a sinusoidal function by the cosinor method, a linear method of least squares.

Technically, the presence of sinusoidally varying circadian trends can be tested by cosinor analysis, which represents a linear reduction of sinusoidal regresión.[35] Parameters of the sinusoidal regression such as MESOR (midline-estimating statistic of rhythm or rhythm-adjusted mean), acrophase [a measure of time, the lag from a defined reference time point (e.g. midnight of the first day of measurement in some analysis; 8:00 AM in others) of the crest time in the cosine-curve-fitted curve to the data], and amplitude (half the extent of rhythmic change, or the difference between the maximum concentration and the MESOR of the fitted curve) can be obtained for circadian, ultradian or infradian periods. For consistency purposes, all cosinor charts included in this chapter present 8:00 AM as defined reference time point and are based on circadian analysis.

Rhythm detection is sought by testing the null hypothesis of zero amplitude with an F test. Finally, the potential role of IL-6 and TNF secretion in mediating daytime fatigue in chronic insomnia has been examined by the evaluation of 24-hour quantitative and temporal pattern of IL-6 and TNF secretion in insomniacs and controls matched for age and body mass index

(BMI). Chronic insomnia is associated with a shift of IL-6 and TNF secretion from nighttime to daytime, which may explain the daytime fatigue and performance impairments associated with this disorder.

Insomniacs are unable to fall asleep compared to normal sleepers during an objective daytime sleep testing [multiple sleep latency test (MSLT)].[9] Many studies have previously attributed this difficulty in falling asleep either at night or during the day to activation of their stress system.

Cosinor analyses, both for the individual and population IL-6 data, indicated a significant circadian rhythm. The pattern typically exhibits a multiple component curve with periods of 12 and 24 hours. These patterns are significant in both insomniac and control subjects ($P \leq .01$). These patterns are distinct, however, between these two groups of subjects. In control subjects, there was a major peak at night (about 2 AM) and a secondary peak in the late afternoon (about 5 PM). Insomniac subjects, compared to control subjects, manifested a significant shift of the major peak from about 2 AM to about 7 PM ($P \leq .001$). The nadir point was temporally and quantitatively the same, however, for both groups, about 9 AM (Figure 2).

Cosinor analyses, both for individual and population TNF data, indicated a significant circadian rhythm, with a multiple component curve, including periods with 12 and 24 hours for control subjects only ($P \leq .05$).

The peak appeared closer and prior to the offset of sleep (about 6 AM), whereas the nadir occurred in mid-afternoon (about 3 PM). Rhythmic pattern was not present in subjects with insomnia (Figure 3). By contrast, TNF daytime secretion showed a periodic rhythmicity of 4 hours with amplitude greater than zero ($P \leq .02$) in insomniac subjects. The estimated amplitude of this pattern in the insomniac group was 0.22 (95% confidence interval [CI^{95}], 0.047 to 0.392; $P \leq .02$). A trend, albeit not significant, for this ultradian rhythm was manifest in normal control subjects, with estimated amplitude of 0.047 (CI^{95}, −0.026 to 0.12; $P \leq .18$) (Figure 4).

5. Influence of Age on Pro-Inflammatory Cytokines

The role of pro-inflammatory cytokines on sleep at different ages remains to be fully elucidated. There is a paucity of studies assessing the role of immune factors at different ages of life, although copious evidence links pro-inflammatory cytokines and sleep. By contrast, consensus emerging from reviews of human studies is that basal corticotropic function is unaffected by aging, suggesting that the negative interaction of stress and aging does not occur in mankind.[36] However, aging has been associated with an evening increase of cortisol levels in a few studies.[37,38] Poor sleep in healthy adults, regardless of age, is associated with elevated circulating cortisol levels. Elevation of endogenous cortisol or exogenous administration of corticosteroids has been associated with sleep disturbance and increased wake time.[12] Aging is associated with 24-h increase of IL-6, with a phase advance of the

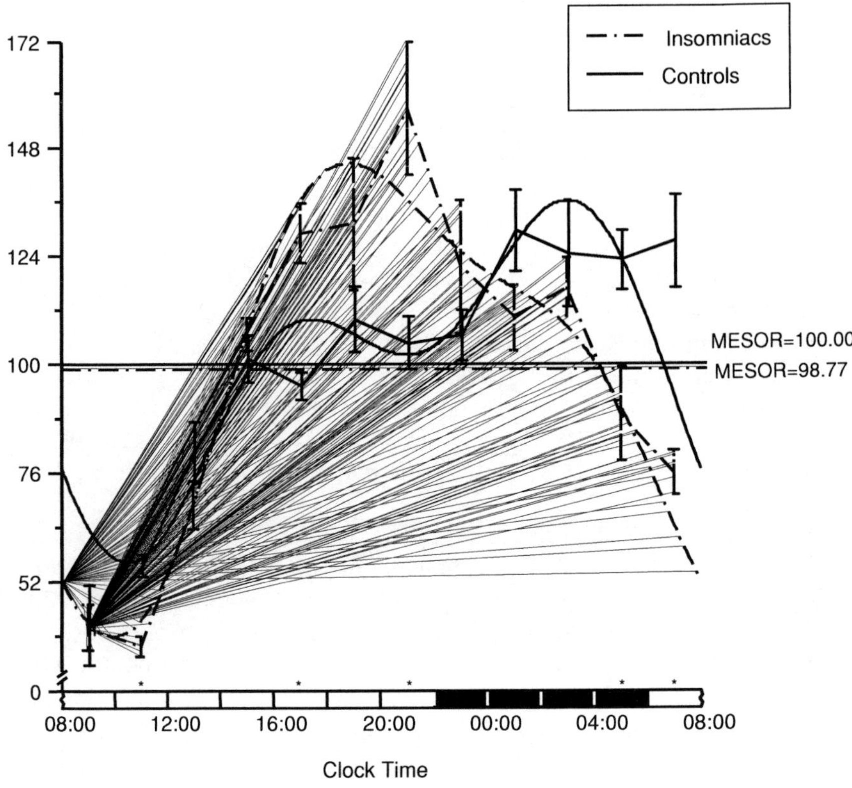

FIGURE 2. Multiple-component (fitted components: 24 and 12 hours) cosinor analysis of 24-hour plasma IL-6 in insomniacs (dotted line) and controls (solid line) expressed as percent variation from the mean. The thick black line on the abscissa represents the sleep-recording period. MESOR, mid-line estimating statistic of rhythm or rhythm-adjusted mean.[*] $P < .05$.

IL-6 circadian wave over that of cortisol by 3–5 h, thus suggesting an increasing over-activity of the HPA axis with increasing age, the latter finding being consistent with a previous report in middle-aged patients with early-untreated rheumatoid artritis.[3] Since higher IL-6 levels are significant correlated with risk of sudden death, this finding may imply that autoimmune diseases may accelerate senescence processes of the immune system.[39,40] Despite its somnogenic properties, IL-6 administration or elevation of its endogenous levels resulted in sleep disturbance when associated with HPA axis activation.[9] The association of IL-6 and cortisol with wake time is stronger in older adults than in young subjects. Middle-aged men show increased vulnerability of sleep to stress hormones, compared with the young.[12,41] A study has already demonstrated that increased depth of sleep

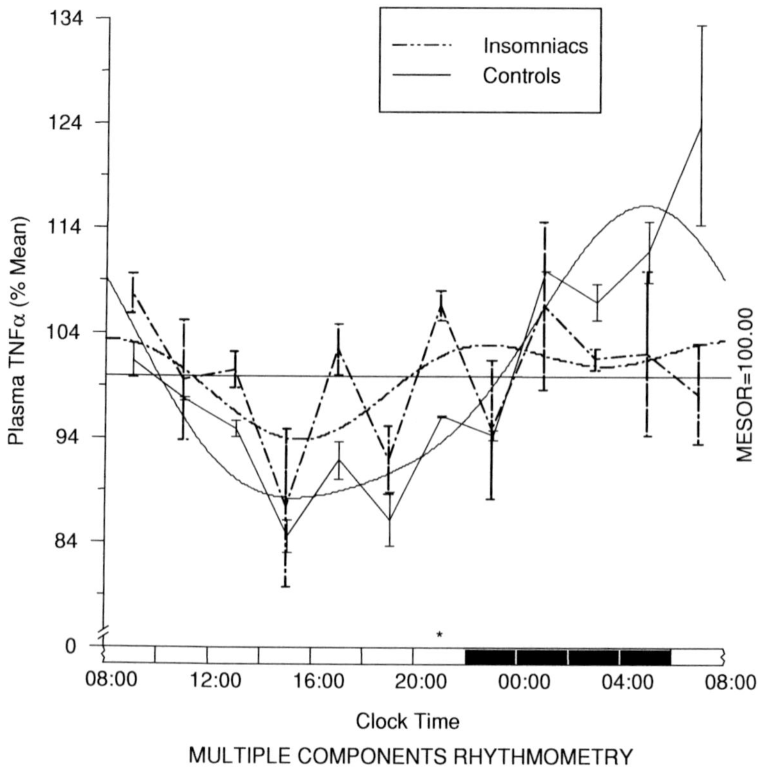

FIGURE 3. Multiple-component (fitted components: 24 and 12 hours) cosinor analysis of 24-hour plasma TNF in insomniacs (dotted line) and controls (solid line) expressed as percent variation from the mean. The thick black line on the abscissa represents the sleep-recording period. MESOR, mid-line estimating statistic of rhythm or rhythm-adjusted mean. *P < .05.

after one night of total sleep deprivation is associated with significantly decreased cortisol levels in young subjects.[9] Increasing age is associated with a sharp decline in Short Wave Sleep (SWS). SWS has an inhibitory effect on the activity of the HPA axis[41].

The potential association of this physiological change with the activity of the HPA axis remains to be elucidated. The HPA axis activity changes little with age in healthy young adult males,[42] but it increases after middle age.[43] A recent study analyzed patterns of sleep in healthy elderly (both men and women), and how those patterns correlate with cytokine secretion.[41] Older, compared to younger adults slept more poorly (wake time after sleep onset and % stage 1 sleep were increased whereas % SWS and % sleep time (ST) were significantly decreased.

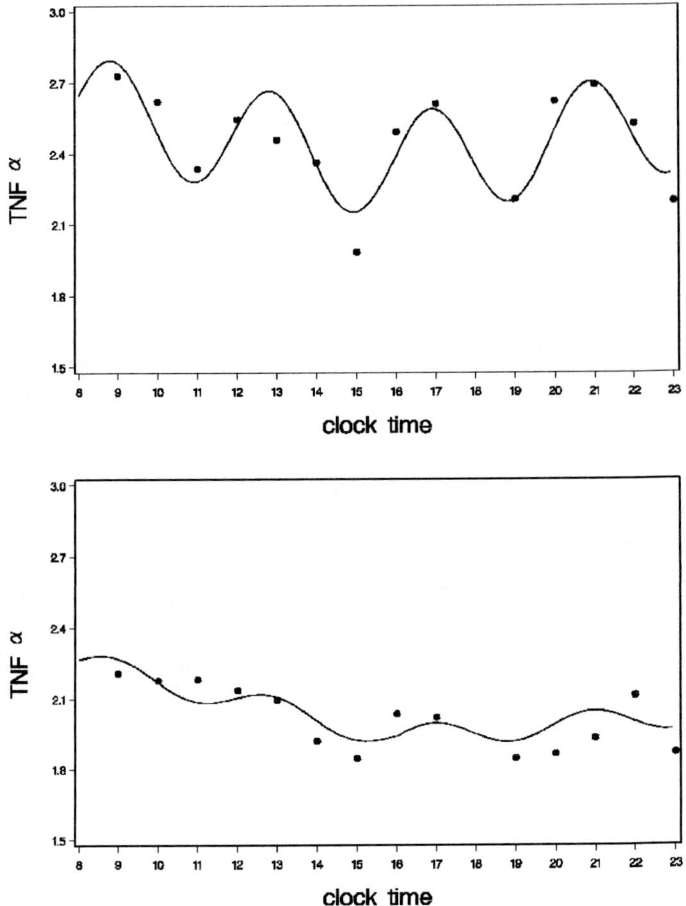

FIGURE 4. Analysis of daytime (8 AM to 11 PM) TNF ultradian rhythmometric pattern by fitting a cosine wave in insomniacs (top) vs. controls (bottom).

The process of aging is associated with increased IL-6 secretion and higher evening cortisol concentrations[41] (Figure 5). The association of the former with age-related sleep changes is unknown. The latter has been associated with lower amount of rapid eye movements (REM) sleep. As a consequence, older adults may be more vulnerable to sleep disturbances during periods of stress compared to young adults.

This might also be confirmed by the absence of significant variations in TNF circadian rhythmicity (i.e., the same study found this in the elderly, but not in younger individuals) (Figure 5). Unfortunately, very few studies are available on TNF role regarding either age or disease and it is still impossible to obtain a clear-cut answer.[41,44,45]

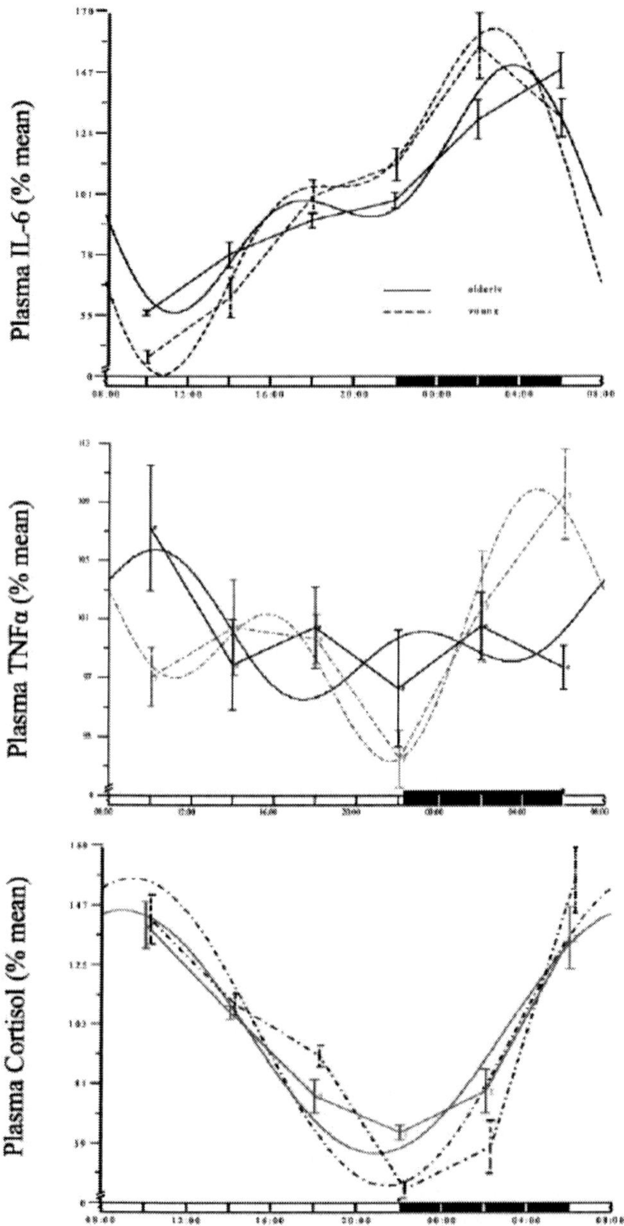

FIGURE 5. Multiple component (fitted components: 24 and 12 h) cosinor analysis of 24-h plasma IL-6 (top), TNF (middle), and cortisol (bottom) in young (dotted line) and old (solid line) healthy subjects expressed as percent variation from the mean. The thick black line on the abscissa represents the sleep-recording period.$^*P$ <.05.

Changes in sleep physiology associated with aging, including elevations of sleep-disturbing hormones and increased sensitivity of the sleep-controlling target organ to the actions of these hormones, play a significant role in the marked increase of insomnia prevalence with aging. IL-6 peripheral levels correlate negatively with sex steroids levels, positively with the amount of adipose tissue, are decreased after a good night's sleep and elevated in chronic pain/inflammatory síndromes.[13]

Old age is associated with decreased sex steroid concentrations, increased proportional body fat, accompanied by decreased quantity and quality of sleep, and frequent chronic pain/inflammatory conditions. Reducing the secretion of IL-6 in the elderly by administration of sex steroids, decreasing body fat through diet and exercise, and controlling adequately chronic pain and inflammation with nonsteroidal anti-inflammatory agents, may improve sleep, daytime alertness, and performance and decrease the risk of common ailments of old age, *e.g.* metabolic and cardiovascular problems, cognitive disorders, and osteoporosis.[13,46] Further studies are needed to enlighten gender differences in circulating pro-inflammatory cytokines and their possible different role in men and women.

6. Concluding Remarks

Several distinct processes are involved in sleep regulation: one can be considered a sleep homeostat, responsible to react to the need of sleep as it depends on prior amounts of sleep and wakefulness. Another is the circadian oscillator, based in the suprachiasmatic nucleus, responsible for the tendency to falling asleep during certain phases of a 24-hour period and sleep and wake time into specific episodes. A third motive, although related to this second one, is represented by circadian rhythmicity of pro-inflammatory cytokines that may as well influence sleepiness and alertness at different phases of the 24-hour cycle.

This new knowledge in the basic mechanisms of the role of sleep alterations, and the underlying importance of emotional stress in modulating biological systems will position us to better enhance health overall. Since IL-6 may contribute to common ailments of the elderly, (i.e., metabolic and cardiovascular problems, cognitive disorders, and osteoporosis) decreasing its hypersecretion through improvement of sleep, decreasing body fat, administering sex hormones, and controlling chronic inflammatory conditions may be associated with improved daytime function and well-being and decreased morbidity and mortality.

Chronic insomnia is associated with a shift of IL-6 and TNF secretion from nighttime to daytime. The daytime shift of IL-6 and TNF secretion, combined with a 24-hour hypersecretion of cortisol, an arousal hormone, may explain the insomniacs' fatigue, which in contrast to disorders of EDS, is not associated with an increased sleep propensity at daytime or nighttime. These findings may lead to either novel approaches in treating chronic insomnia or provide evidence to non-pharmacological, or so-called complementary, interventions.

Chronic activation of the HPA axis in insomnia suggests that insomniacs are at risk not only for mental disorders, like anxiety and depression, but also for significant medical morbidity (e.g., major depression, cardiovascular problems) associated with HPA axis hyperarousal.[47-49] The role of cytokines is intimately intertwined with the HPA axis function. Many studies have addressed the role of HPA axis hormones in sleep disorders, but they did not convey their attention to the immune side. Findings of studies presented here that correlate altered sleep with altered immune function show that sleep disorders must be taken into account as a factor of immune function impairment. However, questions remain unanswered. None of the studies ever employed any disease specific antigen or biologically relevant immunization (e.g. vaccination) in order to evaluate response to an immune challenge in subjects suffering from insomnia or any other sleep disturbance. Given the basic evidence that pro-inflammatory cytokines are somnogenic, further studies should expand on the potential use of these cytokines or their antagonists in evaluation and treatment of disordered sleep in a clinical population. There is a huge need for such improvements in treatment. Sleep disorders cost to the U.S. economy alone was estimated over $ 35 billion in 1994.[50] The least we can do is to encourage research that will eventually give us a good night's sleep in this frantic 24/7 never asleep society.

9. Acknowledgements. The authors thank the students, trainees, co-workers and junior faculty, who have contributed to various aspects of the material presented in this chapter. A very special thank you goes to Dr. Alex Vgontzas. The authors acknowledge support for these studies by the Fondazione Cassa di Risparmio di Cuneo, Italy; the Fondazione Cassa di Risparmio di Saluzzo, Italy; the National Institutes of Health (NIDDKD P50 DK 64539); UCLA School of Dentistry Intramural Seed Grant Program.

10. References

1. Kales and Kales, 1984 Kales A, Kales JD. Evaluation and treatment of insomnia. New York: Oxford University Press 1984.
2. Gold PW, Chrousos GP. The endocrinology of melancholic and atypical depression: relation to neurocircuitry and somatic consequences. *Proc. Assoc. Am. Phys.* 1999 111: 22–34.
3. Chrousos GA, Kattah JC, Beck RW, Cleary PA and the Optic Neuritis Study Group. Side effects of glucocorticoid treatment. *JAMA* 1993 269:2110–2112.
4. Opp M, Obal Jr F, Kreuger JM Corticotropin-releasing factor attenuates interleukin 1-induced sleep and fever in rabbits. *Am J Physiol* 1989 257:R528 R535.
5. Chiappelli F, Abanomy A, Hodgson D, Mazey KA, Messadi DV, Mito RS, Nishimura I, Spigleman, I.. Clinical, experimental and translational psychoneu-

roimmunology research models in oral biology and medicine. In *Psychoneuro-immunology, III*; Robert Ader, Robert Cohen and David Felten eds, Academic Press, Chapter 64, 2001 pp. 645-70.

6. Chiappelli F, Prolo P, Cajulis E, Harper S, Sunga E, Concepcion E. Consciousness, emotional self-regulation, and the psychosomatic network: Relevance to Oral Biology and Medicine. In *Consciousness, Emotional Self-regulation and the Brain*; M. Beauregard, Ed. Advances in Consciousness Research, John Benjamins Publishing Company, Chapter 9 2004 pp. 253-274.

7. Chiappelli F, Cajulis O. Psychobiological Views on "Stress-Related Oral Ulcers". *Quintessence Int.* 2004; 35: 233-237.

8. Prolo P, Chiappelli F, Fiorucci A, Dovio A, Sartori ML, Angeli A. Psychoneuroimmunology: new avenues of research for the 21st century *Ann NY Acad Sci* 2002 966: 400-408.

9. Vgontzas AN, Papanicolaou DA, Bixler EO, Lotsikas A, Zachman K, Kales A, Wong M-L, Licinio J, Prolo P, PW Gold PW, Hermida RC, Mastorakos G, Chrousos GP: Circadian interleukin-6 secretion and quantity and depth of sleep. *Journal of Clinical Endocrinology & Metabolism* 1999 84:2603-2607.

10. Kunz-Ebrecht SR, Mohamed-Ali V, Feldman PJ, Kirschbaum C, Steptoe A. Cortisol responses to mild psychological stress are inversely associated with proin-flammatory cytokines. *Brain Behav Imm* 2003 17:373-383.

11. Steptoe A, Willemsen G, Owen N, Flower L, Mohamed-Ali V. Acute mental stress elicits delayed increases in circulating inflammatory cytokine levels. *Clin Sci (Lond)*. 2001 101:185-192.

12. Vgontzas AN, Bixler EO, Lin H-M, Prolo P, Mastorakos G, Vela-Bueno A, Kales A, Chrousos GP 2001 Chronic insomnia is associated with nyctohemeral activa-tion of the hypothalamic-pituitary-adrenal axis: clinical implications. *J Clin Endocrinol Metab* 2001 86:3787–3794.

13. Irwin M. Effects of sleep and sleep loss on immunity and cytokines. *Brain Behav Immun.* 2002 16:503-512.

14. Drize N, Chertkov J, Samoilina N, Zander A.Effect of cytokine treatment (gran-ulocyte colony-stimulating factor and stem cell factor) on hematopoiesis and the circulating pool of hematopoietic stem cells in mice. *Exp Hematol.* 1996; 24: 816-822.

15. McNiece IK, Briddell RA, Hartley CA, Smith KA, Andrews RG. Stem cell factor enhances in vivo effects of granulocyte colony stimulating factor for stimulating mobilization of peripheral blood progenitor cells. *Stem Cells.* 1993; 11 Suppl 2:36-41.

16. Weber C, Arck P, Mazurek B, Klapp BF. Impact of a relaxation training on psychometric and immunologic parameters in tinnitus sufferers. *J Psychosom Res.* 2002; 52:29-33.

17. <u>Nemunaitis J, Singer JW, Buckner CD, Mori T, Laponi J, Hill R, Storb R, Sullivan KM, Hansen JA, Appelbaum FR.</u> Long-term follow-up of patients who received recombinant human granulocyte-macrophage colony stimulating factor after autologous bone marrow transplantation for lymphoid malignancy. *Bone Marrow Transplant.* 1991;7:49-52

18. Ackerman KD, Martino M, Heyman R, Moyna NM, Rabin BS. Stressor-induced alteration of cytokine production in multiple production sclerosis patients and controls. *Psychosom Med* 1998 60: 484-491.

19. Arck PC, Rose M, Hertwig K, Hagen E, Hildebrandt M, Klapp BF. Stress and immune mediators in miscarriage. *Hum Reprod* 2001 16: 1505-1511.

20. Hauser P, Soler R, Reed S, Kane R, Gulati M, Khosla J, Kling MA, Valentine AD, Meyers CA. Prophylactic treatment of depression induced by interferon-alpha. *Psychosomatics.* 2000;41:439-441.

21. Tilg H, Vogel W, Dinarello CA.Interferon-alpha induces circulating tumor necrosis factor receptor p55 in humans. *Blood.* 1995; 85:433-435.

22. Carlson LE, Speca M, Patel KD, Goodey E. Mindfulness-based stress reduction in relation to quality of life, mood, symptoms of stress and levels of cortisol, dehydroepiandrosterone sulfate (DHEAS) and melatonin in breast and prostate cancer outpatients. *Psychoneuroendocrinology* 2004 29:448-474.

23. Reibel DK, Greeson JM, Brainard GC, Rosenzweig S. Mindfulness-based stress reduction and health-related quality of life in a heterogeneous patient population. *Gen Hosp Psychiatry.* 2001 23:183-192.

24. Solberg EE, Halvorsen R, Sundgot-Borgen J, Ingjer F, Holen A. Meditation: a modulator of the immune response to physical stress? A brief report. *Br J Sports Med.* 1995 29:255-257.

25. Taylor DN. Effects of a behavioral stress-management program on anxiety, mood, self-esteem, and T-cell count in HIV positive men. *Psychol Rep.* 1995 76:451-457.

26. Petry LJ, Weems LB 2nd, Livingstone JN 2nd. Relationship of stress, distress, and the immunologic response to a recombinant hepatitis B vaccine. *J Fam Pract.* 1991 32:481-486.

27. Buijs RM. The anatomical basis for the expression of circadian rhythms: the efferent projections of the suprachiasmatic nucleus. *Prog Brain Res.* 1996;111:229-240.

28. Rosenwasser AM. Alcohol, Antidepressants, and Circadian Rhythms Human and Animal Models 2004 *NIAAA Publications* http://www.niaaa.nih.gov/publications/arh25-2/126-135.htm

29. Kodama H, Nakazawa Y, Kotorii T, Nonaka K, Inanaga K, Ohshima M, Tokoyama T. Biorhythm of core temperature in depressive and non-depressive alcoholics. *Drug and Alcohol Dependence* 1988 21: 1-6.

30. Imatoh N, Nakazawa Y, Ohshima H, Ishibashi M, Yokoyama T. Circadian rhythm of REM sleep of chronic alcoholics during alcohol withdrawal. *Drug and Alcohol Dependence* 1986 18:77-85.

31. Sano H, Suzuki Y, Yazaki R, Tamefusa K, Ohara K, Yokoyama T, Miyasato K, Ohara K. Circadian variation in plasma 5-hydroxyindoleacetic acid level during and after alcohol withdrawal: phase advances in alcoholic patients compared with normal subjects. *Acta Psychiatr Scand.* 1993;87:291-296.

32. Sano H, Suzuki Y, Ohara K, Miyasato K, Yokoyama T, Ohara K. Circadian variations in plasma monoamine metabolites level in alcoholic patients: a possible predictor of alcohol withdrawal delirium.Prog *Neuropsychopharmacol Biol Psychiatry.* 1994;18:741-752.

33. Iranmanesh A, Veldhuis JD, Johnson ML, Lizarralde G. 24-hour pulsatile and circadian patterns of cortisol secretion in alcoholic men. *Journal of Andrology* 1989 10: 54-63.

34. Wong M-L, Kling MA, Munson PJ, Listwak S, Licinio J, Prolo P, Chrousos GP, Oldfield EH, McCann SM, Gold PW: Stress system dysfunction in major depression: Serial cerebrospinal fluid sampling of corticotropin releasing hormone and norepinephrine. *Proceedings of the National Academy of Sciences U.S.A.* 2000 97; 325-330.

35. Fernandez JR, Hermida RC. Inferential statistical method for analysis of nonsinusoidal hybrid time series with unequidistant observations. *Chronobiol Int* 1998 15:191–204.

36. Prinz PN, Bailey SL, Woods DL. Sleep impairments in healthy seniors: roles of stress, cortisol, and interleukin-1 beta. *Chronobiol Int.* 2000;17:391-404.
37. Ferrari E, Cravello L, Muzzoni B, Casarotti D, Paltro M, Solerte SB, Fioravanti M, Cuzzoni G, Pontiggia B, Magri FAge-related changes of the hypothalamic-pituitary-adrenal axis: pathophysiological correlates. *Eur J Endocrinol.* 2001 144:319-329.
38. Van Cauter E, Leproult R, Plat L. Age-related changes in slow wave sleep and REM sleep and relationship with growth hormone and cortisol levels in healthy men. *JAMA* 2000 284:861-868.
39. Luc G, Bard JM, Juhan-Vague I, Ferrieres J, Evans A, Amouyel P, Arveiler D, Fruchart JC, Ducimetiere P. PRIME Study Group. C-reactive protein, inter-leukin-6, and fibrinogen as predictors of coronary heart disease: the PRIME Study. Arterioscler *Thromb Vasc Biol* 2003 23:1255-1261.
40. Singh RB, Kartik C, Otsuka K, Pella D, Pella J. Brain-heart connection and the risk of heart attack. *Biomed Pharmacother* 2002;56 Suppl 2:257s-265s.
41. Vgontzas AN, Zoumakis M, EBixler EO, Lin HM, Prolo P, Vela-Bueno A Kales A, Chrousos GP. Impaired Nighttime Sleep In Healthy Old Vs. Young Adults is Associated with Elevated Plasma IL-6 and Cortisol Levels: Physiologic and Therapeutic Implications. *J Clin Endocrinol Metab* 2003 88: 2087-2095.
42. Pavlov EP, Harman SM, Chrousos GP, Loriaux DL, Blackman MR. Responses of plasma adrenocorticotropin, cortisol, and dehydroepiandrosterone to ovine corticotropin-releasing hormone in healthy aging men. *J Clin Endocrinol Metab.* 1986;62:767-772.
43. Van Cauter E, Leproult R, Kupfer DJ. Effects of gender and age on the levels and circadian rhythmicity of plasma cortisol. *J Clin Endocrinol Metab.* 1996;81: 2468-2473.
44. Imagawa S, Yamaguchi Y, Ogawa K, Obara N, Suzuki N, Yamamoto M, Nagasawa T. Interleukin-6 and tumor necrosis factor-alpha in patients with obstructive sleep apnea-hypopnea syndrome. *Respiration* 2004; 71:24-29.
45. Irwin M, Rinetti G, Redwine L, Motivala S, Dang J, Ehlers C Nocturnal proin-flammatory cytokine-associated sleep disturbances in abstinent African American alcoholics. Brain Behav Immun. 2004;18:349-360.
46. Chrousos GP, Gold PW. A healthy body in a healthy mind—and vice versa—the damaging power of "uncontrollable" stress. *J Clin Endocrinol Metab.* 1998 83: 1842-1845.
47. Gold PW, Wong ML, Chrousos GP, Licinio J. Stress system abnormalities in melancholic and atypical depression: molecular, pathophysiological, and thera-peutic implications. *Mol Psychiatry.* 1996;1:257-264.
48. Prolo P, Chiappelli F, Cajulis E, Bauer J, Spackman S, Romeo H, Carrozzo M, Gandolfo S, Christensen R. Psychoneuroimmunology in Oral Biology and Medicine: The Model of Oral Lichen Planus. *Ann NY Acad Sci* 2002 966:429-440.
49. Prolo P, Licinio J. Molecular neuroendocrinology and its impact on behavior. In: *Genes, Behavior, and Health.* (C.L. Bolis and J. Licinio eds.) World Health Organization, Geneva, Swizerland 1999 pp. 37-60.
50. Chilcott LA, Shapiro CM. The socioeconomic impact of insomnia. An overview. *Pharmacoeconomics.* 1996;10 Suppl 1:1-14.

Sleep and Circadian Neuroendocrine Function in Seasonal Affective Disorder

A. A. Putilov, S. R. Pandi-Perumal and T. Partonen

1. Introduction

The research of winter depression (also known as winter type of seasonal affective disorder or SAD) has a unique history. Unlike other affective diseases, it has been discovered by the investigations in the field of chronobiology[1,2] rather than throughout medical practice. Since the first systematic description of winter depression[2] its biological basis has been extensively studied. A number of hypotheses have been proposed to explain the underlying biological mechanisms of SAD and its successful treatment by 2-hr daily administration of bright light for 1-2 weeks. However, none of these hypotheses is fully compatible with existing empirical results (see, i.e., the reviews by[3,4])

Sleep disturbances are a hallmark of SAD, as they are of the remaining mood disorders. Winter depression is characterized by the so-called atypical depressive symptoms that in particular, include hypersomnia, difficulty waking up in the morning and daytime sleepiness. These abnormal features disappear or even the opposite symptoms of hyposomnia and insomnia can appear throughout summer remission.[2] Sleep in SAD patients was found to have little resemblance to sleep in the majority of patients with non-seasonal major depression. In particular, as compared to the general findings in depressive disorder (i.e.[5]), winter depressives show the opposite trend in sleep continuity measures.[6-10] However, in general, the electroencephalographic (EEG) studies failed to support strong clinical and epidemiological evidence for the dramatic effect of season and bright light treatment on the sleep-wake pattern in SAD.

2. Sleep and Seasonality in Winter Depression

2.1. Contradictive Results on EEG Sleep

The inconclusiveness of sleep EEG findings in winter depression might be exemplified by the contradictive results of studies conducted by several groups soon after SAD discovery. There were both negative and positive findings on the enlarged total sleep time in winter depressives. The differences were significant or absent when depressives were compared to 1) healthy subjects (i.e.[6,11] vs.[7,8], respectively) or 2) themselves during summertime remission (i.e.[7] vs.[6,8], respectively) or 3) themselves after bright light treatment (i.e.[6,8] vs.[7,12], respectively). Similarly, both negative and positive findings were reported on wintertime reduction of slow wave sleep in SAD patients compared to 1) controls (i.e.[7,11] vs.[6,8], respectively), or 2) themselves either in summer (i.e.[7] vs.[6,8], respectively) or 3) in post-treatment condition (i.e.[7,8,11] vs.[6,13], respectively). Hence, we have to be cautious with interpretation of the findings suggesting the significant sleep EEG differences between SAD patients and healthy controls.

Even the results reported by the same group illustrate this notion. For example, earlier findings of the Basel group on the normo- or hyposomnic SAD patients suggested that the characteristics of sleep EEG in SAD patients and controls differ neither during baseline sleep nor during the extended sleep following sleep deprivation.[7] However, more recent report suggested the attenuation of buildup of waking EEG power density during daytime in winter depressives compared to healthy subjects.[14] The results of Siberian group dealing with hypersomnic patients are also contradictive, and, besides, they, in general, do not fit the reports of the Basel group. In the study of nighttime sleep in SAD, the reduction of baseline slow wave sleep and increased total sleep time were found in winter depressives as compared to healthy controls.[8] However, another study applying the 24-hr multiple sleep latency test, failed to provide evidence of patients' increased sleep propensity in day- and nighttime hours[15] and of abnormality of slow-wave sleep pressure in the first 20-min of the subsequent recovery sleep.[16,17] By contrast, the third study[17] supported the finding of the Basel group on the attenuation of buildup of waking EEG power density in SAD.

Several reasons might be suggested to explain so often reported disagreement between the EEG results. The small sample sizes appear to be the most plausible explanation. It has to be noted that most typical EEG sleep abnormalities in depressed subjects, such as reduced rapid eye movement (REM) sleep latency, night sleep interruptions, reduced slow wave sleep, and the like, are similar to those of healthy elderly people. Both depressed and control subjects show the expected age trends, but these trends seem to be somewhat advanced in depressed subjects compared to normals (see, i.e.,[18,19]). Hence, when the subjects of different age are included in the same sample, the intra-group variation of EEG characteristics could mimic inter-group differences.

2.2. *Objective and Subjective Sleep Reports*

Moreover, it has to be emphasized that, despite association of SAD with sleep complains, the neuro-vegetative symptoms do not necessary reflect the deviations of the objective sleep measures from the normal range (i.e. these deviations sometimes could be surprisingly small, if any). Patients with

SAD seem to exhibit the tendency to overestimate their winter sleep duration in retrospective reports and clinical interviews.

In particular, this discrepancy was revealed by comparison of the objectively measured amplitudes of seasonal variations in sleep length (i.e.[8]) with those retrospectively reported (i.e.[20,21]). When retrospective reports suggest 1-to-3-hr difference between sleep duration in summer and winter, the average winter-summer difference in EEG sleep duration did not exceed several minutes (although it often was statistically significant). Earlier published data on systematically recorded sleep timing also indicate that the deviation of SAD sleep from the normal range is far from being as dramatic as retrospectively reported estimates (i.e.[22]).

The reason for exaggeration of sleep problems might be rooted in SAD psychology. Not only physiological functioning, but, most importantly, the psychosocial functioning in the depressed with SAD suffers from the wintertime neuro-vegetative symptoms. The psychosocial problems caused by such habits as late awakening, oversleeping, overeating and the like can contribute to development of both depressed mood state and high beneficial expectations from bright light therapy. Hence, further replication studies are needed to confirm the reported differences between SAD patients and controls on EEG sleep measures.

2.3. Seasonality of Mood

It is naturally to suggest that seasonal variations in sleep in SAD are the extreme cases of the response of sleep regulated mechanisms to annual variations in photoperiod. Although humans are not considered to be very sensitive to the changes in the length of day or photoperiod,[23] their physiology, in general, still reacts on illumination conditions in the same way, as does physiology of other mammals. The discovery of SAD was stimulated by the finding that the exposure of bright light does suppress melatonin secretion from the pineal gland and changes the phase position of the rhythm generated by the circadian pacemaker.[24]

Moreover, seasonal variation in many bodily functions has been acknowledged in the literature. The seasonal variations in mood are of a special interest, because statistics of hospital admission for depression and mania shows significant seasonal patterns, with the incidence of depression being greatest in winter. Specific seasonal effect for affective illness is evidenced by the non-existence of seasonality in other psychiatric disorders (i.e. [25]). In the light of numerous findings of triggering and termination of affective illness by environmental factors, Wehr et al. proposed that these conditions are likely to be the disorders of systems that mediate the organism's adaptation to physical changes in the habitat.[26] The decreasing daylight period as winter approaches is thought to trigger a depressive episode in subjects predisposed to SAD, in particular. However, no causal relationship can be drawn between the incidence of SAD and the relative shortage of light. Although bright-light exposure has been used in treatment, the cause of SAD is not inevitably a lack of light.

2.4. Epidemiology of SAD

Winter type of SAD is an affective illness with recurrent depressive episodes in winter plus hyperthymic periods in spring or summer, as first systematically described in the literature by Rosenthal et al.[2]. About 90% of patients with SAD meet criteria of Diagnostic and Statistical Manual of Mental Disorders for bipolar disorder type II,[26,27] and approximately 10% of affective disorders have the seasonal pattern.[27] Epidemiological studies have shown that there is a greater incidence of SAD or atypical symptoms of depression at higher latitudes and cold climate.[28,29] In particular, over 10% of the Siberian population shows clear seasonal variations in mood and behavior.[20]

Clinical reports pointed on the relationship between SAD and photoperiod. In patients with SAD, depressive symptoms tend to worsen during the winter along with the decreasing hours of daylight.[30] In fact, the onset of depressive symptoms coincides with the most rapid changes in the intensity, spectrum and hours of sunlight that takes place around the autumn equinox. No association of the clinical picture with other environmental variables, such as rainfall, humidity or atmospheric barometric pressure has been reported. A model proposed for SAD is that of a multi-factorial illness in which the genetic effect interacts with the seasonally changing effect, such as light exposure or ambient temperature.[30,31] The genetic effect seems to contribute 29% to the onset of SAD.

There is likely a molecular basis for these unique responses to photoperiod in winter depression. At present, two gene polymorphisms in association with SAD are discovered, both of which are located in the so-called circadian clock genes, showing that the *NPAS2* and *ARNTL* genes take part in the pathogenesis of SAD.[32] A polymorphism (647 Val/Gly) of the *PER3* gene has been linked to not only delayed sleep phase syndrome, but now also to the self-reported circadian preference, those with at least one glycine allele showing preference to display more activity in the morning hours. In addition, there are some interesting discoveries that have importance to the accuracy and stability of the circadian clockwork in general and need to be analyzed in SAD patients as well.

2.5. Bright Light Treatment of SAD

The effect of bright light therapy on SAD was first predicted in theory and has thereafter been demonstrated in many clinical trials (for reviews, see[33,34]). Symptoms may disappear after just one week of daily exposures to bright light (2500 lux or more) for 1 to 2 hours.[2]

However, despite extensive research, the mechanism of beneficial response to bright light remains unknown. The majority of the proposed explanations link it with the regulation of circadian rhythms or serotonin in brain. It has been noted that most of these hypotheses are not mutually exclusive.[3,4,35,36] Some of these hypotheses justified as follows.

3. Photoperiodic Response in Winter Depression

3.1. Timing of Light Treatment

SAD and the antidepressant effect of bright light on depressive symptoms[1,37] were predicted on the basis of two lines of evidence. First, response to the length of day was known to be mediated by the actions of melatonin in animals (i.e.[38]). Second, the exposure to bright light was shown to suppress melatonin secretion in humans as well. These findings led to suggestion that the suprachiasmatic nuclei (SCN) of the anterior hypothalamus play a role as a mediator of changes in mood, and that this response may be mirrored in the seasonal variation in photoperiod and melatonin profile.

Since then, the predicted effect of bright light therapy to reverse the symptoms of SAD has been demonstrated in numerous studies (for reviews, see[39-44]). Soon after discovery of SAD in 1984, the comparison of the light treatments at different time of the day has been proposed as a test of the mechanisms underlying clinical effect of phototherapy. In particular, light in the middle of the day was used to examine at least two chronobiological explanations. The first and earliest explanation suggested the normalization of the timing of melatonin secretion by extension of day length with morning plus evening bright light.[1] The next hypothesis postulated the possibility of correction of abnormally delayed circadian rhythms with early morning bright light.[45] Both explanations suggest only weak, if any, therapeutic response to bright light in the middle of the day, since such a treatment fails to extend day length or reset the abnormally phased circadian rhythms.

However, the efficacy of bright light was found to be rather high at various time of day, including hours when melatonin is not secreted by the pineal gland.[46-50] In particular, bright light even at midday still reverses the symptoms of SAD effectively.[51-55]

3.2. Effect of Melatonin Administration

Oral administration of melatonin was not found to reverse completely the effects of bright light, although it does reproduce the atypical depressive symptoms of SAD.[56] In another study[57] this hormone given to SAD patients either in the morning or in the evening has neither positive nor negative effect on depressive symptoms, while bright light has a positive effect.

The afternoon treatment with melatonin in physiological doses was thought to be an effective antidepressant, because such treatment needed to produce a phase advance. Indeed, the preliminary finding suggested that the afternoon administration alleviates symptoms of SAD.[58] In another study[17] the afternoon melatonin treatment prevented relapse after total sleep deprivation, although no difference in the antidepressant response between placebo and melatonin was seen.

The hypothesis of the involvement of melatonin in SAD has not been supported by data on the effect of antidepressant treatments on the levels of this hormone. In particular, it was noted that the antidepressants reduced depressive symptoms irrespective of their effect on melatonin levels that could be either increased (tricyclics, fluvoxamine) or decreased (fluoxetine) after medication.[59]

In contrast, some other facts point on a possible link between melatonin alterations and the pathogenesis of SAD. For example, the administration of high doses of melatonin to healthy individuals induces drowsiness, decreased attention and prolonged reaction times, and melatonin administration to SAD patients induces a worsening of depressive symptoms.[60]

3.3. *Seasonal Changes in Melatonin Profile*

Seasonal changes in melatonin or other hormones were not detected in a study by Avery et al.[61] Negative results were also presented by Wehr et al. in their early report.[62]

However, more recent studies of healthy individuals, however, have renewed interest in the original hypothesis for SAD. Wehr et al. showed that the duration of melatonin secretion in healthy subjects responded to changes in the photoperiod in a way that was very similar to the response seen in animals.[63] Moreover, Hashimoto et al. reported that midday exposure to bright light might change the duration of melatonin excretion due to an advance of phase of the secretion onset.[64] Spiegel et al. demonstrated that the duration of melatonin secretion might be shortened or lengthened by sleep curtailment or extension, respectively.[65]

Finally, Wehr et al. showed that patients with SAD rather than healthy controls responded to the lengthening of day by shortening of the duration of melatonin secretion.[66,67] Thus, the original assumption that the onset of SAD resembles the mechanism of action in response to the photoperiod, as seen in mammals, cannot be rejected completely and needs further elaboration.

3.4. *Melatonin Supression*

The relationship between suppression of melatonin by light and amelioration of SAD symptoms was reported.[68] Several groups[20,45,50,69,70] have noted that patients with SAD do not differ much from controls in the baseline melatonin phase position, while there are more pronounced differences concerning the extent of advance of the phase position of melatonin circadian rhythm following exposure to morning bright light.

The symptoms of hypersomnia and late awakening may be a sign of both a delayed phase and a long duration of melatonin excretion. Since recent findings show a link between the timing of melatonin secretion and sleep onset,[71,72] the use of the onset of melatonin secretion as a marker of the circadian phase position[73] may lead to underestimation of phase differences between depressed SAD patients in winter and healthy individuals.[20] Possibly, more prominent phase differences between patients and controls exist in the time of offset rather than onset of melatonin secretion.

Since several reports suggest that in SAD patients the melatonin secretion is more sensitive to light than in healthy individuals,[62,74,75]) it is the increased sensitivity to light rather than the initial phase delay that may explain the extensive phase shifts in SAD patients. However, Partonen et al. did not find any difference between patients with SAD and controls in the degree of suppression of melatonin by late evening bright light.[76]

The experiments with two β-blockers, which suppress melatonin secretion, yielded conflicting results. First, a long-acting β-blocker, atenolol, did not reproduce the antidepressant effect of bright light in SAD patients treated in the afternoon, although several individuals responded very well .[77] In contrast, Schlager reported the efficacy of a short-acting β-blocker, propranolol, in SAD patients treated in the early morning.[78] A subsequent study provides additional support for beneficial effect of well-timed β-adrenergic blockade.[79]

3.5. Daytime Melatonin

Several clinical studies have suggested that the daytime secretion of melatonin may be enhanced in SAD patients. In a case study[80] abnormally high levels of this hormone were found in a patient with SAD in the morning and afternoon hours, and early morning bright light resulted in the reduced amplitude and phase advance of melatonin circadian rhythm. One of ten patients with SAD[52] secreted abnormally much amount of melatonin in daytime at baseline, and showed a rise in plasma levels after oral administration of 5-hydroxytryptophan. In some other studies, dietary L-tryptophan, the precursor of 5-hydroxytryptophan, has been as effective as light exposure in patients with SAD.[81] In a case study, a good response to tryptophan treatment was associated with the increased nighttime levels of a melatonin metabolite (aMT6s) excretion.[82]

Diurnal differences in urinary excretion of MLT were low in a sample of Siberian SAD patients, whereas the signs of normalization of the excretion pattern were seen after remission of symptoms followed by light therapy, change of seasons, or after travel to a southern region.[83-85] Further research showed that the daytime circulating melatonin levels were higher in patients with SAD compared with controls during the winter. This difference disappeared after administration of bright light and in summer.[86,87] In addition, the study of the seasonal changes in the diurnal pattern of melatonin levels in an Alaskan population showed that the elevated daytime levels of this hormone in winter are linked to SAD-like symptoms.[88]

By contrast, the abnormalities in the daytime levels of melatonin in SAD have not routinely been seen by several other groups.[51,70,89,90] The abnormally high daytime of melatonin may have extra-pineal origin, and further experimental studies are necessary to understand whether the daytime levels of this hormone have a role in the clinical picture of SAD.[83]

It is not clear whether excessive levels of circulating melatonin due to a prolonged secretion or a compromised elimination might cause SAD symptoms, or whether they are merely an epiphenomenon of other factors, such as dysfunctions of neurotransmission. The latter were suggested by several hypotheses. In particular, according to Depue et al. there is a state-dependent reduction of dopaminergic activity in SAD.[91,92] Abnormal regulation of the serotonergic system has been also suggested. In this regard, the seasonal fluctuations of serotonin metabolism that is also present in healthy individuals, was suggested to be of greater magnitude in patients with SAD.[89]

3.6. Photosensitivity

Some recent evidence gives support for the suggestion of abnormal photosensitivity in SAD. This may explain some circadian rhythm alterations noted in patients with SAD. For example, the decrease in plasma melatonin levels that normally occurs in the early morning is delayed by 2 hours[93] and the rest-activity rhythm is delayed by up to 70 minutes with respect to healthy controls.[94] As discussed in detail below, the importance of the circadian rhythm of melatonin in the pathogenesis of SAD is addressed by the hypothesis that the treatment of choice for SAD is morning light therapy, which acts by inducing a phase advance of the circadian rhythms.[45,95]

Abnormal photosensitivity may also be due to retinal dysfunction.[96] Exposure to light sets into motion a cascade of biochemical reactions in the retina that brings about the amplification and transmission of the signal towards the cerebral cortex, and influences the loss of the retinal cell membranes. The number of cells stimulated is proportional to the intensity of light exposure.[97] Another suggestion concerns the mechanism of light adaptation. In prolonged exposure to light, photoreceptor cells normally adapt at different levels of illumination, according to a mechanism known as photostasis. In SAD, the mechanism of photoreceptive adaptation would be compromised at low levels of illumination in particular. This suggests that there is reduced sensitivity to light, agreeing with the data obtained from the recordings of electrooculography in patients with SAD.[98]

4. Circadian Phase in Winter Depression

4.1. Circadian Phase Delay

The association of winter SAD with the reduced length of day, and with the response to bright light suggests that this disorder might be related not only to changes in the duration of melatonin secretion, but also to changes in the phase of melatonin and other circadian rhythms.[99] When early experiments did not confirm the assumption that the extension of day length is critical for the antidepressant effect of bright light, another chronophysiological responses to light were proposed. In particular, the normalization of abnormal phase positions of circadian rhythms was thought to be of significance for seasonal course of manifestation of depressive symptoms, and therefore the phase shifts were tested as a possible mechanism of action of bright light for SAD. According to the phase-shift hypothesis,[95] the timing of endogenous circadian rhythms in most SAD patients are abnormally delayed with respect to the actual time or their sleep time, and bright light in the morning can correct this abnormality by advancing the phase position and reducing phase angles between the circadian rhythms and sleep-wake cycle.

In studies of masked body temperature rhythms in SAD patients, neither the baseline delays of phase position, nor advances produced by bright light therapy have been observed.[50,85,89,100-102] Under the constant routine, however, both the prior delay and the advance following morning bright light have been reported for such key markers of the circadian phase position as melatonin, body temperature and cortisol.[61,103] In another study using constant routine[54,104] the finding for two markers – melatonin and temperature – were not always positive, but, anyway, the circadian rhythm of body temperature was found to tend to be delayed in winter, and parameters of the rhythm were advanced after midday light exposure.

Studies of pretreatment melatonin phase and the subsequent shift following bright light treatment in winter depression have yielded inconsistent results. A phase delay of this rhythm in depressed SAD patients has been noted in a number of studies.[90,105-107] However, there are also many reports of a normal phase position.[54,70,89,108,109] Moreover, a phase advance of the rhythm following bright light treatment has been reported[45,70,90,95,103,106,110] but not unanimously.[54,89,107-109] Therefore, the significance of the pre-treatment phase delay and the post-treatment phase advance of the melatonin rhythm by morning light for the clinical response can be challenged in general.

As an alternative, Thompson et al. suggested that there may be instability of circadian rhythms in SAD that is mediated by a high-amplitude phase response curve (PRC), rather than a fixed phase abnormality.[70] Another explanation suggests that bright light does not shift the phase of circadian pacemaker, but shifts only the phase of the overt rhythms by simple strengthening of the pacemaker's effect on them.[111] Thus, the phase shift models proposed to explain the mechanisms of SAD and light therapy remains controversial.

4.2. Phase Delay and Hypersomnia

Hypersomnia and morning fatigue have been regarded as signs of the delayed circadian rhythms.[24,112,113] In a number of studies, the symptom of hypersomnia has shown to be a predictor of the clinical response to bright light therapy in SAD.[114-120] Besides, the marked correlation between difficulties in awakening and the severity of depression was reported.[121] In some study samples, however, the symptom of hypersomnia has not been very common, and SAD patients with hypersomnia had responded well to evening or nighttime bright light, although this does not result in advance of the circadian phase.[112]

Avery et al. assumed that a phase delay of circadian rhythms relative to sleep might explain why SAD patients experience hypersomnia.[61] Indeed, under conditions of constant routine which lead to the internal desynchronization, or mismatch, the minimum of core body temperature tend to be delayed relative to sleep onset, and subsequently the sleep duration is often longer than the usual.[122,123]

4.3. Comparison of Morning and Evening Light

The findings of possible differential response to bright light in the morning and evening hours are still inconclusive. Recently, the earlier findings of the superiority of morning light over evening light have been replicated by several groups.[124-126] However, no marked differences in the antidepressant response have been discovered in most studies using a parallel design with random assignment to time of morning and evening treatment.[48,86,107,127] In general, bright light has efficacy at most times of day (for review, see[41,112])

The meta-analysis of light treatment data[44] also did not provide evidence for phase-shifting hypothesis, because the antidepressant effect of a single pulse of light was found to be similar for morning, midday, and evening light, whereas antidepressant effect of the combination of morning-evening light regime was superior to a single pulse of light administered at other times of day.

There could be individual differences between patients with SAD in response to morning or evening light. Lewy et al. reported of a marked correlation between the decrease in symptom ratings and the degree of phase advance, and concluded that early morning bright light cannot be recommended for those patients with early morning awakenings and crushing evening tiredness.[128]

Although there is a positive association between the advanced phase position and antidepressant response in several studies,[45,127,129] there are also reports of a negative association.[61,102] Hence, the causal link between the phase shifts and antidepressant response in SAD still needs verification,[9,70,108,110,117,130,131] and finally elucidation of molecular mechanisms of action.

4.4. Order Effect

It was noted that evening light did not augment the following response to morning light, whereas morning light inhibited the following response to evening light.[41,132] Therefore, Rosenthal and Wehr suggested that there is a circadian variation in sensitivity to the antidepressant responses to light.[3] Morning bright light may advance the sensitive phase in a way that subsequently administered evening light may fall upon the inert portion of the PRC.

Although the typical PRC for mammals suggests that any response to light in the middle of the day is weak, the phase-shifting effect of daytime exposure to light was reported for healthy individuals.[133] Therefore, it seems that the human circadian pacemaker keeps up the sensitivity to light throughout the subjective day, and that midday bright light still able to advance the circadian phase. However, the positive findings of clinical efficacy of schedules of bright light varying day by day disagree with the hypothesis of compromised circadian clock function in SAD.[47-49]

Taking together, these results are arguing against the hypotheses postulating the important contribution of chronobiological mechanisms to the antidepressant action of bright light. The findings concerning the effect of bright light on the circadian phase position in SAD provide little evidence for the view that phase shifts are the key to the pathogenesis of winter depression or the efficacy of light therapy. Phase shifts, however, may play a partial role in the emergence of depressive symptoms in winter and remission following bright light treatment.

5. Physiology of Winter Depression

5.1. Physiological Responses to Bright Light

The physiological changes in SAD seem to be more variable than just alterations of circadian rhythms and sleep. The system of body temperature regulation might be also involved in the symptoms often seen in clinical picture of SAD, and, in some cases, the circadian regulation of sleep might be normal, but alterations in sleep related events, such as regulation of core body temperature, may influence the occurrence of disordered sleep.

In the depressed patients with SAD, the band-specific electroencephalogram of non-REM sleep resembles those who have been sleep deprived.[134] Among SAD patients, there might be normal production but slow elimination of circulating melatonin and, in addition to this, poor cooling activity of the brain during NREM sleep.[134] Whether these two phenomena are truly linked to one another is not clear.

The signs of improvements in several physiological systems were found to be associated with therapeutic benefits treatment of SAD with bright light (i.e.[20,127,135]) and physical exercise.[55] In the study of 4 physiological responses to light, namely, 1) advance of circadian phase, 2) increase in energy expenditure, 3) activation of sympatho-adrenal system, and 4) intensification of non-rapid eye movement sleep, each of these responses was found to be positively associated with remission of the depressive symptoms.[20,127] However, the results provided little evidence of strong association between clinical response and any single physiological response. Moreover, the association between these 4 physiological responses was either weak or insignificant. It was concluded that none of the responses plays a critical role in antidepressant action of light. In the case of favorable response to the treatment, several additive responses (multi-component physiological response) are observed.[127]

In the discussion of the study findings,[127] the notion was made that the results of correlational analysis could not prove that SAD is mainly produced by physiological abnormalities. It remains possible that the link between therapeutic and biological effects is not casual, or that the physiological changes are consequences rather than causes of the psychic disturbances. Thus, correlational evidence leaves open the question of causality, although, in general, the findings pointed to a certain pathogenic and therapeutic significance of the mechanisms of physiological regulation in SAD.

5.2. Seasonality of Neurovegetative Symptoms

The results epidemiological study of seasonal variations in depressive symptoms[20] were similar to the physiological results in that they show neither close associations between the responses of mood and physiological systems to changes in illumination, nor close interrelations between the responses of separate physiological systems. The seasonal variations in mood and well-being were synchronized with variations of ambient temperature averaged over 10-year period, but do not follow closely seasonal variations of photoperiod. Moreover, they significantly lag behind the variations of the majority of neuro-vegetative depressive symptoms (i.e. those reflecting such physiology as sleep length, sleep onset/termination, and metabolism), being synchronized only with arousal/energy symptoms.

Besides, it was found that annual curves of some other psychic symptom also can run ahead compared to mood and energy cycles and behind the photoperiodic cycle. For example, the curve for the symptom of social withdrawal reminds the curves of metabolic symptoms (i.e. the changes in weight, appetite, and alcohol consumption). Thus, although annual variation in the length of day seems to be one of very important seasonal timers for some depressive symptoms (i.e. such as problems with sleep onset/termination), the majority of most common SAD symptoms appear to be under more or less marked modulating influence of ambient temperature and, possibly, some other environmental factors.[20]

5.3. Physiological and Psychological Responses

In general, the study of seasonal variation of SAD symptoms suggest that seasonality of the phase and duration of melatonin secretion may have a role in the emergence of some but not all vegetative symptoms of SAD. Hence, photoperiod appears not to be the only important determinant for seasonal variations in mental and vegetative symptoms. It is likely that some symptoms oscillate more in phase with ambient temperature rather than with photoperiod. Further, these symptoms may originate from seasonal deviation in systems that are not involved in the photoperiodic and circadian time measurement.[17]

Since both the photoperiod and ambient temperature may independently trigger changes in several systems of physiological regulation, this may partly explain the contradicting results of the pathogenesis of SAD. Most experiments have been designed to test only one physiological abnormality, while the action of bright light may be mediated not in a single, but in several different ways, such as suppression of melatonin secretion, advance of the circadian phase position, increase in non-REM pressure, elevation of brain temperature, increase in metabolic rate and adrenal activity, etc. Each of these changes may not be observable in all SAD patients.

6. Psychology of Winter Depression

6.1. Placebo Effect

Because winter SAD patients associate their depression with a dark season and related neuro-vegetative problems, the positive expectations for beneficial action of bright light appear to be much stronger for this form of depression compared to non-SAD depression. Although the parallelism was reported in dynamics of mood state and physiological well-being, not all changes are running in synch, and, besides, there is no solid evidence of causal role of physiological disturbances in development of depression. Additionally, physiological influence cannot explain several facts reported by investigators of the efficacy of bright light as antidepressant (i.e. such as higher efficacy of the open bright light trials compared to efficacy of pharmacological antidepressants in trials designed as double blind cross-over comparison of placebo with active treatment). It might be suggested that the contribution of patients' physiology in development of winter depression and effects of its antidepressant treatment is of less importance compared to placebo response.

Placebo response remains the main challenge for the investigational bright light trails, because patients cannot be "blind" to such interventions as bright light therapy, physical exercise treatment and sleep deprivation. The placebo effect is hardly controlled even in comparisons of presumably less and more effective treatments due to their visibility (i.e. bright vs. dim light).

Moreover, it is impossible to minimize or control the placebo effect even when active and placebo treatments are undistinguishable. The placebo patients are not really left untreated. For example, in the study of Putilov

et al.[17] half of SAD group was blindly assigned to placebo after sleep deprivation, while another half was on melatonin treatment. Unlike melatonin, the
placebo substance was pharmacologically inert. However, it had symbolic
value for those depressed subjects who hoped to improve. Additionally, being
the participants of sleep deprivation and other research procedures, the
patients received information on their symptoms and possibility to verbalize
their problems. They also obtained psychological support from physicians,
research assistants and other participated subjects. Thus, the mood stabiliz-
ing effect in both placebo and melatonin groups may be attributed to the psy-
chotherapeutic action of the experimental trails, and we have to be cautious
with interpretation of clinical benefits of such novel treatments as melatonin
or bright light.

6.2. Expectations for Bright Light

Meta-analysis literature on any antidepressant treatment shows that in the
vast majority of investigational trials the depressed patients assigned to
placebo exhibit substantial symptom reduction (i.e.[136]). It is known that placebo
response is dose-dependent (i.e. see examples in[137]). Many results on comparison of light treatment effects point on the importance of pre-treatment
expectations as modulators of antidepressant response to different non-drug
treatment in "placebo-counting" manner.

Both in seasonal and non-seasonal depression bright light produce
faster antidepressant benefits than pharmacological treatment (see
review[138]). This may be considered as an argument for importance of therapeutic expectations in both patients and researches. Bright light trials
seem to produce more positive expectations compared to the pharmacological trials with implementation of the randomized, double-blind,
placebo-controlled design. In other words, light therapy can provide a better antidepressant response compared to active or placebo pills, because it
is easy for patients to recognize what is, in accordance to their beliefs, a
real antidepressant and what is a placebo or weaker antidepressant (i.e.
treatment that is weaker due to lower intensity or shorter duration or inappropriate timing).

6.3. Expectations for Light from Lamp and Visor

The findings of crossover comparison of the effects of different light timing
and intensity on mood in winter depression suggest that patients' and
researchers expectations are important modulators of the treatment outcomes. They point on correlation between initial expectancy score and magnitude of the antidepressant response. The pre-treatment expectations in
general were found to be rather high. Moreover, for the benefits from morning light they were higher than for those of evening light (i.e.[106]). Similarly,
the initial expectancy scores for both dim and bright light treatments were

rather high, and, additionally, bright light was expected to be more effective than dim light (i.e.[2]). In agreement with the pattern of patients' expectations, these studies showed the superiority of morning light over evening light and superiority of bright light over dim light, respectively.

In contrast to the studies with light emitted by lamps[41] when it is rather easy for a patient to make conscious or unconscious prediction on which of the treatments would be most effective, in the studies of the antidepressant effects of visor the prediction seems not to be so simple. It might be expected that both bright and dim light or green and red light are similarly effective due to closeness of light source to the eyes. This uncertainty appears to complicate distinguishing between different lights in "placebo-counting" manner. The evaluations of initial expectancy scores in the visor studies indicated that, indeed, patients had the same expectations for therapeutic effects of proposed active and placebo treatments (i.e.[139]). Again, the clinical outcome in the studies with head-mounted devices agreed with the pattern of patients' expectations: no relationship to intensity, color or duration was found.

6.4. Expectation and Study Design

The literature on timing of light treatment for SAD also shows that the answer on the question on importance of the treatment timing depends upon the design of clinical trial. For example, rather strong and similar beneficial responses to morning and evening light were found in those studies where patients were not cross over for another treatment (i.e.[49,107,127,140]).

By contrast, the preferential response for morning light was mostly found in those trials where thoughts were concerned by a cross-over design aimed at comparing light exposure in the morning with that in the evening (i.e.[42,61,95,106,141,142]).

6.5. Expectations and Order Effect

In the trials applying a parallel design, inter-individual differences in therapeutic response embarrass the statistical comparison of the treatment groups. However, the use of a cross-over design for bright light investigational trials is also limited, because the reports suggest a substantial order effect (the studies of morning and evening bright light in SAD were reviewed by[41]).

As it has been noted[132] the meta-analysis based on results of several groups, suggests that superiority of morning light in the studies with cross-over design was noted when morning light is given first and evening light is given second. The evening light, when given first, is comparable to morning light given first. Similarly, red light was almost as powerful antidepressant as green light when it was the first treatment, while green light given first was superior over second treatment with red light.[143] Besides, the superiority of morning bright light was not seen when the treatment consisted of two days of morning plus two days of evening bright light and vice versa.[140]

Possibly, the placebo effect of any first treatment mimics the differences between more and less optimal treatment (if, of cause, they are really exist). The success of the first treatment with morning or green light would encourage the patients to make an unconscious conclusion that the first effect differ from the effect of the following, supposedly less effective, treatment. If the first treatment is evening or red light, they also expected to be effective, and no reason to belief that following morning or green light is worse than the first treatment. Again, the reason for the difference between morning and evening light or between green and red light would arise mostly from the differences in the pre-treatment expectations. The first treatment with morning or green light provides patient with knowledge of their high efficacy. After this, the second treatment is evaluated more critically to support pre-treatment expectations for superiority of morning or green light over evening or red light.

6.6. Expectations of Researchers and Patients

It is very natural for the researchers and physicians unconsciously stimulate a placebo response in the patients when their own believes in the treatment efficacy coincide with patients' hopes. One of contrast examples are the results of Eastman and Siberian groups due to difference between them on the attention they paid to the placebo issue. The Eastman was most concerned by the placebo action of bright light.[137] In the investigations of her group the extremely careful attention was devoted to controlling placebo effects in assessing treatment benefits of morning and evening lights. Several years of regular trials passed before this group finally found the significant benefit of morning bright light.[102,124] However, compared to this benefit, much bigger and faster improvements were reported by the majority of other (as we can guess, less skeptical) investigators.

Unlike the Eastman group, most of the Siberian studies[20,50,55,86,127,144] did not focus on the differences between clinical effects of active and placebo treatments. This very dubious for patients' mind issue was not mentioned in the treatment consents. Instead, the main question of the investigation was whether the objective physiological measures correlate with treatment response, whether this correlation depends upon treatment timing, etc. (i.e. [20,127]). Consequently, the antidepressant responses in Siberia were found to be among the highest.

6.7. Effect of Bright light for Non-Seasonal Depression

In contrast to consensus on the usefulness of bright light for winter depression, its benefits for patients with non-seasonal depression are not generally accepted (see[39,138]). There were reports of both positive[145-147] and negative findings[148,149] as well as the reports indicating that the clinical response in non-seasonally depressed patients is modest compared to that in patients with winter depression.[55,108,150] The differences between the seasonally and non-seasonally affected may be attributed, at least in part, to differences between these diagnostic groups in heterogeneity of clinical symptoms, rate of spontaneous remission, and vulnerability to have comorbid psychic and somatic illnesses. However, another important reason would be the differences in psychological aspects of the responses to non-drug treatments in SAD and non-SAD.

6.8. Expectations in Seasonal and Non-Seasonal Depression

It is likely that the positive expectations for antidepressant action of bright light are much stronger in seasonal depressives who, unlike non-seasonal depressives, associate their depression with winter season, and, therefore, light might have a symbolic value for SAD. Indeed, data of all trials are agreed in that SAD patients exhibit better response to bright light compared to those with non-seasonal mood disorder.

Bright light treatment contrasts with sleep deprivation and physical exercise in this respect. It is not reasonable to suggest a dramatic difference in expectations for the effects of these treatments in SAD and non-SAD. Indeed, the comparison of the effects of bright light with the effects of physical exercise or sleep deprivation support this suggestion. It was found that sleep deprivation was equally effective for SAD and non-seasonal mood disorder.[21] Besides, physical exercise was also found to be an effective treatment for both SAD and non-SAD patients. The exercise treatments differed from bright light treatments in that only former showed rather high efficacy for both SAD and non-SAD.[55] Nevertheless, the results indicate that, in general, non-SAD patients tended to respond worse to any treatment compared to SAD patients.

6.9. SAD Personality

Again, the fact that patients with SAD benefit more from non-drug therapy than those with non-seasonal mood disorder might be, at least partly, explained by the difference between them in the magnitude of placebo response. The published reports of the double-blind pharmacological studies support this explanation, because they suggest very high responsiveness to drug placebo in SAD patients (i.e.[151]).

The comparison of SAD patients with patients without seasonality on the personality characteristics[152] revealed that they significantly differ on only one of five dimensions of the five-factor model of personality. Namely, SAD patients scored higher than non-SAD patients on the openness dimension. Based on these results, SAD was suggested to be a psychologically distinct subgroup of depressed patients. They seem to be more imaginative, more emotionally sensitive and they more likely to entertain unconventional ideas.[152] It is reasonable to suggest that this personality trait may account, at least in part, for higher responsiveness of the individuals with SAD to innovative light treatments. Such a treatment provides a novel and interactive experience that increases positive pre-treatment expectations and stimulates a strong placebo effect.

6.10. Some Other Pieces of Evidence for Role of Expectations

The investigators of light treatment effects for which the contribution of psychological rather than chronophysiological factors in antidepressant action might be considered as the best explanation collected a set of paradoxical facts. Several of them are listed below.

It was noted (i.e.[153]) that the extent of reduction of atypical depressive symptoms after the first hour of treatment predicts the final therapeutic outcome. It is unlikely that such a short exposure to bright light may produce immediate effect on mood via alteration of the parameters of physiological rhythms. More plausible explanation of the rapid antidepressant response to light is the impact of positive pre-treatment expectations in the therapeutic outcome.

The placebo explanation looks reasonable for paradoxical fact reported in the study by Richter et al., who compared the effects of real light treatment and imaginary light treatment (i.e. during the hypnotic session patients were made to imagine that they perceived bright light).[154] No statistical differences in treatment outcomes were found. Such a difference was detected only at day 10 after the treatment: the effects of the imaginary light were gone, while the effects of real light treatment were still observed.[154]

The conflicting data on the duration of relapse after successful light treatment is another paradoxical fact of light treatment studies. Some researchers stated that relapse occurs within 3 to 4 day after withdrawal (i.e.[155,156]), while other groups observed remission that lasted longer[157]. Besides, solid data were provided on the increase in response percentages at 10th day after treatment compared to the 3rd day.[49] These discrepancies resemble the above-mentioned contradicting reports on treatment outcomes for bright light. To our minds, they argue against the suggestions of a certain strong physiological background for the antidepressant response.

In general, the remission rate for bright light is rather high compared to that reported in pharmacological studies. The most reasonable explanation for this is a high level of therapeutic hopes, especially in a study with parallel design that, albeit a cross-over controlled trial, does not include a placebo or, at least, less active treatment for a given patient. This conclusion is complemented by the reports on close correlation between positive pre-treatment expectations and beneficial antidepressant action of bright light for winter depression. Thus, placebo effect seems to account for a large portion of the therapeutic response.

7. Explanation of Winter Depression and its Treatment with Light

7.1. Evolutionary Psychology of Depression

The facts suggesting that the efficacy of elaborative bright light treatments might be to a large extent attributed to the placebo response agree with the hypotheses that conceptualize depression as an evolved psychological feature. In brief, one of the particular explanations (i.e.[16,21] is based on idea that SAD, as any other depression, might be mainly considered as a general emotional response to negative psychosocial factors. However, the seasonal form of depression has some specific features that require a more complex explanation. In particular, it could results from the combined action of negative psychosocial factors, from one hand, and seasonal variations in physical environment, from another hand. However, the former could be of the most importance than the latter. The effects of psychosocial and physical environments would meet on the level of biochemical mechanisms, because they would be either the same or related or interact in additive manner to produce similar brain reactions in response to these distinct external factors

In more detail, it is known[158] that different situations people living in different cultures all show very similar facial, gesture, postural, and physiological signs of pain and sadness. The behavior signals of bad feeling, such as physical and psychic pain, grief, despair, and sadness, are universal (cause- and culture-independent). They remind those of infants and children. The response of the members of kin group to these nonspecific signals of suffering pain and sadness is similar to a parental response. This response is usually aimed on termination of the painful stimulus or, if it is not possible, on consoling, comforting and giving help and support.[158]

More specifically, the emotions as sadness, grief and despair might be understood as the evolved non-verbal signals aimed on alarming social environment about subjective experience of considerable lose, failure or other events associated with decrease or threat of decrease in inclusive fitness. The mental state that we call "depression" might be conceptualized as a response on insufficiency of feedback signals on this signal from the kin group. In other words, the suffering person, exposing his/her emotion, unconsciously expects appropriate reaction from his/her close social surrounding (i.e. extended family). The development of full-blown depression might be the secondary response on luck of unconsciously recognized signs of support and protection along with calming, consoling, expression of empathy, sympathy, etc.

The disruption of extended family in modern industrial societies led to deficiency of psychosocial mechanisms that evolved to protect the social individuals from depression. As a result, different forms of depression increase in their rate in these societies compared to the traditional societies or the same societies just several decades ago. The psychosocial components of modern therapy sometimes partly counteract to the effects of this deficiency. These are mostly the components that we call "placebo effect".[21] Thus, the evolutionary psychological explanations predict that psychosocial risk factors, especially those related to social support, play the critical role in manifestation of depression. In particular, findings on postpartum depression[159]) fit well in this prediction: this condition seems to be closely associated with poor social support.

7.2. Psychosocial Risk Factors

The hypothesis suggests that seasonal depression is not an exception in the respect of its dependency from social environment. However, up to now, SAD has been understood as condition that develops in response to periodical changes of physical environment, and, therefore, the influence of these periodicities on bodily functions has been considered to be the primary and most important cause of seasonality of mood. Following this paradigm, the researches have not focused their studies upon non-periodic psychosocial risk factors as possible causes of seasonal depression.

Only recently, the contribution of psychosocial risk factors in SAD has been tested and confirmed to be significant. These reports provide evidence for association of poor social support with seasonality of mood[160,161] or with diagnosis of winter type of SAD.[21,162-164]

For biological theories of seasonal depression, the association of mood seasonality with poor social support is unpredicted finding, but it is in agreement with evolutionary psychological explanations of depression.

7.3. Patient-Physician Pretreatment Interactions

Some facts collected by light treatment researches provide further support for these explanations. For example, Geerts et al. reported that non-verbal behavior and communication between physician and patient during pretreatment interview has a predictable value for future therapeutic benefits of bright light therapy.[165]

In another study[166] a high-pitched voice with small variation in this pitch was found to be a predictor of benefits from light therapy. It was suggested that bright light treatment gives extra comfort in "tense" patients, who become rapid responders to this treatment. Thus, psychosocial component of physical treatments for SAD and non-SAD seems to play an import role in clinical response and, therefore, it requires further elaboration.

7.4. Psychological Correlates of Seasonality

Seasonal psychosocial risk factors are excluded from the diagnostic of SAD. Therefore, it is necessary to explain how non-periodical psychosocial factors could provoke such seasonally manifesting form of affective illnesses as winter depression.

Sometimes, the contribution of psychological factors in seasonality of depression might be unconsciously hidden by a patient. In particular, it was found[167] that the depressive episodes could be anniversary reactions associated with intense traumatic experiences in childhood, adolescence or adulthood. The time and sometimes the place of the traumatic event acted as triggers eliciting the clinical symptoms. Thus, anniversary reactions may constitute a subgroup of seasonal mood disorders, which are precipitated primarily by psychological factors rather than climatic conditions. Besides, patients might unconsciously replace the real but traumatic psychosocial cause of their depression by more neutral but less traumatic physical cause, such as winter darkness.

7.5. *Neurochemical Correlates of Seasonality*

Most cases of SAD, however, can not be explained by seasonality of psychologically traumatic events, because, unlike statistics for other mental disorders, hospital admission statistics for depression and mania that shows significant seasonal patterns with the incidence of depression being highest in winter. The suggestion for most cases would be that the brain biochemistry is influenced in additive way by negative social information, from one hand, and season of the year, from another.[21]

Negative social input associated with unpleasant feelings, such as sadness, loneliness, luck of interest, withdrawal from participation in normal activity and the like, directly alter brain neurotransmitter regulation. Additive seasonal change in these neurotransmitter systems in the same direction can become a triggering event for full-blown depression, although these changes as such do not exceed normal or subclinical range.

7.6. *Serotonergic System and Social Interactions*

Despite possible positive relation between serotonergically induced changes in mood, from one hand, and serotonergically induced changes in chrono-physilogical functioning, from another, they need not be causally related. It is likely that mood disturbances are mostly induced by direct effect of psychosocial factors on brain biochemistry.

The idea that social information can influence the brain functioning is not new. The studies of neurotransmission in monkeys[168,169] might serve as a well-studied example. They showed that central (brain) serotonin sensitivity and peripheral serotonin levels are positively related to social status and prosocial behavior (i.e. grooming). These indexes of serotonergic neurotransmission directly react on alterations in social status. They are dropping down with its lowering and rising up with the opposite change. The number of submissive displays from low-status monkeys serves as social information that primarily responsible for initiating such biochemical changes.

7.7. Serotonergic System and Circadian Physiology

Among reported seasonal fluctuations of the major monoamine transmitters implicated in mood disorders (i.e., serotonin, dopamine and norepinephrine), the magnitude of changes seems to be the greatest for the serotonergic system (i.e.[170]). This fact provided a rationale for hypotheses considering serotonergic dysfunction as the major cause of SAD. Hence, the important progress has been made in a search for evidence of involvement of this neurotransmitter in winter depression (see, i.e.[4]).

Moreover, recent findings have highlighted important relations between serotonin and regulation of circadian rhythms and sleep-wake cycle. Serotonergic system participates in modulation of photic and non-photic responses of the SCN (i.e.[171,172]). In particular, serotonergic projections from the midbrain raphe nuclei to the SCN regulate the photic entrainment of circadian clocks (i.e.[173,174]). It was hypothesized that some seasonal alterations of circadian phase and its response to bright light in SAD may be secondary to impaired serotonergic function in the afferent pathways to the SCN.[70]

7.8. Serotonergic System and Sleep

Earlier, we suggested[8] that the reported seasonal differences in EEG sleep between winter depressed and controls might be understood as mediated by the mechanisms of serotonergic regulation of sleep. Such an explanation predicts that mood state and sleep regulation in winter depression might be linked, because they both are related to the same neurochemical mechanism of action (i.e. serotonergic neurotransmission).

Recent findings highlighted important relations between serotonin and regulation of circadian rhythms and sleep-wake cycle. Serotonergic pathways are likely involved in the SCN projections to effector systems controlling sleep and wake functions. In particular, the data suggest that interleukin-1-induced enhancement of non-rapid eye movement sleep is mediated, in part, by the serotonergic system (i.e.[175]), and that certain serotonin receptors subtypes are implicated in regulation of slow-wave sleep.[176]) Besides, serotenergic regulatory mechanism appears to contribute not only to the regulation of the sleep process, but also, in an opposite manner, to regulation of the waking process.[177] Thus, serotonergic system can contribute to the observed alteration of sleep structure in patients with major depression[19]) and SAD.[8,20,127]

7.9. Social Interactions and Sleep

Despite possible involvement of the same neurotransmitter system (or systems) in seasonal changes of both mood and sleep, the latter is unlikely to be the primary cause of the former. By contrast, the possibility of the opposite causal relationship is not excluded. Low brain serotonergic activity in wintertime could trigger depression in the presence of negative psychosocial factors, although in the absence of these factors, it still could not disrupt the regulation of the sleep-wake cycle. The disruption, however, could become possible in response to the cumulative effect of low brain serotonergic activity over the winter, from one hand, and social input associated with depressogenic life events, from another.

The evidence for influence of psychosocial factors on sleep regulation has been, in particular, provided by Meerlo et al in their experiment with slow-wave sleep regulation in mice.[178] The results of this study supported the notion that slow-wave sleep pressure and subsequent non-REM sleep intensity not only depend on the duration of prior wakefulness but also on the experience during waking (i.e. social defeat stress was found to accelerate the build up of sleep debt).

Thus, the effects of psychosocial and physical environments on mood and physiology, respectively, would meet on the level of biochemical mechanisms in winter depression (i.e. they would be either the same or interrelated or interact in additive manner to produce similar brain reactions in response to these distinct environmental factors). Additionally, the alteration of patients' psychic state might accelerate the changes in their physiology (i.e. sleep-wake pattern). In turn, change in the function of circadian pacemaker might have the same direction effect on mental functions and health. In general, however, the extent of similarity of the primary biochemical mechanisms underlying variations in depression symptom and sleep regulation in SAD cannot be clearly understood right now and it requires future research.

7.10. Dualistic Hypotheses

A dual-vulnerability hypothesis has been proposed in which SAD is suggested to result from separate depression and seasonality factors, each of which may have different pathophysiological mechanisms.[179,180] Another "dualistic" explanation[163] suggests that, by contrast, the primary biochemical mechanism underlying depression and seasonality might be the same (i.e. related to serotonergic system), whereas major environmental causes of depression and seasonality seem to be different (i.e. non-periodical psychosocial situation and seasonal changes in physical factors).

The antidepressant response to bright light therapy was suggested to results mostly from interaction between mind and psychosocially significant components of treatment procedure. By contrast, chronophysiological response is mostly caused by physical components of this procedure. Both responses, however, might be mediated by the same biochemical mechanism (i.e. serotonergic regulation). Besides, the improvement might accelerate normalization of functions of the circadian clock, and this in turn might have a stimulating effect on mood. However, these interactions appear to be of less importance compared to the direct effect of psychosocially significant components of the treatment on brain biochemistry.[163]

In other words, two separate environmental inputs (psychosocial risk factors and abiotic periodicities, respectively) and two mostly separate outputs (mood and somatic changes, respectively) might be related to the same brain processes (i.e. serotonin neurotransmission). These psychic and somatic responses would not to be clear separated on the level of brain biochemistry. Besides, in the certain limits, two separate outputs could interact in mutually accelerating way on higher (i.e. physiological and psychosocial) levels.

7.11. Summary on the Explanation of the Light Treatment Effects

In sum, it is uneasy to find out which of treatment effects - physiological or psychological – mainly contributes to clinical response to bright light. Most likely, the main contributor is patients' psychology rather than their physiology. The regulation of brain neurotransmitters is the first candidate to the biological background for both chronophysiology and psycho-sociology of winter depression. This means that any pure biological theory that stresses the disturbances in circadian rhythm, sleep regulation or neurotransmitter systems is too limited for theoretical understanding of SAD. There is, therefore, a necessity to include psychosocial aspects in the theoretical framework of seasonal depression and light therapy.

Bright light applied in the morning or midday or afternoon or evening markedly reverses the symptoms of SAD. In other words, the clinical response is achieved irrespective of the time of the treatment and proposed phase shifting effect. Other "visible" treatments, such as physical exercise in the middle of the day, are also effective treatment for SAD. Besides, any treatment including placebo, being combined with total night sleep deprivation, prevents relapse for, at least, one treatment week in seasonal depression.[21] The finding, in general, suggests that a contribution of placebo effect to the antidepressant response to a non-pharmacological treatment seems to be both practically and theoretically important issue. Although certain physiological effects of non-drug therapy might be significant for clinical improvement, they seem not to be critical. The depressed patients would improve preferentially to a particular treatment, since it causes a better feeling of the expected normalizing changes in physiological functioning than other treatments. However, psychological factors, such as patients' positive pretreatment expectations and researches' enthusiasm appear to contribute more considerably to antidepressant response. They can explain, in particular, why a non-pharmacological intervention is a more beneficial therapy compared to the pharmacological treatments in the studies with double blind cross-over design.

To understand better the mechanism of mood and sleep regulation in SAD, this depression might be considered in evolutionary perspective - as an evolved feature of general emotional response to negative psychosocial factors. It could results from the combined action of these psychosocial factors, from one hand,

and seasonal variations in physical environment, from another hand, with former being of the most importance. The effects of psychosocial and physical environments would meet on the level of biochemical mechanisms. They would be either the same or related or interact in additive manner to produce similar brain reactions in response to these distinct external factors.

8. References

1. A. J. Lewy, H. A. Kern, N. E. Rosenthal, and T. A. Wehr, Bright artificial light treatment of a manic-depressive patient with a seasonal mood cycle, *Am J Psychiat*.139, 1496-1498 (1982).
2. N. E. Rosenthal, D. A. Sack, J. C. Gillin et al, Seasonal affective disorder: a description of the syndrome and preliminary findings with light therapy, *Arch Gen Psychiatr*.41, 72-80 (1984).
3. N. E. Rosenthal and T. A. Wehr, Towards understanding the mechanism of action of light in seasonal affective disorder, *Pharmacopsychiat*.25, 56-60 (1992).
4. R. W. Lam and R. D. Levitan, Pathophysiology of seasonal affective disorder: a review, *J Psychiatry Neurosci*.25, V469-V480 (2000).
5. R. M. Benca, W. H. Obermeyer, R. A. Thisted, and J. C. Gillin, Sleep and Psychiatric Disorders, *Arch Gen Psychiatry*.49, 651-668 (1992).
6. J. L. Anderson, L. N. Rosen, W. B. Mendelson et al, Sleep in fall/winter seasonal affective disorder: effects of light and changing seasons, *J Psychosom Res*.38, 323-337 (1994).
7. D. P. Brunner, K. Krauchi, D.-J. Dijk, G. Leonhardt, H.-J. Haug, and A. Wirz-Justice, Sleep electroencephalogram in seasonal affective disorder and in control women: effects of midday light treatment and sleep deprivation, *J Biol Psychiatry*.40, 485-496 (1996).
8. V. E. Palchikov, D. Y. Zolotarev, K. V. Danilenko, and A. A. Putilov, Effects of season and of bright light administered at different times of day on sleep EEG and mood in patients with seasonal affective disorder, *Biol Rhythm Res*.28, 166-184 (1997).
9. K. M. Koorengevel, D. G. M. Beersma, J. A. Den Boer, and R. H. Van den Hoofdakker, A forced desynchrony study of circadian pacemaker characteristics in seasonal affective disorder, *J Biol Rhythms*.17, 463-475 (2002).
10. K. M. Koorengevel, D. G. Beersma, J. A. Den Boer, and R. H. Van den Hoofdakker, Sleep in seasonal affective disorder patients in forced desynchrony: an explorative study, *J Sleep Res*.11, 347-356 (2002).
11. T. Partonen, B. Appelberg, and M. Partonen, Effects of light treatment on sleep structure in seasonal affective disorder, *Euro Arch Psychiat Clin Neurosci*.242, 310-313 (1993).
12. M. Kohsaka, H. Honma, N. Kukuda, R. Kobayashi, and K. Honma, Does bright light change sleep structures in seasonal affective disorder?, *Soc Light Treatment Biol Rhythms, 6th Meet Abstr.*, 32 (1994).
13. T. Endo, Morning bright light effects on circadian rhythms and sleep structure of SAD, *Jikeikai Med J*.40, 295-307 (1993).
14. C. Cajochen, D. P. Brunner, K. Krauchi, P. Graw, and A. Wirz-Justice, EEG and subjective sleepiness during extended wakefulness in seasonal affective disorder: circadian and homeostatic influences, *Biol Psychiatry*.47, 610-617 (2000).

15. A. A. Putilov, K. V. Danilenko, V. E. Palchikov, and S. M. Schergin, The multiple sleep latency test in seasonal affective disorder: No evidence for increased sleep propensity, *Soc Light Treatment Biol Rhythms, Abstr, 1995*.7, 30 (1995).

16. A. A. Putilov, Chronobiology and evolutionary psychology of seasonal depression. Evolution, Behavior and Society: Human Ethology. Abstracts of papers presented at the 2nd Summer School (Pushchino, 20th June-7th July, 2002), 36-37 (2002).

17. A. A. Putilov, O. G. Donskaya, O. A. Jafarova et al, Waking EEG power density in hypersomnic winter depression. Daytime and nighttime naps in winter depression: association with chronophysiological and psychiatric ratings, *Soc Light Treatment Biol Rhythms, Abstract*.12, 24 (2000).

18. D. E. Giles, H. P. Roffwarg, A. J. Rush, and D. S. Guzik, Age-adjusted threshold values for reduced REM latency in unipolar depression using ROC analysis, *Biol Psychiatry*.27, 841-853 (1990).

19. K. L. Benson, K. F. Kaull, and V. P. J. Zarcone, The effects of age and serotonergic activity on slow wave sleep in depressive illness, *Biol Psychiatry*.33, 842-844 (1994).

20. A. A. Putilov, G. S. Russkikh, and K. V. Danilenko, Phase of melatonin rhythm in winter depression, *Adv Exp Med Biol 1999*.460, 441-458 (1999).

21. A. A. Putilov, Depressive symptoms as exaggerated signs of physiological and behavior changes in response to extremes of physical environment, *Int J Behavioral Med*.11 (September), 281-282 (2004).

22. C. M. Shapiro, G. M. Devins, B. Feldman, and A. J. Levitt, Is hypersomnolence a feature of seasonal affective disorder?, *J Psychosom Res*.38 (suppl. 1), 49-54 (1994).

23. J. Arendt, Melatonin, circadian rhythms and sleep, *N Eng J Med*.34, 1114-1116 (2000).

24. A. J. Lewy and R. L. Sack, Light therapy and psychiatry, *Proc Soc Exp Biol Med*.183, 11-18 (1986).

25. K. Suhail and R. Cochrane, Seasonal variations in hospital admissions for affective disorders by gender and ethnicity, *Soc Psychiatry Psychiatr Epidemiol 1998*.33, 211-217 (1998).

26. T. A. Wehr, N. E. Rosenthal, and D. A. Sack, Environmental and behavioral influences on affective illness, *Acta Psychiatr Scand*.(Suppl. 77), 44-52 (1988).

27. G. L. Faedda, L. Tondo, M. H. Teicher, R. J. Baldessarini, H. A. Gelbard, and G. F. Floris, Seasonal mood disorders: Patterns of seasonal recurrence in mania and depression, *Arch Gen Psychiatry*.50, 17-23 (1993).

28. J. M. Booker and C. J. Hellekson, Prevalence of seasonal affective disorder in Alaska, *Am J Psychiatry*.149, 1176-1182 (1992).

29. L. A. Palinkas, M. Cravalho, and D. Browner, Seasonal variation of depressive symptoms in Antarctica, *Acta Psychiatr Scand*.91, 423-429 (1995).

30. P. A. Madden, A. C. Heath, N. E. Rosenthal, and N. G. Martin, Seasonal changes in mood and behavior. The role of genetic factors, *Arch Gen Psychiatry*.53, 47-55 (1996).

31. J. Axelsson, J. G. Stefannson, A. Magnusson, H. Sigvaldason, and M. M. Karlsson, Seasonal affective disorders: relevance of icelanding and icelandic-canadian evidence to etiologic hypotheses, *Can J Psychiatry*.47, 153-158 (2002).

32. C. Johansson, M. Willeit, C. Smedh et al, Circadian clock-related polymorphism in seasonal affective disorder and their relevance to diurnal preference, *Neuropsychopharmacology*.28, 734-739 (2003).

33. T. A. Wehr and N. E. Rosenthal, Seasonality and affective illness, *Am J Psychiatry*.146, 829-839 (1989).

34. T. Partonen and J. Lonnqvist, Seasonal affective disorder, *Lancet*.352, 1369-1374 (1998).

35. M. C. Blehar and N. E. Rosenthal, Seasonal affective disorders and phototherapy. Report of a National Institute of Mental Health-sponsored workshop, *Arch Gen Psychiatry*.46, 469-474 (1989).

36. A. Wirz-Justice, P. Graw, K. Krauchi et al, Natural light treatment of seasonal affective disorder, *J Affective Dis*.37, 109-120 (1996).

37. D. F. Kripke, Photoperiodic mechanisms for depression and its treatment,0, 1248-1252 (1981).

38. J. A. Elliott and B. D. Goldman, Seasonal reproduction: photoperiodism and biological clocks,0, 377-423 (1981).

39. R. W. Lam, D. F. Kripke, and J. C. Gillin, Phototherapy for depressive disorders: a review, *Can J Psychiatry*.34, 140-147 (1989).

40. R. W. Lam, M. Terman, and A. Wirz-Justice, Light therapy for depressive disorders: Indications and efficacy,0, 215-234 (1997).

41. M. Terman, J. S. Terman, F. M. Quitkin, P. J. McGrath, J. W. Stewart, and B. Rafferty, Light therapy for seasonal affective disorder: a review of efficacy, *Neuropsychopharmacol*.2, 1-22 (1989).

42. M. Terman, J. S. Terman, and B. W. Williams, Seasonal affective disorder and its treatments, *J Prac Psychiatry Behav Health*.5, 287-303 (1998).

43. E. M. Tam, R. W. Lam, and A. J. Levitt, Treatment of seasonal affective disorder: a review, *Can J Psychiatry 1995*.40, 457-466 (1995).

44. T. M. Lee, C. A. Blashko, H. L. Janzen, J. G. Paterson, and C. C. Chan, Pathophysiological mechanism of seasonal affective disorder, *J Affect Disord*.46, 25-38 (1997).

45. A. J. Lewy, R. L. Sack, C. M. Singer, D. M. White, and T. M. Hoban, Winter depression and the phase-shift hypothesis for bright light/Es therapeutic effects: History, theory, and experimental evidence, *J Biol Rhythms*.3, 121-134 (1988).

46. S. P. James, T. A. Wehr, D. A. Sack, B. L. Parry, and N. E. Rosenthal, Treatment of seasonal affective disorder with light in the evening, *Br J Psychiatry*.147, 424-428 (1985).

47. J. L. Anderson, R. G. Vasile, K. L. Bloomingdale, and J. J. Schildkraut, SAD: Varied schedule phototherapy and catecholamines, *Amer Psychiatr Assoc 141st Annual Meet Program and Abs.*, 166 (1988).

48. B. Lafer, G. S. Sachs, L. A. Labbate, A. Thibault, and J. F. Rosenbaum, Phototherapy for seasonal affective disorder: a blind comparison of three different schedules, *Am J Psychiatry*.151, 1081-1083 (1994).

49. Y. Meesters, The timing of light therapy and response assessment in winter depression, *Acta Neuropsychiatrica*.7, 61-63 (1995).

50. A. A. Putilov, K. V. Danilenko, G. S. Russkikh, and L. K. Duffy, Phase typing of patients with seasonal affective disorder: a test for the phase shift hypothesis, *Biol Rhythm Res*.27, 431-451 (1996).

51. T. A. Wehr, D. A. Sack, F. Jacobsen, L. Tamarkin, J. Arendt, and N. E. Rosenthal, Phototherapy of seasonal affective disorder: time of the day and suppression of melatonin are not critical for antidepressant effects, *Arch Gen Psychiatry*.145, 52-56 (1986).

580   A. A. Putilov, S. R. Pandi-Perumal and T. Partonen

52. F. M. Jacobsen, E. A. Mueller, R. G. Skwerer, D. A. Sack, and N. E. Rosenthal, Morning versus midday phototherapy of seasonal affective disorder, *Am J Psychiatry*.144, 1301-1305 (1987).
53. G. Isaacs, D. S. Stainer, T. E. Sensky, S. Moor, and C. Thompson, Phototherapy and its mechanism of action in seasonal affective disorder, *J Affect Disord*.14, 13-19 (1988).
54. A. Wirz-Justice, K. Krauchi, P. Graw et al, Circadian rhythms of core body temperature and salivary melatonin in winter SAD before and after midday light, *Soc Light Treatment Biol Rhythms 6th Meet Abs.*, 12 (1994).
55. B. B. Pinchasov, A. M. Shurgaja, O. V. Grischin, and A. A. Putilov, Mood and energy regulation in seasonal and non-seasonal depression before and after midday treatment with physical exercise or bright light, *Psychiatry Res*.94, 29-42 (2000).
56. N. E. Rosenthal, D. A. Sack, F. M. Jacobsen et al, Melatonin in seasonal affective disorder and phototherapy, *J Neural Trans*.(Suppl.) 21, 257-267 (1986).
57. A. Wirz-Justice, P. Graw, K. Krauchi et al, Morning or nighttime melatonin is ineffective in seasonal affective disorder, *J Psychiat Res*.24, 129-137 (1990).
58. A. J. Lewy, V. K. Bauer, N. L. Cutler, and R. L. Sack, Melatonin treatment of winter depression: a pilot study, *Psychiatry Res*.77, 57-61 (1998).
59. P. A. Childs, I. Rodin, N. J. Martin et al, Effect of fluoxetine on melatonin in patients with seasonal affective disorder and matched controls, *Br J Psychiatry*.166, 196-198 (1995).
60. S. G. Potkin, M. Zetin, V. Stamenkovic, D. Kripke, and W. E. J. Bunney, Seasonal affective disorder: prevalence varies with latitude and climate, *Clin Neuropharmacol*.9, 181-183 (1986).
61. D. H. Avery, K. Dahl, M. V. Savage et al, Circadian temperature and cortisol rhythms during a constant routine are phase-delayed in hypersomnic winter depression, *Biol Psychiatry*.41, 1109-1123 (1997).
62. T. A. Wehr, P. J. Schwartz, E. H. Turner, C. L. Drake, and N. E. Rosenthal, Summer-Winter differences in duration of nocturnal melatonin secretion in SAD patients and healthy controls, *Soc Light Treatment Biol Rhythms, 7th Meet Abstract.*, 9 (1995).
63. T. A. Wehr, P. J. Schwartz, E. H. Turner, S. Feldman-Naim, C. L. Drake, and N. E. Rosenthal, Bimodal patterns of human melatonin secretion consistent with a two-oscillator model of regulation, *Neuroscience Lett*.194, 105-108 (1995).
64. S. Hashimoto, M. Kohsaka, K. Nakamura, H. Honma, S. Honma, and K. I. Honma, Midday exposure to bright light changes the circadian organization of plasma melatonin rhythm in human, *Neurosci Lett*.221, 89-92 (1997).
65. K. Spiegel, R. Leproult, M. LÆHermite-Baleriaux, and E. Van Cauter, Effect of sleep restriction and sleep extension on the 24-h profiles of melatonin and body temperature, *6th Meet Soc Res Biol Rhythms Amelia Island Plantation and Conf Center, Jacksonville, FL, USA.*, 36 (1988).
66. T. A. Wehr, W. C. J. Duncan, L. Sher et al, SCN signal change of season in seasonal affective disorder Conservation of phtoperiod-responsive mechanisms in humans, *Soc Light Treatment Biol Rhythms, 12th Meet Abs*.265, 2-R857 (2000).
67. T. A. Wehr, W. C. J. Duncan, L. Sher et al, A circadian signal change of season in patients with seasonal affective disorder, *Arch Gen Psychiatry*.58, 1108-1114 (2001).
68. B. F. Kjellman, B. E. Thalen, and L. Wetterberg, Light treatment of depressive states: Swedish experiences at latitude 59 N,0, 1-20 (1993).

69. A. J. Lewy, R. L. Sack, V. K. Bauer, and N. L. Cutler, Melatonin as treatment for winter depression, *Amer Psychiatr Assoc 150th Annual Meet San Diego, Syllabus & Proc.*, 74B (1997).

70. C. Thompson, P. A. Childs, N. J. Martin, I. Rodin, and P. J. Smythe, Effects of morning phototherapy on circadian markers in seasonal affective disorder, *Br J Psychiatry*.170, 431-15 (1997).

71. H. Nakagawa, R. L. Sack, and A. J. Lewy, Sleep propensity free-runs with the temperature, melatonin and cortisol rhythms in a totally blind person, *Sleep*.15, 330-336 (1992).

72. T. Shochat, R. Luboshitzky, and P. Lavie, Nocturnal melatonin onset is phase locked to primary sleep gate, *Am J Physiol*.273 (1 pt 2), R364-R370 (1997).

73. A. J. Lewy and R. L. Sack, The dim light melatonin onset as a marker for circadian phase position, *Chronobiol Int*.6, 93-102 (1989).

74. J. R. Gaddy, K. T. Stewart, B. Byrne, K. Doghramji, M. D. Rollag, and G. C. Brainard, Light-induced plasma melatonin suppression in seasonal affective disorder, *Prog Neuropsychopharmacol Biol Psychiatry*.14, 563-568 (1990).

75. C. Thompson, D. Stinson, and A. Smith, Seasonal affective disorder and seasonal-dependent abnormalities of melatonin suppression by light, *Lancet*.336, 703-706 (1990).

76. T. Partonen, O. Vakkuri, and J. Lonnqvist, Suppression of melatonin secretion by bright light in seasonal affective disorder, *Biol Psychiatry*.42, 509-513 (1997).

77. N. E. Rosenthal, F. M. Jacobsen, D. A. Sack et al, Atenolol in seasonal affective disorder: a test of melatonin hypothesis, *Am J Psychiatry*.145, 52-56 (1988).

78. D. Schlager, Early-morning administration of short-acting beta-blockers for treatment of winter depression, *Am J Psychiat*.151, 1383-1385 (1994).

79. C. C. Norman and D. S. Schlager, Early morning, short-acting beta-blockers as treatment for winter depression, *Amer Psychiatr Assoc 150th Annual Meet San Diego, Syllabus & Proc.*, 210 (1997).

80. J. Prasko, Light therapy in Czechoslovakia. Light Treatment, *Biol Rhythms*.4, 50-51 (1992).

81. R. E. McGrath, B. Buckwald, and E. V. Resnick, The effect of L-tryptophan on seasonal affective disorder, *J Clin Psychiatry*.51, 162-163 (1990).

82. A. J. Levitt, G. M. Brown, S. H. Kennedy, and K. Stern, Tryptophan treatment and melatonin response in a patient with seasonal affective disorder, *J Clin Psychopharmacol*.11, 74-75 (1991).

83. K. V. Danilenko, A. A. Putilov, G. S. Russkikh, L. K. Duffy, and S. O. E. Ebbesson, Diurnal and seasonal variations in melatonin and serotonin in women with seasonal affective disorder, *Arctic Med Res*.53, 137-145 (1994).

84. V. A. Cherepanova, K. V. Danilenko, and A. A. Putilov, The effects of phototherapy (PT) on diurnal rhythms of physiological parameters and melatonin excretion in subjects with seasonal affective disorder (SAD), *Arctic Med Res*.(suppl.), 327-329 (1991).

85. A. A. Putilov, K. V. Danilenko, N. V. Volf et al, Chronophysiological aspects of light treatment for seasonal affective disorder: Siberian studies, *Light Treatment Biol Rhythms*.3, 43-45 (1991).

86. K. V. Danilenko and A. A. Putilov, Diurnal and seasonal variations in cortisol, prolactin, TSH, and thyroid hormones in women with and without seasonal affective disorder, *J Interdis Cycle Res*.24, 185-196 (1993).

87. A. A. Putilov, K. V. Danilenko, G. S. Russkikh, and L. K. Duffy, Daytime melatonin in seasonal affective disorder,0, 241-246 (1994).

88. M. E. Levine, A. N. Milliron, and L. K. Duffy, Diurnal and seasonal rhythms of melatonin, cortisol, and testosterone in interior Alaska, *Arctic Med Res*.53, 25-34 (1994).

89. R. G. Skwerer, F. M. Jacobson, C. C. Dunkan et al, Neurobiology of seasonal affective disorder and phototherapy, *J Biol Rhythms*.3, 135-154 (1988).

90. M. Terman, J. S. Terman, F. M. Quitkin et al, Response of the melatonin cycle to phototherapy for seasonal affective disorder, *J Neural Transm*.72, 147-165 (1988).

91. R. A. Depue, W. G. Iacono, R. Muir, and P. Arbisi, Effect of phototherapy on spontaneous eye blink rate in subjects with seasonal affective disorder, *Am J Psychiatry*.145, 1457-1459 (1988).

92. R. A. Depue, P. Arbisi, M. R. Spoont, S. Krauss, A. Leon, and B. Ainsworth, Seasonal and mood independence of low basal prolactin secretion in premenopausal women with seasonal affective disorder, *Am J Psychiatry 1989*.146, 989-995 (1989).

93. M. Terman, F. M. Quitkin, J. S. Terman, J. W. Stewart, and P. J. McGrath, The timing of phototherapy: effects on clinical response and the melatonin cycle, *Psychopharmacol Bull*.23, 354-357 (1987).

94. M. H. Teicher, C. A. Glod, E. Magnus et al, Circadian rest-activity disturbances in seasonal affective disorder, *Arch Gen Psychiatry*.54, 124-130 (1997).

95. A. J. Lewy, R. L. Sack, L. S. Miller, and T. M. Hoban, Antidepressant and circadian phase-shifting effect of light, *Science*.235, 352-354 (1987).

96. C. Reme, M. Terman, and A. Wirz-Justice, The visual input stage of the mammalian circadian pacemaking system. I.Evidence for a self-sustained ocular oscillator. II. The effect of natural and artifical light schedules on retinal integrity and consequences for circadian rhythm research, *J Biol Rhytms*.6, 31-48 (1990).

97. C. Reme, A. Wirz-Justice, A. Rhyner, and S. Hofmann, Circadian rhythm in the light response of rat retinal disk-shedding and autophagy, *Brain Res*.369, 356-360 (1986).

98. N. Ozaki, N. E. Rosenthal, F. Myers, P. J. Schwartz, and D. A. Oren, Effects of season on electrooculographic ratio in winter seasonal affective disorder, *Psychiatry Res 1995*.59, 151-155 (1995).

99. P. H. Desan, D. A. Oren, R. Malison et al, Gnetic polymorphism at the CLOCK gene locus and major depression, *Am J Med Genet*.96, 418-421 (2000).

100. N. E. Rosenthal, A. A. Levendosky, R. G. Skwerer et al, Effects of light treatment on core body temperature in seasonal affective disorder, *Biol Psychiatry*.27, 39-50 (1990).

101. A. A. Levendosky, J. R. Joseph-Vanderpool, T. Hardin, E. Sorek, and N. E. Rosenthal, Core body temperature in patients with seasonal affective disorder and normal controls in summer and winter, *Biol Psychiatry*.29, 524-534 (1991).

102. C. I. Eastman, L. C. Gallo, H. W. Lahmeyer, and L. F. Fogg, The circadian rhythm of temperature during light treatment for winter depression, *Biol Psychiatry*.34, 210-220 (1993).

103. K. Dahl, D. H. Avery, A. J. Lewy et al, Dim light melatonin onset and circadian temperature during a constant routine in hypersomnic winter depression, *Acta Psychiatr Scand*.88, 60-66 (1993).

104. A. Wirz-Justice, Seasonal affective disorder and light therapy,0, 189-200 (1996).

105. T. Winton, T. Corn, L. Huson, C. Franey, J. Arendt, and S. Checkley, Effects of light treatment upon mood and melatonin in patients with seasonal affective disorder, *Psychol Med*.19, 585-590 (1989).

106. R. L. Sack, A. J. Lewy, D. M. White, C. M. Singer, M. V. Fireman, and R. Vandiver, Morning vs evening light treatment for winter depression: evidence that the therapeutic effects of light are mediated by circadian phase shifting, *Arch Gen Psychiatry*.47, 343-351 (1990).

107. A. Wirz-Justice, P. Graw, K. Krauchi et al, Light therapy in seasonal affective disorder is independent of time of the day or circadian phase, *Arch Gen Psychiatry*.50, 929-937 (1993).

108. B. E. Thalen, B. F. Kjellman, L. Morkrid, and L. Wetterberg, Melatonin in light treatment of patients with seasonal and nonseasonal depression, *Acta Psychiatr Scand*.92, 274-284 (1995).

109. T. Partonen, O. Vakkuri, C. Lamberg-Allardt, and J. Lonnqvist, Effects of bright light on sleepiness, melatonin, and 25-hydorxyvitamin D(3) in winter seasonal affective disorder, *Biol Psychiatry*.39, 865-872 (1996).

110. J. Rice, J. Mayor, A. Tucker, and R. J. Bielski, Effect of light therapy on salivary melatonin in seasonal affective disorder, *Psychiatry Res*.56, 221-228 (1995).

111. J. L. Andersonand A. Wirz-Justice. Biological rhythms in the pathophysiology and treatment of affective disorders, in: *Biol. Aspects of Affective Disorders,* edited by R. Horton and C. Katona. 0 ed. (Academic Press, London, 1991) pp. 223-269.

112. A. Wirz-Justice and J. L. Anderson, Morning light exposure for the treatment of winter depression: The one true light therapy?, *Psychopharmacol Bull*.26, 511-520 (1990).

113. M. Terman, J. S. Terman, A. J. Lewy, and D. J. Dijk, Light treatment for sleep phase and duration disturbances, *Light Treatment Biol Rhythms*.7, 7-14 (1994).

114. D. H. Avery, A. Khan, S. R. Dager, S. Cohen, G. B. Cox, and D. L. Dunner, Morning or evening bright light treatment of winter depression? The significance of hypersomnia, *Biol Psychiatry*.29, 117-126 (1991).

115. D. A. Oren, F. M. Jacobsen, D. L. Cameron, and N. E. Rosenthal, Predictors of response to phototherapy in seasonal affective disorder, *Compr Psychiatry*.33, 111-114 (1992).

116. K. Krauchi, A. Wirz-Justice, and P. Graw, High intake of sweets late in the day predicts a rapid and persistent response to light therapy in winter depression, *Psychiatry Res*.46, 107-117 (1993).

117. M. Terman, Problems and prospects for use of bright light as a therapeutic intervention,0, 421-436 (1993).

118. R. W. Lam, Morning light therapy for winter depression: predictors of response, *Acta Psychiatr Scand*.89, 97-101 (1994).

119. T. Partonen, Effects of morning light treatment on subjective sleepiness and mood in winter depression, *J Affect Dis*.30, 47-56 (1994).

120. M. Terman, L. Amira, J. S. Terman, and D. C. Ross, Predictors of response and nonresponse to light treatment for winter depression, *Am J Psychiatry*.153, 1423-1429 (1996).

121. D. H. Avery, M. A. Bolte, and D. Eder, Difficulty awakening as a symptom of winter depression, 21 (1994).

122. C. A. Czeisler, E. D. Weitzman, M. C. Moore-Ede, J. C. Zimmerman, and R. S. Kronauer, Human Sleep: its duration and organization depend on its circadian phase, *Science*.210, 1264-1267 (1980).

123. J. Zulley, R. Wever, and J. Aschoff, The independence of onset and duration of sleep on the circadian rhythm of rectal temperature, *Pflugers Arch.*391, 314-318 (1981).

124. C. I. Eastman, M. A. Young, L. F. Fogg, L. Liu, and P. M. Meaden, Bright light treatment of winter depression: a placebo-controlled trial, *Arch Gen Psychiatry.*55, 883-889 (1998).

125. A. J. Lewy, V. K. Bauer, N. L. Cutler et al, Morning vs evening light treatment of patients with winter depression, *Arch Gen Psychiatry.*55, 890-896 (1998).

126. M. Terman, J. S. Terman, and D. C. Ross, A controlled trial of timed bright light and negative air ionization for treatment of winter depression, *Arch Gen Psychiatry.*55, 875-882 (1998).

127. A. A. Putilov, Multi-component physiological response mediates therapeutic benefits of bright light in winter seasonal affective disorder, *Biol Rhythm Res.*29, 367-386 (1998).

128. A. J. Lewy, V. K. Bauer, H. A. Bish et al, Antidepressant response correlates with the phase advance in winter depressives, *Soc Light Treatment Biol Rhythms 12th Meet Abstr.*, 22 (2000).

129. M. Terman and J. S. Terman, Morning vs. evening light: Effects on the melatonin rhythm and antidepressant response in winter depression, *Soc Light Treatment Biol Rhythms (abstract).*12, 1 (2000).

130. M. Terman and J. S. Terman, Phase shifts in melatonin and sleep under light therapy for winter depression, *Soc Light Treatment Biol Rhythms, 7th Meet Abstr.*, 15 (1995).

131. J. Kurtz, M. S. Bauer, and R. Poland, A test of the phase-shift hypothesis of SAD, *Amer Psychiatr Assoc 150th Annual Meet San Diego, Syllabus & Proc.*, 453 (1997).

132. B. Rafferty, M. Terman, J. S. Terman, and C. E. Reme, Does morning light prevent evening light effect? A statistical model for morning/evening crossover studies, *Soc Light Treatment Biol Rhythms, 2nd Meet Abstr.*, 18 (1990).

133. M. E. Jewett, D. W. Rimmer, J. F. Duffy, E. B. Klerman, R. E. Kronauer, and C. A. Czeisler, Human circadian pacemaker is sensitive to light throughout subjective day without evidence of transients, *Am J Physiol.*273 (5Pt 2), R1800-R1809 (1997).

134. P. J. Schwartz, N. E. Rosenthal, and T. A. Wehr, Band-specific electroencephalogram and brain cooling abnormalities during NREM sleep in patients with winter depression, *Biol Psychiatry.*50, 627-632 (2001).

135. S. M. Schergin, K. V. Danilenko, and A. A. Putilov, Biological and psychic effects of bright light in seasonal affective disorder,0, 409-411 (1996).

136. A. Khan, H. A. Warner, and W. A. Brown, Symptom reduction and suicide risk in patients treated with placebo in antidepressant clinical trials: an analysis of the food and drug administration database, *Arch Gen Psychiatry.*57, 311-317 (2000).

137. C. I. Eastman, What the placebo literature can tell us about phototherapy for SAD, *Psychopharmacol Bull.*26, 495-504 (1990).

138. D. F. Kripke, Light treatment for nonseasonal depression: speed, efficacy, and combined treatment, *J Affect Disord.*49, 109-117 (1998).

139. M. H. Teicher, C. A. Gold, D. A. Oren et al, The phototherapy light visor: more to it than meets the eye, *Am J Psychiatry.*152, 1197-1202 (1995).

140. Y. Meesters, J. H. C. Jansen, D. G. M. Beersma, A. L. Bouhuys, and R. H. Van den Hoofdakker, Light therapy for seasonal affective disorder: the effect of timing, *Br J Psychiatry.*166, 607-612 (1995).

141. T. M. Hoban, R. L. Sack, A. J. Lewy, L. S. Miller, and C. M. Singer, Entrainment of a free-running human with bright light?, *Chronobiol Int.*6, 347-353 (1989).

142. D. H. Avery, A. Khan, S. R. Dager, G. B. Cox, and D. L. Dunner, Bright light treatment of winter depression: morning versus evening light, *Acta Psychiatr Scand.*82, 335-338 (1990).

143. D. A. Oren, G. C. Brainard, S. H. Johnston, J. R. Joseph-Vanderpool, E. Sorek, and N. E. Rosenthal, Treatment of seasonal affective disorder with green light and red light, *Am J Psychiatry.*148, 509-511 (1991).

144. B. B. Pinchasov, O. V. Grischin, and A. A. Putilov, Rate of oxygen consumption in seasonal and nonseasonal depression, *World J Biol Psychiatry 2002.*3, 101-104 (2002).

145. D. F. Kripke, S. C. Risch, and D. Janowsky, Bright white light alleviates depression, *Psychiatry Res.*10, 105-111 (1983).

146. D. F. Kripke, D. J. Mullaney, M. R. Klauber, S. C. Risch, and J. C. Gillin, Controlled trial of bright light for nonseasonal major depressive disorders, *Biol Psychiatry.*31, 119-134 (1992).

147. N. Yamada, M. T. Martin-Uiverson, K. Daimon, T. Tsukimoto, and S. Takahashi, Clinical and chronobiological effects of light therapy on nonsesonal affective disorders, *Biol Psychiatry.*37: 866-873 (1995).

148. D. Stinson and C. Thompson, Clinical experience with phototherapy, *J Affect Disord.*18, 129-135 (1990).

149. A. Mackert, H. P. Volz, R. D. Stieglitz, and B. Muller-Oerlinghausen, Phototherapy in nonseasonal depression, *Biol Psychiatry 1991.*37, 866-873 (1991).

150. J. A. Deltito, M. Moline, C. Pollak, L. Y. Martin, and I. Maremmani, Effects of phototherapy on non-seasonal unipolar and bipolar depressive spectrum disorders, *J Affect Disord.*23, 231-237 (1991).

151. D. A. Oren, D. E. Moul, P. J. Schwartz, T. A. UER, and N. E. Rosenthal, A controlled trial of levodopa plus carbidopa in the treatment of winter seasonal affective disorder: a test of the dopamine hypothesis, *J Clin Psychopharmacol.*14, 196-200 (1994).

152. R. M. Bagby, D. R. Schuller, A. J. Levitt, R. T. Joffe, and K. L. Harkness, Seasonal and non-seasonal depression and the five-factor model of personality, *J Affect Disord.*5, 89-95 (1996).

153. L. Sher, J. R. Matthews, E. H. Turner, T. T. Postolache, K. S. Katz, and N. E. Rosenthal, Early response to light therapy partially predicts long-term antidepressant effects in patients with seasonal affective disorder, *J Psychiatry Neurosci.*26, 336-338 (2001).

154. P. Richter, A. L. Bouhuys, H. R. H. Van den et al, Imaginary versus real light for winter depression, *Biol Psychiatry.*31, 534-536 (1992).

155. N. E. Rosenthal, D. A. Sack, C. J. Carpenter, B. L. Parry, W. B. Mendelson, and T. A. Wehr, Antidepressant effect of light in seasonal affective disorder, *Am J Psychiatry.*142, 606-608 (1985).

156. M. Terman and J. S. Terman, Light therapy for winter depression,0, 171-187 (1992).

157. L. J. Grota, B. I. Yerevanian, K. Gupta, J. Kruse, and L. Zborowski, Phototherapy for seasonal major depressive disorder: effectiveness of bright light of high or low intensity, *Psychiatry Res.*29, 29-35 (1989).

158. W. Schiefenh^vel, Perception, expression, and social function of pain: a human ethological view, *Science in Context.*8, 31-46 (1995).

159. H. Hagen, The function of postpartum depression, *Evol Human Behav.*20, 325-359 (1999).
160. E. E. Michalak, C. Wilkinson, C. Dowrick, and K. Hood, Seasonal affective disorder, negative life events and social support: evidence of an association in a community sample. Soc Light Treatment, *Biol Rhythms, (abstract)*.13 (2001).
161. E. E. Michalak, C. Wilkinson, K. Hood, C. Dowrick, and G. Wilkinson, Seasonality, negative life events and social support in a community sample, *Br J Psychiatry.*182, 434-438 (2003).
162. A. A. Putilov, B. B. Pinchasov, K. V. Danilenko, and E. Y. Poljakova, Open Trials of mono- and combined non-drug treatments for seasonal and non-seasonal depression: Effect of patientsÆ expectations, *Int J Behavioral Med.*11, 182 (2004).
163. L. V. Pylkova, K. V. Danilenko, and A. A. Putilov, Determinants of seasonal affective disorder: Biological or Psychological?, *Acta Psychiatr Scand.*(Suppl.) (submitted) (2004).
164. L. V. Pylkova, K. V. Danilenko, and A. A. Putilov, Association of seasonal affective disorder with poor social support: A fact supporting evolutionary psychological explanation of winter depression, *Int J Behavioral Med, 2004b.*11 (September), 154 (2004).
165. E. Geerts, N. Bouhuys, and R. H. Van den Hoofdakker, Nonverbal attunement between depressed patients and an interviewer predicts subsequent improvement, *J Affect Disord.*40, 15-21 (1996).
166. A. D. Boenink, A. L. Bouhuys, D. G. Beersma, and Y. Meesters, Prediction of acute and late responses to light therapy from vocal (pitch) and self-rated activation in seasonal affective disorder, *J Affect Disord.*42, 117-126 (1997).
167. S. Beratis, P. Gourzis, and J. Gabriel, Psychological factors in the development of mood disorders with a seasonal pattern, *Psychopathology.*29, 331-339 (1996).
168. M. J. Raleigh, G. I. Brammer, M. T. McGuire, and A. Yuweiler, Dominant social status facilitates the behavioral effects of serotonergic agonists, *Brain Res.*348, 274-282 (1985).
169. M. J. Raleigh, M. T. McGuire, G. I. Brammer, and A. Yuweiler, Social and environmental influences on blood serotonin concentrations in monkeys, *Arch Gen Psychiatry.*41, 405-410 (1884).
170. G. W. Lambert, C. Reid, D. M. Kaye, G. L. Jennings, and M. D. Esler, Effect of sunlight and season on serotonin turnover in the brain, *The Lancet.*360, 9348 (2002).
171. L. P. Morin, Serotonin and the regulation of mammalian circadian rhythmicity, *Ann Med.*31, 12-33 (1999).
172. E. M. Mintz, C. F. Gillespie, C. L. Marvel, K. L. Huhman, and H. E. Albers, Serotonergic regulation of circadian rhythms in Syrian hamsters, *Neuroscience.*79, 563-569 (1997).
173. T. Moriya, Y. Yoshinobu, M. Ikeda, S. Yokota, M. Akiyama, and S. Shibata, Potentiating action of MKC-242, a selective 5-HT1A receptor agonist, on the photic entrainment of the circadian activity rhythm in hamsters, *Br J Pharmacol.*125, 1281-1287 (1998).
174. J. D. Glass, L. A. DiNardo, and J. C. Ehlen, Dorsal raphe nuclear stimulation of SCN serotonin release and circadian phase-resetting, *Brain Res.*859, 224-232 (2000).
175. L. Imeri, M. Mancia, and M. R. Opp, Blockade of 5-hydoxytryptamine (serotonin)-2-receptors alters interleukin-1-induced changes in rat sleep, *Neuroscience.*92, 745-749 (1999).

176. R. Kirov and S. Moyanova, Ritanserin-induced changes in sleep-waking phases in rats, *Acta Physiol Pharmacol Bulg.*21, 87-92 (1995).
177. S. Kantor, R. Jakus, R. Bodizs, P. Halasz, and G. Bagdy, Acute and long-term effects of the 5-HT(2) receptor antagonist ritanserin on EEG power spectra, motor activity, and sleep: changes at the light-dark phase shift, *Arch Gen Psychiatry.*55, 890-896 (1998).
178. P. Meerlo, E. A. de Bruin, A. M. Strijkstra, and S. Daan, A social conflict increases EEG slow-wave activity during subsequent sleep, *Physiol Behav.*73, 331-335 (2001).
179. M. A. Young, P. M. Maeden, L. F. Fogg, E. A. Cherin, and C. I. Eastman, Which environmental variables are related to the onset of seasonal affective disorder?, *J Abnormal Psychol.*106: 554-562 (1997).
180. R. W. Lam, E. M. Tam, L. N. Yatham, and A. P. Zis, Seasonal depression: the dual-vulnerability hypothesis revisited, *J Affect Disord.*63, 123-132 (2001).

Index